Applied Mathematical Sciences
Volume 104

Editors
F. John J.E. Marsden L. Sirovich

Advisors
M. Ghil J.K. Hale J. Keller
K. Kirchgässner B.J. Matkowsky
J.T. Stuart A. Weinstein

Applied Mathematical Sciences

(continued following index)

Miklós Farkas

Periodic Motions

With 105 Illustrations

Springer-Verlag

New York Berlin Heidelberg London Paris
Tokyo Hong Kong Barcelona Budapest

Miklós Farkas
Department of Mathematics
Budapest University of Technology
Budapest H-1521
Hungary

Editors

F. John
Courant Institute of
  Mathematical Sciences
New York University
New York, NY 10012
USA

J.E. Marsden
Department of
  Mathematics
University of California
Berkeley, CA 94720
USA

L. Sirovich
Division of
  Applied Mathematics
Brown University
Providence, RI 02912
USA

Mathematics Subject Classification (1991): 34C25, 58F22, 92D25

Library of Congress Cataloging-in-Publication Data
Farkas, Miklós.
    Periodic motions/Miklós Farkas.
      p.   cm. – (Applied mathematical sciences; v. 104)
    Includes bibliographical references (p.   -   ) and index.
    ISBN 978-1-4419-2838-2
  DM98.00
    1. DIfferential equations – Numerical solutions.   I. Title.
  II. Series: Applied mathematical sciences (Springer-Verlag New York
  Inc.); v. 104.
  QA371.F27   1994   93-50623
  515′.352 – dc20

Printed on acid-free paper.

Photocomposed using the author's LaTex files.

Printed in the United States of America.

9 8 7 6 5 4 3 2 1

To the two Katies of my life and to little Anna

Preface

"The task is done; the Maker rests. And lo!
The engine turns. A million years shall flow,
Ere round its axle shall the wheel run slow
And a new cog be needed. ... "

Madách: The Tragedy of Man
J.C.W. Horne's translation

In this book I tried to sum up the facts and results I considered most
important concerning periodic solutions of ordinary differential equations
(ODEs) produced by this century from Henri Poincaré up to the youngest
mathematician appearing in the list of references. I have included also some
results of my own that did not find their way into monographs in the past.
I have done research in this direction for more than 25 years and have given
graduate courses about some of the topics covered for many years at the
Budapest University of Technology and also at the Universidad Central de
Venezuela in Caracas. I hope that people interested in differential equations
and applications may use this experience. Some may say that periodic
solutions of ODEs has been a closed chapter of mathematics for some time.
My answer to this objection is twofold. First, I hope that the text will
show that there is a large amount of open problems in this field, problems
that are deep and interesting from the point of view of the mathematician
and important in applications at the same time. Secondly, I believe that
it is impossible to go into research in the fields of functional differential
equations, evolution equations, abstract dynamical systems, chaos, etc.,
which are now more in fashion without having a thorough knowledge of

what is going on in the context of ODEs. Most researchers working in more abstract structures gain their intuition secretly from ODEs where "one can see things".

The book is intended for graduate and Ph.D students in mathematics, physics, engineering, biology, etc., and I hope that it can be used by people doing research in the field of dynamical systems and their applications as a standard reference if and when they encounter periodic solutions. For most parts of the book a background in advanced calculus and linear algebra is a sufficient prerequisite. Some knowledge of ODEs is also useful; however, the basic material needed from the general theory of the latter field, existence, uniqueness of solutions, dependence on initial values and parameters, stability, etc., is summed up in the first chapter. In the introduction, the theorems are given without proofs. In the rest of the chapters, proofs are provided for almost all the theorems; the exceptions being some statements in lateral branches of the theory. In each case when the proof is not included, references are given where it can be found.

Chapters 2, 3, 4 and 5 deal with the existence and stability problems of periodic solutions of linear, two dimensional autonomous, periodic and higher dimensional autonomous systems, respectively. Besides the general theory, Chapter 2 treats Hill's and Mathieu's equation, Chapter 3 presents the Liénard, the Van der Pol, the Duffing and the Lotka-Volterra models and introduces the reader to Hilbert's Sixteenth Problem. Chapter 4 deals with non-autonomous periodic systems applying mainly topological methods. Chapter 5 contains the definition of orbital stability and the Andronov-Witt Theorem which is fundamental in the study of non-constant periodic solutions of dynamical systems. Some results here concerning the isolation and isochronism of periodic solutions have not been published yet in monographs. Chapter 6 presents the most important results achieved by perturbation methods. The method of controllably periodic perturbations and results about aperiodic perturbations expounded here have not been published yet in monographs. This chapter provides a concise insight into the methods of averaging and of singular perturbations too. Chapter 7 is about bifurcations with emphasis on bifurcations generating periodic solutions; thus, the Andronov-Hopf bifurcation plays a central role here. The phenomenon of the zip bifurcation is presented here for the first time in a book. Although this book is about ODEs, the last two sections of this chapter extend beyond the actual topic, giving some insight into related material in the theory of retarded functional differential equations on the one hand and chaos on the other.

There are three Appendices. Appendix 1 contains knowledge from advanced matrix theory. Here most theorems are given with proofs. Appendix 2 gives a concise treatment of topological degree and fixed point theorems and provides some basic concepts of functional analysis that are needed in some places. Appendix 3 deals with differentiable manifolds, the Riemannian metric and the invariant manifolds of dynamical systems. In the last two appendices, proofs on functional analysis and on manifolds had to be omitted because they would have filled a book.

Although the classical systems of non-linear oscillations (Liénard, Van der Pol, Duffing, etc.) are also treated, most concrete models illustrating the theory come from biomathematics, more exactly, from population dynamics (different predator-prey, competitive and cooperative systems including the Lotka-Volterra one). In some cases because of didactic reasons not the sharpest available statement is proved, but if this happens, it is noted. At the end of each chapter some problems are given; their solution may help the understanding of the text. Formulae, definitions, theorems (lemmata and corollaries considered as synonyms), examples, figures and problems are numbered by their respective chapter, section and place in the sequence of their kind Ak denoting the kth "section" of the appendix ($k = 1, 2, 3$). Thus, Definition 1.2.3, resp. Theorem 1.2.3 is the third definition, resp. theorem (lemma, corollary) in Chapter 1, Section 2.

The writing of this book was partially supported by the Hungarian National Foundation for Scientific Research (grant numbers 1186 and 1994).

I wish to express my gratitude to those friends and colleagues who helped me in the preparation of this volume. Barna Garay, Viktor Kertész and Gábor Stépán of the Budapest University of Technology and Dana Schlomiuk of l'Université de Montreal read parts of the manuscript, corrected many errors and made invaluable suggestions that improved the text considerably. The participants of the Stability Theory Seminar at the Budapest University of Technology contributed much by their active criticism when listening to talks about some of the topics. My student Tamás Kalmár-Nagy corrected many misprints. Zoltán Farkas drew the figures with great care and much skill. Gábor Salfer solved with ease all the problems caused by our computer dependence in preparing the LATEXversion of the manuscript. Miss Flóra Géczy has done a fine job bravely learning TEXand typing the text. Karen Kosztolnyik of Springer-Verlag New York organized and arranged everything with utmost efficiency when, finally, the publication of the book gained momentum. I thank them all. Finally, I thank my wife, Kati, for her encouragement, moral support and patience, that is, for playing the role of the Muse superbly; without this, the book probably would have not been written.

Hegyesd, March 1994   Miklós Farkas

Contents

1
Introduction

We observe *periodic motions*, periodic variations in every field of science and everywhere in real life. We observe the revolution of the Moon around the Earth, the rotation of the Earth around its axis, the swinging movement of the pendulum of a clock, the wheel of a moving car, an engine in working order, the effect of alternating electric current; we observe the periodic ups and downs of economy, heart beat, respiration, etc. To be sure, some of these phenomena can better be described as *"almost periodic motions"*; still, even these can be approximated in a satisfactory way by periodic variations. If the phenomenon is characterised by some parameters, the so-called, *phase coordinates*, then the periodicity of the phenomenon is represented by the periodic variation of these phase coordinates, i.e. by a periodic motion in the *phase space*. A periodic motion in the phase space is a path that after some fixed *time* returns to its starting point and is continuing to do this in the whole future. This means that a periodic motion is represented by a *closed path* in the phase space. Behind such a periodic motion, usually there stands a natural (mechanical, physical, biological, etc.) or an economic law. This natural or economic, etc., law is, usually, represented by a *differential equation* which determines how the rate of change of the phase coordinates depends on their actual or past values, on the actual time and on other parameters. In this book we shall consider and *call* the "independent variable " or one of the independent variables, *time*. We shall be concerned with variations periodic *in time*, though naturally there are important phenomena that show *spatial periodicity* or periodicity with respect to some other variable. We shall denote time by t if possible, and assume that either $t \in \mathbf{R}$ or $t \in \mathbf{R}_+ = [0, \infty]$ or $t \in I$ where I is some interval on the real line.

In this introductory chapter the most important prerequisites are summed up from the *theory of ordinary differential equations*. At some places of the book, delay differential equations are also considered; basic facts about these are summed up in Section 7.5. In this chapter in Section 1 basic concepts and theorems about the existence, the uniqueness and some properties of solutions are treated. In Section 2 the most important properties of *linear systems* are presented since the study of non-linear problems is reduced often in some way also to linear systems. Autonomous systems are treated in Section 3. In real situations only solutions that are *stable* in some sense can be observed directly. Therefore, stability theory is presented in a concise way in Sections 4 and 5.

1.1   Existence, Uniqueness and Analytic Properties of Solutions

Differential equations will be treated in $\mathbf{R}^n$. Most results can be extended to Banach spaces or to differentiable manifolds. The proofs of the theorems of this section can be found in Hartman [1964] or in Hirsch-Smale [1974]. The elements or points in $\mathbf{R}^n$ will be denoted by lowercase Roman letters, and their coordinates, if possible, by the same letter with indices, so $x = (x_1, ..., x_n) \in \mathbf{R}_n$. In matrix algebraic calculations, x will be considered as a column vector, *transpose* will be denoted by a prime. The derivative of a function $\varphi : \mathbf{R} \to \mathbf{R}^n$ with respect to "time" $t \in \mathbf{R}$ will be denoted by *dot* : $\dot{\varphi}$, the derivative matrix of a function $f : \mathbf{R}^n \to \mathbf{R}^n$ by f'_x, i.e. $f'_x(x)$ is the matrix whose entries are the first order partial derivatives of the coordinate functions of f at $x \in \mathbf{R}^n : f'_{ix_k}(x)$. The derivative of a scalar function $F : \mathbf{R}^n \to \mathbf{R}$ is the vector grad $F(x) = (F'_{x_1}(x), ..., F'_{x_n}(x))$.

Let the subset $X \subset \mathbf{R}^n$ be open and connected, and consider a continuous function $f : \mathbf{R} \times X \to \mathbf{R}^n$, i.e. $f \in C^0(\mathbf{R} \times X)$. The equation

$$\dot{x} = f(t, x) \qquad (1.1.1)$$

is called an *n-dimensional system of first order ordinary differential equations* for the unknown function x. Let $I \subset \mathbf{R}$ be an interval; the function $\varphi : I \to \mathbf{R}^n$ is called a *solution* of system (1.1.1) if for $t \in I$: $\varphi(t) \in X$, $\varphi \in C^1(I)$, and if it satisfies (1.1.1), i.e.

$$\dot{\varphi}(t) \equiv f(t, \varphi(t)).$$

We have an *initial value problem* attached to (1.1.1) if the value of the solution to be found is prescribed at a given moment $t_0 \in \mathbf{R}$, i.e. if we require that

$$x(t_0) = x^0 \in X. \qquad (1.1.2)$$

The following theorem is a weakened version of the Cauchy-Lipschitz-Picard-Lindelöf Theorem about the existence and uniqueness of the

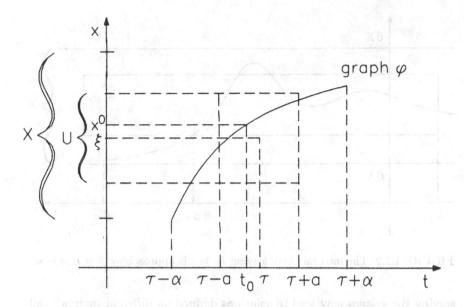

FIGURE 1.1.1. The integral curve belonging to the solution of initial value problem (1.1.1)-(1.1.2).

solution of the initial value problem (1.1.1) - (1.1.2).

Theorem 1.1.1. *Let $\tau \in \mathbf{R}$, $a > 0$, $\xi \in X$ and U be a neighbourhood of ξ such that the closure $\overline{U}$ of U is in X : $\overline{U} \subset X$, assume that $f \in C^0(\mathbf{R} \times X)$ and $f'_x \in C^0(\mathbf{R} \times X)$; then an $\alpha > 0$ exists such that for every $t_0 \in (\tau - \alpha, \tau + \alpha)$ and $x^0 \in U$ there exists a solution $\varphi : (\tau - \alpha, \tau + \alpha) \to X$ of the initial value problem (1.1.1), (1.1.2) and every solution of the problem defined on $(\tau - \alpha, \tau + \alpha)$ (or on a subinterval of this) is identical to φ. If $t_0 = \tau$, $x^0 = \xi$, the radius of U is $\rho > 0$ and*

$$M := \sup |f(t,x)| = \sup \left(\sum_{i=1}^{n} |f_i(t,x)|^2 \right)^{1/2} \quad \text{when} \quad (t,x) \in [\tau - a, \tau + a] \times \overline{U}$$

for some $a > 0$, then $\alpha \geq \min(a, \rho/(M\sqrt{n}))$.

The theorem says that if the right-hand side of (1.1.1) is not only continuous but also continuously differentiable with respect to x, then the initial value problem has one and only one solution. Moreover, it says that to a closed neighbourhood of any point $(\tau, \xi) \in \mathbf{R} \times X$ there belongs an interval containing τ such that the solutions of all the initial value problems with initial values in this neighbourhood are defined on this common interval; see Figure 1.1.1.

If the initial values, t_0, $x^0 \in X$, are fixed, then different ways of

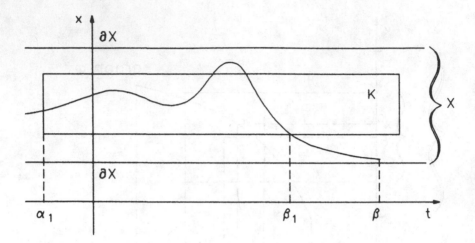

FIGURE 1.1.2. The integral curve leaving K as t is approaching β or $\alpha = -\infty$.

solving the system may lead to solutions defined on different intervals (all containing, naturally, t_0). Because of the uniqueness part of the theorem, these solutions must coincide on the intersection of their respective domains. There exists a solution of this initial value problem with *maximal domain* (α, β), say where α may be $-\infty$, β may be ∞ and $t_0 \in (\alpha, \beta)$. This means that the domain of any solution of this initial value problem is a subinterval of (α, β). Because of uniqueness, any solution of the same initial value problem is a restriction (to its domain) of the solution with the maximal domain. The solution with the maximal domain is called the *maximal solution*. However, in the future when we say solution we mean the maximal solution for sake of brevity.

Let $\varphi : (\alpha, \beta) \to X$ be a solution; the graph of this function in $\mathbf{R} \times X$ is called an *integral curve*; the parametrically given curve in $X \subset \mathbf{R}^n$, the set

$$\gamma = \{x \in X : x = \varphi(t), t \in (\alpha, \beta)\}$$

is called the *path (trajectory or orbit)* of this solution; the function φ itself is sometimes called a *motion*.

If $\varphi : (\alpha, \beta) \to X$ is a maximal solution, and t tends to $\beta - 0$ or $\alpha + 0$, then the integral curve of this solution approaches the boundary of the set $\mathbf{R} \times X$ in $\mathbf{R} \times \mathbf{R}^n$; more exactly the following theorem holds.

Theorem 1.1.2. *Let $\varphi : (\alpha, \beta) \to X$ be a maximal solution of (1.1.1) and $K \subset \mathbf{R} \times X$ a compact set; there exist α_1, β_1 satisfying $\alpha < \alpha_1 < \beta_1 < \beta$ such that for $t \in (\alpha, \alpha_1)$ and for $t \in (\beta_1, \beta) : (t, \varphi(t)) \in \mathbf{R} \times X \setminus K$ (see Figure 1.1.2).*

This may mean that either $\beta = \infty$ or $\lim_{t \to \beta} \text{dist}\,(\varphi(t), \partial X) = 0$, or

$\lim_{t \to \beta} | \varphi(t) | = \infty$, and similarly, either
$\alpha = -\infty$ or $\lim_{t \to \alpha} \text{dist}(\varphi(t), \partial X) = 0$ or $\lim_{t \to \alpha} |\varphi(t)| = \infty$; if $\beta < \infty$
(resp. $\alpha > -\infty$), then other more complex situations may also arise.

The solution of (1.1.1) satisfying the initial condition (1.1.2) will be
denoted by $\varphi(t, t_0, x^0)$, i.e. for all $t_0 \in \mathbf{R}$ and $x^0 \in X$ the corresponding
solution φ satisfies

$$\varphi(t_0, t_0, x^0) = x^0. \tag{1.1.3}$$

Let us denote the maximal domain of definition of φ by
$(\alpha(t_0, x^0), \beta(t_0, x^0))$; the end points of the interval, clearly, depend on
the initial values. One can prove (see Hartman [1964]) that if the
conditions of Theorem 1.1.1 hold, then β is a *lower semicontinuous*
and α an *upper semicontinuous* function of (t_0, x^0). This means that
to any $\alpha(t_0, x^0) < \alpha_1 < \beta_1 < \beta(t_0, x^0)$ there belongs a neighbourhood
$W \subset \mathbf{R} \times X$ of (t_0, x^0) such that for all $(t_1, x^1) \in W$ we have $\beta(t_1, x^1) > \beta_1$
and $\alpha(t_1, x^1) < \alpha_1$.

Often we are confronted with problems in which the right-hand side of
the differential equation depends on parameters, i.e. we have to consider
systems of the form

$$\dot{x} = f(t, x, \mu) \tag{1.1.4}$$

with initial condition $x_0 = x(t_0)$ where $f \in C^0(\mathbf{R} \times X \times \mathbf{R}^m, \mathbf{R}^n)$,
$f'_x \in C^0$, $f'_\mu \in C^0$ the subset $X \subset \mathbf{R}^n$ being open and connected, the
parameters being denoted by $\mu \in \mathbf{R}^m$. The solutions and their maximal
domains of definition depend now, naturally, on the parameters too. The
solution of the initial value problem attached to (1.1.4) will be denoted
by $\varphi(t, t_0, x^0, \mu)$ and its maximal domain by $(\alpha(t_0, x^0, \mu), \beta(t_0, x^0, \mu))$. The
functions β and α are again lower and upper semicontinuous functions,
respectively. Thus, if $\alpha(t_0, x^0, \mu) < \alpha_1 < \beta_1 < \beta(t_0, x^0, \mu)$ are fixed, we may
choose a neighbourhood $W \subset \mathbf{R} \times X$ of (t_0, x^0) and a neighbourhood
$M \subset \mathbf{R}^m$ of the actual value of μ such that the solution φ considered as a
function of the time t, the initial values (t_0, x^0), and the parameters μ is
a function defined on $[\alpha_1, \beta_1] \times W \times M$:

$$\varphi : [\alpha_1, \beta_1] \times W \times M \to X.$$

For this function we have the result:

Theorem 1.1.3. *The solution of (1.1.4) as a function of time,
the initial values and the parameters is continuously differentiable:*
$\varphi \in C^1([\alpha_1, \beta_1] \times W \times M, X)$; *moreover,* $\varphi'_{t_0}, \varphi'_{x^0}, \varphi'_\mu$ *are continuously
differentiable with respect to t.*

In this introductory chapter we are not supplying the proofs that can
be found in standard textbooks. However, in this particular case, we note
that the proof of the fact that $\varphi'_{x^0_k}$ exists and is continuous ($k = 1, ..., n$;

x_k^0 is the k-th coordinate of the initial value x^0) goes by showing that the corresponding difference ratio tends to the solution of the system

$$\dot{y} = f_x'(t, \varphi(t, t_0, x^0, \mu), \mu) y \qquad (1.1.5)$$

that satisfies the initial condition $y_i(t_0) = 0$, $i \neq k$, $y_k(t_0) = 1$. In fact, if we knew already that φ has the differentiability properties expressed in the theorem, we may substitute it into equation (1.1.4), and differentiate the identity

$$\dot{\varphi}(t, t_0, x^0, \mu) = f(t, \varphi(t, t_0, x^0, \mu), \mu)$$

with respect to x_k^0. This yields

$$\dot{\varphi}_{x_k^0}'(t, t_0, x^0, \mu) = f_x'(t, \varphi(t, t_0, x^0, \mu), \mu) \varphi_{x_k^0}'(t, t_0, x^0, \mu),$$

showing that $\varphi_{x_k^0}'$ satisfies (1.1.5) indeed. Differentiating the identity $\varphi(t_0, t_0, x^0, \mu) = x^0$ with respect to x_k^0 we get that the k-th coordinate of $\varphi_{x_k^0}'(t_0, t_0, x^0, \mu)$ is 1 and all the other coordinates are equal to zero. System (1.1.5) plays an important role in the theory. It is called the *variational system* of (1.1.4) with respect to the solution $\varphi(t, t_0, x^0, \mu)$. The reason for this is that if we fix this solution, denote an arbitrary solution of system (1.1.4) by $x(t)$ and the difference of the two solutions, the *"variation of the solution"* by

$$z(t) = x(t) - \varphi(t, t_0, x^0, \mu),$$

then z, obviously, satisfies the system

$$
\begin{aligned}
\dot{\varphi}(t, t_0, x^0, \mu) + \dot{z}(t) &= f(t, \varphi(t, t_0, x^0, \mu) + z(t), \mu) \\
&= f(t, \varphi(t, t_0, x^0, \mu), \mu) + f_x'(t, \varphi(t, t_0, x^0, \mu), \mu) z(t) \\
&\quad + o(|z(t)|),
\end{aligned}
$$

i.e.

$$\dot{z} = f_x'(t, \varphi(t, t_0, x^0, \mu), \mu) z + o(|z|).$$

Neglecting the higher order terms we get (1.1.5) as the approximating system of the variation of the solution. Equation (1.1.5) is a homogeneous linear system (see Section 1.2), and therefore is also called the *linearization of system* (1.1.4) *around the solution* φ. Its study is essential if we want to say something about how close/far other solutions get to/from φ.

The previous theorem can be generalized.

Theorem 1.1.4. *If, besides the conditions imposed above, the right hand side of (1.1.4) is in the C^k class in x and μ, then the solution φ is also in* *the C^k class in x^0 and μ, $(k = 1, 2, 3, ...)$, and the k-th derivatives of φ are yet continuously differentiable with respect to t.*

The theory summed up to now concerns the system (1.1.1) or, if parameters are taken into account, the system (1.1.4) of first order explicit differential equations. It is easy to see that a system of explicit higher order differential equations is equivalent to a system of first order equations of type (1.1.4), the former meaning a system where the highest order derivatives of the unknown functions are expressed as functions of the rest of the data. The initial conditions attached to a higher order system are transformed into initial conditions for the equivalent first order system. We are to show this only in the case of a single higher order differential equation. Let $F \in C^0(\mathbf{R} \times X, \mathbf{R})$ where $X \subset \mathbf{R}^n$ is open and connected, and consider the equation

$$y^{(n)} = F(t, y, \dot{y}, \ddot{y}, ..., y^{(n-1)}) \tag{1.1.6}$$

which is called an *n-th order explicit ordinary differential equation*. A function $\eta : I \to \mathbf{R}$ where I is an open interval is said to be a *solution* of (1.1.6) if $\eta \in C^n(I, \mathbf{R})$, for $t \in I : (\eta(t), \dot{\eta}(t), ..., \eta^{(n-1)}(t)) \in X$, and if it satisfies (1.1.6) . An *initial value problem* attached to (1.1.6) is to find a solution that satisfies

$$\eta(t_0) = y_0, \quad \dot{\eta}(t_0) = \dot{y}_0, ..., \eta^{(n-1)}(t_0) = y_0^{(n-1)} \tag{1.1.7}$$

where $y_0, \dot{y}_0, ..., y_0^{(n-1)}$ are given constants such that $(y_0, \dot{y}_0, ..., y_0^{(n-1)}) \in X$. Now, introducing the notations $x_1 = y, x_2 = \dot{y}, ..., x_n = y^{(n-1)}$, (1.1.6) is transformed into

$$\dot{x}_1 = x_2, \quad \dot{x}_2 = x_3, \quad ..., \quad \dot{x}_{n-1} = x_n,$$

$$\dot{x}_n = F(t, x_1, x_2, ..., x_n). \tag{1.1.8}$$

Obviously, if η is a solution of (1.1.6), then the vector $(\eta, \dot{\eta}, ..., \eta^{(n-1)})$ is a solution of (1.1.8), and if φ is a solution of (1.1.8), then its first coordinate φ_1 is a solution of (1.1.6). In this equivalence the solution of (1.1.6) corresponding to the initial conditions (1.1.7) is equivalent to the solution φ of (1.1.8) that corresponds to the initial conditions

$$\varphi_1(t_0) = y_0, \qquad \varphi_2(t_0) = \dot{y}_0, \quad ..., \quad \varphi_n(t_0) = y_0^{(n-1)}. \tag{1.1.9}$$

System (1.1.8) is already of type (1.1.1). It is called the *Cauchy normal form* of equation (1.1.6). Results concerning systems given in Cauchy normal form transform in an obvious way into results concerning the higher order equation (1.1.6).

Definition 1.1.1. A function $V \in C^1(\mathbf{R} \times X, \mathbf{R})$ is called a *first integral* of system (1.1.1) if it is constant along any solution, i.e. if for every solution φ of (1.1.1) we have

$$V(t, \varphi(t)) \equiv \text{const}.$$

Definition 1.1.2. A differentiable manifold (see Definition A3.1) $M \subset \mathbf{R} \times X$ is said to be an *integral manifold* of system (1.1.1) if $(t_0, x^0) \in M$ implies that $(t, \varphi(t, t_0, x^0)) \in M$ for all t in the domain of φ.

Intuitively this definition says that motions belonging to initial conditions prescribed on the surface M stay inside M in the whole future and past. Clearly, if V is a *regular* first integral, i.e. $\operatorname{grad}_x V \neq 0$, and the constant c is taken from the range of V, then $V(t, x) = c$ is an invariant manifold of system (1.1.1). The possession of first integrals simplifies the study of the system. If we have a regular first integral, we may reduce the system to an $(n-1)$ dimensional one. Geometrically this means that we restrict the differential system to the n dimensional hypersurfaces $V = c$ in the $n + 1$ dimensional space $\mathbf{R} \times \mathbf{R}^n$. If we know n *independent* first integrals $V_1, V_2, ..., V_n$ (independent meaning that their Jacobian with respect to x is non-zero), then, in principle, we may determine any solution by solving the system of equations

$$V_k(t, x) = c_k \qquad (k = 1, 2, ..., n)$$

for x.

Theorem 1.1.5. $V \in C^1(\mathbf{R} \times X, \mathbf{R})$ *is a first integral of system (1.1.1) if and only if it satisfies the first order partial differential equation*

$$V_t' + \sum_{k=1}^{n} V_{x_k}' f_k = 0 \qquad (1.1.10)$$

(or, in concise form, $V_t' + < \operatorname{grad}_x V, f > = 0$ *where* $< .,. >$ *is the Euclidean scalar product).*

Condition (1.1.10) has a clear geometric interpretation. Assume that $n = 2$, consider the system

$$\dot{x}_1 = f_1(t, x_1, x_2), \quad \dot{x}_2 = f_2(t, x_1, x_2) \qquad (1.1.11)$$

and assume that V is a regular first integral, $V_{x_2}' \neq 0$, say. Then the normal vector of the level surface $V = c$ in the 3 dimensional space $\mathbf{R} \times \mathbf{R}^2$ is $(V_t', V_{x_1}', V_{x_2}')$, and the tangent vector of the integral curve $(t, \varphi_1(t), \varphi_2(t))$ belonging to the solution φ is $(1, f_1, f_2)$. This integral curve belongs to the level surface $V = c$ if and only if its tangent is orthogonal to the normal vector of the surface everywhere; this is expressed by (1.1.10). Because of the regularity condition, in a neighbourhood of one of its points $V = c$ determines a unique continuously differentiable function $\varphi_2(t, x_1, c)$ such that $V(t, x_1, \varphi_2(t, x_1, c)) = c$, and the given point satisfies $x_2 = \varphi(t, x_1, c)$.

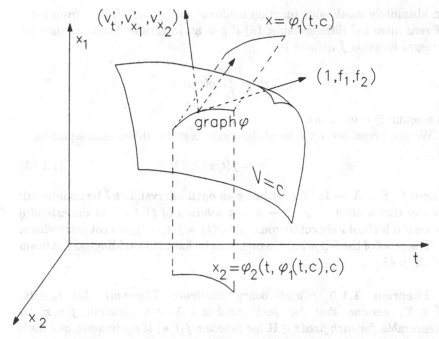

FIGURE 1.1.3. The projection of the integral curve belonging to the solution φ_1 of (1.1.12) upon the level surface $V = c$ of a first integral is the integral curve belonging to the solution φ of the system (1.1.11).

If we substitute the function φ into the first differential equation, we get a single scalar differential equation for the unknown function x_1 :

$$\dot{x}_1 = f_1(t, x_1, \varphi_2(t, x_1, c)). \tag{1.1.12}$$

If a solution of this equation is denoted by φ_1, then $\varphi(t) = (\varphi_1(t, c), \varphi_2(t, \varphi_1(t, c), c))$ is a solution of (1.1.11) such that the corresponding integral curve belongs to the level surface $V = c$; see Figure 1.1.3.

In a few cases we shall need the generalization of the concept of "solution" and a generalized existence theorem due to Carathéodory. First of all the concept of an "absolutely continuous" function is needed.

Definition 1.1.3. The function $f : [a, b] \to \mathbf{R}$ where $a < b$ are real numbers is said to be *absolutely continuous* if for arbitrary $\varepsilon > 0$ there exists a number $\delta > 0$ such that for any finite set of pairwise disjoint subintervals $(a_k, b_k) \subset [a, b]$ $(k = 1, 2, ..., n)$ with total length less than $\delta : \sum_{k=1}^{n}(b_k - a_k) < \delta$ we have $\sum_{k=1}^{n} |f(b_k) - f(a_k)| < \varepsilon$.

This Definition and the properties of absolutely continuous functions can be found in any textbook on functional analysis, for instance, in Kolmogorov-Fomin [1972]. For us the most important properties are: (i) an absolutely continuous function is almost everywhere (apart from a set of zero measure) differentiable; (ii) if g is an integrable function, then its integral function f defined by

$$f(x) := \int_a^x g(t)dt$$

is absolutely continuous.

We shall consider, for sake of simplicity, a scalar differential equation

$$\dot{x} = f(t, x) \tag{1.1.13}$$

where $f : \mathbf{R} \times X \to \mathbf{R}$, $X \subset \mathbf{R}$ being an open interval. Let I be an interval; we say that a function $\varphi : I \to X$ is *a solution of (1.1.13) in the extended sense* if it is absolutely continuous, and $\dot{\varphi}(t) = f(t, \varphi(t))$ almost everywhere.

The proof of the following Theorem can be found in Coddington-Levinson [1955] p.43.

Theorem 1.1.6 *(Carathéodory Existence Theorem). Let $t_0 \in \mathbf{R}$, $x^0 \in X$, assume that for each fixed $x \in X$ the function $f(\bullet, x)$ is measurable, for each fixed $t \in \mathbf{R}$ the function $f(t, \bullet)$ is continuous, and there exists a Lebesgue integrable function m defined on some neighbourhood of t_0 such that $|f(t, x)| \leq m(t)$ in their common domain of definition in t and for all $x \in X$. Then (1.1.13) has a solution φ in the extended sense defined on a neighbourhood of t_0 such that $\varphi(t_0) = x^0$.*

A similar theorem is valid in the n dimensional case, but obvious modifications are to be made.

Occasionally *analytic systems* will be considered, i.e. systems of the form

$$\dot{x} = f(t, x) \tag{1.1.14}$$

where $f : \mathbf{R} \times X \to \mathbf{R}^n$, $X \subset \mathbf{R}^n$ being an open and connected subset, and $f \in C^0$ is assumed to be a real analytic function in x. Theorems 1.1.1-1.1.4 are, obviously, valid in this case; moreover, one may prove (see Coddington-Levinson [1955] p.36) that if $\varphi(t, t_0, x^0)$ denotes the solution that assumes the value $x^0 \in X$ at $t = t_0$, then this is an analytic function of x^0.

1.2   Linear Systems

Linear systems are important not just for their own sake but also because often approximations and stability conditions for solutions of non-linear systems are formulated in terms of the linearization of the original system (see (1.1.5)). The proofs of the theorems in this Section can be found in Hartman [1964] and in Rouche-Mawhin [1973]. For $\alpha < \beta$ let the matrix function $A : (\alpha, \beta) \to \mathbf{R}^{n^2}$ and the vector $b : (\alpha, \beta) \to \mathbf{R}^n$ be continuous, i.e. $A \in C^0((\alpha, \beta), \mathbf{R}^{n^2})$ and $b \in C^0((\alpha, \beta), \mathbf{R}^n)$. The general form of an n dimensional inhomogeneous linear system of differential equations is

$$\dot{y} = A(t)y + b(t) \qquad (1.2.1)$$

where $y = (y_1, ..., y_n)$ and b are considered as column vectors. The system

$$\dot{y} = A(t)y \qquad (1.2.2)$$

is the *homogeneous linear system* corresponding to (1.2.1).

We shall state most of the results for systems with real valued coefficient matrix A and "forcing term" b. However, these results remain true if A and b are complex valued functions, and in some cases we have to consider them as such. The "*complexification*" is carried out in detail in Hirsch-Smale [1974] and in Rouche-Mawhin [1973]. The conditions of the existence and uniqueness Theorem 1.1.1 are, obviously, satisfied by system (1.2.1) (and (1.2.2)) in $(\alpha, \beta) \times \mathbf{R}^n$ (or $\mathbf{C}^n$). Moreover, the following Theorem holds true.

Theorem 1.2.1 *All maximal solutions of (1.2.1) (and (1.2.2)) are defined over the whole interval (α, β) of continuity of the right-hand side.*

The set of solutions of the homogeneous system (1.2.2) has a particularly simple structure.

Theorem 1.2.2. *If A is real, then the set of real solutions of (1.2.2) forms an n dimensional linear subspace of the linear space $C^1((\alpha, \beta), \mathbf{R}^n)$.*

Let the functions $y^k : (\alpha, \beta) \to \mathbf{R}^n$ form a *basis* of the space of solutions to (1.2.2), i.e. assume that $y^k \in C^1((\alpha, \beta), \mathbf{R}^n)$ satisfies (1.2.2) for $k = 1, 2, ..., n$, and that $y^1, y^2, ..., y^n$ are linearly independent. The previous Theorem means that an arbitrary linear combination of the functions y^k $(k = 1, 2, ..., n)$ is a solution of (1.2.2) and that any solution y of (1.2.2) can be expressed uniquely as a linear combination of the y^ks:

$$y = \sum_{k=1}^{n} c_k y^k \qquad (1.2.3)$$

where $(c_1, ..., c_n) \in \mathbf{R}^n$ is uniquely determined.

Let $y^k \in C^1((\alpha, \beta), \mathbf{R}^n)$ $(k = 1, 2, ..., n)$ be arbitrary solutions of (1.2.2). Their *Wronskian* is, by definition, the determinant of the $n \times n$ matrix whose column vectors are y^k :

$$W(y^1, ..., y^n) := \det[y^1, ..., y^n].$$

Theorem 1.2.3. *The solutions y^k $(k = 1, 2, ..., n)$ of (1.2.2) form a basis of the space of solutions if and only if their Wronskian is nowhere zero:*

$$W(y^1(t), ..., y^n(t)) \neq 0, \qquad t \in (\alpha, \beta). \tag{1.2.4}$$

We may consider along with (or instead of) (1.2.2) the *matrix differential equation*

$$\dot{Y} = A(t)Y. \tag{1.2.5}$$

A matrix function $Y : (\alpha, \beta) \to \mathbf{R}^{n^2}$ is a solution of this equation if $Y \in C^1$ and makes an identity out of (1.2.5) if substituted. The following Theorem is known as *Liouville's formula.*

Theorem 1.2.4. *Let Y be a solution of (1.2.5), and $t_0 \in (\alpha, \beta)$. Then for $t \in (\alpha, \beta)$*

$$\det Y(t) = \det Y(t_0) \exp \int_{t_0}^{t} \operatorname{Tr} A(s) ds \tag{1.2.6}$$

where if the entries of A are denoted by a_{ik}, then $\operatorname{Tr} A(s) := \sum_{i=1}^{n} a_{ii}(s)$.

A solution of (1.2.5) is called a *matrix solution* of (1.2.2). From (1.2.6) we have that if a matrix solution is regular at some $t_0 \in (\alpha, \beta)$, then it is regular for all $t \in (\alpha, \beta)$. Clearly, columns of a matrix solution are solutions of (1.2.2). A regular matrix solution is called a *fundamental matrix.* It is easy to see that if Φ is a fundamental matrix, then any solution Y of (1.2.5) can be expressed uniquely as

$$Y(t) = \Phi(t)C \tag{1.2.7}$$

where C is a constant $n \times n$ matrix. n solutions $y^1, y^2, ..., y^n$ of (1.2.2) form a basis if and only if the matrix $Y = [y^1, ..., y^n]$ is a fundamental matrix. Thus, we see that if (1.2.4) holds at a $t_0 \in (\alpha, \beta)$, then it holds everywhere in (α, β) . We may put now (1.2.3) in the following equivalent form. If y is an arbitrary solution of (1.2.2) and Φ is a fundamental matrix, then there is a unique constant vector $c = (c_1, ..., c_n)$ such that

$$y(t) = \Phi(t)c. \tag{1.2.8}$$

The fundamental matrix that is equal to the $n \times n$ unit matrix I at $t_0 \in (\alpha, \beta)$ will be denoted by $\Phi(t, t_0)$, i.e.

$$\dot{\Phi}(t, t_0) \equiv A(t)\Phi(t, t_0), \qquad \Phi(t_0, t_0) = I. \tag{1.2.9}$$

If this fundamental matrix is known, then the solution of (1.2.2) satisfying the initial condition $y(t_0) = y^0$, denoted by $\varphi(t, t_0, y^0)$ as before, is given by

$$\varphi(t, t_0, y^0) = \Phi(t, t_0)y^0.$$

Applying this notation and considering t_0 and $s \in (\alpha, \beta)$ fixed for a moment, we have

$$\Phi(t, t_0) = \Phi(t, s)\Phi(s, t_0) \qquad (1.2.10)$$

since by (1.2.7) both sides are matrix solutions, and at $t = s$ they assume the same value.

We turn now to the inhomogeneous system (1.2.1). The so-called "method of the variation of constants" yields the next result.

Theorem 1.2.5. *The solution φ of (1.2.1) satisfying the initial condition $y(t_0) = y^0$ is given by*

$$\varphi(t, t_0, y^0) = \Phi(t, t_0) \left(y^0 + \int_{t_0}^{t} \Phi^{-1}(s, t_0)b(s)ds \right) \qquad (1.2.11)$$

where $\Phi(t, t_0)$ is defined by (1.2.9).

Applying (1.2.10) we may put the last formula into the equivalent form

$$\varphi(t, t_0, y^0) = \Phi(t, t_0)y^0 + \int_{t_0}^{t} \Phi(t, s)b(s)ds. \qquad (1.2.12)$$

Now we are going to treat *linear systems with constant coefficients*, i.e. the important special case when the coefficient matrix A of system (1.2.1) resp. (1.2.2) is constant. Consider first the homogeneous linear system

$$\dot{y} = Ay \qquad (1.2.13)$$

where A is an $n \times n$ constant matrix. The fundamental matrix that assumes the unit matrix I at $t = 0$ of this system is

$$\Phi(t, 0) = \exp(At) := \sum_{k=0}^{\infty} (1/k!)A^k t^k \qquad (1.2.14)$$

where $A^0 = I$. The series on the right-hand side is convergent for every matrix A and for all $t \in \mathbf{R}$, the sum is differentiable, and

$$\frac{d}{dt} \exp(At) = A \exp(At)$$

(see Hirsch-Smale [1974] and Yakubovich-Starzhinskij [1972]). The exponential of a square matrix satisfies the properties

$$\exp O = I, \qquad \det \exp A \neq 0, \qquad (\exp A)^{-1} = \exp(-A);$$

if the square matrices A and B commute, then

$$\exp A \exp B = \exp B \exp A = \exp(A + B). \qquad (1.2.15)$$

In Appendix 1 we show how $\exp(At)$ can be calculated with the help of the Hermite interpolation polynomial of $\exp(\lambda t)$ corresponding to the spectrum of A. Denote the different eigenvalues of A by $\lambda_1, \lambda_2, ..., \lambda_r$, and assume that their respective multiplicities in the *minimal polynomial* of A are $m_1, ..., m_r$ where $\sum_{k=1}^r m_k = m \le n$. From (A1.7) we see that the elements of the matrix $\exp(At) = [e_{ij}(t)]$ are of the type

$$e_{ij}(t) = \sum_{k=1}^r p_{ijk}(t)e^{\lambda_k t} \qquad (i, j = 1, 2, ..., n) \qquad (1.2.16)$$

where p_{ijk} is a polynomial of degree $(m_k - 1)$ at most $(k = 1, 2, ..., r)$. Let A be real. Since complex eigenvalues may occur even in this case, the elements of the matrix $\exp(At)$ may, seemingly, be complex valued functions. However, as it is shown in Corollary A1.7, they are, in fact, real functions due to the fact that in case of a real matrix A the complex eigenvalues occur in conjugate pairs whose multiplicities are equal. In case A is real, and a complex valued solution turns up due to complex eigenvalues, we can always find an independent conjugate complex solution, and by linear combination we can generate two independent real solutions out of them. (The conjugate of a solution is also a solution, and they are linearly independent except when the original solution is a real one multiplied by a constant complex factor.)

Example 1.2.1. Find the fundamental matrix $\exp(At)$ of the system

$$\begin{aligned} \dot{y}_1 &= 3y_1 - 3y_2 + 2y_3, \\ \dot{y}_2 &= -y_1 + 5y_3 - 2y_3, \\ \dot{y}_3 &= -y_1 + 3y_2 \end{aligned}$$

where we denote the coefficient matrix by A. In Example A1.1 we show that the minimal polynomial of this matrix is $\Delta(\lambda) = \lambda^2 - 6\lambda + 8$. Its roots are 4 and 2 (the latter is a double root of the characteristic polynomial). The Hermite interpolation polynomial of $\exp(\lambda t)$ corresponding to the spectrum of A is denoted by $h(\lambda) = h_1 \lambda + h_0$, and the coefficients are to be determined from the conditions $4h_1 + h_0 = \exp(4t)$, $2h_1 + h_0 = \exp(2t)$. They are $h_1(t) = (\exp(4t) - \exp(2t))/2$, $h_0(t) = -\exp(4t) + 2\exp(2t)$. Thus,

$$e^{At} = h(A) = \begin{pmatrix} (e^{4t} + e^{2t})/2 & -3(e^{4t} - e^{2t})/2 & e^{4t} - e^{2t} \\ -(e^{4t} - e^{2t})/2 & (3e^{4t} - e^{2t})/2 & e^{4t} - e^{2t} \\ -(e^{4t} - e^{2t})/2 & 3(e^{4t} - e^{2t})/2 & -e^{4t} + 2e^{2t} \end{pmatrix}.$$

Example 1.2.2. Find the fundamental matrix $\exp(At)$ of the system

$$\dot{y}_1 = -7y_1 + y_2, \qquad \dot{y}_2 = -2y_1 - 5y_2$$

where we denote the coefficient matrix by A. The characteristic polynomial of A is $D(\lambda) = \lambda^2 + 12\lambda + 37$. The eigenvalues are $\lambda_{1,2} = -6 \pm i$. Thus, the minimal polynomial is identical to D. The Hermite interpolation polynomial of $\exp(\lambda t)$ corresponding to the spectrum of A is $h(\lambda) = h_1\lambda + h_0$. Its coefficients satisfy the conditions

$$h_1(-6+i) + h_0 = \exp((-6+i)t), \quad h_1(-6-i) + h_0 = \exp((-6-i)t).$$

Hence, $h_1(t) = e^{-6t}(e^{it} - e^{-it})/(2i) = e^{-6t}\sin t$,
$h_0(t) = e^{-6t}(6\sin t + \cos t)$.

$$e^{At} = h(A) = \begin{pmatrix} -\sin t + \cos t & \sin t \\ -2\sin t & \sin t + \cos t \end{pmatrix} e^{-6t}.$$

Clearly, for arbitrary $t, t_0 \in \mathbf{R}$, the matrices At and At_0 commute. Thus, by (1.2.15) and (1.2.7) the fundamental matrix of system (1.2.13) that assumes the unit matrix I at t_0 is

$$\Phi(t, t_0) = \exp(A(t - t_0)),$$

and the solution satisfying the initial condition $y(t_0) = y^0$ is

$$\varphi(t, t_0, y^0) = \exp(A(t - t_0))y^0. \tag{1.2.17}$$

Consider now the inhomogeneous linear system with constant coefficients

$$\dot{y} = Ay + b(t) \tag{1.2.18}$$

where A is an $n \times n$ constant matrix as before, and $b \in C^0(\alpha, \beta)$. Applying the general formula (1.2.12), the solution of (1.2.18) that satisfies the initial condition $y(t_0) = y^0$ for some $t_0 \in (\alpha, \beta)$ is

$$\varphi(t, t_0, y^0) = \exp(A(t - t_0))y^0 + \int_{t_0}^{t} \exp(A(t - s))b(s)ds. \tag{1.2.19}$$

There is an easier way of determining a fundamental matrix solution of system (1.2.13) in the important special case when each eigenvalue of the coefficient matrix A has as many independent eigenvectors as its *multiplicity in the characteristic polynomial*. Denote the eigenvalues of A by $\lambda_1, \lambda_2, ..., \lambda_n$ where now they may be equal, and assume that one may choose non-zero vectors $s^1, s^2, ..., s^n$ in such a way that s^k is an eigenvector corresponding to λ_k and the system $s^1, ..., s^n$ is linearly independent. In other words, this means that the matrix A is diagonalizable. This is the case, e.g. when all eigenvalues are simple roots of the characteristic polynomial, i.e. $\lambda_i \neq \lambda_k$, $i \neq k$. Then it is easy to see that the system

$$s^1 \exp(\lambda_1 t), \quad s^2 \exp(\lambda_2 t), \quad ..., \quad s^n \exp(\lambda_n t) \tag{1.2.20}$$

is a basis of the space of solutions of system (1.2.13) . We note, finally, that the Cauchy normal form (see (1.1.8)) of an n-th order linear differential equation

$$y^{(n)} + a_{n-1}(t)y^{(n-1)} + \cdots + a_1(t)\dot{y} + a_0(t)y = b(t) \qquad (1.2.21)$$

is an n-dimensional linear system. The results concerning systems can be interpreted easily for the scalar differential equation (1.2.21). In particular, n solutions $y^k \in C^n(\alpha, \beta)$ $(k = 1, 2, ..., n)$ $((\alpha, \beta)$ being the interval of continuity of the coefficients a_i and the right-hand side b) form a *basis* of the space of solutions of (1.2.21) if and only if their *Wronskian* (cf. (1.2.4))) is non-zero. The Wronskian is now

$$W(y^1, y^2, ..., y^n) = \det \begin{pmatrix} y^1 & y^2 & \cdot & \cdot & \cdot & y^n \\ \dot{y}^1 & \dot{y}^2 & \cdot & \cdot & \cdot & \dot{y}^n \\ \cdot & \cdot & & & & \cdot \\ \cdot & \cdot & & & & \cdot \\ \cdot & \cdot & & & & \cdot \\ y^{1(n-1)} & y^{2(n-1)} & \cdot & \cdot & \cdot & y^{n(n-1)} \end{pmatrix} . \qquad (1.2.22)$$

If in (1.2.21) the coefficients a_i $(i = 0, 1, ..., n-1)$ are constants, then the *characteristic polynomial* of the corresponding homogeneous differential equation

$$y^{(n)} + a_{n-1}y^{(n-1)} + \cdots + a_1\dot{y} + a_0 y = 0 \qquad (1.2.23)$$

is

$$D(\lambda) = \lambda^n + a_{n-1}\lambda^{n-1} + \cdots + a_1\lambda + a_0$$

as one can check easily by taking into consideration that the coefficient matrix of the Cauchy normal form is

$$\begin{pmatrix} 0 & 1 & 0 & \cdots & 0 \\ 0 & 0 & 1 & \cdots & 0 \\ \cdot & \cdot & \cdot & \cdots & \cdot \\ -a_0 & -a_1 & -a_2 & \cdots & -a_{n-1} \end{pmatrix} .$$

1.3 Dynamical Systems

A *dynamical system* or a *flow* (in the mathematical sense) is modelling the stationary flow of an incompressible fluid. If we look at a slow river for a not too long period of time, we see that the motion of fluid particles is "smooth", and that if a certain position and a certain duration of time is fixed, then any fluid particle occupying that position at any moment will get to a fixed position independent of the starting moment after the fixed duration has passed. In this case, clearly, the velocity field of the flow does not change in time, i.e. it is independent of time. Most Laws of Nature are independent of time: if they act, they act identically in any moment of time. Therefore, if they are considered deterministic and are modelled by differential equations, the result is, mostly, a dynamical system.

Let the subset $X \subset \mathbf{R}^n$ be open and connected. We recall that a *diffeomorphism* of X is a one-to-one (invertible) mapping of X into itself which as well as its inverse is at least once continuously differentiable.

Definition 1.3.1. Let the subset $X \subset \mathbf{R}^n$ be open and connected; the map $\varphi : \mathbf{R} \times X \to X$ is called a *dynamical system* (or a *flow*) *on* X if

(i) $\varphi \in C^1$, and for any fixed $t \in \mathbf{R}$ the mapping $\varphi_t := \varphi(t, .)$ is a diffeomorphism of X into itself;

(ii) φ_0 is the identity, i.e. $\varphi_0(x) = \varphi(0, x) = x$ for all $x \in X$;

(iii) $\varphi_t \circ \varphi_s = \varphi_{t+s}$, i.e. $\varphi(t, \varphi(s, x)) = \varphi(t + s, x)$ for all $t, s \in \mathbf{R}$ and $x \in X$.

Note that the conditions imply $\varphi_t \circ \varphi_{-t} = \varphi_{-t} \circ \varphi_t = I$, the identity. This means that φ_t is a smooth one parameter group of diffeomorphisms of X with respect to composition.

An example of a dynamical system is the solution of the homogeneous linear system with constant coefficients (1.2.13) as a function of the initial value x for fixed initial time, say $t_0 = 0$:

$$\varphi(t, x) = \varphi(t, 0, x) = e^{At} x. \tag{1.3.1}$$

This function, obviously, satisfies the conditions of the previous Definition: (iii) follows from the identity $\exp(At) \exp(As) = \exp(A(t + s))$ which is true because At and As commute.

A few expressions will be introduced in connection with flows now. The set $\{\varphi(t, x) \in X : t \in \mathbf{R}\}$ is called the *path (trajectory, orbit)* of $x \in X$. For fixed $x \in \mathbf{R}$ the function $\varphi(., x)$ is called the *motion* of x. If for some $x^0 \in X : \varphi(t, x^0) = x^0$ for all $t \in \mathbf{R}$, we call x^0 a *fixed point* (or an *equilibrium*) of the flow. If for some $x^p \in X$ there exists a $T > 0$ such that $\varphi(T, x^p) = x^p$, then we say that x^p is a *periodic point*, and T is a *period*. If x^p is a periodic point with period T, then the motion of x^p is a periodic function. This follows easily from (iii) of Definition 1.3.1, as $\varphi(t + T, x^p) = \varphi(t, \varphi(T, x^p)) = \varphi(t, x^p)$ for all $t \in \mathbf{R}$. A subset M of X is called an *invariant set* of the flow if $x \in M$ implies that its path remains in M, i.e. $\varphi(t, x) \in M$ for all $t \in \mathbf{R}$. The subset $P \subset X$ is called *positively invariant* if $x \in P$ implies $\varphi(t, x) \in P$ for all $t > 0$. A *negatively invariant* set is defined analogously. The set $\{\varphi(t, x) \in X : t \in \mathbf{R}_+\}$ is called the *positive semitrajectory* of x. The *negative semitrajectory* is defined analogously.

Our first example (1.3.1) of a dynamical system was the solution of a homogeneous linear system of differential equations with constant coefficients as a function of the initial value x assumed at $t = 0$. Dynamical systems, in general, are generated by *autonomous differential equations*, the general form of which is

$$\dot{x} = f(x) \tag{1.3.2}$$

where $f \in C^1(X, \mathbf{R}^n)$, $X \subset \mathbf{R}^n$ being an open and connected subset. This system is called autonomous because the right-hand side does not depend on time, i.e. it does not depend on external conditions which may vary in time. The velocity $\dot{x}$ of the motion governed by the differential equation (1.3.2) is determined completely by the state x of the system. Let us denote the solution of (1.3.2) that assumes the initial value $x^0 \in X$ at $t_0 = 0$ by $\varphi(t, x^0) := \varphi(t, 0, x^0)$. By substitution into the differential equation one can see that for arbitrary $t_0 \in \mathbf{R}$ the function $\varphi(t - t_0, x^0)$ is also a solution, and this latter solution assumes the value x^0 at $t = t_0 : \varphi(t - t_0, x^0)|_{t=t_0} = \varphi(0, x^0) = x^0$. Because of uniqueness,

$$\varphi(t - t_0, x^0) = \varphi(t, t_0, x^0). \tag{1.3.3}$$

Fix $x^0 \in X$ in the phase space. All the trajectories that pass though x^0 belong to solutions with initial values $t_0 \in \mathbf{R}$ and x^0. But we see that the parametric equations of these trajectories are $x = \varphi(t - t_0, x^0)$, $t_0 \in \mathbf{R}$. These equations (for different values of t_0), clearly, describe the same curve in the phase space. Thus, we got that *through each point $x^0 \in X$ there passes one and only one path* though this path belongs to an infinite number of solutions (1.3.3) where x^0 is fixed and $t_0 \in \mathbf{R}$. Phrasing the same fact in a geometric way: for fixed $x^0 \in X$ and varying $t_0 \in \mathbf{R}$, all the integral curves of the solutions $\varphi(t, t_0, x^0)$ project onto the same path in the phase space; see Figure 1.3.1. Assume that *all solutions of* (1.3.2) *are defined for all $t \in \mathbf{R}$*. Then the solutions as functions of the initial values assumed at $t = 0$, i.e. the function $\varphi : \mathbf{R} \times X \to X$, satisfies the requirements of Definition 1.3.1. (i) is clear because of Theorem 1.1.3 and the uniqueness of the trajectories established above. Condition (ii) is obviously true. In order to see (iii) note that for fixed $s \in \mathbf{R}$, the solution $\varphi(t + s, x)$ assumes the value $\varphi(s, x)$ at $t = 0$, and so does $\varphi(t, \varphi(s, x))$. Because of uniqueness of the solutions of initial value problems,

$$\varphi(t + s, x) = \varphi(t, \varphi(s, x))$$

for all $t, s \in \mathbf{R}$ and $x \in X$. This way system (1.3.2) generates a flow provided that all its solutions are defined on $\mathbf{R}$.

The restrictive assumption that all solutions must be defined on the whole real line is not essential in the case $X = \mathbf{R}^n$. If the solutions of (1.3.2) have bounded maximal domains, we introduce the "arc length" of the integral curve as a new parameter:

$$s = \int_0^t \left(1 + f^2(x(\tau))\right)^{1/2} d\tau. \tag{1.3.4}$$

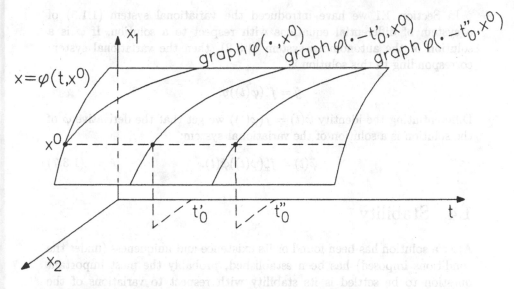

FIGURE 1.3.1. Integral curves and path of solutions belonging to the same initial value x^0 in case of an autonomous system.

This way system (1.3.2) transforms into

$$\frac{dx}{ds} = f(x)\big(1 + f^2(x)\big)^{-1/2}. \tag{1.3.5}$$

This is a different system of autonomous differential equations, but it is equivalent to (1.3.2) in the sense that both have the *same trajectories* including the direction of motion along the trajectories (s is increasing with t). Now, the right-hand side of (1.3.5) is bounded in the whole space $X = \mathbf{R}^n : |f(x)(1 + f^2(x))^{-1/2}| < 1$. By a theorem of Wintner (see Hartman [1964] p.29) all solutions are defined on the whole real line. Thus, allowing a possible transformation (1.3.4) into (1.3.5) we may say that every autonomous system (1.3.2) whose right-hand side is continuously differentiable in the whole space $\mathbf{R}^n$ generates a flow.

Conversely, let a dynamical system be given in the sense of Definition 1.3.1 with the additional assumption $\varphi \in C^2$ and define

$$f(x) := \dot{\varphi}(t, x)|_{t=0}. \tag{1.3.6}$$

Then the autonomous system $\dot{x} = f(x)$ generates the flow φ in the sense described above because

$$\dot{\varphi}(t, x) = \dot{\varphi}(\tau + t, x)|_{\tau=0} = \dot{\varphi}(\tau, \varphi(t, x))|_{\tau=0} = f(\varphi(t, x)).$$

Thus, every C^2 flow generates a C^1 autonomous system whose solutions are the motions of the flow.

In Section 1.1 we have introduced the variational system (1.1.5) of a system of differential equations with respect to a solution. If φ is a solution of the autonomous system (1.3.2), then the variational system corresponding to this solution is

$$\dot{y} = f'_x(\varphi(t))y.$$

Differentiating the identity $\dot{\varphi}(t) = f(\varphi(t))$ we get that the derivative $\dot{\varphi}$ of the solution is a solution of the variational system:

$$\ddot{\varphi}(t) = f'_x(\varphi(t))\dot{\varphi}(t). \tag{1.3.7}$$

1.4   Stability

After a solution has been found or its existence and uniqueness (under the conditions imposed) has been established, probably the most important question to be settled is its stability with respect to variations of the initial conditions. A solution corresponding to prescribed initial conditions is "stable" if small changes in these latter cause only small changes in the solution in the whole future, or even if solutions corresponding to varied initial conditions tend to the desired one as time tends to infinity. We are going to give here, first, the most important definitions, and then the basic facts about "linearized stability". Liapunov's direct method will be surveyed in the next Section.

Consider the system

$$\dot{x} = f(t, x) \tag{1.4.1}$$

where $f \in C^0(\mathbf{R}_+ \times X)$, $f'_x \in C^0(\mathbf{R}_+ \times X)$ and $X \subset \mathbf{R}^n$ is an open and connected subset. (Instead of $\mathbf{R}_+ = [0, \infty)$ we could use any interval not bounded from above.) Let $\psi : \mathbf{R}_+ \to X$ be a solution of (1.4.1) and denote its initial value at an arbitrary $t_0 \geq 0$ by $\psi^0 = \psi(t_0)$. Often the problem of stability of an arbitrary solution ψ defined in the whole future is reduced to the stability of a constant solution, an *equilibrium* of a system. Though we are not following this road, we are showing the procedure of reduction. Introduce new coordinates by $y = x - \psi(t)$. If a solution is substituted for x, then the function y is the *variation of the solutions* with respect to the fixed one ψ. If x satisfies (1.4.1), then we have for y,

$$\dot{y} = f(t, \psi(t) + y) - f(t, \psi(t)), \tag{1.4.2}$$

i.e. y satisfies this differential system which has $y \equiv 0$ as a constant solution. The zero solution of (1.4.2), clearly, corresponds to the solution ψ of (1.4.1). The absolute value of a solution of (1.4.2) is equal to the absolute value of the difference of the corresponding solution x and $\psi : |y| = |x - \psi|$. Therefore the stability of the zero solution of (1.4.2) takes care of the stability of ψ. We are not applying this reduction because in the

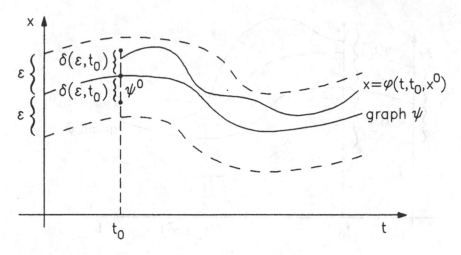

FIGURE 1.4.1. Stable solution ψ in the Liapunov sense.

important special case when (1.4.1) is autonomous, this method reduces the problem of stability of a non-constant solution ψ to the stability of the zero solution of the *non-autonomous system* (1.4.2). However, since equilibria are important in applications, and they are the simplest solutions with a *compact path*, their stability will be treated separately where it is appropriate.

Definition 1.4.1. The solution $\psi : \mathbf{R}_+ \to X$ of (1.4.1) is said to be *stable in the Liapunov sense* if to every $\varepsilon > 0$ and $t_0 \geq 0$ there belongs a $\delta(\varepsilon, t_0) > 0$ such that for all $|x^0 - \psi^0| < \delta(\varepsilon, t_0)$ the solution $\varphi(t, t_0, x^0)$ is defined in $[t_0, \infty)$, and for all $t > t_0$

$$|\varphi(t, t_0, x^0) - \psi(t)| < \varepsilon \qquad (1.4.3)$$

(see Figure 1.4.1).

Definition 1.4.2. We say that the solution ψ is *uniformly stable* if it is stable in the Liapunov sense, and δ of the previous definition does not depend on t_0.

In the following definitions $B(a, \delta)$ is the *ball* with centre at a and radius δ, i.e.

$$B(a, \delta) = \{x \in \mathbf{R}^n : |x - a| < \delta\}. \qquad (1.4.4)$$

Definition 1.4.3. We say that the solution $\psi : \mathbf{R}_+ \to X$ of (1.4.1) is *attractive* if to every $t_0 \geq 0$ there belongs an $\eta(t_0) > 0$ such that $|x^0 - \psi^0| < \eta(t_0)$ implies

$$\lim_{t \to \infty} |\varphi(t, t_0, x^0) - \psi(t)| = 0, \qquad (1.4.5)$$

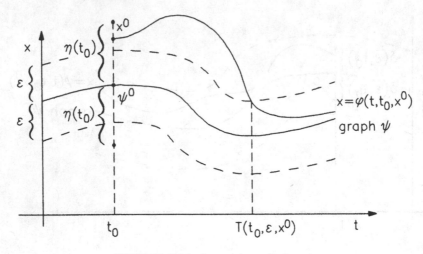

FIGURE 1.4.2. Attractive solution ψ.

i.e. to every $\varepsilon > 0$ and $x^0 \in B(\psi^0, \eta(t_0))$ there belongs a $T(t_0, \varepsilon, x^0) > 0$ such that for all $t \geq t_0 + T(t_0, \varepsilon, x^0)$ we have $|\varphi(t, t_0, x^0) - \psi(t)| < \varepsilon$ (see Figure 1.4.2).

Definition 1.4.4. We say that the solution ψ is *uniformly attractive* if it is attractive, and in the previous definition η does not depend on t_0 and T does not depend on t_0 and $x^0 \in B(\psi^0, \eta)$.

Definition 1.4.5. We say that the solution ψ is *asymptotically stable* if it is stable in the Liapunov sense and it is attractive.

Definition 1.4.6. We say that the solution ψ is *uniformly asymptotically stable* if it is uniformly stable and uniformly attractive.

From Definition 1.4.3 it is clear that the solution ψ "attracts" the solutions from a certain distance, from a neighbourhood, and this neighbourhood may vary in time. The maximal open set from which at time t_0 the attractive solution attracts is called the *basin* (or the *domain of attractivity*) of ψ at time t_0:

$$A(t_0) = \left\{ x^0 \in X : \lim_{t \to \infty} |\varphi(t, t_0, x^0) - \psi(t)| = 0 \right\}. \qquad (1.4.6)$$

If for all $t_0 \geq 0$ the basin $A(t_0) = X = \mathbf{R}^n$, then we say that ψ is *globally attractive*. If ψ is stable in the Liapunov sense and globally attractive, it is said to be *globally asymptotically stable*.

Though the stability concepts introduced above may seem far too many, their number could be increased almost arbitrarily. For instance, they are not enough for the study of the stability of periodic solutions belonging to autonomous systems. We are going to introduce further necessary stability concepts in Chapter 5.

We are turning now to the problem of stability in case of linear systems. First, we show that it is reasonable to speak about the stability of a linear *system*. In case of a non-linear system one solution may be stable, and the other not. This cannot happen to linear systems.

Consider the system

$$\dot{y} = A(t)y + b(t) \tag{1.4.7}$$

where $A \in C^0(\mathbf{R}_+, \mathbf{R}^n)$, $b \in C^0(\mathbf{R}_+, \mathbf{R}^n)$.

Theorem 1.4.1. *If one solution of (1.4.7) is stable in the Liapunov sense (resp. asymptotically stable), then all the solutions are stable (resp. asymptotically stable). In this case we say that the system is stable in the Liapunov sense (resp. asymptotically stable).*

Proof. We know (see Theorem 1.2.1) that all solutions are defined over $\mathbf{R}_+$. Let us fix a solution ψ, and transform the coordinate system by $x = y - \psi(t)$. If y satisfies (1.4.7), then x clearly is a solution of the corresponding homogeneous system

$$\dot{x} = A(t)x, \tag{1.4.8}$$

and vice versa. ψ is stable in the Liapunov sense (resp. asymptotically stable) if and only if the solution $x = 0$ of (1.4.8) has the corresponding property. $\square$

Corollary 1.4.2. *If system (1.4.8) is stable in the Liapunov sense (resp. asymptotically stable), then for arbitrary $b \in C^0(\mathbf{R}_+, \mathbf{R}^n)$, system (1.4.7) is stable (resp. asymptotically stable).*

Corollary 1.4.3. *If system (1.4.7) is asymptotically stable, then it is globally asymptotically stable.*

Proof. This follows from the fact that the solutions of (1.4.8) form an n dimensional linear space, i.e. every solution can be expressed as a linear combination of a basis (see (1.2.3)). Thus, if "small solutions" of (1.4.8) tend to zero then "large solutions" do the same. $\square$

It is not easy at all to determine the stability of a general linear system. We are supplying a sufficient criterion of asymptotic stability in the next section. The situation is different in the case of a linear system with constant coefficients. In this important particular case the problem reduces to an algebraic problem, namely, to the determination of the signs of the real parts of the eigenvalues.

The following terminology will be used. A *polynomial* is said to be *stable* if all its roots have negative real parts. A *square matrix* is said to be *stable* if all its eigenvalues have negative real parts. The proofs of the following theorems can be found in Rouche-Mawhin [1973].

Theorem 1.4.4. *The linear system with constant coefficients*

$$\dot{x} = Ax \qquad (1.4.9)$$

is asymptotically stable if and only if the coefficient matrix A is stable.

Recall that Corollaries 1.4.2 and 1.4.3 remain, naturally, valid in this special case.

There is an important estimate related to the previous Theorem which will be used occasionally. If the matrix A is stable, and the number $\lambda > 0$ is such that for all the eigenvalues λ_k of A we have $\operatorname{Re} \lambda_k < -\lambda$, then a constant $K > 0$ exists such that for $t \geq 0$

$$|\exp(At)| \leq K \exp(-\lambda t). \qquad (1.4.10)$$

Here $|\exp(At)|$ is the *norm* of this matrix (see (A1.19)). The inequality (1.4.10) is a simple consequence of (1.2.16). Because of (1.4.10) the solutions of the system (1.4.9) tend to zero exponentially as t tends to infinity. Therefore, we say that the system or the trivial solution is *"exponentially asymptotically stable"*.

Theorem 1.4.5. *The linear system with constant coefficients (1.4.9) is stable in the Liapunov sense (but not asymptotically) if and only if all the eigenvalues of the coefficient matrix A have non-positive real parts, there are eigenvalues with zero real parts, but their multiplicity in the minimal polynomial of A is one (see Appendix 1).*

Theorem 1.4.6. *If the coefficient matrix A of system (1.4.9) either has at least one eigenvalue with a positive real part or at least one eigenvalue has a zero real part, and its multiplicity in the minimal polynomial is greater than one, then the system is unstable.*

Thus, we see that the asymptotic stability of system (1.4.9) is determined by the stability of the matrix A, i.e. by the stability of its characteristic polynomial.

We are now supplying some criteria for the stability of a polynomial. These can be used well, especially for a relatively low dimensional system. Proofs can be found in Demidovich [1967$_a$]. Consider the n-th degree polynomial

$$p(\lambda) = a_n \lambda^n + a_{n-1} \lambda^{n-1} + \cdots + a_1 \lambda + a_0 \qquad (1.4.11)$$

where $n \geq 1$, $a_n > 0$, and the coefficients are real.

Theorem 1.4.7. *If the polynomial p is stable, then all its coefficients are positive: $a_k > 0 (k = 0, 1, ..., n)$.*

Note that in the $n = 2$ case this necessary condition is also sufficient for the stability. If $n \geq 3$, then this is no longer true. For the general case we have the basic *Routh-Hurwitz criterion*.

Theorem 1.4.8. *Assume that every coefficient a_k of the polynomial p is positive, and consider the Hurwitz matrix of the polynomial*

$$H_p = \begin{pmatrix} a_1 & a_0 & 0 & 0 & ... & 0 \\ a_3 & a_2 & a_1 & a_0 & ... & 0 \\ . & . & . & . & ... & . \\ a_{2n-1} & a_{2n-2} & a_{2n-3} & a_{2n-4} & ... & a_n \end{pmatrix}$$

where $a_k = 0$ if $k > n$; the polynomial is stable if and only if all the leading principal minors of H_p are positive, i.e. (in view of the positivity of all the coefficients, in particular, a_1 and a_n)

$$\Delta_2 = \begin{vmatrix} a_1 & a_0 \\ a_3 & a_2 \end{vmatrix} > 0, ..., \Delta_{n-1} > 0$$

where Δ_{n-1} is the minor of the last element of the last row.

In particular, the cubic polynomial

$$a_3\lambda^3 + a_2\lambda^2 + a_1\lambda + a_0, \qquad a_3 > 0,$$

is stable if and only if a_2, a_1, $a_0 > 0$, and

$$a_1a_2 - a_0a_3 > 0. \qquad (1.4.12)$$

In order to illustrate the behaviour of the solutions of a linear system with constant coefficients in case of stability, resp. instability and for the sake of future reference we are presenting a complete classification of two dimensional linear systems according to the character of the *trivial equilibrium point* $(0,0)$. Consider the system with real coefficients

$$\dot{x}_1 = a_{11}x_1 + a_{12}x_2, \qquad \dot{x}_2 = a_{21}x_1 + a_{22}x_2. \qquad (1.4.13)$$

The constant coefficient matrix will be denoted by $A = [a_{ik}]$ $(i, k = 1, 2)$. The characteristic polynomial of this system is

$$D(\lambda) = \lambda^2 - (a_{11} + a_{22})\lambda + a_{11}a_{22} - a_{12}a_{21}$$

or

$$D(\lambda) = \lambda^2 - \lambda \operatorname{Tr} A + \det A. \qquad (1.4.14)$$

According to the remark following Theorem 1.4.7 *this system is asymptotically stable if and only if the trace of the matrix A is negative, and its determinant is positive:*

$$\text{Tr } A < 0, \qquad \det A > 0. \qquad (1.4.15)$$

According to the possible values of the two eigenvalues λ_1 and λ_2 of the matrix A, the following cases may be distinguished. A has (i) real eigenvalues of different signs, (ii) real non-zero eigenvalues of equal signs, (iii) a pair of complex conjugate eigenvalues with non-zero real part, (iv) a pair of pure imaginary eigenvalues, (v) a zero and a non-zero eigenvalue and (vi) a double zero eigenvalue. If $\lambda_1 \neq \lambda_2$, then applying formula (1.2.20) the "general solution" of system (1.4.13) can be written in the form

$$x(t) = c_1 s^1 \exp(\lambda_1 t) + c_2 s^2 \exp(\lambda_2 t)$$

where s^i is an eigenvector belonging to the eigenvalue λ_i $(i = 1, 2)$, and c_1, c_2 are arbitrary constants. The same form applies if $\lambda_1 = \lambda_2$, but it has two independent eigenvectors s^1, s^2. If we denote the coordinates of the system based on the basis s^1, s^2 by (ξ_1, ξ_2), i.e.

$$x = \xi_1 s^1 + \xi_2 s^2$$

(this is a not necessarily orthogonal Cartesian system of coordinates), then the parametric form of the solution is

$$\xi_1(t) = c_1 \exp(\lambda_1 t), \qquad \xi_2(t) = c_2 \exp(\lambda_2 t). \qquad (1.4.16)$$

If A has a double eigenvalue $\lambda = \lambda_1 = \lambda_2$, and it is not diagonalizable, then by linear transformation it is similar to the matrix

$$A \sim \begin{pmatrix} \lambda & 0 \\ 1 & \lambda \end{pmatrix}.$$

Denoting the coordinates of the system in which A assumes this form again by (ξ_1, ξ_2), an easy calculation yields the "general solution"

$$\xi_1(t) = c_1 \exp(\lambda t), \qquad \xi_2(t) = (c_1 t + c_2) \exp(\lambda t). \qquad (1.4.17)$$

Case (i). $\lambda_1 < 0 < \lambda_2$: *Saddle.* Dividing ξ_2 of (1.4.16) by $(\xi_1)^{\lambda_2/\lambda_1}$ the equation of the trajectories assumes the form $\xi_2 = c\xi_1^{\lambda_2/\lambda_1}$ where the exponent λ_2/λ_1 is negative. So the trajectories are hyperbolae and the half coordinate axes ξ_1, ξ_2; see Figure 1.4.3 (i).

Case (ii). $\text{sgn } \lambda_1 = \text{sgn } \lambda_2 \neq 0$: *Node.* (ii/a) In case $\lambda_1 \neq \lambda_2$, then proceeding as in the previous case the equation of the trajectories becomes $\xi_2 = c\xi_1^{\lambda_2/\lambda_1}$ where, now, the exponent λ_2/λ_1 is positive (*improper node*). (ii/b) If $\lambda_1 = \lambda_2 = \lambda$ but (1.4.16) applies, then the equation of the trajectories is $\xi_2 = c\xi_1$ (*proper node*, a path comes in or starts off the origin in each direction). (ii/c) If $\lambda_1 = \lambda_2 = \lambda$ and (1.4.17) applies, then the equation of the trajectories becomes

$$\xi_2 = \xi_1(c_3 \ln \xi_1 + c_4)$$

(*improper node*); see Figures 1.4.3 (ii/a,b,c).

Case (iii). $\lambda_1 = \lambda = \alpha + i\beta$, $\lambda_2 = \overline{\lambda} = \alpha - i\beta$, $\alpha \neq 0$, $\beta \neq 0$: *Spiral point*. Applying what has been said following formula (1.2.16) and denoting the coordinates of the system whose basis consists of the real and of the imaginary parts of the eigenvector belonging to λ, by (ξ_1, ξ_2), the parametric equation of the trajectories assumes the real form

$$\xi_1(t) = c_3 e^{\alpha t} \cos(\beta t + c_4), \qquad \xi_2(t) = c_3 e^{\alpha t} \sin(\beta t + c_4);$$

see Figure 1.4.3 (iii).

Case (iv). $\lambda_1 = \lambda = i\beta$, $\lambda_2 = \overline{\lambda} = -i\beta$, $\beta \neq 0$: *Centre*. Proceeding as in the previous case we have

$$\xi_1(t) = c_3 \cos(\beta t + c_4), \qquad \xi_2(t) = c_3 \sin(\beta t + c_4),$$

all trajectories are closed curves (ellipses); see Figure 1.4.3 (iv).

Case (v). $\lambda_1 = 0$, $\lambda_2 \neq 0$. The origin is no longer an isolated equilibrium point. As one can see from (1.4.16) each point of the axis ξ_1 is an equilibrium. Besides these, the trajectories are half lines parallel to the axis ξ_2; see Figure 1.4.4 (v).

Case (vi). $\lambda_1 = \lambda_2 = 0$. (vi/a) If A is non-diagonalizable, then (1.4.17) applies with $\lambda = 0$. Clearly, each point of the axis ξ_2 is an equilibrium, and the other trajectories are straight lines parallel to the axis ξ_2; see Figure 1.4.4 (vi/a). (vi/b) If A is diagonalizable, then it cannot be but the zero matrix, so all points of the phase plane are equilibria; see Figure 1.4.4 (vi/b).

In case (i) the origin is unstable. In case (ii) if the common sign of the two eigenvalues is negative, the origin is asymptotically stable; if the sign is positive, then it is unstable. In case (iii) if $\alpha = \text{Re } \lambda$ is negative, the origin is asymptotically stable; if the real part is positive, it is unstable. In case (iv) the origin is stable in the Liapunov sense. In case (v) if $\lambda_2 < 0$, the origin and all the equilibria are stable in the Liapunov sense; if $\lambda_2 > 0$, they are unstable. In case (vi/a) all the equilibria are unstable (zero is a double root of the minimal polynomial as well). In case (vi/b) all points are stable in the Liapunov sense (zero is a simple root of the minimal polynomial). In Figures 1.4.3 and 1.4.4 we have shown the stable versions everywhere.

The notions of saddle, node, spiral point and centre generalize to non-linear systems. Consider the system

$$\dot{x}_1 = f_1(x_1, x_2), \qquad \dot{x}_2 = f_2(x_1, x_2) \qquad (1.4.18)$$

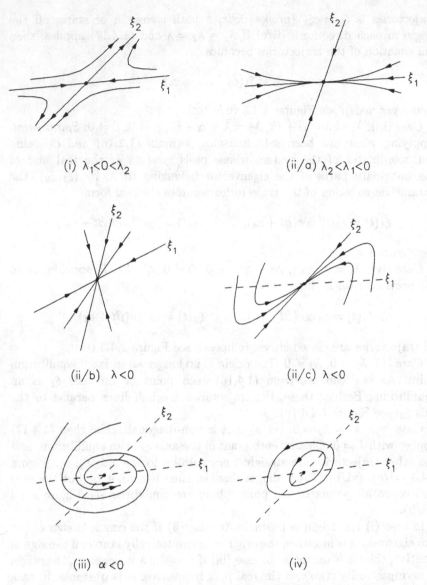

FIGURE 1.4.3. The trivial solution of (1.4.13) is a saddle (i), a stable improper node (ii/a, ii/c), a stable proper node (ii/b), a stable spiral point (iii), a centre (iv). The arrows refer to the motion along the trajectories. (i) is unstable, (ii/a,b,c) and (iii) are asymptotically stable and (iv) is stable in the Liapunov sense.

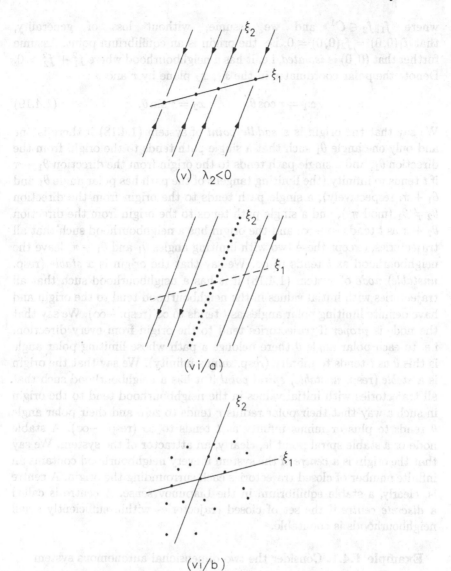

FIGURE 1.4.4. Phase plane of system (1.4.13) in the degenerate cases when there is a straight line of equilibria (v, vi/a), resp. all points of the plane are equilibria (vi/b). In case (v) and (vi/b) the origin and all equilibria are stable in the Liapunov sense; in case (vi/a) all equilibria are unstable.

where $f_1, f_2 \in C^1$ and we assume, without loss of generality, that $f_1(0,0) = f_2(0,0) = 0$, i.e. the origin is an equilibrium point. Assume further that $(0,0)$ is *isolated*, i.e. it has a neighbourhood where $f_1^2 + f_2^2 > 0$. Denote the polar coordinates in the x_1, x_2 plane by r and θ :

$$x_1 = r\cos\theta, \qquad x_2 = r\sin\theta. \qquad (1.4.19)$$

We say that the origin is a *saddle point* of system (1.4.18) if there is one and only one angle θ_1 such that a single path tends to the origin from the direction θ_1, and a single path tends to the origin from the direction $\theta_1 + \pi$ if t tends to infinity (the limiting tangent of the path has polar angle θ_1 and $\theta_1 + \pi$, respectively), a single path tends to the origin from the direction $\theta_2 \neq \theta_1$ (mod π), and a single path tends to the origin from the direction $\theta_2 + \pi$ as t tends to $-\infty$, and the origin has a neighbourhood such that all trajectories, except those two with limiting angles θ_1 and $\theta_1 + \pi$, leave the neighbourhood as t tends to ∞. We say that the origin is a *stable* (resp. *unstable*) *node* of system (1.4.18) if it has a neighbourhood such that all trajectories with initial values in the neighbourhood tend to the origin and have definite limiting polar angles as t tends to ∞ (resp. $-\infty$). We say that the node is *proper* if trajectories tend to the origin from every direction, i.e. to each polar angle θ there belongs a path whose limiting polar angle is this θ as t tends to infinity (resp. minus infinity). We say that the origin is a *stable* (resp. *unstable*) *spiral point* if it has a neighbourhood such that all trajectories with initial values in the neighbourhood tend to the origin in such a way that their polar radius r tends to zero and their polar angle θ tends to plus or minus infinity as t tends to ∞ (resp. $-\infty$). A stable node or a stable spiral point is, clearly, an attractor of the system. We say that the origin is a *centre* of the system if every neighbourhood contains an infinite number of closed trajectories each surrounding the origin. A centre is, clearly, a stable equilibrium in the Liapunov sense. A centre is called a *discrete centre* if the set of closed trajectories within sufficiently small neighbourhoods is countable.

Example 1.4.1. Consider the two dimensional autonomous system

$$\dot{x}_1 = -x_2 + x_1(x_1^2 + x_2^2)\sin\left(1/\sqrt{x_1^2 + x_2^2}\right),$$
$$\dot{x}_2 = x_1 + x_2(x_1^2 + x_2^2)\sin\left(1/\sqrt{x_1^2 + x_2^2}\right)$$

where the right-hand side is, naturally, defined as zero at $(0,0)$. It is easy to see that the right-hand side is in the C^1 class in the whole plane. Introducing polar coordinates by (1.4.19), the polar coordinates of the solutions satisfy the system

$$\dot{r} = r^3\sin(1/r), \qquad \dot{\theta} = 1.$$

Hence, the system has a countably infinite number of closed trajectories, the circles $r = 1/(k\pi)$ $(k = 1, 2, ...)$. The rest of the trajectories are spirals that wind upon every second closed path as t tends to infinity.

A basic question about being a saddle, a node, a spiral point or a centre is how "*robust*" these properties are. There are two ways of approaching this problem. First, one may ask whether the linearization of the system determines the character of the equilibrium point. We expand system (1.4.18) in the form

$$
\begin{aligned}
\dot{x}_1 &= a_{11}x_1 + a_{12}x_2 + g_1(x_1, x_2), \\
\dot{x}_2 &= a_{21}x_1 + a_{22}x_2 + g_2(x_1, x_2)
\end{aligned}
\qquad (1.4.20)
$$

where $a_{ik} = f'_{ix_k}(0,0)$, $g_i \in C^0$ and

$$
g_i\left(\sqrt{x_1^2 + x_2^2}\right) = o\left(\sqrt{x_1^2 + x_2^2}\right) \qquad (i = 1, 2).
$$

The linearized system is $\dot{x} = Ax$ where $A = [a_{ik}]$. In order to prove some of the following results, one has to assume that $g_i \in C^1$. The proofs can be found in Coddington-Levinson [1955]. If the origin is a saddle point, an improper node of type (ii/a) or a spiral point of the linear system, then it is the same for the non-linear system (1.4.20). If it is a proper node or an improper node of type (ii/c) of the linear system then it is is either a node or a spiral point of system (1.4.20). If the origin is a centre of the linear system, then it is either a centre or a spiral point of the non-linear system. Finally, if the origin is an *attractor* of the linear system, i.e. it is attractive (Definition 1.4.3), then it is an attractor of the non-linear system as well. We have seen that if an equilibrium point of a linear system is attractive, then it is globally attractive. This last property, naturally, is not inherited by the non-linear system. The basin of the attractive equilibrium point of system (1.4.20), in general, is no longer the whole plane. If the origin is stable in the Liapunov sense for the linear system, i.e. it is a centre or falls into one of the stable degenerate cases (v) and (vi), then concerning the non-linear system anything may happen: the origin may stay stable in the Liapunov sense, it may become asymptotically stable or it may be unstable.

There is a second, probably, more important and more up-to-date approach to the robustness of these properties. In the previous approach we considered systems with common linearization, or in other words, systems which differed in the higher order terms of the right-hand sides (the g_is in (1.4.20)) only. The difference of two such systems tends to zero as $|x| \to 0$ in a higher order, and this means that if we take the derivatives of the right-hand sides, *their* difference tends to zero too as $|x| \to 0$. In other words these systems are such that the "C^1-*norm*" of their differences tends to zero as $|x| \to 0$ (see Definition 7.1.1). For instance, we may reformulate one of the statements above saying that if the C^1-norm of the difference of a non-linear system and a linear system tends to zero as $|x| \to 0$ and the linear system has a stable spiral point at the origin, then the non-linear system has the same. The *structural stability* of a property (see Definition 7.1.3) requires that all systems that are *close* to the given one in the C^1-norm also have the property. The C^1-norm of the difference of the systems must *not* tend to zero as $|x| \to 0$ in this context. If we are interested in the character of an equilibrium point and say that two systems are close in the C^1-norm, then we mean that they are close in a sufficiently small neighbourhood of the equilibrium. Putting things still in an other form, in the previous paragraph we considered *"perturbations"* of the right-hand side which affected the higher order terms only, leaving the linear part unchanged. Now we allow small perturbations of the linearization as well, i.e. we consider systems of the form

$$
\begin{aligned}
\dot{x}_1 &= \tilde{a}_{11}x_1 + \tilde{a}_{12}x_2 + \tilde{g}_1(x_1, x_2), \\
\dot{x}_2 &= \tilde{a}_{21}x_1 + \tilde{a}_{22}x_2 + \tilde{g}_2(x_1, x_2)
\end{aligned}
\tag{1.4.21}
$$

where the coefficient matrix $\tilde{A} = [\tilde{a}_{ik}]$ is close to the coefficient matrix A of system (1.4.20) in some matrix norm and $\tilde{g}_1$, $\tilde{g}_2$ satisfy conditions similar to those imposed upon g_1, g_2. The following propositions hold. If the origin is a saddle point, an improper node of type (ii/a) or a spiral point of system $\dot{x} = Ax$, then it is the same for system (1.4.21). If it is a proper node or an improper node of type (ii/c) of the system, then it is either a node or a spiral point of system (1.4.21). If it is a centre of $\dot{x} = Ax$, then it is either a centre or a spiral point of (1.4.21). These propositions follow immediately from the respective propositions in the previous paragraph, and from the fact that the eigenvalues are continuous functions of the entries of a matrix (or equivalently the roots are continuous functions of the coefficients of a polynomial with leading coefficient one, see Problem 1.6.1 at the end of the chapter). Clearly, if A has two different real eigenvalues, then $\tilde{A}$ has also two different real eigenvalues that have the same signs as those of A. If A has a pair of complex conjugate eigenvalues with non-zero real parts, then so does $\tilde{A}$. If A has a double real eigenvalue, then this may "bifurcate" into a pair of complex conjugate eigenvalues for $\tilde{A}$. Finally, if A has a pair of pure imaginary eigenvalues, then $\tilde{A}$ will have a pair of complex conjugate eigenvalues with positive, negative or zero real part. All these hold, naturally, only if $\tilde{A}$ is close enough to A.

We return now to general n dimensional systems treating the problem whether the linearization determines the stability of the equilibrium of the non-linear system.

The following theorem due to Perron (see Coddington-Levinson [1955]) is rather a "perturbation theorem" since the linearization of a general n dimensional system (cf. (1.1.5)) does not lead, usually, to a system with constant coefficients.

Theorem 1.4.9 *(Perron's Theorem). Assume that A is a stable $n \times n$ matrix, $g \in C^0(\mathbf{R}_+ \times X, \mathbf{R}^n)$ where $X \subset \mathbf{R}^n$ is a neighbourhood of the origin, and $g(t, x) = o(|x|)$ uniformly in $t \in \mathbf{R}_+$; then the origin is an asymptotically stable equilibrium point of system*

$$\dot{x} = Ax + g(t, x). \tag{1.4.22}$$

There holds also

Theorem 1.4.10. *If the matrix A has at least one eigenvalue with positive real part, and g satisfies the conditions of the previous theorem, then the origin is an unstable equilibrium of system (1.4.22).*

It is equally important to remember that if the matrix A is *critical*, i.e. all its eigenvalues have non-positive real parts, and at least one real part is zero, then we cannot tell anything about the stability with respect to the non-linear system (1.4.22).

A simple Corollary of Perron's Theorem is its specialization to autonomous systems which goes back to Liapunov. The importance of this is due to the fact that the linearization of an autonomous system at an equilibrium point is, clearly, always a system with constant coefficients.

Theorem 1.4.11. *Assume that $f \in C^1(X, \mathbf{R}^n)$ where X is an open and connected subset of $\mathbf{R}^n$, and $f(a) = 0$ for some $a \in X$; if the matrix $f'_x(a)$ is stable, then $x = a$ is an asymptotically stable equilibrium of the system*

$$\dot{x} = f(x). \tag{1.4.23}$$

As it was mentioned above, this result follows easily from Perron's Theorem. If we displace the origin into a and expand system (1.4.23) there, denoting the new coordinates again by x we get

$$\dot{x} = f'_x(a)x + g(x),$$

and this system satisfies the conditions of Theorem 1.4.9. Finally, we quote the following fairly often used

Theorem 1.4.12 *(Yoshizawa [1966] p.30). If in (1.4.1) f does not depend on t or is periodic in t, and $f(t, 0) \equiv 0$, then the Liapunov stability, resp. the asymptotic stability, of the origin $x = 0$ implies its uniform stability, resp. uniform asymptotic stability.*

1.5   Liapunov's Direct Method

Establishing asymptotic stability by linearization (as treated in the previous Section) is unquestionably the simplest way, especially for autonomous systems. The problem is reduced to the determination of the signs of the real parts of the eigenvalues. However, the method has two drawbacks, at least. First, if the coefficient matrix of the linearized system is critical, i.e. besides eigenvalues with negative real parts it has also some with real part zero, then the method cannot decide the stability with respect to the non-linear system. Secondly, even if this method establishes asymptotic stability, it cannot tell anything about the extent of the basin of the attractive equilibrium. Besides these difficulties, linearization leaves open the stability problem for non-autonomous systems because the linearization of a non-autonomous system at an equilibrium point is no longer a system with constant coefficients. A powerful method has been established in treating stability problems by Liapunov [1892] in his famous paper published the first time in Russian 100 years ago. The essence of this method, called *Liapunov's direct* (or *second*) *method*, will be treated now. Proofs of the theorems can be found in Demidovich [1967$_a$] and in Rouche-Habets-Laloy [1977]. The method has its origin in Mechanics, in Lagrange's theorem announced in 1788 but proved completely by Dirichlet only about 60 years later. According to this theorem if the potential energy of a conservative mechanical system has a strict minimum at a point, then this point is stable. First, we shall give an idea about the method by an example.

Example 1.5.1. Consider the two dimensional autonomous system

$$\dot{x}_1 = -5x_2 - 2x_1^3, \qquad \dot{x}_2 = 5x_1 - 3x_2^3. \qquad (1.5.1)$$

It can be seen that $(0,0)$ is the only equilibrium point of this system, and it is a centre for the linearized system $\dot{x}_1 = -5x_2$, $\dot{x}_2 = 5x_1$. So that Theorem 1.4.12 does not solve the problem of the stability for the non-linear system. Let us see how the trajectories of the system intersect the concentric circles with centre at the origin, i.e. the level curves of the function $V(x_1, x_2) = x_1^2 + x_2^2$. In order to see this we substitute the solutions $(\varphi_1(t), \varphi_2(t))$ into V and differentiate the composite function by t:

$$\frac{d}{dt}V(\varphi_1(t), \varphi_2(t)) = \;\; < \text{grad}\, V(\varphi_1(t), \varphi_2(t)), (\dot{\varphi}_1(t), \dot{\varphi}_2(t)) >$$

$$= \;\; -(4\varphi_1^4(t) + 6\varphi_2^4((t))$$

$$(1.5.2)$$

where we made use of the fact that (φ_1, φ_2) satisfies (1.5.1). What we got means that at every point (x_1, x_2) of the plane this derivative is

$$\dot{V}(x_1, x_2) = -(4x_1^4 + 6x_2^4) < 0, \qquad (x_1, x_2) \neq (0,0).$$

The geometric meaning of this result is that the tangent vector of the path forms an obtuse angle with the outward normal grad V of the level curve

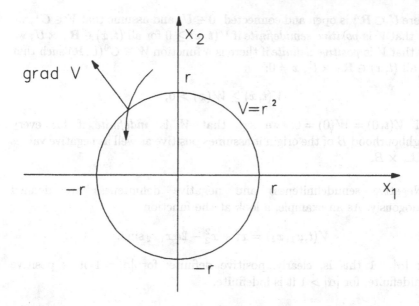

FIGURE 1.5.1. A path of system (1.5.1) cutting a level curve of $V(x_1, x_2) = x_1^2 + x_2^2$.

(the concentric circle through (x_1, x_2)), i.e. the motion on the path is from the outside of the circle into the inside (see Figure 1.5.1). Going further, and by an abuse of notation denoting $V(t) = V(\varphi_1(t), \varphi_2(t))$ from (1.5.2) we get the following estimate

$$\begin{aligned}
\dot{V}(t) &\leq -4(\varphi_1^4(t) + \varphi_2^4(t)) = -4(\varphi_1^2(t) + \varphi_2^2(t))^2 + 8\varphi_1^2(t)\varphi_2^2(t) \\
&\leq -4(\varphi_1^2(t) + \varphi_2^2(t))^2 + 2(\varphi_1^2(t) + \varphi_2^2(t))^2 = -2V^2(t).
\end{aligned}$$

Hence, $\dot{V}/V^2 \leq -2$, and integrating this differential inequality, $0 \leq V(t) \leq 1/(1/V(0) + 2t) \to 0$, as $t \to \infty$. This, clearly, means that $(0,0)$ is asymptotically stable. Moreover, since in $V(0) = V(\varphi_1(0), \varphi_2(0))$ one may take any point of the plane for initial value $(\varphi_1(0), \varphi_2(0)) \neq (0,0)$ the origin is globally asymptotically stable.

The solution of the stability problem in this Example was achieved by finding a function that had a minimum at the equilibrium point of the system and that was decreasing along the trajectories. This led to the conclusion that the motion along the trajectories tended to the equilibrium. This method can be applied to general differential systems.

Definition 1.5.1. Consider the function

$$V : \mathbf{R}_+ \times U \to \mathbf{R}$$

where $U \subset \mathbf{R}^n$ is open and connected, $0 \in U$, and assume that $V \in C^1$; we say that V is *positive semidefinite* if $V(t, x) \geq 0$ for all $(t, x) \in \mathbf{R}_+ \times U$; we say that V is *positive definite* if there is a function $W \in C^0(U, \mathbf{R})$ such that for all $(t, x) \in \mathbf{R}_+ \times U$, $x \neq 0$:

$$V(t, x) \geq W(x) > 0,$$

and $V(t, 0) = W(0) = 0$; we say that V is *indefinite* if for every neighbourhood B of the origin it assumes positive as well as negative values in $\mathbf{R}_+ \times B$.

Negative semidefiniteness and negative definiteness are defined analogously. As an example, a look at the function

$$V(t, x_1, x_2) = x_1^2 + x_2^2 - 2\alpha x_1 x_2 \sin t.$$

For $|\alpha| < 1$ this is, clearly, positive definite; for $|\alpha| = 1$ it is positive semidefinite; for $|\alpha| > 1$ it is indefinite.

Definition 1.5.2. We say that the function $h : \mathbf{R}^+ \to \mathbf{R}^+$ belongs to the *function class* $\mathcal{H}$ if $h(0) = 0$, and it is strictly increasing and continuous.

Let $X \subset \mathbf{R}^n$ be open and connected, $x = 0 \in X$, $f \in C^0(\mathbf{R}_+ \times X)$, $f'_x \in C^0(\mathbf{R}_+ \times X)$, $f(t, 0) = 0$, and consider the system

$$\dot{x} = f(t, x). \tag{1.5.3}$$

We shall be concerned with the stability of the equilibrium $x = 0$ of this system.

Definition 1.5.3. Let $V \in C^1(\mathbf{R}_+ \times U)$ where $U \subset X \subset \mathbf{R}^n$ is open and connected; the *derivative of V with respect to the system* (1.5.3) at $(t_0, x^0) \in \mathbf{R}_+ \times U$ is

$$
\begin{aligned}
\dot{V}_{(1.5.3)}(t_0, x^0) &= \frac{d}{dt} V(t, \varphi(t, t_0, x^0))\Big|_{t=t_0} \\
&= V'_t(t_0, x^0) + \sum_{k=1}^{n} V'_{x_k}(t_0, x^0) f_k(t_0, x^0) \\
&= V'_t(t_0, x^0) + < \operatorname{grad} V(t_0, x^0), f(t_0, x^0) >
\end{aligned}
$$

where φ is the solution of (1.5.3) assuming the initial value x^0 at $t = t_0$.

Clearly, $\dot{V}_{(1.5.3)} \in C^0(\mathbf{R}_+ \times U, \mathbf{R})$. If no confusion arises we shall drop the index referring to the system and shall denote this derivative, simply, by $\dot{V}$.

Now, we are able to announce the three basic theorems of the method which in their original form are due to Liapunov. In the following theorems,

U denotes always an open and connected subset of X which contains the origin.

Theorem 1.5.1 *(Liapunov's First Theorem). If there exists a function* $V \in C^1(\mathbf{R}_+ \times U, \mathbf{R})$ *where* $0 \in U \subset X$, *and a function* $h \in \mathcal{H}$ *such that for* $(t,x) \in \mathbf{R}_+ \times U : V(t,x) \geq h(|x|)$, $V(t,0) = 0$, *and* $\dot{V}_{(1.5.3)}$ *is negative semidefinite, then the origin is stable in the Liapunov sense.*

Note that the conditions imply that V is positive definite. Inversely, one may prove (see Rouche-Mawhin [1973]) that if V is positive definite, then such an $h \in$ exists. One may also prove that if besides the condition of this theorem there exists an $h_2 \in \mathcal{H}$ such that $V(t,x) \leq h_2(|x|)$, then 0 is uniformly stable.

Example 1.5.2. Consider the motion of a point of unit mass in a *potential force field*. The equation of motion is by Newton's axiom

$$\ddot{x} = -\operatorname{grad} W(x)$$

where W is the potential function and $x \in \mathbf{R}^3$ the physical space. Assume that $x = 0$ is an equilibrium, i.e. $\operatorname{grad} W(0) = 0$. The Cauchy normal form of this system is

$$\dot{x}_k = x_{3+k} \qquad \dot{x}_{3+k} = -W'_{x_k}(x_1, x_2, x_3) \qquad (k = 1, 2, 3). \tag{1.5.4}$$

Consider the *"energy integral"*

$$V(x_1, x_2, x_3, x_4, x_5, x_6) := W(x_1, x_2, x_3) + (x_4^2 + x_5^2 + x_6^2)/2.$$

The derivative of this function with respect to system (1.5.4) is

$$\dot{V}_{(1.5.4)} = \sum_{k=1}^{3} (W'_{x_k} x_{3+k} - x_{3+k} W'_{x_k}) \equiv 0,$$

and this means that $\dot{V}$ is (negative) semidefinite. Now, if V, i.e. W is positive definite, then $(x_1, x_2, ..., x_6) = (0, 0, ..., 0)$ is stable in the Liapunov sense by the previous theorem. W is positive definite if and only if $W(0) = 0$, and this value is a strict minimum. What we have got is, as a matter of fact, Lagrange's theorem.

Theorem 1.5.2 *(Liapunov's Second Theorem). Assume that there exist a function* $V \in C^1(\mathbf{R} \times U, \mathbf{R})$ *where* $0 \in U \subset X$ *and functions* $h_1, h_2, h_3 \in \mathcal{H}$ *such that* $h_1(|x|) \leq V(t,x) \leq h_2(|x|)$, *and* $\dot{V}_{(1.5.3)}(t,x) \leq -h_3(|x|)$; *then the origin is uniformly asymptotically stable.*

Note that the conditions imply that V is positive definite and tends to zero uniformly in $t \in \mathbf{R}_+$ as $x \to 0$ and that $\dot{V}$ is negative definite; cf. also the remark that follows the previous theorem.

Example 1.5.3. Consider a *"gradient system"* in $\mathbf{R}^3$ that may be considered as a model of a (hydrodynamic) flow with a *velocity potential*:

$$\dot{x} = -\operatorname{grad} V(x), \qquad x \in \mathbf{R}^3. \tag{1.5.5}$$

In other words, this is the equation of a flow with a curl-free velocity field. We assume that $V \in C^2$ and that it is positive definite, $V(0) = V'_{x_k}(0) = 0$ $(k = 1, 2, 3)$. The differential system (with these assumptions) means that the particles of the fluid move from the places of high velocity potential towards the regions of low potential, and that $x = 0$ is the point of minimal potential. The derivative of the velocity potential V with respect to system (1.5.5) is

$$\dot{V}_{(1.5.5)}(x) = -\operatorname{grad}^2 V(x) \leq 0.$$

If $\operatorname{grad} V(x) \neq 0$, $x \neq 0$, then, obviously, the assumptions of the previous theorem hold, and $x = 0$ is uniformly asymptotically stable. A sufficient condition of the existence of a neighbourhood of the origin in which $\operatorname{grad} V(x) \neq 0$ for $x \neq 0$ is

$$\det[V''_{x_i x_k}(0)] \neq 0$$

because of the Inverse Function Theorem. Clearly,

$$V(x) = (1/2) \sum_{i,k=1}^{3} V''_{x_i x_k}(0) x_i x_k + o(|x|^2).$$

The positive definiteness of V implies that the matrix $\left[V''_{x_i x_k}(0)\right]$ is positive definite. This means that all the eigenvalues of this matrix are positive. Hence, $\operatorname{Tr}\left[V''_{x_i x_k}(0)\right] = \sum_{i=1}^{3} V''_{x_i x_i}(0) > 0$, or using the notations of vector analysis $-\operatorname{div} \operatorname{grad} V(0) < 0$, i.e. $x = 0$ is a *sink*.

Corollary 1.5.3. *If the conditions of the previous theorem hold in a ball $B(0, \delta)$, then every solution φ of system (1.5.3) that stays in $B(0, \delta) : \varphi(t, t_0, x^0) \in B(0, \delta)$ for $t \geq t_0$ tends to zero as $t \to \infty$.*

This Corollary helps us to give a lower estimate of the basin of $x = 0$ if its asymptotic stability has been established by Liapunov's Second Theorem; cf. Problem 1.6.3.

Theorem 1.5.4 *(Liapunov's Third Theorem). Assume that there exist a function $V \in C^1(\mathbf{R} \times U, \mathbf{R})$ where $0 \in U \subset X$ and functions $h_2, h_3 \in \mathcal{H}$ such that $|V(t, x)| \leq h_2(|x|)$, $\dot{V}_{(1.5.3)}(t, x) \leq -h_3(|x|)$, $(t, x) \in \mathbf{R}_+ \times U$, and*

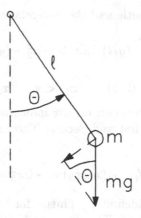

FIGURE 1.5.2. The mathematical pendulum.

that a $t_0 \in \mathbf{R}_+$ exists such that in every neighbourhood $B(0, \delta)$ of the origin there is an $x \in B(0, \delta)$ for which $V(t_0, x) < 0$; then the origin is unstable.

The functions denoted by V in Theorems 1.5.1 - 1.5.4 are called, loosely, *Liapunov function*. This is not a rigorous concept, though in some texts one may find attempts to a formal definition. A Liapunov function is a scalar valued function in the C^1 class having some definiteness property together with its derivative with respect to the system. The method sketched in this series of theorems is often called the *method of Liapunov functions*. To find an appropriate Liapunov function that may solve the problem of stability in a particular case is not an easy task at all. Useful ideas and methods for constructing Liapunov functions can be found, e.g. in Barbashin [1968]. In the following Example we try to show an often applicable possibility, namely, the modification of the energy integral if the latter does not work.

Example 1.5.4. Consider the *damped mathematical pendulum* of length l and mass m whose equation of motion is

$$ml\ddot{\theta} + b\dot{\theta} + mg\sin\theta = 0$$

where $\theta(t)$ is the angle of displacement from the vertical (hanging) position at time t, g is the gravitational acceleration and $b > 0$ is the damping coefficient (see Figure 1.5.2). The latter is, normally, positive. Introducing the notations $x_1 = \theta, x_2 = \dot{\theta}, B = b/(ml)$, the Cauchy normal form of the system will be

$$\dot{x}_1 = x_2, \qquad \dot{x}_2 = -(g/l)\sin x_1 - Bx_2. \qquad (1.5.6)$$

If there was no damping, this would be a conservative system. Its energy integral, the sum of the kinetic and the potential energies, would be

$$W(x_1, x_2) = x_2^2/2 + \int_0^{x_1} (g/l) \sin s \, ds = x_2^2 + (g/l)(1 - \cos x_1) > 0,$$

$$(x_1, x_2) \neq (0,0), \quad |x_1| < \pi, \quad x_2 \in \mathbf{R}.$$

This is a positive definite function on the indicated domain satisfying the conditions of Liapunov's First and Second Theorems. Its derivative with respect to the system is

$$\dot{W}_{(1.5.6)}(x_1, x_2) = -Bx_2^2$$

which is negative semidefinite. Thus, for the time being, we may apply Liapunov's First Theorem, and consider the equilibrium $(\theta, \dot{\theta}) = (x_1, x_2) = (0,0)$ to be stable in the Liapunov sense. However, if we modify the function W and make a try with the Liapunov function

$$V(x_1, x_2) = x_2^2 + (Bx_1 + x_2)^2 + 4(g/l)(1 - \cos x_1),$$

we see that this is also positive definite, and its derivative with respect to the system is

$$\dot{V}_{(1.5.6)}(x_1, x_2) = -2B(x_2^2 + (g/l)x_1 \sin x_1) < 0,$$

$$(x_1, x_2) \neq (0,0), \quad |x_1| < \pi, \quad x_2 \in \mathbf{R}.$$

Thus, V satisfies the conditions of Liapunov's second theorem, i.e. the equilibrium $(0,0)$ is uniformly asymptotically stable. We got this result under the natural assumption that $B > 0$. It is worthwhile to note that if $B < 0$ ("negative damping"), then the Liapunov function $-V$ and its derivative $-\dot{V}_{(1.5.6)}$ satisfy the conditions of Liapunov's third theorem, so $(0,0)$ is unstable.

Liapunov's theorems are, naturally, valid for autonomous systems in the case of that, usually, time independent Liapunov functions are applied. There is an important, often used theorem about the asymptotic stability of the equilibrium of an autonomous system that relaxes the requirements in Theorem 1.5.2 (see Barbashin-Krasovskij [1952]). Consider the autonomous system

$$\dot{x} = f(x) \tag{1.5.7}$$

where $f \in C^1(X, \mathbf{R}^n)$, $0 \in X$ is an open and connected subset of $\mathbf{R}^n$ and $f(0) = 0$.

Theorem 1.5.5 *(Barbashin-Krasovskij Theorem). Assume that there exist a function $V \in C_1(U, \mathbf{R})$ where $0 \in U \subset X$ and functions $h_1, h_2 \in \mathcal{H}$ such that $h_1(|x|) \leq V(x) \leq h_2(|x|)$, $\dot{V}_{(1.5.7)}(x) \leq 0$, and the set $M = \{x \in$*

$U : \dot{V}_{(1.5.7)}(x) = 0\}$ does not contain any positive semitrajectory apart from the origin; then the equilibrium $x = 0$ is asymptotically stable.

This Theorem can be generalized for non-autonomous periodic systems (see Theorem 4.2.4).

If we apply the Barbashin-Krasovskij Theorem for the previous example, the Liapunov function W works. W is positive definite on $U = \{(x_1, x_2) : |x_1| < \pi, x_2 \in \mathbf{R}\}$, and its derivative $\dot{W}_{(1.5.6)}(x_1, x_2) = -Bx_2^2 = 0$ for $x_2 = 0$, i.e. $M = \{(x_1, x_2) : |x_1| < \pi, x_2 = 0\}$. In this set there is no positive semitrajectory apart from the origin.

In the rest of this Section we are looking at some implications of Liapunov's direct method on linear systems. The easiest functions that can be used as Liapunov functions are the quadratic forms (provided that they work). This is due to the practically well applicable criteria of their definiteness. As we shall see, in the case of linear systems they can be used indeed. Consider, first, a homogeneous linear system with constant coefficients

$$\dot{x} = Ax \tag{1.5.8}$$

where A is an $n \times n$ real constant matrix. Let $V(x) = x'Bx$ be a quadratic form with coefficient matrix B which is, naturally, symmetric. The derivative of V with respect to the system is

$$\begin{aligned} \dot{V}_{(1.5.8)}(x) &= \dot{x}'Bx + x'B\dot{x} = x'(A'B + BA)x \\ &= -x'Cx \end{aligned}$$

where

$$A'B + BA = -C. \tag{1.5.9}$$

This is called *Liapunov's matrix equation*. The proof of the following theorem can be found in Barbashin [1967$_a$].

Theorem 1.5.6. *If the matrix A is stable, then for an arbitrary positive definite (symmetric) matrix C, Liapunov's matrix equation (1.5.9) has one and only one solution B, and this is a (symmetric) positive definite matrix.*

If C is positive definite then $-C$ is negative definite. According to this theorem, if we prescribe arbitrarily a negative definite quadratic form $-x'Cx$, then we may find a unique positive definite quadratic form $x'Bx$ by solving (1.5.9) for B such that the derivative of the latter with respect to the system (1.5.8) is the former (provided, of course, that (1.5.8) is an asymptotically stable system); cf. Problem 1.6.4. So if A is a stable matrix, then in view of Theorem 1.5.2 and Corollary 1.5.3 this theorem implies the global asymptotic stability of system (1.5.8) that has been stated in the previous Section.

In the case that the linear system is non-autonomous, i.e. the coefficient matrix is not a constant one, the stability problem is by no means trivial. We are going to prove the following:

Theorem 1.5.7. *Consider the system*

$$\dot{x} = A(t)x \tag{1.5.10}$$

where $A \in C^0(\mathbf{R}_+, \mathbf{R}^{n^2})$, *and denote the (real) eigenvalues of the symmetric matrix* $A(t) + A^T(t)$ *by* $\lambda_i(t)$ $(i = 1, 2, ..., n)$; *if there exists an* $\alpha > 0$ *such that for* $t \in \mathbf{R}_+$ $(i = 1, ..., n)$ *we have* $\lambda_i(t) \leq -\alpha$, *then the system (1.5.10) is asymptotically stable.*

Proof. Assume the positive definite Liapunov function $V(x) = x'x = \sum_{i=1}^{n} x_i^2$. Its derivative with respect to the system is

$$\dot{V}_{(1.5.10)}(x) = \dot{x}'x + x'\dot{x} = x'(A(t) + A'(t))x.$$

We know that by an orthogonal transformation $x = M(t)\tilde{x}$ where $M(t)$ is an orthogonal matrix for all t the last quadratic form can be brought into diagonal form:

$$\dot{V}_{(1.5.10)}(M(t)\tilde{x}) = \sum_{i=1}^{n} \lambda_i(t)\tilde{x}_i^2 \leq -\alpha \sum_{i=1}^{n} \tilde{x}_i^2.$$

This, clearly, means that $\dot{V}$ is negative definite, and this proves the theorem. □

Liapunov's direct method has been established originally for equilibria, but it can be generalized to more complicated invariant sets of differential systems. The essence of the method is to find a scalar function that has its minimum at the equilibrium point and decreases along the trajectories. In the case of an invariant set consisting of more than a single point, one may try to find a function that is, say, zero on the manifold, positive outside it and decreasing along the trajectories. This way one may hope to establish the *attractivity of invariant manifolds*. We shall return to this point when it is necessary in Section 6.7.

1.6 Problems

Problem 1.6.1. Show that the roots of the polynomial $p(\lambda) = \lambda^n + a_{n-1}\lambda^{n-1} + \cdots + a_1\lambda + a_0$ are continuous functions of the coefficients $a_0, a_1, ..., a_{n-1}$. (Hint: apply Rouché 's theorem to $p(\lambda)$ and its "δ perturbation" on small circles of radius ε with centres in the roots.)

Problem 1.6.2. Show that the origin is a spiral point of system

$$\dot{x}_1 = -x_1 + x_2/\ln(x_1^2 + x_2^2)^{1/2}, \qquad \dot{x}_2 = x_1 - x_2 - x_1/\ln(x_1^2 + x_2^2)^{1/2}$$

while it is an improper node of type (ii/c) of the linearized system. (Hint: use polar coordinates and differential inequalities.)

Problem 1.6.3. Show that $(x_1, x_2) = (0, 0)$ is an asymptotically stable equilibrium of the system

$$\dot{x}_1 = 2x_1^3 x_2^2 - x_1, \qquad \dot{x}_2 = -x_2,$$

and that the ball $B(0, 1)$ is a subset of the basin. (Hint: use the Liapunov function $x_1^2 + x_2^2$.)

Problem 1.6.4. Let A be a stable 2×2 matrix, and $C = I$ the 2×2 unit matrix. Solve Liapunov's matrix equation (1.5.9) explicitly for B, and show that B is a stable matrix indeed. (Hint: use the criterion that a 2×2 matrix is stable if and only if its trace is negative and its determinant positive (1.4.15).)

Problem 1.6.5. (*Gronwall's Lemma*). Let $u, f \in C^0[t_0, t_1]$, where t_1 may be ∞, $u(t), f(t) > 0$, $t \in [t_0, t_1]$, and $k > 0$; if for $t \in [t_0, t_1]$

$$u(t) \leq k + \int_{t_0}^{t} f(\tau)u(\tau)d\tau,$$

then

$$u(t) \leq k \exp \int_{t_0}^{t} f(\tau)d\tau.$$

(Hint: multiply the first inequality by $f(t)$.)

Problem 1.6.6 (*Gronwall's Lemma 2*). Let $u, f, t_0 < t_1$ and k be as in the previous Problem and $m > 0$; if for $t \in [t_0, t_1]$

$$u(t) \leq k + \int_{t_0}^{t} (f(\tau)u(\tau) + m)d\tau,$$

then

$$u(t) \leq (k + m(t - t_0)) \exp \int_{t_0}^{t} f(\tau)d\tau.$$

2
Periodic Solutions of Linear Systems

In this chapter the existence and stability problems of periodic solutions of linear systems will be treated. For centuries people used linear systems as models of phenomena in Nature, in Mechanics, in Physics, etc. So that in the theories of elasticity, heat propagation, the propagation of waves, electromagnetic phenomena, etc. basic differential equations are linear ones. Besides that, as this can be seen from (1.1.5) and (1.3.7), the variational system with respect to a periodic solution of a periodic or an autonomous system is a linear system with periodic coefficients. As a consequence, results obtained in this chapter will be used extensively in the study of periodic solutions of *non-linear* systems. In the first Section we treat linear systems with constant coefficients though they are special cases of linear systems with periodic coefficients. In Section 2 we are treating homogeneous linear systems with periodic coefficients and show that they are *"reducible"* to systems with constant coefficients. In Section 3 *"forced oscillations"* will be dealt with, i.e. inhomogeneous linear systems with periodic "forcing term". Stability problems will be treated in the fourth Section, and in the last one we shall study second order linear differential equations with periodic or harmonic coefficients: Hill's and Mathieu's equations.

2.1   Linear Systems with Constant Coefficients

Linear systems with constant coefficients are, naturally, special cases of linear systems with periodic coefficients. Nevertheless, it is worthwhile to treat them separately because they occur more frequently in applications, and because the results for such systems are more explicit.

As we can see from (1.2.16) and (1.2.20) the homogeneous linear system

$$\dot{y} = Ay \tag{2.1.1}$$

with constant coefficient matrix A has periodic solutions if and only if A has a pure imaginary eigenvalue $\lambda = i\beta$, $\beta \neq 0$. If $s \neq 0$ is a corresponding eigenvector, then $\varphi^1(t) = s\exp(\lambda t)$ is a (complex valued) periodic solution. If A is real, then $\overline{\lambda} = -i\beta$ is also an eigenvalue with $\overline{s}$ as an eigenvector. So $\varphi^2(t) = \overline{s}\exp(\overline{\lambda}t)$ is another solution. We shall perform here the replacement of φ^1 and φ^2 by two independent real valued solutions described after (1.2.16). Let us denote the real part of the vector s by $u/2$, and the imaginary part by $-v/2$, i.e. $u = 2\,\mathrm{Re}\,s$, $v = -2\,\mathrm{Im}\,s$. Then

$$\varphi^1(t) = e^{i\beta t}(1/2)(u - iv), \qquad \varphi^2(t) = e^{-i\beta t}(1/2)(u + iv).$$

It is easy to see that neither u nor v is zero, and that these two solutions are linearly independent. By linear combination, the following two solutions are formed:

$$\begin{aligned}
\varphi^3(t) &= \varphi^1(t) + \varphi^2(t) = u\cos\beta t + v\sin\beta t, \\
\varphi^4(t) &= -i\varphi^1(t) + i\varphi^2(t) = u\sin\beta t - v\cos\beta t
\end{aligned}$$

which are, again, linearly independent. Since an arbitrary linear combination of φ^3 and φ^4 is also a solution, we get that for arbitrary $c \in \mathbf{R}$ and $\alpha \in [0, 2\pi]$

$$\varphi(t) = c(u\cos(\beta t - \alpha) + v\sin(\beta t - \alpha)) \tag{2.1.2}$$

is a periodic solution of (2.1.1). If for fixed c we vary α, then we get different solutions having the same path (see (1.3.3)). By varying c and keeping α fixed we get all the ellipses in the plane spanned by the independent vectors u and v. This means that the trajectories corresponding to these periodic solutions cover the two dimensional subspace of $\mathbf{R}^n$ spanned by u and v completely. Thus, we see that *if* (2.1.1) *has a non-constant periodic solution, then it has a whole family of closed trajectories corresponding to periodic solutions of the same period $2\pi/\beta$ depending on one parameter (c) at least.* Clearly, $i\beta$ might be a multiple root of the characteristic equation (together with $-i\beta$), and the subspace filled by closed trajectories corresponding to periodic solutions of period $2\pi/\beta$ may have higher dimension than two. All this implies that a real autonomous linear system of differential equations (2.1.1) cannot have an "*isolated*" closed path (which is not a single point of equilibrium), i.e. a closed path with a neighbourhood that does not contain any other closed path. A by-product of this result is that a non-constant periodic solution of (2.1.1) cannot be attractive, which is, naturally, clear considering that A has eigenvalues with zero real part, so the system cannot be asymptotically stable (cf. Theorems 1.4.4-1.4.5).

If $i\beta$ is a multiple root (of multiplicity m) of the characteristic polynomial but a simple root of the minimal polynomial, then the subspace filled by closed trajectories corresponding to periodic solutions of period $2\pi/\beta$ is $2m$ dimensional (the multiplicity of $-i\beta$ is also m). If $i\beta$ is a multiple root of the minimal polynomial too, then besides families of periodic solutions of period $2\pi/\beta$ the system has solutions whose coordinates are periodic functions $(\cos\beta t, \sin\beta t)$ multiplied by polynomials of degree one at least. These solutions are oscillating, but their amplitude tends to infinity as t tends to $\pm\infty$. These unbounded solutions occur even if the rest of the eigenvalues have negative real parts. The system is said to have *internal resonance* in this case.

We turn now to the more intricate problem of the existence of periodic solutions in case we have an inhomogeneous system with constant coefficients and a *periodic forcing term* (or *excitation*):

$$\dot{y} = Ay + b(t). \tag{2.1.3}$$

Here A is a constant $n \times n$ matrix, the function $b \in C^0(\mathbf{R})$, and it is periodic with least positive period T, i.e. $b(t+T) = b(t)$. According to (1.2.19) an arbitrary solution is

$$\varphi(t) = \exp(At)\varphi^0 + \int_0^t \exp(A(t-s))b(s)ds \tag{2.1.4}$$

where $\varphi^0 = \varphi(0)$. Now, φ *is a periodic solution of (2.1.3) if and only if* $\varphi(T) = \varphi(0) = \varphi^0$. The "only if" part is obvious. The "if" part follows from the fact that $\psi(t) := \varphi(t+T)$ is also a solution of (2.1.3), and $\psi(0) = \varphi(T) = \varphi(0)$. Because of uniqueness, $\psi(t) = \varphi(t+T) = \varphi(t)$. Thus, *the necessary and sufficient condition of the periodicity of φ is*

$$\varphi(T) = \exp(AT)\varphi^0 + \int_0^T \exp(A(T-s))b(s)ds = \varphi^0. \tag{2.1.5}$$

It is easy to prove the following:

Theorem 2.1.1. *If the homogeneous system corresponding to (2.1.3) has no periodic solution of period T apart from the trivial one, then (2.1.3) has one and only one periodic solution of period T.*

Proof. The periodicity condition (2.1.5) has one and only one solution φ^0 if and only if the coefficient matrix of the linear algebraic system of equations

$$(\exp(-AT) - I)\varphi^0 = \int_0^T \exp(-As)b(s)ds \tag{2.1.6}$$

is non-singular, i.e. if and only if

$$\det(\exp(-AT) - I) \neq 0.$$

This condition is, clearly, equivalent to

$$\det(I - \exp(AT)) \neq 0.$$

Now, $\det(I - \exp(AT)) = 0$ if and only if the number 1 is an eigenvalue of the matrix $\exp(AT)$. By Corollary A1.8 this happens if and only if $2k\pi i/T$ is an eigenvalue of A for some $k = 0, 1, 2, \dots$. The homogeneous system has a periodic solution of period T if and only if $2k\pi i/T$ is an eigenvalue of A for some k (if $k \geq 1$, then the *least* positive period is T/k; if $k = 0$, then A is singular, and the homogeneous system has non-trivial constant solutions). Thus, if the homogeneous system has no periodic solution of period T apart from the trivial one, then (2.1.6) can be solved uniquely for φ^0. $\square$

In order to proceed to the case when the homogeneous system has non-trivial periodic solutions, we have to introduce the "adjoint system" of a linear system of differential equations. We shall apply this concept also in cases when the coefficients of the system are complex numbers. So let $A \in \mathbf{C}^{n^2}$ an $n \times n$ matrix with complex entries, the *adjoint system* of $\dot{y} = Ay$ is

$$\dot{\eta} = -A^* \eta \tag{2.1.7}$$

where A^* is the conjugate transpose of A. (In the case A is real, then $A^* = A'$, the transpose.) We note that if $y(t)$ is a solution of $\dot{y} = Ay$, and $\eta(t)$ a solution of (2.1.7), then their scalar product is constant:

$$< y(t), \eta(t) > = y'(t)\overline{\eta}(t) = \text{const}, \tag{2.1.8}$$

because differentiating the left-hand side,

$$\begin{aligned}(y'(t)\overline{\eta}(t))^{\cdot} &= \dot{y}'(t)\overline{\eta}(t) + y'(t)\dot{\overline{\eta}}(t) = y'(t)A'\overline{\eta}(t) \\ &- y'(t)A'\overline{\eta}(t) = 0.\end{aligned}$$

There holds also the following:

Lemma 2.1.2. *If the subspace of T-periodic solutions of system $\dot{y} = Ay$ is of dimension k (i.e. the maximal number of linearly independent T-periodic solutions of this system is k, cf. Theorem 1.2.2), then the subspace of T-periodic solutions of (2.1.7) is also of dimension k.*

Proof. The periodicity condition (2.1.5) applied to the homogeneous system $\dot{y} = Ay$ (i.e. $b = 0$) is

$$(\exp(AT) - I)\varphi^0 = 0. \tag{2.1.9}$$

According to the assumption this system has k (and not more) linearly independent solutions, or, in other words, the kernel of $(\exp(AT) - I)$ is of dimension k. Multiplying (2.1.9) by $\exp(-AT)$ and taking the conjugate transpose we get

$$\varphi^{0*}(I - \exp(-A^*T)) = 0.$$

This means, obviously, that the kernel of $(\exp(-A^*T) - I)$ is also of dimension k. However, for system (2.1.7) the periodicity condition is

$$(\exp(-A^*T) - I)v = 0, \qquad (2.1.10)$$

and this proves the Lemma. $\square$

Now, we are able to establish the conditions under which the inhomogeneous system (2.1.3) has periodic solutions even if the corresponding homogeneous system has some.

Theorem 2.1.3. *Assume that the subspace of T-periodic solutions of the homogeneous system corresponding to (2.1.3) is of dimension $k \geq 1$, and denote k linearly independent T-periodic solutions of the adjoint system (2.1.7) by $\psi^1, ..., \psi^k$; system (2.1.3) has T-periodic solutions if and only if the "orthogonality conditions"*

$$\int_0^T \psi^{i\prime}(t)\bar{b}(t)dt = 0 \qquad (i = 1, 2, ..., k) \qquad (2.1.11)$$

hold.

Proof. First note that a linear system $Bx = c$ where B is an $n \times n$ matrix and c an n dimensional column vector has a solution x if and only if for $v \in \mathbf{C}^n$ the equation $v'\overline{B} = 0$ implies $v'\bar{c} = 0$. The "only if part" is clear. The proposition that $v'\overline{B} = 0$ implies $v'\bar{c} = 0$ means that every vector that is orthogonal to the column-space of B is also orthogonal to the vector c. But this implies that c is a vector in the subspace spanned by the columns of B, i.e. the system has a solution.

Let us apply this criterion to the periodicity condition (2.1.6). φ^0 is the initial value of a T-periodic solution of (2.1.3) if and only if it is a solution of this equation. Now, the coefficient matrix on the left-hand side of (2.1.6) is singular. Equation (2.1.6) has solutions if and only if every vector v that satisfies

$$v'(\exp(-\overline{A}T) - I) = 0 \qquad (2.1.12)$$

also satisfies

$$v'\int_0^T \exp(-\overline{A}s)\bar{b}(s)ds = 0. \qquad (2.1.13)$$

Transposing (2.1.12) we get that it is equivalent to (2.1.10). So we may reformulate the previous condition stating that (2.1.6) has solutions if and only if every initial value v belonging to a T-periodic solution of (2.1.7) satisfies also (2.1.13). Transposing the latter condition yields

$$\int_0^T b^*(s) \exp(-A^* s) v \, ds = 0.$$

But $\exp(-A^* t)v$ is exactly the (T-periodic) solution of the adjoint equation (2.1.7) belonging to the initial value v. Thus, we got that the inhomogeneous system (2.1.3) has T-periodic solutions if and only if the vector function $b(t)$ is orthogonal to every T-periodic solution of (2.1.7) on the interval T. Taking into consideration that under the assumptions $\psi^1, ..., \psi^k$ span the subspace of periodic solutions of (2.1.7) this completes the proof. $\square$

The previous Theorem is a special case of an analogous general theorem about linear systems with periodic coefficients which will be presented in Section 3. If the orthogonality conditions (2.1.11) do not hold, then system (2.1.3) has no periodic solution. In the general setting of Section 3 we shall prove Massera's Theorem according to which if an inhomogeneous linear system with periodic coefficients and periodic forcing term has a bounded solution on $t \in [0, \infty)$, then it has a periodic solution too. Thus, if the homogeneous system corresponding to (2.1.3) has T-periodic solutions apart from the trivial one, and the orthogonality conditions do not hold, then every solution of (2.1.3) is *unbounded* on $[0, \infty)$. This is called the case of *resonance*. Note that instead of saying that conditions (2.1.11) do not hold, one may say, equivalently, that the adjoint equation (2.1.7) has at least one T-periodic solution that is not orthogonal to the function $b(t)$ on the interval T.

Example 2.1.1 *The linear spring.* Consider a body of mass $m > 0$ attached to the end of a spring that may move in a horizontal tube (the gravitational force is not to be taken into account). Assume that Hooke's law applies, i.e. the elastic force is proportional to the elongation of the spring and that we may neglect the mass of the spring itself. Then according to Newton's axiom the equation of motion is

$$m\ddot{y} = -b\dot{y} - ky + F(t) \tag{2.1.14}$$

where $y(t)$ is the displacement of the body from its rest point, $b > 0$ is the *damping coefficient* (the damping being proportional to the velocity), $k > 0$ is the *restoring force coefficient* and F is the *periodic forcing term*: it is continuous and periodic with period $T > 0$. First, the free system will be treated, i.e. the homogeneous equation

$$m\ddot{y} + b\dot{y} + ky = 0.$$

The eigenvalues of the characteristic equation $m\lambda^2 + b\lambda + k = 0$ are

$$\lambda_{1,2} = -b/(2m) \pm (b^2 - 4mk)^{1/2}/(2m).$$

If $b^2 > 4mk$, resp. $b^2 = 4mk$ (the damping is relatively large compared to the restoring force and the mass), λ_1 and λ_2 are negative real numbers (in the second case they are equal), and the "general solution" is

$$y(t) = c_1 \exp(\lambda_1 t) + c_2 \exp(\lambda_2 t),$$

resp.

$$y(t) = (c_1 + c_2 t) \exp(\lambda_1 t) \qquad (\lambda_1 = \lambda_2).$$

In both cases the displacement tends to zero exponentially along with the velocity when $t \to \infty$. The origin of the plane $y, \dot{y}$ is a stable node (cf. (1.4.16) - (1.4.17)). If $b^2 < 4mk$, then $\lambda_{1,2} = \alpha \pm i\beta$ where $a = -b/(2m)$ and $\beta = (4mk - b^2)^{1/2}/(2m)$. In this case the "general solution" is

$$y(t) = ae^{\alpha t} \cos(\beta t + \delta)$$

where $a \geq 0$ and δ are arbitrary constants. When $t \to \infty$, the displacement is oscillating with exponentially decreasing amplitude. The origin of the plane $y, \dot{y}$ is a stable spiral point.

If a periodic force is applied on the body, i.e. if equation (2.1.14) is considered, then Theorem 2.1.1 can be applied, and that means that the equation has a periodic solution of period T. According to Corollaries 1.4.2 - 1.4.3 this periodic solution is globally asymptotically stable.

Now assume that the damping is so small that it can be neglected and consider the equation of motion

$$\ddot{y} + (k/m)y = 0.$$

This is sometimes called the one dimensional *harmonic oscillator*. The eigenvalues are now $\lambda_{1,2} = \pm i\beta$ where $\beta = (k/m)^{1/2}$ is called the *eigenfrequency* of the system. The "general solution" is

$$y(t) = a \cos(\beta t + \delta) = c_1 \cos(\beta t) + c_2 \sin(\beta t)$$

where $a \geq 0$, δ, c_1, c_2 are arbitrary constants. In this case, clearly, all trajectories in the plane $y, \dot{y}$ are closed curves, and the origin is a centre. The equation of the periodically forced harmonic oscillator is

$$\ddot{y} + (k/m)y = F(t)/m. \qquad (2.1.15)$$

If for the period T of F we have $T \neq l2\pi(k/m)^{-1/2}$ $(l = 1, 2, ...)$, then this equation has a T-periodic solution, and it is stable in the Liapunov sense (see Corollary 1.4.2). If a positive integer l exists such that $T = l2\pi(k/m)^{-1/2}$, then (2.1.15) has a T-periodic solution if and only if

$$\int_0^T F(t) \cos(2\pi l t/T)dt = \int_0^T F(t) \sin(2\pi l t/T)dt = 0. \qquad (2.1.16)$$

(One may see transforming (2.1.15) into Cauchy normal form that the scalar second order equation corresponding to the adjoint equation of the homogeneous system has $\cos(\beta t)$ and $\sin(\beta t)$ also as periodic solutions.) The orthogonality conditions (2.1.11) boil down to (2.1.16). The latter conditions mean that in the Fourier series of F

$$F(t) \sim a_0/2 + \sum_{j=1}^{\infty} (a_j \cos(2\pi j t/T) + b_j \sin(2\pi j t/T))$$

the terms with $j = l$ do not turn up, i.e. $a_l = b_l = 0$. If (2.1.16) does not hold, then (2.1.15) has no periodic solution, and all solutions are unbounded.

2.2 Homogeneous Linear Systems with Periodic Coefficients

In this Section the system

$$\dot{y} = A(t)y \tag{2.2.1}$$

will be treated where $A \in C^0(\mathbf{R}_+)$, and it is periodic with period $T > 0 : A(t+T) = A(t)$, $t \in \mathbf{R}_+$. We shall, usually, tacitly assume that T is the *least positive period* of the coefficient matrix. However, in most of the results this assumption does not play any role, and if the system can have arbitrarily small positive periods, then it contains the case of systems with constant coefficients.

Before proceeding, we need a result which is valid for general (not only linear) periodic systems. Consider the system

$$\dot{x} = f(t,x) \tag{2.2.2}$$

where $f, f_x' \in C^0(\mathbf{R}_+ \times X)$ and f is periodic in t with period $T > 0 : f(t+T,x) = f(t,x)$, $t \in \mathbf{R}_+$, $x \in X$, X being an open and connected subset of $\mathbf{R}^n$ (or $\mathbf{C}^n$).

Lemma 2.2.1. *The solution $\varphi : \mathbf{R}_+ \to X$ of system (2.2.2) is T-periodic if and only if a $t_0 \in \mathbf{R}_+$ exists for which*

$$\varphi(t_0 + T) = \varphi(t_0). \tag{2.2.3}$$

Proof. If φ is T-periodic, then (2.2.3) is, obviously, true. Assume that (2.2.3) holds. The function $\psi(t) := \varphi(t+T)$ is also a solution of (2.2.2) since

$$\dot{\psi}(t) = \dot{\varphi}(t+T) = f(t+T, \varphi(t+T)) = f(t, \psi(t))$$

because of the periodicity of f in t. However, $\psi(t_0) = \varphi(t_0 + T) = \varphi(t_0)$ because of condition (2.2.3). Hence, because of uniqueness, $\psi(t) = \varphi(t)$, i.e. $\varphi(t+T) = \varphi(t)$, $t \in \mathbf{R}_+$. $\square$

Returning now to (2.2.1) let $\Phi(t)$ be a fundamental matrix solution (see (1.2.7)) of this system. Just as in the previous proof one may show that $\Phi(t+T)$ is also a fundamental matrix solution. Therefore a constant matrix C exists for which

$$\Phi(t+T) = \Phi(t)C, \qquad (2.2.4)$$

or

$$C = \Phi^{-1}(t)\Phi(t+T) = \Phi^{-1}(0)\Phi(T).$$

If we assume that Φ is the fundamental matrix that assumes the unit matrix I at $t = 0$, i.e. with the notation of (1.2.9) $\Phi(t) = \Phi(t,0)$, then also $\Phi^{-1}(0) = I$; hence

$$C = \Phi(T,0). \qquad (2.2.5)$$

Substituting this value back into (2.2.4) we get

$$\Phi(t+T,0) = \Phi(t,0)\Phi(T,0). \qquad (2.2.6)$$

The (regular) matrix C defined by (2.2.5) is called the *principal matrix* of system (2.2.1).

Observe that if we used (2.2.4) as the definition of C without assuming that Φ is the fundamental matrix with value I at $t = 0$, then C would depend on the fundamental matrix chosen. Nevertheless, if another fundamental matrix $\tilde{\Phi}$ is chosen, and $\tilde{C}$ is the matrix defined by $\tilde{\Phi}(t+T) = \tilde{\Phi}(t)\tilde{C}$, then taking into account that a regular matrix U exists satisfying $\tilde{\Phi}(t) = \Phi(t)U$ one gets

$$\tilde{C} = \tilde{\Phi}^{-1}(0)\tilde{\Phi}(T) = U^{-1}\Phi^{-1}(0)\Phi(T)U = U^{-1}CU.$$

This means that $\tilde{C}$ is *similar* to C; consequently the spectrum of all these matrices C belonging to different fundamental matrices is invariant and determined by the system.

The eigenvalues of the principal matrix C defined by (2.2.5) are called the *characteristic multipliers* of system (2.2.1). Since Φ is a regular matrix, the number zero cannot be a characteristic multiplier.

Theorem 2.2.2. *If λ is a characteristic multiplier of system (2.2.1), then a non-trivial solution φ exists satisfying*

$$\varphi(t+T) = \lambda\varphi(t); \qquad (2.2.7)$$

inversely, if for a non-trivial solution φ of (2.2.1)

$$\varphi(T) = \lambda\varphi(0) \qquad (2.2.8)$$

holds, then λ is a characteristic multiplier; moreover, $\varphi(0)$ is a corresponding eigenvector.

Proof. Let λ be a characteristic multiplier, and $s \neq 0$ a corresponding eigenvector. Consider the solution φ of system (2.2.1) that assumes the value s at $t = 0$: $\varphi(0) = s$. Then $\varphi(t) = \Phi(t, 0)s$, and

$$\varphi(t + T) = \Phi(t + T, 0)s = \Phi(t, 0)Cs = \lambda\Phi(t, 0)s = \lambda\varphi(t)$$

where (2.2.6) and (2.2.5) were applied. Inversely, assume that (2.2.8) holds for some non-zero solution φ. Since $\varphi(T) = \Phi(T, 0)\varphi(0)$, this means that

$$(C - \lambda I)\varphi(0) = 0.$$

Hence, $\det(C - \lambda I) = 0$, i.e. λ is a characteristic multiplier, and $\varphi(0)$ an eigenvector corresponding to it. $\square$

Corollary 2.2.3. *System (2.2.1) has a non-trivial T-periodic, resp. $2T$-periodic solution if and only if the number 1, resp. -1 is a characteristic multiplier.*

Proof. The "if part" is clear from (2.2.7). (In the case $\lambda = -1$, $\varphi(t + 2T) = -\varphi(t + T) = \varphi(t)$). If φ^1 is a non-trivial T-periodic solution, then (2.2.8) holds with $\lambda = 1$. If φ^2 is a non-trivial periodic solution with *least* positive period $2T$, then applying the definition (2.2.5) of the principal matrix C and (2.2.6) :

$$\varphi^2(2T) = C^2\varphi^2(0) = \varphi^2(0),$$

i.e. 1 is an eigenvalue of C^2 with eigenvector $\varphi^2(0) \neq 0$. Because of Corollary A 1.8 either the number 1 or -1 (or both) is a characteristic multiplier, of the system. Clearly,

$$(C^2 - I)\varphi^2(0) = (C + I)(C - I)\varphi^2(0) = 0.$$

If -1 were not a characteristic multiplier, then the first matrix in this product would be regular, and we would get $(C - I)\varphi^2(0) = 0$, i.e. $\varphi^2(T) = C\varphi^2(0) = \varphi^2(0)$ which would mean that T would be a period of φ^2 contrary to the assumption that $2T$ is the least positive period. $\square$

Another simple consequence of Theorem 2.2.2 is

Corollary 2.2.4. *If system (2.2.1) has k linearly independent T-periodic (resp. $2T$-periodic) solutions, then the multiplicity of the characteristic multiplier 1 (resp. -1) is at least k in the characteristic polynomial of the principal matrix.*

Proof. Indeed, (2.2.8) holds for $\lambda = 1$ (resp. $\lambda = -1$) with the initial vectors of the k linearly independent T-periodic (resp. $2T$-periodic) solutions. This means that to the characteristic multiplier 1 (resp. -1)

there belongs k linearly independent eigenvectors, i.e. its eigenspace is of dimension k at least. $\square$

Now we are in the position to prove the basic theorem of *"Floquet's theory"*.

Theorem 2.2.5 *(Floquet [1883]). The fundamental matrix $\Phi(t,0)$ of system (2.2.1) can be written in the form*

$$\Phi(t,0) = P(t)\exp(Bt) \qquad (2.2.9)$$

where $P \in C^1(\mathbf{R}_+, \mathbf{C}^{n^2})$ is a regular T-periodic matrix: $P(t+T) = P(t)$, $P(0) = I$, and B is a constant matrix (with possibly complex entries); if $A(t)$ is real and (2.2.1) is considered as a periodic system with period $2T$, then

$$\Phi(t,0) = P_1(t)\exp(B_1 t) \qquad (2.2.10)$$

where $P_1 \in C^1(\mathbf{R}_+, \mathbf{R}^{n^2})$ is a regular $2T$-periodic matrix: $P_1(t+2T) = P_1(t)$, $P_1(0) = I$, and B_1 is a real constant matrix.

Proof. Let B be a matrix satisfying

$$\exp(BT) = C = \Phi(T,0) \qquad (2.2.11)$$

(see (A 1.9)). We may write

$$\Phi(t,0) = \Phi(t,0)\exp(-Bt)\exp(Bt).$$

Consider the factor $P(t) := \Phi(t,0)\exp(-Bt)$. We obtain, applying (2.2.6) and (2.2.11),

$$
\begin{aligned}
P(t+T) &= \Phi(t+T,0)\exp(-B(t+T)) \\
&= \Phi(t,0)\Phi(T,0)\exp(-Bt)\exp(-BT) \\
&= \Phi(t,0)\exp(BT)\exp(-BT)\exp(-Bt) \\
&= \Phi(t,0)\exp(-Bt) = P(t).
\end{aligned}
$$

The definition of P and the last relation imply that P meets the requirements of the Theorem. Now, if $A(t)$ is real, then applying Corollary A 1.10, a *real* matrix B_1 exists satisfying

$$\exp(B_1 2T) = C^2 = \Phi(T,0)\Phi(T,0) = \Phi(2T,0) \qquad (2.2.12)$$

where (2.2.6) was used also. Introducing $P_1(t) := \Phi(t,0)\exp(-B_1 t)$ we may repeat the first part of the proof, writing P_1, B_1, and $2T$ for P, B, and T, respectively, everywhere. Since the fundamental matrix $\Phi(t,0)$ is real, the matrix function $P_1(t)$ meets the requirement of the Theorem. $\square$

The eigenvalues of the matrix B defined by (2.2.11) are called the *characteristic exponents* of system (2.2.1). Obviously, if ν is a characteristic exponent, then $\lambda = \exp(\nu T)$ is a characteristic multiplier of the system (cf. Corollary A 1.8), and vice versa. Note, however, that the characteristic exponents are not uniquely determined (since the matrix B is not either). If λ is a characteristic multiplier, then the corresponding characteristic exponents are among

$$\nu_k = \frac{1}{T}\ln\lambda = \frac{\ln|\lambda|}{T} + \frac{i(\arc\lambda + 2k\pi)}{T} \qquad (k = 0, \pm 1, \pm 2, ...). \quad (2.2.13)$$

Thus, a multiple and several different characteristic exponents may correspond to the same characteristic multiplier. It is also worthwhile to observe that the eigenvalues of the real matrix B_1 defined by (2.2.12) are also in the set of the ν_k's given by (2.2.13) (λ running through the set of the characteristic multipliers). Indeed, if ν^1 is an eigenvalue of B_1, then a characteristic multiplier λ exists such that

$$\exp(\nu^1 2T) = \lambda^2.$$

Hence

$$\nu^1 = \frac{1}{2T}\ln\lambda^2 = \frac{\ln|\lambda|}{T} + \frac{i(\arc\lambda + k\pi)}{T} = \nu_k$$

for some $k = 0, \pm 1, \pm 2, ...$.

In stability investigations it will be important to remember that according to (2.2.13) the following two statements are equivalent: *the real part of a characteristic exponent is negative* ($\operatorname{Re}\nu_k < 0$); *the modulus of the corresponding characteristic multiplier is less than one* ($|\lambda| < 1$).

Floquet's Theorem implies that to each characteristic exponent ν there corresponds a solution φ of system (2.2.1) of the form

$$\varphi(t) = \exp(\nu t)p(t) \qquad (2.2.14)$$

where the vector p is periodic in t with period T: $p(t + T) = p(t)$. In order to see this, denote an eigenvector of the matrix B (resp. B_1) corresponding to the eigenvalue ν by s, i.e. $Bs = \nu s$ (resp. $B_1 s = \nu s$), and consider the solution φ with initial value $\varphi(0) = s$. Then

$$\varphi(t) = P(t)\exp(Bt)s = P(t)e^{\nu t}s = e^{\nu t}p(t)$$

where $p(t) = P(t)s$ is, obviously, periodic with period T. We applied here Corollary A 1.8. The reasoning is similar with B_1.

Applying Floquet's theory, we are going to show that every homogeneous linear system with periodic coefficients can be transformed into a linear system with constant coefficients by a linear t-dependent transformation. This is the problem of *reducibility* due to Liapunov which is treated in a more general setting in Demidovich [1967a].

Theorem 2.2.6 *(Liapunov [1892]). If $A(t)$ is real, then there exists a regular T-periodic matrix $P \in C^1(\mathbf{R}_+, \mathbf{C}^{n^2})$ (resp. a regular $2T$-periodic matrix $P \in C^1(\mathbf{R}_+, \mathbf{R}^{n^2})$) satisfying $P(0) = I$ such that the coordinate transformation $y = P(t)z$ carries system (2.2.1) into a homogeneous linear system with constant (resp. real constant) coefficients.*

Proof. We are going to show that the matrix $P(t) = \Phi(t, 0)\exp(-Bt)$ of formula (2.2.9) where B is given by (2.2.11) serves the purpose. Let us substitute $y = P(t)z$ into (2.2.1), and see the differential equation satisfied by z:

$$\dot{P}z + P\dot{z} = APz$$

(we suppress the writing out of the argument t where possible). Reordering the last expression yields

$$\dot{z} = P^{-1}(AP - \dot{P})z.$$

The coefficient matrix here is

$$
\begin{aligned}
P^{-1}(AP - \dot{P}) &= \exp(Bt)\Phi^{-1}(A\Phi\exp(-Bt) \\
&\quad - \dot{\Phi}\exp(-Bt) + \Phi B\exp(-Bt)) \\
&= \exp(Bt)\Phi^{-1}\Phi B\exp(-Bt) = B,
\end{aligned}
$$

i.e. z satisfies

$$\dot{z} = Bz \qquad (2.2.15)$$

where B is given by (2.2.11). An analogous reasoning (with $P_1(t)$ instead of $P(t)$) carries the system into $\dot{z} = B_1 z$ where B_1 is defined by (2.2.12). $\square$

We have defined the characteristic exponents of system (2.2.1) as the eigenvalues of the coefficient matrix B of system (2.2.15). If we consider the latter system as a periodic one with period T, then its principal matrix is, clearly, $\exp(BT)$. So that (2.2.15) considered as a T-periodic system has the same characteristic multipliers as system (2.2.1). This is a particular case of a more general property. If we transform system (2.2.1) into another one by a periodic transformation, then the characteristic multipliers of the new system will be the same as those of the old one.

Theorem 2.2.7 *(Liapunov [1892]). Let $S \in C^1(\mathbf{R}_+, \mathbf{C}^{n^2})$ be regular and T-periodic; the transformation $y = S(t)z$ carries system (2.2.1) into a homogeneous linear system with T-periodic coefficient matrix whose characteristic multipliers coincide with those of (2.2.1).*

Proof. Substitute $y = S(t)z$ into (2.2.1) :

$$\dot{S}(t)z + S(t)\dot{z} = A(t)S(t)z;$$

thus, z satisfies the system

$$\dot{z} = S^{-1}(t)(A(t)S(t) - \dot{S}(t))z. \qquad (2.2.16)$$

The coefficient matrix of this system is, clearly, continuous and T-periodic. Let $\Phi(t,0)$ be the fundamental matrix of (2.2.1) satisfying $\Phi(0,0) = I$. Then $\Psi(t) := S^{-1}(t)\Phi(t,0)$ is a fundamental matrix of (2.2.16). According to what has been said after (2.2.6) the characteristic multipliers of (2.2.16) are the eigenvalues of $\Psi^{-1}(0)\Psi(T)$. But

$$\begin{aligned} \Psi^{-1}(0)\Psi(T) &= \Phi^{-1}(0,0)S(0)S^{-1}(T)\Phi(T,0) \\ &= IS(0)S^{-1}(0)C = C \end{aligned}$$

where we used (2.2.5) and the fact that S^{-1} is also T periodic. $\square$

We have seen that the characteristic multipliers form an invariant set of a homogeneous linear system with periodic coefficients in the sense that their set is invariant under periodic coordinate transformation. The properties of this set determine the behaviour of the solutions, the existence of a periodic solution and, as we shall see, the stability. However in order to determine the characteristic multiplier, we need a fundamental matrix, i.e. we need to know all the solutions. To find the transformation that reduces the periodic system to a system with constant coefficients is not easier than to find a fundamental matrix solution. This vicious circle can be broken in some special cases, especially in two dimensions when Liouville's formula (Theorem 1.2.4) may help. (See also Problem 2.6.2.) Thus, we need a method producing approximations of the characteristic multipliers.

We are going to construct an algorithm that leads to the approximate determination of the characteristic multipliers (see Demidovich [1967a], also Kotsis [1976] and Hsu [1974]). For sake of simplicity assume that $A(t)$ is real. Divide the interval $[0, T]$ into N equal parts by $0 = t_0 < t_1 < \cdots < t_N = T$, i.e. $h = t_{k+1} - t_k = T/N$, $k = 0, 1, ..., N-1$, and replace the coefficient matrix of the matrix differential equation

$$\dot{Y} = A(t)Y \qquad (2.2.17)$$

corresponding to (2.2.1) by a piecewise constant matrix $A_h(t)$ defined in the following way:

$$A_h(t) = A_{hk} \qquad \text{for} \qquad t \in [t_k, t_{k+1}) \qquad (k = 0, 1, ..., N-1)$$

where A_{hk} is a constant matrix satisfying

$$\min A(t) \le A_{hk} \le \max A(t), \qquad t \in [t_k, t_{k+1}]. \qquad (2.2.18)$$

(Here $\min A(t)$ is the matrix whose elements are the minima of the corresponding elements of $A(t)$ on the interval; the definition of $\max A(t)$

is similar; a matrix A is less than a matrix B of the same dimensions if all its elements are less than the respective elements of B.) Thus, e.g. one may set $A_{hk} = A(t_k)$. Let $Y_h : [0, T] \to \mathbf{R}^{n^2}$ be the continuous matrix that satisfies the system

$$\dot{Y}_h = A_h(t)Y_h \qquad (2.2.19)$$

at points of continuity of A_h, i.e. for $t \in \cup_{k=0,...,N-1}(t_k, t_{k+1})$, and $Y_h(0) = I$, the unit matrix. It is easy to see that

$$Y_h(t) = \exp((t - t_k)A_{hk}) \exp(hA_{h,k-1}) \cdots \exp(hA_{h0}), \qquad (2.2.20)$$

$$t \in [t_k, t_{k+1}] \qquad (k = 0, ..., N - 1);$$

thus,

$$Y_h(T) = \exp(hA_{h,N-1}) \exp(hA_{h,N-2}) \cdots \exp(hA_{h0}). \qquad (2.2.21)$$

By an abuse of notation denote the solution of (2.2.17) that satisfies $Y(0) = I$ by $Y(t)$, i.e. $Y(T)$ is the principal matrix of system (2.2.1). Its eigenvalues are the characteristic multipliers of the system. We are going to show that $Y_h(T)$ tends to $Y(T)$ as h tends to zero. Integrating the identities (2.2.17) and (2.2.19) we get

$$Y(t) = I + \int_0^t A(\tau)Y(\tau)d\tau,$$

$$Y_h(t) = I + \int_0^t A_h(\tau)Y_h(\tau)d\tau.$$

Hence, for $t \in [0, T)$,

$$Y_h(t) - Y(t) = \int_0^t (A_h(\tau) - A(\tau))Y_h(\tau)d\tau$$

$$+ \int_0^t A(\tau)(Y_h(\tau) - Y(\tau))d\tau.$$

Taking any usual matrix norm we get the inequality

$$\|Y_h(t) - Y(t)\| \leq \int_0^t \|A_h(\tau) - A(\tau)\| \ \|Y_h(\tau)\|d\tau$$

$$+ \int_0^t \|A(\tau)\| \ \|Y_h(\tau) - Y(\tau)\|d\tau. \qquad (2.2.22)$$

Let $\|A(t)\| \leq M$ for $t \in [0, T]$; then because of (2.2.18), $\|A_{hk}\| \leq M$ also $(k = 0, 1, ..., N - 1)$, and because of (2.2.20)

$$\|Y_h(t)\| \leq \exp(h\|A_{hk}\|) \exp(h\|A_{h,k-1}\|) \cdots \exp(h\|A_{h0}\|)$$
$$\leq \exp(hNM) = \exp(TM), \qquad t \in [0, T].$$

On the other hand, because of the uniform continuity of A on the interval $[0, T]$, to every $\varepsilon > 0$ there belongs a $\delta(\varepsilon) > 0$ such that for every $\tau_1, \tau_2 \in [0, T]$, $|\tau_1 - \tau_2| < \delta(\varepsilon)$ we have $\|A(\tau_1) - A(\tau_2)\| < \varepsilon$. As a consequence, because of (2.2.18), $\|A_h(\tau) - A(\tau)\| < \varepsilon$ also for $\tau \in [0, T]$ if $h < \delta(\varepsilon)$. Thus, inequality (2.2.22) can be continued in the following way:

$$\|Y_h(t) - Y(t)\| \leq \varepsilon T \exp(TM) + \int_0^t M \|Y_h(\tau) - Y(\tau)\| d\tau$$

if $h < \delta(\varepsilon)$. Applying Gronwall's Lemma (Problem 1.6.5) we get

$$\|Y_h(t) - Y(t)\| \leq \varepsilon T \exp(TM + Mt), \qquad t \in [0, T],$$

hence,

$$\|Y_h(T) - Y(T)\| \leq \varepsilon T \exp(2TM),$$

and this means that

$$Y(T) = \lim_{h \to 0} Y_h(T).$$

Since the eigenvalues of a matrix are continuous functions of the entries, this implies that the eigenvalues of $Y_h(T)$ tend to the characteristic multipliers of system (2.2.1). The matrix $Y_h(T)$ can be determined numerically for arbitrary positive h by applying formula (2.2.21).

2.3   Forced Linear Oscillations

In this Section we treat inhomogeneous linear periodic systems of the form

$$\dot{y} = A(t)y + b(t) \tag{2.3.1}$$

where the coefficient matrix $A(t)$ and the forcing term $b(t)$ are continuous and T-periodic functions: $A \in C^0(\mathbf{R}_+)$, $b \in C^0(\mathbf{R}_+)$, $A(t + T) = A(t)$, $b(t + T) = b(t)$ for some $T > 0$ and all $t \in \mathbf{R}_+$. We are going to extend Theorems 2.1.1-2.1.3 to system (2.3.1). This will be done by applying Theorem 2.2.6 which establishes the reducibility of periodic systems.

Theorem 2.3.1. *If the homogeneous system*

$$\dot{y} = A(t)y \tag{2.3.2}$$

corresponding to (2.3.1) has no periodic solution of period T apart from the trivial one, then (2.3.1) has one and only one periodic solution of period T.

Proof. By Theorem 2.2.6 a T-periodic regular matrix function $P(t)$ exists that carries (2.3.1) by the coordinate transformation $y = P(t)z$ into (2.2.15), i.e.

$$\dot{z} = Bz \tag{2.3.3}$$

where B is a constant matrix. The same transformation carries (2.3.1) into

$$\dot{z} = Bz + P^{-1}(t)b(t). \tag{2.3.4}$$

Now, if (2.3.2) has no periodic solution apart from the trivial one, then, clearly, (2.3.3) does not have one either. By Theorem 2.1.1 this implies that (2.3.4) has a T-periodic solution, and this means that (2.3.1) has one too. The uniqueness of this solution, follows from the fact that if (2.3.1) had two different T-periodic solution then their difference would be a non-trivial T-periodic solution of (2.3.2) contrary to the assumption. $\square$

Corollary 2.2.3 implies:

Corollary 2.3.2. *If* 1 *is not a characteristic multiplier of system (2.3.2), then the inhomogeneous system (2.3.1) has one and only one T-periodic solution.*

We turn now to the more intricate problem of the existence of a periodic solution in the case the homogeneous system that has non-trivial periodic solutions. The results will be similar to those of the case of a constant coefficient matrix. This is again a consequence of the reducibility of a periodic system. We introduce, first, the *adjoint system*

$$\dot{\eta} = -A^*(t)\eta \tag{2.3.5}$$

of (2.3.2), and note that if y is a solution of (2.3.2) and η a solution of (2.3.5), then

$$y'(t)\overline{\eta}(t) = \text{const}. \tag{2.3.6}$$

This can be proved exactly the same way as (2.1.8). The analogue of Lemma 2.1.2 is valid:

Lemma 2.3.3. *For system (2.3.2) and its adjoint (2.3.5) the dimensions of the subspaces of T-periodic solutions are equal.*

Proof. The proof proceeds by reducing both systems to systems with constant coefficients. We know by Theorem 2.2.6 that a continuously differentiable, regular, T-periodic matrix $P(t)$ exists that transforms system (2.3.2) to the form (2.3.3). We are going to show that the transformation

$$\eta = P^{*-1}(t)\zeta \tag{2.3.7}$$

reduces the adjoint system (2.3.5) to

$$\dot{\zeta} = -B^*\zeta \tag{2.3.8}$$

which is the adjoint of system (2.3.3). Indeed, if η satisfies (2.3.5), then ζ satisfies

$$\dot{\zeta} = P^*(t)(-A^*(t)P^{*-1}(t) - \dot{P}^{*-1}(t))\zeta. \tag{2.3.9}$$

The coefficient of the last system is

$$-P^*(t)A^*(t)P^{*-1}(t) - P^*(t)\dot{P}^{*-1}(t) = -P^*(t)A^*(t)P^{*-1}(t) + \dot{P}^*(t)P^{*-1}(t)$$
$$= -(P^{-1}(t)(A(t)P(t) - \dot{P}(t)))^* = -B^*$$

where we used the identity $(P^*(t)P^{*-1}(t))^{\bullet} = I^{\bullet} = 0$. Thus, (2.3.9) is the adjoint (2.3.8) of system (2.3.3) indeed. Now, if (2.3.2) has k linearly independent T-periodic solutions, then, clearly, the same is true for system (2.3.3). But then by Lemma 2.1.2 its adjoint system (2.3.8) has k linearly independent T-periodic solutions also, and this implies that the same is true for system (2.3.5). $\square$

It might be useful to look at the following scheme illustrating this Lemma:

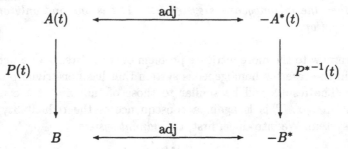

We turn now to the analogue of Theorem 2.1.3.

Theorem 2.3.4. *Assume that the subspace of T-periodic solutions of the homogeneous system (2.3.2) is of dimension $k \geq 1$, and denote k linearly independent T-periodic solutions of the adjoint system (2.3.5) by $\eta^1, \eta^2, ..., \eta^k$; system (2.3.1) has T-periodic solutions if and only if the "orthogonality conditions"*

$$\int_0^T \eta^{i\prime}(t)\bar{b}(t)dt = 0 \qquad (i = 1, 2, ..., k) \qquad (2.3.10)$$

hold.

Proof. By the previous Lemma, (2.3.5) and (2.3.8) have k (and not more) linearly independent T-periodic solutions. Denote such a set for system (2.3.5) by $\eta^1, ..., \eta^k$ and the *corresponding* set for system (2.3.8) by $\zeta^1, \zeta^2, ..., \zeta^k$, i.e. let $\eta^i(t) = P^{*-1}(t)\zeta^i(t)$ $(i = 1, 2, ..., k)$.

System (2.3.1) has T-periodic solutions if and only if its "reduced system" (2.3.4) has. By Theorem 2.1.3 the latter has one, at least, if and only if the orthogonality conditions

$$\int_0^T \zeta^{i\prime}(t)\bar{P}^{-1}(t)\bar{b}(t)dt = 0 \qquad (i = 1, 2, ..., k)$$

hold. But

$$\int_0^T \zeta^{i'} \overline{P}^{-1}(t)\overline{b}(t)dt = \int_0^T \eta^{i'}(t)\overline{P}(t)\overline{P}^{-1}(t)\overline{b}(t)dt$$

$$= \int_0^T \eta^{i'}(t)\overline{b}(t)dt,$$

and this proves the Theorem. $\square$

Condition (2.3.10) has an intuitive interpretation. The subspace spanned by the periodic solutions of the homogeneous system (2.3.2) is closely related to the subspace of periodic solutions of the adjoint system (2.3.5). This intrinsic relation is expressed by (2.3.6). If the system is one dimensional, then the periodic solution of the adjoint system is a constant times the reciprocal of the periodic solution of (2.3.2). Now, if the external excitation, the periodic forcing term b, has no projection on this subspace in which the system is "self-oscillating", i.e. (2.3.10) holds, then the forced system may oscillate periodically. If this projection is non-zero, then a "*resonance*" occurs, the forced system has no periodic solutions, and as we shall see, all solutions are unbounded.

Example 2.3.1. Consider the two scalar periodic inhomogeneous linear differential equations

$$\dot{y} = y\cos t + \cos t \qquad\qquad (2.3.11)$$

and

$$\dot{y} = y\cos t + \sin t. \qquad\qquad (2.3.12)$$

Both belong to the same homogeneous equation $\dot{y} = y\cos t$. The general solution of the latter is $y_H(t) = c\exp\sin t$. This, clearly, means that the homogeneous equation has one (linearly independent) periodic solution, say, $y_0(t) = \exp\sin t$. The adjoint equation is $\dot{\eta} = -\eta\cos t$. It has also one (linearly independent) periodic solution that is the reciprocal of y_0, namely $\eta_0(t) = \exp(-\sin t)$. The function $\cos t$ is orthogonal to η_0 :

$$\int_0^{2\pi} e^{-\sin t}\cos t\, dt = 0,$$

and it is easy to see that the general solution of (2.3.11) is $y_1(t) = -1 + c\exp\sin t$. Thus, this inhomogeneous equation has a one parameter family of periodic solutions. On the other hand, the function $\sin t$ is *not* orthogonal to η_0 :

$$\int_0^{2\pi} e^{-\sin t}\sin t\, dt < 0. \qquad\qquad (2.3.13)$$

(This is clear because $\exp(-\sin t) < 1$ on $(0,\pi)$, and it is greater than 1 on $(\pi, 2\pi)$.) So (2.3.12) has no periodic solution. In fact, the general solution of this inhomogeneous equation is

$$y_2(t) = \int_0^t e^{-\sin\tau} \sin\tau \ d\tau \ e^{\sin t} + ce^{\sin t}.$$

By (2.3.13) all solutions tend to $-\infty$ as t tends to infinity.

Corollary 2.3.5. *If the homogeneous system (2.3.2) has k linearly independent T-periodic solutions, and the orthogonality conditions (2.3.10) hold, then the inhomogeneous system (2.3.1) has a k parameter family of T-periodic solutions.*

Proof. This follows from the fact that if (2.1.9) has k linearly independent solutions, then (2.1.6), if solvable, has a k parameter family of solutions. □

We turn now to the *resonance* case, i.e. to the case when the forcing term has a non-zero projection to the subspace of the periodic solutions of the adjoint system; in other words, to the case when (2.3.10) does not hold. We are going to show that in this case all solutions of the inhomogeneous system are unbounded. This result is a consequence of the following:

Theorem 2.3.6 *(Massera [1950]). If system (2.3.1) has a bounded solution on $t \in [0, \infty)$, then it has also a T-periodic solution.*

Proof. Denote the solution of (2.3.1) bounded on $[0, \infty)$ by φ, and the fundamental matrix solution of (2.3.2) that satisfies $\Phi(0) = I$ by Φ. Then applying formula (1.2.11)

$$\varphi(t) = \Phi(t)\varphi^0 + \int_0^t \Phi(t)\Phi^{-1}(s)b(s)ds$$

where $\varphi^0 = \varphi(0)$, and

$$\varphi(T) = \Phi(T)\varphi^0 + \int_0^T \Phi(T)\Phi^{-1}(s)b(s)ds,$$

or denoting the integral on the right-hand side by c :

$$\varphi(T) = \Phi(T)\varphi^0 + c.$$

We know that the function $\varphi(t+T)$ is also a solution, and it assumes the value $\varphi(T)$ at $t = 0$. So

$$\varphi(t+T) = \Phi(t)\varphi(T) + \int_0^t \Phi(t)\Phi^{-1}(s)b(s)ds$$

and
$$\varphi(2T) = \Phi(T)\varphi(T) + c = \Phi^2(T)\varphi^0 + (\Phi(T) + I)c.$$

Proceeding this way we get for arbitrary $m = 1, 2, \dots$,

$$\varphi(mT) = \Phi^m(T)\varphi^0 + \sum_{k=0}^{m-1} \Phi^k(T)c. \qquad (2.3.14)$$

Assume now, contrary to the statement, that (2.3.1) has no periodic solution of period T. Applying formula (1.2.11) again, this means that the periodicity condition

$$y(T) = \Phi(T)y^0 + c = y^0$$

has no solution, i.e. the linear algebraic system

$$(I - \Phi(T))y^0 = c$$

has no solution for y^0. From the idea used at the beginning of the proof of Theorem 2.1.3 this implies that a non-zero vector v exists satisfying $v'(I - \overline{\Phi}(T)) = 0$ such that $v'\overline{c} \neq 0$, or, taking transposes,

$$(I - \Phi^*(T))v = 0, \qquad c^*v \neq 0.$$

The first condition means that $\Phi^*(T)v = v$; hence,

$$\Phi^{*^k}(T)v = v \qquad (k = 0, 1, 2, \dots). \qquad (2.3.15)$$

Take the scalar product of $\varphi(mT)$ with v :

$$
\begin{aligned}
\varphi'(mT)\overline{v} &= \varphi^{0'}\Phi^{m'}(T)\overline{v} + \sum_{k=0}^{m-1} c'\Phi^{k'}(T)\overline{v} \\
&= \varphi^{0'}(\overline{\Phi^{*m}(T)v}) + \sum_{k=0}^{m-1} \overline{c^*\Phi^{*k}(T)v} \\
&= \varphi^{0'}\overline{v} + \sum_{k=0}^{m-1} \overline{c^*v} = \varphi^{0'}\overline{v} + m\overline{c^*v}
\end{aligned}
$$

where (2.3.14) and (2.3.15) were used. If m tends to infinity, the modulus of the right-hand side tends to infinity, and this means that $|\varphi(mT)|$ tends to infinity contrary to the assumption that φ is bounded, a contradiction. $\square$

Corollary 2.3.7. *If system (2.3.1) has no periodic solution, then every solution is unbounded on $t \in [0, \infty)$.*

2.4   Stability of Linear Systems

If a linear system has periodic coefficients (and forcing term), then one may add something to the general results about the stability of linear systems expressed in Theorems 1.4.1-1.4.3 and 1.5.6. In this Section we deal with these additional criteria.

Consider first the homogeneous linear system with periodic coefficients

$$\dot{y} = A(t)y \qquad (2.4.1)$$

where $A \in C^0(\mathbf{R}_+)$, $A(t+T) = A(t)$ for some $T > 0$ and all $t \in \mathbf{R}_+$. Theorems 1.4.1-1.4.3 imply that if (2.4.1) has a non-trivial periodic solution, then the system (and this solution) cannot be asymptotically stable.

The stability conditions for system (2.4.1) are simple consequences of the corresponding conditions for linear systems with constant coefficients. We sum them up in the following:

Theorem 2.4.1. *(a) System (2.4.1) is asymptotically stable if and only if all its characteristic multipliers are in modulus less than one.*

(b) The system is stable in the Liapunov sense if and only if all its characteristic multipliers are in modulus less than or equal to one, and those with modulus one are simple in the minimal polynomial.

(c) The system is unstable if it has either a characteristic multiplier with modulus greater than one or a characteristic multiplier with modulus one and multiplicity greater than one in the minimal polynomial.

Proof. These statements follow from Theorems 1.4.4-1.4.6, from Theorem A 1.12 and from Theorem 2.2.6. According to this last Theorem, system (2.4.1) can be transformed into a system with constant coefficients $\dot{z} = Bz$ by a periodic coordinate transformation, i.e. the solutions $y(t)$ of (2.4.1) and the solutions $z(t)$ of this system with constant coefficients are related by $y(t) = P(t)z(t)$ where P is a regular, periodic, continuous matrix function. So that if $z(t)$ tends to zero when t tends to infinity, then so does $y(t)$; if $z(t)$ stays bounded, then the same holds for $y(t)$; and if there are unbounded z's, the corresponding y's are also unbounded. The eigenvalues of B are the characteristic exponents of system (2.4.1) related to the multipliers by (2.2.13). $\square$

Note that the conditions for (a) , (b) and (c) could be expressed in an obvious way also in terms of the characteristic exponents. System (2.4.1) is asymptotically stable, stable in the Liapunov sense, unstable iff all the characteristic exponents have negative real parts, all the characteristic exponents have non-positive real parts, and those with zero real part are simple in the minimal polynomial, there is at least one exponent either with positive real part or with zero real part and with multiplicity greater than one in the minimal polynomial, respectively. If we tried to simplify the criteria and applied the characteristic polynomial instead of the minimal one, we could run into trouble as shown in the following Example. This is because different characteristic exponents may correspond to the same multiplier as this has been pointed out around formula (2.2.13).

Example 2.4.1. Consider the 5-periodic homogeneous linear system

$$
\begin{aligned}
\dot{y}_1 &= -\frac{13\pi}{30} y_2 + y_3 \frac{2\pi}{5} \cos^2\left(\frac{2\pi}{5}t\right) + y_4 \frac{\pi}{5} \sin\left(\frac{4\pi}{5}t\right) \\
\dot{y}_2 &= \frac{13\pi}{30} y_1 \qquad - y_3 \frac{\pi}{5} \sin\left(\frac{4\pi}{5}t\right) + y_4 \frac{2\pi}{5} \cos^2\left(\frac{2\pi}{5}t\right) \\
\dot{y}_3 &= -\frac{13\pi}{30} y_4 \\
\dot{y}_4 &= \frac{13\pi}{30} y_3.
\end{aligned} \tag{2.4.2}
$$

The 5-periodic linear transformation $y = P(t)z$ where

$$
P(t) = \begin{pmatrix}
\cos\left(\frac{2\pi}{5}t\right) & -\sin\left(\frac{2\pi}{5}t\right) & \sin\left(\frac{2\pi}{5}t\right) & 0 \\
\sin\left(\frac{2\pi}{5}t\right) & \cos\left(\frac{2\pi}{5}t\right) & 0 & \sin\left(\frac{2\pi}{5}t\right) \\
0 & 0 & \cos\left(\frac{2\pi}{5}t\right) & -\sin\left(\frac{2\pi}{5}t\right) \\
0 & 0 & \sin\left(\frac{2\pi}{5}t\right) & \cos\left(\frac{2\pi}{5}t\right)
\end{pmatrix}
$$

carries system (2.4.2) into the system with constant coefficients

$$
\dot{z}_1 = -\frac{\pi}{30} z_2, \quad \dot{z}_2 = \frac{\pi}{30} z_1, \quad \dot{z}_3 = -\frac{13\pi}{30} z_4, \quad \dot{z}_4 = \frac{13\pi}{30} z_3 \tag{2.4.3}
$$

(cf. Theorem 2.2.6). The coefficient matrix

$$
B = \begin{pmatrix}
0 & -\pi/30 & 0 & 0 \\
\pi/30 & 0 & 0 & 0 \\
0 & 0 & 0 & -13\pi/30 \\
0 & 0 & 13\pi/30 & 0
\end{pmatrix}
$$

of this system has the following eigenvalues: $\pm i\pi/30$, $\pm i13\pi/30$, i.e. system (2.4.2) has four simple characteristic exponents all with real part zero. Hence, (2.4.2) is a stable system in the Liapunov sense. A fundamental matrix of (2.4.3) is

$$
\exp(Bt) = \begin{pmatrix}
\cos\left(\frac{\pi}{30}t\right) & -\sin\left(\frac{\pi}{30}t\right) & 0 & 0 \\
\sin\left(\frac{\pi}{30}t\right) & \cos\left(\frac{\pi}{30}t\right) & 0 & 0 \\
0 & 0 & \cos\left(\frac{13\pi}{30}t\right) & -\sin\left(\frac{13\pi}{30}t\right) \\
0 & 0 & \sin\left(\frac{13\pi}{30}t\right) & \cos\left(\frac{13\pi}{30}t\right)
\end{pmatrix}.
$$

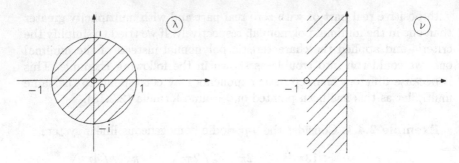

FIGURE 2.4.1. The transformation $\lambda = (\nu + 1)/(\nu - 1)$.

The corresponding fundamental matrix of system (2.4.2) is $\Psi(t) = P(t)\exp(Bt)$, and, clearly, $\Psi(0) = I$. The corresponding principal matrix is

$$C = \Psi(5) = I\exp(5B) = \begin{pmatrix} \sqrt{3}/2 & -1/2 & 0 & 0 \\ 1/2 & \sqrt{3}/2 & 0 & 0 \\ 0 & 0 & \sqrt{3}/2 & -1/2 \\ 0 & 0 & 1/2 & \sqrt{3}/2 \end{pmatrix}.$$

The characteristic polynomial of C is

$$\det(C - \lambda I) = (\lambda^2 - \lambda\sqrt{3} + 1)^2;$$

thus, the system has two characteristic multipliers $\exp(\pm i\pi/6)$, both double roots of the characteristic polynomial. However, it is easy to see that the minimal polynomial is $\lambda^2 - \lambda\sqrt{3} + 1$, i.e. the characteristic multipliers with modulus one are *simple in the minimal polynomial.*

As we have seen, the asymptotic stability of a periodic linear system depends on whether the characteristic multipliers lie in the interior of the unit circle or not. Criteria for that, i.e. conditions guaranteeing that all the roots of a polynomial have modulus less than 1, can be deduced from the Routh-Hurwitz criterion (Theorem 1.4.8), applying the linear rational complex transformation

$$\lambda = \frac{\nu + 1}{\nu - 1}. \tag{2.4.4}$$

This transformation carries the interior of the unit circle of the "λ-plane" into the interior of the left half plane of the "ν-plane" (see Figure 2.4.1).

Thus, the polynomial

$$p(\lambda) = \lambda^n + a_{n-1}\lambda^{n-1} + \cdots + a_1\lambda + a_0$$

has all its roots in the interior of the unit circle if and only if the polynomial

$$P(\nu) = (\nu - 1)^n p\left(\frac{\nu+1}{\nu-1}\right)$$

is a stable one, i.e. if the latter (or its -1 multiple) satisfies the Routh-Hurwitz criterion. How these conditions look like in the case of low degrees can be seen from Problems 2.6.3 -2.6.4; for an application, see e.g. Stépán, Steven, Maunder [1990]; for the more sophisticated "Jury's criterion," see Kuo [1977].

We know (see Corollary 1.4.2) that the stability property of the homogeneous system (2.4.1) settles the stability of the inhomogeneous system

$$\dot{y} = A(t)y + b(t) \tag{2.4.5}$$

where $A \in C^0(\mathbf{R}_+)$, $b \in C^0(\mathbf{R}_+)$, $A(t+T) = A(t)$, $b(t+T) = b(t)$ for some $T > 0$ and all $t \in \mathbf{R}_+$. Still, it may be worthwhile to consider some implications.

In the case that all the characteristic multipliers of system (2.4.1) are in modulus less than 1, i.e. (2.4.1) is asymptotically (globally) stable, clearly, it has no periodic solution apart from the trivial one. In this case, (2.4.5) has a single T-periodic solution (see Theorem 2.3.1), and it is globally asymptotically stable, i.e. all solutions tend to the periodic one as t tends to infinity.

If (2.4.1) is stable in the Liapunov sense but not asymptotically (e.g. if it has non-trivial periodic solutions), then such is the inhomogeneous system (2.4.5) too. In this case there are two possibilities, (2.4.5) falling in exactly one of these categories. Either all solutions are unbounded as t tends to infinity or (2.4.5) has T-periodic solutions (see Corollary 2.3.7). In the latter case each periodic solution is stable in the Liapunov sense.

At this point we note that system (2.4.1), resp. (2.4.5) may have periodic solutions of period different from T, and it may happen that a T-periodic solution does not exist at all (apart from the trivial one in case of the homogeneous system). This follows easily from Floquet's theory, and from (2.2.9) in particular. If $\exp(Bt)$ does not have a T-periodic column but has a column with period a rational multiple of T, then system (2.4.1) (i.e. (2.2.1)) has no T-periodic solution (apart from the trivial one) but has a periodic solution with period an integer multiple of T. We know from Corollary 2.2.3 that (2.4.1) has a non-trivial T-periodic solution iff the number 1 is a characteristic multiplier. Now, if (2.4.1) is considered to be an mT-periodic system ($m \in \mathbf{N}$), then (2.2.4) implies that $\Phi(t+mT) = \Phi(t)C^m$, i.e. if C denotes the principal matrix of system

(2.4.1) considered as a T-periodic system, then C^m is the principal matrix of the same system if it is considered an mT-periodic one. Since, if λ is an eigenvalue of C, then λ^m is an eigenvalue of C^m, we get to the conclusion that (2.4.1) *has an mT periodic solution if and only if it has an m-th root of unity as a characteristic multiplier* (if considered as a T-periodic system). We draw attention to the fact that although the *characteristic multipliers* λ of (2.4.1) is raised to integer powers, $\lambda^2, \lambda^3, ..., \lambda^m, ...$ as the system is considered a $2T$-, a $3T$-, ... an mT-, ... periodic one, the *characteristic exponents* do not change. The formula defining the matrix B whose eigenvalues are the characteristic exponents is (2.2.11). Now, if the system is considered to be mT-periodic, then we have to replace T by mF and C by C^m, respectively, in this formula. But, clearly, with the old matrix B defined by (2.2.11) we have

$$\exp(BmT) = C^m.$$

In Example 2.4.1 system (2.4.2) is 5-periodic but it has no non-trivial 5-periodic solution. However, as this can be seen from the form of $\exp(Bt)$, the system has periodic solutions of period 60. In fact, the characteristic multipliers, $\exp(\pm i\pi/6)$, are 12th roots of unity: $60 = 12 \cdot 5$. In other words, if (2.4.2) is considered to be a 60-periodic system, then its principal matrix is C^{12}, and the characteristic multipliers are $(\exp(\pm i\pi/6))^{12} = \exp(\pm 2\pi i) = 1$.

In the rest of this Section we are going to deal with the important special case when the periodic linear system is "close" to an asymptotically stable one with constant coefficients. Many situations in real life can be modelled by linear systems with constant coefficients, some parameters of which undergo small periodic *perturbations*. The problem is then how a large perturbation can the system bear without losing its stability? We present a solution to this problem due to Chetaev [1955] (cf. also Yakubovich-Starzhinskij [1972]). Besides the importance of the problem on its own account, this is the first instance in this book where we treat a perturbation result.

Consider the *perturbed linear system*

$$\dot{y} = (G + \mu H(t))y \tag{2.4.6}$$

where G is a real stable $n \times n$ matrix, $H \in C^0(\mathbf{R}_+)$ is a periodic matrix function with period $T > 0$: $H(t + T) = H(t)$, and $\mu \in \mathbf{R}$ is a "small parameter". We treat first the unperturbed system ($\mu = 0$)

$$\dot{y} = Gy \tag{2.4.7}$$

which is, by assumption, asymptotically stable. If we choose a positive definite (symmetric) matrix C, then according to Theorem 1.5.5 there exists a positive definite matrix B that solves Liapunov's matrix equation

$$G'B + BG = -C.$$

This means that if we use the positive definite Liapunov function $V(y) = y'By$, then its derivative with respect to system (2.4.7)

$$\dot{V}_{(2.4.7)}(y) = -y'Cy$$

is negative definite. Now, we take the derivative of this Liapunov function found for the unperturbed system (2.4.7) with respect to the perturbed system (2.4.6):

$$\dot{V}_{(2.4.6)}(t, y) = -y'(C + \mu C_1(t))y \qquad (2.4.8)$$

where $C_1(t) = -(H'(t)B + BH(t))$. It is well known that the quadratic form $y'(C + \mu C_1(t))y$ is positive definite if and only if each principal minor of the matrix $C + \mu C_1(t)$ is positive. Denote these principal minors by $\Delta_k(t, \mu)$ $(k = 1, 2, ..., n;$ $\Delta_n(t, \mu) = \det(C + \mu C_1(t)))$, and those of the positive definite constant matrix C by δ_k $(k = 1, 2, ..., n)$. Then $\Delta_k(t, 0) = \delta_k > 0$ $(k = 1, 2, ..., n)$, and because of the continuity and the periodicity with respect to t of $\Delta_k(t, \mu)$, a $\mu_1 > 0$ exists such that for $t \in [0, \infty)$ (which is the same as for $t \in [0, T]$) and $|\mu| \leq \mu_1$ we have

$$\Delta_k(t, \mu) > 0 \qquad (k = 1, 2, ..., n). \qquad (2.4.9)$$

This means that for $|\mu| \leq \mu_1$ and $t \in [0, T]$ the quadratic form $y'(C + \mu C_1(t))y$ is positive definite. Denote the least and the largest eigenvalue of the matrix $C + \mu C_1(t)$ by $\lambda_{\min}(t, \mu)$ and $\lambda_{\max}(t, \mu)$, respectively:

$$0 < \lambda_{\min}(t, \mu) \leq \lambda_{\max}(t, \mu), \qquad |\mu| \leq \mu_1, \qquad t \in [0, T].$$

It is well known (see e.g. Demidovich [1967a] p.34) that the following inequality holds:

$$\lambda_{\min}(t, \mu)|y|^2 \leq y'(C + \mu C_1(t))y \leq \lambda_{\max}(t, \mu)|y|^2. \qquad (2.4.10)$$

Since $\lambda_{\min}$ is continuous and positive on the compact quadrangle $[0, T] \times [-\mu_1, \mu_1]$, it has a positive minimum, $\lambda_1 > 0$. Thus,

$$\dot{V}_{(2.4.6)}(t, y) = -y'(C + \mu C_1(t))y \leq -\lambda_1|y|^2, \qquad t \in [0, \infty], \qquad |\mu| \leq \mu_1,$$

i.e. the derivative of the positive definite quadratic form V with respect to system (2.4.6) is negative definite for $|\mu| \leq \mu_1$. By Theorem 1.5.2 the origin, i.e. the system (2.4.6), is uniformly asymptotically stable. The value of μ_1 can be determined by solving the set of inequalities (2.4.9) for μ on $t \in [0, T]$. This settles the problem, in principle, of how large perturbations are permitted.

2.5   Hill's and Mathieu's Equations

As an application of the previous results we shall now discuss Hill's differential equation and then its special case: Mathieu's equation. Both equations have widespread applications and a vast literature (see Hill [1886] , Mathieu [1868] and, for instance, Yakubovich-Starzhinskij [1972] and Meixner-Schäfke [1954]). Here we shall restrict ourselves to the most instructive part of the theory and to some characteristic applications. We shall rely on Demidovich [1967$_a$] and Farkas [1964].

Hill's equation is the scalar differential equation

$$\ddot{y} + p(t)y = 0 \tag{2.5.1}$$

where $p \in C^0(\mathbf{R})$, $p(t+T) = p(t)$ for some $T > 0$. Note that an arbitrary second order homogeneous linear differential equation with periodic and continuously *differentiable* coefficients can be reduced to (2.5.1). More exactly let $a_1 \in C^1(\mathbf{R})$, $a_0 \in C^0(\mathbf{R})$ and consider the equation

$$\ddot{z} + a_1(t)\dot{z} + a_0(t)z = 0 \tag{2.5.2}$$

where also $a_1(t+T) = a_1(t)$, $a_0(t+T) = a_0(t)$ for some $T > 0$. Introducing the new unknown function y by the transformation

$$z = y \exp\left(-\frac{1}{2}\int_0^t a_1(s)ds\right) \tag{2.5.3}$$

we get that z satisfies (2.5.2) if and only if y satisfies (2.5.1), where now

$$p(t) = a_0(t) - a_1^2(t)/4 - \dot{a}_1(t)/2.$$

Returning to equation (2.5.1) we shall present a method due to Liapunov [1902] for the determination of the general solution and the conditions of stability. The stability of equation (2.5.1) means, according to what has been said after formula (1.1.6), the stability of the equivalent Cauchy normal form

$$\dot{y}_1 = y_2, \qquad \dot{y}_2 = -p(t)y_1. \tag{2.5.4}$$

Since $y = y_1$, $\dot{y} = y_2$, the stability of this system means that all solutions of (2.5.1) stay bounded on $[0, \infty)$ along with their first derivatives.

First the fundamental matrix solution $\Phi(t)$ of (2.5.4) satisfying $\Phi(0) = I$, will be determined in the form of a series. For this purpose, instead of (2.5.1) consider first the one parameter family of differential equations:

$$\ddot{y} = \mu p(t)y \tag{2.5.5}$$

$\mu \in \mathbf{R}$; we obtain (2.5.1) again by substituting $\mu = -1$. The solution of (2.5.5) satisfying $y(0) = 1$, $\dot{y}(0) = 0$ will be denoted by $\varphi_\mu(t)$, and we let $\varphi = \varphi_{-1}$. Assume that this solution can be determined as the sum of a convergent power series of the parameter μ:

$$\varphi_\mu(t) = \sum_{k=0}^{\infty} \varphi_k(t)\mu^k. \tag{2.5.6}$$

Substituting this into the differential equation (2.5.5) we get

$$\sum_{k=0}^{\infty} \ddot{\varphi}_k(t)\mu^k = \sum_{k=0}^{\infty} p(t)\varphi_k(t)\mu^{k+1}.$$

Equating equal powers of μ yields

$$\ddot{\varphi}_0(t) = 0, \qquad \ddot{\varphi}_k(t) = p(t)\varphi_{k-1}(t) \qquad (k = 1, 2, ...).$$

In order to satisfy the initial conditions, we impose the requirements that $\varphi_0(0) = 1$, $\dot{\varphi}_0(0) = 0$, $\varphi_k(0) = \dot{\varphi}_k(0) = 0$, $k \geq 1$. Hence,

$$\varphi_0(t) = 1, \qquad \varphi_k(t) = \int_0^t \left(\int_0^{t_1} p(t_2)\varphi_{k-1}(t_2)dt_2 \right) dt_1, \qquad k \geq 1.$$

Considering the last integral as a value of a double integral and interchanging the order of integration we get

$$\varphi_k(t) = \int_0^t \left(\int_{t_2}^t p(t_2)\varphi_{k-1}(t_2)dt_1 \right) dt_2 = \int_0^t (t-\tau)p(\tau)\varphi_{k-1}(\tau)d\tau. \quad (2.5.7)$$

A simple mathematical induction yields that $\varphi_k \in C^2(\mathbf{R})$, and the estimate

$$|\varphi_k(t)| \leq \frac{M^k t^{2k}}{(2k)!} \qquad (k = 1, 2, ...)$$

holds where M is an upper bound of the periodic function p, i.e. $|p(t)| \leq M$, $t \in \mathbf{R}$. Thus, for arbitrary $\mu_0 > 0$ and $t_0 > 0$ the convergent numerical series

$$\sum_{k=0}^{\infty} \mu_0^k M^k t_0^{2k}/(2k)! = \cosh(t_0(\mu_0 M)^{1/2})$$

is a majorant series of (2.5.6) for $|\mu| < \mu_0$, $|t| < t_0$, so that the latter series is absolute and uniformly convergent, and, as a consequence, its sum is the required solution of (2.5.5) indeed. The solution of (2.5.1) satisfying the initial conditions $\varphi(0) = 1$, $\dot{\varphi}(0) = 0$ is

$$\varphi(t) = \varphi_{-1}(t) = \sum_{k=0}^{\infty} (-1)^k \varphi_k(t).$$

In a similar way one may determine that solution of (2.5.1) that satisfies $y(0) = 0$, $\dot{y}(0) = 1$. Denoting this solution by $\psi(t)$ we have

$$\psi(t) = t + \sum_{k=1}^{\infty} (-1)^k \psi_k(t)$$

where $\psi_0(t) = t$, and

$$\psi_k(t) = \int_0^t (t - \tau)p(\tau)\psi_{k-1}(\tau)d\tau, \qquad k \geq 1.$$

The fundamental matrix Φ of (2.5.4) that satisfies $\Phi(0) = I$ is then

$$\Phi(t) = \begin{pmatrix} \varphi(t) & \psi(t) \\ \dot{\varphi}(t) & \dot{\psi}(t) \end{pmatrix}.$$

The characteristic equation determining the characteristic multipliers is

$$\begin{vmatrix} \varphi(T) - \lambda & \psi(T) \\ \dot{\varphi}(T) & \dot{\psi}(T) - \lambda \end{vmatrix} = 0.$$

By Liouville's formula (Theorem 1.2.4)

$$\det \Phi(T) = \exp \int_0^T \text{Tr} \begin{pmatrix} 0 & 1 \\ -p(t) & 0 \end{pmatrix} dt = 1,$$

so that the characteristic equation is

$$\lambda^2 - a\lambda + 1 = 0 \qquad (2.5.8)$$

where $a = \varphi(T) + \dot{\psi}(T) = \text{Tr } \Phi(T)$ is called the *Liapunov constant*. Since the product of the two characteristic multipliers is 1, system (2.5.4) cannot be asymptotically stable (cf. also Problem 2.6.3). The Liapunov stability of the system depends on the value of the Liapunov constant. In order to determine this value we are going to find an explicit expression for $\varphi(T)$ and $\dot{\psi}(T)$. Applying the recursion formula (2.5.7) k times,

$$\varphi_k(t) = \int_0^t \int_0^{t_1}$$
$$\cdots \int_0^{t_{k-1}} (t - t_1)(t_1 - t_2) \cdots (t_{k-1} - t_k)p(t_1)p(t_2) \cdots p(t_k)dt_k...dt_2dt_1.$$

Similarly

$$\psi_k(t) = \int_0^t \int_0^{t_1}$$
$$\cdots \int_0^{t_{k-1}} (t - t_1)(t_1 - t_2) \cdots (t_{k-1} - t_k)t_kp(t_1)p(t_2) \cdots p(t_k)dt_k...dt_2dt_1.$$

Differentiating the last expression, and substituting into the respective series we get

$$a = \varphi(T) + \dot{\psi}(T) = 2 - T \int_0^T p(t)dt + \sum_{k=2}^{\infty}(-1)^k \int_0^T \int_0^{t_1}$$
$$\cdots \int_0^{t_{k-1}} (T - t_1 + t_k)(t_1 - t_2) \cdots (t_{k-1} - t_k)p(t_1)p(t_2) \cdots p(t_k)dt_k...dt_2dt_1.$$

$$(2.5.9)$$

Theorem 2.5.1. *If the Liapunov constant is in modulus greater than 2, then system (2.5.4) is unstable; if the modulus of the Liapunov constant is less than 2, then the system is stable in the Liapunov sense.*

Proof. The roots of (2.5.8) are

$$\lambda_{1,2} = (1/2)(a \pm (a^2 - 4)^{1/2}),$$

so that the conclusions are implied by Theorem 2.4.1. $\square$

If the Liapunov constant is in modulus equal to 2, then we are confronted with a critical case because $|a| = 2$ implies that either 1 or -1 is a characteristic multiplier with multiplicity 2 *in the characteristic polynomial.* According to Theorem 2.4.1 the stability of the system depends on whether the multiplicity of 1, resp. -1 is one or two in *the minimal polynomial.* The following holds:

Theorem 2.5.2. *Let the Liapunov constant be $a = 2$, resp. $a = -2$; system (2.5.4) is stable in the Liapunov sense if and only if all its solutions are periodic with period T, resp. with least positive period $2T$.*

Proof. If $a = 2$, resp. $a = -2$, then the characteristic polynomial of the system is $(\lambda - 1)^2$, resp. $(\lambda + 1)^2$. (We know by Corollary 2.2.3 that in the first case a non-trivial T-periodic, in the second a non-trivial $2T$-periodic solution exists.) The system is stable in the Liapunov sense if and only if the minimal polynomial is $\lambda - 1$, resp. $\lambda + 1$ (and not $(\lambda - 1)^2$, resp. $(\lambda + 1)^2$). Now, according to the definition of the minimal polynomial (see Theorem A 1.3) $\lambda - 1$, resp. $\lambda + 1$ is the minimal polynomial iff it annuls the principal matrix

$$C = \Phi(T) = \begin{pmatrix} \varphi(T) & \psi(T) \\ \dot{\varphi}(T) & \dot{\psi}(T) \end{pmatrix},$$

i.e. iff $C - I = 0$, resp. $C + I = 0$. In the first case this means that $\varphi(T) = 1 = \varphi(0)$, $\dot{\varphi}(T) = 0 = \dot{\varphi}(0)$, and $\psi(T) = 0 = \psi(0)$, $\dot{\psi}(T) = 1 = \dot{\psi}(0)$, implying that both these independent solutions, and with them all the solutions, are T-periodic. In the second case this means that $\varphi(T) = -1 = -\varphi(0)$, $\dot{\varphi}(T) = 0 = \dot{\varphi}(0)$, i.e $[\varphi(T), \dot{\varphi}(T)] = -[\varphi(0), \dot{\varphi}(0)]$, and $\psi(T) = 0 = \psi(0)$, $\dot{\psi}(T) = -1 = -\dot{\psi}(0)$, i.e. $[\psi(T), \dot{\psi}(T)] = -[\psi(0), \dot{\psi}(0)]$. This, obviously, implies that $[\varphi(2T), \dot{\varphi}(2T)] = -[\varphi(T), \dot{\varphi}(T)] = [\varphi(0), \dot{\varphi}(0)]$, i.e. that φ is $2T$-periodic, and similarly that ψ is $2T$-periodic too. $\square$

Summing up the results of the last two Theorems we have the following complete classification of the possible cases:

(i) If $|a| > 2$, then the system has a characteristic multiplier in modulus greater than 1, and another in modulus less than 1 (both real).

This means that some solutions of Hill's equation may oscillate with amplitudes tending exponentially to infinity (cf. Floquet's theory, Theorem 2.2.5), and some tend to zero as t tends to infinity.

(ii) If $|a| < 2$, then the system has two different complex conjugate characteristic multipliers with modulus equal to one. All solutions of Hill's equation are bounded and oscillating but none (apart from the trivial one) is periodic with period T or $2T$.

(iii) If $a = 2$, resp. $a = -2$, then either all solutions are periodic with period T, resp. period $2T$, or one non-trivial solution is periodic with period T, resp. $2T$, and every solution independent of this one oscillates with amplitudes tending to infinity.

It is worthwhile to consider two important special cases: equation (2.5.1) with a non-positive, resp. non-negative coefficient p.

Theorem 2.5.3. *If the function p in (2.5.1) is non-positive and not identically zero, then Hill's equation is unstable (falling into class (i) of the previous classification).*

Proof. If $p(t) \leq 0$, $t \in [0,T]$ but is not identically zero, then $\int_0^T p(t)dt < 0$, and the multiple integral in the k-th term of the infinite series of (2.5.9) has sign equal to $(-1)^k$; thus, each term on the right-hand side of expression (2.5.9) is positive. Hence, $a > 2$. $\square$

Theorem 2.5.4. *If the function p in (2.5.1) is non-negative, and*

$$0 < T \int_0^T p(t)dt \leq 4, \qquad (2.5.10)$$

then Hill's equation is stable (falling into class (ii) in the previous classification).

Proof. Denoting the multiple integral occurring in the k-th term of the series in (2.5.9) by M_k ($k = 2, 3, ...$), and introducing the notation

$$M_1 = T \int_0^T p(t)dt,$$

the expression for the Liapunov constant a becomes

$$a = 2 - \sum_{k=1}^{\infty} (-1)^{k-1} M_k. \qquad (2.5.11)$$

We are going to show that the series on the right-hand side is of Leibniz's type. By the assumptions of the Theorem $M_k > 0$ ($k = 1, 2, ...$), and since we know that the series is convergent, we have $M_k \to 0$, as $k \to \infty$:

$$M_{k+1} = \int_0^T \int_0^{t_1} \cdots \int_0^{t_{k-1}} \int_0^{t_k} (T - t_1 + t_{k+1})(t_1 - t_2) \cdots (t_{k-1} - t_k)$$
$$\times (t_k - t_{k+1})p(t_1)p(t_2) \cdots p(t_k)p(t_{k+1})dt_{k+1}dt_k...dt_2dt_1.$$

Applying the inequality between the geometric and the arithmetic means we get

$$(T - t_1 + t_{k+1})(t_k - t_{k+1}) \leq (1/4)(T - t_1 + t_k)^2$$
$$< (T/4)(T - t_1 + t_k), \qquad 0 \leq t_k < t_1.$$

Hence,

$$M_{k+1} < (T/4)\int_0^T \int_0^{t_1} \cdots \int_0^{t_{k-1}} (T - t_1 + t_k)(t_1 - t_2) \cdots (t_{k-1} - t_k)$$

$$\times \ p(t_1)p(t_2) \cdots p(t_k)\int_0^{t_k} p(t_{k+1})dt_{k+1}dt_k...dt_2dt_1$$

$$< (T/4)M_k \int_0^T p(t_{k+1})dt_{k+1} \leq M_k \qquad (k = 1, 2, ..)$$

by (2.5.10). We have established that the series on the right-hand side of (2.5.11) is of Leibniz's type. As a consequence, using also inequality (2.5.10),

$$-2 \leq -T \int_0^T p(t)dt + 2 < a < 2,$$

and this proves the Theorem. $\square$

It is to be noted that though the inequality (2.5.10) gives a fairly conservative estimate of the stability region, in the case that it is translated into inequalities concerning the parameters in the continuous function p, the number 4 is sharp (see Demidovich [1967a]).

In the rest of this Section, applications, and an important special case, of Mathieu's equation will be presented.

First consider the phenomenon of wave propagation in a plane. The waves may be elastic ones in a membrane, electromagnetic ones, etc. Their propagation is governed by the *two dimensional wave equation*

$$\partial^2 U/\partial t^2 = c^2 \Delta U$$

where t is the time, U is the amplitude of the wave, $c > 0$ is a constant and $\Delta = \partial^2/\partial x^2 + \partial^2/\partial y^2$ is the Laplacian in Cartesian coordinates x, y. If we assume that the oscillation is harmonic and apply the method of the separation of variables by writing $U(t, x, y) = \exp(i\omega t)u(x, y)$, then u satisfies

$$\Delta u + k^2 u = 0 \tag{2.5.12}$$

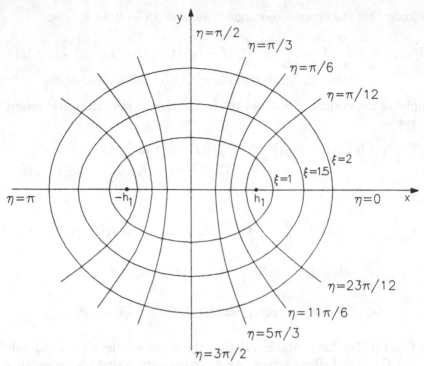

FIGURE 2.5.1. Elliptic coordinates in the plane.

where $k = \omega/c$. We treat this problem assuming that *the boundary conditions are given along an ellipse*. In this case it is useful to introduce orthogonal elliptic coordinates ξ and η by the transformation

$$x = h_1 \cosh \xi \cos \eta, \qquad y = h_1 \sinh \xi \sin \eta$$

where $h_1 > 0$ is a constant. Figure 2.5.1 shows the $\xi = \text{const}$ and $\eta = \text{const}$ coordinate curves that are confocal ellipses and hyperbolas, respectively, with common foci at $x = \pm h_1$, $y = 0$. Applying the form of the Laplacian in orthogonal curvilinear coordinates, equation (2.5.12) goes into

$$\partial^2 u/\partial \xi^2 + \partial^2 u/\partial \eta^2 + (k^2 h_1^2/2)(\cosh 2\xi - \cos 2\eta)u = 0$$

where, by an abuse of notation, u denotes the function depending on ξ and η again. Separating the variables once again by assuming u in the form $u(\xi, \eta) = A(\xi)B(\eta)$ we get

$$A''/A + 2h^2 \cosh 2\xi = -B''/B + 2h^2 \cos 2\eta = r$$

where prime denotes differentiation of a function by its own variable, $h^2 = k^2 h_1^2/4$, and r is an arbitrary constant. Thus, the functions A and B satisfy the differential equations

$$A'' - (r - 2h^2 \cosh 2\xi)A = 0, \qquad B'' + (r - 2h^2 \cos 2\eta)B = 0,$$

respectively. By a simple complex transformation these equations can be transformed into each other. The equation, using now notations we are accustomed to,

$$d^2y/dt^2 + (r - 2q\cos 2t)y = 0 \qquad (2.5.13)$$

is called *Mathieu's equation*. It comes into the picture almost every time in mathematical physics (wave propagation, electromagnetic, elastic phenomena, heat conduction) when there is some elliptic symmetry in the problem.

Mathieu's equation is a special case of Hill's equation with coefficient $p(t) = r - 2q\cos 2t$ periodic in t with period $T = \pi$. A basic problem is to determine the π- or 2π-periodic solutions of this differential equation. From the point of view of the wave propagation problem by which Mathieu's differential equation has been introduced, only these solutions may have physical meaning since η is a "cyclical coordinate," a kind of polar angle (see Figure 2.5.1), and, therefore, solutions have to assume the same value at t, and at $t + 2\pi$. The value of h^2 (q in (2.5.13)) is determined by the physical and geometrical data of the problem, by the parameters in the differential equation and by the shape of the ellipse along which the boundary conditions are given. If $q \in \mathbf{R}$ is fixed, there is only a discrete set of values of r for which (2.5.13) has a π- or 2π-periodic solution; these are the values at which $|a| = 2$ in (2.5.9). For the details and the proofs of this and the following propositions, see Meixner-Schäfke [1954]. Those values of $r(q)$ for which (2.5.13) has a non-trivial π- or 2π-periodic solution are called *characteristic values*; the π- or 2π-periodic solutions are the *elliptic cylinder functions* or *Mathieu functions of the first kind*. It can be proved that to each $m = 1, 2, \ldots$ there belong two characteristic values $\rho_m(q)$ and $\sigma_m(q)$ such that if these values are substituted for r into (2.5.13), then the equation has a π- or 2π-periodic even, resp. odd solution which has exactly m zeroes in the interval $[0, \pi)$ (to $m = 0$ there belong just one characteristic value and an even Mathieu function); these Mathieu functions are denoted by $ce_m(t, q)$ and $se_m(t, q)$, respectively. ce_m and se_m are 2π-periodic if m is odd and π-periodic if m is even. In the particular case $q = 0$ the Mathieu equation assumes the form

$$\ddot{y} + ry = 0 \qquad (2.5.14)$$

This equation has π- or 2π-periodic solutions if and only if $r = m^2$ ($m = 0, 1, 2, \ldots$), so $ce_m(t, 0) = \cos mt$, $se_m(t, 0) = \sin mt$, $\rho_m(0) = \sigma_m(0) = m^2$, and these functions have m zeroes in $[0, \pi)$ indeed. Note that if $q \neq 0$, then $\rho_m(q) \neq \sigma_m(q)$, and if (2.5.13) has a non-trivial π- or 2π-periodic solution, then no other linearly independent solution may be π- or 2π-periodic. This means that in the case of the Mathieu equation with $q \neq 0$ the necessary and sufficient condition of stability expressed in Theorem 2.5.2 cannot hold when for the Liapunov constant $|a| = 2$ holds.

Having established the properties of equation (2.5.14) it is reasonable to assume that the characteristic values of (2.5.13) and the corresponding

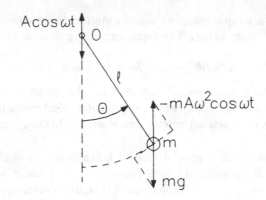

FIGURE 2.5.2. Mathematical pendulum with vertically oscillating point of suspension.

Mathieu functions can be developed into power series of the parameter q:

$$\rho_m(q) = m^2 + \sum_{k=1}^{\infty} \alpha_k q^k, \qquad ce_m(t,q) = \cos mt + \sum_{k=1}^{\infty} q^k c_k(t) \quad (2.5.15)$$

$$(m = 0, 1, 2, ...),$$

$$\sigma_m(q) = m^2 + \sum_{k=1}^{\infty} \beta_k q^k, \qquad se_m(t,q) = \sin mt + \sum_{k=1}^{\infty} q^k s_k(t) \quad (2.5.16)$$

$$(m = 1, 2, 3, ...).$$

The coefficients α_k, β_k and the functions c_k, s_k in the expansion can be determined by substituting into the differential equation; cf. Problem 2.6.6.

Another typical important problem of applied mathematics leading to Mathieu's equation is the motion of the *mathematical pendulum with oscillating point of suspension*. A material point of mass m is suspended on a weightless rod of length l which rotates around a suspension point O. The latter is assumed to undergo vertical harmonic oscillations of amplitude $A > 0$ and circular frequency $\omega > 0$. Friction is supposed to produce a damping effect with coefficient $b > 0$ proportional to the angular velocity. The equation of motion assumes the form (see Figure 2.5.2)

$$lm\ddot{\theta} = -mg\sin\theta + mA\omega^2 \cos(\omega t)\sin\theta - b\dot{\theta}.$$

(Compare Example 1.5.4.) This time we assume that the angular displacement θ is small and replace $\sin\theta$ in the equation of motion by

$\theta \approx \sin \theta$. We get

$$\ddot{\theta} + \frac{b}{lm}\dot{\theta} + \left(\frac{g}{l} - \frac{A\omega^2}{l}\cos(\omega t)\right)\theta = 0. \tag{2.5.17}$$

Performing the transformation (2.5.3),

$$\theta = \psi \exp(-bt/(2lm)), \tag{2.5.18}$$

the differential equation for ψ is obtained in the form

$$\ddot{\psi} + \left[\frac{g}{l} - \frac{A\omega^2}{l}\cos\omega t - \left(\frac{b}{2lm}\right)^2\right]\psi = 0.$$

Introducing the new "dimensionless time" $\tau = \omega t/2$ and the notations

$$r = \frac{4g}{\omega^2 l} - \left(\frac{b}{\omega lm}\right)^2, \qquad q = \frac{2A}{l}, \tag{2.5.19}$$

we obtain Mathieu's equation

$$d^2\psi/d\tau^2 + (r - 2q\cos 2\tau)\psi = 0. \tag{2.5.20}$$

Applying Theorem 2.5.4 we get that the last differential equation is stable in the Liapunov sense if

$$\pi\left(g/l - (b/(2lm))^2\right)^{1/2} \le \omega \le (l/A)^{1/2}\left(g/l - (b/(2lm))^2\right)^{1/2}.$$

Here we have to assume, naturally, that $g/l - (b/(2lm))^2 > 0$. Now, it is well known that the eigenfrequency of the damped pendulum with fixed suspension point is $\omega_0(b) = \left(g/l - (b/2lm)^2\right)^{1/2}$. (If there is no damping, then $\omega_0(0) = (g/l)^{1/2}$.) So that a sufficient condition for the stability of our pendulum with oscillating point of suspension is

$$\pi \le \omega/\omega_0(b) \le (l/A)^{1/2}. \tag{2.5.21}$$

If this condition is satisfied, then every solution of (2.5.20) is bounded, and if $\psi(0)$ and $\dot{\psi}(0)$ are small, $\psi(t)$ and $\dot{\psi}(t)$ will stay small for $t > 0$. However, even if (2.5.21) is not satisfied, and the parameter values of equation (2.5.20) fall into class (i) or (iii) in the classification following Theorem 2.5.2 (and in case (iii), Mathieu's equation with $q \ne 0$ is always unstable), the original differential equation (2.5.17) will be asymptotically stable provided that (in case (i)) the real part of the characteristic exponent corresponding to the larger characteristic multiplier is less than $b/(2lm)$. This follows immediately from (2.5.18) since every solution of (2.5.20) is multiplied by $\exp(-bt/2lm)$.

As we have mentioned earlier the stability region is, in general, much larger than that provided by Theorem 2.5.4. Using these sharper estimates one may even show that by vertical oscillations of the point of suspension one may stabilize the unstable upper ($\theta = \pi$) equilibrium of the pendulum. (In this case one has to linearize the equation, naturally, in the neighbourhood of this equilibrium.) For the details see Meixner-Schäfke [1954].

2.6 Problems

Problem 2.6.1. Solve the differential equation (2.1.15) if $F(t) = A \cos \omega t$. Consider the cases $\omega \neq \beta$, $\omega = \beta$.

Problem 2.6.2. Consider the scalar second order homogeneous linear differential equation

$$\ddot{x} + P(t)\dot{x} + Q(t)x = 0$$

where P and Q are T-periodic continuous functions. Show that if the characteristic exponents of the Cauchy normal form of this equation are chosen appropriately and denoted by ν_1 and ν_2, then

$$\nu_1 + \nu_2 = -(1/T) \int_0^T P(t)dt.$$

(Hint: apply Liouville's formula, Theorem 1.2.4.)

Problem 2.6.3. Show that the polynomial $p(\lambda) = \lambda^2 + a_1\lambda + a_0$ has its roots in the interior of the unit circle if and only if $-1 + |a_1| < a_0 < 1$.

Problem 2.6.4. Show that the polynomial $p(\lambda) = \lambda^3 + a_2\lambda^2 + a_1\lambda + a_0$ has its roots in the interior of the unit circle if and only if $1 + a_1 > |a_0 + a_2|$, $3 - a_1 > |3a_0 - a_2|$ and $1 - a_1 > a_0(a_0 + a_2)$.

Problem 2.6.5. Every solution of the homogeneous system corresponding to

$$\dot{y}_1 = -y_2 2\pi/3 + \cos(2\pi t), \qquad \dot{y}_2 = y_1 2\pi/3 + \sin(2\pi t)$$

is periodic with least positive period 3. Determine the single 1-periodic solution of the inhomogeneous system.

Problem 2.6.6. Show that for $m = 1$ the characteristic values and the corresponding Mathieu functions are

$$\rho_1(q) = 1 + q - q^2/8 - q^3/64 - q^4/1536 + 11q^5/36864 + O(q^6),$$

$$\begin{aligned} \text{ce}_1(t,q) = {} & \cos t - (q/8)\cos 3t + (q^2/64)(-(1/2)\cos t - \cos 3t + (1/3)\cos 5t) \\ & -(q^3/512)(\cos t - (1/6)\cos 3t - (4/9)\cos 5t + (1/18)\cos 7t) \\ & +O(q^4), \end{aligned}$$

$$\sigma_1(q) = 1 - q - q^2/8 + q^3/64 - q^4/1536 - 11q^5/36864 + O(q^6),$$

$$\begin{aligned} \text{se}_1(t,q) = {} & \sin t - (9/8)\sin 3t + (q^2/64)((-1/2)\sin t + 3\sin 3t + (1/3)\sin 5t) \\ & -(q^3/512)(-\sin t - (1/6)\sin 3t + (4/q)\sin 5t + (1/18)\sin 7t) \\ & +O(q^4). \end{aligned}$$

3
Autonomous Systems in the Plane

There is an old joke well known, probably, in most countries of the world. In the middle of the night the policeman sees a drunkard on all fours at the foot of a lamppost on the main street and asks him: 'What are you doing here, friend'? The man replies: 'I am looking for my lost purse, officer'. 'Have you lost it here at this lamp?' 'No, I have lost it in that side-street'. 'Then why aren't you looking for it there?' 'I can't, officer, it is too dark over there'. ... No doubt, this is one of the reasons why two dimensional systems have been treated so extensively: there is some clarity in the two dimensional plane which disappears as the dimension of the system is increased. The clarity is mainly due to "*Jordan's Theorem*" according to which a simple closed Jordan curve divides the plane into two disconnected components. Jordan curves do not generate a similar division of three or higher dimensional spaces, and that is the main reason why the existence problem of closed trajectories, i.e. periodic solutions, is much more difficult in dimensions higher than two than on the plane. Therefore, one can say more about periodic solutions of two dimensional systems than about those of higher dimensional ones, and one may illustrate general situations relatively easily on the former. One of the purposes of this chapter is the introduction of those classical two dimensional autonomous systems (Van der Pol, Liénard, Duffing, Volterra) that will serve as standard references in the general theory.

However, there are other reasons why two dimensional systems are important and are to be treated separately in some detail. Because of the character of Newton's axiom concerning the relation between force and the derivative of the impulse, one degree of freedom mechanical systems have second order differential equations as equations of motion. Also such equations govern the behaviour of single electric circuits. In population dynamics and in economics predator-prey, resp. consumer-supplier, competitive, and cooperative situations may already be studied on two dimensional models.

In this chapter we start the study of periodic solutions of non-linear systems. In Section 1 the *Poincaré-Bendixson theory* is presented; in Section 5 the *Poincaré index* is introduced. Sections 2-4 treat important classical models. In Section 6 the reader will be led to the very edge of the jungle consisting of results concerning *Hilbert's sixteenth problem*: the existence and the number of closed trajectories of polynomial systems. We shall be concerned with the existence problem of periodic solutions throughout this chapter. Stability will be treated in Chapter 5 where autonomous systems of arbitrary finite dimension will be studied.

3.1   The Poincaré-Bendixson Theory

The theory to be expanded in this Section is due to H. Poincaré [1886] and I. Bendixson [1901]. It is the most powerful geometrical-topological method enabling us to prove the existence of a closed path, and even to determine its relative position in two dimensions. The exposition is based on Andronov, Leontovich, Gordon, Mayer [1966], Coddington, Levinson [1955] and Cesari [1963].

Let $D \subset \mathbf{R}^2$ be an open and connected subset of the plane, and $f \in C^1(D, \mathbf{R}^2)$. We shall be concerned with the system

$$\dot{x} = f(x) \quad \text{or} \quad \dot{x}_1 = f_1(x_1, x_2), \quad \dot{x}_2 = f_2(x_1, x_2). \qquad (3.1.1)$$

Let $L \subset D$ be a compact straight line segment. We say that L is a *transversal* of the system (3.1.1) (or of the flow generated by the system) if for all $x \in L$, the vector $f(x)$ is *not* parallel to L. In particular, if L is a transversal, then there is no equilibrium point on L, i.e. if $x \in L$, then $f(x) \neq 0$. The basic properties of a transversal are expressed in the following two lemmata.

Lemma 3.1.1. *(a) Every non-equilibrium point x of D is an interior point of a transversal that may have any direction except $f(x)$. (b) If a path has a common point with a transversal then crosses it, and all trajectories that cross a transversal do that in the same direction (i.e. from left to right, say, if the transversal is directed appropriately). (c) If x^0 is an interior point of the transversal L and $\varepsilon > 0$, then there is a $\delta(x^0, \varepsilon) > 0$ such that all trajectories corresponding to solutions with initial condition in the δ-neighbourhood of x^0 cross L at some time t, where $|t| < \varepsilon$, i.e. if $|x - x^0| < \delta(x^0, \varepsilon)$, then there is a $|t| < \varepsilon$ such that $\varphi(t, x) \in L$ (here, naturally, φ is the solution that assumes the value x at $t = 0$). (d) If γ is*

*a closed path of the system (3.1.1), then it may have at most one common
point with any transversal.*

Proof. (a) is trivial. (b) is a consequence of the continuity of f and of
the requirement that it can nowhere be parallel to the transversal. (c) is a
consequence of the continuous dependence of solutions on initial conditions
on a compact interval. (For these, see Figure 3.1.1 (A).) For (d), consult
Figure 3.1.1 (B). □

Lemma 3.1.2. *If L is a transversal and γ a compact arc of a path,
then $L \cap \gamma$ is a finite set, and the points of intersection of γ with L form
a directionally monotone finite sequence on L if considered in the order of
increasing moments of time to which they belong. (See Figure 3.1.2.)*

Proof. Assume first that there is an infinite number of points of
intersection. Let φ denote the solution of (3.1.1) to whose path γ
belongs and assume that γ is the image of the compact interval I. Since
there is an infinite sequence of moments at which the corresponding
point of γ is on L, we may choose a sequence $t_n \in I$, $n \in \mathbf{N}$ that
converges to a moment in I, $\lim t_n = \bar{t} \in I$ such that $\varphi(t_n) \in L$. Then
$\lim \varphi(t_n) = \varphi(\bar{t}) \in L$, and $\lim(\varphi(t_n) - \varphi(\bar{t}))/(t_n - \bar{t}) = \dot{\varphi}(\bar{t}) = f(\varphi(\bar{t})) \neq 0$.
At the same time, $(\varphi(t_n) - \varphi(\bar{t}))/(t_n - \bar{t})$ is parallel to L for all n; hence,
$f(\varphi(\bar{t}))$ is parallel to L, a contradiction. So $L \cap \gamma$ is finite. For the proof of
the second statement consult Figure 3.1.2 and note that after the crossing
has taken place at $\varphi(t_2)$ the arc γ cannot cross L on the part that is towards
$\varphi(t_1)$ since it cannot get out of (into) the set bounded by the segment $\varphi(t_1)$,
$\varphi(t_2)$ and the arc $\varphi(t_1)$, $\varphi(t_2)$. So that if $t_3 > t_2$, and $\varphi(t_3) \in L$, then $\varphi(t_3)$
must be away from $\varphi(t_1)$. □

We need some basic concepts of topological dynamics. Let $\varphi : \mathbf{R} \to \mathbf{R}^2$
be a solution (defined for all t) of (3.1.1), and γ its path. We say that
$\omega \in \mathbf{R}^2$ is an *omega limit point* of the path γ if there is an infinite
sequence of moments t_n tending to infinity, $t_n \to \infty$ as $n \to \infty$, such that
$\varphi(t_n) \to \omega$, $n \to \infty$. Similarly, $\alpha \in \mathbf{R}^2$ is called an *alpha limit point* of γ if
there is a sequence $t'_n \to -\infty$, $n \to \infty$ such that $\varphi(t'_n) \to \alpha$, $n \to \infty$. The
set of all omega limit points (resp. alpha limit points) of γ is called its
omega limit set and denoted by $\Omega(\gamma)$ (resp. *alpha limit set*, $A(\gamma)$). The
omega, resp. alpha limit points and sets of solutions which are defined on
$[0, \infty)$, resp. $(-\infty, 0]$ are defined similarly. If φ is a periodic (in particular
a constant) solution, then its path γ is, obviously, a closed curve (in
particular a single point). In this case, clearly $\Omega(\gamma) = A(\gamma) = \gamma$. In the
following lemmata we state and prove some properties of the omega limit
set of the positive semitrajectory $\gamma^+ = \{x \in \mathbf{R}^2 : x = \varphi(t), \ t \in [0, \infty)\}$.
The analogous propositions are true for the alpha limit set of the negative
semitrajectory $\gamma^- = \{x \in \mathbf{R}^2 : x = \varphi((t), \ t \in (-\infty, 0]\}$.

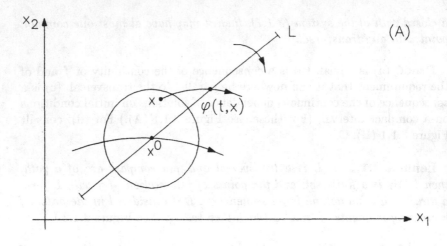

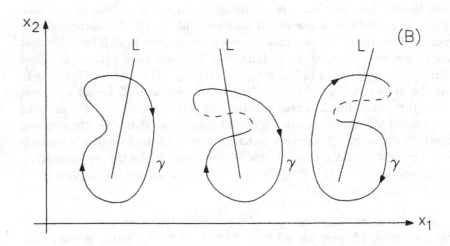

FIGURE 3.1.1. (A) L: a transversal of system (3.1.1) illustrating Lemma 3.1.1 (b) and (c); (B) L: a transversal, γ: a closed path; possible and impossible cases illustrating Lemma 3.1.1 (d).

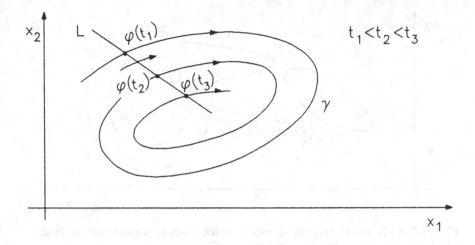

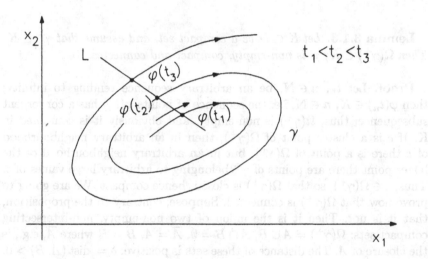

FIGURE 3.1.2. L: a transversal, γ: a compact arc of a path illustrating Lemma 3.1.2; two possible cases.

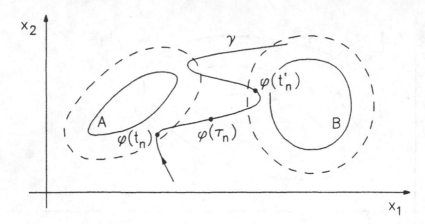

FIGURE 3.1.3. Illustrating the situation which leads to a contradiction in the proof of Lemma 3.1.3: $\Omega(\gamma^+) = A \cup B$.

Lemma 3.1.3. *Let $K \subset D$ be a compact set, and assume that $\gamma^+ \subset K$. Then $\Omega(\gamma^+) \subset K$; it is non-empty, compact and connected.*

Proof. Let t_n, $n \in \mathbf{N}$, be an arbitrary sequence tending to infinity; then $\varphi(t_n) \in K$, $n \in \mathbf{N}$, i.e. the sequence of points $\varphi(t_n)$ has a convergent subsequence; thus, $\Omega(\gamma^+)$ is non-empty, and obviously it is contained in K. If c is a cluster point of $\Omega(\gamma^+)$, then in an arbitrary neighbourhood of c there is a point of $\Omega(\gamma^+)$, but in an arbitrary neighbourhood of the latter point there are points of γ^+ belonging to arbitrary large values of t. Thus, $c \in \Omega(\gamma^+)$, so that $\Omega(\gamma^+)$ is closed, hence compact. We are going to prove now that $\Omega(\gamma^+)$ is connected. Suppose, contrary to the proposition, that it is not. Then it is the union of two non-empty, non-intersecting compact sets: $\Omega(\gamma^+) = A \cup B$, $A \cap B = \emptyset$, $\overline{A} = A$, $\overline{B} = B$ where $\overline{A}$, e.g., is the closure of A. The distance of these sets is positive: $\delta = \text{dist}\,(A, B) > 0$. Consider the $\delta/4$-neighbourhood of A, and of B. Obviously, there is a sequence $t_n \to \infty$, $n \in \mathbf{N}$, and a sequence $t'_n \to \infty$, $n \in \mathbf{N}$, such that $\varphi(t_n)$ is in the $\delta/4$-neighbourhood of A, and $\varphi(t'_n)$ is in the $\delta/4$-neighbourhood of B for all $n \in \mathbf{N}$. Because of the continuity of φ in each interval $[t_n, t'_n]$, there is a moment $\tau_n \in (t_n, t'_n)$ such that $\text{dist}\,(\varphi(\tau_n), A) = \delta/2$, $\text{dist}\,(\varphi(\tau_n), B) = \delta/2$. Clearly, $\tau_n \to \infty$, and there is a subsequence $\varphi(\tau'_n)$ of $\varphi(\tau_n)$ that is converging to a point in K, i.e. $\varphi(\tau'_n) \to \omega \in \Omega(\gamma^+)$, $\tau'_n \to \infty$, $n \to \infty$. But $\text{dist}\,(\omega, A) = \text{dist}\,(\omega, B) = \delta/2$; hence, $\omega \notin \Omega(\gamma^+)$, a contradiction (see Figure 3.1.3). $\square$

Lemma 3.1.4. *If γ^+ and $\Omega(\gamma^+)$ have a common point, then γ is a closed orbit.*

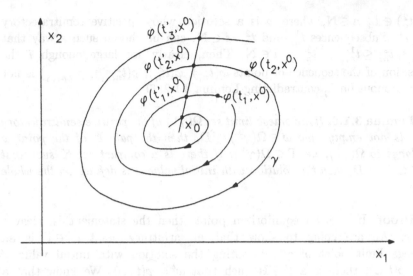

FIGURE 3.1.4. Illustrating the proof of Lemma 3.1.4.

Proof. Let $x^0 \in \gamma^+ \cap \Omega(\gamma^+)$. If x^0 is an equilibrium point, then $\gamma^+ = \{x^0\} = \Omega(\gamma^+)$, and the proposition is trivially true. So we may assume that $f(x^0) \neq 0$. According to Lemma 3.1.1 (a) a transversal L can be chosen through x^0. Applying Lemma 3.1.1 (c) let $\delta = \delta(x^0, 1) > 0$ be the radius of the neighbourhood of x^0 belonging to $\varepsilon = 1$. Since $x^0 \in \Omega(\gamma^+)$, a positive sequence t_n, $n \in \mathbf{N}$, can be chosen such that $t_{n+1} > t_n + 2$, $\varphi(t_n, x^0) \in B(x^0, \delta)$ (the disc of radius δ around x^0), and $\varphi(t_n, x^0) \to x^0$, $n \to \infty$. According to the Lemma quoted, for every positive integer n a t'_n can be found such that $|t_n - t'_n| < 1$, and $\varphi(t'_n, x^0) \in L$, i.e. in every interval (t_{n-1}, t_{n+1}) there is at least one t'_n at which the orbit γ^+ crosses L. Assume that $\varphi(t'_1, x^0) \neq x^0$; then according, to Lemma 3.1.2, $\varphi(t'_2, x^0)$ is away from x^0 on L, and so on. This means that the sequence $\varphi(t'_n, x^0)$ is tending away from x^0, and with it the sequence $\varphi(t_n, x^0)$ is tending away too, a contradiction (see Figure 3.1.4). So $\varphi(t'_1, x^0) = x^0$, $t'_1 \neq 0$ and this means that γ is a closed path corresponding to a periodic solution (see Section 1.3). $\square$

A similar argument proves the following:

Lemma 3.1.5. *The intersection of a transversal L and an omega limit set $\Omega(\gamma^+)$ consists of a single point at most.*

Proof. Assume, contrary to the statement, that $\Omega(\gamma^+)$ has two common points $\omega^1 \neq \omega^2$ with the transversal L. Repeating the construction in the proof of the previous Lemma, two sequences $t_n^1 \to \infty$, $t_n^2 \to \infty$ can be found such that $\varphi(t_n^1) \to \omega^1$, $\varphi(t_n^2) \to \omega^2$, $n \to \infty$, and $\varphi(t_n^1) \in L$,

$\varphi(t_n^2) \in L$, $n \in \mathbf{N}$, where φ is a solution whose positive semitrajectory is γ^+. Subsequences $t_{n_i}^1$ and $t_{n_i}^2$, $i \in \mathbf{N}$, can be chosen such a way that $t_{n_i}^1 \le t_{n_i}^2 \le t_{n_{i+1}}^1 \le t_{n_{i+1}}^2$, $i \in \mathbf{N}$. Then, clearly, for large enough i the position of the sequence of points $\varphi(t_{n_i}^1)$, $\varphi(t_{n_i}^2)$, $\varphi(t_{n_{i+1}}^1)$, $\varphi(t_{n_{i+1}}^2)$ is not monotonous on L, contradicting Lemma 3.1.2. □

Lemma 3.1.6. *If the omega limit set $\Omega(\gamma^+)$ of the positive semitrajectory γ^+ is not empty, and $\omega \in \Omega(\gamma^+) \subset D$, then the path Γ of the point ω belongs to $\Omega(\gamma^+)$, i.e. $\Gamma \subset \Omega(\gamma^+)$; if there is a compact set K such that $\gamma^+ \subset K \subset D$, then the solution with initial value ω is defined on the whole $\mathbf{R}$.*

Proof. If ω is an equilibrium point, then the statement is, clearly, true. We are going to show that an arbitrary point $\omega' \in \Gamma$ is an omega limit point of γ^+. Denoting the solution with initial value ω by $\varphi(t, \omega)$, there is a $t' \in \mathbf{R}$ such that $\omega' = \varphi(t', \omega)$. We know that a sequence $t_n \to \infty$, $n \to \infty$, exists such that $\varphi(t_n, x^0) \to \omega$ where $\varphi(t, x^0)$ is the solution whose path is γ. Now, applying Definition 1.3.1 (iii), $\varphi(t_n, x^0) = \varphi(0, \varphi(t_n, x^0))$. Because of the continuous dependence of solutions on initial conditions, $\varphi(t', \varphi(t_n, x^0)) \to \varphi(t', \omega)$, as $n \to \infty$, i.e. $\varphi(t' + t_n, x^0) \to \omega'$, and $t' + t_n \to \infty$ as $n \to \infty$, so that $\omega' \in \Omega(\gamma^+)$. This proves that $\Gamma \subset \Omega(\gamma^+)$. If $\gamma^+ \subset K$, then $\Gamma \subset \Omega(\gamma^+) \subset K \subset D$, and this implies that $\varphi(t, \omega)$ is defined on the whole real line $\mathbf{R}$. □

The previous lemmata yields our last

Lemma 3.1.7. *If $\gamma^+ \subset K \subset D$ where K is a compact set, and $\Omega(\gamma^+)$ contains a closed path Γ which is not an equilibrium point, then $\Omega(\gamma^+) = \Gamma$.*

Proof. Suppose, contrary to the statement, that $\Omega(\gamma^+) \setminus \Gamma \ne \emptyset$. Since by Lemma 3.1.3, $\Omega(\gamma^+)$ is connected, Γ contains a point $\omega \in \Gamma$ which is a cluster point of $\Omega(\gamma^+) \setminus \Gamma$. Consider a transversal L through ω (Lemma 3.1.1 (a)), let $B(\omega, \delta)$ be a disc of radius δ around ω such that all trajectories with initial value in $B(\omega, \delta)$ cross L (Lemma 3.1.1 (c)), and let $\omega^1 \in (\Omega(\gamma^+) \setminus \Gamma) \cap B(\omega, \delta)$. Then (by Lemma 3.1.6) the path Γ^1 through ω^1 is contained completely in $\Omega(\gamma^+)$, and it crosses L at some point that is different from ω since, clearly, $\Gamma \cap \Gamma^1 = \emptyset$. This means that $\Omega(\gamma^+)$ crosses L in two distinct points which contradicts Lemma 3.1.5. □

With the help of these lemmata we may prove now

Theorem 3.1.8 *(Poincaré-Bendixson Theorem).* *If $\gamma^+ \subset K \subset D$ where K is a compact set, and $\Omega(\gamma^+)$ does not contain any equilibrium point, then $\Omega(\gamma^+)$ is a closed path.*

Proof. If γ^+ is a closed path, then $\Omega(\gamma^+) = \gamma^+$, and the statement holds.

Assume that γ^+ is not a closed path. Then by Lemmata 3.1.3 and 3.1.6, $\Omega(\gamma^+)$ contains a path Γ corresponding to a solution defined over $\mathbf{R}$. The path Γ has an omega limit point ω which belongs also to $\Omega(\gamma^+)$ since the latter set is closed by Lemma 3.1.3. By the assumption of the Theorem, ω is not an equilibrium, so that we may consider a transversal L through ω. Since $\omega \in \Omega(\gamma^+)$, this set has no point common with L besides ω (Lemma 3.1.5). Γ^+ gets arbitrarily close to ω, so that by Lemma 3.1.1 (c), Γ^+ intersects L too. Since $\Gamma^+ \subset \Omega(\gamma^+)$, this point of intersection must be ω. So that Γ^+ and *its* omega limit set have a common point, and by Lemma 3.1.4 this implies that Γ is a closed orbit. Finally, Lemma 3.1.7 yields that $\Omega(\gamma^+) = \Gamma$ is a closed orbit. □

The Poincaré-Bendixson theory is, probably, the most widely used method for settling the problem of the existence of a non-constant periodic solution, i.e. a non-trivial closed path in two dimensions. The following Corollary is applied directly in most cases.

Corollary 3.1.9. *If $\gamma^+ \subset K \subset D$ where K is a compact set, and K does not contain any equilibrium point, then it contains a closed path.*

Now, the demonstration of the existence of a closed path proceeds in the following way. By a "*phase plane analysis*" the course of a path is followed until it can be shown that it crosses a transversal twice. The arc of the path and the segment of the transversal between the two points of intersection form the boundary of a compact set, out of which no trajectory may *escape* (see Figure 3.1.5). Sometimes this set is called a *Bendixson sack*. According to the previous Corollary a Bendixson sack contains at least one closed path; however, this may be just a trivial one, an equilibrium point. The trouble is, as we shall see later, that a non-trivial closed path always contains an equilibrium point in its interior, and this means that a Bendixson sack contains an equilibrium point necessarily. If a Bendixson sack has been constructed, then two roads can lead to the existence of a non-trivial closed path. First, one may show that the sack contains a *single* equilibrium, and this is "*repelling*", i.e. attracting for $t \to -\infty$. In this case it cannot be the omega limit set of any orbit, so that –since the omega limit set of an orbit entering the sack cannot be the empty set – the repelling equilibrium must be surrounded by a closed path. Secondly, one may construct a second smaller Bendixson sack in the interior of the former, bounded by an arc of a path and by a segment of a transversal having two common points with the arc such that no trajectory can *enter* it. If one can show that there is no equilibrium point in the annular domain between the boundaries of the two Bendixson sacks, then this annular region contains a non-trivial closed path.

In order to make the previous description of the possibilities precise we need one more theorem, and for its proof we need

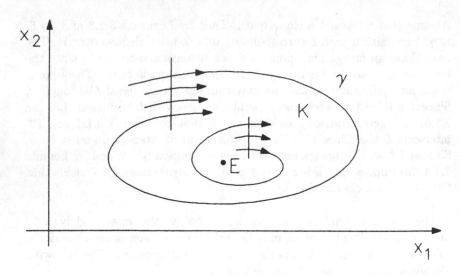

FIGURE 3.1.5. A Bendixson sack K out of which no trajectory can escape and an inner Bendixson sack into which no trajectory can enter, E: an equilibrium point.

Lemma 3.1.10. *Let $\gamma^+ \subset K \subset D$ where K is a compact set; if $\Gamma \subset \Omega(\gamma^+)$ is a path and it has an omega limit point that is not an equilibrium, then Γ is a closed path.*

Proof. The proof is a repetition of that of Theorem 3.1.8. □

Theorem 3.1.11. *Let $\gamma^+ \subset K \subset D$ where K is a compact set, and assume that K contains only a finite number of equilibria; then the omega limit set $\Omega(\gamma^+)$ of γ^+ falls into exactly one of the following three categories: (i) it is a single equilibrium and γ^+ tends to it as $t \to \infty$; (ii) it is a nontrivial closed path; (iii) it consists of a finite number of equilibria, and of orbits corresponding to solutions defined on the whole real line joining the equilibria in the sense that if Γ is such an orbit, then both its alpha limit set $A(\Gamma)$ and its omega limit set $\Omega(\Gamma)$ are single equilibria.*

Proof. If $\Omega(\gamma^+)$ consists of equilibria only, then it is a single equilibrium since it is connected (Lemma 3.1.3). If $\Omega(\gamma^+)$ contains a non-trivial closed path, then it is equal to it (Lemma 3.1.7). If none of the previous possibilities prevail, then $\Omega(\gamma^+)$ consists of equilibria and of orbits that correspond to solutions defined over the whole real line, and none of the latter is a closed path. If $\Gamma \subset \Omega(\gamma^+)$ is an orbit, then its alpha and omega limit sets are not empty, and by Lemma 3.1.10 both consist of equilibria only. Because of the connectedness of the omega (alpha) limit set of Γ both these sets consist of a single equilibrium each. Since $\Omega(\gamma^+)$ is connected, each equilibrium point belonging to it must be connected by an orbit to the set of the rest of the equilibria. (The possible situations are shown on Figure 3.1.6.) □

We note that a non-trivial closed path is called a *limit cycle* if it is the omega or alpha limit set of some orbit. A path joining two different points of equilibria is called a *heteroclinic orbit*. A path whose alpha and omega limit set is the *same* equilibrium point is called a *homoclinic orbit*.

Example 3.1.1. *A predator-prey model.* The following two dimensional system of differential equations occurred in Hsu, Hubbel, Waltman [1978$_b$] in the context of a more complicated model. Since then, it has been studied extensively (see e.g. Cheng [1981]). $N(t)$ and $P(t)$ denote the quantity of prey, resp. of predator at time t, $\varepsilon > 0$ is the *intrinsic birth rate of prey*, $K > 0$ is the *carrying capacity of the environment*, $m > 0$ denotes the *limiting birth rate of predator* when the quantity of prey tends to infinity, $a > 0$ is the so-called "*half saturation constant*", and $\gamma > 0$ denotes the *mortality of predator*. It is assumed that in the absence of predation $(P = 0)$ the growth of the prey quantity is governed by a logistic differential equation, and that the "*functional response*", i.e. the effect of consuming the prey on the predator's growth rate, is of "Holling's type": it is increasing with the increase of the prey quantity but is bounded above. These assumptions lead to the system

$$
\begin{aligned}
\dot{N} &= \varepsilon N(1 - N/K) - mNP/(a + N), \\
\dot{P} &= mNP/(a + N) - \gamma P.
\end{aligned}
\tag{3.1.2}
$$

If $N = a$, then the birth rate of the predator is equal to $m/2$, i.e. half of the maximal "saturation" value, hence the name for a.

It is clear that – (3.1.2) being a *Kolmogorov system* (see Kolmogoroff [1936])–the positive quadrant of the N, P-plane is an invariant set of the system. First we are going to show that *all solutions with non-negative initial conditions are bounded for $t > 0$* . This is clear in the case when either $N(0)$ or $P(0)$ is zero; since the positive half coordinate axes are also invariant sets, the solution on the positive P-axis tends towards zero, and the solutions on the positive N-axis tend towards $K > 0$ as t tends to infinity. If $N(t)$ is the first coordinate of an arbitrary solution with $N(0) > 0$, then from the first equation

$$
\dot{N}(t) \leq \varepsilon N(t)(1 - N(t)/K),
$$

so that for $t > 0$ the function $N(t)$ can be estimated from above by the corresponding solution of the logistic differential equation $\dot{z} = \varepsilon z(1 - z/K)$. The solutions of the latter equation with non-negative initial values are bounded and (apart from the solution $z = 0$) tend to K as t tends to infinity. Therefore, an $\tilde{N} > 0$ exists such that $N(t) \leq \tilde{N}$ for $t \geq 0$. If $m \leq \gamma$

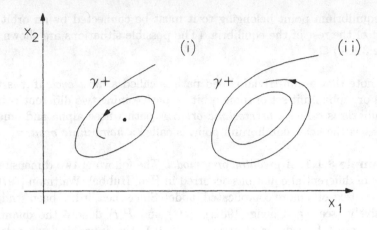

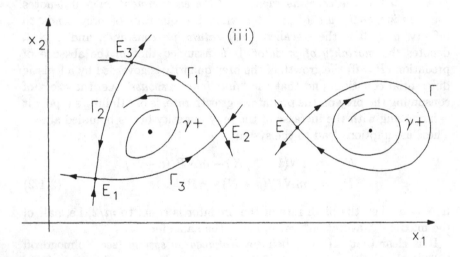

FIGURE 3.1.6. Different omega limit sets of positive semitrajectories; (i) a single equilibrium; (ii) a single closed path: a limit cycle; (iii) on the left the omega limit set is the union of the equilibria E_i and the heteroclinic orbits Γ_i ($i = 1, 2, 3$), on the right it is the union of the equilibrium E and the homoclinic orbit Γ.

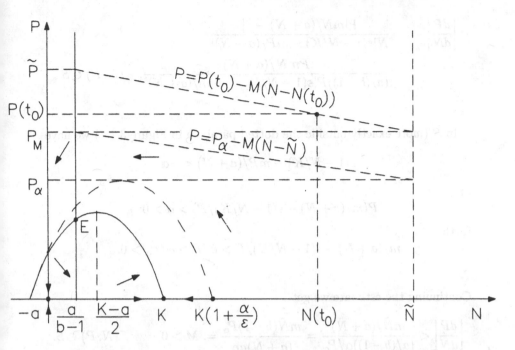

FIGURE 3.1.7. Phase plane of system (3.1.2). The solid parabola is the $\dot{N} = 0$ isocline; the solid vertical line is the $\dot{P} = 0$ isocline; the arrows indicate the direction of the flow; the equilibria are $(0,0)$, $(K,0)$ and E, the latter is left to the maximum point of the solid parabola according to condition (3.1.3).

(i.e. the maximal birth rate of the predator is less than or equal to its death rate), then $\dot{P} < 0$, and $P(t)$ is obviously bounded for $t > 0$. *Assume that* $b := m/\gamma > 1$. From the second equation, $\dot{P} = 0$ if $N = a/(b-1) > 0$. In the strip $0 \leq N < a/(b-1)$ we have $\dot{P} < 0$, i.e. when the path gets into this strip, P cannot grow. For the rest of the proof it might be useful to study Figure 3.1.7. In this Figure the case when $a/(b-1) < K$ is shown. This is the more interesting case since the system has an equilibrium point E in the interior of the positive quadrant. However, the proof goes the same way in the case $0 < K < a/(b-1)$. The $\dot{N} = 0$ isocline is the parabola $\varepsilon(1 - N/K) - mP/(a+N) = 0$ (apart from the $N = 0$ axis). Let $\alpha > 0$ be an arbitrarily fixed positive number, consider the parabola

$$\varepsilon(1 - N/K) - mP/(a+N) = -\alpha,$$

i.e.

$$P = (a+N)(K(1+\alpha/\varepsilon) - N)\varepsilon/(mK),$$

and denote its maximum by $P_\alpha > 0$. If $P(t) \leq P_\alpha$ for $t \geq 0$, then the proposition holds. The set to be studied is the strip $S = \{(N,P) : a/(b-1) < N < \tilde{N}, P > P_\alpha\}$. We are going to estimate the slope of the trajectories in this strip:

$$\left| \frac{dP}{dN} \right| = \frac{P|mN/(a+N)-\gamma|}{N|\varepsilon(1-N/K)-mP/(a+N)|}$$

$$< \frac{Pm\tilde{N}/(a+\tilde{N})}{(a/(b-1))P|\varepsilon(1-N/K)/P-m/(a+N)|}, \quad (N,P) \in S.$$

In S (more exactly outside the dotted parabola in Figure 3.1.7) we have

$$\varepsilon(1-N/K)-mP/(a+N) < -\alpha$$

or

$$P(m/(a+N)-\varepsilon(1-N/K)/P) > \alpha > 0;$$

further

$$m/(a+N)-\varepsilon(1-N/K)/P > \alpha/P > \alpha/P_\alpha > 0.$$

Continuing the estimate we get

$$\left| \frac{dP}{dN} \right| < \frac{m\tilde{N}/(a+\tilde{N})}{(a/(b-1))\alpha/P_\alpha} = \frac{m\tilde{N}(b-1)P_\alpha}{(a+\tilde{N})a\alpha} =: M > 0, \qquad (N,P) \in S.$$

If $P(t) \le P_M := P_\alpha - M(a/(b-1) - \tilde{N})$ for $t \ge 0$, then the proposition holds. Assume that at some moment $t_0 > 0, P(t_0) > P_M$. Since the absolute value of the slope (which is negative in the strip) of the path is less than M, the path hits the vertical line $N = a/(b-1)$ below $\tilde{P} = P(t_0) - M(a/(b-1) - N(t_0))$. After this crossing $P(t)$ is decreasing until it does not cross the line $N = a/(b-1)$ again. But then it cannot rise above $P_M < \tilde{P}$ any more, so that $P(t) \le \tilde{P}$ for $t > t_0$. This, naturally, implies that the solution $(N(t), P(t))$ is defined on the whole interval $[0, \infty)$ (see Theorem 1.1.2).

Now, it is easy to prove the following proposition. *If $b := m/\gamma > 1$ and*

$$K > a + 2a/(b-1), \tag{3.1.3}$$

then system (3.1.2) has a limit cycle in the interior of the positive quadrant. The assumptions imply that $K > a/(b-1)$, i.e. the system has an equilibrium point E in the interior. Condition (3.1.3) is equivalent to $(K - a)/2 > a/(b-1)$, meaning that E is left of the maximum point of the solid parabola in Figure 3.1.7. Besides E the system has two equilibria: the origin $(0,0)$ and the point $(K,0)$. If we consider, for instance, the solution with initial value $(N(t_0), P(t_0))$ in Figure 3.1.7, then, clearly, its positive semitrajectory stays inside the compact set $\{(N, P) : 0 \le N \le \tilde{N}, \ 0 \le P \le \tilde{P}\}$. The equilibria $(0,0)$ and $(K,0)$ are saddle points, as can be seen by linearizing the system at these points. The "ingoing trajectories" for $(0,0)$ are the positive and negative P-axes;

the "ingoing trajectories" for $(K, 0)$ are the two parts of the positive
N-axis. So none of these two equilibria can be an omega limit point
of our positive semitrajectory. Denote the coordinates of E by $(\overline{N}, \overline{P})$.
Here $\overline{N} = a/(b-1)$, and $\overline{P} = \varepsilon(1 - \overline{N}/K)(a + \overline{N})/m$. The characteristic
polynomial of the system linearized at E is

$$D(\lambda) = \lambda^2 + \lambda\overline{N}(\varepsilon/K - m\overline{P}/(a + \overline{N})^2) + m^2 a\overline{N}\,\overline{P}/(a + \overline{N})^3.$$

The constant term is positive; however, the coefficient of λ is

$$\overline{N}\left(\frac{\varepsilon}{K} - \frac{\varepsilon(1 - \overline{N}/K)(a + \overline{N})}{(a + \overline{N})^2}\right) = \frac{\overline{N}\varepsilon}{K(a + \overline{N})}(a - K + 2\overline{N})$$

$$= \frac{\overline{N}\varepsilon}{K(a + \overline{N})}\left(a + \frac{2a}{b - 1} - K\right) < 0$$

because of our assumption. By (1.4.15) this means that E is unstable;
moreover, since the real parts of *both* eigenvalues are negative, it is a
repeller, i.e. all neighbouring solutions tend to it as t tends to $-\infty$! As
a consequence, E cannot be an omega limit point either. Then by Theorem
3.1.11 the omega limit set of our trajectory is a limit cycle.

In the following three Sections further important applications will be
treated.

3.2   Liénard's Equation

First, a model governing the behaviour of *the electrical circuit with a triode
and with inductive feedback* will be set up following Cesari [1963] (see also
Farkas-Pidal [1981]). The circuit is shown in Figure 3.2.1. Today it might
be considered interesting only from the point of view of science history;
however, its differential equation is one of the most widely used models in
the study of non-linear oscillations.

The triode has a plate (anode), a grid and a cathode. The potential
difference between anode and cathode is V_a, and between grid and cathode
V_g. The behaviour of the triode depends on the linear combination
$v = V_g + DV_a$ of these potentials where $0 < D < 1$ is a constant depending
on the triode. The cathode is heated by a small battery and emits electrons.
When these electrons reach the anode, an anode current i_p can be measured.
The combination potential v shall stay below a negative number in order
to prevent all electrons to reach the anode, i.e. if $v \le v_0 < 0$ then $i_p = 0$.
As v is increased above v_0 the electrons reach the anode, and the anode
current is increasing with v until v reaches a saturation value $v_1 > 0$. If v is
increased beyond v_1, the anode current does not increase any further but
stays at a constant level. The *characteristic* of the triode is the function
describing the dependence of i_p on v. A typical characteristic is shown

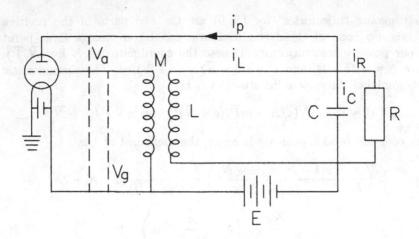

FIGURE 3.2.1. Electrical circuit with a triode and with inductive feedback modelled by Van der Pol's equation.

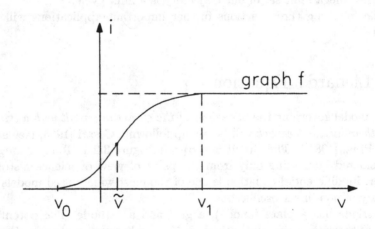

FIGURE 3.2.2. Characteristic of a triode.

on Figure 3.2.2. If it is denoted by f, then a good approximation can be reached by expanding it into a Taylor series at the point of inflection $\tilde{v}$ and taking into consideration terms up to third degree:

$$f(v) = f(\tilde{v}) + a(v - \tilde{v}) - b(v - \tilde{v})^3/3, \quad a > 0, \quad b > 0.$$

The equations determining the behaviour of the whole circuit are as follows. By Kirchoff's law

$$i_p = i_L + i_c + i_R, \tag{3.2.1}$$

the feedback regulating the potential of the grid by mutual inductance is (denoting time by τ)

$$V_g = M di_L/d\tau. \tag{3.2.2}$$

The drop of the potential through the solenoid L, the capacitance C and the resistance R is the same and is equal to the difference of the potential of the main battery E and the potential of the plate V_a, i.e.

$$L di_L/d\tau = q_c = R i_R = E - V_a \tag{3.2.3}$$

where q_c is the charge of the capacitor. The current through the capacitor is

$$i_c = dq_c/d\tau,$$

so that differentiating the first equation of (3.2.3) making use of (3.2.1) and taking into consideration that $i_p = f(v)$ we have

$$\begin{aligned} L d^2 i_L/d\tau^2 &= i_c/C = (1/C)(i_p - i_L - i_R) \\ &= (1/C)(f(v) - i_L - (L/R)di_L/d\tau) \end{aligned}$$

or

$$\begin{aligned} LC d^2 i_L/d\tau^2 + (L/R)di_L/d\tau + i_L &= f(M di_L/d\tau + D(E - L di_L/d\tau)) \\ &= f(DE + (M - DL)di_L/d\tau) \tag{3.2.4} \end{aligned}$$

where also (3.2.2) and (3.2.3) have been applied. We choose the physical parameters of the triode in such a way that $\tilde{v} = DE$ shall hold, and we introduce

$$V = (M - DL)di_L/d\tau$$

as an "unknown function". Differentiating (3.2.4) and multiplying it by $(M - DL)$ we get

$$LC d^2 V/d\tau^2 + (L/R)dV/d\tau + V = (M - DL)f'(\tilde{v} + V)dV/d\tau$$

where f' denotes the derivative of f with respect to its argument. Taking into consideration the expression for f, the last equation assumes the form

$$LC d^2 V/d\tau^2 + (L/R)dV/d\tau + V = (M - DL)(a - bV^2)dV/d\tau.$$

Introducing the new independent variable t by $\tau = \sqrt{LC}t$ we get

$$\frac{d^2V}{dt^2} + (LC)^{-1/2}(\frac{L}{R} + DLa - Ma + (M - DL)bV^2)\frac{dV}{dt} + V = 0$$

or

$$\frac{d^2V}{dt^2} + \frac{Ma - DLa - L/R}{(LC)^{1/2}}\left(-1 + \frac{(M - DL)b}{Ma - DLa - L/R}V^2\right)\frac{dV}{dt} + V = 0.$$

Assume that the feedback is sufficiently strong, i.e. the mutual inductance M is large enough. Then

$$m := (Ma - DLa - L/R)(LC)^{-1/2} > 0,$$

and the coefficient of V^2 in the brackets of the last differential equation is positive. Finally, introducing the new "unknown function" u by

$$u = ((M - DL)b/(Ma - DLa - L/R))^{1/2}V$$

we get *Van der Pol's differential equation* (see Van der Pol [1922,1926])

$$\ddot{u} + m(u^2 - 1)\dot{u} + u = 0 \tag{3.2.5}$$

where $m > 0$, the dot denotes differentiation with respect to t, and summing up all the transformations performed

$$u(t) = \left(\frac{(M - DL)^{3/2}b^{1/2}}{(Ma - DLa - L/R)^{1/2}}\right)\frac{di_L}{d\tau}(t\sqrt{LC}).$$

The main feature of the differential equation (3.2.5) is that although it has a simple linear restoring force equal to the displacement u and of opposite direction, the damping, i.e. the term containing the first derivative, is *negative* for small values of u. This means that the "damping" is actually amplifying the displacement from equilibrium until the displacement reaches the value 1 and only above this value does it play the role of a true damping. So one may expect that Van der Pol's differential equation exhibits oscillations in spite of the fact that it does not contain any periodic forcing term. The system corresponding to (3.2.5) is

$$\dot{u} = v, \qquad \dot{v} = -u + m(1 - u^2)v. \tag{3.2.6}$$

Linearizing this system in the only equilibrium point $u = 0$, $\dot{u} = v = 0$, we get that the characteristic polynomial is

$$D(\lambda, m) = \lambda^2 - m\lambda + 1.$$

For $m > 0$ the equilibrium is, clearly, unstable. One may consider van der Pol's equation for $m \leq 0$. If $m = 0$, one gets the (undamped) harmonic oscillator for which the point $(u, v) = (0, 0)$ is a centre. If $m < 0$, then the equilibrium is, obviously, asymptotically stable.

Van der Pol's equation is a special case of *Liénard's differential equation* which we assume in its generalized form

$$\ddot{x} + f(x)\dot{x} + g(x) = 0 \qquad (3.2.7)$$

where f and g satisfy conditions to be specified in the following two theorems. A. Liénard [1928] actually studied the case $g(x) = x$. The essence of his method formulated for (3.2.7) is the introduction of the variables (x, η) on the "*Liénard plane*" by writing (3.2.7) in the equivalent form

$$\dot{x} = \eta - F(x), \qquad \dot{\eta} = -g(x) \qquad (3.2.8)$$

where $F(x) := \int_0^x f(s)ds$. Because of the equivalence of (3.2.8) and (3.2.7) the conditions $f \in C^0(\mathbf{R})$, $g \in C^1(\mathbf{R})$ ensure existence and uniqueness for arbitrary initial conditions.

There is a classical result about the existence of a non-constant periodic solution for (3.2.7) due to Levinson and Smith [1942] which we are presenting here without proof (see also Cesari [1963]).

Theorem 3.2.1. *Assume that $f \in C^0(\mathbf{R})$, $g \in C^1(\mathbf{R})$ and let $G(x) := \int_0^x g(s)ds$; if f is even, there exists an $x_0 > 0$ such that $F(x) < 0$ for $0 < x < x_0$, and $F(x) > 0$ for $x > x_0$, g is odd, $xg(x) > 0$ for $x \neq 0$, F is monotone increasing in (x_0, ∞), and $F(x) \to \infty$, $G(x) \to \infty$ as $x \to \infty$ then (3.2.7) has a unique non-constant periodic solution (apart from shifts of the time t).*

Van der Pol's equation (3.2.5), obviously, satisfies all the requirements.

We are going to prove a more recent theorem due to Gabriele Villari [1982], the main advantage being that it does not require symmetry properties from the functions f and g. On the other hand, it does not guarantee the uniqueness of the periodic solution.

Theorem 3.2.2 *(Villari [1982]). Assume that*

(i) $f \in C^0(\mathbf{R})$, $g \in C^1(\mathbf{R})$;

(ii) $f(0) < 0$, and there exists an $x_0 > 0$ such that $f(x) > 0$ for $|x| > x_0$;

(iii) $xg(x) > 0$, for $x \neq 0$;

(iv) $\min(\limsup_{x \to \infty}(g(x)/f(x)), \limsup_{x \to -\infty}(-g(x)/f(x))) < \infty$;

(v) *there exist positive constants $\bar{x} > x_0$ and $b > 0$ such that $f(x) + |g(x)| > b > 0$ for $|x| > \bar{x}$.*

Then (3.2.7) has at least one non-constant periodic solution.

Proof. We introduce the system

$$\dot{x} = y, \qquad \dot{y} = -f(x)y - g(x) \tag{3.2.9}$$

equivalent to (3.2.7). Existence and uniqueness of solutions corresponding to arbitrary initial conditions hold for this system because it is also equivalent to (3.2.8). Our aim is the construction of a Bendixson sack (see the description after Corollary 3.1.9) in the phase plane x, y.

It is clear from assumption (iii) that the only equilibrium of system (3.2.9) is the origin $(0,0)$. We claim that it cannot be the omega limit point of any trajectory (except, of course, itself). This will be proved by showing that the origin is a repeller. Reverse the time in system (3.2.9), i.e. introduce the new independent variable τ by $t = -\tau$, then (3.2.9) assumes the form

$$dx/d\tau = -y, \qquad dy/d\tau = f(x)y + g(x). \tag{3.2.10}$$

Consider the Liapunov function

$$V(x,y) = G(x) + y^2/2$$

where $G(x) = \int_0^x g(s)ds$. Because of (iii), clearly, V is positive definite. Its derivative with respect to system (3.2.10) is

$$\dot{V}_{(3.2.10)}(x,y) = f(x)y^2.$$

Because of (ii), $x = 0$ has a neighbourhood where $f(x) < 0$, so that in this neighbourhood $\dot{V}_{(3.2.10)}(x,y) \le 0$. It is zero if and only if $y = 0$, i.e. on the axis x. Because of (iii) there is no other positive semitrajectory on the axis x but the origin. Thus, by the Barbashin-Krasovskij Theorem 1.5.5 the point $(0,0)$ is an asymptotically stable equilibrium of system (3.2.10), and this means that the solutions of the original system (3.2.9) tend to it as t tends to *minus* infinity.

Now, consider the $\dot{y} = 0$ isocline of the system. Its equation is

$$y = -g(x)/f(x).$$

The function on the right-hand side is defined everywhere except at zeroes of f. According to condition (ii) there is a least and a largest zero of f: $x_1 < 0 < x_2$, i.e. $f(x_1) = f(x_2) = 0$, and $f(x) > 0$ for $x < x_1$ and for $x > x_2$. (This means that $x_0 = \max(|x_1|, x_2)$.) Introduce the notation $u(x) = -g(x)/f(x)$. We have

$$\lim_{x \to x_1 - 0} u(x) = \infty, \qquad \lim_{x \to x_2 + 0} u(x) = -\infty,$$

and by condition (iv), u is bounded either as $x \to -\infty$ or as $x \to \infty$. Assume that the first case applies. The proof in the second case is similar. Choose an abscissa $\xi < x_1$ such that $u(\xi) > u(x) > 0$ for $x < \xi$, and denote the solution with initial values $(\xi, u(\xi))$ by $(\varphi(t), \psi(t))$. In the upper half plane $(y > 0)$ we have $\dot\varphi(t) > 0$, and introducing the notations

$$a_1 = \max_{[\xi, \bar{x}]} |f(x)|, \qquad a_2 = \max_{[\xi, \bar{x}]} |g(x)|,$$

the estimate $\dot\psi \le a_1 \psi(t) + a_2$ holds while $\xi \le \varphi(t) \le \bar{x}$. This implies that ψ is increasing less than exponential, and that the path cannot escape into infinity in the strip $\xi \le x \le \bar{x}$ in finite time. All trajectories that cross the negative x-axis cross it in the upward direction. If our path did not cross the positive y-axis, then $\varphi(t) \to c \le 0$ as $t \to \infty$ would hold. In this case $\psi(t) \to \infty$ as $t \to \infty$ should hold since if ψ did not tend to infinity the vertical line $x = c$ would contain the omega limit set of our trajectory which is impossible (all trajectories intersect this line for $y > 0$ transversally). But if $\psi(t) = \dot\varphi(t) \to \infty$ were true, then $\varphi(t)$ would tend to infinity, contrary to the assumption. We conclude that our path crosses the positive y-axis. The same estimate yields that the path either crosses the positive x-axis between 0 and $\bar{x}$ (all trajectories that cross the positive x-axis cross it in the downward direction) or crosses the vertical line $x = \bar{x}$ above the x-axis. Take the second possibility. If $\varphi(t) > \bar{x}$, then $g(\varphi(t)) > 0$, $f(\varphi(t)) > 0$, and the following estimates hold:

$$\dot\psi = f(\varphi(t))(1 - \psi(t)) - \big(f(\varphi(t)) + g(\varphi(t))\big) < f(\varphi(t))(1 - \psi(t)) - b$$

and

$$\begin{aligned} \dot\psi(t) &= -\big(f(\varphi(t)) + g(\varphi(t))\big)\psi(t) + g(\varphi(t))(\psi(t) - 1) \\ &< -b\psi(t) + g(\varphi(t))(\psi(t) - 1) \end{aligned}$$

where condition (v) has been used. This means that $\dot\psi(t) < -b$ if $\psi(t) > 1$, and $\dot\psi(t) < -b\psi(t)$ if $0 < \psi(t) < 1$. In the latter case the slope of the path is $\dot\psi(t)/\dot\varphi(t) < -b$. These estimates imply that the path crosses the x-axis at some point beyond $\bar{x}$. No matter whether the crossing has taken place in the interval $(0, \bar{x})$ or to the right of $\bar{x}$, in the lower half plane $(y < 0)$ where $\dot\varphi(t) < 0$ similar arguments apply and result in our path having to cross the negative x-axis either in the interval $(\xi, 0)$ or left to ξ. If the crossing takes place in the interval $(\xi, 0)$, then again in the upper half plane the path has to move to the right and stay below its earlier self, so that it has to cross the positive y-axis at a point below the former crossing. This way a Bendixson sack has been formed around the origin, the transversal being the segment of the y-axis between the two crossings. If the crossing of the negative x-axis takes place to the left of the point ξ, then because of the choice of ξ the path stays below the horizontal line $y = u(\xi)$ (because in the case it cuts into the graph of u, this is necessarily below $u(\xi)$, and above the graph of u the ordinate $\psi(t)$ of the path is decreasing). As a consequence,

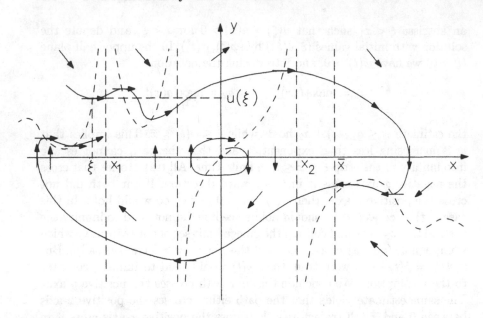

FIGURE 3.2.3. The large solid curve shows a path of system (3.2.9) forming a Bendixson sack; the dotted curves show the graph of the function u in the proof of Theorem 3.2.2, i.e. the locus of $\dot{y} = 0$; the origin is repelling as shown by the small path around it.

the path has to cross the vertical line $x = \xi$ at some point below $u(\xi)$, forming a Bendixson sack again. The latter situation is shown on Figure 3.2.3. Having taken into consideration all the possibilities, a Bendixson sack has been constructed containing a single equilibrium point, the origin which is a repeller. By Theorem 3.1.11 this implies that system (3.2.9) has a non-constant periodic solution with path inside the Bendixson sack. In case u is bounded as $x \to \infty$ we start at a properly chosen point $\xi > x_2$, $u(\xi) < 0$. The argumentation is analogous. $\square$

Van der Pol's equation (3.2.5) satisfies all the conditions of this Theorem too, and this means that it has a closed path in the phase plane $u, \dot{u}$ around the origin. The previous Theorem 3.2.1 ensures the uniqueness of this path also. We conclude that the electrical circuit shown in Figure 3.2.1 is capable to "self-excited oscillations", i.e. though the energy source provides DC the derivative $di_L/d\tau$ and, because of (3.2.2), (3.2.3), the grid and anode potentials V_g and V_a are periodic functions provided that the initial conditions are chosen appropriately. Whether this oscillation is observable experimentally depends on its stability. We shall return to the problem of stability of the periodic solution in Chapter 5. Now, we are going to generalize Liénard's equation (3.2.7) and prove one more theorem about the existence of a non-constant periodic solution of the generalized equation.

The following Theorem is due to P.J. Ponzo - N. Wax [1984] and to Zheng-Zuo-Huan [1990].

Consider the differential equation

$$\ddot{x} + f(x, \dot{x})\dot{x} + g(x) = 0 \tag{3.2.11}$$

where we assume that $f \in C^1(\mathbf{R}^2)$, $g \in C^1(\mathbf{R})$. As we see, (3.2.11) differs from (3.2.7) in that that the damping coefficient depends also on the velocity. This is a more realistic assumption in modelling many real world phenomena. Equation (3.2.11) is rewritten in the equivalent Cauchy normal form

$$\dot{x} = y, \qquad \dot{y} = -f(x, y)y - g(x). \tag{3.2.12}$$

Theorem 3.2.3. *Assume that*

(i) $f(0,0) < 0$, *there are* $a < 0 < b$ *such that*
   $f(a, y) = f(b, y) = 0$ *for all* $y \in \mathbf{R}$,
   $f(x, y) > 0$ *if* $x \in (-\infty, a) \cup (b, \infty)$ *for all* $y \in \mathbf{R}$,
   for $y \geq 0$ *and for every* $x < a$ *the function* $yf(x, y)$ *is increasing in*
      y, *and* $\lim_{y \to \infty} yf(x, y) = \infty$,
   there is an $M > 0$ *such that for* $x \in [a, b] : f(x, y) \geq -M$;

(ii) $xg(x) > 0$ *for* $x \neq 0$,
   for $G(x) = \int_0^x g(s)ds$, $\lim_{x \to \pm\infty} G(x) = \infty$;

(iii) *for the function* $u : (-\infty, a) \to \mathbf{R}_+$ *defined implicitly by*

$$uf(x, u) + g(x) = 0, \qquad x < a \tag{3.2.13}$$

   assume that $\bar{u}(x) := \max_{s \leq x} u(s)$ *exists for* $x < a$.

Then (3.2.12) has at least one non-constant periodic solution.

Note that Van der Pol's equation satisfies all the conditions of this Theorem too, the function u defined in (iii) being $u(x) = -x/(m(x^2 - 1))$, $x \in (-\infty, -1)$ and $\bar{u}(x) = u(x)$.

It is easy to see that conditions (i) and (ii) imposed upon f and g imply that (3.2.13) determines a unique positive valued function u, and that $u \in C^0(-\infty, a)$. The function $\bar{u}$ is also continuous and it is non-decreasing. We are going to show that

$$\lim_{x \to a-0} u(x) = \infty. \tag{3.2.14}$$

Indeed, if this was not the case, a sequence $x_n \to a - 0$ as $n \to \infty$ would exist such that $u(x_n) < K$ for some $K > 0$. Because of (i), $0 < u(x_n)f(x_n, u(x_n)) < Kf(x_n, K) \to 0$ as $n \to \infty$. But this would mean that $-g(x_n) \to 0$ as $n \to \infty$, a contradiction since g is continuous and $-g(a) > 0$.

Proof. System (3.2.12) has a single equilibrium, the origin. An annular region will be constructed around the origin from which no path may escape. Consider the Liapunov function (the "energy integral")

$$V(x,y) = y^2/2 + G(x).$$

Assumptions (ii) imply that its level curves $V(x,y) = c$, for $c > 0$, are nested closed "ovals" (the level c increasing along any ray emanating from the origin). The derivative of V with respect to the system is

$$\dot{V}_{(3.2.12)}(x,y) = -f(x,y)y^2.$$

Condition $f(0,0) < 0$ implies that the origin has a neighbourhood in which $f(x,y) < 0$. For a level curve which lies in this neighbourhood completely, $\dot{V} \geq 0$, i.e. the trajectories cannot cross this level curve inwardly. We choose such a level curve, $V = c_1 > 0$, say, for the inner boundary of the annulus. In order to construct the other boundary we obverse that the curve $y = u(x)$, $x \in (-\infty, a)$ is the zero isocline $\dot{y} = 0$. This means that above this curve $\dot{y} < 0$; below it $\dot{y} > 0$. Clearly, $\bar{u}(x) \geq u(x)$, and we have

$$\lim_{x \to \substack{-\infty \\ a-0}} (\bar{u}^2(x)/2 + G(x)) = \infty$$

because of (ii) and (3.2.14), so that the function has an absolute positive minimum in $(-\infty, a)$:

$$0 < \alpha := \min(\bar{u}^2(x)/2 + G(x)), \qquad x \in (-\infty, a),$$

i.e. there is an $x_0 \in (-\infty, a)$ such that

$$\alpha = \bar{u}^2(x_0)/2 + G(x_0). \tag{3.2.15}$$

In $x < x_0$, $y > 0$ we use the graph $\bar{u}$ as part of the outer boundary of the annular region; in $[x_0, a]$ we use the level curve $y^2/2 + G(x) = \alpha$, i.e. the graph of the function $y = \sqrt{2}\,(\alpha - G(x))^{1/2}$. Because of (3.2.15) we have $\bar{u}(x_0) = \sqrt{2}\,(\alpha - G(x_0))^{1/2}$. These curves are crossed by the orbits of the system inwardly. Our level curve belonging to the value α meets the vertical line $x = a$ at $y_a = \sqrt{2}\,(\alpha - G(a))^{1/2} > 0$. Consider now the solution of the system corresponding to the initial value (a, y_a) (see Figure 3.2.4). We repeat part of the argument in the proof of Theorem 3.2.2. Because of (i) in the strip $x \in [a, b]$ for $y > 0$ we have

$$\dot{y} < My + \bar{g}$$

where $\bar{g} > 0$ is an upper bound of $|g(x)|$, $x \in [a, b]$. This means that the y coordinate of the solution is increasing less than exponential so that the

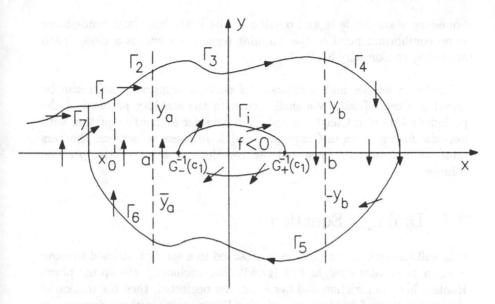

FIGURE 3.2.4. Annular region containing a limit cycle under the conditions of Theorem 3.2.3. Γ_i: inner boundary $V = c_1$; $\Gamma_o = \Gamma_1 \cup \Gamma_2 \cup \Gamma_3 \cup \Gamma_4 \cup \Gamma_5 \cup \Gamma_6 \cup \Gamma_7$ outer boundary, Γ_1: graph $\bar{u}$, $\Gamma_2 : V = \alpha$, Γ_3: a path, $\Gamma_4 : V = c_2$, Γ_5: a path, $\Gamma_6 : V = c_3$, Γ_7: a vertical straight line segment.

path cannot escape into infinity in finite time while x stays in $[a, b]$. Just like in the proof of Theorem 3.2.2 it follows that the path either crosses the x-axis between $G_+^{-1}(c_1) > 0$ and b (where G_+^{-1} is the inverse of the function G restricted to $[0, \infty)$) or it crosses the vertical line $x = b$ at some $y_b > 0$. In the former case we follow this path until it enters the $y > 0$ half plane. In the latter case let the level curve through (b, y_b) be $V = c_2$. This level curve will be the next part of the outer boundary until it meets the vertical line $x = b$ again at $-y_b < 0$. The next section of the outer boundary will be the path that starts at $(b, -y_b)$. In both cases we have a path that enters the strip $x \in [a, b]$, $y < 0$. A similar consideration to that which was used in the $y > 0$ half plane yields that such a path either crosses the x-axis between a and $G_-^{-1}(c_1) < 0$ (where G_-^{-1} is the inverse of the function G restricted to $(-\infty, 0]$) or crosses the vertical line $x = a$ at some $\bar{y}_a < 0$. In the former case the same path shall cross the y-axis somewhere above the crossing of the inner boundary $V = c_1$ and below the crossing of the outer boundary; this way a Bendixson sack is formed. In the latter case we continue the outer boundary with the level curve through $(a, \bar{y}_a)$ whose equation is $V = c_3$, say. This crosses the negative x-axis left of a, and we complete the outer boundary with a vertical line to the graph of $\bar{u}$ or to $V = \alpha$ depending on whether the crossing of the negative x-axis has taken place to the left or to the right of x_0. Any trajectory that crosses the outer boundary stays inside it and outside of the inner boundary. Since there is no equilibrium point in this annular region, it contains a closed path according to Corollary 3.1.9. □

Further existence and uniqueness of periodic solutions results can be found in Cesari [1963]. We shall return to the stability problem of the periodic solution in Chapter 5 and shall consider Liénard's equation with periodic forcing term in Chapters 4 and 6. In these subsequent chapters methods will also be presented for the determination of the periodic solution.

3.3 Duffing's Equation

It is well known that if a mass m is attached to a spring restricted to move along a horizontal straight line (gravitation excluded), the spring obeys Hooke's law, and friction and hysteresis are neglected, then the motion of this one degree of freedom system, the *linear spring without damping*, is governed by the differential equation

$$m\ddot{x} = -kx \qquad (3.3.1)$$

where x is the displacement from the equilibrium (no elastic force) position, $k > 0$ is the *stiffness* of the spring and $-kx$ is the *restoring force* (see Figure 3.3.1). The general solution of the *harmonic oscillator* (3.3.1), $\ddot{x} + xk/m = 0$, is $x(t) = c_1 \cos \omega t + c_2 \sin \omega t$ where $\omega = (k/m)^{1/2}$. This means that all non-trivial solutions are periodic with the same (least positive) period $T = 2\pi/\omega$, the trajectories are concentric circles around the origin of the x, $\dot{x}/\omega$ plane.

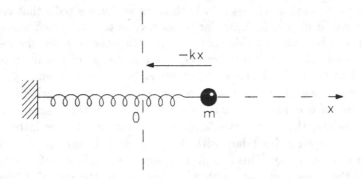

FIGURE 3.3.1. The linear spring without damping.

We may consider a spring that does not obey Hooke's law, i.e. in it the elastic restoring force is not proportional to the displacement. In other words, the stiffness is not a (positive) constant but depends on the displacement. If the stiffness is increasing with the displacement, then we say that the spring is *hard*; if the stiffness is a decreasing function, we have a *soft spring*. Since we require that the restoring force should behave symmetrically for displacements to the left and to the right of the equilibrium position, the stiffness of the simplest *non-linear spring* can be assumed in the form $k(x) = \alpha \pm \beta x^2$ with $\alpha, \beta > 0$, keeping the plus or minus sign according to whether the spring is hard or soft. Replacing k in (3.3.1) by this stiffness function, dividing by m and introducing new notations, we get *Duffing's differential equation without damping*

$$\ddot{x} + ax + bx^3 = 0 \qquad (3.3.2)$$

where $a > 0$, $b \in \mathbf{R}$ (Duffing [1918]).

Note that the differential equation of the mathematical pendulum without damping is $\ddot{\theta} + (g/l)\sin\theta = 0$ where g is the gravitational acceleration and l is the length (cf. Example 1.5.4). Here θ is the angle formed with the vertical direction. If the sin function is developed into a Taylor series, and the series is truncated at the third order term, we have $\sin\theta \approx \theta - \theta^3/6$ so that the differential equation governing the pendulum for *small displacements* becomes

$$\ddot{\theta} + (g/l)(\theta - \theta^3/6) = 0,$$

identical to (3.3.2) with $a = g/l$, $b = -g/(6l) < 0$ (the soft spring case). It turns out that this equation describes the motion of the pendulum relatively well if the maximal displacement θ is less than 1 radian.

Writing (3.3.2) in Cauchy normal form

$$\dot{x} = y, \qquad \dot{y} = -x(a + bx^2) \qquad (3.3.3)$$

we see that the energy integral $V(x, y) = (1/2)(ax^2 + bx^4/2 + y^2)$ is a first integral of the system (see Theorem 1.1.5), since $\dot{V}_{(3.3.3)}(x, y) \equiv 0$. This means that the level curves

$$ax^2 + bx^4/2 + y^2 = c, \qquad c \in \mathbf{R}, \qquad (3.3.4)$$

are the trajectories of the system. We may determine the time dependence of the phase variables x and y by solving (3.3.4) for y and substituting the result into the first differential equation of the system: $\dot{x} = \pm\sqrt{c - ax^2 - bx^4/2}$, according to whether $y \gtrless 0$. From this equation we obtain

$$\int_{x_0}^{x} \frac{ds}{\pm(c - as^2 - bs^4/2)^{1/2}} = t \qquad (3.3.5)$$

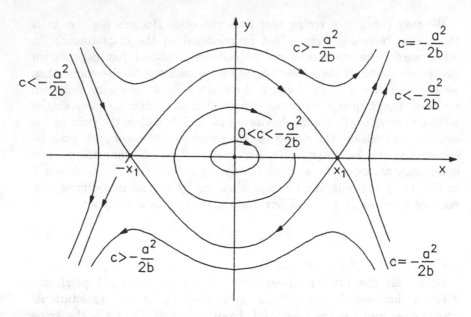

FIGURE 3.3.2. Phase portrait of Duffing's equation without damping (3.3.3) in the soft spring case $(a > 0, b < 0)$.

where $x(0) = x_0$. On the left-hand side we have an elliptic integral of the first kind which can be evaluated from tables, and we may even express x as a function of t in terms of elliptic functions (see e.g. Farkas [1964]). However, we may draw the phase picture of the system without the actual determination of the solutions. We have to distinguish the $b > 0$ hard spring and the $b < 0$ soft spring cases.

In the first case $(b > 0)$ we get real level curves from (3.3.4) only for $c \geq 0$, and these are, obviously, simple closed curves around the origin. The system has a single equilibrium point $(0,0)$, and all the other solutions are periodic.

In the second $(b < 0)$ case the system has three equilibria: $(0,0)$, $(x_1, 0)$ and $(-x_1, 0)$ where $x_1 = (-a/b)^{1/2}$. Simple calculations show that $(0,0)$ is a centre for the linearized system, and $(\pm x_1, 0)$ are saddle points. These latter equilibria lie on the level curve corresponding to the value $c = -a^2/(2b)$ in (3.3.4); the equation of this curve is $y = \pm(2|b|)^{-1/2}(a + bx^2)$. In the set

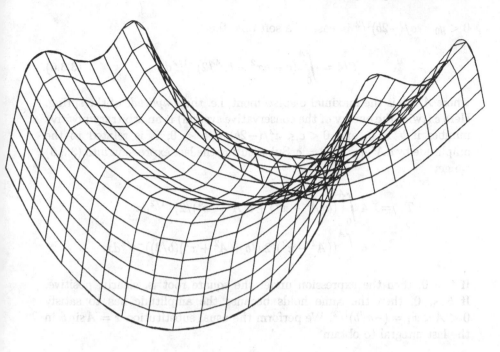

FIGURE 3.3.3. Graph of the energy integral belonging to Duffing's equation without damping.

$U = \{(x, y) : -(2|b|)^{-1/2}(a + bx^2) < y < (2|b|)^{-1/2}(a + bx^2), \ |x| < x_1\}$ the level curves are closed "ovals" around the origin; they belong to values in the interval $0 < c < -a^2/(2b)$. This implies that $(0, 0)$ is a centre also of the non-linear system (3.3.3). Hence, in the soft spring case we have periodic motion only if the initial energy is relatively small, i.e., the initial value $(x_0, y_0) \in U$. The two saddles $(\pm x_1, 0)$ are connected by two heteroclinic orbits. The rest of the solutions are unbounded (see Figure 3.3.2; the graph of the energy integral is shown in Figure 3.3.3).

We may determine the period T of the periodic solutions in both cases from equation (3.3.4). Since the flow is symmetric to both the x- and y- axes (with directions reversed at reflections), if the initial values of a periodic solution are $(x_0, y_0) = (0, y_0)$, $y_0 > 0$ in case of a hard spring, and

$0 < y_0 < a/(-2b)^{1/2}$ in case of a soft one, then

$$T/4 = \int_0^A (c - ax^2 - bx^4/2)^{-1/2} dx \qquad (3.3.6)$$

where $A > 0$ is the maximal displacement, i.e. the *amplitude* of the spring. Here c (twice the energy of the conservative system) is an arbitrary positive number if $b > 0$, and $0 < c < a^2/(-2b)$ if $b < 0$; it is related to the amplitude by $aA^2 + bA^4/2 = c$. Substituting the last expression into (3.3.6) we get

$$
\begin{aligned}
T &= 4 \int_0^A (a(A^2 - x^2) + b(A^4 - x^4)/2)^{-1/2} dx \\
&= 4 \int_0^A ((A^2 - x^2)(2a/b + A^2 + x^2)(b/2))^{-1/2} dx.
\end{aligned}
$$

If $b > 0$, then the expression under the square root is, clearly, positive. If $b < 0$, then the same holds because the amplitude has to satisfy $0 < A < x_1 = (-a/b)^{1/2}$. We perform the usual substitution $x = A\sin\theta$ in the last integral to obtain

$$T(A) = 4\sqrt{2} \int_0^{2\pi} (2a + A^2 b(1 + \sin^2\theta))^{-1/2} d\theta. \qquad (3.3.7)$$

If $b = 0$, i.e. in case of the linear spring $T = 2\pi/\sqrt{a}$, or interpreting the system as a mathematical pendulum with small displacements, $T = 2\pi\sqrt{l/g}$, since as we have seen $a = g/l$, l being the length of the pendulum.

In the case $b \neq 0$, the periodic solutions are no longer "*isochronous*". If $b > 0$, then, clearly, the period T is decreasing, the frequency increasing with the amplitude, i.e. a hard spring's oscillations are wilder if its energy is larger. If $b < 0$, then the period T is increasing as the amplitude A is increased in the interval $0 < A < (-a/b)^{1/2}$. As $A \to (-a/b)^{1/2} - 0$ the period $T(A)$ tends to infinity:

$$\lim_{A \to (-a/b)^{1/2} - 0} T(A) = 4(2/a)^{1/2} \int_0^{\pi/2} d\theta/\cos\theta = \infty.$$

This is clear from the phase portrait (Figure 3.3.2), since in this case the upper half, say, of the closed orbits tends to the heteroclinic trajectory connecting the equilibria $(-x_1, 0)$ and $(x_1, 0)$; to travel on this trajectory from one equilibrium to the other "takes infinite time".

We turn now to the *damped Duffing equation*

$$\ddot{x} + d\dot{x} + ax + bx^3 = 0 \qquad (3.3.8)$$

where the damping coefficient $d > 0$, and $a > 0$, $b \in \mathbf{R}$ as in (3.3.2). We assume that the damping is proportional to the velocity $\dot{x}$; this is sometimes called *viscous damping* (see Stoker [1950]). This differential equation could be considered to be of Liénard's type (cf. (3.2.7)), and for $b > 0$, at least, $g(x) = ax + bx^3$ satisfies the conditions of Theorem 3.2.1; however, here the coefficient of $\dot{x}$ is $f(x) = d$ which is positive everywhere, so that it does not satisfy the conditions of Theorems 3.2.1-3.2.3. And, in fact, in Section 5 we shall see that (3.3.8) does not have any non-constant periodic solution. Writing the differential equation (3.3.8) in Cauchy normal form

$$\dot{x} = y, \qquad \dot{y} = -ax - bx^3 - dy, \qquad (3.3.9)$$

and using the same energy integral as in the case of (3.3.3) as a Liapunov function,

$$V(x, y) = (1/2)(ax^2 + bx^4/2 + y^2),$$

we see that its derivative with respect to the damped system (3.3.9) is no longer zero but

$$\dot{V}_{(3.3.9)}(x, y) = -dy^2 \leq 0.$$

We have to distinguish again the hard spring $(b > 0)$ and the soft spring $(b < 0)$ cases.

If $b > 0$, then the system (3.3.9) has no equilibrium but $(0,0)$. In this case the set in which $\dot{V} = 0$, i.e. the x-axis does not contain any positive semitrajectory apart from the origin. Therefore, by the Barbashin-Krasovskij Theorem 1.5.5 the origin is asymptotically stable. Since V is a global positive definite function, it follows that $(0,0)$ is globally asymptotically stable. As is easy to see, the characteristic polynomial of the system linearized at $(0,0)$ is $\lambda^2 + d\lambda + a$, so that the origin is either a spiral point or a node according to whether $d^2 - 4a$ is negative or positive. This means that if the stiffness is large compared to the damping, $a > d^2/4$, then the spring returns to equilibrium by damped oscillations. If the stiffness is small, and damping large, $0 < a \leq d^2/4$, then the return to equilibrium happens in one move. Since the characteristic polynomial does not depend on b, these last statements are true also for soft springs.

If $b < 0$, then system (3.3.9) has three equilibria just like in the undamped case: $(0,0)$, $(x_1, 0)$ and $(-x_1, 0)$ where $x_1 = (-a/b)^{1/2}$. From what has been said about the characteristic polynomial above it is clear that $(0,0)$ is again asymptotically stable (but it is no longer globally asymptotically stable). The characteristic polynomial at the equilibria $(\pm x_1, 0)$ is $\lambda^2 + d\lambda - 2a$, i.e. these equilibrium points are saddles.

We shall return to Duffing's equation with periodic forcing term in Chapter 4.

3.4   The Lotka-Volterra Predator-Prey Model and Generalizations

Assume that two species form an ecological system, one serving as the only food for the other. We denote the quantity of the nutrient (prey) species at time t by $N(t)$, and the quantity of the predator species by $P(t)$. Let the prey have the constant $\varepsilon > 0$ as *specific* (per capita) *growth rate* (birth rate minus death rate) in absence of predation, and assume that predators decrease the growth rate proportionally to their quantity: one unit of predator decreasing the growth rate of prey by $\alpha > 0$. Then the dynamics of the prey quantity is governed by the differential equation

$$\dot{N}(t)/N(t) = \varepsilon - \alpha P(t).$$

Let the *mortality* of the predator in the absence of food be the constant $\gamma > 0$, and assume that the prey increases the growth rate of predator proportionally to prey quantity, the rate of proportionality, the *conversion rate* (of prey biomass into predator biomass) being $\beta > 0$. Then the differential equation governing the dynamics of predator quantity is

$$\dot{P}(t)/P(t) = -\gamma + \beta N(t).$$

Thus, the predator-prey system is modelled by the two dimensional autonomous system of differential equations

$$\dot{N} = N(\varepsilon - \alpha P), \qquad \dot{P} = P(-\gamma + \beta N). \qquad (3.4.1)$$

This is called the *Lotka-Volterra predator-prey system*. The model (3.4.1) was suggested by V. Volterra [1931] to explain the change in the composition of catch observed by fishermen on the Adriatic Sea after World War One during which fishing had been reduced considerably. We shall return to this problem after having treated system (3.4.1). The same model occurred in A.J. Lotka [1924].

It is clear that for system (3.4.1) the positive quadrant $\mathbf{R}_+^2$ of the N, P plane, and the N- and P- axes are invariant sets. We are going to study the system in $\mathbf{R}_+^2$ since only non-negative values of N and P may be interpreted ecologically. In $\mathbf{R}_+^2$ the system has two equilibria: $(0,0)$ and the point $E = (\gamma/\beta, \varepsilon/\alpha)$. The eigenvalues of the system linearized at the origin are ε and $-\gamma$ so that the origin is a saddle point. The eigenvalues of the system linearized at E are $\pm i(\varepsilon\gamma)^{1/2}$. Thus, for the linearized system E is a centre. The period of every periodic solution of the linearized system is

$$T = 2\pi/(\varepsilon\gamma)^{1/2}. \qquad (3.4.2)$$

We are going to show that (3.4.1) has a first integral. Multiply the first equation by β, the second by α and add them:

$$\beta\dot{N} + \alpha\dot{P} = \varepsilon\beta N - \alpha\gamma P. \qquad (3.4.3)$$

Multiply the first equation by γP, the second by εN and add them:

$$\gamma P \dot{N} + \varepsilon N \dot{P} = -\alpha \gamma N P^2 + \varepsilon \beta N^2 P$$

or

$$\gamma \dot{N}/N + \varepsilon \dot{P}/P = -\alpha \gamma P + \varepsilon \beta N.$$

Subtracting the last equation from (3.4.3) we obtain

$$\frac{d}{dt}(\beta N - \gamma \ln N + \alpha P - \varepsilon \ln P) = 0.$$

Thus, the function

$$F(N,P) := \beta N - \gamma \ln N + \alpha P - \varepsilon \ln P, \quad (N,P) \in \text{Int } \mathbf{R}_+^2, \qquad (3.4.4)$$

is a first integral of system (3.4.1). Easy calculations show that $F \to \infty$ if $N \to 0+0$ or ∞ or $P \to 0+0$ or ∞, and that F has an absolute minimum at $E = (\gamma/\beta, \varepsilon/\alpha)$. The value of F at E is

$$F(\gamma/\beta, \varepsilon/\alpha) = \gamma(1 - \ln(\gamma/\beta)) + \varepsilon(1 - \ln(\varepsilon/\alpha)).$$

Deducting this from F, the function

$$V(N,P) = F(N,P) - F(\gamma/\beta, \varepsilon/\alpha) \qquad (3.4.5)$$

becomes positive definite with respect to the point E, i.e. $V(N,P) > 0$ for $N > 0, P > 0, (N,P) \neq (\gamma/\beta, \varepsilon/\alpha)$ and $V(\gamma/\beta, \varepsilon/\alpha) = 0$. Since $\dot{V}_{(3.4.1)} = 0$, the equilibrium E is stable in the Liapunov sense. Easy calculations show that the graph of $V : z = V(N,P)$ is a convex down surface in N, P, z space, thus, the level curves that are the orbits of system (3.4.1) are all closed ovals. As a consequence, the equilibrium E is a centre also for the non-linear system (3.4.1), and all the solutions are periodic functions. If their amplitude is small, i.e. their path stays near to E, then the period is approximately equal to (3.4.2). See Figure 3.4.1.

Let $(N(t), P(t))$ be an arbitrary solution of (3.4.1), and denote its period by $T > 0$. We are going to calculate the time average of the quantity of prey and predator. Dividing the first equation by N, the second by P, and integrating from 0 to T we get

$$0 = \ln(N(T)/N(0)) = \int_0^T \dot{N}(t)/N(t)\,dt = \varepsilon T - \alpha \int_0^T P(t)\,dt$$

and

$$0 = \ln(P(T)/P(0)) = \int_0^T \dot{P}(t)/P(t)\,dt = -\gamma T + \beta \int_0^T N(t)\,dt.$$

Hence

$$(1/T)\int_0^T N(t)\,dt = \gamma/\beta, \qquad (1/T)\int_0^T P(t)\,dt = \varepsilon/\alpha. \qquad (3.4.6)$$

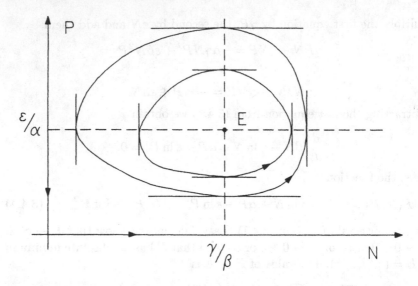

FIGURE 3.4.1. Phase plane of the Lotka-Volterra predator-prey system (3.4.1). E is the single equilibrium with positive coordinates. The closed curves are the orbits; they are the level curves of the function (3.4.4).

Regardless of the positive initial values, we have obtained that the time average of the prey, resp. the predator quantity is always the same and equal to the equilibrium value, i.e. the respective coordinate of the equilibrium point E.

It is easy to see that the closed trajectories have vertical tangents iff $P = \varepsilon/\alpha$, and horizontal ones iff $N = \gamma/\beta$. This means that the amount of predator is decreasing when the amount of prey is below the average γ/β, and increasing when this amount is above the average. Similarly, the amount of prey is decreasing, resp. increasing according to whether the amount of predator is above or below its average ε/α (see Figure 3.4.1).

There are several results in the literature concerning the dependence of the period of a solution on the amplitude, or, rather, the "energy" of the corresponding path. One of the most recent ones is that of F. Rothe [1985]. As it is often done, he transforms system (3.4.1) applying the transformation

$$u = \ln(\beta N/\gamma), \qquad v = \ln(\alpha P/\varepsilon)$$

into

$$\dot{u} = \varepsilon(1 - \exp v), \qquad \dot{v} = -\gamma(1 - \exp u);$$

the latter is already a Hamiltonian system with the Hamiltonian function

$$H(u, v) = \gamma(\exp u - u - 1) + \varepsilon(\exp v - v - 1),$$

since, clearly,

$$\partial H/\partial u = -\gamma(1 - \exp u), \qquad \partial H/\partial v = -\varepsilon(1 - \exp v).$$

Our first integral V of the original system is related to H by

$$V(N, P) = H(\ln(\beta N/\gamma), \ln(\alpha P/\varepsilon)).$$

Applying the theory of Hamiltonian systems and Laplace transform technique, Rothe obtains the following results which we are presenting here without proof. For arbitrary $h > 0$ the equation $V(N, P) = h$ determines a closed path of system (3.4.1). The positive constant h is called the *energy* of this path. The least positive period of a solution corresponding to this path will be denoted by $T(h)$.

Theorem 3.4.1 *(Rothe [1985]). The energy-period function $T : \mathbf{R}^+ \to \mathbf{R}^+$ is strictly increasing.*

This means that the Lotka-Volterra system behaves like a soft spring.

Theorem 3.4.2 *(Rothe [1985]). For small values of the energy $h > 0$, the period can be expanded into a convergent Taylor series, and*

$$T(h) = \sum_{k=0}^{3} T^{(k)}(0)h^k/k! + o(h^3)$$

where

$$T(0) = 2\pi/(\varepsilon\gamma)^{1/2}, \qquad T'(0) = \pi(1/\varepsilon + 1/\gamma)/(6(\varepsilon\gamma)^{1/2}),$$
$$T''(0) = T'(0)^2/(2T(0)) = \pi(1/\varepsilon + 1/\gamma)^2/(144(\varepsilon\gamma)^{1/2}),$$
$$T'''(0) = -139\pi(1/\varepsilon + 1/\gamma)(1/\varepsilon^2 + 1/\gamma^2$$
$$-154/(139\varepsilon\gamma))/(25920(\varepsilon\gamma)^{1/2}) < 0.$$

According to an earlier result due to J.S. Frame [1974] the energy-period function can be approximated well by the modified Bessel function of the first kind I_0. Frame's formula is

$$T(h) = 2\pi/(\varepsilon\gamma)^{1/2}I_0\big((h(\varepsilon + \gamma)/(3\varepsilon\gamma))^{1/2}\big) \qquad (3.4.7)$$

where

$$I_0(z) = J_0(iz) = \sum_{k=0}^{\infty} z^{2k}/(2^k k!)^2.$$

The fairly crude model (3.4.1) describes an intuitively plausible phenomenon: the quantity of prey is increasing. This results in the increase of the predator population; if the quantity of predators becomes too large, prey is diminishing rapidly which, in turn, results in the decrease of the predator population which has no food. After that, the prey population may increase again, beginning a new cycle. This oscillation has been observed in nature. As we have seen, the system behaves like a soft spring (see Section 3.3), the cycle time increasing with the amplitude, i.e. the farther the initial values are from the equilibrium point E, the larger the period of the oscillation is. The model gives an explanation also of the change in the composition of catch as a result of the change in the intensity of fishing. If we assume that "harvesting" is proportional to the quantities of the different species, then the quantity δN, resp. δP has to be subtracted from the right-hand side of the first, resp. the second equation of (3.4.1) where $\delta > 0$ measures the intensity of fishing. This means that (3.4.1) is replaced by

$$\dot{N} = N(\varepsilon - \delta - \alpha P), \qquad \dot{P} = P(-\gamma - \delta + \beta N). \qquad (3.4.8)$$

The equilibrium, and with it the time average of the solutions of (3.4.8), is now $((\gamma + \delta)/\beta, (\varepsilon - \delta)/\alpha)$. Since $(\gamma + \delta)/\beta > \gamma/\beta$, and $(\varepsilon - \delta)/\alpha < \varepsilon/\alpha$, it is apparent that the increase of the intensity of fishing is beneficial to the prey species, and detrimental to the predator. This was exactly what statistical data on the Adriatic Sea showed. Another often quoted instance for which the same effect was observed was the following. The "cottony cushion scale insect" (*Icerya purchasi*) had accidentally been introduced in 1868 to the west coast of the United States from Australia, and threatened to destroy the citrus plantations. Its natural enemy, a ladybird beetle (*Rodolia cardinalis*), had then been introduced from Australia, and this kept the scale insect population under control. When, much later, DDT was discovered it was applied at the citrus plantations in order to bring down the quantity of scale insect population still more. The effect was disastrous: the scale insect population grew rapidly while the predator, the ladybird beetle, went down.

In spite of the successes in describing some observable phenomena, the Lotka-Volterra model cannot be considered a satisfactory one, primarily, because it is not structurally stable (see Definition 7.1.3). Arbitrary small perturbations may destroy its phase portrait, and replace the centre with a stable or unstable spiral point.

A generalization of the Lotka-Volterra model is the following one which goes back, partly, to Gause [1934] and Gause-Smaragdova-Witt [1936]; see also Freedman [1980] p.66:

$$\dot{N} = Nf(N) - PNg(N), \qquad \dot{P} = P(-\gamma + \beta Ng(N)), \qquad (3.4.9)$$

where N and P denote again the quantities of prey and predator, respectively, the *mortality* γ of the predator and the *conversion rate* β

are positive constants, $f, g \in C^1(\mathbf{R}_+)$

$$f(0) > 0, \qquad f'(N) < 0, \qquad N \in \mathbf{R}_+, \qquad f(K) = 0 \qquad \text{for some} \qquad K > 0, \tag{3.4.10}$$

ensuring a saturation effect in the growth of prey at $N = K$,

$$g(N) > 0, \qquad (Ng(N))' > 0, \qquad N \in \mathbf{R}_+, \tag{3.4.11}$$

ensuring that prey is beneficial to predator, and that the *"functional response function"* $Ng(N)$ grows with the increase of prey quantity.

Note that a special case of (3.4.9) satisfying all the conditions has been treated as an Example in Section 3.1.

We give a simple graphical criterion for the existence of a limit cycle of system (3.4.9) in the positive quadrant of the N, P plane. First the equilibria will be surveyed. The equilibria in $\mathbf{R}_+^2$ are $(0,0)$, $(K,0)$, and, possibly, an equilibrium point in the interior of the positive quadrant. Linearization shows that $(0,0)$ is a saddle, $(K,0)$ is an asymptotically stable equilibrium, namely, a stable node if $\gamma - \beta Kg(K) > 0$, and it is a saddle if

$$\gamma - \beta Kg(K) < 0. \tag{3.4.12}$$

The zero isocline of the prey, apart from the P-axis, is the curve with equation $P = f(N)/g(N)$; the zero isocline of the predator is the vertical line $N = \overline{N}$ where $\overline{N}$ is the solution of the equation $Ng(N) = \gamma/\beta$. Now, because of (3.4.11), this last equation has one solution at most. If such a solution exists, it satisfies $0 < \overline{N} < K$ if and only if $\gamma/\beta = \overline{N}g(\overline{N}) < Kg(K)$, i.e. if and only if (3.4.12) holds. The vertical line $N = \overline{N}$ intersects the prey isocline in the positive quadrant exactly when this last condition is satisfied (see Figure 3.4.2). We conclude that an equilibrium point with positive coordinates $(\overline{N}, \overline{P} = f(\overline{N})/g(\overline{N}))$ exists if and only if (3.4.12) holds, and in this case the other two equilibria $(0,0)$ and $(K,0)$ are saddle points. Condition (3.4.12) will be assumed in the sequel. The stability of $(\overline{N}, \overline{P})$ will be studied now. The characteristic polynomial of the system linearized at $(\overline{N}, \overline{P})$ is

$$D(\lambda) = \lambda^2 + \lambda \overline{N}(\overline{P}g'(\overline{N}) - f'(\overline{N})) + \gamma \overline{P}(g(\overline{N}) + \overline{N}g'(\overline{N})).$$

The constant term is positive because of (3.4.11), so that the stability depends on the sign of the coefficient of λ, i.e. on the sign of

$$
\begin{aligned}
\overline{P}g'(\overline{N}) - f'(\overline{N}) &= g'(\overline{N})f(\overline{N})/g(\overline{N}) - f'(\overline{N}) \\
&= (f(\overline{N})g'(\overline{N}) - g(\overline{N})f'(\overline{N}))/g(\overline{N}) \\
&= -g(\overline{N})(f(N)/g(N))'|_{N=\overline{N}}.
\end{aligned}
$$

Thus, if $(f/g)'$ is negative at $N = \overline{N}$, then the interior equilibrium is asymptotically stable; if $(f/g)'$ is positive at $N = \overline{N}$, then it is unstable,

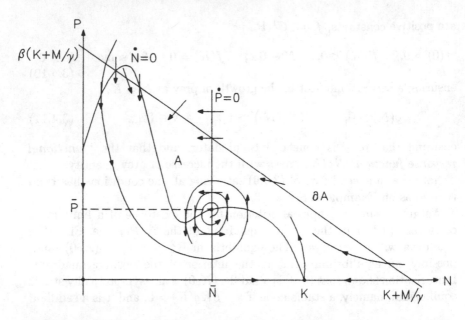

FIGURE 3.4.2. Phase portrait of system (3.4.9) when (3.4.12) holds, and the interior equilibrium is on an ascending branch of the prey isocline. The boundary of the positively invariant set A is the straight line $N + P/\beta = K + M/\gamma$.

namely, it is an unstable spiral point or node, in both cases a repeller. In other words, *if the interior equilibrium is on an ascending branch of the prey isocline, it is unstable; if it is on a descending branch, it is asymptotically stable.* This graphical criterion is, naturally, not a rigorous one, the exact criterion being the positivity, resp. the negativity of $(f/g)'$. Figure 3.4.2 shows the unstable situation.

We are going to show now that if the interior equilibrium is a repeller, then the system has a limit cycle in the interior of the positive quadrant. This result will be obtained through the construction of a positively invariant compact set. In the construction, a method due to Freedman and So [1985] will be used. Let

$$M := \max_{N \in [0,K]} N f(N).$$

Because of (3.4.10), clearly, $M > 0$. We are to prove that the set

$$A = \{(N, P) \in \mathbf{R}_+^2 : 0 \le N \le K, \ 0 \le N + P/\beta \le K + M/\gamma\} \quad (3.4.13)$$

is positively invariant, i.e. if $(N(0), P(0)) \in A$, then $(N(t), P(t)) \in A$ for $t > 0$ where, naturally, $(N(t), P(t))$ is an arbitrary solution of (3.4.9). From the first differential equation $\dot{N}(t) \le N(t)f(N(t))$, and $f(K) = 0$, $f(N) < 0$ for $N > K$; thus, if $0 \le N(0) \le K$, then $N(t)$ cannot grow above the bound K for $t > 0$. Dividing the second differential equation of the system by β and adding the equations

$$\dot{N}(t) + \dot{P}(t)/\beta \leq N(t)f(N(t)) - \gamma P(t)/\beta \leq -\gamma(N(t) + P(t)/\beta)$$
$$+ \gamma N(t) + N(t)f(N(t)) \leq -\gamma(N(t) + P(t)/\beta) + \gamma K + M.$$

Compare the function $N(t) + P(t)/\beta$ with the function $z(t)$ that assumes the same initial value $z(0) = N(0) + P(0)/\beta$ and satisfies the differential equation

$$\dot{z} = -\gamma z + \gamma K + M.$$

This last function is, clearly,

$$z(t) = K + M/\gamma + (z(0) - (K + M/\gamma)) \exp(-\gamma t).$$

Assuming that $z(0) = N(0) + P(0)/\beta \leq K + M/\gamma$ we obtain that $z(t) \leq K + M/\gamma$ for $t \geq 0$. Because of the differential inequality satisfied by $N(t) + P(t)/\beta$ we get, finally, that

$$N(t) + P(t)/\beta \leq z(t) \leq K + M/\gamma, \qquad t \geq 0,$$

and this proves the proposition. A simple calculation shows, by the way, that the equilibrium point $(\overline{N}, \overline{P})$ is in the interior of the set A, i.e.

$$\overline{P} = f(\overline{N})/g(\overline{N}) < -\beta\overline{N} + \beta(K + M/\gamma),$$

$P = -\beta N + \beta(K + M/\gamma)$ being the equation of the "upper" boundary of the set A (see Figure 3.4.2).

Let $(N(0), P(0)) \in \text{Int } A$ and $(N(0), P(0)) \neq (\overline{N}, \overline{P})$. Neither $(0,0)$ nor $(K, 0)$ nor $(\overline{N}, \overline{P})$ can be an omega limit point of the corresponding solution since $(0,0)$ and $(K, 0)$ are saddles with ingoing trajectories on the axes, and $(\overline{N}, \overline{P})$ is a repeller by assumption. Since the positive semitrajectory of $(N(0), P(0))$ stays in A, its omega limit set is not empty, and by Theorem 3.1.11 it is a limit cycle. Thus, we have proved that *if* $(\overline{N}, \overline{P})$ *is a repeller,* e.g. $(f/g)'(\overline{N}) > 0$, *then A contains a non-trivial closed path.* In case we knew that there is a *single* non-trivial closed path in A, then the proof would imply that this is the limit cycle of every solution with initial value in Int A except, of course, the equilibrium point.

The results deduced above can be interpreted intuitively from the point of view of ecology. First we have to note that the shape of the zero prey isocline on Figure 3.4.2 is not usual or typical in ecological applications. It has been assumed in that form in order to show that the assumptions allow fairly general curves. Usually a "unimodal" graph is assumed, one with a single maximum in the interval $[0, K]$ like the one in Figure 3.1.7. The interval between 0 and the maximum point is called the *Allée effect zone* (see Allée-Emerson [1949]). Some populations do not multiply fast if their density is low. Up to a certain value the increase of their density or quantity is beneficial to their growth rate. When the density is already sufficiently high, then the saturation effect of overcrowding occurs, and a further increase of quantity reduces the per capita growth rate. The interval in which the overcrowding does not yet affect the population growth is the Allée effect zone. If the equilibrium of the system is in the Allée effect zone, i.e. on the ascending branch of the prey isocline, then an *increase of prey* has to be compensated by an *increase of predator* in order to keep the growth rate of prey zero. If the equilibrium is on the descending branch of the prey isocline, then an *increase of prey* has to be compensated by a *decrease of predator* to keep the growth rate of prey on the zero level. If the equilibrium is in the Allée effect zone, and the system is driven out from the equilibrium by an increase of prey, say, then the system gets into that part of the phase plane where both prey and predator are growing, i.e. the dynamics is driving the system still further away from the equilibrium. This means that the equilibrium must be unstable, and since there is no other equilibrium and the solutions are bounded, the system has to oscillate in accordance with our results. To the contrary, if the equilibrium is in the "saturation effect zone", i.e. on the descending branch of the prey curve, and the system is driven out from the equilibrium by an increase of prey, say, then it gets into a region where prey is decreasing and predator increasing, i.e. the dynamics is driving the system back into the equilibrium position. This means that the equilibrium must be asymptotically stable as it has been proved.

We shall return to this model in Chapter 7.

3.5 The Poincaré Index and Non-existence of Cycles

If a continuous vector field is given on the plane, an integer may be made to correspond to each smooth closed curve and to each point where the value of the vector field is zero. These numbers, the "indices", carry some information about the flow generated by the vector field. First we shall introduce the "Poincaré index" and establish some of its properties. Then we are going to prove *Dulac's Criterion* about the non-existence of periodic solutions.

Let D be a simply connected open set in $\mathbf{R}^2$, and let $f \in C^0(D, \mathbf{R}^2)$ be a continuous vector field defined on D. The coordinates of f will be denoted by f_1, f_2. When we are to apply the theory to two dimensional autonomous systems of differential equations we shall assume more, $f \in C^1$, say, in order to ensure the existence and uniqueness of solutions (cf. (3.1.1)); however, to build up the theory of the Poincaré index, the continuity of f is sufficient. If at $x \in D : f(x) = 0$, we say that x is a *critical point* of the field f. If f is the right-hand side of a system of differential equations, then a critical point is, clearly, an equilibrium point of the system or, in other words, a fixed point of the flow induced by the system.

In the following construction we shall omit the proofs of the geometrically more obvious facts. Rigorous proofs of these elementary statements can be found in Andronov, Leontovich, Gordon, Mayer [1966]. A rotation angle is measured positive if the rotation of a vector is counterclockwise, and negative in the opposite direction. The *angle* $\theta(f^1, f^2)$ *of vectors* f^1 and f^2 (in this order) is the rotation angle, in modulus less than π, that carries the vector f^1 into the direction of f^2. If the rotation angle is $\pm\pi$, then we may choose either value. Clearly, $\theta(f^2, f^1) = -\theta(f^1, f^2)$, and $-\pi \leq \theta(f^1, f^2) \leq \pi$ for any two vectors. A *polar angle of the vector* f is the angle between the directional vector of the positive axis x_1 and f (in this order) plus any integer multiple of 2π. Let $c \subset D$ be a simple arc (a continuous curve without self-intersection, in particular, not closed) with finite arc length, and assume that there is no critical point of the vector field $f \in C^0(D, \mathbf{R}^2)$ on c. *An angular function of the vector field* f *restricted to* c is the function $F : c \to \mathbf{R}$ if $F \in C^0$, and if for every $x \in c$ the value $F(x)$ is a polar angle of the field vector $f(x)$. If F_1 and F_2 are two angular functions of the vector field f restricted to c, then their difference is an integer multiple of 2π. Let $x^1, x^2 \in c$, the *rotation of the field* f *along* c *from* x^1 *to* x^2 is, by definition,

$$w(f, x^1, x^2) := F(x_2) - F(x_1) \tag{3.5.1}$$

where F is an arbitrary angular function of f restricted to c. If c is a *directed* arc, then the *rotation of* f *along* c from the starting point to the end point is denoted, simply, by $w(f, c)$. The rotation of the field along a directed curve does not depend on the choice of the angular function, nor does it depend on the particular coordinate system. The rotation satisfies the relations

$$w(f, x^2, x^1) = -w(f, x^1, x^2),$$

and if $x^1, x^2, x^3 \in c$, then

$$w(f, x^1, x^3) = w(f, x^1, x^2) + w(f, x^2, x^3).$$

Now, let c be a positively directed simple, closed curve, and assume that the vector field f has no critical point on c. Choose two points x^1 and x^2 on c, and denote by c_1 the part of c going from x^1 to x^2 according to the orientation of c, and by c_2 the other part which goes from x^2 to x^1 (see Figure 3.5.1).

Definition 3.5.1. The *Poincaré index of the closed curve* c *with respect to the field* f is

$$i(f, c) := (w(f, c_1) + w(f, c_2))/2\pi. \tag{3.5.2}$$

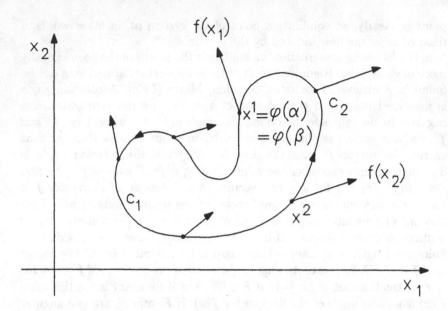

FIGURE 3.5.1. A positively directed, simple, smooth closed curve split up to parts c_1 and c_2.

It is easy to see that the index does not depend on the choice of x^1 and x^2, i.e. on how c is split up to c_1 and c_2. If the field is fixed, and no misunderstanding may occur, then the index of a closed curve c will be denoted by $i(c)$. Let F_1 and F_2 be angular functions on c_1 and c_2, respectively. Then applying (3.5.1) we have

$$i(f, c) = (F_1(x^2) - F_1(x^1) + F_2(x^1) - F_2(x^2))/2\pi. \tag{3.5.3}$$

However, $F_2(x^1)$ and $F_1(x^1)$ are polar angles of the same vector $f(x^1)$, and therefore their difference is an integer multiple of 2π, and similarly $F_1(x^2) - F_2(x^2)$ is also an integer multiple of 2π. Hence, *the index of a closed curve is an integer*. We may introduce a parametrization on the closed curve, $c = \{x \in \mathbf{R}^2 : x = \varphi(t), \ \alpha \le t \le \beta\}$ where, naturally, $\varphi(\alpha) = \varphi(\beta)$. We may also introduce an angular function F along c considering the curve as one with starting point $\varphi(\alpha)$ and end point $\varphi(\beta)$ and parametrize F by $t \in [\alpha, \beta]$. Then, clearly, $F : [\alpha, \beta] \to \mathbf{R}$ is continuous but, in general, $F(\alpha) \ne F(\beta)$. Using the restriction of this function as F_1 on c_1 and as F_2 on c_2, and choosing $x^1 = \varphi(\alpha) = \varphi(\beta)$, we obtain

$$i(f, c) = (F(\beta) - F(\alpha))/2\pi \tag{3.5.4}$$

where $F(\beta)$ and $F(\alpha)$ are polar angles of the same vector $f(\varphi(\beta)) = f(\varphi(\alpha))$; see Figure 3.5.1.

In the sequel, the expression "closed curve" will mean, always, a positively oriented, simple closed curve, and it will be assumed that the vector fields have no critical point on it.

Definition 3.5.2. Let c be a closed curve and f^0 and f^1 two continuous vector fields defined on c, and assume that f^0 and f^1 have no critical point on c; we say that f^0 *and* f^1 *are homotopic* if there is a one parameter family of vector fields $f : [0, 1] \times c \to \mathbf{R}^2$ such that $f \in C^0$, $f(0, x) = f^0(x)$, $f(1, x) = f^1(x)$ for all $x \in c$, and $f(\mu, x) \neq 0$ for $\mu \in [0, 1]$, $x \in c$.

"*Homotopy*" means that one vector field can continuously be deformed into the other in such a way that in the process no intermediate vector field has a critical point on the curve. Homotopy is, obviously, an equivalence relation; it is reflexive, symmetric and transitive. We shall prove now two lemmata that guarantee the invariance of index by homotopy.

Lemma 3.5.1. *Let g and h be two continuous vector fields defined on the closed curve c and having no critical point on c; if there is no point $x \in c$ at which the vectors $g(x)$ and $h(x)$ have opposite directions, then $i(g, c) = i(h, c)$.*

Proof. Let $c = \{x \in \mathbf{R}^2 : x = \varphi(t), \quad t \in [\alpha, \beta]\}$ be a parametrization of the curve, $G : [\alpha, \beta] \to \mathbf{R}$ an angular function of the field g, and $\theta(t) := \theta(g(\varphi(t)), h(\varphi(t)))$ the angle of the vectors of the two fields at points of the curve. Since, by assumption, $|\theta(t)| < \pi$ for all $t \in [\alpha, \beta]$ it follows that $\theta : [\alpha, \beta] \to (-\pi, \pi)$ is a continuous function. Clearly, the function $H(t) := G(t) + \theta(t)$ is an angular function of h. Applying formula (3.5.4) we obtain

$$i(h, c) = (H(\beta) - H(\alpha))/2\pi - (G(\beta) - G(\alpha) + \theta(\beta) - \theta(\alpha))/2\pi$$
$$= (G(\beta) - G(\alpha))/2\pi = i(g, c)$$

since $\theta(\beta) = \theta(\alpha)$. ($\theta(\beta) = \theta(\alpha) \pm 2k\pi$, $k \neq 0$, is impossible because $|\theta(t)|$ is less than π.) $\square$

Lemma 3.5.2. *If f^0 and f^1 are continuous, homotopic vector fields on the closed curve c, then the indices of c with respect to these two fields are the same: $i(f^0, c) = i(f^1, c)$.*

Proof. Let $f : [0, 1] \times c \to \mathbf{R}^2$ be the one parameter family of vector fields that establishes the homotopy between f^0, and f^1, so that $f^0(x) = f(0, x)$, $f^1(x) = f(1, x)$ for $x \in c$. Since f is continuous on the compact set $[0, 1] \times c$ and has no critical point, the minimum of its absolute

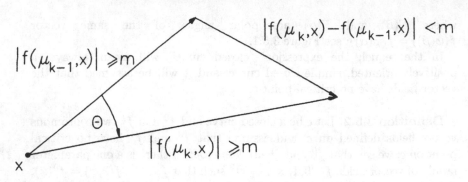

FIGURE 3.5.2. Proof of Lemma 3.5.2: θ is an acute angle since it is opposite the shortest side of the triangle.

value is positive:

$$m := \min_{\mu \in [0,1],\ x \in c} |f(\mu, x)| > 0.$$

Since f is uniformly continuous on $[0,1] \times c$, given the number $m > 0$, a $\delta > 0$ exists such that if $\mu', \mu'' \in [0,1]$ and $|\mu' - \mu''| < \delta$, then $|f(\mu', x) - f(\mu'', x)| < m$ for all $x \in c$. Divide the interval $[0,1]$ into subintervals of length less than δ, i.e. let $0 = \mu_0 < \mu_1 < \cdots < \mu_{k-1} < \mu_k < \cdots < \mu_n = 1$, and $\max |\mu_k - \mu_{k-1}| < \delta$ $(k = 1, 2, ..., n)$, then for every $x \in c$ and any $1 \le k \le n$ the vectors $f(\mu_{k-1}, x)$, $f(\mu_k, x)$ form an acute angle (see Figure 3.5.2), i.e. for $x \in c$ we have $|\theta(f(\mu_{k-1}, x), f(\mu_k, x))| < \pi/2 < \pi$; thus, by the previous Lemma $i(f_{\mu_{k-1}}, c) = i(f_{\mu_k}, c)$ $(k = 1, 2, ..., n)$ where $f_{\mu_k}(x) := f(\mu_k, x)$. Hence $i(f^0, c) = i(f^1, c)$. $\square$

Now we are in the position to establish some basic facts about the indices of closed curves with respect to dynamical systems. In the sequel, $D \subset \mathbf{R}^2$ will denote a simply connected, open set, $f : D \to \mathbf{R}^2$ a continuous vector field, and $c \subset D$ a positively directed simple closed curve as before; however, when we treat the implications of the results for dynamical systems, we shall consider the system

$$\dot{x} = f(x) \qquad\qquad (3.5.5)$$

where $f \in C^1(D, \mathbf{R}^2)$ will be assumed (in order to ensure the existence and uniqueness of solutions and also of trajectories passing through a given point).

Theorem 3.5.3. *Let $c \subset D$ be a simple closed curve, and assume that the vector field f has no critical point on c and in its interior, then $i(f,c) = 0$.*

Proof. We prove the statement, first, for the unit circle k with centre at the origin. We assume that $k \subset D$ and that f has no critical point on k and in the disc bounded by k. Introducing polar coordinates by $x_1 = r\cos\theta$, $x_2 = r\sin\theta$ we shall denote the value of f at (x_1, x_2) by $f(r, \theta)$, an abuse of notation. By the assumption, $f(r, \theta) \neq 0$, $0 \leq r \leq 1$, $0 \leq \theta \leq 2\pi$. Define now two vector fields on the circle k (whose equation is $r = 1$ in polar coordinates). $f^0 : k \to \mathbf{R}^2$ shall have the value $f(0, \theta)$ at the point of the circle given by the polar angle θ, i.e. $f^0(\theta) := f(0, \theta)$. The other field, $f^1 : k \to \mathbf{R}^2$, will be defined by $f^1(\theta) := f(1, \theta)$. The field f^0 is constant, since at each point of the circle k, its value is equal to the value of f at the origin. We may consider the vector field $f(r, \theta)$ defined on the unit disc as a one parameter family of vector fields defined on the circle k; for each fixed $0 < r < 1$ the function $f(r, \bullet)$ gives the vector field on the circle k that is parametrized by the polar angle $0 \leq \theta \leq 2\pi$. Clearly, $f(r, \theta)$ establishes a homotopy between f^0 and f^1 on the circle k. By Lemma 3.5.2, $i(f^1, k) = i(f^0, k)$. However, f^1 is the restriction of f onto k, and any angular function of f^0 on k is constant since f^0 is constant. Hence, $i(f, k) = i(f^1, k) = 0$.

Turning to the general case, denote the interior of the closed curve c by C and the interior of the unit circle k with centre an the origin by K. Let $H : C \cup c \to K \cup k$ be a homeomorphism, and define a vector field g on $K \cup k$ by $g(y) = f(H^{-1}(y))$ for $y \in K \cup k$ (see Figure 3.5.3). The field g is nowhere zero because f is nowhere zero. Since the values of g and f are equal at corresponding points, any angular function of f on c gives rise to an angular function of g on k, the two angular functions having the same values at corresponding points of c, resp. k. This implies that the index of c with respect to f is equal to the index of k with respect to g. But the latter is zero as has been proved above. $\square$

Corollary 3.5.4. *Let $c \subset D$ be a simple closed curve, and assume that the system (3.5.5) has no equilibrium point on c and in its interior, then $i(f, c) = 0$.*

Theorem 3.5.5. *Let c be a simple, smooth, closed curve, and denote by $v : c \to \mathbf{R}^2$ the field of its tangent vectors; the index of c with respect to v is 1, i.e. $i(v, c) = 1$.*

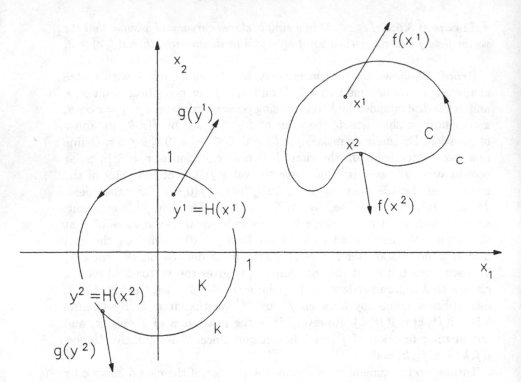

FIGURE 3.5.3. Homeomorphism of $C \cup c$ onto $K \cup k$, carrying the vector field f into g.

The assumption "*smooth*" means that c admits a continuously differentiable parametrization such that the derivative of the vector function describing c, i.e. the tangent vector, is nowhere zero.

Proof. Denote the (or a) point of c with least x_2 coordinate by x^0, and introduce arc length parametrization on c starting at x^0 in positive direction. Let $\bar{s}$ be the arc length of c :

$$c = \{x \in \mathbf{R}^2 : x = \varphi(s), \quad 0 \le s \le \bar{s}\},$$

and $x^0 = \varphi(0) = \varphi(\bar{s})$; see Figure 3.5.4 (a). Let OAB be the triangle formed by the straight lines $x_2 = x_1$, $x_2 = \bar{s}$, and the x_2-axis in the x_1, x_2 plane; see Figure 3.5.4 (b). We define a vector field f on the triangular domain bounded by OAB the following way:

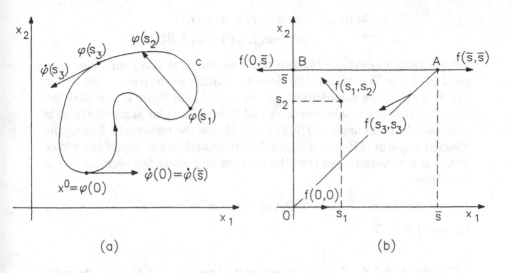

FIGURE 3.5.4. Proof of Theorem 3.5.5.

$$
\begin{aligned}
f(s_1, s_2) &:= (\varphi(s_2) - \varphi(s_1))/|\varphi(s_2) - \varphi(s_1)|, \qquad 0 \le s_1 < s_2 < \bar{s}, \\
f(s, s) &:= \dot{\varphi}(s)/|\dot{\varphi}(s)|, \qquad 0 \le s \le \bar{s}, \\
f(s, \bar{s}) &:= (\varphi(\bar{s}) - \varphi(s))/|\varphi(\bar{s}) - \varphi(s)| = (\varphi(0) - \varphi(s))/|\varphi(0) - \varphi(s)|, \\
&\qquad 0 < s < \bar{s}, \\
f(0, \bar{s}) &:= -\dot{\varphi}(0)/|\dot{\varphi}(0)|.
\end{aligned}
$$

An easy inspection yields that f is a continuous field defined on the closed curve $OABO$ and in its interior, and that f is nowhere zero. By Theorem 3.5.3

$$i(f, OABO) = 0.$$

By Definition 3.5.1 the index is the rotation of the field along the curve divided by 2π, and by the additivity of the rotation

$$2\pi i(f, OABO) = w(f, OA) + w(f, AB) + w(f, BO).$$

The definition of f implies that the rotation of f along OA is equal to the rotation of the tangent vector field along the curve c. In the last equation the left-hand side is equal to zero, so that

$$2\pi i(v,c) \; = \; w(v,c) = w(f,OA)$$
$$= \; -w(f,AB) - w(f,BO).$$

We are going to evaluate the two terms on the right-hand side. Since the point $\varphi(\bar{s}) = \varphi(0) = x^0$ is in the "lowest" position on the curve c, the vector $f(s,\bar{s})$ is "pointing downwards", i.e. its second coordinate is non-positive; thus, its polar angle is between $-\pi$ and 0, but the polar angle of $f(\bar{s},\bar{s})$ is zero, and the polar angle of $f(0,\bar{s})$ is $-\pi$. Hence, the rotation of f along the directed segment AB is $-\pi$. Similarly, the second coordinate of the vector $f(0,s)$ is non-negative, so that the rotation of f along the segment BO is $-\pi$. Thus,

$$2\pi i(v,c) = \pi + \pi,$$

i.e. $i(v,c) = 1$. $\square$

Corollary 3.5.6. *If γ is a closed orbit of system (3.5.5), then its index with respect to the system is 1.*

The last Corollary and Corollary 3.5.4 imply the following:

Corollary 3.5.7. *If γ is a closed orbit of system (3.5.5), then its interior contains at least one equilibrium point of the system.*

Theorem 3.5.5 and Lemma 3.5.1 imply the following:

Corollary 3.5.8. *If $c \subset D$ is a simple smooth closed curve "transversal" to the trajectories of system (3.5.5), then its index with respect to the system is 1.*

A smooth curve is said to be *transversal* to the trajectories of a flow or simply to the flow if its tangent vector is at no point parallel to the vector of the field defining the flow, i.e. if the trajectories of the flow intersect the curve without contact. In particular, this implies that there is no equilibrium point on the curve. It is worthwhile to note that if a simple smooth closed curve is transversal to the flow, then either all trajectories that cross it go from the outside to the interior or all are leaving the interior (cf. the definition of a transversal straight line segment after formula (3.1.1); see Figure 3.5.5).

Corollary 3.5.9. *Let $c \subset D$ be a simple closed curve, and $c_1, c_2, ..., c_n$ be*

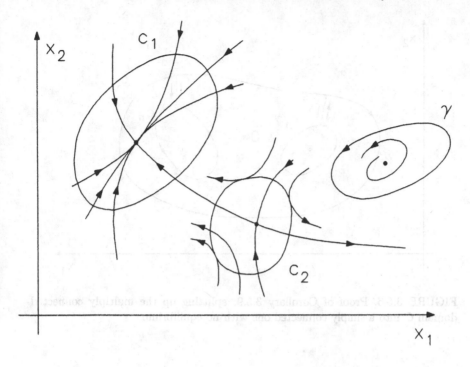

FIGURE 3.5.5. Different closed curves in the phase portrait of a flow. c_1: a transversal, $i(c_1) = 1$. Curve c_2 is not a transversal, $i(c_2) = -1$. γ is a closed path, $i(\gamma) = 1$.

simple closed curves in the interior of c not intersecting each other, each c_k lying in the exterior of all the other c_j's $(j = 1, 2, ... n; \ j \neq k)$; if the system (3.5.5) has no equilibrium point on $c_1, c_2, ..., c_n$, and in the domain C between c and the c_j's, then

$$i(f, c) = \sum_{j=1}^{n} i(f, c_j).$$

Proof. The proof is analogous to the proof of the generalization of Cauchy's Fundamental Theorem in complex analysis (see Ahlfors [1966] and Figure 3.5.6.) □

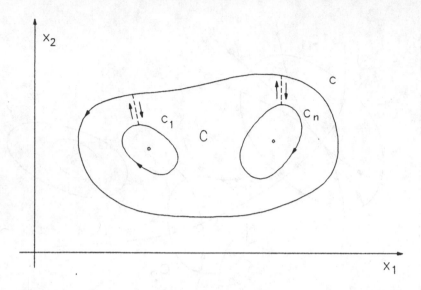

FIGURE 3.5.6. Proof of Corollary 3.5.9: splitting up the multiply connected domain C into a simply connected one with no equilibrium.

We are going to define now the Poincaré index of an equilibrium point. It is easy to prove applying the last Corollary that if $a \in D$ is an equilibrium point of system (3.5.5), $c_1 \subset D$ and $c_2 \subset D$ are simple closed curves that contain a in their interior, and the system has no other equilibrium point on c_1, c_2 and in their interiors, then the indices of these two curves are the same.

Definition 3.5.3. Let $a \in D$ be an equilibrium point of system (3.5.5) and $c \subset D$ an arbitrary simple closed curve that contains a in its interior and does not contain any other equilibrium point of the system on itself and in its interior; then the index of c with respect to the system is called the *Poincaré index of the equilibrium a*, denoted by

$$i(a) := i(f, c).$$

The remarks before the Definition imply that the index of a does not depend on the choice of the surrounding curve c. It is also clear that if a is an *isolated* equilibrium point, i.e. it has an open neighbourhood that does not contain any other equilibrium, then its index is defined. An immediate consequence of the last Corollary is the following:

Corollary 3.5.10. *Let $c \subset D$ be a simple closed curve, assume that system (3.5.5) has no equilibrium point on c but has n equilibria, $a_1, ..., a_n$ in the interior of c; then*

$$i(f, c) = \sum_{j=1}^{n} i(a_j). \tag{3.5.6}$$

There holds also

Corollary 3.5.11. *If γ is a non-trivial closed path of system (3.5.5), and all the equilibria of the system are isolated, then the interior of γ contains a finite number of equilibria, and the sum of their indices is equal to 1.*

We are deducing now an integral formula enabling us to calculate the Poincaré index of a closed curve. This will help us to determine the index of isolated equilibria.

Theorem 3.5.12. *Let $c \subset D$ be a simple, smooth closed curve, and assume that system (3.5.5) has no equilibrium on c; then*

$$i(f, c) = (1/2\pi) \oint_{(C)} |f|^{-2}(f_1 \, \text{grad} \, f_2 - f_2 \, \text{grad} \, f_1) dx \tag{3.5.7}$$

where $|f| = (f_1^2 + f_2^2)^{1/2}$, $\text{grad} \, f_i = (f'_{ix_1}, f'_{ix_2})$ $(i = 1, 2)$, and the line integral is to be taken in positive direction.

Proof. Let $x = \varphi(t)$, $\alpha \leq t \leq \beta$ be a parametrization of c, and $\varphi(\alpha) = \varphi(\beta)$. We have to consider an angular function $F : c \to \mathbf{R}$ and its variation as we move around c once. The parametrization of c induces a parametrization of F; by an abuse of notation $F(t) = (F \circ \varphi)(t)$ is a polar angle of the vector $f(\varphi(t))$. Now, our assumptions imply that at an arbitrary $t_0 \in [\alpha, \beta]$ either $f_1(\varphi(t_0)) \neq 0$ or $f_2(\varphi(t_0)) \neq 0$. If one of these inequalities holds, then t_0 has a whole neighbourhood where the same inequality is valid because of continuity. If $f_1(\varphi(t_0)) \neq 0$, then in a neighbourhood of t_0 we may write

$$F(t) = \tan^{-1}\big(f_2(\varphi(t))/f_1(\varphi(t))\big) + \text{constant} \, ;$$

if $f_2(\varphi(t_0)) \neq 0$, then in a neighbourhood we may write

$$F(t) = -\tan^{-1}\big(f_1(\varphi(t))/f_2(\varphi(t))\big) + \text{constant} \, .$$

Hence, for every $t \in [\alpha, \beta]$ we have

$$F'(t) = \frac{f_1(\varphi(t))df_2(\varphi(t))/dt - f_2(\varphi(t))df_1(\varphi(t))/dt}{|f(\varphi(t))|^2}.$$

Thus, by (3.5.4)

$$2\pi i(f,c) = \int_\alpha^\beta \frac{f_1(\varphi(t))\operatorname{grad} f_2(\varphi(t)) - f_2(\varphi(t))\operatorname{grad} f_1(\varphi(t))}{|f(\varphi(t))|^2}\dot\varphi(t)dt$$

$$= \oint_{(c)} |f(x)|^{-2}(f_1(x)\operatorname{grad} f_2(x) - f_2(x)\operatorname{grad} f_1(x))dx.\ \Box$$

As an application to be used later, the index of the trivial equilibrium of a homogeneous linear system will be determined in case it is isolated (cf. the classification of systems of type (1.4.13)).

Consider the system

$$\dot x_1 = a_{11}x_1 + a_{12}x_2, \qquad \dot x_2 = a_{21}x_1 + a_{22}x_2, \qquad (3.5.8)$$

and introduce the notations

$$A = \begin{pmatrix} a_{11} & a_{12} \\ a_{21} & a_{22} \end{pmatrix},$$

$$l(x) = [a_{11}x_1 + a_{12}x_2, \ a_{21}x_1 + a_{22}x_2].$$

Theorem 3.5.13. *If $\det A \neq 0$, then the Poincaré index of the equilibrium $(0,0)$ of (3.5.8) is plus or minus one according to whether $\det A$ is positive or negative, i.e.*

$$i((0,0)) = \left\{ \begin{matrix} 1 \\ -1 \end{matrix} \right. \qquad \text{if} \qquad \det A \gtrless 0.$$

Proof. The right-hand side of system (3.5.8), the vector $l(x)$, is nowhere zero but at the origin because the coefficient matrix is regular, so that the index of the origin is equal to the index of an arbitrary closed curve around it. We shall use the ellipse E defined by

$$(a_{11}x_1 + a_{12}x_2)^2 + (a_{21}x_1 + a_{22}x_2)^2 = 1.$$

According to Definition 3.5.3

$$i((0,0)) = i(l,E).$$

We are applying formula (3.5.7) for the evaluation of the index of the ellipse E. By the affine coordinate transformation

$$y_1 = a_{11}x_1 + a_{12}x_2, \qquad y_2 = a_{21}x_1 + a_{22}x_2, \qquad (3.5.9)$$

the equation of the ellipse assumes the form $y_1^2 + y_2^2 = 1$. Introducing the parametrization $y_1 = \cos\theta$, $y_2 = \sin\theta$, $0 \le \theta \le 2\pi$, the image of the ellipse in the y_1, y_2 plane, the unit circle is travelled round in the positive direction with the increase of θ. The orientation of the ellipse E is the same, i.e. positive if the determinant of the matrix A of the coordinate transformation is positive, but it is negative if the determinant is negative. From (3.5.9) the direct relation between (x_1, x_2) and θ is

$$x_1 = (a_{22} \cos \theta - a_{12} \sin \theta)/\det A, \qquad x_2(a_{11} \sin \theta - a_{21} \cos \theta)/\det A,$$

and the derivatives which are needed at the integration are

$$
\begin{aligned}
dx_1/d\theta &= (-a_{22} \sin \theta - a_{12} \cos \theta)/\det A, \\
dx_2/d\theta &= (a_{11} \cos \theta + a_{21} \sin \theta)/\det A, \\
2\pi i(l, E) &= \oint_{(E)} (a_{11}x_1 + a_{12}x_2)[a_{21}, a_{22}] - (a_{21}x_1 + a_{22}x_2)[a_{11}, a_{12}])dx \\
&= \pm \frac{1}{\det A} \int_0^{2\pi} [a_{21} \cos \theta - a_{11} \sin \theta, a_{22} \cos \theta - a_{12} \sin \theta] \\
&\quad \bullet [-a_{22} \sin \theta - a_{12} \cos \theta, a_{11} \cos \theta + a_{21} \sin \theta]d\theta \\
&= \pm \frac{1}{\det A} \int_0^{2\pi} \det A d\theta = \pm 2\pi
\end{aligned}
$$

where the plus or the minus sign is to be chosen according to whether $\det A$ is positive or negative. $\square$

The result just achieved is important because it is robust with respect to perturbations. First, it is obvious that if we have a homogeneous linear system "close" to (3.5.8), i.e. the norm of the difference of the two coefficient matrices is small, then the positivity, resp. negativity of $\det A$ implies the same property for the coefficient matrix of this other system. Secondly, in the following Theorem we show that if the condition of the previous Theorem holds for the linearization of a non-linear system at an equilibrium point, then this equilibrium point has the same index with respect to the original non-linear system as with respect to the linear one.

We are writing system (3.5.5) in the form

$$\dot{x} = Ax + g(x), \tag{3.5.10}$$

assuming without loss of generality that the origin $x = 0$ is an equilibrium point. Here

$$A = \begin{pmatrix} a_{11} & a_{12} \\ a_{21} & a_{22} \end{pmatrix},$$

$a_{ik} = f'_{i x_k}(0,0)$ $(i, k = 1, 2)$ and $|g(x)| = o(\sqrt{x_1^2 + x_2^2})$. We shall also use the notation

$$l(x) = [a_{11}x_1 + a_{12}x_2, \quad a_{21}x_1 + a_{22}x_2]$$

for the linearization of the right-hand side.

Theorem 3.5.14. *Assume that det $A \neq 0$; then the index of the origin $x = 0$ with respect to system (3.5.10) is equal to its index with respect to the linearized system $\dot{x} = Ax$.*

Proof. Denote the minimum of the modulus of the vector $l(x)$ on the circle of radius $r > 0$ with centre at the origin by $m(r)$:

$$m(r) := \min_{|x|=r} |l(x)|.$$

Since $|l(x)|$ is continuous and, by the assumption of the Theorem, non-zero on the circle $m(r) > 0$, we have

$$\begin{aligned} |l(x)| &= ((a_{11}x_1 + a_{12}x_2)^2 + (a_{21}x_1 + a_{22}x_2)^2)^{1/2} \\ &= r((a_{11}x_1/r + a_{12}x_2/r)^2 + (a_{21}x_1/r + a_{22}x_2/r)^2)^{1/2}. \end{aligned}$$

Whereas (x_1, x_2) runs through the circle $x_1^2 + x_2^2 = r^2$, the point $(x_1/r, x_2/r)$ runs through the circle $x_1^2 + x_2^2 = 1$; therefore,

$$m(r) = \min_{|x|=r} |l(x)| = r \min_{|x|=1} |l(x)| = rm(1).$$

We know that $|g(x)|/|x| \to 0$ as $x \to 0$, i.e. for $m(1) > 0$ there exists a $\delta > 0$ such that for $0 < |x| < \delta$ we have $|g(x)|/|x| < m(1)$, or

$$|g(x)| < |x|m(1) \qquad \text{for } 0 < |x| < \delta.$$

Let us denote now a circle with center in the origin and radius $0 < r < \delta$ by c. For $x \in c$, i.e. if $|x| = r$, we have

$$|g(x)| < rm(1) = m(r) \leq |l(x)|;$$

hence, for $x \in c$ there holds the inequality

$$|l(x) + g(x)| \geq ||l(x)| - |g(x)|| > 0, \qquad (3.5.11)$$

and *a fortiori*

$$|l(x) + \mu g(x)| \geq ||l(x)| - \mu|g(x)|| > 0$$

for $0 \leq \mu \leq 1$. Thus, we see that the vector fields l and $l+g$ are homotopic on the circle c; the homotopy is realized by the one parameter family $l + \mu g$, $0 \leq \mu \leq 1$ (see Definition 3.5.2). By Lemma 3.5.2 this means that the indices of c with respect to the two fields are the same,

$$i(l + g, c) = i(l, c). \qquad (3.5.12)$$

It is clear that (3.5.11) holds for $0 < |x| < \delta$, i.e. neither the linearized system $\dot{x} = Ax$ nor system (3.5.10) has an equilibrium in the interior of c but the origin. Therefore, the index of c is equal to the index of the origin, i.e. (3.5.12) implies the statement. □

The previous Theorems imply the important

Corollary 3.5.15. *Let the eigenvalues of the linearization of system (3.5.10) be not zero, then the index of the origin is 1 if the origin is a node, a spiral point or a centre, and it is −1 if the origin is a saddle (see the classifications following (1.4.13) and (1.4.18)).*

Proof. The determinant of the coefficient matrix A of the linearized system is the product of the two eigenvalues. This is negative in case of a saddle and positive in the rest of the non-degenerate cases. □

We prove now one of the most often used criteria due to H. Dulac [1923] that guarantees that in a given domain there is no closed path.

Theorem 3.5.16 *(Dulac's Criterion). If a function $h \in C^1(D, \mathbf{R})$ exists such that the divergence of the vector field hf is semidefinite on D, i.e.*

$$\operatorname{div}(hf) = \partial(hf_1)/\partial x_1 + \partial(hf_2)/\partial x_2 \geq 0 \quad or \quad \leq 0, \quad x \in D,$$

and div $(hf) = 0$ only on a subset of measure zero; then system (3.5.5) has no non-trivial closed path in D.

Proof. Assume, contrary to the statement, that $\gamma \subset D$ is a non-trivial closed orbit of system (3.5.5), and denote the line integral of the vector field $[-hf_2, hf_1]$ along γ by

$$L = \oint_{(\gamma)} [-h(x)f_2(x), \ h(x)f_1(x)]dx.$$

Let $\varphi : [0, T] \to D$ be a solution whose path is γ,

$$\gamma = \{x \in D : x = \varphi(t), \ 0 \leq t \leq T\}.$$

Then

$$
\begin{aligned}
L &= \int_0^T h(\varphi(t))(-f_2(\varphi(t))\dot{\varphi}_1(t) + f_1(\varphi(t))\dot{\varphi}_2(t))dt \qquad (3.5.13)\\
&= \int_0^T h(\varphi(t))(-f_2(\varphi(t))f_1(\varphi(t)) + f_1(\varphi(t))f_2(\varphi(t)))dt = 0.
\end{aligned}
$$

On the other hand, by Stokes' theorem

$$L = \int\int_{(\Gamma)} (\partial(h(x)f_1(x))/\partial x_1 + \partial(h(x)f_2(x))/\partial x_2)dx_1 dx_2$$

where Γ is the interior of γ. But the last integral is non-zero since the integrand is non-negative, resp. non-positive everywhere, and it is zero only on a set of measure zero. The contradiction proves the Theorem. □

Note that the same conditions also guarantee that the system has neither homoclinic trajectories nor closed invariant curves consisting of heteroclinic trajectories in the set D provided that the right-hand side f is analytic (see Andronov, Leontovich, Gordon, Mayer [1966]).

Sometimes one may establish the semidefiniteness of $\operatorname{div}(hf)$ in an annular domain. This yields also useful information.

Corollary 3.5.17. *Let $D_0 \subset D$ be an annular (doubly connected) open subset, assume that a function $h \in C^1(D_0, \mathbf{R})$ exists such that the divergence of the vector field hf is semidefinite on D_0, i.e.*

$$\operatorname{div}(hf) = \partial(hf_1)/\partial x_1 + \partial(hf_2)/\partial x_2 \geq 0 \quad or \quad \leq 0, \qquad x \in D_0,$$

and $\operatorname{div}(hf) = 0$ only on a subset of measure zero, then system (3.5.5) has at most one non-trivial closed path in D_0.

Note that if the system has a non-trivial closed path in D_0, then this path must contain the "inner boundary of D_0" in its interior. If it did not the divergence would be semidefinite in its whole interior, contradicting the previous theorem.

Proof. Assume that the system has two non-trivial closed orbits γ_1 and γ_2 in D_0, and that, say, γ_2 contains γ_1 in its interior (see Figure 3.5.7). Denote the domain between γ_1 and γ_2 by Γ. We split up Γ by a smooth curve c joining a point $x^1 \in \gamma_1$ with a point $x^2 \in \gamma_2$, and consider Γ as a simply connected set bounded by $\gamma_1 \cup c_+ \cup \gamma_2 \cup c_-$ where $c_+ = c$ directed from x^1 to x^2, and $c_- = c$ directed from x^2 to x^1. Integrating the vector field $[-hf_2, hf_1]$ along the closed curve $\gamma_1 \cup c_+ \cup \gamma_2 \cup c_-$ where γ_1 is oriented positively and γ_2 negatively, we get zero since (3.5.13) applies to γ_1 and γ_2, and the line integrals on c_+ and c_- cancel out. However, applying Stokes' theorem the same integral is equal to

$$\int\int_{(\Gamma)} (\partial(h(x)f_1(x))/\partial x_1 + \partial(h(x)f_2(x))/\partial x_2)dx_1 dx_2 \neq 0$$

since $\Gamma \subset D_0$ is a set of positive measure, and the integrand is non-negative, resp. non-positive, being zero only on a set of measure zero. The contradiction proves the Corollary. □

As an application, consider system (3.3.9) equivalent to the damped Duffing equation. The divergence of the field is, clearly, $-d < 0$, so that the system does not have any non-trivial closed path.

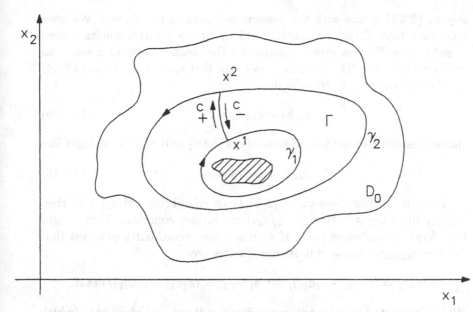

FIGURE 3.5.7. Proof of Corollary 3.5.17: in the doubly connected set D_0 there cannot lie more than one closed orbit.

Example 3.5.1. As an application of Dulac's Criterion *the general two dimensional Lotka-Volterra system*

$$
\begin{aligned}
\dot{x}_1 &= x_1(\varepsilon_1 - a_{11}x_1 - a_{12}x_2), \\
\dot{x}_2 &= x_2(\varepsilon_2 - a_{21}x_1 - a_{22}x_2)
\end{aligned}
\tag{3.5.14}
$$

will be considered. The system is characterized by that that the per capita growth rates $\dot{x}_k/x_k$ are linear functions of the quantities x_1, x_2 ($k = 1, 2$). It describes the interaction of the species 1 and 2 whose quantities (densities) are x_1 and x_2. The terms containing $a_{kk}x_k$ ($k = 1, 2$) express the "*intraspecific competition*" of the respective species. It is, usually, assumed that the increase of the density is disadvantageous for the growth; therefore, $a_{kk} \geq 0$ ($k = 1, 2$). Three basic types of interaction are taken into account. (a) The system is said to be *competitive* if $a_{12} > 0$, $a_{21} > 0$; the terms containing these coefficients express the "*interspecific competition*", i.e. the measure of how much the increase of one of the species affects the other. In this case it is assumed that $\varepsilon_1 > 0$, $\varepsilon_2 > 0$, meaning that the growth rates at zero densities ("at very low densities") are positive. (b) The system is said to be *cooperative* if $a_{12} < 0$, $a_{21} < 0$. In this case, clearly, the increase of a species is advantageous for the other. When the system is cooperative, the ε_k's may be positive or negative but, usually, it is assumed that $\varepsilon_1\varepsilon_2 > 0$. (c) The system is a *predator-prey* system if $a_{12}a_{21} < 0$, say, $a_{12} > 0$, $a_{21} < 0$, and $\varepsilon_1 > 0$, $\varepsilon_2 < 0$. In this case species 1 is the prey and species 2 the predator. We have already considered a special case of this in Section 4, system (3.4.1) where with the present notations $a_{11} = a_{22} = 0$. We have seen there that all the trajectories of (3.4.1) in the positive quadrant were closed curves. We now give a criterion for the absence of periodic solutions in the general case. The zero isoclines of the first species of system (3.5.14) are the x_2-axis and the straight line

$$a_{11}x_1 + a_{12}x_2 = \varepsilon_1; \tag{3.5.15}$$

the zero isoclines of the second species are the x_1-axis and the straight line

$$a_{21}x_1 + a_{22}x_2 = \varepsilon_2. \tag{3.5.16}$$

Thus, in all the three cases $(0,0)$ is an equilibrium. If $a_{11} \neq 0$, then $(\varepsilon_1/a_{11}, 0)$; if $a_{22} \neq 0$, then $(0, \varepsilon_2/a_{22})$ are further equilibria. There might be a fourth equilibrium point E with non-zero coordinates provided that $\det A = a_{11}a_{22} - a_{12}a_{21} \neq 0$. Its coordinates are

$$E : x_{10} = (\varepsilon_1 a_{22} - \varepsilon_2 a_{12})/\det A, \quad x_{20} = (\varepsilon_2 a_{11} - \varepsilon_1 a_{21})/\det A.$$

We are interested in the positive quadrant of the x_1, x_2 plane only (which is invariant); therefore, *we assume that $x_{10} > 0$, $x_{20} > 0$*. If $\det A = 0$, then the two straight lines (3.5.15) and (3.5.16) are parallel, and there is no fourth equilibrium unless the two equations determine the same line. In both cases there can be no closed path by Corollary 3.5.7. Hence, $\det A \neq 0$ is a necessary condition for the existence of a non-trivial closed path. Assume that this condition holds and that the coordinates (x_{10}, x_{20}) of the equilibrium E are positive. An easy calculation yields the characteristic polynomial of the system linearized at E. It is

$$\lambda^2 + \lambda B/\det A + x_{10}x_{20} \det A$$

where

$$B = \varepsilon_2 a_{11}(a_{22} - a_{12}) - \varepsilon_1 a_{22}(a_{21} - a_{11}). \tag{3.5.17}$$

The signs of the coefficients of this polynomial determine the stability of E (see (1.4.12) - (1.4.13)) unless $B = 0$. In the latter case the linearized system may have a centre at E, and the stability of E with respect to the non-linear system (3.5.14) may remain unresolved. Exactly this is the only case when system (3.5.14) may have non-trivial periodic solutions as is shown by the following Proposition. We note that in case of system (3.4.1), $B = 0$.

Proposition. *If $\det A = a_{11}a_{22} - a_{12}a_{21} \neq 0$ and $B \neq 0$, then system (3.5.14) has no non-trivial closed orbit.*

Proof. Dulac's Criterion will be applied with $h(x_1, x_2) = x_1^{r_1} x_2^{r_2}$ where $r_1 = a_{22}(a_{21} - a_{11})/\det A - 1$, $r_2 = a_{11}(a_{12} - a_{22})/\det A - 1$. A simple calculation yields

$$\partial(x_1^{r_1} x_2^{r_2} x_1 (\varepsilon_1 - a_{11} x_1 - a_{12} x_2))/\partial x_1$$
$$+ \quad \partial(x_1^{r_1} x_2^{r_2} x_2 (\varepsilon_2 - a_{21} x_1 - a_{22} x_2))/\partial x_2 = -x_1^{r_1} x_2^{r_2} B / \det A,$$

i.e. the divergence is nowhere zero and of the same sign for $x_1 > 0$, $x_2 > 0$.
□

Dulac's Criterion has several generalizations. A recent one is due to Busenberg and Van den Driessche [1993].

3.6   Hilbert's Sixteenth Problem

A chapter on two dimensional autonomous systems cannot be concluded without treating the problem in the title to some extent. The problem, actually, a part of Hilbert's Sixteenth Problem (see Hilbert [1902]), is the following one. Consider all the two dimensional autonomous systems

$$\dot{x} = P_n(x, y), \qquad \dot{y} = Q_n(x, y) \tag{3.6.1}$$

where P_n and Q_n are polynomials of degree n with real coefficients. *Find the maximal number H_n of limit cycles, i.e. non-trivial, isolated, closed orbits, a system like (3.6.1) may have and their relative position.* This problem has attracted much attention in this century, and the answer still seems to be far away. Under natural conditions Ilyashenko [1984] and, independently, Écalle et al. [1987] have proved that any given two dimensional polynomial system (3.6.1) has a finite number of limit cycles (thus proving *Dulac's Conjecture*; see also Ilyashenko [1991] and Écalle [1993]). In 1989 J.W.Reijn [1989] set up a bibliography of the quadratic case, i.e. $n = 2$, that contains about five hundred titles. In this case the prevailing conjecture (far from being proved) is, for the time being, that $H_2 = 4$. Systems that have on the right-hand sides quadratic polynomials are important in applications too. Such systems occur often in chemical reaction kinetics and in population dynamics. Also, obviously, if we approximate a non-linear system by the Taylor expansion of the right-hand side, a quadratic system is obtained in the first non-trivial step beyond linearization. We are going to give a survey of results concerning these quadratic systems

$$\dot{x} = P_2(x, y), \qquad \dot{y} = Q_2(x, y) \tag{3.6.2}$$

where

$$P_2(x, y) = \sum_{i+k=0}^{2} a_{ik} x^i y^k, \qquad Q_2(x, y) = \sum_{i+k=0}^{2} b_{ik} x^i y^k,$$

$a_{ik}, b_{ik} \in \mathbf{R}$, and P_2 and Q_2 are relative primes. Unlike other sections we are presenting most of the results without proof here since otherwise with all the background material this section would fill up the rest of the book. However, some classical results will be proved following Coppel's [1966] standard reference survey paper that summed up the state of art in 1966.

As we shall see later the problem is related to the "*centre problem*" : to determine whether an equilibrium point of a two dimensional autonomous system is a centre or not if it is a centre of the linearized system. In the discussion following formula (1.4.20) we have mentioned that if the equilibrium is a centre of the linearized system, then it is either a centre or a stable or unstable spiral point of the original non-linear one. To decide which case occurs is by no means a trivial problem. We are presenting here the classical, straightforward method since we have to make reference to this in an other context in the chapter on bifurcation. (see Sansone-Conti [1964] and also the paper of Negrini-Salvadori [1979] where applications can be found too).

If a non-linear *analytic* system has an isolated equilibrium point that is a centre for the system linearized at this point, then we may assume without loss of generality that the equilibrium is the origin, and the coordinate system has been transformed linearly (see Lemma A1.20) in such a way that the system has the form

$$\dot{x} = -\omega y + f(x,y), \qquad \dot{y} = \omega x + g(x,y) \qquad (3.6.3)$$

where $\omega > 0$, and f and g are analytic in a neighbourhood of the origin and contain terms of degree two and higher only. Choose a positive integer $m \geq 3$ and consider the Liapunov function

$$F(x,y) = x^2 + y^2 + \sum_{k=3}^{m} F_k(x,y) \qquad (3.6.4)$$

where F_k is a homogeneous polynomial of degree $k = 3, 4, ..., m$ to be determined. The derivative of F with respect to system (3.6.3) is

$$\begin{aligned} \dot{F}_{(3.6.3)}(x,y) \;=\;& 2xf(x,y) + 2yg(x,y) \\ & - \omega y F'_{3x}(x,y) + \omega x F'_{3y}(x,y) + \text{ terms of order 4 at least}, \end{aligned}$$

i.e. $\dot{F}_{(3.6.3)}$ starts with terms of order 3; we write it in the form

$$\dot{F}_{(3.6.3)}(x,y) = \sum_{k=3}^{\infty} \tilde{F}_k(x,y)$$

where $\tilde{F}_k$ is a homogeneous polynomial of degree $k = 3, 4, ...$ The following Theorem can be proved (for polynomial systems it was proved by Poincaré [1886]).

Theorem 3.6.1 *(Liapunov [1892] p. 361). Assume that F_k ($k =$*

$3, ..., m-1$) *have been chosen in such a way that* $\tilde{F}_3 = \tilde{F}_4 = \cdots = \tilde{F}_{m-1} = 0$; *if* m *is odd, then* F_m *can be determined (uniquely) in such a way that* $\tilde{F}_m = 0$; *if* m *is even, then* F_m *can be determined in such a way that*

$$\tilde{F}_m(x,y) = G_m(x^2 + y^2)^{m/2}$$

where $G_m \in \mathbf{R}$ *is uniquely determined (irrespective of which appropriate* F_m *was chosen).*

Since, in the discussion above m was arbitrary, this Theorem implies that a formal series

$$F(x,y) = x^2 + y^2 + \sum_{k=3}^{\infty} F_k(x,y) \tag{3.6.5}$$

can be determined where F_k is a homogeneous polynomial of degree $k = 3, 4, ...$ such a way that the derivative of F with respect to the system (3.6.3) is the formal series

$$\dot{F}_{(3.6.3)}(x,y) = G_{2k}(x^2 + y^2)^k + \sum_{h=2k+1}^{\infty} \tilde{F}_h(x,y).$$

Definition 3.6.1. The constant G_{2k} is called the k-th *Poincaré-Liapunov coefficient* of the equilibrium $(0,0)$; if $G_4 = \cdots = G_{2k} = 0$ but $G_{2k+2} \neq 0$, then the origin is called a *weak spiral point of order* k (if $G_4 \neq 0$, then it is a *first order weak spiral point*).

It is easy to see by introducing polar coordinates, say, that a weak spiral point is a spiral point in the sense of the definition following formula (1.4.19). If the origin is a weak spiral point of order k, i.e. if the first non-vanishing Poincaré-Liapunov coefficient is G_{2k+2}, then the procedure described in Theorem 3.6.1 can be terminated at $m = 2k + 2$ and we do not have to introduce a formal series. In this case the procedure determines a Liapunov function (3.6.4) which is positive definite in a neighbourhood of the origin whose derivative with respect to the system is

$$\dot{F}_{(3.6.3)}(x,y) = G_{2k+2}(x^2 + y^2)^{k+1} + \text{higher order terms}.$$

Obviously (see Theorems 1.5.2 and 1.5.4), if $G_{2k+2} < 0$, then the origin is asymptotically stable; if $G_{2k+2} > 0$, then it is unstable. It can be proved (see Sansone-Conti [1964]) that if all the Poincaré-Liapunov coefficients are zero, $G_{2k} = 0$ ($k = 2, 3, 4, ...$), then the origin is a centre. (In this case the system admits a first integral of the form (3.6.5).)

Putting the centre problem aside for a while we return to system (3.6.2) now. First we are proving a Lemma that will enable us to establish some basic properties of closed orbits (see Coppel [1966]). This Lemma states,

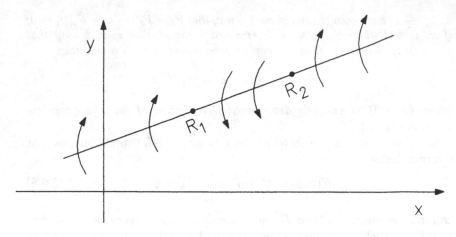

FIGURE 3.6.1. Trajectories of system (3.6.2) intersecting a straight line outside and inside two equilibria, resp. points of contact R_1, R_2.

loosely speaking, that three equilibria can never lie on the same straight line.

Lemma 3.6.2. *On any straight line that is not composed of orbits the maximal number of equilibria and points of contact (i.e. points where the vector field is parallel to the line) is two. If R_1 and R_2 are two such points, then the trajectories intersect the straight line determined by R_1 and R_2 outside the segment $\overline{R_1R_2}$ in the same direction and inside the segment in the direction opposite to the former (see Figure 3.6.1).*

Proof. Consider a straight line $ax + by + c = 0$ where $a, b, c \in \mathbf{R}$ are arbitrary, $a^2 + b^2 > 0$. The isocline of the field (P_2, Q_2) joining the points with the same slope as this line is $aP_2(x, y) + bQ_2(x, y) = 0$ which is a conic section. This isocline contains also the equilibria of the system since the latter are characterised by $P_2 = Q_2 = 0$. If our line has three common points with this conic section, i.e. three points that are either equilibria or points of contact, then it is contained in the conic section, and hence it is composed of orbits contrary to the assumption. If our line intersects the conic section in two points R_1 and R_2, then the unbounded segments ∞R_1 and $R_2\infty$ must belong to the domain where $aP_2 + bQ_2 > 0$ and the segment R_1R_2 to the domain $aP_2 + bQ_2 < 0$ or vice versa. This means that in the first domain the field $[P_2, Q_2]$ forms an acute angle with the normal vector $[a, b]$ of the line, and in the second domain it forms an obtuse angle, or vice versa. $\square$

Theorem 3.6.3. *The interior of a closed path of (3.6.2) is convex.*

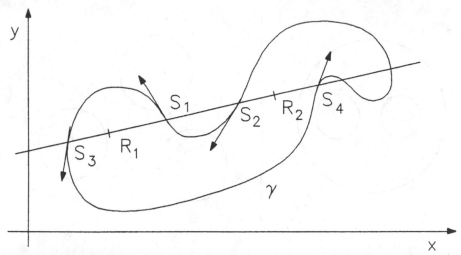

FIGURE 3.6.2. The interior of the closed path γ must be convex. Proof of Theorem 3.6.3.

Proof. Let γ be the closed path in question and assume, contrary to the statement, that there are two points R_1, R_2 in its interior such that the straight line segment $\overline{R_1 R_2}$ contains a point outside γ (see Figure 3.6.2). Then γ must cross the segment between R_1 and R_2 at least twice, at S_1 and S_2, say, and these crossings are in the opposite directions. γ must cross the line $R_1 R_2$ also outside this segment to the "left" and to the "right" at least twice at S_3 and S_4, say. If there are just these four crossings, the direction of the crossing at S_3 must be opposite to that of S_1, and similarly at S_4 it is opposite to that of S_2. Thus, on each segment $\overline{S_3 S_1}$, $\overline{S_1 S_2}$, $\overline{S_2 S_4}$ there lies a point of contact or an equilibrium, at least, and this contradicts the Lemma. If there were more crossings, then there would be more than three points of contact on the straight line. $\square$

We know (see Corollary 3.5.7) that the interior of a closed path contains an equilibrium point at least. In case of a quadratic system we may say more.

Theorem 3.6.4. *The interior of each closed path of system (3.6.2) contains exactly one equilibrium point.*

Proof. Assume, contrary to the proposition, that the interior of the closed path γ contains two equilibria R_1 and R_2. Then γ must cross the straight line $R_1 R_2$ outside the segment $\overline{R_1 R_2}$ in opposite directions, and this contradicts the last part of the Lemma. $\square$

Theorem 3.6.5. *If two closed trajectories of system (3.6.2) are not situated one inside the other, then they are oppositely oriented.*

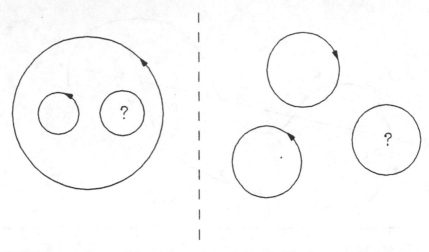

FIGURE 3.6.3. Impossible configurations of closed orbits for a quadratic system.

Proof. Denote by γ_1 and γ_2 the two closed orbits whose interiors have no common point, and the equilibria contained in γ_1 and in γ_2 by R_1 and R_2, respectively. If γ_1 and γ_2 were similarly oriented, then they would cross the segment $\overline{R_1 R_2}$ in opposite directions, and this contradicts the Lemma. $\square$

Theorem 3.6.6. *If a closed path lies in the interior of another closed path, then these two trajectories are similarly oriented.*

Proof. Assume that the closed path γ_1 is contained in the interior of the path γ_2 and denote the equilibrium in the interior of γ_1 by R. If γ_1 and γ_2 were oppositely oriented, then γ_1 and γ_2 would intersect an arbitrary straight line through R "to the left of R" in opposite directions and "to the right of R" also in opposite directions. As a consequence, on this line there would lie two points of contact, one to the left and one to the right of R. But this contradicts the Lemma. $\square$

The previous theorems imply that no configuration of closed orbits shown in Figure 3.6.3 may arise for system (3.6.2). The character of an equilibrium situated in the interior of a closed orbit is established in the following:

Theorem 3.6.7. *An equilibrium in the interior of a closed path of system (3.6.2) must be either a spiral point or a centre.*

Proof. Without loss of generality assume that the equilibrium point in the interior of a closed path is the origin $(0,0)$. Then we have $a_{00} = b_{00} = 0$ in (3.6.2). Introducing polar coordinates by $x = r \cos \theta$, $y = r \sin \theta$, the differential equation for the polar angle θ becomes

$$\dot{\theta} = f(\theta) + rg(\theta)$$

where

$$
\begin{aligned}
f(\theta) &= b_{10}\cos^2\theta + (b_{01} - a_{10})\cos\theta\sin\theta - a_{01}\sin^2\theta, \\
g(\theta) &= b_{20}\cos^3\theta + (b_{11} - a_{20})\cos^2\theta\sin\theta \\
&\quad + (b_{02} - a_{11})\cos\theta\sin^2\theta - a_{02}\sin^3\theta.
\end{aligned}
$$

Assume that the origin is neither a spiral point nor a centre. Then the equation $f(\theta) = 0$ has at least one root $\theta_1 \in [0, 2\pi)$. Since the equation $g(\theta) = 0$ is a cubic equation in $\tan\theta$, say, it has to have at least one root $\theta_2 \in [0, 2\pi)$. If we had $\theta_1 = \theta_2$, then the ray $\theta = \theta_1$ would be composed of orbits, and therefore no closed orbit could surround the origin. So that we may suppose that $0 < \theta_2 - \theta_1 < 2\pi$, say, and that $f(\theta) \neq 0$ for $\theta_1 < \theta \leq \theta_2$, $g(\theta) \neq 0$ for $\theta_1 \leq \theta < \theta_2$. Suppose that $f(\theta)g(\theta) < 0$ for $\theta \in (\theta_1, \theta_2)$. At $\theta = \theta_1$ we have $\dot{\theta} = rg(\theta_1) \neq 0$, and at $\theta = \theta_2$ we have $\dot{\theta} = f(\theta_2) \neq 0$. Since the signs of $rg(\theta_1)$ and $f(\theta_2)$ are opposite, all trajectories that cross the rays $\theta = \theta_1$ and $\theta = \theta_2$ either enter the sector $\theta_1 < \theta < \theta_2$ or leave it with increasing t. Thus, if a solution has a point inside this sector for $t = 0$, then either its positive or its negative semitrajectory has to remain in the sector implying that the trajectory cannot be closed. If $f(\theta)g(\theta)$ is positive in the sector, we consider the sector $\theta_1 + \pi < \theta < \theta_2 + \pi$ instead. Clearly, $f(\theta + \pi) = f(\theta)$, $g(\theta + \pi) = -g(\theta)$. $\square$

This Theorem leads us back to the centre problem. We have to exclude systems with a centre. By the way, it has been proved (see Ye Yanqian [1982,1986]) that system (3.6.2) cannot have a centre and a spiral point simultaneously. Assuming that the origin $(0,0)$ is a centre or a weak spiral point of system (3.6.2) we write the system, applying a linear coordinate transformation and rescaling the independent variable in the form (3.6.3), i.e.

$$\dot{x} = -y + p_2(x, y), \qquad \dot{y} = x + q_2(x, y) \tag{3.6.6}$$

where p_2 and q_2 are homogeneous polynomials of degree two. It has been proved by Bautin [1952] that if the first three "independent" Poincaré-Liapunov coefficients of system (3.6.6) are zero, then the origin is a centre. Performing the calculations, the following criteria are obtained (see Coppel [1966] and the references therein):

Theorem 3.6.8. *The system*

$$
\begin{aligned}
\dot{x} &= -y - bx^2 - (2c + \beta)xy - dy^2, \\
\dot{y} &= x + ax^2 + (2b + \alpha)xy + cy^2
\end{aligned}
\tag{3.6.7}
$$

has a centre at the origin if and only if one of the following three conditions is satisfied:

    (A) $a + c = b + d = 0$;

    (B) $\alpha(a + c) = \beta(b + d)$,

         $a\alpha^3 - (3b + \alpha)\alpha^2\beta + (3c + \beta)\alpha\beta^2 - d\beta^3 = 0$;

    (C) $\alpha + 5(b + d) = \beta + 5(a + c) = ac + bd + 2(a^2 + d^2) = 0$.

This Theorem enables one to exclude systems with a centre, and one may concentrate on systems with one or two spiral points. Already Bautin [1952] has proved the following:

Theorem 3.6.9. *For a quadratic system (3.6.2) a weak spiral point is of order three at most; as a consequence there can be three small amplitude limit cycles at most around any equilibrium point.*

For a long time it was believed that the maximal number of limit cycles of a quadratic system was three. This belief was based on a paper of Petrovskij and Landis [1955] which later turned out to be erroneous. Probably Shi [1980] was the first one to produce an example of a quadratic system with 4 limit cycles. We quote this from Shi [1981].

Consider the system

$$\dot{x} = \lambda x - y + lx^2 + (5a + \delta)xy + ny^2,$$
$$\dot{y} = x + ax^2 + (3l + 5n + (\delta(l + n) + 8\varepsilon)/a)xy. \qquad (3.6.8)$$

Theorem 3.6.10. *If*

$$a(2a^2 + 2n^2 + ln)(a(5l + 6n) - 3(l + 2n)(l + n^2)) \neq 0$$

$$\delta a(2a^2 + 2n^2 + ln) > 0, \qquad \delta\varepsilon < 0, \qquad \varepsilon\lambda < 0,$$

$$0 < |\lambda| << |\varepsilon| << |\delta| << 1,$$

$$3a^2 - l(l + 2n) < 0,$$
$$25a^2 + 12n(l + 2n) < 0,$$
$$l(3l + 5n)^2 - 5a^2(3l + 5n) + na^2 < 0,$$
$$a^2(5l + 8n) - ((2l + 5n)^2 + 15a^2)(25a^2 + 3n(2l + 5n)) > 0,$$

then there are three limit cycles around the origin $(0,0)$, and there is a limit cycle around the equilibrium point $(0, 1/n)$.

A choice satisfying the conditions of the Theorem is

$$l = -10, \quad a = 1, \quad \delta = -10^{-13}, \quad \varepsilon = 10^{-52}, \quad \lambda = -10^{-240}.$$

Since then several examples have been established, all yielding the configuration in Figure 3.6.4. See, e.g., Kayumov, Rozet, Begiev [1981],

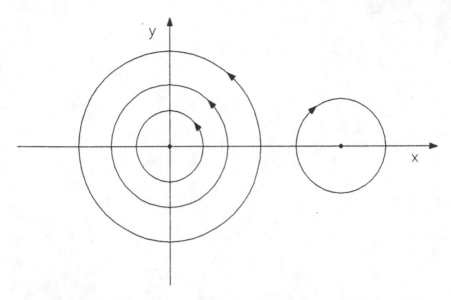

FIGURE 3.6.4. The only configuration of four limit cycles of a quadratic system (?).

Ilyashenko [1985] and Blows-Lloyd [1984]. It is conjectured now that the maximal number of limit cycles of a quadratic system (3.6.2) is four, and this number can be achieved only in the configuration of Figure 3.6.4. However, no proof of this is known yet.

Still less is known about systems with higher degree polynomials on the right-hand side. Some results concerning third and higher degree cases can be found, for instance, in Blows-Lloyd [1984] and in Cima-Llibre [1989].

3.7   Problems

Problem 3.7.1. Let a_n, $n \in \mathbf{N}$ be a bounded sequence of real numbers; show that if each convergent subsequence tends to the same limit, then a_n, $n \in \mathbf{N}$, is convergent. Realize that a version of this proposition is necessary for the completion of the proof of Theorem 3.1.11.

Problem 3.7.2. Prove Theorem 3.2.1. (Hint: use Liénard's plane and the "Liapunov function" $\eta^2/2 + G(x)$.)

Problem 3.7.3. Draw the phase portrait of the damped Duffing equation (system (3.3.9)) for different parameter configurations ($b > 0$, $b < 0$, $a > d^2/4$, $a \le d^2/4$).

Problem 3.7.4. Assume that system (3.5.5) can be written in the form $\dot{x} = h(x) + g(x)$ where $h_1(x)$ and $h_2(x)$ are homogeneous polynomials of degree $n \ge 1$ and $|g(x)| = o(|x|^n)$. Prove that the index of the origin with respect to this system is equal to its index with respect to the system $\dot{x} = h(x)$.

4
Periodic Solutions of Periodic Systems

In this chapter we study the existence, stability and isolation of periodic solutions belonging to n-dimensional systems of periodic nonlinear differential equations of the form $\dot{x} = f(t, x)$ where f is periodic in t with some period $T > 0$: $f(t + T, x) = f(t, x)$. One may believe that an autonomous system of the form $\dot{x} = f(x)$ is a special case since, obviously, here the right-hand side is periodic in t with arbitrary positive period. Though this is true, autonomous systems cannot be treated similarly to periodic non-autonomous ones. This is so because in the case of an autonomous system we do not know *a priori* what may be the period of a periodic solution if there exists any, and also because the *integral curve* belonging to a non-constant periodic solution of an autonomous system can never be "isolated". More will be said about these problems at the appropriate places. We have mentioned these problems here in order to explain why autonomous systems will be treated in the next chapter. The methods developed with the aim of establishing the existence and stability of periodic solutions can be classified in two groups. The first is the group of *topological methods* based on degree theory and fixed point theorems. These methods will be presented in the first Section of this chapter. The background material can be found in Appendix 2. The second group consists of (small) *perturbation methods*. These are more effective but have the disadvantage that they work under the assumption that the given differential equation is a "perturbation" of another one whose periodic solution is known. Both methods have their origin in the works of H. Poincaré [1899]. We shall treat the perturbation methods separately in Chapter 6. In the second section of this chapter we study the stability and isolation problems of periodic solutions. In Sections 3, 4 and 5, applications will be presented.

4.1 Existence of Periodic Solutions

In this section some topological methods will be presented that may enable us to establish the existence of a periodic solution of a periodic system. These methods are based on degree theory and on various fixed point theorems. A concise summary of the topological tools to be applied can be found in Appendix 2. These methods can be applied in situations when very little information is known about the system considered; their disadvantage is that even in case they are successful they do not provide a method for the actual determination of the periodic solution, neither are they of any help in problems of stability of the solution. Still, without them in some important cases we could not prove the existence of a periodic solution.

The system considered will be assumed in the form

$$\dot{x} = f(t, x), \tag{4.1.1}$$

where $f \in C^0(\mathbf{R} \times X, \mathbf{R}^n)$, $f'_x \in (\mathbf{R} \times X, \mathbf{R}^{n^2})$, X being an open and connected subset of $\mathbf{R}^n$, and it will be assumed that there exists a $T > 0$ such that for every $t \in \mathbf{R}$, $x \in X$

$$f(t + T, x) = f(t, x),$$

i.e. f is periodic in t with (not necessarily the least) period T.

A periodic system (4.1.1) may have a periodic solution whose period has nothing to do with T. Consider, for instance, the system

$$\dot{x}_1 = -x_2 + (x_1^2 + x_2^2 - 1)\cos \omega t, \qquad \dot{x}_2 = x_1 + (x_1^2 + x_2^2 - 1)\sin \omega t. \tag{4.1.2}$$

This is a periodic system with period $2\pi/\omega$, and we let ω be an arbitrary non-zero real number. On the other hand, obviously, $x_1(t) = \cos t$, $x_2(t) = \sin t$ is a 2π-periodic solution of the system. This phenomenon is possible because there is a subset of $\mathbf{R}^n$ where the periodicity of the system (with a fixed period) disappears. In the case of system (4.1.2) this subset is $\{(x_1, x_2) \in \mathbf{R}^2 : x_1^2 + x_2^2 = 1\}$, and the path of the periodic solution with a different period lies in this subset.

Such periodic solutions will not be considered. We shall be interested in periodic solutions of period T, i.e. that of the system, or in "*subharmonics*" whose least period is an integer multiple of T or, sometimes, in "*superharmonics*" whose period is a rational fraction of T.

Denote the solution of (4.1.1) that assumes the value $x^0 \in X$ at $t = 0$ by $\varphi(t, 0, x^0)$. For the initial moment we could choose any other $t_0 \in \mathbf{R}$ instead of zero. Suppose that there exists a non-empty subset $X_0 \subset X$ such that for each $x^0 \in X_0$ the solution $\varphi(t, 0, x^0)$ is defined on the closed interval $[0, T]$ at least. The function

$$\varphi_T : X_0 \to X, \quad \varphi_T(x) := \varphi(T, 0, x), \quad x \in X_0 \qquad (4.1.3)$$

is called the *Poincaré-Andronov operator* belonging to the periodic system (4.1.1). The problem of the existence of a periodic solution of period T is reduced to the existence of a fixed point of this operator by the following Theorem which is due to Poincaré and is an immediate consequence of Lemma 2.2.1.

Theorem 4.1.1. *Under the assumptions made above, system (4.1.1) has a periodic solution of period T if and only if the operator φ_T has a fixed point $x^0 \in X_0$, i.e. $\varphi_T(x^0) = x^0$; if x^0 is a fixed point of the operator (4.1.3), then $\varphi(t, 0, x^0)$ is a periodic solution of (4.1.1).*

Thus, the task is to guarantee the existence of a fixed point. This can be done by establishing the conditions of a fixed point theorem. In many cases Brouwer's Fixed Point Theorem (Corollary A2.9) is used, like in the proof of the following:

Theorem 4.1.2. *Assume that a subset $D \subset X_0 \subset X$ exists that is homeomorphic to the closed ball $B = \{x \in \mathbf{R}^n : |x| \leq 1\}$ and for which $x^0 \in D$ implies that the solution $\varphi(t, 0, x^0)$ of (4.1.1) is defined on $[0, T]$ at least; if for every $x^0 \in D$ we have $\varphi(t, 0, x^0) \in D$ for $t > 0$, then (4.1.1) has a periodic solution of period T whose path lies in D.*

Proof. The assumptions and the continuous dependence on initial conditions (Theorem 1.1.3) imply that the operator φ_T defined by (4.1.3) maps D into itself and is continuous. Since D and φ_T satisfy the conditions of Brouwer's Fixed Point Theorem (Corollary A2.9), φ_T has a fixed point in D. $\square$

There are two difficulties with the application of this Theorem. First, it is hard to determine an appropriate set D that has the property that initial conditions belonging to it determine solutions that are defined on sufficiently long intervals. This difficulty can be overcome in some important cases by finding *a priori* estimates for the solutions. The second difficulty is that the Theorem does not distinguish non-constant periodic solutions from equilibria. An equilibrium point is, clearly, a fixed point of the operator φ_T. However, there is an important class of systems at which this problem does not arise. These are the autonomous systems subject to the periodic forcing term:

$$\dot{x} = f(x) + b(t) \qquad (4.1.4)$$

where $f \in C^1(X, \mathbf{R}^n)$, $X \subset \mathbf{R}^n$ an open and connected set, $b \in C^0(\mathbf{R}, \mathbf{R}^n)$, and $b(t + T) = b(t)$ for some $T > 0$ and all $t \in \mathbf{R}$. *If b is non-constant, then system (4.1.4) has no equilibrium point*, as can be seen easily by assuming the contrary.

We illustrate Theorem 4.1.2 by the following one dimensional but non-trivial Example.

Example 4.1.1 (see Gopalsamy, Kulenovic, Ladas [1990]). The time evolution of a single species is often modelled by the logistic differential equation $\dot{N} = rN(1 - N/K)$ where $N(t)$ is the quantity of the population at time t, and the positive constants r and K are the *intrinsic growth rate* and the *carrying capacity*, respectively. However, it has been pointed out (see F.E. Smith [1963]) that there are populations like *Daphnia* (a water flea) for which the logistic equation is not realistic, and the model

$$\dot{N} = rN(K - N)/(K + cN) \tag{4.1.5}$$

has been suggested instead. Here r, K and c are positive constants. It is easy to see that as $N = 0$ is a solution, any solution with positive initial value stays positive for all $t > 0$, and that solutions with positive initial values are defined and stay bounded over $[0, \infty]$ ($\dot{N}$ is negative if $N > K$). Also if $N(0) > 0$, then $N(t) \to K$ as $t \to \infty$, i.e. the solution $N = K$ is globally asymptotically stable with respect to the half line $(0, \infty)$.

The question arises of what happens to the population if due to seasonal changes, say, the parameters in (4.1.5) are replaced by positive periodic functions of the same period. This means that we have to consider the differential equation

$$\dot{N} = r(t)N(K(t) - N)/(K(t) + c(t)N) \tag{4.1.6}$$

where $r, K, c \in C^0$, $r(t) > 0$, $K(t) > 0$, $c(t) \geq 0$ for $t \in \mathbf{R}$; they are periodic with period $T > 0$, i.e. $r(t + T) = r(t)$, $c(t + T) = c(t)$, $K(t + T) = K(t)$, and K, at least, is definitely non-constant, $K(t) \neq$ constant. Note that (4.1.6) contains the periodic logistic equation $\dot{N} = r(t)N(1 - N/K(t))$ as a special case if $c(t) \equiv 0$ is chosen. We are going to show that under these assumptions, *(4.1.6) has a non-constant positive periodic solution, and every solution with positive initial value tends to this periodic solution as t tends to infinity.*

Introduce the notations

$$K_1 = \min_{[0,T]} K(t), \qquad K_2 = \max_{[0,T]} K(t).$$

Clearly, $0 < K_1 < K_2$. If $N(t) > K(t)$, then $\dot{N}(t) < 0$; if $N(t) < K(t)$, then $\dot{N}(t) > 0$. Therefore if $N_0 \in [K_1, K_2]$, then $N(t, 0, N_0) \in [K_1, K_2]$ for $t \geq 0$ where $N(t, 0, N_0)$ is the solution that satisfies the initial condition $N(0, 0, N_0) = N_0$. By Theorem 4.1.2 equation (4.1.6) has a periodic solution

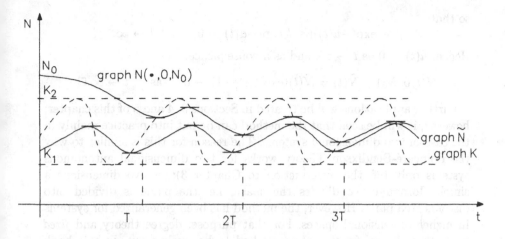

FIGURE 4.1.1. Graph of the periodic solution $\tilde{N}$ of (4.1.6) and a solution $N(t, 0, N_0)$ tending to it.

$\tilde{N}(t)$ of period T such that $K_1 \leq \tilde{N}(t) \leq K_2$ for $t \in \mathbf{R}$. Since (4.1.6) cannot have a positive constant solution, $\tilde{N}(t) \neq$ const.

We are going to show that if $N_0 > 0$, then $\lim_{t \to \infty}(N(t, 0, N_0) - \tilde{N}(t)) = 0$. Write $N(t, 0, N_0)$ in the form $N(t, 0, N_0) = \tilde{N}(t) \exp u(t)$. Obviously, if $N_0 > \tilde{N}(0)$, then $u(0) > 0$ and $u(t) > 0$ for $t > 0$, and if $N_0 < \tilde{N}(0)$, then $u(0) < 0$ and $u(t) < 0$ for $t > 0$ (see Figure 4.1.1). Substituting this expression of $N(t, 0, N_0)$ into the differential equation (4.1.6) we obtain the following differential equation for the function u:

$$\dot{u} = r(t)(K(t) - \tilde{N}(t) \exp u)/(K(t) + c(t)\tilde{N}(t) \exp u)$$
$$-r(t)(K(t) - \tilde{N}(t))/(K(t) + c(t)\tilde{N}(t));$$

or applying the Mean Value Theorem

$$\dot{u} = -r(t)\frac{K(t)(1 + c(t))}{(K(t) + c(t)\nu(t))^2}\tilde{N}(t)(\exp u - 1)$$

where $\nu(t) \in (\tilde{N}(t), \tilde{N}(t) \exp u)$. Integrating the last differential equation by "separating the variables" we obtain

$$|1 - \exp(-u(t))| = k \exp\left(-\int_0^t (r(\tau)K(\tau)\tilde{N}(\tau)(1 + c(\tau))/(K(\tau) + c(\tau)\nu(\tau))^2)d\tau\right).$$

But, clearly, there are positive constants $0 < a_1 < a_2$ such that the integrand can be estimated by

$$0 < a_1 < r(t)K(t)\tilde{N}(t)(1 + c(t))/(K(t) + c(t)\nu(t)) < a_2,$$

so that
$$|1 - \exp(-u(t))| \leq k \exp(-a_1 t) \to 0, \qquad t \to \infty.$$

Hence, $u(t) \to 0$ as $t \to \infty$, and as a consequence,

$$N(t, 0, N_0) - \tilde{N}(t) = \tilde{N}(t)(\exp u(t) - 1) \to 0 \qquad \text{as} \quad t \to \infty. \square$$

Further applications will be treated in Sections 3, 4 and 5 of this chapter; however, it will be seen that the method can be put into practice mainly in the case of two dimensional systems. The reason for this is similar to why the Poincaré-Bendixson Theory works for two dimensional autonomous systems only (cf. the introduction to Chapter 3): in two dimensions a simple Jordan curve divides the space, i.e. the plane is divided into disconnected parts. However, the method has been generalized for systems in higher dimensional spaces. For that purpose, degree theory and *fixed point theorems in functional spaces* had to be worked out. In this book very little functional analysis is used. The most important background material that is needed is summed up in Appendix 2. The basic idea of this generalized method is that an *operator* is made to correspond to the periodic system (4.1.1) that acts on the space of continuous periodic functions of period $T > 0$,

$$C_T = \{x \in C^0(\mathbf{R}, \mathbf{R}^n) : x(t + T) = x(t), \ t \in \mathbf{R}\}, \tag{4.1.7}$$

and fixed points of this operator correspond to periodic solutions of (4.1.1). The space C_T is a Banach space (see Definition A2.5) if the supremum norm is introduced for $x \in C_T$:

$$\|x\|_{C_T} := \sup_{t \in [0,T]} |x(t)|.$$

We are treating this method following, first of all, the presentation in Rouche-Mawhin [1973], relying also on Cronin [1964]. We are going to write system (4.1.1) in the form

$$\dot{x} = A(t)x + g(t, x) \tag{4.1.8}$$

where $A \in C^0(\mathbf{R}, \mathbf{R}^{n^2})$, $g, g'_x \in C^0(\mathbf{R} \times X, \mathbf{R}^n)$, $X \subset \mathbf{R}^n$ being an open and connected set, and both A and g are periodic functions of t with period $T > 0$, i.e. $A(t + T) = A(t)$, $g(t + T, x) = g(t, x)$ for all $t \in \mathbf{R}$ and $x \in X$. The advantage of considering systems in the form (4.1.8) is that if a T-periodic function $p \in C^1$ is substituted into g for x, then the existence problem of a periodic solution of the nonlinear system (4.1.8) is reduced to the problem of whether p is a solution of the inhomogeneous linear system

$$\dot{x} = A(t)x + g(t, p(t)). \tag{4.1.9}$$

We shall consider the special case when *the homogeneous linear system*

$$\dot{x} = A(t)x \tag{4.1.10}$$

corresponding to (4.1.8) has no T-periodic solution apart from the trivial one. Instead of (4.1.9) consider now an arbitrary inhomogeneous system corresponding to (4.1.10),

$$\dot{x} = A(t)x + b(t) \qquad (4.1.11)$$

where $b \in C^0(\mathbf{R}, \mathbf{R}^n)$ and $b(t + T) = b(t)$ for $t \in \mathbf{R}$. By (1.2.11) the solution φ of (4.1.11) that assumes the value φ^0 at $t = 0$ is

$$\varphi(t) = \Phi(t)(\varphi^0 + \int_0^t \Phi^{-1}(s)b(s)ds) \qquad (4.1.12)$$

where Φ is the fundamental matrix of (4.1.10) that assumes the unit matrix at $t = 0$: $\Phi(0) = I$. According to Lemma 2.2.1 the solution φ is periodic with period T if and only if $\varphi(T) = \varphi(0)$, i.e. if and only if

$$(I - \Phi(T))\varphi^0 = \Phi(T) \int_0^T \Phi^{-1}(s)b(s)ds,$$

or if and only if

$$\varphi^0 = (I - \Phi(T))^{-1}\Phi(T) \int_0^T \Phi^{-1}(s)b(s)ds.$$

The matrix $I - \Phi(T)$ is invertible since we have assumed that (4.1.10) has no nontrivial T-periodic solution. If the last expression for φ^0 is substituted into (4.1.12), we get the explicit expression for the unique T-periodic solution of (4.1.11) (see Theorem 2.3.1). Thus, we have defined an operator

$$\mathcal{K} : C_T \to C_T$$

that orders the unique periodic solution φ of (4.1.11) to any $b \in C_T$:

$$
\begin{aligned}
\varphi(t) &= (\mathcal{K}b)(t) \\
&= \Phi(t)(I - \Phi(T))^{-1}\Phi(T) \int_0^T \Phi^{-1}(s)b(s)ds \\
&\quad + \Phi(t) \int_0^t \Phi^{-1}(s)b(s)ds \\
&= (\Phi(t)(I - \Phi(T))^{-1}\Phi(T) + \Phi(t)) \int_0^t \Phi^{-1}(s)b(s)ds \\
&\quad + \Phi(t)(I - \Phi(T))^{-1}\Phi(T) \int_t^T \Phi^{-1}(s)b(s)ds \\
&= \Phi(t)(I - \Phi(T))^{-1} \int_0^t \Phi^{-1}(s)b(s)ds \\
&\quad + \Phi(t + T)(I - \Phi(T))^{-1} \int_t^T \Phi^{-1}(s)b(s)ds
\end{aligned}
$$

where (2.2.6) and the fact that $\Phi(T)$ and $(I - \Phi(T))^{-1}$ commute has been applied. Introducing the *Green matrix function*

$$K(t,s) := \begin{cases} \Phi(t)(I - \Phi(T))^{-1}\Phi^{-1}(s), & \text{for } 0 \leq s \leq t \leq T \\ \Phi(t+T)(I - \Phi(T))^{-1}\Phi^{-1}(s), & \text{for } 0 \leq t \leq s \leq T, \end{cases}$$

we have that

$$\varphi(t) = (\mathcal{K}b)(t) = \int_0^T K(t,s)b(s)ds. \tag{4.1.13}$$

Comparing (4.1.9) with (4.1.11) we see that the unique periodic solution of (4.1.9) is

$$\varphi = \mathcal{K}(g \circ p) = \int_0^T K(t,s)g(s,p(s))ds. \tag{4.1.14}$$

Let C_{TX} be the subset of C_T containing the functions whose range is in X, i.e.

$$C_{TX} = \{x \in C_T : x(t) \in X, \ t \in \mathbf{R}\},$$

and let

$$\mathcal{G} : C_{TX} \to C_T$$

be the operator defined by $(\mathcal{G}p)(t) = g(t, p(t))$ for any $p \in C_{TX}$. This way (4.1.14) can be put in the form

$$\varphi = \mathcal{K}\mathcal{G}p.$$

There holds, obviously, the following:

Theorem 4.1.3. *System (4.1.8) where (4.1.10) has no nontrivial T-periodic solution has a T-periodic solution if and only if the operator*

$$\mathcal{K}\mathcal{G} : C_{TX} \to C_T$$

has a fixed point.

More general situations in which (4.1.10) has nontrivial periodic solutions are treated in Rouche-Mawhin [1973]. Such a study has to rely on Theorem 2.3.4 and requires the introduction of more sophisticated operators; nevertheless, the existence of a periodic solution can be established through a fixed point theorem analogous to Theorem 4.1.3. Because of a lack of space we have to omit this generalization; however, we are going to treat that most degenerate case when $A(t) \equiv 0$, i.e. when (4.1.10) has the form $\dot{x} = 0$, and as a consequence, every solution of it is periodic, constant to be sure.

Keeping the notations established above, (4.1.8) has the form

$$\dot{x} = g(t, x), \tag{4.1.15}$$

and if $p \in C_{TX}$ the question is whether it is a solution of the analogue of (4.1.9)

$$\dot{x} = g(t, p(t)). \tag{4.1.16}$$

The solutions of (4.1.16) are $x(t) = c + \int_0^t g(s, p(s))ds$ where $c \in \mathbf{R}^n$. Unfortunately, the integral of a periodic function is not necessarily periodic. We shall need

Lemma 4.1.4. *For $f \in C_T$ its integral function $F(t) := \int_0^t f(s)ds$ is periodic with period T if and only if the mean value of f is zero:*

$$\frac{1}{T} \int_0^T f(t)dt = 0. \tag{4.1.17}$$

Proof.

$$F(t+T) = \int_0^{t+T} f(s)ds = \int_0^T f(t)dt + \int_T^{T+t} f(s)ds = F(t) + \int_0^T f(t)dt$$

because of the periodicity of f. From here the Lemma obviously follows. $\square$

As we see, the solutions of (4.1.16) are periodic if and only if the mean value of the right-hand side is zero, which is not necessarily the case. However, if the mean of the right-hand side is subtracted from it, then a periodic function with zero mean is obtained, and this implies that for the system

$$\dot{x} = g(t, p(t)) - \frac{1}{T} \int_0^T g(t, p(t))dt \tag{4.1.18}$$

all solutions

$$x(t) = c + \int_0^t g(s, p(s))ds - \frac{t}{T} \int_0^T g(s, p(s))ds$$

where $c \in \mathbf{R}^n$ is an arbitrary constant are periodic with period T. If we replace p by x in the last expression, we obtain that all solutions of the integral equation

$$x(t) = c + \int_0^t g(s, x(s))ds - \frac{t}{T} \int_0^T g(s, x(s))ds \tag{4.1.19}$$

are periodic solutions of the differential system

$$\dot{x} = g(t, x) - \frac{1}{T} \int_0^T g(t, x(t))dt. \tag{4.1.20}$$

In (4.1.19) the constant vector c may be chosen arbitrarily, but, obviously, $c = x(0)$. Now, let H be an arbitrary $n \times n$ regular matrix and replace the integral equation (4.1.19) by the following:

$$x(t) = x(0) + H\left(\frac{1}{T}\int_0^T g(t, x(t))dt\right)$$

$$+ \int_0^t g(s, x(s))ds - \frac{t}{T}\int_0^T g(s, x(s))ds. \qquad (4.1.21)$$

What happened is that c was chosen the following way:

$$c = x(0) + H\left(\frac{1}{T}\int_0^T g(t, x(t))dt\right).$$

But still, from (4.1.21) it is clear that c must be the initial value $x(0)$ of the solution, i.e.

$$x(0) = x(0) + H\left(\frac{1}{T}\int_0^T g(t, x(t))dt\right)$$

or because of the regularity of H

$$\int_0^T g(t, x(t))dt = 0. \qquad (4.1.22)$$

Thus, if x is a C^0 solution of (4.1.21), then it is in the C^1 class, is periodic with period T, and satisfies (4.1.22) and (4.1.20) with the second term on the right-hand side equal to zero, i.e. it is satisfies (4.1.15). It is easy to see that, vice versa, if x is a T-periodic solution of system (4.1.15) with initial value $x(0)$, then it satisfies condition (4.1.22) and it is a solution of the integral equation (4.1.21) which is reduced, because of (4.1.22), to

$$x(t) = x(0) + \int_0^t g(s, x(s))ds.$$

Now, the following notations will be introduced. For $x \in C_{TX}$ the composite function $g \circ x : \mathbf{R} \to \mathbf{R}^n$ is defined by $(g \circ x)(t) = g(t, x(t))$ for all $t \in \mathbf{R}$, and $g \circ x = \mathcal{G}x \in C_T$. For an arbitrary $p \in C_T$

$$\tilde{p} := \frac{1}{T}\int_0^T p(t)dt, \qquad (4.1.23)$$

i.e. $\tilde{p}$ is the mean value of p. For an arbitrary $p \in C_T$ the operator $\mathcal{H} : C_T \to C_T$ is defined by

$$(\mathcal{H}p)(t) = H\tilde{p} + \int_0^t (p(s) - \tilde{p})ds \qquad (4.1.24)$$

where H is an arbitrary $n \times n$ regular matrix, and finally the operator $\mathcal{V} : C_T \to C_T$ orders to every $x \in C_T$ its initial value $x(0)$ which as a constant function is considered to be an element of C_T :

$$(\mathcal{V}x)(t) = x(0), \quad \text{for} \quad x \in C_T, \quad t \in \mathbf{R}.$$

With these notations the integral equation (4.1.21) can be written in the form

$$x = (\mathcal{V} + \mathcal{H}\mathcal{G})x.$$

Thus, we have proved the following

Theorem 4.1.5. *System (4.1.15) has a T-periodic solution if and only if the operator*

$$\mathcal{V} + \mathcal{H}\mathcal{G} : C_{TX} \to C_T$$

has a fixed point.

In Appendix 2 Schauder's and Banach's Fixed Point Theorems are treated. However, it is by no means an easy task to determine whether a given operator satisfies the conditions of a fixed point theorem. We shall return to this problem, presenting some applications, in Section 3 of this chapter.

4.2 Stability and Isolation of Periodic Solutions

Once a periodic solution has been determined or its existence proved, the problem of its stability arises. If it is attracting, then there can be no other periodic solution in its neighbourhood, i.e. it is isolated. If it is not attracting, e.g. it is only stable in the Liapunov sense, then the question arises of whether it is isolated or not. These problems are to be treated in this Section. We consider, first, the stability of a periodic solution of a non-linear system by linearization. The system to be studied is system (4.1.1) which will be written out here again with the assumptions for the sake of convenience. It is

$$\dot{x} = f(t, x) \tag{4.2.1}$$

where $f \in C^0(\mathbf{R} \times X, \mathbf{R}^n)$, $f'_x \in (\mathbf{R} \times X, \mathbf{R}^{n^2})$, X being an open and connected subset of $\mathbf{R}^n$, and it will be assumed that there exists a $T > 0$ such that for every $t \in \mathbf{R}$, $x \in X$

$$f(t + T, x) = f(t, x),$$

i.e. f is periodic in t with (not necessarily the least) period T. Assume that (4.2.1) has a non-constant periodic solution $p : \mathbf{R} \to X$ with period

T. Performing the coordinate transformation $z = x - p(t)$, system (4.2.1) assumes the form

$$\dot{z} = f(t, p(t) + z) - f(t, p(t)),$$

or expanding

$$\dot{z} = f'_x(t, p(t))z + o(|z|). \tag{4.2.2}$$

The linearization of system (4.2.1) at the solution p or, in other words, the variational system with respect to the solution p (cf. (1.1.5)) is

$$\dot{y} = A(t)y \tag{4.2.3}$$

where $A(t) := f'_x(t, p(t))$. The assumptions imply that A is a T-periodic continuous matrix function, and that $o(|z|)/|z|$ in (4.2.2) tends to zero as $z \to 0$ uniformly in t. In many cases one may determine the stability of the periodic solution p of the non-linear system (4.2.1) by this linearization, reducing the problem to the stability of system (4.2.3). The basic tool is the following:

Theorem 4.2.1. *If all the characteristic multipliers (see Section 2.2) of system (4.2.3) are in modulus less than 1, then p is a uniformly asymptotically stable solution of (4.2.1); if (4.2.3) has at least one characteristic multiplier with modulus greater than 1, then p is unstable.*

Proof. This follows from Theorems 1.4.9, 1.4.10 and 1.4.12. By Liapunov's Theorem 2.2.6, system (4.2.3) can be transformed into a system with constant coefficients

$$\dot{v} = Bv$$

by a transformation $y = P(t)v$ where P is a continuously differentiable regular matrix that is periodic with period T or $2T$. The eigenvalues of B are the characteristic exponents of system (4.2.3). If all the characteristic multipliers are in modulus less than 1, then (see (2.2.13)) all the characteristic exponents have negative real parts; if a characteristic multiplier is in modulus greater than 1, then there is a characteristic exponent with positive real part. Now, the coordinate transformation $z = P(t)w$ carries system (4.2.2) into

$$\dot{w} = Bw + o(|w|).$$

The properties of P imply that its norm is bounded, so that the transformation does not alter the stability relations, and the last system is already of the form for which Theorems 1.4.9 and 1.4.10 apply. $\square$

Note that this Theorem enables one to determine the stability in concrete non-critical cases by finding the characteristic multipliers applying the method presented at the end of Section 2.2. This numerical procedure can be applied even when the periodic solution p is given only approximately.

Liapunov's direct method can be applied to stability problems concerning periodic solutions of non-linear periodic systems too. These applications are based on the obvious fact that the Liapunov stability, resp. asymptotic stability of the $z = 0$ solution of system (4.2.2) is equivalent to the Liapunov stability, resp. asymptotic stability of the periodic solution p of (4.2.1).

Theorem 4.2.2. *Let $U \subset \mathbf{R}^n$ be a neighbourhood of the origin $z = 0$ such that for all $t \in [0, T]$ the relation $x - p(t) \in U$ implies $x \in X$, and $V \in C^1(U, \mathbf{R}_+)$ be a positive definite function; if*

$$< \operatorname{grad} V(x - p(t)), f(t, x) - f(t, p(t)) > \ \leq 0 \qquad (4.2.4)$$

for all $t \in [0, T]$ and $x - p(t) \in U$, then the T periodic solution p of system (4.2.1) is stable in the Liapunov sense.

Proof. The derivative of the positive definite function $V(z)$ with respect to system (4.2.2) is

$$\dot{V}_{(4.2.2)}(t, z) = < \operatorname{grad} V(z), f(t, p(t) + z) - f(t, p(t)) >, \qquad (4.2.5)$$

and this is negative semidefinite because of (4.2.4). The statement follows then from Liapunov's First Theorem 1.5.1. $\square$

Theorem 4.2.3. *Let $U \subset \mathbf{R}^n$ be as in the previous theorem, and $V \in C^1(U, \mathbf{R}_+)$ a positive definite function; if*

$$< \operatorname{grad} V(x - p(t)), f(t, x) - f(t, p(t)) > \ < 0 \qquad (4.2.6)$$

for all $t \in [0, T]$, $x - p(t) \in U$, $x \neq p(t)$, then the solution p of (4.2.1) is uniformly asymptotically stable.

Proof. The derivative of the positive definite function $V(z)$ (which does not depend on t and therefore tends to zero as z tends to zero uniformly in t) is given by (4.2.5). By our assumptions it is negative for all $t \in [0, T]$, $z \in U$, $z \neq 0$. Introduce the function $W : U \to \mathbf{R}_+$ by

$$-W(z) := \max_{t \in [0, T]} \dot{V}_{(4.2.2)}(t, z).$$

For fixed $z \in U$, $z \neq 0$ the continuous function $\dot{V}_{(4.2.2)}$ assumes its supremum in $[0, T]$, and it is, clearly, negative, i.e. $-W(z) < 0$, $z \neq 0$. Thus, $-W$ is negative definite and, because of the periodicity of $\dot{V}_{(4.2.2)}$ we have $\dot{V}_{(4.2.2)}(t, z) \leq -W(z)$ for all $t \in \mathbf{R}_+$, $z \in U$. The Theorem follows then from Liapunov's Second Theorem 1.5.2. $\square$

Note that the previous Theorem remains valid even if condition (4.2.6) is replaced by the following milder one: there exists a $t_0 \in [0, T]$ such that in a neighbourhood of $p(t_0)$ we have

$$< \operatorname{grad} V(x - p(t_0)), f(t_0, x) - f(t_0, p(t_0)) > \; < 0,$$

and (4.2.4) holds for $t \in [0, T]$, $x - p(t) \in U$.

In the previous two Theorems Liapunov functions depending on x only were applied in order to establish stability conditions for a periodic solution. We are going to prove now the Barbashin-Krasovskij Theorem for periodic systems (cf. Theorem 1.5.5 and see Krasovskij [1963] and Rouche-Habets-Laloy [1977]). This Theorem concerns the stability of an equilibrium point of a periodic system and applies periodic Liapunov functions. Consider the system

$$\dot{z} = g(t, z) \tag{4.2.7}$$

where $g, g_z' \in C^0(\mathbf{R} \times X)$, X being an open and connected subset of $\mathbf{R}^n$ containing the origin. We assume that a $T > 0$ exists for which

$$g(t + T, z) = g(t, z) \quad \text{for} \quad t \in \mathbf{R}, \quad z \in X,$$

and that $g(t, 0) \equiv 0$, i.e. $z = 0$ is a constant solution of (4.2.7).

Note that system (4.2.2) satisfies all these conditions, and the stability of its equilibrium point $z = 0$ is equivalent to the stability of the periodic solution p of system (4.2.1), so that the following Theorem can be applied in establishing the stability of a general (non-constant) periodic solution of a periodic system.

Theorem 4.2.4 *(Barbashin-Krasovskij Theorem). Assume that a function $V \in C^1(\mathbf{R} \times X, \mathbf{R})$ exists for which for all $t \in \mathbf{R}$ and $z \in X$*

(i)   *there exists a $T > 0$ such that $V(t + T, z) = V(t, z)$;*

(ii) *there exists a function $f \in \mathcal{H}$ (see Definition 1.5.2) such that $V(t, z) \geq h(|z|)$, and $V(t, 0) \equiv 0$;*

(iii) $\dot{V}_{(4.2.7)}(t, z) \leq 0$;

(iv) *the set $M = \{(t, z) \in \mathbf{R} \times X : \dot{V}_{(4.2.7)}(t, z) = 0\}$ does not contain a complete "positive integral curve," i.e. a set of the form $\{(t, \varphi(t, t_0, z^0)) \in \mathbf{R} \times X : t \geq t_0\}$ where φ is the solution of (4.2.7) with initial values $t_0, z^0 \neq 0$ apart from the trivial one $(t, 0)$.*

Let $B(0, \rho)$ be a ball around the origin $z = 0$ with radius $\rho > 0$ such that $\overline{B}(0, \rho) \subset X$, and for any $t \in \mathbf{R}$, set

$$U(t, \rho) := \{z \in X : V(t, z) \leq h(\rho)\};$$

then the origin $z = 0$ is uniformly asymptotically stable, and its basin $A(t)$ (see (1.4.6)) contains $U(t, \rho)$.

Proof. Because of (ii) and (iii), Liapunov's First Theorem 1.5.1 and Theorem 1.4.12 imply that $z = 0$ is uniformly stable. Let $t_0 \in \mathbf{R}$ be chosen arbitrarily, and $z^0 \in U(t_0, \rho)$. Then because of (ii), $|z^0| < \rho$, and as V is non-increasing along solutions, $|\varphi(t, t_0, z^0)| \leq \rho$ in the whole future, i.e. φ is defined on $[t_0, \infty)$.

We are going to show that $\varphi(t, t_0, z^0) \to 0$ as $t \to \infty$. Because of uniform stability it is sufficient to prove that for arbitrary $\delta > 0$ there exists a $t_1 > t_0$ such that $|\varphi(t_1, t_0, z^0)| < \delta$. Since, if this is known, then for arbitrary $\varepsilon > 0$ a $\delta(\varepsilon) > 0$ exists such that $|\varphi(t_1, t_0, z^0)| < \delta(\varepsilon)$ implies $|\varphi(t, t_1, \varphi(t_1, t_0, z^0))| < \varepsilon$ for all $t \geq t_1$ but $\varphi(t, t_1, \varphi(t_1, t_0, z^0)) \equiv \varphi(t, t_0, z^0)$. Assume, contrary to the statement, that a $\delta > 0$ exists such that

$$\delta \leq |\varphi(t, t_0, z^0)| \leq \rho \quad \text{for} \quad t \leq t_0. \tag{4.2.8}$$

Consider the sequence $z^k := \varphi(t_0 + kT, t_0, z^0)$ $(k = 1, 2, ...)$. This, clearly, has a convergent subsequence that will be denoted again by z^k $(k = 1, 2, ...)$. We have that $z^k \to \bar{z}$, as $k \to 0$, and $\delta \leq |\bar{z}| \leq \rho$. Now, $V(t, \varphi(t, t_0, z^0))$ is positive and non-increasing, i.e. it has a limit as t tends to infinity. Because of (i) and the continuity of V

$$\lim_{t \to \infty} V(t, \varphi(t, t_0, z^0)) = \lim_{k \to \infty} V(t_0 + kT, \varphi(t_0 + kT, t_0, z^0))$$
$$= \lim_{k \to \infty} V(t_0, z^k) = V(t_0, \bar{z}). \tag{4.2.9}$$

Consider now the solution $\varphi(t, t_0, \bar{z})$. Because of (iv) there is a $t_2 > t_0$ such that $\dot{V}_{(4.2.7)}(t_2, \varphi(t_2, t_0, \bar{z})) < 0$ and as a consequence

$$V(t_2, \varphi(t_2, t_0, \bar{z})) \neq V(t_0, \bar{z}). \tag{4.2.10}$$

It is easy to see that if a function φ is a solution of a T-periodic system like (4.2.7), then the function ψ defined by $\psi(t) : \varphi(t + T)$ is also a solution. Further, the solution $\varphi(t + T, t_0 + T, z^k)$ of (4.2.7) assumes the initial value z^k at $t = t_0$, so that

$$\varphi(t + T, t_0 + T, z^k) \equiv \varphi(t, t_0, z^k).$$

Hence,

$$\varphi(t_2, t_0, z^k) = \varphi(t_2 + kT, t_0 + kT, z^k)$$
$$= \varphi(t_2 + kT, t_0 + kT, \varphi(t_0 + kT, t_0, z^0))$$
$$= \varphi(t_2 + kT, t_0, z^0) \tag{4.2.11}$$

where in the last step we used the fact that the solution $\varphi(t + kT, t_0 + kT, \varphi(t_0 + kT, t_0, z^0))$ assumes the value $\varphi(t_0 + kT, t_0, z^0)$ at $t = t_0$, so that it is identical to $\varphi(t + kT, t_0, z^0)$.

Because of (4.2.9), (4.2.11) and the periodicity of V we have

$$
\begin{aligned}
V(t_0, \bar{z}) &= \lim_{k \to \infty} V(t_2 + kT, \varphi(t_2 + kT, t_0, z^0)) \\
&= \lim_{k \to \infty} V(t_2, \varphi(t_2, t_0, z^k)) \\
&= V(t_2, \varphi(t_2, t_0, \bar{z}))
\end{aligned}
$$

and this contradicts (4.2.10). We have proved that $U(t, \rho) \subset A(t)$: the basin of $z = 0$ at t. But since $V(t, z) \to 0$ as $z \to 0$ a $\delta(t) > 0$ exists such that if $|z| < \delta(t)$, then $V(t, z) < h(\rho)$, i.e. $U(t, \rho)$ contains a neighbourhood of the origin $z = 0$, and this implies that $z = 0$ is asymptotically stable. Uniform asymptotic stability follows from Theorem 1.4.12. □

Corollary 4.2.5 *If the conditions of the previous Theorem hold with* $X = \mathbf{R}^n$, *and the function h in (ii) is such that* $h(r) \to \infty$, *as* $r \to \infty$, *then* $z = 0$ *is globally asymptotically stable.*

Proof. Indeed, because of the periodicity of V in t for arbitrary $z^0 \in \mathbf{R}^n$ we may find a $\rho > 0$ such that $V(t, z^0) \leq h(\rho)$, i.e. $z^0 \in U(t, \rho)$ for all t, and this implies that with arbitrary t_0 the solution $\varphi(t, t_0, z^0) \to 0$ as $t \to \infty$.
□

We note that Theorem 4.2.4 is not valid for arbitrary systems that are not periodic in t. This is shown by the example $\dot{z} = -b(t)z$ due to Matrosov [1962] where $z \in \mathbf{R}$, $b \in C^0(\mathbf{R}, \mathbf{R})$, $b(t) > 0$, $t \in \mathbf{R}$, and $\int_0^\infty b(t)dt$ is convergent. The solutions are

$$
z(t, 0, z^0) = z^0 \exp(- \int_0^t b(\tau)d\tau),
$$

and because of the convergence of the improper integral, these do not tend to zero as t tends to infinity ($z^0 \neq 0$). On the other hand, if b is a continuous positive *periodic* function, then, clearly, $\int_0^\infty b(t)dt = \infty$ and all solutions tend to zero. In both cases the Liapunov function $V(z) = z^2$ satisfies every condition of Theorem 4.2.4.

As it was mentioned before, the Barbashin-Krasovskij Theorem may be applied to determine the asymptotic stability of the non-constant periodic solution p of system (4.2.1) if the latter is transformed into (4.2.2). However, we draw the attention to the fact that this way cannot be followed in the case of an *autonomous system*. For, if $\dot{x} = f(x)$ where $f \in C^1(X, \mathbf{R}^n)$ has a non-constant periodic solution p with period $T > 0$, and the autonomous system is transformed into the T-periodic system

$$
\dot{z} = f(p(t) + z) - f(p(t)) \tag{4.2.12}
$$

then the $z = 0$ solution of the latter *can never be asymptotically stable*. This is so because (4.2.12) has a one parameter family of periodic

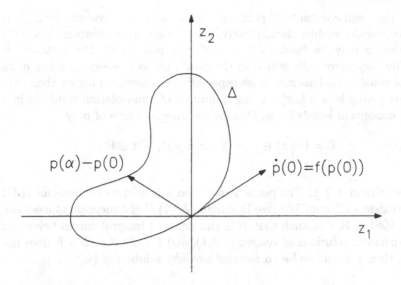

FIGURE 4.2.1. The curve $\Delta = \{z^0 \in \mathbf{R}^2 : z^0 = p(\alpha) - p(0), \alpha \in [0, T]\}$ is the set of initial values that give rise to T-periodic solutions of (4.2.12): if $z^0 \in \Delta$, then $\varphi(t, 0, z^0)$ is a non-constant periodic solution.

solutions in which there are non-constant periodic solutions of arbitrary small amplitude around the origin. Indeed, it is easy to see that for arbitrary $\alpha \in \mathbf{R}$ the T-periodic function $q_\alpha(t) : p(t + \alpha) - p(t)$ is a solution of (4.2.12), and because of the uniform continuity of p on the interval $[0, T]$ for arbitrary $\varepsilon > 0$, there exists an $\alpha_0 > 0$ such that for $|\alpha| < \alpha_0$ we have $|q_\alpha(t)| < \varepsilon$ for all $t \in \mathbf{R}$. This means that for arbitrary $t_0 \in \mathbf{R}$ we have initial values $q_\alpha(t_0) = p(t_0 + \alpha) - p(t_0)$ that are in modulus less than ε, whereas the corresponding solution $q_\alpha(t)$ does not tend to zero as t tends to infinity. Figure 4.2.1 shows the possible set Δ of initial values $p(\alpha) - p(0)$, $\alpha \in [0, T]$ that give rise to the periodic solutions $q_\alpha(t) = \varphi(t, 0, p(\alpha) - p(0)) = p(t + \alpha) - p(t)$ of (4.2.12). The curve Δ is congruent with the path $\gamma = \{x \in \mathbf{R}^n : x = p(t), t \in [0, T]\}$. Clearly, $(p(\alpha) - p(0))/\alpha \to \dot{p}(0)$ as $\alpha \to +0$, $(p(\alpha) - p(0))/(T - \alpha) \to -\dot{p}(0)$ as $\alpha \to T - 0$. We shall return to this point in the next chapter where we introduce new stability definitions applicable to non-constant periodic solutions of autonomous systems.

In the rest of this Section we treat the problem of "isolation" of periodic solutions. The term was used already by H. Poincaré [1899] in a vague sense expressing the property that there is no other periodic solution in some neighbourhood of the given one. Apart from a very much generalized treatment by D.C. Lewis [1961] the concept was rigorously studied in Farkas [1975a]. We are following the last paper.

If the non-constant T-periodic solution p of system (4.2.1) is asymptotically stable, then, clearly, there is no other solution "nearby"; however, p may be "isolated" even if it is not stable. For instance, it may be "asymptotically stable in the past", i.e. as $t \to -\infty$ or other more complicated situations may arise especially in dimensions higher than two. We are giving here a fairly strong definition of non-isolation resulting in a weak concept of isolatedness. Denote the integral curve of p by

$$\Gamma = \{(t,x) \in \mathbf{R} \times X : x = p(t), \quad t \in \mathbf{R}\}.$$

Definition 4.2.1. The periodic solution p is said to be *non-isolated* if there exists a C^1-manifold (see Definition A3.1) P of dimension at least two embedded in $\mathbf{R} \times X$ such that P is the union of integral curves belonging to T-periodic solutions of system (4.2.1), and $\Gamma \subset P$. If such a P does not exist, then p is said to be an *isolated periodic solution* of (4.2.1).

We shall give a sufficient criterion that guarantees the isolatedness of a periodic solution in the sense of the previous Definition. We need the following

Lemma 4.2.6. *If p is a non-isolated periodic solution with period T of system (4.2.1), and the dimension of the differentiable manifold P in Definition 4.2.1 is $\dim P = m$, $2 \le m \le n$, then there exists an open and connected set $Q \subset \mathbf{R}^{m-1}$ and a function $q : \mathbf{R} \times Q \to X$ of class C^1 such that*

$$\{(t,x) \in \mathbf{R} \times X : x = q(t,\alpha), \quad t \in \mathbf{R}, \quad \alpha \in Q\} \subset P,$$

for every fixed $\alpha = (\alpha_2, ..., \alpha_m) \in Q$ the function $q(\bullet, \alpha)$ is a T-periodic solution of (4.2.1); there exists an $\alpha^0 \in Q$ for which $p(t) \equiv q(t, \alpha^0)$, and

$$\operatorname{rank} q'_\alpha(0, \alpha^0) = m - 1. \tag{4.2.13}$$

(Here q'_α denotes the $n \times (m-1)$ matrix with elements $q'_{i\alpha_k}$ $(i = 1, 2, ..., n; \; k = 2, ..., m)$).

Proof. Set $R_0^n = \{(0,x) \in \mathbf{R} \times \mathbf{R}^n : x \in \mathbf{R}^n\}$. The periodic solution p is defined at $t = 0$, and $(0, p(0)) \in R_0^n$. Denote one of the coordinate neighbourhoods covering P and containing the point $(0, p(0))$ by P_0, let $t = \eta_0(\tilde{\alpha})$, $x = \eta(\tilde{\alpha})$ be a regular representation of P_0 where $\tilde{\alpha} = (\alpha_1, \alpha_2, ..., \alpha_m) \in \mathbf{R}^m$, $\eta_0 \in C^1(\mathbf{R}^m, \mathbf{R})$, $\eta \in C^1(\mathbf{R}^m, \mathbf{R}^n)$, and let $\eta_0(\tilde{\alpha}^0) = 0$, $\eta(\tilde{\alpha}^0) = p(0)$, $\tilde{\alpha}^0 = (\alpha_1^0, \alpha_2^0, ..., \alpha_m^0)$. Then

$$\operatorname{rank} \begin{pmatrix} \eta'_{0\alpha_1}(\tilde{\alpha}^0) & \eta'_{0\alpha_2}(\tilde{\alpha}^0) & \cdots & \eta'_{0\alpha_m}(\tilde{\alpha}^0) \\ \eta'_{\alpha_1}(\tilde{\alpha}^0) & \eta'_{\alpha_2}(\tilde{\alpha}^0) & \cdots & \eta'_{\alpha_m}(\tilde{\alpha}^0) \end{pmatrix} = m \tag{4.2.14}$$

where the matrix is of type $(n+1) \times m$, $\eta'_{\alpha_i}(\tilde{\alpha}^0)$ being an n dimensional column vector $(i = 1, 2, ..., m)$. The tangent vector of the integral curve Γ at $(0, p(0))$ is $(1, \dot{p}(0))$. Since this vector is an element of the tangent space to P_0 at $(0, p(0))$, and this tangent space is spanned by the column vectors of the matrix in (4.2.13), at least one of the elements in the first row of this matrix is non-zero. We may assume without loss of generality that

$$\eta'_{0\alpha_1}(\tilde{\alpha}^0) \neq 0.$$

Because of this the Implicit Function Theorem can be applied to $\eta_0(\tilde{\alpha}) = 0$ taking into account that $\eta_0(\tilde{\alpha}^0) = 0$. Thus, an open and connected set $\hat{Q} \subset \mathbf{R}^{m-1}$ and a function $\hat{\alpha}_1 \in C^1(\hat{Q}, \mathbf{R})$ exists such that $\alpha^0 = (\alpha_2^0, ..., \alpha_m^0) \in \hat{Q}$, $\hat{\alpha}_1(\alpha^0) = \alpha_1^0$ and $\eta_0(\hat{\alpha}_1(\alpha), \alpha) \equiv 0$. Clearly,

$$\{(0, x) \in \mathbf{R} \times \mathbf{R}^n : x = \eta(\hat{\alpha}_1(\alpha), \alpha), \ \alpha \in \hat{Q}\} \subset \mathbf{R}_0^n \cap P_0,$$

and $\eta(\hat{\alpha}_1(\alpha^0), \alpha^0) = p(0)$. Denoting the composite function by $\hat{\eta}(\alpha) := \eta(\hat{\alpha}_1(\alpha), \alpha)$ we have

$$\hat{\eta}'_{\alpha_i}(\alpha^0) = \eta'_{\alpha_1}(\hat{\alpha}_1(\alpha^0), \alpha^0)\frac{\partial \hat{\alpha}_1(\alpha^0)}{\partial \alpha_i} + \eta'_{\alpha_i}(\hat{\alpha}_1(\alpha^0), \alpha^0) \quad (i = 2, ..., m),$$

and differentiating the identity $\eta_0(\hat{\alpha}_1(\alpha), \alpha) \equiv 0$ we obtain

$$0 = \eta'_{0\alpha_1}(\hat{\alpha}_1(\alpha^0), \alpha^0)\frac{\partial \hat{\alpha}_1(\alpha^0)}{\partial \alpha_i} + \eta'_{0\alpha_i}(\hat{\alpha}_1(\alpha^0), \alpha^0) \quad (i = 2, ..., m).$$

The last two equalities imply that the matrix

$$\begin{pmatrix} \eta'_{0\alpha_1}(\tilde{\alpha}^0) & 0 & \cdots & 0 \\ \eta'_{\alpha_1}(\tilde{\alpha}^0) & \hat{\eta}'_{\alpha_2}(\alpha^0) & \cdots & \hat{\eta}'_{\alpha_m}(\alpha^0) \end{pmatrix}$$

is obtained from the matrix in (4.2.14) by elementary column operations, so that the rank of this one is also m. Hence, the last $m - 1$ column vectors, and, as a consequence, the vectors $\hat{\eta}'_{\alpha_2}(\alpha^0), ..., \hat{\eta}'_{\alpha_m}(\alpha^0)$ are linearly independent, i.e.

$$\text{rank } \hat{\eta}'_{\alpha}(\alpha^0) = m - 1. \tag{4.2.15}$$

Because of the continuity of the derivative matrix $\hat{\eta}'_{\alpha}$ the point α^0 has a neighbourhood $Q \subset \hat{Q}$ such that the set

$$\{(0, \hat{\eta}(\alpha)) \in \mathbf{R} \times \mathbf{R}^n : \alpha \in Q\}$$

is an $(m - 1)$ dimensional coordinate neighbourhood of the point $(0, p(0))$ contained in $\mathbf{R}_0^n \cap P_0$.

Consider the solution $\varphi(t, 0, x^0)$ of system (4.2.1); the function q defined by

$$q(t, \alpha) := \varphi(t, 0, \hat{\eta}(\alpha)), \qquad t \in \mathbf{R}, \qquad \alpha \in Q$$

satisfies the requirements of the Lemma. It is of the C^1 class. For every $\alpha \in Q$ we have $(0, \hat{\eta}(\alpha)) \in P_0$; therefore, $q(\bullet, \alpha)$ is a T-periodic solution of (4.2.1), and $q(t, \alpha) \in P$ for $t \in \mathbf{R}$. Clearly, $q(t, \alpha^0) = p(t)$. Finally, in order to prove that the rank of q'_α at $t = 0$, $\alpha = \alpha^0$ is $m - 1$ we have to recall that $\varphi'_{x^0}(t, 0, p(0))$ is a fundamental matrix of the variational system

$$\dot{y} = f'_x(t, p(t))y \tag{4.2.16}$$

and $\varphi'_{x^0}(0, 0, p(0)) = I$, the unit matrix (see (1.1.5) and what follows). Applying (4.2.15) we obtain

$$\begin{aligned}
\operatorname{rank} q'_\alpha(0, \alpha^0) &= \operatorname{rank}\left(\varphi'_{x^0}(0, 0, \hat{\eta}(\alpha^0))\hat{\eta}'_\alpha(\alpha^0)\right) \\
&= \operatorname{rank}\left(\varphi'_{x^0}(0, 0, p(0))\hat{\eta}'_\alpha(\alpha^0)\right) \\
&= \operatorname{rank} \hat{\eta}'_\alpha(\alpha^0) = m - 1. \square
\end{aligned}$$

Theorem 4.2.7. *If p is a T-periodic solution of system (4.2.1) and the number 1 is not a characteristic multiplier of the variational system (4.2.16), then p is an isolated periodic solution.*

Proof. Assume that p is non-isolated. Then there exists a function q satisfying the requirements of the Lemma, so that the identity

$$\dot{q}(t, \alpha) \equiv f(t, q(t, \alpha))$$

holds. Differentiate this identity by α_i $(i = 2, ..., m)$:

$$\dot{q}'_{\alpha_i}(t, \alpha) \equiv f'_x(t, q(t, \alpha))q'_{\alpha_i}(t, \alpha) \qquad (i = 2, ..., m).$$

At $\alpha = \alpha^0$ this is

$$\dot{q}'_{\alpha_i}(t, \alpha^0) \equiv f'_x(t, p(t))q'_{\alpha_i}(t, \alpha^0) \qquad (i = 2, ..., m).$$

This means that the T-periodic function $q'_{\alpha_i}(\bullet, \alpha^0)$ is a solution of the variational system (4.2.16) $(i = 2, ..., m)$. Because of (4.2.13) this means that the variational system has $m - 1 \geq 1$ linearly independent (nontrivial) periodic solutions at least, so that 1 is a characteristic multiplier (see Corollaries 2.2.3-2.2.4). $\square$

Two remarks are needed about the limitations of Definition 4.2.1 and Theorem 4.2.7. First, the Definition and the Theorem takes into consideration only those solutions that have T as a period (not necessarily the least positive). According to the Definition a T-periodic solution p is isolated even if there are, e.g., $2T$-periodic solutions arbitrary near it. Also the case when there is but a countable set of T-periodic solutions in an arbitrary neighbourhood of p falls into the isolated class.

Secondly, the Definition and Theorem 4.2.7 do not apply to autonomous system, since a non-constant periodic solution of an autonomous system can never be isolated in this sense. This is so because with $p(t)$, the function $p(t + \alpha)$ is also a T-periodic solution for arbitrary $\alpha \in \mathbf{R}$ (see (1.3.3)), and the set

$$\{(t, p(t + \alpha)) \in \mathbf{R} \times \mathbf{R}^n : t \in \mathbf{R}, \quad \alpha \in \mathbf{R}\}$$

is a two dimensional C^1 manifold containing the integral curve of the solution p and consisting of integral curves of T-periodic solutions (see Figure 1.3.1). The isolation problem in the autonomous case will be treated in the next chapter where other suitable definitions will be introduced.

4.3 Periodically Forced Liénard and Duffing Equations

In this Section we apply fixed point theorems and degree theory to prove the existence of periodic solutions of two dimensional periodic systems. Liénard's and Duffing's equation (cf. Sections 3.2 and 3.3) with a periodic forcing term will be treated.

First, a fairly general second order differential equation with bounded forcing term will be considered that covers both the Liénard and the Duffing equation under some conditions. A closed curve will be constructed in the phase plane such that all solutions that have a common point with this curve at any time cross it inward.

Consider the second order differential equation

$$\ddot{x} + f(x, \dot{x})\dot{x} + g(x) = b(t) \tag{4.3.1}$$

where $f \in C^1(\mathbf{R}^2, \mathbf{R})$, $g \in C^1(\mathbf{R}, \mathbf{R})$, $b \in C^0(\mathbf{R}, \mathbf{R})$, and the last function is supposed to be bounded. The Cauchy normal form of (4.3.1) is the system

$$\dot{x} = y \qquad \dot{y} = -f(x, y)y - g(x) + b(t). \tag{4.3.2}$$

The following Theorem is due to N. Levinson [1943]. We follow the presentation of L. Cesari [1963], filling in some gaps there.

Theorem 4.3.1. *Assume that there are constants $a, m, M > 0$ such that for $|x| \geq a$ and $|y| \geq a$ there holds $f(x, y) > m$, for $(x, y) \in \mathbf{R}^2$ we have $f(x, y) \geq -M$, for $|x| \geq a$ there holds $xg(x) > 0$, and in $(-\infty, -a)$ and in (a, ∞) the function g is monotone increasing, assume further that $|g(x)| \to \infty$ as $|x| \to \infty$, and denoting the integral function of g by $G(x) := \int_0^x g(u)du$, that $g(x)/G(x) \to 0$ as $|x| \to \infty$; then there exists a closed curve C in the x, y plane containing the square*

$$\{(x, y) \in \mathbf{R}^2 : |x| \leq a, \quad |y| \leq a\}$$

in its interior such that each path of (4.3.2) having a common point with C crosses it from the outside into the inside.

Proof. We shall make use of the "energy function"

$$V(x,y) = y^2/2 + G(x) \tag{4.3.3}$$

(which would be the energy integral if f and b were equal to zero). For $|x|$ sufficiently large, i.e. outside an interval containing $x = 0$, the function G is positive, increasing with the increase of $|x|$, and $G(x) \to \infty$ as $|x| \to \infty$. This implies that for sufficiently large positive constants c the level curves $V(x,y) = c$ are closed, containing the origin $(x,y) = (0,0)$ in their interior, and for such large $0 < c_1 < c_2$ the curve $V(x,y) = c_2$ contains the curve $V(x,y) = c_1$ in its interior. The derivative of the function V with respect to system (4.3.2) is

$$\dot{V}(t,x,y) := \dot{V}_{(4.3.2)}(t,x,y) = -f(x,y)y^2 + b(t)y. \tag{4.3.4}$$

Denote an upper bound of $|b|$ by B, i.e. let $B \geq |b(t)|$ for $t \in \mathbf{R}$, choose an $a > 0$ so large that

$$ma > 2B \tag{4.3.5}$$

and denote

$$g_a := \max |g(x)|, \qquad |x| \leq a.$$

We shall construct the closed curve $C = P_1 P_2 \cdots P_{13}$ from thirteen parts P_i denoting the vertex joining the i-th and the $(i+1)$-st part. The notations $P_i = (x_i, y_i)$, $V_i = V(x_i, y_i)$ $(i = 1, 2, ..., 13)$ will be used (see Figure 4.3.1). We shall choose an $\overline{x} > a$ such that

$$|g(x)| > 2(Ma + B) \quad \text{for} \quad |x| > \overline{x}. \tag{4.3.6}$$

Consider the curve $y = -(g(x) + B)/m$ and choose a point with abscissa $x_4 > \overline{x}$ on this curve, i.e. $P_4 = (x_4, y_4)$ where $y_4 = -(g(x_4) + B)/m$. Later we shall dispose of $\overline{x}$ and x_4, making them sufficiently large to meet new requirements.

For $P_4 P_3$ we choose the arc on the level curve $V(x,y) = V_4$ that ends in the point $P_3 = (x_3, y_3)$ where $y_3 = -a$. On $P_4 P_3$ we have $a < \overline{x} < x_4 \leq x$, $y \leq -a$, and applying (4.3.4) and (4.3.5)

$$\begin{aligned} \dot{V}(t,x,y) \quad &< \quad -my^2 + B|y| = my^2(B/(m|y|)) - 1) \\ &\leq \quad my^2(B/(ma) - 1) < -my^2/2 < 0, \end{aligned}$$

so that any solution crossing $P_4 P_3$ at any time t goes inward. We have also $V_4 - V_3 = 0$.

For $P_2 P_3$ the straight line $x = x_3$ is chosen with end point $P_2 = (x_2, y_2)$ on the x-axis, i.e. $y_2 = 0$. Since $\dot{x} = y < 0$ on this segment, the solutions cross $P_2 P_3$ inward. We have also $x_2 = x_3$, and as a consequence $V_3 - V_2 = a^2/2$.

For $P_1 P_2$ the portion of the curve

$$y^2/2 + G(x) - (Ma + B)x = G(x_2) - (Ma + B)x_2 \tag{4.3.7}$$

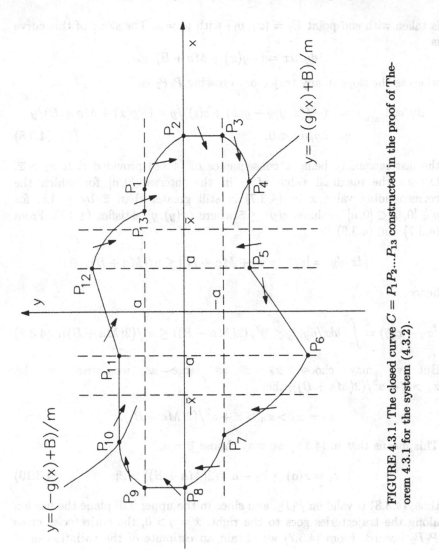

FIGURE 4.3.1. The closed curve $C = P_1 P_2 ... P_{13}$ constructed in the proof of Theorem 4.3.1 for the system (4.3.2).

is taken with end point $P_1 = (x_1, y_1)$ with $y_1 = a$. The slope of this curve is

$$dy/dx = (-g(x) + Ma + B)/y,$$

whereas the slope of any trajectory crossing $P_1 P_2$ is

$$
\begin{aligned}
dy/dx\big|_{\text{path}} &= (-f(x,y)y - g(x) + b(t))/y < (-g(x) + Ma + B)/y \\
&= dy/dx < 0, \hspace{4cm} (4.3.8)
\end{aligned}
$$

the last inequality being a consequence of (4.3.6) provided that $x_1 > \bar{x}$. Denote the maximal value of y in the interval $[0, a]$ for which the corresponding value x in (4.3.7) is still greater than $\bar{x}$ by $\bar{y}$, i.e. for $y \in [0, \bar{y}] \subset [0, a]$ we have $x(y) \geq \bar{x}$ where $(x(y), y)$ satisfies (4.3.7). From (4.3.7) and (4.3.6)

$$|dx/dy| = |y/(-g(x) + Ma + B)| \leq y/(Ma + B);$$

hence,

$$x_2 - x(\bar{y}) = \int_0^{\bar{y}} |dx/dy|\, dy \leq \bar{y}^2/(2(Ma+B)) \leq a^2/(2(Ma+B)). \quad (4.3.9)$$

But we may choose $x_4 > \bar{x}$ as large as we wish, so let $x_4 > \bar{x} + a^2/(2(Ma + B))$; then,

$$x_2 = x_3 > x_4 > \bar{x} + a^2/(2(Ma + B)).$$

This means that in (4.3.9) we may choose $\bar{y} = a$:

$$x_1 = x(a) \geq x_2 - a^2/(2(Ma + B)) > \bar{x}; \hspace{2.5cm} (4.3.10)$$

thus, (4.3.8) is valid on $P_1 P_2$, and since in the upper half plane the motion along the trajectories goes to the right, $\dot{x} = y > 0$, the trajectories cross $P_1 P_2$ inward. From (4.3.7) we obtain an estimate of the variation of V along $P_1 P_2$:

$$
\begin{aligned}
V_2 - V_1 &= V(x_2, 0) - V(x_1, a) = G(x_2) - (G(x_1) + a^2/2) \\
&= (Ma + B)(x_2 - x_1) \leq a^2/2
\end{aligned}
$$

where in the last step, (4.3.10) was applied.

Returning to the point P_4 the arc $P_4 P_5$ will be that section of the straight line $y = y_4$ which extends to the point $P_5 = (x_5, y_5)$ with $x_5 = a$. Thus, $y_5 = y_4$, and on $P_4 P_5$ we have for the solutions

$$
\begin{aligned}
\dot{y}(t) &= -f(x, y_4)y_4 - g(x) + b(t) > -g(x) - my_4 - B \\
&= -g(x) - B + g(x_4) + B = g(x_4) - g(x) \geq 0
\end{aligned}
$$

because g is monotone increasing. Hence, all solutions crossing P_4P_5 move inward, and we have

$$V_5 - V_4 = V(a, y_4) - V(x_4, y_4) = G(a) - G(x_4).$$

For P_5P_6 we take the straight line of slope $2M$ extending to $P_6 = (x_6, y_6)$ with $x_6 = -a$, i.e. $y_6 = y_5 - 4Ma$. The slope of any path crossing P_5P_6 is

$$\begin{aligned}
dy/dx &= (-f(x,y)y - g(x) + b(t))/y \\
&< M - (g_a + B)/y_5 = M - (g_a + B)/y_4 \\
&= M + m(g_a + B)/(g(x_4) + B).
\end{aligned}$$

Now assume that $\bar{x}$ has been chosen so large that

$$|g(x)| > 2m(g_a + B)/M \qquad \text{for} \qquad |x| > \bar{x}. \qquad (4.3.11)$$

Then continuing the previous estimate,

$$dy/dx < M + (M/2)g(x_4)/(g(x_4) + B) < 3M/2 < 2M,$$

i.e. the slope is less than the slope of the straight line P_5P_6, and since the motion in the lower half plane is to the left, $\dot{x} = y < 0$, all solutions crossing P_5P_6 move inward. We have also

$$\begin{aligned}
V_6 - V_5 &= y_6^2 + G(-a) - y_5^2/2 - G(a) \\
&= (y_5 + y_6)(-4Ma)/2 + G(-a) - G(a) \\
&= -4May_4 + 8M^2a^2 + G(-a) - G(a).
\end{aligned}$$

For P_6P_7 the level curve $V(x,y) = V_6$ is chosen up to the point $P_7 = (x_7, y_7)$ with $y_7 = -a$. Proceeding analogously to the P_3P_4 case we may prove that the trajectories cross this arc inward. We have also $V_7 - V_6 = 0$.

For P_7P_8 the curve

$$y^2/2 + G(x) + (Ma + B)x = a^2/2 + G(x_7) + (Ma + E)x_7 \qquad (4.3.12)$$

is chosen up to the point $P_8 = (x_8, 0)$. We may proceed as in the P_1P_2 case proving that all trajectories that cross this arc go inward and that $V_8 - V_7 < a^2/2$ provided that $x_7 < \bar{x}$. If this is not the case, we increase x_4 until this requirement is met. If the function V is taken along the curve (4.3.12), we get

$$V(x,y)\big|_{y^2/2 = -G(x) - (Ma+B)x+c} = -(Ma + B)x + c$$

where the constant on the right-hand side has been denoted by c. Differentiating with respect to x,

$$\frac{d}{dx}V(x,y)\big|_{(4.3.12)} = -(Ma + B)$$

is obtained. This means that V is decreasing along this curve with the increase of x, i.e. $V_8 - V_7 > 0$.

For $P_8 P_9$ the straight line $x = x_8$ is chosen up to the point $P_9 = (x_9, a)$. The case is similar to the case $P_2 P_3$. We conclude that every path crosses inward, and that $V_9 - V_8 = a^2/2$.

For $P_9 P_{10}$ the level curve $V(x, y) = V_9$ is chosen up to the point where it meets the curve

$$y = (-g(x) + B)/m \qquad (4.3.13)$$

in a point $x \leq -\bar{x}$. If such an intersection does not take place, x_4 is increased again until it does. The curve (4.3.13) is the graph of a function which is positive, monotone decreasing in $(-\infty, -a]$ and tending to infinity as $x \to -\infty$, so that $P_{10} = (x_{10}, y_{10})$ where $y_{10} = (-g(x_{10}) + B)/m$, $x_9 < x_{10} \leq \bar{x}$. We may proceed again as in the case $P_3 P_4$, obtaining that every path crosses inward, and that $V_{10} - V_9 = 0$.

For $P_{10} P_{11}$ we take the straight line $y = y_{10}$ ending in the point $P_{11} = (-a, y_{11})$ where $y_{11} = y_{10}$. On this straight line segment we have for the solutions

$$\begin{aligned}\dot{y}(t) &= -f(x, y_{10})y_{10} - g(x) + b(t) \leq -my_{10} - g(x) + B \\ &= -g(x) + B + g(x_{10}) - B = g(x_{10}) - g(x) < 0\end{aligned}$$

because g is monotone increasing. Thus, the trajectories that meet $P_{10} P_{11}$ cross it inward (downward):

$$V_{11} - V_{10} = V(-a, y_{10}) - V(x_{10}, y_{10}) = G(-a) - G(x_{10}).$$

For $P_{11} P_{12}$ the straight line of slope $2M$ is chosen up to the point $P_{12} = (a, y_{12})$. We proceed as in the case $P_5 P_6$ using (4.3.11). The slope of any path crossing $P_{11} P_{12}$ is

$$\begin{aligned}dy/dx &= (-f(x, y)y - g(x) + b(t))/y < M + (g_a + B)/y_{10} \\ &< M + (g_a + B)m/(-g(x_{10}) + B) \\ &< M + (M/2)|g(x_{10})|/(|g(x_{10})| + B) < 3M/2 < 2M,\end{aligned}$$

i.e. the slope is less than the slope of the straight line $P_{11} P_{12}$, and since the motion in the upper half plane goes to the right, $\dot{x} = y > 0$, all trajectories crossing $P_{11} P_{12}$ move inward. We have also

$$V_{12} - V_{11} = 4May_{10} + 8M^2 a^2 + G(a) - G(-a).$$

For $P_{12} P_{13}$ the level curve $V(x, y) = V_{12}$ is chosen to the end point $P_{13} = (x_{13}, y_{13})$ with $y_{13} = a$. Applying (4.3.4) and (4.3.5) again we get that the trajectories cross $P_{12} P_{13}$ inward, and that $V_{13} - V_{12} = 0$.

On the straight line $y = a$ on the segment $P_{13} P_1$ we have for the solutions

$$\begin{aligned}\dot{y} &= -f(x, y)y - g(x) + b(t) < -ma - g(x) + B \\ &< -ma + B < -2B + B = -B < 0\end{aligned}$$

where (4.3.5) was applied. So that if P_{13} is at the left of P_1, then the trajectories cross $P_{13}P_1$ inward.

We are proving now that if $\bar{x}$ is chosen sufficiently large, then P_{13} is at the left of P_1, i.e. $x_{13} < x_1$. Since $y_{13} = y_1 = a$, we have $V_1 - V_{13} = G(x_1) - G(x_{13})$, so that $V_1 > V_{13}$ would imply the inequality to be proven. Combining the estimates of the differences $V_{i+1} - V_i$ $(i = 1, ..., 13)$ we get

$$
\begin{aligned}
V_{13} - V_1 &= \sum_{i=1}^{13} (V_{i+1} - V_i) \\
&\leq 2a^2 + 16M^2a^2 + G(a) + G(-a) \\
&\quad + 4May_{10} - G(x_{10}) - 4May_4 - G(x_4).
\end{aligned}
$$

But $y_4 = -(g(x_4) + B)/m$ and $y_{10} = (-g(x_{10}) + B)/m$, so that

$$
\begin{aligned}
4May_{10} - G(x_{10}) &= 4MaB/m - G(x_{10})(1 + (4Ma/m)g(x_{10})/G(x_{10})) \\
-(4May_4 + G(x_4)) &= 4MaB/m - G(x_4)(1 - (4Ma/m)g(x_4)/G(x_4)).
\end{aligned}
$$

Substituting into the previous estimate yields

$$
\begin{aligned}
V_{13} - V_1 \leq\ & 2a^2 + 16M^2a^2 + 8MaB/m + G(a) + G(-a) \\
& -G(x_4)(1 - (4Ma/m)g(x_4)/G(x_4)) \\
& -G(x_{10})(1 + (4Ma/m)g(x_{10})/G(x_{10})).
\end{aligned}
$$

But $G(x) \to \infty$ as $|x| \to \infty$, $g(x)/G(x) \to 0$ as $|x| \to \infty$, and $x_4 > \bar{x}$, $x_{10} < -\bar{x}$. Finally, choose $\bar{x}$ so large that for $|x| > \bar{x}$ the following inequalities hold:

$$
|g(x)|/G(x) < m/(8Ma),
$$

$$
G(x) > 2a^2 + 16M^2a^2 + 8MaB/m + G(a) + G(-a).
$$

Then the estimate can be continued in the following way:

$$
\begin{aligned}
V_{13} - V_1 <\ & 2a^2 + 16M^2a^2 + 8MaB/m + G(a) + G(-a) \\
& -(G(x_4) + G(x_{10}))/2 < 0.
\end{aligned}
$$

As we have said before, this implies $x_{13} < x_1$. Thus, the curve $C = P_1 P_2 \cdots P_{13}$ constructed above meets the requirements of the Theorem. $\square$

All the portions of the curve C depend continuously on P_4 which is tending to infinity on the curve $y = -(g(x) + B)/m$ as $x_4 \to \infty$. Also, every point of C tends to infinity in this case. This implies

Corollary 4.3.2. *If the conditions of the previous theorem hold, then through every point of the plane x, y sufficiently far from the origin there*

passes a curve C satisfying the requirements of the Theorem.

Theorem 4.3.1 combined with Brouwer's Fixed Point Theorem (Corollary A2.9) makes it is easy to prove the existence of a periodic solution of equation (4.3.1) provided that the forcing term is periodic.

Theorem 4.3.3. *Suppose that the functions f, g and b satisfy the conditions of Theorem 4.3.1 and that b is a non-constant periodic function with period $T > 0$, $b(t + T) = b(t)$, $b(t) \not\equiv const$; then (4.3.1) has a non-constant T-periodic solution.*

Proof. By Theorem 4.3.1 there is a simple closed curve C enclosing the origin such that every path having a common point with C moves into the interior of C. Denote by $\overline{D}$ the union of the interior of C with C; $\overline{D}$ is, obviously, homeomorphic to the complete unit disc $\{(x, y) \in \mathbf{R}^2 : x^2 + y^2 \leq 1\}$. For every point $(x^0, y^0) \in \overline{D}$ consider the solution $\varphi(t, 0, (x^0, y^0))$ of the equivalent system (4.3.2), and define the mapping $F : \overline{D} \to \overline{D}$ by

$$F(x^0, y^0) = \varphi(T, 0, (x^0, y^0)) \quad \text{for} \quad (x^0, y^0) \in \overline{D}.$$

F is, clearly, a continuous mapping of $\overline{D}$ into itself, so that by Corollary A2.9 it has a fixed point $(\overline{x}, \overline{y}) \in \overline{D}$, i.e.

$$\varphi(T, 0, (\overline{x}, \overline{y})) = (\overline{x}, \overline{y}) = \varphi(0, 0, (\overline{x}, \overline{y})).$$

By Lemma 2.2.1, $\varphi(t, 0, (\overline{x}, \overline{y}))$ is a T-periodic solution, and according to what has been said about (4.1.4), it is not constant. $\square$

The conditions of the Theorem cover a wide class of Liénard's equations (see Section 3.2) with a periodic forcing term. For instance *Van der Pol's equation* (3.2.5) *with a periodic forcing term*

$$\ddot{u} + m(u^2 - 1)\dot{u} + u = b(t) \tag{4.3.14}$$

is covered if $m > 0$, and $b \in C^0(\mathbf{R}, \mathbf{R})$, $b(t + T) = b(t)$ for some $T > 0$, $b(t) \neq const$. So that the existence of a non-constant T-periodic solution of (4.3.14) is guaranteed. If $m < 0$, the conditions imposed upon the function f of (4.3.1) are not fulfilled. However, reversing the time by the transformation $\tau = -t$, equation (4.3.14) goes into

$$d^2 u / d\tau^2 - m(u^2 - 1) du / d\tau + u = \tilde{b}(\tau)$$

where $\tilde{b}(\tau) := b(-\tau)$. If $m < 0$, then this equation has all the properties required, and as a consequence, it has a non-constant T-periodic solution. This means, naturally, that (4.3.14) has a non-constant T-periodic solution also when $m < 0$. If $m = 0$, then those possibilities arise which were discussed in connection with the periodically forced harmonic oscillator (2.1.15).

The *damped Duffing equation (3.3.8) with periodic forcing term*

$$\ddot{x} + d\dot{x} + ax + bx^3 = h(t) \qquad (4.3.15)$$

is covered too if $h \in C^0(\mathbf{R}, \mathbf{R})$, $h(t + T) = h(t)$ for some $T > 0$, $h(t) \neq$ const, and $d > 0$, $b > 0$ (this is the *"hard spring case"*; see what has been written preceding (3.3.2) and following (3.3.9)). Thus, in this case, (4.3.15) has a non-constant T-periodic solution.

For the *"soft spring case"* $b < 0$, we quote a theorem from Rouche-Mawhin [1973] without proof. The proof makes use of the Brouwer degree (see the Definitions A2.1-A2.3).

Theorem 4.3.4. *If in Duffing's equation (4.3.15), $h(t) = H\sin(2\pi t/T)$, $t > 0$, $d \in \mathbf{R}$, $a \in \mathbf{R}$, $H \in \mathbf{R}$ are arbitrary, and $b < 0$, then there exists a T-periodic solution $p : \mathbf{R} \to [-M, M]$ where $M > 0$ is such that $aM + bM^3 < |H|$.*

Note that Theorem 4.3.1 is valid in the autonomous case too when $b(t) \equiv B \in \mathbf{R}$. Since this case is reducible to

$$\ddot{x} + f(x, \dot{x})\dot{x} + g(x) = 0 \qquad (4.3.16)$$

by replacing the function $g(x)$ by $g(x) - B$, we shall speak about (4.3.1) with $b(t) \equiv 0$. If the conditions of Theorem 4.3.1 hold, then applying Brouwer's Fixed Point Theorem we may establish the existence of a periodic solution of an arbitrary period $T > 0$. The problem is that we are unable to tell *a priori* whether the periodic solution whose existence is established is a constant solution or not. The conditions of Theorem 4.3.1 imply that the function g has a zero x^0 in the interval $(-a, a) : g(x^0) = 0$, $-a < x^0 < a$. Then, clearly, $(x, \dot{x}) = (x^0, 0)$ is an equilibrium of (4.3.16), i.e. a periodic solution of arbitrary period. In order to establish the existence of a *non-constant* periodic solution besides the curve C of Theorem 4.3.1 we need a smaller curve around the equilibrium point such that every solution with a path crossing this inner curve moves outward. If we can show that there is no equilibrium point in the annular domain between the two curves, the existence of a closed path that is not a point follows (cf. the Poincaré-Bendixson Theory and Corollary 3.1.9 in particular). A theorem established on these lines can be found among the Problems at the end of this chapter: Problem 4.6.7.

In the rest of this Section the *Duffing type equation*

$$\ddot{x} + d\dot{x} + g(t, x) = 0 \qquad (4.3.17)$$

will be considered where $d > 0$ is a constant, $g, g'_x \in C^0$ and g is T-periodic in the variable t for some $T > 0 : g(t + T, x) = g(t, x)$, $t \in \mathbf{R}$, $x \in \mathbf{R}$. The

differential equation (4.3.17) is said to be of Duffing type because the
damping is linear in the velocity $\dot{x}$ with constant coefficient and the
restoring force is non-linear. The conditions and the methods to be applied
now are completely different from those that have been applied in this
Section up to this point. The main condition will be that lower and upper
bounds will be imposed upon the x derivative of the function g, i.e. we shall
assume that

$$a \leq g'_x(t, x) \leq b, \qquad (t, x) \in \mathbf{R}^2 \qquad (4.3.18)$$

where the constants a and b satisfy certain conditions. The consequence is
that Duffing's equation (4.3.15) is *not* a special case of (4.3.17) if g satisfies
(4.3.18). Clearly (4.3.18) does not allow a cubic (or quadratic or higher
degree) term in x. On the other hand, the uniqueness and the asymptotic
stability of the periodic solution will also be established. The results we
present here are due to Lazer and McKenna [1990] and references in this
paper.

Theorem 4.3.5. *If there are numbers $a \leq b$, $r > (b - a)/2$ such that
(4.3.18) holds, and the complex numbers*

$$z_m := (2\pi m/T)^2 - id2\pi m/T \qquad (m = 0, \pm 1, \pm 2, ...) \qquad (4.3.19)$$

do not belong to the closed disc

$$D := \{z \in \mathbf{C} : |z - (a + b)/2| \leq r\}$$

*of the complex plane, then (4.3.17) has a unique periodic solution with
period T.*

Note that the conditions of this Theorem may hold if a and b have the
same sign: $ab > 0$. If $ab < 0$, i.e. g'_x assumes positive as well as negative
values, then the disc D contains necessarily the origin $z_0 = 0$ ($m = 0$ in
(4.3.19)).

Proof. Denote the Hilbert space (see Definition A2.6) of the T-periodic
real valued functions whose restriction to the interval $[0, T]$ is square
integrable with the usual inner product $< \bullet, \bullet >$ by E, i.e. for $u, v \in E$

$$< u, v > := \int_0^T u(t)v(t)dt.$$

Consider the linear operator L defined by

$$Lu = -\ddot{u} - d\dot{u} - cu.$$

Theorem 2.3.1 and the Carathéodory Existence Theorem 1.1.6 imply that
if $h \in E$, then there exists a unique T-periodic function $u \in C^1$ for which

$\dot{u}$ is absolutely continuous with $\ddot{u} \in E$ such that $Lu = h$. In fact, if the complex Fourier series of h is

$$h(t) = \sum_{m=-\infty}^{\infty} c_m \exp(i2\pi mt/T)$$

where $c_{-m} = \bar{c}_m$ and the equation is to be understood according to the norm in E,

$$\lim_{N \to \infty} \left\| h - \sum_{m=-N}^{N} c_m \exp(i2\pi m \bullet /T) \right\| = 0,$$

then a simple calculation shows that

$$u(t) = (L^{-1}h)(t) = \sum_{m=-\infty}^{\infty} c_m \exp(i2\pi mt/T)/(4\pi^2 m^2/T^2 - i2\pi mdT - c)$$

where $c := (a + b)/2$, and the equation is to be understood again in E. From the theory of Fourier series it follows that L^{-1} is a compact linear mapping of E into E (see Definition 2.7). Moreover, because of the assumptions of the Theorem for $m = 0, \pm 1, \pm 2, ...$,

$$|4\pi^2 m^2/T^2 - i2\pi md/T - c| > r > (b - a)/2,$$

and applying Bessel's identity concerning Fourier series we obtain the estimate

$$\|L^{-1}h\| \le \|h\|/r \tag{4.3.20}$$

for $h \in E$.

Now, u is a periodic solution with period T of (4.3.17) if and only if

$$Lu = g(t, u) - cu$$

or if and only if

$$u = L^{-1}Gu \tag{4.3.21}$$

where the operator $G : E \to E$ is defined by

$$(Gu)(t) := g(t, u(t)) - cu(t), \qquad \text{for} \qquad u \in E.$$

For $u_1, u_2 \in E$ we have

$$\begin{aligned}
(Gu_1)(t) - (Gu_2)(t) &= g(t, u_1(t)) - g(t, u_2(t)) - c(u_1(t) - u_2(t)) \\
&= (g'(t, \nu(t)) - c)(u_1(t) - u_2(t))
\end{aligned}$$

where $u_1(t) < \nu(t) < u_2(t)$. Hence, by (4.3.18) and taking into account that $c = (a + b)/2$ we obtain the estimate

$$\|Gu_1 - Gu_2\| \le \|u_1 - u_2\|(b - a)/2. \tag{4.3.22}$$

The estimates (4.3.20), (4.3.22) yield that

$$\|L^{-1}Gu_1 - L^{-1}Gu_2\| \leq \|u_1 - u_2\|(b-a)/(2r)$$

where, according to the assumptions, $(b-a)/2r < 1$. Thus, $L^{-1}G$ is a contraction mapping so that by Theorem A2.17 it has a unique fixed point, i.e. (4.3.21) has a unique solution implying that (4.3.17) has a unique T-periodic solution. $\square$

The asymptotic stability of the periodic solution can be proved under similar but somewhat more restrictive conditions.

Theorem 4.3.6. *If there are positive numbers* $0 < a \leq b$, $r > (b-a)/2$ *such that (4.3.18) holds, and the complex numbers*

$$\tilde{z}_m := (\pi m/T)^2 - id\pi m/T \qquad (m = \pm 0, \pm 1, \pm 2, ...)$$

do not belong to the closed disc

$$D = \{z \in \mathbf{C} : |z - c| \leq r\}$$

where $c = (a+b)/2$, *then (4.3.17) has a unique T-periodic solution, and it is asymptotically stable.*

Proof. The assumption that $\tilde{z}_m \notin D$, clearly, implies that $z_m \notin D$ $(m = 0, \pm 1, \pm 2, ...)$; hence, by the previous Theorem the equation (4.3.17) has a unique T-periodic solution. Denote this solution by $p(t)$ and consider the variational system of (4.3.17) with respect to this solution:

$$\dot{y}_1 = y_2, \qquad \dot{y}_2 = -g'_x(t, p(t))y_1 - dy_2 \tag{4.3.23}$$

or in equivalent scalar second order form,

$$\ddot{y} + d\dot{y} + g'_x(t, p(t))y = 0. \tag{4.3.24}$$

A homotopy type argument will be applied showing that the characteristic multipliers of (4.3.23) are in modulus less than 1. By Theorem 4.2.1 this will prove the asymptotic stability of p.

Since $0 < a \leq c = (a+b)/2 \leq b$ and (4.3.18) holds, we have for all $0 \leq s \leq 1$ that

$$a \leq (1-s)c + sg'_x(t, p(t)) \leq b. \tag{4.3.25}$$

Consider the one parameter family of second order linear differential equations

$$\ddot{y} + d\dot{y} + \big((1-s)c + sg'_x(t, p(t))\big)y = 0. \tag{4.3.26}$$

For $s = 1$ this reduces to (4.3.24); for $s = 0$ it is

$$\ddot{y} + d\dot{y} + cy = 0.$$

Since both d and c are positive, the latter is an asymptotically stable differential equation with constant coefficients, i.e. its eigenvalues have negative real parts; hence, the characteristic multipliers of the equivalent system

$$\dot{y}_1 = y_2, \qquad \dot{y}_2 = -cy_1 - dy_2 \qquad (4.3.27)$$

are in modulus less than 1. By (4.3.25) the differential equation (4.3.26) considered as a $2T$-periodic one satisfies all the conditions of Theorem 4.3.5; hence, it has a unique $2T$-periodic solution for every $s \in [0,1]$ but the trivial solution $y \equiv 0$ is such, so that (4.3.26) has no non-trivial $2T$-periodic solution. Since the fact that $\tilde{z}_m \notin D$ implies that $z_m \notin D$ ($m = 0, \pm 1, \pm 2, ...$), (4.3.26) considered as a T-periodic equation satisfies also the conditions of Theorem 4.3.5; hence, it has no non-trivial T-periodic solution either. By Corollary 2.2.3 we conclude that neither 1 nor -1 is a characteristic multiplier of the system equivalent to (4.3.26):

$$\dot{y}_1 = y_2, \qquad \dot{y}_2 = -((1-s)c + sg'_x(t, p(t)))y_1 - dy_2 \qquad (4.3.28)$$

for $s \in [0,1]$. Neither can the characteristic multipliers $\lambda_1(s)$ and $\lambda_2(s)$ be complex conjugates with modulus equal to 1 since their product is the determinant of the principal matrix

$$\lambda_1(s)\lambda_2(s) = \exp \int_0^T (-d)dt = \exp(-dT) < 1$$

by Liouville's formula (Theorem 1.2.4). Thus, for $s \in [0,1]$ the characteristic multipliers of system (4.3.28) which are, obviously, continuous functions of s, cannot fall on the unit circle of the complex plane. Since for $s = 0$ (system (4.3.27)) they lie in the interior, they do the same for $s = 1$, i.e. in the case of system (4.3.23). $\square$

4.4 Two Competing Species in a Periodically Changing Environment

In this Section we study a *two dimensional periodic competing species problem*. The presentation is based on the results of Alvarez and Lazer [1986]. The model to be studied is the Lotka-Volterra system with periodic coefficients

$$\begin{aligned} \dot{u} &= u(a(t) - b(t)u - c(t)v), \\ \dot{v} &= v(d(t) - e(t)u - f(t)v) \end{aligned} \qquad (4.4.1)$$

where $a, b, ..., f \in C^0(\mathbf{R})$; all these functions are positive everywhere and periodic with period $T > 0$. This is a model of two competing species whose quantities at time t are $u(t)$ and $v(t)$, respectively, in a periodic environment. The functions a and d are the respective *intrinsic growth rates*, b and f measure the respective *intraspecific competition* within species u and v, and the functions c and e measure the *interspecific competition* between the two species.

If $a, b, ..., f$ are positive constants, then an elementary calculation and stability analysis shows that *the system has an asymptotically stable positive equilibrium if and only if*

$$af > cd \qquad and \qquad bd > ae. \tag{4.4.2}$$

A simple phase plane analysis shows that *this equilibrium point is globally asymptotically stable* with respect to the open positive quadrant. Conditions (4.4.2) have an intuitive interpretation. The first inequality means that "u" has an advantage over "v": the product of its birth rate with the saturation coefficient of "v" is higher than the product of the birth rate of "v" with the measure of the adverse effect "v" has on "u". The second one expresses a similar advantage on the part of "v" over "u". The two inequalities together express a certain balance between the two species. In some respect, one is better; in another respect, the other is better. This is needed (and sufficient) for a stable coexistence of the two competing species. If one of the inequalities were reversed, then one of the species would be fitter in every respect and would outcompete the other.

Returning to the general system (4.4.1) where the coefficients are not necessarily constants, we shall show that inequalities analogous to (4.4.2) are sufficient for the existence of a unique positive asymptotically stable T-periodic solution. For a function $g \in C^0(\mathbf{R})$ such that $g(t) > 0$, $g(t + T) = g(t)$, for $t \in \mathbf{R}$, the following notations will be used:

$$g_M := \max g(t), \qquad g_L; = \min g(t), \qquad t \in \mathbf{R}.$$

Clearly, both quantities are positive. The inequalities that replace (4.4.2) in the general case are

$$a_L f_L > c_M d_M \qquad and \qquad b_L d_L > a_M e_M. \tag{4.4.3}$$

First, the existence of a T-periodic solution and then its uniqueness and stability will be proved. Both theorems need considerable preparation, and the uniqueness problem is much more difficult to settle.

It is clear that the first quadrant of the u, v plane is invariant with respect to system (4.4.1) in the sense that if for some $t_0 \in \mathbf{R}$ the initial values $(u(t_0), v(t_0))$ are in this quadrant, then the corresponding path lies entirely in it. The coordinate axes u and v are invariant too. Moreover, every solution with non-negative initial values $(u(t_0), v(t_0))$ is defined in the whole future $t \in [t_0, \infty)$. This follows easily from the obvious inequalities satisfied by every solution with a path in the positive quadrant:

$$\dot{u} \leq u(a_M - b_L u - c_L v) \leq 0,$$
$$\dot{v} \leq v(d_M - e_L u - f_L v) \leq 0$$

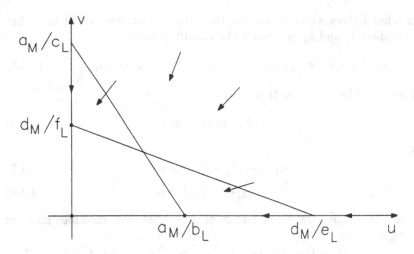

FIGURE 4.4.1. The flow of (4.4.1) "outside the straight lines" $b_L u + c_L v = a_M$, $e_L u + f_L v = d_M$ in the positive quadrant. The Figure shows the situation when (4.4.3) holds.

if $a_M - b_L u - c_L v \leq 0$ and $d_M - e_L u - f_L v \leq 0$ (see Figure 4.4.1).

Lemma 4.4.1. *If $(u_k(t), v_k(t))$ $(k = 1, 2)$ are two solutions of (4.4.1) such that $u_1(0) \geq u_2(0) \geq 0$ and $v_2(0) \geq v_1(0) \geq 0$, then $u_1(t) \geq u_2(t)$ and $v_2(t) \geq v_1(t)$ for $t \geq 0$; if the inequalities are strict in the assumption, then they are strict in the conclusion too.*

Proof. Since the solutions are continuous functions of the initial values, it is sufficient to prove the statement concerning the strict inequalities only, so that we suppose that the inequalities are strict in the assumption of the Lemma. By continuity

$$u_1(t) > u_2(t) \qquad \text{and} \qquad v_2(t) > v_1(t) \qquad (4.4.4)$$

for $t > 0$ and sufficiently small. If the inequalities do not hold for every positive t in the common domain of the two solutions, then there is a first $\bar{t} > 0$ such that either $u_1(\bar{t}) = u_2(\bar{t})$ or $v_1(\bar{t}) = v_2(\bar{t})$, and (4.4.4) is valid in $[0, \bar{t})$. Assume, e.g., that the first equality applies. Because of continuity, $v_1(\bar{t}) \leq v_2(\bar{t})$, and because of uniqueness of solutions, $v_1(\bar{t}) < v_2(\bar{t})$. The relations $u_1(t) - u_2(t) > 0$, $t \in [0, \bar{t})$ and $u_1(\bar{t}) - u_2(\bar{t}) = 0$ imply that $\dot{u}_1(\bar{t}) - \dot{u}_2(\bar{t}) \leq 0$; but from (4.4.1)

$$\dot{u}_1(\bar{t}) - \dot{u}_2(\bar{t}) = u_1(\bar{t}) c(\bar{t})(v_2(\bar{t}) - v_1(\bar{t})) > 0$$

because $u_1(\bar{t}) > 0$ by the invariance of the open positive quadrant. This contradiction proves the statement in the first case. If v_1 becomes equal to v_2 first, the proof is similar. $\square$

In what follows we shall assume that the inequalities (4.4.3) hold. Let the numbers k_1 and k_2 be chosen the following way:

$$a_M/b_L < k_1 < d_L/e_M, \qquad d_M/f_L < k_2 < a_L/c_M; \qquad (4.4.5)$$

and let $\delta > 0$ be chosen so that

$$0 < \delta < \min(k_1, k_2) \qquad (4.4.6)$$

and

$$a_L - c_M k_2 - b_M \delta > 0, \qquad (4.4.7)$$

$$d_L - e_M k_1 - f_M \delta > 0. \qquad (4.4.8)$$

Because of (4.4.3) such k_1, k_2 and δ, clearly, exist. We introduce also the rectangle

$$D_\delta = \{(u, v) \in \mathbf{R}^2 : \delta < u < k_1, \quad \delta < v < k_2\}$$

where most of the events will take place. If a $\delta > 0$ satisfies (4.4.6) - (4.4.8), then, obviously, any positive number smaller than this δ satisfies these requirements too. Later we shall use also the rectangle

$$D_0 = \{(u, v) \in \mathbf{R}^2 : 0 < u < k_1, \quad 0 < v < k_2\}.$$

Lemma 4.4.2. *Suppose that (4.4.3) holds, and k_1, k_2, δ have been chosen according to (4.4.5) - (4.4.8); if $(u_1(t), v_1(t))$ and $(u_2(t), v_2(t))$ denote the respective solutions that satisfy the initial conditions*

$$(u_1(0), v_1(0)) = (k_1, \delta), \qquad (u_2(0), v_2(0)) = (\delta, k_2),$$

then these solutions satisfy the inequalities

$$u_2(t) < u_1(t), \qquad v_1(t) < v_2(t), \qquad (4.4.9)$$

$$u_1(0) > u_1(t) > u_1(t+T), \qquad (4.4.10)$$

$$v_1(0) < v_1(t) < v_1(t+T), \qquad (4.4.11)$$

$$u_2(0) < u_2(t) < u_2(t+T), \qquad (4.4.12)$$

$$v_2(0) > v_2(t) > v_2(t+T) \qquad (4.4.13)$$

for $t \in [0, \infty)$.

Proof. The inequalities (4.4.9) are consequences of the previous Lemma. From the differential equation applying (4.4.5) we obtain

$$\begin{aligned}
\dot{u}_1(0) &= u_1(0)(a(0) - b(0)u_1(0) - c(0)v_1(0)) \\
&< u_1(0)(a(0) - b(0)a_M/b_L) \leq 0,
\end{aligned}$$

and applying (4.4.8)

$$\dot{v}_1(0) = v_1(0)(d(0) - e(0)u_1(0) - f(0)v_1(0))$$
$$\geq v_1(0)(d_L - e_M k_1 - f_M \delta) > 0.$$

(Similarly, one may show that $\dot{v}_2(0) < 0$, $\dot{u}_2(0) > 0$.) These inequalities imply that for small positive t

$$u_1(0) > u_1(t), \qquad v_1(0) < v_1(t) \qquad (4.4.14)$$

(and, similarly, $u_2(0) < u_2(t)$, $v_2(0) > v_2(t)$). If the inequalities (4.4.14) did not hold for the whole domain of definition, $t > 0$, there would be a $\bar{t} > 0$ such that (4.4.14) holds on $[0, \bar{t})$, and either $u_1(\bar{t}) = u_1(0) = k_1$ or $v_1(\bar{t}) = v_1(0) = \delta$. If the first equality held, then $v_1(\bar{t}) \geq v_1(0) > 0$, and we would have $\dot{u}_1(\bar{t}) \geq 0$, and from (4.4.1) and (4.4.5)

$$\dot{u}_1(\bar{t}) = u_1(\bar{t})(a(\bar{t}) - b(\bar{t})u_1(\bar{t}) - c(\bar{t})v_1(\bar{t}))$$
$$< u_1(\bar{t})(a(\bar{t}) - b(\bar{t})a_M / b_L) \leq 0,$$

a contradiction. If the second equality held, then $u_1(\bar{t}) \leq u_1(0) = k_1$, and $\dot{v}_1(\bar{t}) \leq 0$. But from (4.4.1) and (4.4.8) we have

$$\dot{v}_1(\bar{t}) = v_1(\bar{t})(d(\bar{t}) - e(\bar{t})u_1(\bar{t}) - f(\bar{t})v_1(\bar{t}))$$
$$\geq v_1(0)(d_L - e_M k_1 - f_M \delta) > 0,$$

another contradiction. Hence, (4.4.14) holds for all $t > 0$. The proof of the first parts of inequalities (4.4.12), (4.4.13) is analogous. The periodicity of system (4.4.1) implies that

$$(\tilde{u}_1(t), \tilde{v}_1(t)) := (u_1(t + T), v_1(t + T))$$

is also a solution. By (4.4.9), $u_1(0) > u_1(T) = \tilde{u}_1(0)$ and $v_1(0) < v_1(T) = \tilde{v}_1(0)$. Therefore, by the previous Lemma, $u_1(t) > \tilde{u}_1(t) = u_1(t + T)$ and $v_1(t) < \tilde{v}_1(t) = v_1(t + T)$ for all $t > 0$. This proves the second parts of inequalities (4.4.10), (4.4.11). The proof of the second parts of (4.4.12), (4.4.13) is analogous. $\square$

Now we are able to establish the existence of a T-periodic solution.

Theorem 4.4.3. *Suppose that the inequalities (4.4.3) hold and that k_1, k_2 and δ have been chosen according to (4.4.5) - (4.4.8); then system (4.4.1) has a T-periodic solution with initial values in the rectangle D_δ.*

Proof. For arbitrary $(r, s) \in \overline{D}_\delta$ denote the solution with initial values (r, s) at $t = 0$ by $(u(t, r, s), v(t, r, s))$, i.e. $(u(0, r, s), v(0, r, s)) = (r, s)$. Consider the time T map $\varphi : \overline{D}_\delta \to \mathbf{R}_+^2$ defined by

$$\varphi(r, s) := (u(T, r, s), v(T, r, s)) \qquad \text{for} \qquad (r, s) \in \overline{D}_\delta.$$

The mapping φ is, clearly, continuous. We show that it maps $\overline{D}_\delta$ into itself. Obviously,

$$u_2(0) = \delta \leq r \leq k_1 = u_1(0),$$
$$v_1(0) = \delta \leq s \leq k_2 = v_2(0)$$

where (u_1, v_1) and (u_2, v_2) are the solutions in Lemma 4.4.2. Applying Lemma 4.4.1 and the inequalities (4.4.10) - (4.4.13) we obtain

$$\delta = u_2(0) < u_2(T) \leq u(T, r, s) \leq u_1(T) < u_1(0) = k_1,$$
$$\delta = v_1(0) < v_1(T) \leq v(T, r, s) \leq v_2(T) < v_2(0) = k_2$$

for $(r, s) \in \overline{D}_\delta$, i.e. $(u(T, r, s), v(T, r, s)) \in \overline{D}_\delta$. Applying Brouwer's Fixed Point Theorem (Corollary A2.9) we get that φ has a fixed point $(r_0, s_0) \in D_\delta, (r_0, s_0) = \varphi(r_0, s_0)$, i.e.

$$(u(T, r_0, s_0), v(T, r_0, s_0)) = (u(0, r_0, s_0), v(0, r_0, s_0)),$$

and this, by Lemma 2.2.1, implies that the solution $(u(t, r_0, s_0), v(t, r_0, s_0))$ is periodic with period T. $\square$

Let us denote the T-periodic solution whose existence has been established in the previous Theorem by

$$(u_0(t), v_0(t)) = (u(t, r_0, s_0), v(t, r_0, s_0)), \quad (r_0, s_0) \in D_\delta,$$

and consider the variational system with respect to this solution (see (1.1.5)),

$$\dot{x} = A(t)x \tag{4.4.15}$$

where

$$A(t) = \begin{pmatrix} a(t) - 2b(t)u_0(t) - c(t)v_0(t) & -c(t)u_0(t) \\ -e(t)v_0(t) & d(t) - e(t)u_0(t) - 2f(t)v_0(t) \end{pmatrix}.$$

As it has been shown following (1.1.5) the matrix

$$\Phi(t) = \begin{pmatrix} u'_r(t, r_0, s_0) & u'_s(t, r_0, s_0) \\ v'_r(t, r_0, s_0) & v'_s(t, r_0, s_0) \end{pmatrix}$$

is a solution matrix of (4.4.15), and since $u(0, r, s) \equiv r$, $v(0, r, s) \equiv s$ we have $\Phi(0) = I$, the unit matrix. We recall (see the paragraph preceding Theorem 2.2.2) that the eigenvalues of the matrix $\Phi(T)$ are the characteristic multipliers of system (4.4.15). There holds

Theorem 4.4.4. *If the inequalities (4.4.3) hold, then the T-periodic solution $(u_0(t), v_0(t))$ is asymptotically stable.*

Proof. We shall prove that the characteristic multipliers of system (4.4.15) are in modulus less than 1. By Theorem 4.2.1 this will prove the statement. This will be done by transforming the variational system (4.4.15) into a system whose principal matrix is a matrix with positive elements and by applying Perron's Theorem A1.13 for positive matrices.

Substituting the solution $(u_0(t), v_0(t))$ into system (4.4.1) and dividing the first equation by $u_0(t)$ and the second by $v_0(t)$, we get

$$\dot{u}_0(t)/u_0(t) = a(t) - b(t)u_0(t) - c(t)v_0(t),$$
$$\dot{v}_0(t)/v_0(t) = d(t) - e(t)u_0(t) - f(t)v_0(t)$$

$(u_0(t) \neq 0, v_0(t) \neq 0$ because of the invariance of the open positive quadrant). Thus, system (4.4.15) can be written in the form

$$\dot{x}_1 = x_1\dot{u}_0(t)/u_0(t) - b(t)u_0(t)x_1 - c(t)u_0(t)x_2,$$
$$\dot{x}_2 = x_2\dot{v}_0(t)/v_0(t) - e(t)v_0(t)x_1 - f(t)v_0(t)x_2$$

or

$$(\dot{x}_1 u_0(t) - \dot{u}_0(t)x_1)/u_0^2(t) = -b(t)u_0(t)x_1/u_0(t) - c(t)v_0(t)x_2/v_0(t),$$
$$(\dot{x}_2 v_0(t) - \dot{v}_0(t)x_2)/v_0^2(t) = -e(t)u_0(t)x_1/u_0(t) - f(t)v_0(t)x_2/v_0(t).$$

This means that performing the coordinate transformation $y_1 = x_1/u_0(t)$, $y_2 = x_2/v_0(t)$, system (4.4.15) assumes the form

$$\dot{y} = B(t)y \tag{4.4.16}$$

in the new coordinates where

$$B(t) = \begin{pmatrix} -b(t)u_0(t) & -c(t)v_0(t) \\ -e(t)u_0(t) & -f(t)v_0(t) \end{pmatrix}.$$

Applying the matrix of the coordinate transformation

$$P(t) = \begin{pmatrix} 1/u_0(t) & 0 \\ 0 & 1/v_0(t) \end{pmatrix}$$

to the fundamental matrix $\Phi(t)$ of system (4.4.15), a fundamental matrix $P(t)\Phi(t)$ of (4.4.16) is obtained. Multiplying this by the constant matrix $P(0)^{-1}$ from the right, the matrix $\Psi(t) := P(t)\Phi(t)P^{-1}(0)$ is also a fundamental matrix of (4.4.16), and clearly, $\Psi(0) = I$. Moreover, because of the periodicity of $(u_0(t), v_0(t))$ we have $P(T) = P(0)$, and as a consequence

$$\Psi(T) = P(T)\Phi(T)P^{-1}(T),$$

i.e. $\Psi(T)$ is similar to $\Phi(T)$, and this means that it has the same eigenvalues. We perform one more coordinate transformation, $z_1 = y_1$, $z_2 = -y_2$ or in matrix form $z = Ry$ where

$$R = \begin{pmatrix} 1 & 0 \\ 0 & -1 \end{pmatrix} = R^{-1}.$$

By this transformation, system (4.4.16) goes into

$$\dot{z} = C(t)z \tag{4.4.17}$$

where

$$C(t) = RB(t)R^{-1} = \begin{pmatrix} -b(t)u_0(t) & c(t)v_0(t) \\ e(t)u_0(t) & -f(t)v_0(t) \end{pmatrix}.$$

$R\Psi(t)$ and also $\Theta(t) := R\Psi(t)R^{-1}$ is, clearly, a fundamental matrix of (4.4.17) and $\Theta(0) = I$. This means that $\Theta(T) = R\Psi(T)R^{-1}$ is also similar to $\Psi(T)$ and to $\Phi(T)$ so that its eigenvalues are the characteristic multipliers.

We are going to show that every element of $\Theta(t)$ is positive for $t \in (0, T]$. Indeed, the off diagonal elements of $C(t)$ are positive; therefore, we may choose a $\gamma > 0$ so large that every entry of the matrix $C(t) + \gamma I$ is positive for $t \in [0, T]$. If $\tilde{\Theta}(t) = \Theta(t)\exp(\gamma t)$, then

$$\dot{\tilde{\Theta}}(t) = (C(t) + \gamma I)\tilde{\Theta}(t), \qquad \tilde{\Theta}(0) = I,$$

so that

$$\tilde{\Theta}(t) \equiv I + \int_0^t (C(s) + \gamma I)\tilde{\Theta}(s)ds. \tag{4.4.18}$$

Let $N \in \mathbf{N}$ be such that

$$\frac{T}{N} \max_{s \in [0,T]} |C(s) + \gamma I| < 1.$$

Then the method of successive approximation applies to the matrix sequence $\Theta_0(t) = I$,

$$\Theta_{m+1}(t) = I + \int_0^t (C(s) + \gamma I)\Theta_m(s)ds \qquad (m = 0, 1, 2, ...)$$

which tends uniformly to $\tilde{\Theta}(t)$ on $(0, T/N]$. A simple mathematical induction shows that every entry of each element of the matrix sequence $\Theta_m(t)$ $(m = 1, 2, ...)$ is non-negative. Hence, the same is true for the limit $\tilde{\Theta}(t)$. Since $\tilde{\Theta}(t)$ is a regular matrix for every t, the identity (4.4.18) implies that all its entries are strictly positive on $(0, T/N]$. This implies that the same is true for $\Theta(t) = \tilde{\Theta}(t)\exp(-\gamma t)$. Repeating this procedure on $[kT/N, (k+1)T/N]$ $(k = 1, 2, ..., N-1)$ replacing I in (4.4.18) by $\tilde{\Theta}(kT/N)$, we obtain the positivity of each entry on $[0, T]$.

Now, Perron's Theorem A1.13 says that the positive matrix $\Theta(T)$ has a simple positive eigenvalue that is strictly greater than the other eigenvalue and has a corresponding eigenvector with positive coordinates. Denoting the characteristic multipliers by λ_1 and λ_2 by Liouville's formula (Theorem 1.2.4)

$$\lambda_1\lambda_2 = \det \Theta(T) = \exp \int_0^T \operatorname{Tr} C(t)dt > 0,$$

so that $0 < \lambda_1 < \lambda_2$.

We shall prove now that the larger (largest) positive eigenvalue λ_2 is less than 1. Multiplying the right-hand sides, resp. the left-hand sides of the inequalities (4.4.3) and dividing by $a_L d_L$ we obtain

$$f_L b_L > c_M e_M. \tag{4.4.19}$$

Let $w = (w_1, w_2)$ be an eigenvector corresponding to the largest positive eigenvalue of the matrix $\Theta(T)$, i.e. $\Theta(T)w = \lambda_2 w$, $w_1 > 0$, $w_2 > 0$ according to Perron's Theorem, and consider the solution of (4.4.17) defined by $\theta(t) = \Theta(t)w$. Because of the positivity of the entries of $\Theta(t)$ on $(0, T]$ and of the coordinates of w, and taking into consideration also $\Theta(0) = I$, we have that the coordinates of the solution $\theta(t) = (\theta_1(t), \theta_2(t))$ are positive on the closed interval $[0, T]$. Substituting this solution into (4.4.17) we get the estimates

$$\dot{\theta}_1(t) \leq -b_L u_0(t)\theta_1(t) + c_M v_0(t)\theta_2(t),$$
$$\dot{\theta}_2(t) \leq e_M u_0(t)\theta_1(t) - f_L v_0(t)\theta_2(t).$$

Multiplying the first inequality by e_M and the second by b_L, adding them and taking (4.4.19) into consideration we obtain

$$e_M \dot{\theta}_1(t) + b_L \dot{\theta}_2(t) \leq (c_M e_M - b_L f_L)\theta_2(t) < 0$$

for $t \in [0, T]$; thus,

$$e_M \theta_1(T) + b_L \theta_2(T) < e_M \theta_1(0) + b_L \theta_2(0).$$

But $\theta(T) = \Theta(T)w = \lambda_2 w = \lambda_2 \theta(0)$, and substituting this into the left-hand side of the last inequality we get

$$\lambda_2(e_M \theta_1(0) + b_L \theta_2(0)) < e_M \theta_1(0) + b_L \theta_2(0);$$

hence, $\lambda_2 < 1$. $\square$

Applying degree theory we are able to prove also the uniqueness of the periodic solution whose existence has been established in Theorem 4.4.3.

Theorem 4.4.5. *If the inequalities (4.4.3) hold, then system (4.4.1) has a unique T-periodic solution with positive coordinates $u_0(t) > 0$, $v_0(t) > 0$.*

Proof. First we are proving that there is a unique periodic solution with initial values (r_0, s_0) at $t = 0$ in the rectangle

$$D_0 = \{(u, v) \in \mathbf{R}^2 : 0 < u < k_1, \ 0 < v < k_2\}$$

where k_1 and k_2 satisfy (4.4.5). In the second phase of the proof we shall show that the path of every solution with positive initial values enters the rectangle D_0 eventually.

Out of the time T map φ applied in the proof of Theorem 4.4.3 and the identity map I, we construct the mapping $F := I - \varphi : \mathbf{R}_+^2 \to \mathbf{R}_+^2$ defined by

$$F(r, s) = (r - u(T, r, s), s - v(T, r, s)) \qquad \text{for} \qquad (r, s) \in \mathbf{R}_+^2$$

where $(u(t, r, s), v(t, r, s))$ is the solution of (4.4.1) with initial values (r, s) at $t = 0$. Let $\delta > 0$ be chosen according to (4.4.6), and consider the rectangle

$$D_\delta = \{(u, v) \in \mathbf{R}^2 : \delta < u < k_1, \ \delta < v < k_2\}.$$

We shall prove first that the topological degree of F with respect to the set D_δ and to zero (see Definitions A2.1-A2.2) is 1: $d[F, D_\delta, 0] = 1$. In the proof of Theorem 4.4.3 we have shown that the time T mapping φ maps $\overline{D}_\delta$ into itself and ∂D_δ into the interior of D_δ so that $F(r, s) \neq 0$ for $(r, s) \in \partial D_\delta$. For $\lambda \in [0, 1]$ and a fixed point $(r_1, s_1) \in D_\delta$, define the mapping N by

$$N(r, s, \lambda) := (r_1 + \lambda(u(T, r, s) - r_1), s_1 + \lambda(v(T, r, s) - s_1))$$

$$\text{for} \qquad (r, s) \in \overline{D}_\delta.$$

Since D_δ is convex, $(r, s) - N(r, s, \lambda) \neq 0$ for $(r, s) \in \partial D_\delta$. Consequently, $I - N$ establishes an admissible homotopy (see Theorem A2.5) between $F(r, s) = (r, s) - N(r, s, 1)$ and $(r - r_1, s - s_1) = (r, s) - N(r, s, 0)$. Because of the invariance of the degree by homotopy, $d[F, D_\delta, 0]$ is equal to the degree of the mapping $(r - r_1, s - s_1)$. But this last mapping has a single zero (r_1, s_1) in D_δ and its Jacobian is identically equal to 1, its degree with respect to D_δ and zero is 1, and as a consequence, $d[F, D_\delta, 0] = 1$. We show next that if $(r_0, s_0) \in D_\delta$ is a zero of F, then the Jacobian of F at (r_0, s_0) is positive:

$$JF(r_0, s_0) = \det \begin{pmatrix} 1 - u_r'(T, r_0, s_0) & -u_s'(T, r_0, s_0) \\ -v_2'(T, r_0, s_0) & 1 - v_s'(T, r_0, s_0) \end{pmatrix} > 0.$$

But $F(r_0, s_0) = 0$ implies that $(u(t, r_0, s_0), v(t, r_0, s_0))$ is a T-periodic solution of system (4.4.1); hence, the eigenvalues λ_1, λ_2 of the matrix

$$\begin{pmatrix} u_r'(T, r_0, s_0) & u_s'(T, r_0, s_0) \\ v_r'(T, r_0, s_0) & v_s'(T, r_0, s_0) \end{pmatrix}$$

are the characteristic multipliers of the variational system (4.4.15) with respect to this periodic solution. However, we have shown in the proof of Theorem 4.4.4 that these characteristic multipliers satisfy $0 < \lambda_1 < \lambda_2 < 1$. Obviously, $JF(r_0, s_0) = (1 - \lambda_1)(1 - \lambda_2) > 0$, and this proves the statement. We conclude that F is non-degenerate in D_δ (see the paragraph preceding Lemma A2.1) for arbitrary small $\delta > 0$. Applying Definition A2.1, taking into consideration that the degree of F is 1 and its Jacobian is positive in every possible zero in D_0, we get that F has a unique zero in D_0. As a consequence, system (4.4.1) has a unique T-periodic solution with initial values in D_0 at $t = 0$.

In the second part of the proof we show that every solution with positive initial values enters D_0 as t tends to infinity. We recall that from Example 4.1.1 it follows that the scalar differential equations

$$\dot{U} = U(a(t) - b(t)U), \tag{4.4.20}$$

$$\dot{V} = V(d(t) - f(t)V), \tag{4.4.21}$$

special cases of (4.1.6), have a unique positive periodic solution each, $U_0(t)$, and $V_0(t)$, respectively and every solution of (4.4.20), resp. (4.4.21) with positive initial value tends to U_0, resp. V_0 as t tends to infinity. It is easy to see that the inequalities

$$U_0(t) \leq a_M/b_L, \qquad V_0(t) \leq d_M/f_L \tag{4.4.22}$$

hold. Indeed, if U_0 attains its maximum at $t_1 \geq 0$, then $\dot{U}_0(t_1) = 0$; consequently, $U_0(t_1) = a(t_1)/b(t_1) \leq a_M/b_L$. The proof is similar for V_0.

Let $(u(t), v(t))$ be an arbitrary solution of system (4.4.1) with positive initial values $u(0) > 0$, $v(0) > 0$, and let $U(t)$, resp. $V(t)$ be the solution of (4.4.20), resp. (4.4.21) with initial value $U(0) = u(0)$, $V(0) = v(0)$, respectively. Clearly, the vector functions $(U(t), 0)$ and $(0, V(t))$ are both solutions of system (4.4.1). Applying Lemma 4.4.1 we conclude that $u(t) \leq U(t)$, $v(t) \leq V(t)$ for $t \geq 0$. Since $U(t) - U_0(t) \to 0$ and $V(t) - V_0(t) \to 0$ as $t \to \infty$, and $U_0(t) \leq a_M/b_L < k_1$, $V_0(t) \leq d_M/f_L < k_2$ for $t \geq 0$ by (4.4.22), it follows that for sufficiently large t we have $u(t) \leq U(t) < k_1$, $v(t) \leq V(t) < k_2$. In particular, there is an integer $m \geq 1$ such that $(u(mT), v(mT)) \in D_0$. Because of the periodicity of system (4.4.1),

$$(\tilde{u}(t), \tilde{v}(t)) := (u(t + mT), v(t + mT))$$

is also a solution and $(\tilde{u}(0), \tilde{v}(0)) \in D_0$. If $(u(t), v(t))$ were T-periodic, then $(\tilde{u}(t), \tilde{v}(t))$ would be T-periodic too, and because of the uniqueness of a periodic solution with initial values in D_0 established in the first part of the proof, it would coincide with $(u(t, r_0, s_0), v(t, r_0, s_0))$. But because of the periodicity, $(\tilde{u}(t), \tilde{v}(t))$ coincides also with $(u(t), v(t))$ and this proves the uniqueness of a T-periodic solution in the whole positive quadrant. $\square$

As we have mentioned at the start the previous study of system (4.4.1) has been based on the results due to Alvarez and Lazer [1986] who have proven also that the unique positive T-periodic solution of the system is globally asymptotically stable with respect to the open positive quadrant. The periodic two dimensional competition model has also been studied by Cushing [1980], de Mottoni and Schiaffino [1981] and Gopalsamy [1982].

4.5   Applications in Higher Dimensions

In this Section two problems will be treated: first, an n-dimensional competitive Lotka-Volterra system in a periodically varying environment; secondly, a periodically excited n-dimensional conservative system.

First, the *Lotka-Volterra system*

$$\dot{x}_i = x_i\left(b_i(t) - \sum_{j=1}^n a_{ij}(t)x_j \right) \tag{4.5.1}$$

will be considered where $b_i, a_{ij} \in C^0(\mathbf{R})$, $b_i(t) > 0$, $a_{ij}(t) > 0$, and there exists a $T > 0$ such that $b_i(t+T) = b_i(t)$, $a_{ij}(t+T) = a_{ij}(t)$ for $t \in \mathbf{R}$ $(i,j = 1,2,...,n)$. This system is modelling the competition of n species in a periodically changing environment that comes into the picture through the T-periodicity of the intrinsic growth rates b_i and of the coefficients a_{ij}. System (4.5.1) is, clearly, the n-dimensional version of the two dimensional system (4.4.1). We shall prove the existence of a T-periodic solution under conditions analogous to (4.4.3), but we shall need additional assumptions in order to prove uniqueness and global asymptotic stability. We shall denote the minimum and the maximum of a function $g : [0,T] \to \mathbf{R}$ by g_L and g_M, respectively, so that

$$b_{iL} = \min b_i(t), \qquad b_{iM} = \max b_i(t),$$

$$a_{ijL} = \min a_{ij}(t), \qquad a_{ijM} = \max a_{ij}(t),$$

$$t \in [0,T] \qquad (i,j = 1,2,...,n).$$

Our assumptions imply that

$$0 < b_{iL} \le b_{iM}, \qquad 0 < a_{ijL} \le a_{ijM}. \tag{4.5.2}$$

The following results are due to Gopalsamy [1985]. The basic assumptions will be (cf. (4.4.3))

$$b_{iL} > \sum_{j=1,j\neq i}^n a_{ijM}b_{jM}/a_{jjL} \qquad (i = 1,2,...,n), \tag{4.5.3}$$

and the following notations will be introduced:

$$x_{iL} := \left(b_{iL} - \sum_{j=1,j\neq i}^n a_{ijM}b_{jM}/a_{jjL} \right)/a_{iiM} \tag{4.5.4}$$

$$x_{iM} := b_{iM}/a_{iiL} \qquad (i = 1,2,...,n). \tag{4.5.5}$$

Inequalities (4.5.2) and (4.5.3), clearly, imply that

$$0 < x_{iL} < x_{iM} \qquad (i = 1,2,...,n).$$

Lemma 4.5.1. *If (4.5.3) holds, then the set*

$$S = \{x \in \mathbf{R}_+^n : x_{iL} \leq x_i \leq x_{iM} \qquad (i = 1, 2, ..., n)\}$$

is positively invariant with respect to system (4.5.1).

Proof. If $\varphi(t, 0, x^0)$ denotes the solution of (4.5.1) with initial value x^0 at time 0, and the coordinates of x^0 are positive, $x_i^0 > 0$ $(i = 1, 2, ..., n)$, then $\varphi_i(t, 0, x^0) > 0$ in the whole domain since the positive orthant of the space is invariant. As a consequence

$$\dot{\varphi}_i(t, 0, x^0) \leq \varphi_i(t, 0, x^0)(b_{iM} - a_{iiL}\varphi_i(t, 0, x^0)).$$

If $0 < x_i^0 \leq b_{iM}/a_{iiL}$, then the previous inequality implies that

$$0 < \varphi_i(t, 0, x^0) \leq b_{iM}/a_{iiL} \qquad (i = 1, 2, ..., n) \qquad (4.5.6)$$

for $t \geq 0$, but because of this last inequality the domain of the solution contains $\mathbf{R}_+$. On the other hand, because of (4.5.6)

$$\dot{\varphi}_i(t, 0, x^0) \geq \varphi_i(t, 0, x^0)\left(b_{iL} - \sum_{j=1, j\neq i}^{n} a_{ijM}b_{jM}/a_{jjL} - a_{iiM}\varphi_i(t, 0, x^0)\right)$$

$$(4.5.7)$$

if $0 < x_i^0 < b_{iM}/a_{iiL}$ $(i = 1, 2, ..., n)$. Thus, if

$$\left(b_{iL} - \sum_{j=1, j\neq i}^{n} a_{ijM}b_{jM}/a_{jjL}\right)/a_{iiM} \leq x_i^0 \leq b_{iM}/a_{iiL},$$

then the same inequality is valid for $\varphi_i(t, 0, x^0)$, i.e. with the notations (4.5.4) - (4.5.5)

$$x_{iL} \leq \varphi_i(t, 0, x^0) \leq x_{iM} \qquad (i = 1, 2, ..., n), \qquad t \in \mathbf{R}_+. \square$$

Applying this Lemma, the existence of a periodic solution follows readily.

Theorem 4.5.2. *If (4.5.3) holds, then the Lotka-Volterra system (4.5.1) has a T-periodic solution with positive coordinates.*

Proof. Consider the Poincaré-Andronov operator (the time T map, see (4.1.3)) defined on the set S of the Lemma,

$$\varphi_T : S \to \mathbf{R}^n, \qquad \varphi_T(x) := \varphi(T, 0, x), \qquad x \in S.$$

By the Lemma this operator maps the compact convex set S into itself, and, naturally, it is a continuous function of the initial value $x \in S$. By Brouwer's Fixed Point Theorem (Corollary A2.9) it has a fixed point $x^0 \in S : \varphi_T(x^0) = x^0$. According to Theorem 4.1.1 this means that system

(4.5.1) has a T-periodic solution $\varphi(t, 0, x^0)$. Since S is positively invariant, $\varphi(t, 0, x^0) \in S$ for $t \in \mathbf{R}$, in particular all the coordinates of φ are positive everywhere. $\square$

In the two dimensional case applying the Alvarez-Lazer [1986] results we have proved the uniqueness and asymptotic stability of the positive periodic solution under conditions corresponding to (4.5.3). Alvarez and Lazer have proved also the global asymptotic stability under the same conditions. Gopalsamy [1985] has proved the uniqueness and the global asymptotic stability under additional conditions; however, we shall make somewhat stronger assumptions and provide a different proof.

Theorem 4.5.3. *Assume that (4.5.3) holds, and also*

$$a_{iiL} > \sum_{j=1, j \neq i}^{n} a_{jiM}, \qquad a_{iiL} > \sum_{j=1, j \neq i}^{N} a_{ijM} x_{jM} / x_{iL} \qquad (4.5.8)$$

where x_{iL} and x_{jM} are given by (4.5.4) and (4.5.5), respectively; then system (4.5.1) has one and only one T-periodic solution with positive coordinates and this solution is globally asymptotically stable with respect to the positive orthant.

Proof. Denote a T-periodic solution guaranteed by the previous Theorem by ψ. We know that $\psi(t) \in S$ of Lemma 4.5.1 for $t \in \mathbf{R}$, and this implies that all the coordinates of the periodic function ψ are positive everywhere. Performing the coordinate transformation

$$y_i = (x_i - \psi_i(t)) / \psi_i(t) \qquad (i = 1, 2, ..., n),$$

system (4.5.1) is transformed into

$$\dot{y}_i = -(y_i + 1) \sum_{j=1}^{n} a_{ij}(t) \psi_j(t) y_j \qquad (i = 1, 2, ..., n). \qquad (4.5.9)$$

The orthant $Q := \{y \in \mathbf{R}^n : y_i > -1 \ (i = 1, 2, ..., n)\}$ is, clearly, positively invariant with respect to the last system, and the global asymptotic stability of the $y = 0$ solution of (4.5.9) with respect to Q is equivalent to the global asymptotic stability of the periodic solution ψ of (4.5.1) with respect to the open positive orthant $\mathrm{Int} \mathbf{R}^n_+$. The global asymptotic stability of ψ with respect to $\mathrm{Int} \ \mathbf{R}^n_+$ implies its uniqueness. In order to prove the global asymptotic stability of the origin with respect to system (4.5.9), the Liapunov function

$$V(y) := \sum_{i=1}^{n} (y_i - \ln(y_i + 1)), \qquad y \in Q$$

is introduced. This function is positive definite: $V(y) > 0$ for $y \in Q \setminus \{0\}$, $V(0) = 0$, and $V(y) \to \infty$ as $|y| \to \infty$ or one or more $y_i \to -1 + 0$. The derivative of V with respect to system (4.5.9) is

$$\dot{V}_{(4.5.9)}(t, y) = -\sum_{i,j=1}^{n} a_{ij}(t)\psi_j(t)y_i y_j.$$

The quadratic form on the right-hand side can be written in the form

$$\sum_{i,j=1}^{n} a_{ij}(t)\psi_j(t)y_i y_j = \sum_{i,j=1}^{n} h_{ij}(t)y_i y_j$$

where the matrix $H = [h_{ij}]$ is already symmetric:

$$h_{ij}(t) = (a_{ij}(t)\psi_j(t) + a_{ji}(t)\psi_i(t))/2$$

$$(i, j = 1, 2, ..., n).$$

Now, since all the functions a_{ij}, ψ_i are positive continuous and periodic $h_{ii}(t) \geq a_{iiL}\psi_i(t)$, and applying (4.5.8)

$$a_{iiL}\psi_i(t)/2 > \sum_{j=1, j \neq i}^{n} a_{jiM}\psi_i(t)/2 \geq \sum_{j=1, j \neq i}^{n} a_{ji}(t)\psi_i(t)/2$$

and

$$a_{iiL}\psi_i(t)/2 \ > \ a_{iiL}x_{iL}/2 > \sum_{j=1, j \neq i}^{n} a_{ijM}x_{jM}/2$$

$$\geq \sum_{j=1, j \neq i}^{n} a_{ij}(t)\psi_j(t)/2.$$

Hence,

$$h_{ii}(t) > \sum_{j=1, j \neq i}^{n} h_{ij}(t), \qquad t \in \mathbf{R} \qquad (i = 1, 2, ..., n);$$

thus, by the Corollary A1.16 of Hadamard's Theorem, $\dot{V}$ is negative definite for $t \in \mathbf{R}$, $y \in Q$. Because of the continuity, periodicity and strict positivity of all the functions h_{ij}, the function

$$h(r) := \min(-\dot{V}(t, y)), \qquad t \in \mathbf{R}, \qquad |y| = r > 0, \qquad y \in Q$$

is continuous and positive for $r \neq 0$, and

$$\dot{V}(t, y) \leq -h(|y|) \qquad \text{for} \qquad t \in \mathbf{R} \qquad y \in Q.$$

By Liapunov's Second Theorem 1.5.2 this implies that $y = 0$ is uniformly asymptotically stable, and since the domain of attractivity is the whole set Q, the origin is globally asymptotically stable with respect to Q. $\square$

The intuitive meaning of the inequalities (4.5.3) has been explained in the two dimensional case after the corresponding conditions (4.4.2). Inequalities (4.5.8) express an ecological rule of thumb: *in order to expect stability, the intraspecific competition has to be stronger than the interspecific competition among the different species* (see, e.g., Farkas [1990]). Indeed, a_{ii} measures the intraspecific competition: how much a unit increase of x_i decreases the specific growth rate of species i; a_{ij} and a_{ji} $(j \neq i)$ measure the interspecific competition: how much a unit increase of x_j, resp. x_i decreases the specific growth rate of species i, resp. j. According (4.5.8) a_{ii} dominates the off diagonal elements in its row and column.

We note that the uniqueness and the global asymptotic stability of the periodic solution has been proved under conditions (4.5.3) without conditions (4.5.8) by Tineo and Alvarez [1991].

In the rest of this Section the *periodically excited conservative system*

$$\ddot{x} + \operatorname{grad} G(x) = p(t, x) \qquad (4.5.10)$$

will be treated where $G \in C^2(\mathbf{R}^n, \mathbf{R})$, $p \in C^0(\mathbf{R} \times \mathbf{R}^n, \mathbf{R}^n)$, and p is periodic in t with period 2π: $p(t + 2\pi, x) = p(t, x)$ for $t \in \mathbf{R}$, $x \in \mathbf{R}^n$. The results to be presented are due to Ward [1978]. The case in which (4.5.10) contains also a dissipative term $C\dot{x}$ has been treated in Rouche-Mawhin [1973]. See also the references in the Ward paper. The Hessian matrix of the scalar function G will be denoted by $G''(x) = [g''_{x_i x_k}(x)]$, and the following Theorem will be proved.

Theorem 4.5.4. *If there is an $m > 0$ such that $|p(t, x)| \leq m$ for $(t, x) \in \mathbf{R} \times \mathbf{R}^n$, there are an integer N and positive numbers q_1, q_2 and r such that $N^2 < q_1 < q_2 < (N + 1)^2$, and for $|x| \leq r$ the matrices $G''(x) - q_1 I$, $q_2 I - G''(x)$ are positive semidefinite, then (4.5.10) has a 2π-periodic solution.*

Proof. Hilbert space methods (see Definition A2.6) and Schauder's Fixed Point Theorem will be used. We shall assume, without loss of generality, that $\operatorname{grad} G(0) = 0$. (We may subtract $\operatorname{grad} G(0)$ from both sides of equation (4.5.10) without changing its character and the conditions.)

The Hilbert space of square integrable functions on the interval $[0, 2\pi]$ will be denoted by $E := L^2([0, 2\pi], \mathbf{R}^n)$. The usual scalar product will be defined:

$$< u, v > := \int_0^{2\pi} u'(t)v(t)dt, \qquad u, v \in E$$

where u' is the transpose of the column vector $u \in \mathbf{R}^n$. The norm of a function $u \in E$ is generated by this scalar product : $||u|| := < u, u >^{1/2}$. The set of absolutely continuous functions will be denoted by C^A (see Definition 1.1.3), and the following subset of E will be used:

$$D := \{u \in E : u, \dot{u} \in C^A, \quad \ddot{u} \in E, \quad u(0) = u(2\pi), \quad \dot{u}(0) = \dot{u}(2\pi)\}.$$

We convert (4.5.10) into an operator equation.

First, the linear operator $L : D \to E$ is defined by $Lu := \ddot{u}$ for $u \in D$. The assumptions on the Hessian G'' of the Theorem imply that a positive number $a > 0$ exists such that $| \operatorname{grad} G(x)| \leq a|x|$ for $x \in \mathbf{R}^n$. Therefore, the operator $N : E \to E$ defined by $Nu := \operatorname{grad} G(u)$ is continuous. Also the operator $P : E \to E$ defined by $(Pu)(t) := p(t, u(t))$, $t \in [0, 2\pi]$, is continuous. With these operators we may write (4.5.10) in the form

$$Lx + Nx = Px, \tag{4.5.11}$$

and it is clear that x is a 2π-periodic solution of (4.5.10) if and only if its restriction to $[0, 2\pi]$ is a solution of (4.5.11). The spectrum of the linear operator L is $\{-k^2 : k \in \mathbf{N}\}$, i.e. $Lx = \lambda x$ has a solution in D if and only if $\lambda = -k^2$, a non-positive square number (the corresponding eigenfunctions are $\cos kt, \sin kt$ $(k = 0, 1, 2, ...)$). Let $q := (q_1 + q_2)/2$; because of the assumptions of the Theorem, $-q$ is not an element of L's spectrum, and as a consequence, the inverse $(L + qI)^{-1}$ exists and is a bounded linear operator on E. It can also be shown that $(L + qI)^{-1}$ is a compact operator (see Definition A2.7 and Coddington-Levinson [1955] p.194). The operator equation (4.5.11) can be written in the form

$$x = (L + qI)^{-1}(P - N + qI)x. \tag{4.5.12}$$

The operator $P - N + qI$ maps bounded subsets of E into bounded subsets; hence, the composite operator $T := (L + qI)^{-1}(P - N + qI)$ is completely continuous.

We are going to show now that T maps a closed ball of E into itself. It is easy to see by partial integration that the operator L on D is *self-adjoint*, i.e. $< u, Lv >=< Lu, v >$ for $u, v \in D$; hence, the (in modulus) maximal eigenvalue of the inverse operator $(L + qI)^{-1}$ is the reciprocal of the (in modulus) minimal eigenvalue of $L + qI$ (see Coddington-Levinson [1955]). The spectrum of $L + qI$ is $\{q - k^2 : k \in \mathbf{N}\}$; hence,

$$\|(L + qI)^{-1}\| = \text{the modulus } of \text{ the maximal eigenvalue}$$
$$= 1/\min(q - N^2, (N + 1)^2 - q). \tag{4.5.13}$$

On the other hand, for $u \in E$,

$$\|(P - N + qI)u\| \leq \|Pu\| + \|(N - qI)u\|$$
$$\leq m\sqrt{2\pi} + \|(N - qI)u\|. \tag{4.5.14}$$

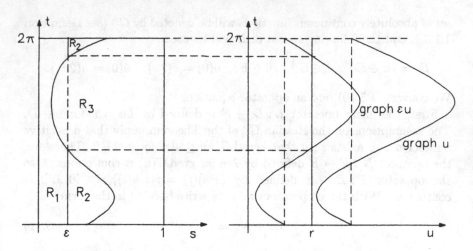

FIGURE 4.5.1. Division of the rectangle R in the proof of Theorem 4.5.4 according to the values of the function $su(t)$.

The estimation of the last norm requires more work. Since $\operatorname{grad} G(0) = 0$, clearly,

$$\operatorname{grad} G(x) = \int_0^1 (d \operatorname{grad} G(sx)/ds) ds = \int_0^1 G''(sx) x \, ds;$$

hence,

$$
\begin{aligned}
\|Nu - qu\|^2 &= \int_0^{2\pi} |\operatorname{grad} G(u(t)) - qu(t)|^2 dt \\
&= \int_0^{2\pi} \left| \int_0^1 (G''(su(t)) - qI) ds \, u(t) \right|^2 dt \\
&\leq \int_0^{2\pi} \left(\int_0^1 |G''(su(t)) - qI| ds \right)^2 |u(t)|^2 dt.
\end{aligned}
$$

The norm of the symmetric matrix $G''(su(t)) - qI$ is equal to the modulus of its (in modulus) maximal eigenvalue. The eigenvalues of this matrix are $\lambda_i - q$ $(i = 1, 2, ..., n)$ where λ_i are the eigenvalues of the symmetric matrix $G''(su(t))$. Now, the assumptions of the Theorem imply that if $|su(t)| \geq r$, then the eigenvalues of $G''(su(t))$ are between q_1 and q_2, i.e.

$$q_1 \leq \lambda_i \leq q_2 \qquad (i = 1, 2, ..., n).$$

Thus, if $|su(t)| \geq r$, we have

$$|G''(su(t)) - qI| = \max_i |\lambda_i - q| \leq (q_2 - q_1)/2.$$

For an arbitrary $\varepsilon > 0$ we divide the rectangle $R := [0,1] \times [0, 2\pi]$ of the s, t plane into three parts:

$$R_1 := \{(s,t) \in R : |su(t)| \leq r, \quad 0 \leq s \leq \varepsilon\},$$
$$R_2 := \{(s,t) \in R : |su(t)| \leq r, \quad \varepsilon \leq s \leq 1\}$$
$$R_3 := \{(s,t) \in R : |su(t)| > r\}.$$

(See Figure 4.5.1.) Continuing the estimate of the norm of $Nu - qu$ we get

$$\|Nu - qu\|^2 \leq \int\int_R |G''(su(t)) - qI|^2 |u(t)|^2 ds\, dt$$

$$= \int\int_{R_1} + \int\int_{R_2} + \int\int_{R_3}$$

$$\leq \varepsilon M^2 \|u\|^2 + 2\pi M^2 r^2 / \varepsilon^2 + \|u\|^2 (q_2 - q_1)^2 / 4$$

where $M := \max |G''(x) - qI|$, $|x| \leq r$. If $\|u\| \geq Mr\sqrt{2\pi}/\varepsilon^{3/2}$, then from the previous inequality we obtain

$$\|Nu - qu\| \leq (\varepsilon(M^2 + 1) + (q_2 - q_1)^2 / 4)^{1/2} \|u\|. \tag{4.5.15}$$

Now, (4.5.13) and the inequalities $N^2 < q_1 < q < q_2 < (N+1)^2$ with $q = (q_1 + q_2)/2$ imply that $\min(q - N^2, (N+1)^2 - q) > (q_2 - q_1)/2$; hence,

$$l := \|(L + qI)^{-1}\| < 2/(q_2 - q_1). \tag{4.5.16}$$

Thus, for $\|u\| \geq MR\sqrt{2\pi}/\varepsilon^{3/2}$ we have

$$\|(L + qI)^{-1}\| \ \|Nu - qu\| \leq (\varepsilon l^2 (M^2 + 1) + l^2 (q_2 - q_1)^2 / 4)^{1/2} \|u\|.$$

By (4.5.16), $l^2 (q_2 - q_1)^2 / 4 < 1$, so that an $\varepsilon_1 > 0$ can be found such that

$$\delta := (\varepsilon_1 l^2 (M^2 + 1) + l^2 (q_2 - q_1)^2 / 4)^{1/2} < 1,$$

and, consequently, for $\|u\| \geq Mr\sqrt{2\pi}/\varepsilon_1^{3/2}$ we obtain

$$\|(L + qI)^{-1}\| \ \|Nu - qu\| \leq \delta \|u\|, \qquad 0 < \delta < 1. \tag{4.5.17}$$

We are able to show now that a positive integer k exists such that the operator $T = (L + qI)^{-1}(P - N + qI)$ maps the closed ball

$$\overline{B}(0, k) = \{u \in E : \|u\| \leq k\}$$

into itself. Assume that, contrary to the statement, such a k does not exist. Then we can find a sequence $x_i \in E$ $(i = 1, 2, ...)$ such that $x_i \in \overline{B}(0, i)$ and $\|Tx_i\| > i$. We must have $\|x_i\| \to \infty$ as $i \to \infty$, since otherwise T would map a bounded set onto an unbounded one. Thus, an index $j \in \mathbf{N}$ exists such that for all $i > j$ we have $||x_i|| > Mr\sqrt{2\pi}/\varepsilon_1^{3/2}$, and applying (4.5.14) and (4.5.17),

$$||x_i|| \le i < ||Tx_i|| \le lm\sqrt{2\pi} + \delta||x_i||;$$

hence,

$$||x_i|| \le lm\sqrt{2\pi}/(1 - \delta), \qquad i > j,$$

a contradiction. Thus, a positive integer k exists such that T maps the closed ball $\overline{B}(0, k)$ into itself. We may now apply Schauder's Fixed Point Theorem A2.16 and conclude that an $x^0 \in \overline{B}(0, k)$ exists such that $x^0 = Tx^0$. Hence, $x^0 \in D$ and satisfies (4.5.11), i.e.

$$\ddot{x}^0(t) \equiv - \operatorname{grad} G(x^0(t)) + p(t, x^0(t)), \qquad t \in [0, 2\pi].$$

The last identity implies that $x^0 \in C^2([0, 2\pi], \mathbf{R}^n)$. We know also that $x^0(0) = x^0(2\pi)$, $\dot{x}^0(0) = \dot{x}^0(2\pi)$, so that x^0 can be periodically extended to the whole real line, and the resulting 2π-periodic function is a solution of (4.5.10). □

4.6 Problems

Problem 4.6.1. Consider the *periodic logistic differential equation*

$$\dot{N} = r(t)N(1 - N/K(t)) \tag{4.6.1}$$

where $r, K \in C^0$, $r(t) > 0$, $K(t) > 0$ for $t \in \mathbf{R}$ and there is a $T > 0$ such that $r(t + T) = r(t)$, $K(t + T) = K(t) \ne \text{const}$ for $t \in \mathbf{R}$. Show that (4.6.1) has a non-constant periodic solution of period T, and that all solutions with positive initial value tend to this periodic solution as t tends to infinity. (Hint: cf. Example 4.1.1.)

Problem 4.6.2. Let $D \subset \mathbf{R}^n$ be a bounded open set and $f, g \in C^0(\overline{D}, \mathbf{R}^n)$; if $f(x) = g(x) \ne 0$ for $x \in \partial D$, then $d[f, D, 0] = d[g, D, 0]$. Prove this statement which means that the topological degree of a mapping depends only on the values assumed on the *boundary* of the domain. (Hint: Use homotopy and cf. the proof of Corollary A2.6.)

Problem 4.6.3. Let f be a continuous real valued function defined on the compact interval $[a, b]$, and assume that $f(a) \ne 0 \ne f(b)$. Prove that

$$d[f, (a, b), 0] = \begin{cases} 1 & \text{if} \quad f(a) < 0 < f(b) \\ -1 & \text{if} \quad f(a) > 0 > f(b) \\ 0 & \text{if} \quad f(a)f(b) > 0. \end{cases}$$

(Hint: Join f by homotopy to the linear function $l(x) = f(a) + (x - a)(f(b) - f(a))/(b - a)$ and determine the degree of the latter by Definition A2.1.)

Problem 4.6.4. Consider the polynomial of degree n

$$p_n(x) = a_n x^n + a_{n-1} x^{n-1} + \cdots + a_1 x + a_0$$

where n is a positive integer and $a_n \neq 0$; show that if $[a, b]$ is a sufficiently large interval (in both directions) containing the origin in its interior $(a < 0 < b)$, then

$$d[p_n, (a, b), 0] = \begin{cases} 1 & \text{if} \quad n \text{ is odd and } a_n > 0 \\ -1 & \text{if} \quad n \text{ is odd and } a_n < 0 \\ 0 & \text{if} \quad n \text{ is even .} \end{cases}$$

(Hint: Assume that (a, b) contains all the real roots of p_n and make use of the previous Problem.)

Problem 4.6.5. Assume that every condition but (ii) of Theorem 4.2.4 holds, and (ii) is replaced by (ii'), there exists a $t_0 \in \mathbf{R}$ such that for all $\delta > 0$, and there is $z^0 \in B(0, \delta)$ such that $V(t_0, z^0) < 0$, and $V(t, 0) = 0$. Prove that in this case the z=0 solution of (4.2.7) is unstable.

Problem 4.6.6. Prove that a non-trivial T-periodic solution p of the homogeneous linear system $\dot{x} = A(t)x$ where $A \in C^0$ and $A(t + T) = A(t)$, $(T > 0)$ can never be isolated. Construct a two dimensional C^1 manifold containing the integral curve of p and consisting of the integral curves of T-periodic solutions.

Problem 4.6.7. Assume that $f, g \in C^1$, these functions satisfy the conditions of Theorem 4.3.1 and also the conditions $f(0, 0) < 0$, $xg(x) > 0$ for every $x \neq 0$; prove that the differential equation

$$\ddot{x} + f(x, \dot{x})\dot{x} + g(x) = 0$$

has a non-constant periodic solution whose path surrounds the origin $(0, 0)$ in the phase plane x, $\dot{x}$.

(Hint: Apply Theorem 4.3.1, Corollary 3.1.9 and the level curves $y^2/2 + G(x) = c$ for small $c > 0$ where $y = \dot{x}$.)

Problem 4.6.8. Consider the *damped mathematical pendulum with periodic forcing* (cf. Example 1.5.4)

$$\ddot{x} + d\dot{x} + (g/l) \sin x = b(t)$$

where $d > 0$, $g/l > 0$, $b \in C^0$ and $b(t + T) = b(t)$ for some $T > 0$. Show that this differential equation cannot satisfy the conditions of Theorem 4.3.5.

Problem 4.6.9. Consider the *damped mathematical pendulum with periodic forcing that is proportional to the displacement*

$$\ddot{x} + d\dot{x} + (g/l) \sin x = -xb(t)$$

where d, g/l and the function b are as in the previous problem. Establish conditions for b such that the equation shall satisfy the conditions of Theorem 4.3.6.

5
Autonomous Systems of Arbitrary Dimension

In this chapter we continue the study of periodic solutions of autonomous systems, this time, of arbitrary finite dimension. The tools that are on hand to prove the existence of periodic solutions are, in principle, the same as in the case of non-autonomous periodic systems. However, as this has been pointed out in the introduction to the previous chapter, the application of fixed point theorems is more difficult in the present case because an autonomous system does not contain any *a priori* information about the period of a possible periodic solution: the period T in the Poincaré-Andronov operator (4.1.3) is not known. We shall postpone the treatment of some methods that may help in overcoming this difficulty until Section 4. (Perturbation methods will be treated in the next chapter.) In Section 1 we introduce the stability concept which is the most suitable to be applied in case of non-constant periodic solutions of autonomous systems: orbital stability. Here we shall prove the Andronov-Witt Theorem which is fundamental in stability investigations. In Section 2 the isolation problem will be treated and in case a periodic solution is not isolated we shall discuss also the problem of isochronism: whether all the periodic solutions have the same period or not ("a wheel or not a wheel?"). In Section 3 we extend the discussion of periodic solutions to "D-periodic solutions" of cylindrical systems, i.e. to systems that are periodic in some of the state variables. In Sections 4 and 5, methods will be shown that help to prove the existence of periodic solutions in dimensions higher than two. This will be done by presenting applications that are of interest on their own account. We note that there is a now classical text of Pliss [1966] concerning these problems and a more recent survey by Li [1981]. Section 6 deals with manifolds of periodic orbits and invariant manifolds related to periodic solutions.

5.1   Orbital Stability

Though an autonomous system can be considered as a special case of a periodic one (any number is a period of the system), the stability problem of a non-constant periodic solution of the former is much more intricate than that of the latter. A non-constant T-periodic $(T > 0)$ solution $p(t)$ of the autonomous system $\dot{x} = f(x)$, $f \in C^1(\mathbf{R}^n)$, *cannot be asymptotically stable*. This is clear because of (1.3.3), since for arbitrary $t_0 \in \mathbf{R}$ the function $p(t - t_0)$ is also a non-constant T-periodic solution, the initial values $p(0)$ and $p(-t_0)$ can be arbitrarily close (if $|t_0|$ is small enough), and still $p(t) - p(t - t_0)$ does not tend to zero as t tends to infinity. One may say in a somewhat picturesque way that the point $p(t - t_0)$ is chasing the point $p(t)$ along their common closed trajectory, but is always lagging behind by t_0 units of time, the difference being a non-constant T-periodic function. This means, naturally, that for a non-constant periodic solution of an autonomous system the conditions of the first part of Theorem 4.2.1 cannot hold. Indeed, with the previous notations $\dot{p}(t)$ is a non-trivial T-periodic solution of the variational system $\dot{y} = f'(p(t))y$, and this implies (Corollary 2.2.3) that 1 is a characteristic multiplier.

There are problems with Liapunov stability too. Consider, e.g., the simplest model of the solar system assuming that a large mass M (the Sun) is located at the origin, and a small mass m (a planet, the Earth) is revolving around it. The system is considered to be conservative; Newton's equation of motion being $m\ddot{r} = -\gamma Mmr/|r|^3$ where $r = [x_1, x_2, x_3]$ is the radius vector of the planet and $\gamma > 0$ is the gravitational constant. The right-hand side of the differential system $\ddot{r} = -\gamma Mr/|r|^3$ is the gradient of the potential $\gamma M/|r|$. We know that the *orbits* of the planets, i.e. the trajectories corresponding to initial conditions (position and velocity) from a fairly large open domain of the phase space are ellipses, and if we vary the initial position and velocity just a little, then the corresponding ellipse varies only a little. This means that some kind of stability prevails; nevertheless the motion of the planet is *not stable in the Liapunov sense*. This can be seen, for instance, remembering Kepler's law, according to which the ratio of the squares of the revolution times (i.e. the periods) of two planets is equal to the ratio of the cubes of the "mean distances from the Sun". (More exactly, one has to consider the half great axes of the orbital ellipse instead of the mean distances.) This means that the slightest change in the initial conditions leading to a small change of the mean distance results in a change in the period of the motion. The consequence of the least non-zero difference in the periods is that in the arbitrary far future moments will occur at which the two planets, the original one and the perturbed one, will be in opposition, i.e. their distance will be large.

This model of the "two-body problem" serves as an example of an autonomous system of differential equations whose non-constant periodic solutions are not stable in the Liapunov sense but the closed trajectories of which show some kind of "*orbital stability*". There are, of course, autonomous systems with non-constant periodic solutions that are stable in the Liapunov sense. Such is, for instance, a homogeneous linear system with constant coefficients satisfying the requirements of Theorem 1.4.5 and having a pair of pure imaginary conjugate eigenvalues. However, there is something "unnatural" about such linear systems: a system like this has at least one one-parameter family of closed trajectories filling in a two dimensional subspace corresponding to periodic solutions of the *same period*. In the case of the simple "harmonic oscillator" $\dot{y}_1 = -y_2$, $\dot{y}_2 = y_1$, the whole plane y_1, y_2 is rotating around the origin like a wheel with the same angular velocity, implying that points far from the origin have very large speeds; the speed tends to infinity as the distance from the origin tends to infinity. This is an instance of "*isochronism*", characteristic of autonomous linear systems, which will be dealt with in the next Section. Typically one would expect that if a system has a family of periodic solutions (with different orbits), then the period is not constant, depends on the orbit and this is the case indeed, usually, when a *non-linear system* has a family of closed trajectories. This is another basic difference between linear and non-linear autonomous systems besides the fact that a non-linear autonomous system may have, generically, a non-constant periodic solution with isolated path (see, e.g., Theorem 3.2.1), in the case of a linear autonomous system this is impossible.

In order to grasp the situation presented by the "two body problem", we are going to introduce a new stability concept. Consider the autonomous system of differential equations

$$\dot{x} = f(x) \tag{5.1.1}$$

where $f \in C^1(X)$, $X \subset \mathbf{R}^n$ being an open and connected subset. Let $\psi : \mathbf{R}_+ \to X$ be a solution of (5.1.1) and denote its path by

$$\gamma := \{x \in X : x = \psi(t), \quad t \in \mathbf{R}_+\}.$$

Occasionally we may consider solutions which are defined on the whole real line, i.e. $\psi : \mathbf{R} \to X$. In this case the path is $\gamma := \{x \in X : x = \psi(t), \ t \in \mathbf{R}\}$, and we speak about the *positive semitrajectory* $\gamma^+ := \{x \in X : x = \psi(t), \ t \in \mathbf{R}_+\}$, resp. *negative semitrajectory* $\gamma^- := \{x \in X : x = \psi(t), \ t \in \mathbf{R}_-\}$.

Definition 5.1.1. The solution $\psi : \mathbf{R}_+ \to X$ of (5.1.1) is said to be *orbitally stable* if for every $\varepsilon > 0$ there exists a $\delta(\varepsilon) > 0$ such that if the distance of the initial value x^0 from the path γ of ψ is less than $\delta(\varepsilon) :$ dist $(x^0, \gamma) < \delta(\varepsilon)$, then the solution $\varphi(t, x^0)$ that assumes the value

x^0 at $t = 0$ is defined for $t \in [0, \infty)$, and for $t \geq 0$:

$$\text{dist} \, (\varphi(t, x^0), \gamma) < \varepsilon.$$

Note, first, that if the solution ψ is orbitally stable, then each solution with the same path γ, i.e. every solution $\psi(t + \alpha)$ for $\alpha \in \mathbf{R}$, is orbitally stable too. Therefore orbital stability of a solution means, in fact, the stability of its orbit.

Secondly, this Definition says roughly that if a motion gets near enough to the path γ at time $t = 0$, then it stays arbitrarily close to it in the whole future $t > 0$. Now, the moment $t = 0$ has no importance in this Definition. Indeed, assume that for some t_0 in the domain of $\varphi(t, x^0)$ we have dist $(\varphi(t_0, x^0), \gamma) < \delta(\varepsilon)$ where $\delta(\varepsilon)$ is the delta that belongs to ε in the Definition. Then, clearly, dist $(\varphi(t, \varphi(t_0, x^0)), \gamma) < \varepsilon$ for all $t \geq 0$, i.e. dist $(\varphi(t + t_0, x^0), \gamma) < \varepsilon$ for $t \in [0, \infty]$. This means that $\varphi(t, x^0)$ is defined for $t \in [t_0, \infty)$, and that dist $(\varphi(t, x^0), \gamma) < \varepsilon$ for $t \in [t_0, \infty)$.

Thirdly, it is clear that stability in the Liapunov sense (Definition 1.4.1) implies orbital stability. By the way, an argument similar to that in the previous paragraph yields that if a solution of an autonomous system is stable in the Liapunov sense, then it is uniformly stable (see Definition 1.4.2). On the other hand, orbital stability does not imply Liapunov stability. Besides referring to the model of the "two body problem" having been treated intuitively, we are going to show this by the following:

Example 5.1.1. We have referred to the simple harmonic oscillator $\dot{y}_1 = -y_2$, $\dot{y}_2 = y_1$ in this Section already. We are considering now a non-linear system with the same orbits, i.e. concentric circles around the origin, namely

$$\dot{x}_1 = -x_2(x_1^2 + x_2^2)^{1/2}, \qquad \dot{x}_2 = x_1(x_1^2 + x_2^2)^{1/2}. \qquad (5.1.2)$$

The right-hand sides are continuously differentiable in the whole plane except the origin. Introducing polar coordinates by $x_1 = r \cos \theta$, $x_2 = r \sin \theta$, the system assumes the form

$$\dot{r} = 0, \qquad \dot{\theta} = r,$$

i.e. in polar coordinates, the general solution is $r = r_0$, $\theta(t) = r_0 t + \theta_0$ where $r_0 \geq 0$, and $\theta_0 \in \mathbf{R}$ can be chosen arbitrarily. Consider two solutions with initial conditions $r_1 > 0$ $\theta_1 = 0$, and $r_2 > 0$, $\theta_2 = 0$, respectively. In the original Cartesian coordinates they are

$$\varphi^1(t) = (r_1 \cos r_1 t, r_1 \sin r_1 t) \quad \text{and} \quad \varphi^2(t) = (r_2 \cos r_2 t, r_2 \sin r_2 t).$$

Their difference is

$$|\varphi^1(t) - \varphi^2(t)| = (r_1^2 + r_2^2 - 2r_1 r_2 \cos(r_1 - r_2)t)^{1/2}.$$

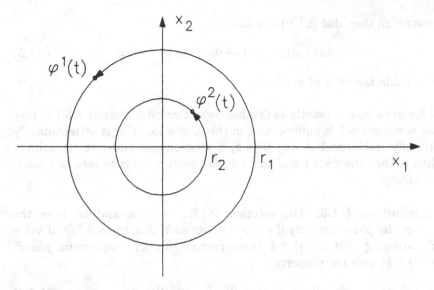

FIGURE 5.1.1. Motions governed by system (5.1.2). The angular velocity of φ^1 is r_1/r_2 times as big as that of φ^2.

Assume without loss of generality that $r_1 > r_2$. This difference assumes the value $r_1 + r_2$ at

$$t_k = (2k+1)\pi/(r_1 - r_2) \to \infty, \quad \text{as} \quad k \to \infty, \quad k \in \mathbf{N}.$$

This means that the solution $\varphi^2(t)$, say, is *not stable in the Liapunov sense*, since if we choose a $0 < \varepsilon \le r_2$, then no matter how small the difference of the initial values $r_1 - r_2$ is, the difference of the solutions will assume values greater than $2\varepsilon < r_1 + r_2$ in the arbitrary far future. Thus, no solution of (5.1.2) (except the equilibrium point $(x_1, x_2) = (0,0)$) is stable in the Liapunov sense, though obviously, *all solutions are orbitally stable* (see Figure 5.1.1). This is in contrast to the harmonic oscillator where we have isochronism (the solutions are $(r_0 \cos(t + \theta_0), r_0 \sin(t + \theta_0))$, $r_0 \ge 0$, $\theta_0 \in \mathbf{R}$), and as a consequence, all solutions are stable in the Liapunov sense. (The heedful reader may observe that system (5.1.2) is "still less natural" than the harmonic oscillator since in it the angular velocity is increasing with the distance from the origin. He/she is right.)

There is an important stability concept related to orbital stability, loosely speaking, as asymptotic stability is related to stability in the Liapunov sense.

Definition 5.1.2. The solution $\psi : \mathbf{R}_+ \to X$ of system (5.1.1) is said to be *asymptotically orbitally stable* if it is orbitally stable, and if a $\delta > 0$

exists such that dist $(x^0, \gamma) < \delta$ implies

$$\text{dist } (\varphi(t, x^0), \gamma) \to 0, \quad \text{as} \quad t \to \infty \quad (5.1.3)$$

(γ is again the orbit of ψ).

It can be shown, exactly as this has been done after Definition 5.1.1, that the moment $t = 0$ is unimportant in this Definition. If ψ is asymptotically orbitally stable, and at *any time* t_0 a solution gets closer to the orbit γ than δ, then the distance of this solution from γ tends to zero as t tends to infinity.

Definition 5.1.3. The solution $\psi : \mathbf{R}_+ \to X$ is said to have the *asymptotic phase property* if a $\delta > 0$ exists such that to each initial value x^0 satisfying dist $(x^0, \gamma) < \delta$ there corresponds an *"asymptotic phase"* $\alpha(x^0) \in \mathbf{R}$ with the property

$$\lim_{t \to \infty} |\varphi(t + \alpha(x^0), x^0) - \psi(t)| = 0. \quad (5.1.4)$$

Note that the requirement (5.1.4) is, obviously, equivalent to

$$\lim_{t \to \infty} |\varphi(t, x^0) - \psi(t - \alpha(x^0))| = 0. \quad (5.1.5)$$

Thus, we see that the asymptotic phase property, just like orbital stability and asymptotic orbital stability, is, in fact, a property of the orbit γ. Equation (5.1.5) means that given an initial value x^0 near enough to γ, a solution having γ as its path can be chosen in such a way that the difference of the latter and of $\varphi(t, x^0)$ tends to zero as t tends to infinity. If asymptotic stability were possible in case of a non-constant solution of an autonomous system, then it would, clearly, imply asymptotic orbital stability with the asymptotic phase property. For each initial value near enough to the given solution, the asymptotic phase would be zero. At the beginning of this Section we have shown that this, i.e. asymptotic stability, is impossible for a non-constant solution of an autonomous system. On the other hand, if a non-constant solution ψ of an autonomous system is asymptotically orbitally stable, and we consider an initial value x^0 *on the path γ of ψ* : dist $(x^0, \gamma) = 0$, then a t_0 exists such that $\psi(t_0) = x^0$, so $\varphi(t, x^0) = \psi(t + t_0)$. The asymptotic phase belonging to this initial value $x^0 \in \gamma$ is then $\alpha(x^0) = -t_0$, since

$$\varphi(t - t_0, x^0) - \psi(t) = \psi(t - t_0 + t_0) - \psi(t) \equiv 0.$$

We are going to establish now a fundamental criterion, *Andronov-Witt's Theorem* which enables one to determine the asymptotic orbital stability of a periodic solution. We shall need the following:

Lemma 5.1.1. *Let $A : \mathbf{R} \to \mathbf{R}^{n^2}$ be a real, continuous, periodic matrix function, $A \in C^0(\mathbf{R})$, $A(t+T) = A(t)$ for some $T > 0$, and assume that the homogeneous linear system*

$$\dot{y} = A(t)y \tag{5.1.6}$$

has the number 1 as a simple characteristic multiplier (in the characteristic polynomial), and the remaining $n - 1$ characteristic multipliers are in modulus less than one; then system (5.1.6) has a real fundamental matrix Φ of the form

$$\Phi(t) = P(t) \begin{pmatrix} 1 & 0 \\ 0 & \exp(B_1 t) \end{pmatrix} \tag{5.1.7}$$

where P is a regular, continuously differentiable, T-periodic matrix, and B_1 is an $(n-1) \times (n-1)$ stable matrix.

Proof. According to (1.2.9) we denote by $\Phi(t,0)$ the real fundamental matrix of (5.1.6) that assumes the unit matrix I at $t = 0 : \Phi(0,0) = I$. The principal matrix (see (2.2.5) et seq.) $C = \Phi(T,0)$ has 1 as a simple eigenvalue, and $n - 1$ eigenvalues with modulus less than one. Therefore (see Gantmacher [1954]), a real non-singular matrix M exists that brings C to the form

$$\begin{pmatrix} 1 & 0 \\ 0 & C_1 \end{pmatrix} = M^{-1}CM$$

by a similarity transformation where C_1 is a real, non-singular $(n-1) \times (n-1)$ matrix all the eigenvalues of which are in modulus less than one (cf. (A3.15) et seq.). Let us take now the real fundamental matrix $\Phi(t) = \Phi(t,0)M$. By (2.2.6)

$$\begin{aligned} \Phi(t+T) &= \Phi(t,0)\Phi(T,0)M \\ &= \Phi(t) \begin{pmatrix} 1 & 0 \\ 0 & C_1 \end{pmatrix}. \end{aligned} \tag{5.1.8}$$

Let B_1 be an $(n-1) \times (n-1)$ matrix satisfying

$$\exp(B_1 T) = C_1 \tag{5.1.9}$$

(see Theorem A1.9). We know that every eigenvalue of B_1 has a negative real part. If the $n \times n$ matrix B is defined by

$$B := \begin{pmatrix} 0 & 0 \\ 0 & B_1 \end{pmatrix},$$

then it is easy to see that

$$B^k = \begin{pmatrix} 0 & 0 \\ 0 & B_1^k \end{pmatrix} \qquad (k = 1, 2, 3, ...),$$

and therefore

$$\exp(Bt) = \begin{pmatrix} 1 & 0 \\ 0 & \exp(B_1 t) \end{pmatrix}, \qquad (5.1.10)$$

in particular

$$\exp(BT) = \begin{pmatrix} 1 & 0 \\ 0 & \exp(B_1 T) \end{pmatrix} = \begin{pmatrix} 1 & 0 \\ 0 & C_1 \end{pmatrix}. \qquad (5.1.11)$$

Let $P(t) := \Phi(t) \exp(-Bt)$, then applying (5.1.8) and (5.1.11) we have

$$P(t + T) = \Phi(t + T)e^{-BT}e^{-Bt} = \Phi(t)e^{-Bt} = P(t),$$

hence (5.1.7), follows, $P(t)$ and B_1 satisfying all the requirements. $\square$

Let us return now to the non-linear autonomous system (5.1.1), assume that this system has a non-constant periodic solution $p(t)$ with period $T > 0$, and consider the variational system with respect to this solution

$$\dot{y} = f'(p(t))y. \qquad (5.1.12)$$

We know from (1.3.7) that $\dot{p}(t)$ is a non-trivial T-periodic solution of this periodic homogeneous linear system, and, therefore, by Corollary 2.2.3 the number 1 is a characteristic multiplier of (5.1.12).

Theorem 5.1.2 *(Andronov-Witt [1933], Demidovich [1967$_a$, 1967$_b$] see also Coddington-Levinson [1955]). If 1 is a simple characteristic multiplier of the variational system (5.1.12), and the remaining $n - 1$ characteristic multipliers are in modulus less than one, then the periodic solution $p(t)$ of (5.1.1) is asymptotically orbitally stable, having the asymptotic phase property.*

Proof. Denote the path of the T-periodic solution p by

$$\gamma = \{x \in X : x = p(t), \ 0 \le t \le T\}.$$

The essence of the proof is that a hypersurface crossing the orbit γ "transversally" will be constructed such that the difference between solutions with initial values in this surface on the one hand and $p(t)$, on the other, tends to zero exponentially as t tends to infinity. Then we are going to show that all solutions with initial conditions near enough to γ cross this surface at some moment (see Figure 5.1.2).

Without loss of generality we assume that the origin has been displaced to the point $p(0)$, i.e. $p(0) = 0$, and that the coordinate system has been rotated such a way that the tangent vector $\dot{p}^0 = \dot{p}(0)$ points in the direction of the positive x_1, axis, i.e. the coordinates of this vector are $\dot{p}^0 = (\dot{p}_1^0, 0, ..., 0)$, $\dot{p}_1^0 > 0$. Next we introduce the variation of solutions as a

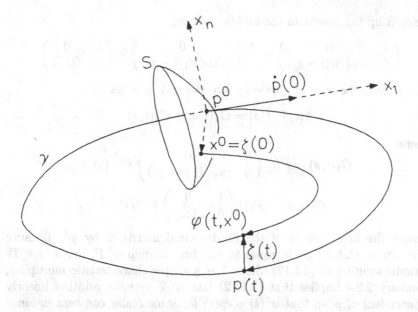

FIGURE 5.1.2. The path γ of the asymptotically orbitally stable periodic solution p with the manifold S of initial values determining solutions φ that tend to p as t tends to infinity.

new "unknown function". Let $\varphi(t)$ be an arbitrary solution of (5.1.1), and note that the differential equation satisfied by $z(t) = \varphi(t) - p(t)$ is

$$\dot{z}(t) = f(p(t) + z(t)) - f(p(t)) = f'(p(t))z(t) + r(t, z(t)) \qquad (5.1.13)$$

where

$$r(t, z) := f(p(t) + z) - f(p(t)) - f'(p(t))z. \qquad (5.1.14)$$

Clearly, r is a continuous function; it is also continuously differentiable with respect to z:

$$r'_z(t, z) = f'(p(t) + z) - f'(p(t)),$$

so that $r(t, 0) \equiv 0$, $r'_z(t, 0) \equiv 0$, and it is T-periodic in t. Further $r(t, z) \to 0$, $r'_z(t, z) \to 0$, as $z \to 0$ uniformly in $t \in [0, \infty]$.

Now, an integral equation will be constructed whose solutions satisfy system (5.1.13). By the assumptions of the Theorem the variational system (5.1.12) satisfies the conditions of the previous Lemma. Let $\Phi(t)$ be the fundamental matrix of (5.1.12) of the form (5.1.7). The fundamental matrix that is equal to the unit matrix I at $t = s$ (s fixed) is

$$
\Phi(t)\Phi^{-1}(s) = P(t) \begin{pmatrix} 1 & 0 \\ 0 & \exp(B_1 t) \end{pmatrix} \begin{pmatrix} 1 & 0 \\ 0 & \exp(-B_1 s) \end{pmatrix} P^{-1}(s)
$$

$$
= P(t) \begin{pmatrix} 1 & 0 \\ 0 & \exp(B_1(t - s)) \end{pmatrix} P^{-1}(s).
$$

We split up the matrix in the middle, writing

$$\begin{pmatrix} 1 & 0 \\ 0 & \exp(B_1(t-s)) \end{pmatrix} = \begin{pmatrix} 0 & 0 \\ 0 & \exp(B_1(t-s)) \end{pmatrix} + \begin{pmatrix} 1 & 0 \\ 0 & O_{n-1} \end{pmatrix}$$

(O_{n-1} being the $(n-1) \times (n-1)$ zero matrix), and get

$$\Phi(t)\Phi^{-1}(s) = G_1(t,s) + G_2(t,s)$$

where

$$G_1(t,s) = P(t) \begin{pmatrix} 0 & 0 \\ 0 & \exp(B_1(t-s)) \end{pmatrix} P^{-1}(s),$$

$$G_2(t,s) = P(t) \begin{pmatrix} 1 & 0 \\ 0 & O_{n-1} \end{pmatrix} P^{-1}(s).$$

Denote the first column of the fundamental matrix Φ by φ^1. Because of structure (5.1.7), φ^1 is equal to the first column of P, i.e. it is a T-periodic solution of (5.1.12). Since 1 is a simple characteristic multiplier, Corollary 2.2.4 implies that (5.1.12) has no T-periodic solution linearly independent of $\dot{p}$, so that $\varphi^1(t) = c\dot{p}(t)$ for some scalar constant c. Since $\dot{p}^0 = \dot{p}(0) = (\dot{p}_1^0, 0, ..., 0)$, we may assume without loss of generality that $\varphi^1(0) = [1, 0, ..., 0]$. Now, $G_2(t,s)$ is the product of the matrices

$$P(t) \begin{pmatrix} 1 & 0 \\ 0 & O_{n-1} \end{pmatrix} \quad \text{and} \quad \begin{pmatrix} 1 & 0 \\ 0 & O_{n-1} \end{pmatrix} P^{-1}(s).$$

The first column of the first matrix is the first column of P, i.e. φ^1, and the rest of the elements are zero; hence, this matrix is real (because Φ is real) and a solution of (5.1.12). The first row of the second matrix is the first row of $P^{-1}(s)$, and, again, because of the structure of (5.1.7) this is equal to the first row of the real matrix $\Phi^{-1}(s)$; the rest of the elements are zero. Hence, the second matrix is real. As a consequence, for fixed $s \in \mathbf{R}$ the matrix $G_2(t,s)$ is real and a solution of (5.1.12). Since $G_1 + G_2$ is a real matrix solution, we get that G_1 is also a real matrix solution. Denote the eigenvalues of the matrix B_1 by $\nu_2, \nu_3, ..., \nu_n$. According to construction (5.1.9), $\mathrm{Re}\,\nu_k < 0$ $(k = 2, ..., n)$. Let a $\nu > 0$ be chosen such that $\mathrm{Re}\,\nu_k < -\nu$ for all $k = 2, 3, ..., n$. Applying inequality (1.4.10) and taking into account that P and P^{-1} are regular, periodic, continuously differentiable matrices, a constant $K > 0$ exists such that

$$|G_1(t,s)| \le K \exp(-\nu(t-s)), \qquad t \ge s, \qquad (5.1.15)$$

$$|G_2(t,s)| \le K, \qquad t, s \in \mathbf{R}. \qquad (5.1.16)$$

Let $a = [0, a_2, ..., a_n]$ be a vector orthogonal to $\dot{p}^0$, i.e. with first coordinate zero. Then because of structure (5.1.7) of Φ and by (1.4.10), the following estimate holds

$$|\Phi(t)a| \le K_1|a| \exp(-\nu t), \qquad t \ge 0 \qquad (5.1.17)$$

where K_1 is a positive constant. Using the elements introduced above, the following integral equation is setup:

$$\zeta(t) = \Phi(t)a + \int_0^t G_1(t,s)r(s,\zeta(s))ds - \int_t^\infty G_2(t,s)r(s,\zeta(s))ds. \quad (5.1.18)$$

First, we are going to show that if $|a|$ is small enough, then (5.1.18) has a solution $\zeta(t,a)$ defined for $t \in [0,\infty]$, and this solution tends to zero exponentially as t tends to infinity. Then we are going to show that this solution is a solution of (5.1.13).

In establishing the existence of a solution to the integral equation we shall make use of the estimates achieved above and of the fact that the properties of the function r imply that for an arbitrary "Lipschitz constant" $K_2 > 0$, there is a $\delta(K_2) > 0$ such that for $|z^1| < \delta(K_2)$, $|z^2| < \delta(K_2)$ we have

$$|r(t,z^1) - r(t,z^2)| \leq K_2|z^1 - z^2| \quad (5.1.19)$$

uniformly in $t \in [0,\infty]$. The appropriate value of K_2 will be specified later. We shall show that if $|a| < \delta(K_2)/2K_1$, then a solution $\zeta(t,a)$ defined in $[0,\infty]$ exists satisfying the inequality

$$|\zeta(t,a)| \leq 2K_1|a|\exp(-\nu t/2), \qquad t \geq 0. \quad (5.1.20)$$

Successive approximation will be applied. Let $\zeta^0(t,a) = 0$, and

$$\zeta^{k+1}(t,a) := \Phi(t)a + \int_0^t G_1(t,s)r(s,\zeta^k(s))ds - \int_t^\infty G_2(t,s)r(s,\zeta^k(s))ds.$$

We are going to show by mathematical induction that $\zeta^k(t,a)$ is defined for $t \in [0,\infty]$ and satisfies

$$|\zeta^{k+1}(t,a) - \zeta^k(t,a)| \leq K_1|a|\exp(-\nu t/2)/2^k \qquad (k = 0,1,2,...). \quad (5.1.21)$$

From the last inequality it, clearly, follows that

$$|\zeta^{k+1}(t,a)| \leq 2K_1|a|\exp(-\nu t/2), \quad t \geq 0 \quad (k = 1,2,...). \quad (5.1.22)$$

Obviously, because of (5.1.17)

$$|\zeta^1(t,a)| = |\zeta^1(t,a) - \zeta^0(t,a)| \leq K_1|a|\exp(-\nu t/2).$$

If we assume that (5.1.21) holds with k replaced by $1, 2, ..., k-1$, then this implies (5.1.22) with $(k+1)$ replaced by k, which, in turn, because of the properties of the function r, guarantees the convergence of the improper integral in the integral equation with ζ^k substituted for ζ, and the existence of $\zeta^{k+1}(t,a)$ for $t \in [0,\infty]$. Applying (5.1.15), (5.1.16), (5.1.17), (5.1.19) and the assumption of the induction we get that

$$|\zeta^{k+1}(t,a) - \zeta^k(t,a)| \leq 4KK_2K_1|a|\exp(-\nu t/2)/(2^{k-1}\nu).$$

Let us choose now $K_2 = \nu/(8K)$. This yields (5.1.21). Thus, the sequence $\zeta^{k+1}(t,a)$ converges uniformly on $t \in [0, \infty]$, $|a| < \delta(\nu/(8K))/2K_1$, and each term of the sequence satisfies (5.1.22). So that a limit function $\zeta(t,a)$ exists on the same domain, it is continuous and satisfies (5.1.18) and (5.1.20). The latter inequality means that $\zeta(t,a) \to 0$, as $t \to \infty$ uniformly in a.

Now, it will be shown that $\zeta(t,a)$ is a solution of (5.1.13). The properties of the integrands in (5.1.18) and the continuity of ζ imply that ζ is, in fact, continuously differentiable with respect to t. Substituting the solution $\zeta(t,a)$ into (5.1.18), differentiating and taking into account that $G_1(t,s)$ and $G_2(t,s)$ are solutions of (5.1.12) for fixed s and that $G_1(t,t) + G_2(t,t) = \Phi(t)\Phi^{-1}(t) = I$ we get that $\zeta(t,a)$ is a solution of (5.1.13) indeed.

We are going to determine the initial values of ζ in terms of the $n-1$ parameters $a_2, ..., a_n$.

$$
\begin{aligned}
\zeta(0,a) &= \Phi(0)a - \int_0^\infty G_2(0,s)r(s,\zeta(s,a))ds \\
&= P(0)a - P(0)\begin{pmatrix} 1 & 0 \\ 0 & O_{n-1} \end{pmatrix}\int_0^\infty P^{-1}(s)r(s,\zeta(s,a))ds
\end{aligned}
$$

where (5.1.7) and the definition of G_2 has been used. The first column of the matrix $P(0)$ is $[1, 0, ..., 0]$; therefore, the second vector on the right-hand side is of the form $\tilde{h}(a) = [\tilde{h}_1(a), 0, ..., 0]$ (the last $n-1$ coordinates are zero). The properties of the function r and inequality (5.1.20) imply that $\tilde{h}_1 \in C^1$ and that $\tilde{h}_1(a) = o(|a|)$ as a tends to zero. Denote the coordinates of the initial vector $\zeta(0,a)$ by $\zeta_1^0, ..., \zeta_n^0$, and the coordinates of the matrix $P(0)$ by p_{ik} $(i, k = 1, 2, ..., n)$. Then the equation $\zeta(0,a) = P(0)a - \tilde{h}(a)$ can be written out in coordinates in the following way:

$$
\zeta_1^0 = \sum_{k=2}^n p_{1k}a_k - \tilde{h}_1(a_2, ..., a_n), \tag{5.1.23}
$$

$$
\zeta_i^0 = \sum_{k=2}^n p_{ik}a_k \qquad (i = 2, 3, ..., n). \tag{5.1.24}
$$

Since $\det P(0) \neq 0$ and the first column of $P(0)$ is the vector $[1, 0, ..., 0]$, the coefficient matrix of the last system is regular, i.e. the last system represents a linear, regular one-to-one mapping of the set of $(n-1)$-tuples $(a_2, ..., a_n)$ onto the set of initial values $(\zeta_2^0, ..., \zeta_n^0)$. Expressing the a_k's $(k = 2, ..., n)$ from this system as linear functions of the ζ_i^0's $(i = 2, ..., n)$, and substituting these expressions into (5.1.23), we get an equation

$$
\zeta_1^0 + \sum_{i=2}^n q_i\zeta_i^0 - h(\zeta_2^0, ..., \zeta_n^0) = 0 \tag{5.1.25}
$$

where, clearly, the q_i's are constants determined by the elements of $P(0)$, $h \in C^1$, and

$$h(\zeta_2^0, ..., \zeta_n^0) = o((\zeta_2^{0^2} + \cdots + \zeta_n^{0^2})^{1/2}).$$

Because of the construction, equation (5.1.25) determines those initial conditions that define solutions tending to zero exponentially according to (5.1.20) as t tends to infinity. Since $p(0) = 0$, the transformation that led to equation (5.1.13) implies that $\zeta(0, a) = z(0) = \varphi(0) - p(0)$ where φ is the corresponding solution of the original system (5.1.1). Thus, if the coordinates of the initial value $\varphi(0) \in X$ satisfy the equation

$$x_1 + \sum_{i=2}^{n} q_i x_i - h(x_2, ..., x_n) = 0, \qquad (5.1.26)$$

then $\zeta(t, a) = \varphi(t) - p(t) \to 0$ exponentially as t tends to infinity. (Here the coordinates $a_2, ..., a_n$ of a are determined by the invertible system (5.1.24) with the ζ_i^0's equal to the corresponding coordinates of the initial value $\varphi(0)$ $(i = 2, ..., n)$.)

Equation (5.1.26) determines an $(n - 1)$ dimensional C^1 manifold (a hypersurface) $S \subset X$ in a neighbourhood of the origin $x = p(0) = 0$. The properties of the function h imply that the equation of the tangent space of this manifold (at $x = 0$) is $x_1 + \sum_{i=2}^{n} g_i x_i = 0$, and the first coordinate of the gradient of the left-hand side in (5.1.26) is 1. Thus, the scalar product of the normal vector of S at $x = 0$ with the tangent vector $\dot{p}^0 = (\dot{p}_1^0, 0, ..., 0)$ of the orbit γ is non-zero, i.e. the path γ intersects S transversally (see Figure 5.1.2). This, clearly, implies than an $\varepsilon_1 > 0$ exists such that the path of every solution φ that differs from p less than ε_1 over the interval $[0, 2T]$, say, intersects S at some $t_1 \in [0, 2T]$.

Now, let an arbitrary $0 < \varepsilon \le \varepsilon_1$ be given. Because of the continuous dependence of solutions on the initial values (on finite intervals), a $\delta(\varepsilon) > 0$ exists such that if $\operatorname{dist}(x^0, \gamma) < \delta(\varepsilon)$, then the solution $\varphi(t, x^0)$ is defined on $[0, 2T]$, and $\operatorname{dist}(\varphi(t, x^0), \gamma) < \varepsilon \le \varepsilon_1$ for $t \in [0, 2T]$. Hence, the path of $\varphi(t, x^0)$ intersects S at some $t_1 \in [0, 2T] : \varphi(t_1, x^0) \in S$. The solution $\varphi(t, \varphi(t_1, x^0)) = \varphi(t + t_1, x^0)$ has its initial value in S, so that φ is defined in $[0, \infty]$, and $\varphi(t + t_1, x^0) - p(t) \to 0$ exponentially as $t \to \infty$. Thus, $t_1 = \alpha(x^0)$, the asymptotic phase, and this completes the proof. $\square$

The Andronov-Witt Theorem can be applied even if the periodic solution is known only approximately, for instance if it has been determined numerically. In this case, one may apply the algorithm for the approximate determination of the characteristic multipliers presented at the end of Section 2.2.

The application of the Theorem is particularly simple in the *two dimensional case*. If (5.1.1) is two dimensional, and it has a non-constant periodic solution p with period $T > 0$, then as we know the number one is a characteristic multiplier of the variational system (5.1.12). Thus, asymptotic orbital stability depends on the modulus of the other characteristic multiplier: λ. But by Liouville's formula (1.2.6) the product of the two characteristic multipliers, i.e. the determinant of the principal matrix, is

$$1 \cdot \lambda = \exp \int_0^T (f'_{1x_1}(p(t)) + f'_{2x_2}(p(t)))dt.$$

Hence, *if*

$$\int_0^T (f'_{1x_1}(p(t)) + f'_{2x_2}(p(t)))dt < 0, \tag{5.1.27}$$

then p is orbitally asymptotically stable with asymptotic phase.

Applying this criterion to *Van der Pol's differential equation* (3.2.5) we recall that for $m > 0$ it has a non-constant periodic solution. Let us denote this solution by u_m and its period by $T_m > 0$. The variational system of system (3.2.6) that is equivalent to van der Pol's equation is

$$\dot{y}_1 = y_2, \quad \dot{y}_2 = -(1 + 2mu_m(t)\dot{u}_m(t))y_1 + m(1 - u_m^2(t))y_2. \tag{5.1.28}$$

According to (5.1.27), if

$$m \int_0^{T_m} (1 - u_m^2(t))dt < 0,$$

then the periodic solution u_m is orbitally asymptotically stable with asymptotic phase. For $m > 0$ the last inequality is equivalent to

$$(1/T_m) \int_0^{T_m} u_m^2(t)dt > 1, \tag{5.1.29}$$

i.e. for asymptotic orbital stability, it is sufficient if the average of the squared periodic solution is greater than 1. It is to be noted that by Dulac's Criterion (Theorem 3.5.16) the system (3.2.6) cannot have a non-constant closed path in the strip $|u| \leq 1$ because the divergence of the vector field is $m(1 - u^2) \geq 0$. Therefore, u_m must assume values in modulus greater than one. By numerical computation, Urabe [1967] has indicated that (5.1.29) holds for all $m > 0$, i.e. *the nonconstant periodic solution of Van der Pol's equation is asymptotically orbitally stable with asymptotic phase for all $m > 0$*. We shall return to this topic in Chapter 6.

There holds the following instability result completing the Andronov-Witt Theorem in a natural way.

Theorem 5.1.3. *Assuming the conditions preceding Theorem 5.1.2 if at least one of the characteristic multipliers of the variational system (5.1.12) is in modulus greater than 1, then the periodic solution $p(t)$ of system (5.1.1) is orbitally unstable.*

Proof. First a Lemma analogous to Lemma 5.1.1 is to be proved and then the proof is similar to that of Theorem 5.1.2. It is left to the reader. □

5.2 Poincaré Map, Isolation and Isochronism

The problem of isolation of non-constant periodic solutions of non-autonomous differential systems has been treated in Section 4.2. At the end of that Section we have shown that a non-constant periodic solution of an *autonomous* system could never be isolated in the sense applied there. This was so because the concept applied there (Definition 4.2.1) concerned, in fact, the isolation or non-isolation of the *integral curve*, and the integral curve of a non-constant periodic solution of an autonomous system lies always in a two dimensional "cylinder surface" that consists of integral curves of periodic solutions that arise by a shift of the independent variable. Also, according to Definition 4.2.1 a periodic solution of a non-autonomous periodic system is non-isolated if it lies in a manifold of integral curves belonging to solutions of the *same period*. In the case of an autonomous system if a periodic solution is "surrounded by periodic solutions" in some sense, the latter ones cannot be expected to have the same period. Normally, periodic solutions that have different paths will have different periods. If this is *not* the case, we speak about *isochronism* being treated in the last part of this Section. Obviously, if the important concept of isolation is to be extended to the autonomous case, it has to concern the *path* and must exclude neighbouring periodic solutions with different periods as well. In order to simplify the language, the trajectory of a *non-constant* periodic solution will be called a *periodic orbit*, i.e. a periodic orbit is a closed Jordan curve (in the phase space) that is *not a point*. We could proceed by accepting a definition analogous to Definition 4.2.1 (see Farkas [1975a, 1978b]); however, this would declare isolated a periodic orbit that is a cluster set of a countable set of periodic orbits. Such situations occur in "*strange attractors*" of "*chaotic dynamical systems*" (see, e.g., Wiggins [1988, 1990]), and one would miss the point if one declared such a periodic solution isolated. Therefore, the definition given by Moson [1976] will be adopted. (See also Moson [1977] where the relation of different possible definitions is analysed.) However, before embarking upon the study of isolation we shall introduce the concept of the *Poincaré mapping*, which plays an important role also in problems of existence, stability, perturbations and bifurcations of periodic solutions. The construction that follows is, in some respect, a generalization of what has been done in the Poincaré-Bendixson Theory (Section 3.1) on a transversal. Confer also the Poincaré-Andronov operator introduced in (4.1.3) and the construction in the proof of the Andronov-Witt Theorem 5.1.2.

Assume that the autonomous system (5.1.1) has a non-constant solution

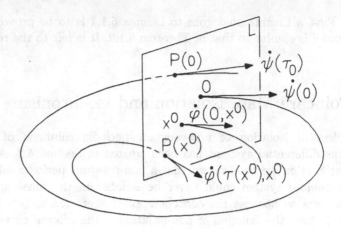

FIGURE 5.2.1. The Poincaré map.

$\psi : \mathbf{R}_+ \to X$, and set $\psi^0 = \psi(0)$. Consider an $n-1$ dimensional hyperplane L passing through the point ψ^0 and transversal to the path of ψ. (It may be, for instance, orthogonal to the tangent vector $\dot\psi(0) = f(\psi^0)$.) Assume that the path of ψ hits the hyperplane L transversally at some positive moment τ_0 again: $\psi(\tau_0) \in L$, $\tau_0 > 0$, and the tangent vector $\dot\psi(\tau_0)$ of the path at $\psi(\tau_0)$ points into the same half space as $\dot\psi(0)$. (If L is orthogonal to $\dot\psi(0)$, then $\dot\psi(\tau_0)$ and $\dot\psi(0)$ form an acute angle.) Because of the continuous dependence of the solution on the initial values, the point ψ^0 has a neighbourhood U in L such that every solution $\varphi(t, x^0)$ with initial value $x^0 \in U$ hits the hyperplane L again transversally at some moment $\tau(x^0) > 0$ close to τ_0, i.e. for $x^0 \in U \subset L$ there exists a $\tau(x^0) > 0$ such that $\varphi(\tau(x^0), x^0) \in L$, and $\dot\varphi(\tau(x^0), x^0)$ is not parallel to L. This implies, in particular, that U does not contain any equilibrium point of the system.

Definition 5.2.1. The mapping $P : U \to L$ defined by $P(x^0) := \varphi(\tau(x^0), x^0)$ for $x^0 \in U$ where U is a sufficiently small neighbourhood of ψ^0 in L is called the *Poincaré map* of system (5.1.1) (see Figure 5.2.1).

We shall establish now a few properties of the Poincaré map.

Theorem 5.2.1. *The Poincaré map is a diffeomorphism of U into L.*

Proof. The properties of flows generated by autonomous systems of differential equations (see Section 1.3) imply that P establishes a one-to-one correspondence between points of U and L.

We shall prove that $P \in C^1(U, L)$. Clearly, it is sufficient to prove that the function $\tau : U \to \mathbf{R}_+$ belongs to the C^1 class. We may assume without loss of generality that the origin has been moved into the point ψ^0 and that the coordinate system has been rotated such a way that the x_1-axis is parallel to the vector $\dot\psi(0) = f(\psi^0)$. This means that $\dot\psi(0) = f(\psi^0) = [f_1(\psi^0), 0, ..., 0]$, and L is the coordinate plane $x_1 = 0$. The point x^0 belongs to L if and only if its first coordinate is zero: $x_1^0 = 0$. Thus, the trajectory of the solution $\varphi(t, x^0)$ corresponding to the initial value $x^0 \in L$ intersects L again at $t > 0$ if $\varphi_1(t, x^0) = 0$. By assumption, this equation has a solution, namely, $\varphi_1(\tau_0, 0) = \psi_1(\tau_0) = 0$. Since the path of ψ intersects the plane L transversally at $t = \tau_0$, it follows that $\dot\varphi_1(\tau_0, 0) = \dot\psi_1(\tau_0) \neq 0$. Thus, by the Implicit Function Theorem, $x^0 = 0 = \psi^0$ has a neighbourhood $U \subset L$, and there exists a unique function $\tau : U \to \mathbf{R}_+$ such that $\tau \in C^1$, $\tau(0) = \tau_0$, and $\varphi_1(\tau(x^0), x^0) \equiv 0$, i.e. $\varphi(\tau(x^0), x^0) \in L$.

Having proved the continuous differentiability of P we note that the inverse P^{-1} is obtained by following the same trajectories in the opposite direction from $P(x^0)$ to x^0; in other words, if $y^0 := P(x^0)$, then $P^{-1}(y^0) = \varphi(\sigma(y^0), y^0)$ where $\sigma(y^0) = -\tau(x^0)$. The continuous differentiability of the inverse mapping can be proved exactly as has been done for P. $\square$

We note the obvious fact that system (5.1.1) has a periodic solution in the neighbourhood of the solution ψ if and only if the Poincaré map or one of its iterates has a fixed point, i.e. iff there is a positive integer k such that $P^k(x^0) = x^0$ for some $x^0 \in U$ (see Fig. 5.2.2.) The fact that the motion of x^0 may intersect the hyperplane several times in the same direction as at x^0 before closing in at x^0, i.e. that we may have $P^j(x^0) \neq x^0$ for $j = 1, 2, ..., k - 1$ and $P^k(x^0) = x^0$ is a basic difference between the two dimensional and the higher dimensional case (cf. Lemma 3.1.1 (d)). If the k-th iterate is indeed the first that returns to x^0, and the notation $P^j(x^0) = P(P^{j-1}(x^0)) = \varphi(\tau(P^{j-1}(x^0)), P^{j-1}(x^0))$ is used ($j = 1, 2, ..., k$), then because of Definition 1.3.1 (iii) we have $P^k(x^0) = \varphi(T, x^0) = x^0$ where

$$T = \sum_{j=1}^{k} \tau(P^{j-1}(x^0))$$

is the (least positive) period of the periodic solution $\varphi(t, x^0)$. In the argumentation above we had to assume, tacitly, that all iterates $P^j(x^0)$ ($j = 1, ..., k - 1$) are in the domain of the Poincaré map, and, naturally, $P^0(x^0) \equiv x^0$.

We turn now to the case when the solution on whose transversal the Poincaré map is defined is a periodic one. Assume that $p(t)$ is a non-constant periodic solution of (5.1.1) with period $T > 0$, and again without loss of generality that $p(0) = 0$, and $\dot p(0) = [\dot p_1(0), 0, ..., 0] \neq 0$, and consider the Poincaré map in the neighbourhood of O on the transversal $L = \{x \in \mathbf{R}^n : x_1 = 0\}$. By what has been established above, $O \in L$ has a neighbourhood $U \subset L$ such that the Poincaré map $P : U \to L$ defined by

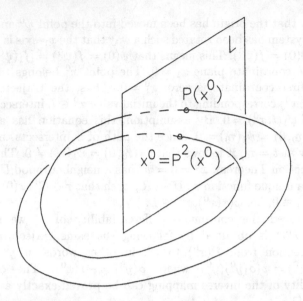

FIGURE 5.2.2. A periodic orbit of system (5.1.1) which intersects the transversal plane L twice in the same direction.

$P(x^0) = \varphi(\tau(x^0), x^0) \in L, << \dot{\varphi}(\tau(x^0), x^0), \dot{p}(0) >> 0$ is a diffeomorphism. Here $\varphi(t, x^0)$ is the solution of system (5.1.1) with initial value $x^0 = \varphi(0, x^0) = (0, x_2^0, ..., x_n^0) \in U$, $\varphi(t, 0) = p(t)$, and $\tau : U \to \mathbf{R}^+$ is a C^1 function with $\tau(0) = T$. Now the matrix of the linearization of P at $x^0 = 0$ will be determined. The linearization DP_0 is a linear operator of the $n - 1$ dimensional subspace L into itself. Its matrix is

$$DP_0 = \left[\dot{\varphi}_i \tau'_{x_k^0} + \partial \varphi_i / \partial x_k^0 \right]_{t=T, x^0=0}. \tag{5.2.1}$$

This is an $n \times n$ matrix but, since $\varphi_1(\tau(x^0), x^0) \equiv 0$, its first row ($i = 1$) is zero, and since P does not depend on $x_1^0 \equiv 0$, the first column ($k = 1$) is also zero. Further, $\dot{\varphi}_i(\tau(0), 0) = \dot{\varphi}_i(T, 0) = \dot{p}_i(T) = \dot{p}_i(0) = 0$ ($i = 2, ..., n$) because of the periodicity of p. Hence,

$$DP_0 = \begin{pmatrix} 0 & 0 & \cdots & 0 \\ 0 & \partial \varphi_2 / \partial x_2^0 & \cdots & \partial \varphi_2 / \partial x_n^0 \\ \vdots & \vdots & \ddots & \vdots \\ 0 & \partial \varphi_n / \partial x_2^0 & \cdots & \partial \varphi_n / \partial x_n^0 \end{pmatrix}_{t=T, x^0=0}.$$

Thus, by an abuse of notation, DP_0 is represented by the $(n-1) \times (n-1)$ matrix $\left[\partial \varphi_i / \partial x_k^0 \right]_{t=T, x^0=0}$ ($i, k = 2, ..., n$), and acts on L by

$$DP_0 \tilde{x}^0 = \left[\partial \varphi_i / \partial x_k^0 \right]_{t=T, x^0=0} \tilde{x}^0 \tag{5.2.2}$$

where $\tilde{x}^0$ is the $n-1$ dimensional column vector $\tilde{x}^0 = \text{col}\,[x_2^0, ..., x_n^0]$. We are to prove the following:

Theorem 5.2.2. *The eigenvalues of the linearization DP_0 of the Poincaré map attached to the T-periodic solution p of system (5.1.1) considered as an $n-1$ dimensional linear operator are equal to the characteristic multipliers of the variational system with respect to the solution p provided that the number 1 is deleted once from the set of the multipliers.*

By Corollary 2.2.3 we know that 1 is a characteristic multiplier of the variational system. This Theorem had to be formulated such an intricate way because 1 may be a multiple characteristic multiplier, in which case it will be an eigenvalue of DP_0 with multiplicity one less than in the spectrum of the principal matrix of the variational system.

Proof. The variational system of (5.1.1), with respect to p is

$$\dot{y} = f_x'(p(t))y. \tag{5.2.3}$$

As it has been shown following (1.1.5), $\Phi(t) := \varphi_{x^0}'(t, 0)$ is the fundamental matrix of (5.2.3) that assumes the unit matrix at $t = 0$, $\Phi(0) = I$. The first column of $\Phi(t)$ is that solution of (5.2.3) that assumes col $[1, 0, ..., 0]$ at $t = 0$. But we know from (1.3.7) that $\dot{p}(t)$ is a solution, hence, $\dot{p}(t)/\dot{p}_1(0)$ is also a solution and this is the one that satisfies the initial condition $\dot{p}(0)/\dot{p}_1(0) = \text{col}\,[1, 0, ..., 0]$. By (2.2.5) the principal matrix of (5.2.3) is $C = \Phi(T)$. But the first column of $\Phi(T)$ is $\dot{p}(T)/\dot{p}_1(0) = \dot{p}(0)/\dot{p}_1(0) = \text{col}\,[1, 0, ..., 0]$ because of the periodicity of p. Hence,

$$C = \begin{pmatrix} 1 & \varphi_{1x_2^0}' & \cdots & \varphi_{1x_n^0}' \\ 0 & \varphi_{2x_2^0}' & \cdots & \varphi_{2x_n^0}' \\ \vdots & \vdots & \ddots & \vdots \\ 0 & \varphi_{nx_2^0}' & \cdots & \varphi_{nx_n^0}' \end{pmatrix}_{t=T, x^0=0},$$

and this proves the Theorem. $\square$

Intuitively it is to expected that there is a correlation between the asymptotic orbital stability of the periodic solution on the one hand and the contractivity (cf. Definition A2.10) of the corresponding Poincaré map on the other. And, indeed, the previous Theorem provides another way for proving the Andronov-Witt Theorem 5.1.2. If 1 is a simple characteristic multiplier of the variational system (5.2.3) and the remaining $n-1$ characteristic multipliers are in modulus less than 1, then by the previous Theorem all $n-1$ eigenvalues of the Poincaré map linearized at $x^0 = 0$, (5.2.2), are in modulus less than 1. This implies that in a

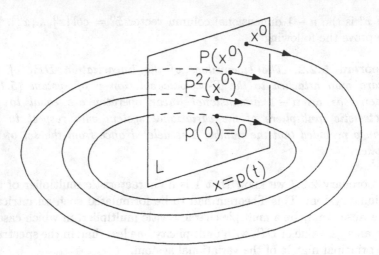

FIGURE 5.2.3. Contractive Poincaré map around an orbitally asymptotically stable periodic solution p. $|P^k(x^0)| \leq \alpha^k|x^0|$, $0 < \alpha < 1$.

neighbourhood of $\tilde{x}^0 = 0 \in L$ the Poincaré map is a contraction mapping. Indeed,

$$P(\tilde{x}^0) - P(0) = P(\tilde{x}^0) = DP_0\tilde{x}^0 + o(|\tilde{x}^0|); \qquad (5.2.4)$$

hence, for arbitrary $\delta > 0$

$$|P(\tilde{x}^0)| \leq (|DP_0| + \delta)|\tilde{x}^0|$$

if $|\tilde{x}^0|$ is small enough. Let $0 < \alpha < 1$ be strictly greater than the modulus of the maximal eigenvalue of DP_0; then by Theorem A1.17 there is a linear coordinate transformation $\tilde{x}^0 \to \hat{x}^0$, $DP_0 \to \hat{D}P_0$ such that $|\hat{D}P_0| + \delta \leq \alpha < 1$ provided that $\delta > 0$ is small enough. Hence,

$$|P(\hat{x}^0)| \leq \alpha|\hat{x}^0|;$$

thus, P is contractive, i.e. taking the iterates of P we get that $P^k(\hat{x}^0) \to 0$ as $k \to \infty$. Since by a shift of the independent variable t every point on the path of the periodic solution p can be taken as $p(0)$, the construction above can be performed along the whole periodic orbit resulting in the conclusion that all the solutions of (5.1.1) that have initial values $\hat{x}^0$ close to the periodic orbit tend to it as t tend to infinity. (The Andronov-Witt Theorem is proved this way in Hartman [1964] p.254. See Figure 5.2.3.)

We turn now to the problem of when shall we call a periodic solution or rather the corresponding periodic orbit isolated, and how this isolation can be guaranteed. Let p be a non-constant T-periodic ($T > 0$) solution of (5.1.1), and γ its orbit. The following Definition will be adopted (see Moson [1976]).

Definition 5.2.2. We say that the non-constant periodic solution p with period $T > 0$ of system (5.1.1) is *non-isolated* if to each $\varepsilon > 0$ one may find an $x^0 \in \mathbf{R}^n$ such that

(i) $0 < \text{dist}(x^0, \gamma) < \varepsilon$;

(ii) $\varphi(t, x^0)$ is a periodic solution with period $\tau(x^0)$;

(iii) $|\tau(x^0) - T| < \varepsilon$.

If these conditions do not hold, then the solution p is said to be *isolated*.

It is to be noted that this isolation concept differs from the one applied in the case of *non-autonomous* periodic systems (cf. Definition 4.2.1) and that it depends upon the period of the periodic solution too. If the periodic orbit has a neighbourhood where there are no periodic orbits *with periods near the given one*, then it is isolated. This presents no problem if the dimension of system (5.1.1) is 2. In this case any periodic orbit may have but one point of intersection with a transversal (see Lemma 3.1.1 (d)) and cannot revolve around our periodic orbit γ several times before closing in. The continuous dependence on the initial conditions implies then that any periodic solution near to γ must have a period close to the period of p. In dimensions higher than 2 the situation is different; this is illustrated by the following:

Example 5.2.1 (*Moson* [1977]). In the Cartesian system of coordinates (x, y, z) "*toroidal coordinates*" will be introduced by the transformation formulae

$$
\begin{aligned}
x &= (1 + r\cos\eta)\cos\theta, \\
y &= r\sin\eta, \\
z &= (1 + r\cos\eta)\sin\theta.
\end{aligned}
$$

Here $r \geq 0$ is the distance from the circle $\gamma = \{(x, y, z) : x^2 + z^2 = 1, \ y = 0\}$, θ is the polar angle along the circle γ in the plane x, z, and η is a polar angle in any plane perpendicular to γ with origin in the point of intersection of γ and the plane (see Figure 5.2.4). Consider the autonomous system given in toroidal coordinates by

$$
\dot{r} = 0, \qquad \dot{\theta} = 1, \qquad \dot{\eta} = r^2.
$$

We shall look at this system, but think about the corresponding system in the Cartesian coordinates x, y, z which will not be written out. The system has a periodic solution $(x(t), y(t), z(t)) = (\cos t, 0, \sin t)$ (which corresponds to $r = 0$, $\theta = t$, $\eta = \text{const}$) with period $T = 2\pi$. The first differential equation $\dot{r} = 0$ implies that all the orbits of the system lie on torus surfaces

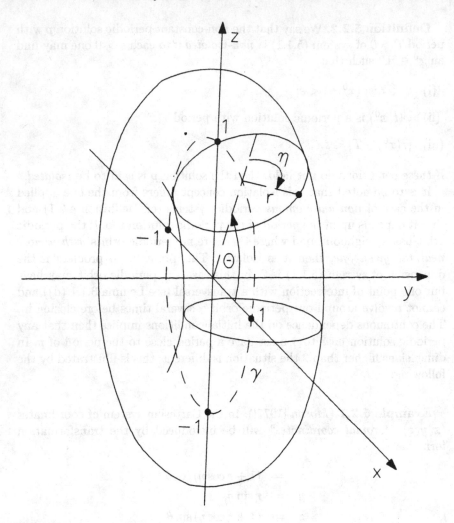

FIGURE 5.2.4. Toroidal coordinates r, θ, η around the circle $\gamma : x^2 + z^2 = 1$ in the x, z plane.

given by $r = $ const. If $r^2 = m/q$ where m and q are relative prime positive integers, then the solutions in the x, y, z coordinates corresponding to $r \equiv (m/q)^{1/2}$, $\theta = t$, $\eta = (m/q)t$ are periodic with least positive period $2\pi q$. If $r^2 = $ an irrational number, then no solution on this torus surface is periodic. Thus, considering arbitrarily small rational values of r^2 we conclude that there are periodic orbits arbitrarily near $(\cos t, 0, \sin t)$. Nevertheless, this periodic solution is *isolated* because as $m/q \to 0$, we must have $q \to \infty$, so that the least positive period of the solutions on the torus $r^2 = m/q$ tends to infinity and will not be near 2π.

The following Theorem is due to Moson [1976] sharpening a theorem in Farkas [1975$_a$].

Theorem 5.2.3. *If the number 1 is a simple characteristic multiplier of the variational system (5.2.3), then the T-periodic solution p of system (5.1.1) is isolated.*

Proof. Suppose, contrary to the statement, that p is *not* isolated. Applying the construction preceding Theorem 5.2.2, this implies that there is a sequence of points $x^k \in L$ ($x_1^k = 0$) such that $x^k \neq 0 = p(0)$, $x^k \to 0$ as $k \to \infty$; $\varphi(t, x^k)$ is periodic with period $\tau(x^k)$ and $\tau(x^k) \to T$ as $k \to \infty$. If $\tilde{x}^k$ denotes the $n - 1$ dimensional vector consisting of the last $n - 1$ coordinates of x^k, then by (5.2.4)

$$\tilde{x}^k = P(\tilde{x}^k) = DP_0\tilde{x}^k + o(|\tilde{x}^k|) \qquad (k = 1, 2, ...)$$

or

$$(I - DP_0)\tilde{x}^k = o(|\tilde{x}^k|)$$

where I is the $n - 1$ dimensional unit matrix. Since, by assumption, 1 is not an eigenvalue of DP_0, the matrix $I - DP_0$ is invertible, so that

$$\tilde{x}^k = (I - DP_0)^{-1}o(|\tilde{x}^k|).$$

Taking the norm of both sides and dividing by $|\tilde{x}^k|$ we obtain

$$1 \leq |(I - DP_0)^{-1}| \ |o(|\tilde{x}^k|)|/|\tilde{x}^k| \to 0 \qquad k \to \infty,$$

a contradiction . $\square$

The condition that 1 be a simple characteristic multiplier is only sufficient for the isolation of the periodic solution; it is not necessary. This is seen also from the previous Example (cf. also Problem 5.7.1). The following Example shows another instance for this and at the same time shows that the Andronov-Witt Theorem 5.1.2 gives also only a sufficient condition for the asymptotic orbital stability.

Example 5.2.2 (*Farkas [1975$_a$]*). Consider a two dimensional system

$$\dot{x}_1 = f_1(x_1, x_2), \qquad \dot{x}_2 = f_2(x_1, x_2) \qquad (5.2.5)$$

where $\tilde{f} = (f_1, f_2) \in C^1$ which has a non-constant periodic solution $\tilde{p} = (p_1, p_2)$ with period $T > 0$; suppose that the variational system (5.2.5) with respect to $\tilde{p}$ has a characteristic multiplier in modulus less than 1 (the other multiplier is 1 as we know). We may take, for instance, Van der Pol's equation (3.2.5) (cf. Theorems 3.2.1, 3.2.2 and formulae (5.1.28)-(5.1.29)). Assume further that $\dot{p}_1(0) = 1$, $\dot{p}_2(0) = 0$. Consider also the system

$$\dot{x}_3 = -(2\pi/T)x_4 - x_3^3, \qquad \dot{x}_4 = (2\pi/T)x_3 - x_4^3. \qquad (5.2.6)$$

The origin $(x_3, x_4) = (0, 0)$ is an asymptotically stable equilibrium point of the latter system which can be shown applying the Liapunov function $V(x_3, x_4) = x_3^2 + x_4^2$ whose derivative with respect to system (5.2.6) is negative definite:

$$\dot{V}_{(5.2.6)}(x_3, x_4) = -2x_3^4 - 2x_4^4.$$

If (5.2.5) and (5.2.6) are glued together into the four dimensional system

$$\dot{x} = f(x) \qquad (5.2.7)$$

where $x = [x_1, x_2, x_3, x_4] \in \mathbf{R}^4$ and

$$f(x) = [f_1(x_1, x_2), f_2(x_1, x_2), -(2\pi/T)x_4 - x_3^3, (2\pi/T)x_3 - x_4^3],$$

then, clearly, $p = [p_1, p_2, 0, 0]$ is a T-periodic solution of (5.2.7). The corresponding variational system is $\dot{y} = f_x'(p(t))y$ where

$$f_x'(p(t)) = \begin{pmatrix} f_{1x_1}'(\tilde{p}(t)) & f_{1x_2}'(\tilde{p}(t)) & 0 & 0 \\ f_{2x_1}'(\tilde{p}(t)) & f_{2x_2}'(\tilde{p}(t)) & 0 & 0 \\ 0 & 0 & 0 & -2\pi/T \\ 0 & 0 & 2\pi/T & 0 \end{pmatrix}.$$

The variational system has three linearly independent T-periodic solutions, namely, $\dot{p}(t) = (\dot{p}_1(t), \dot{p}_2(t), 0, 0)$, $(0, 0 \cos 2\pi t/T, \sin 2\pi t/T)$ and $(0, 0, -\sin 2\pi t/T, \cos 2\pi t/T)$. By Corollary 2.2.4 this implies that 1 *is a triple characteristic multiplier* of the system. The fourth multiplier is identical to the characteristic multiplier of the variational system of (5.2.5) which is in modulus less than one. We show that in spite of this, *p is isolated and orbitally asymptotically stable*. The distance in $\mathbf{R}^2$, resp. $\mathbf{R}^4$ will be denoted by dist$_2$ and dist$_4$, respectively. If $\varphi = (\varphi_1, \varphi_2, \varphi_3, \varphi_4)$ is a solution of (5.2.7), then (φ_1, φ_2) and (φ_3, φ_4) are solutions of (5.2.5) and (5.2.6), respectively. If dist$_4(\varphi(0), p(0))$ is small, then, obviously, dist$_2((\varphi_1(0), \varphi_2(0)), (p_1(0), p_2(0)))$ and dist$_2((\varphi_3(0), \varphi_4(0)), (0, 0))$ are small too. Denote the path of $\tilde{p}$ and of p by $\tilde{\gamma}$ and γ, respectively; $\tilde{\gamma} \subset \mathbf{R}^2$, $\gamma \subset \mathbf{R}^4$ but the curves $\tilde{\gamma}$ and γ are the same in fact. If $\text{dist}_2((\varphi_1(0), \varphi_2(0)), (p_1(0), p_2(0)))$ is sufficiently small, then $\text{dist}_2((\varphi_1(t), \varphi_2(t)), \tilde{\gamma}) \to 0$ as $t \to \infty$ because of the asymptotic orbital stability of $\tilde{p}$. If $\text{dist}_2((\varphi_3(0), \varphi_4(0)), (0, 0))$ is sufficiently small (in fact arbitrary), then $\text{dist}_2((\varphi_3(t), \varphi_4(t)), (0, 0)) \to 0$ as $t \to \infty$ because of the (global) asymptotic stability of $(0, 0)$. Applying the triangle inequality we have for all $t > 0$

$$
\begin{aligned}
\text{dist}_4(\varphi(t), \gamma) &\leq \text{dist}_4(\varphi(t), (\varphi_1(t), \varphi_2(t), 0, 0)) \\
&+ \text{dist}_4((\varphi_1(t), \varphi_2(t), 0, 0), \gamma) \\
&= \text{dist}_2((\varphi_3(t), \varphi_4(t)), (0, 0)) \\
&+ \text{dist}_2((\varphi_1(t), \varphi_2(t)), \tilde{\gamma}).
\end{aligned}
$$

Therefore, if $\text{dist}_4(\varphi(0), p(0))$ is sufficiently small, then $\text{dist}_4(\varphi(t), \gamma) \to 0$ as $t \to \infty$, i.e. φ is not periodic, and p is orbitally asymptotically stable.

The isolation problem is to be treated in some detail separately in the *analytic case*. There is a classical result due to Poincaré which ensures that in the case of a *two dimensional analytic system* no "discrete clustering of periodic orbits" such as in Example 5.2.1, may occur. Consider the two dimensional autonomous system

$$
\dot{x}_1 = f_1(x_1, x_2), \qquad \dot{x}_2 = f_2(x_1, x_2) \tag{5.2.8}
$$

where f_1 and f_2 are analytic functions in an open and connected set $X \subset \mathbf{R}^2$, and assume that $p = (p_1, p_2)$ is a non-constant periodic solution of (5.2.8).

Theorem 5.2.4 *(Poincaré's Isolation Theorem, see Sansone-Conti [1964] p. 173). The periodic solution p of the analytic system (5.2.8) is either isolated or its path lies in the interior of an open annular set which is the union of periodic orbits (a continuum of periodic orbits).*

Proof. Let L be a transversal to the path of p at $p(0)$ and consider the Poincaré map of a neighbourhood of $p(0)$ in L into L defined by $P(x^0) = \varphi(\tau(x^0), x^0)$ (see Definition 5.2.1). Since $\varphi(t, x^0)$ is an analytic function by the remark following (1.1.14), and the function τ is obtained by the Implicit Function Theorem applied, this time, to analytic functions so that it is analytic too, it follows that the Poincaré map is analytic in a neighbourhood of $p(0)$. Suppose that in each neighbourhood U_k of $p(0)$ there is an initial value $x^k \in L$ of a periodic solution, $x^k \neq p(0)$ $(k = 1, 2, ...)$. We may assume that $x^k \to p(0)$ as $k \to \infty$. Since the x^k's are fixed points of the Poincaré map, this means that the analytic function $F(x) : x - P(x)$ has an infinite number of zeroes, $F(x^k) = 0$ $(k = 1, 2, ...)$, clustering to $p(0)$, but this implies that $F(x) \equiv 0$, i.e. every point of L in a neighbourhood of $p(0)$ is an initial value of a periodic solution. $\square$

This classical result can be extended to a certain class of n dimensional analytic systems. We mention this in order to show that it is not completely hopeless to carry over some results in the plane to higher dimensions. The following Theorem which will be presented without a proof is due to R.A. Smith [1984]. Consider the system

$$\dot{x} = f(x) \tag{5.2.9}$$

where $f : X \to \mathbf{R}^n$ is an analytic function on $X \subset \mathbf{R}^n$, an open and connected set.

Theorem 5.2.5 *(Smith's Theorem). Assume that there exists a real symmetric $n \times n$ matrix M having two negative and $n - 2$ positive eigenvalues, there are positive constants λ, ε such that for all $x, y \in X$ the inequality*

$$2(y - x)'M(f(y) - f(x) + \lambda(y - x)) \le -\varepsilon|y - x|^2 \tag{5.2.10}$$

holds, and (5.2.9) has a periodic orbit $\gamma \subset X$; then either γ is isolated or it belongs to a continuum of periodic orbits in X.

Smith proves also that there is a two dimensional plane $\mathbf{R}^2 \subset \mathbf{R}^n$ such that the projection of the continuum of periodic orbits that occur in the Theorem to this plane is an open annular set that contains the projection of γ in its interior. We draw the attention to the fact that condition (5.2.10) would be a strengthened version of condition (4.2.6) if the quadratic form $x'Mx$ was positive definite, which is not the case.

We turn our attention now towards the classical problem of *isochronism*. The problem has already been treated by Galilei in the 16th century in connection with the study of the pendulum. The basic problem was to find that curve in a vertical plane that has the following property: a weighing body (in the gravitational field) released from rest (with zero initial velocity) at any point of the curve and moving along the curve should consume always the same amount of time to reach a fixed horizon. The problem was solved by Huygens in 1673 who proved that the *cycloid* is the curve with this "isochronous" property. He applied this result in the construction of clocks with cycloidal pendulums.

In Section 3.6 around formula (3.6.3) we have discussed the centre problem in some detail. Now *we assume* in the sequel that the system

$$\dot{x} = -y + f(x, y), \qquad \dot{y} = x + g(x, y) \tag{5.2.11}$$

where f and g are analytic in a neighbourhood of the origin $(x, y) = (0, 0)$ and their Taylor series expansion does not contain terms of degree less than two *has a centre in the origin*, i.e. the origin has a neighbourhood that is the union of periodic orbits (plus of course the origin which is an equilibrium point). We have seen that this is the case with Duffing's equation (3.3.2) without damping, which is a model for the undamped mathematical pendulum. This is also the case with the original pendulum equation $\ddot{\theta} + (g/l)\sin\theta = 0$ without damping (cf. Example 1.5.4), as this can be seen easily considering the energy integral $V(x_1, x_2) = x_2^2/2 + (g/l)(1 - \cos x_1)$ defined in the phase plane $x_1 = \theta$, $x_2 = \dot{\theta}$. We have also seen in the case of Duffing's equation (see (3.3.7)) that apart from the linear spring case ($b = 0$) the origin is *not* an isochronous centre, i.e. the period of the periodic solution depends on the amplitude. Also in the case of the Lotka-Volterra predator-prey system (3.4.1) the positive equilibrium point is a non-isochronous centre: the period depends on the amplitude or rather on the "energy" (see Theorems 3.4.1-3.4.2). The question arises of whether there are systems at all apart from the harmonic oscillator that are isochronous? We say that a centre of a two dimensional system is *isochronous* if it has a neighbourhood in which all periodic solutions have the same (least) period. The problem has a vast literature. A good survey paper of this topic is that of Amel'kin [1977]. We are presenting here only the simplest results (cf. also Rudenok [1975] and Amel'kin-Lukashevich [1977]). System (5.2.11) will be studied where the origin $(x, y) = (0, 0)$ is *supposed to be a centre*. The fact that the linearized system has period 2π, i.e. eigenvalues $\pm i$ is not a restriction of generality; any system with a centre can be transformed into this form, as we shall show this in Example 5.2.3. The main tool of the study will be the transformation of system (5.2.11) into polar coordinates by

$$x = r\cos\theta, \qquad y = r\sin\theta. \tag{5.2.12}$$

Multiplying the first equation of (5.2.11) by $-y$, the second by x and adding them up, we obtain

$$x\dot{y} - y\dot{x} = x^2 + y^2 + xg(x, y) - yf(x, y). \tag{5.2.13}$$

On the other hand, using the polar transformation formulae (5.2.12), we see that

$$
\begin{aligned}
x\dot{y} - y\dot{x} &= r\cos\theta(\dot{r}\sin\theta + \dot{\theta}r\cos\theta) - r\sin\theta(\dot{r}\cos\theta - \dot{\theta}r\sin\theta) \\
&= r^2\dot{\theta}.
\end{aligned}
\tag{5.2.14}
$$

There holds the following:

Theorem 5.2.6. *If*

$$G(x, y) := xg(x, y) - yf(x, y) \equiv 0, \tag{5.2.15}$$

then the centre at the origin is isochronous; if $G(x, y) \geq 0$ (resp. ≤ 0) in a neighbourhood of the origin, then the period of the periodic solutions is less (resp. greater) than 2π.

Note the geometric significance of condition (5.2.15). It expresses the fact that the vector of the higher order terms of the field of (5.2.11), $[f,g]$, is perpendicular to the linear part of the field, $[-y,x]$, i.e. the vector of the higher order terms has no "tangential component", or still in other words, nothing is added to the field of the harmonic oscillator in the "tangential direction" ("tangential" meant here in the sense: orthogonal to the radius vector).

Proof. Substituting (5.2.14) into (5.2.13) and taking (5.2.15) into account, we obtain that $\dot{\theta} = 1$, i.e. the angular velocity is constant, in other words the angular frequency is constant: it does not depend on the periodic orbit on which the point moves. Now, if $G \geq 0$ (resp. ≤ 0), then $\dot{\theta} = 1 + G(x,y)/r^2 \geq 1$ (resp. ≤ 1), i.e. the angular velocity is greater (resp. smaller) than 1; hence, the period is less (resp. greater) than 2π. $\square$

In particular, we have the obvious:

Corollary 5.2.7. *If $f(x,y) = xh(x,y)$, $g(x,y) = yh(x,y)$ where h is analytic and its Taylor expansion does not contain a constant term then the centre at the origin is isochronous.*

Example 5.2.3. Duffing's equation without damping (3.3.2) will be treated whose Cauchy normal form is (cf. (3.3.3))

$$\dot{x} = y, \qquad \dot{y} = -x(a + bx^2). \qquad (5.2.16)$$

Here $a > 0$, $b \in \mathbf{R}$. First we transform this system into the form (5.2.11); the calculation will be made in unnecessary generality in order to illustrate the method. We introduce new coordinates (u,v) and "new time" τ by writing

$$u(\tau) := Ax(H\tau), \qquad v(\tau) := By(H\tau),$$

i.e. $t = H\tau$ where A, B and H are real constants to be determined. If (x,y) satisfies (5.2.16), then the corresponding (u,v) satisfies

$$
\begin{aligned}
du/d\tau &= A\dot{x}(H\tau)H = AHv/B, \\
dv/d\tau &= B\dot{y}(H\tau)H = -BHau/A - BHbu^3/A^3,
\end{aligned}
$$

where dot denotes differentiation with respect to t. Now, we want to have

$$AH/B = -1, \qquad -aBH/A = 1,$$

or $A/B = -1/H$, $A/B = -aH$, i.e. $H^2 = 1/a$, $H = 1/\sqrt{a}$. Only $A/B = -\sqrt{a}$ is defined. We may choose, say, $A = 1$, $B = -1/\sqrt{a}$. Thus, the system for (u,v) assumes the form

$$du/d\tau = -v, \qquad dv/d\tau = u + u^3 b/a. \qquad (5.2.17)$$

We know that (5.2.16) has a centre at $(x, y) = (0, 0)$; hence, the last system has a centre at $(u, v) = (0, 0)$, and if this last system has a periodic solution with period $S > 0$, then the period of the corresponding solution of (5.2.16) will be $T = HS = S/\sqrt{a}$. The function G of Theorem 5.2.6 for (5.2.17) is $G(u, v) = u^4 b/a \geq 0$ (resp. ≤ 0) provided that $b > 0$ (resp. $b < 0$). This is in accordance with formula (3.3.7) et seq. If $b > 0$ (resp. < 0), then the periods of the periodic solutions of the original system (5.2.16) are less (resp. greater) than $T = 2\pi/\sqrt{a}$.

From the great number of necessary and sufficient conditions of isochronism we quote but two theorems without proofs. Although these theorems are important from a theoretical point of view, it is very difficult to apply them.

Theorem 5.2.8 *(Vorobev [1963$_1$, 1963$_2$], Urabe [1967]). The centre of system (5.2.11) at the origin is isochronous if and only if there exists an analytic coordinate transformation of the form*

$$u = x + \sum_{i+k=2}^{\infty} a_{ik} x^i y^k, \qquad v = y + \sum_{i+k=2}^{\infty} b_{ik} x^i y^k$$

where the series are convergent in a neighbourhood of the origin that carries the system over to $\dot{u} = -v$, $\dot{v} = u$.

Theorem 5.2.8 *(Rudenok [1975]). The centre of system (5.2.11) at the origin is isochronous if and only if there exists an analytic coordinate transformation of the form*

$$u = x + \sum_{i=1}^{\infty} a_{2i} y^{2i}, \qquad v = y + \sum_{i+k=2}^{\infty} b_{ik} x^i y^k$$

where the series are convergent in a neighbourhood of the origin that carries the system over to a similar one with $G(u, v) \equiv 0$ where G is defined by the formula analogous to (5.2.15) for the new system.

Note that isochronism does not imply necessarily that the flow generated by the two dimensional autonomous system acts on the plane as a rotation of a rigid body, as a wheel (in a neighbourhood of the centre, at least). The angular velocity $\dot{\theta}$ need not be constant in order to have isochronism; it may depend even on the polar radius r. What is needed is that the equation

$$2\pi = \int_0^T \dot{\theta}(r(t), \theta(t)) dt$$

holds with a *fixed* $T > 0$ if an arbitrary periodic solution $(r(t), \theta(t))$ is substituted into the expression for $\dot{\theta}$. This T is then the common period of all the solutions in a neighbourhood of the origin. A neighbourhood of the origin rotates around it like a wheel if $\dot{\theta}$ is constant, and like a single spoke of a wheel if $\dot{\theta}$ depends on θ only (this last condition is sufficient for isochronism). There are stronger isochronism definitions (see Rudenok [1975]) that model this requirement.

Note also that isochronism is related to "area preserving transformations" too (see Birkhoff [1927], Arnold [1974,1984]) which is an important field in topological dynamics but would take us far from the topic of this book.

5.3   D-periodic Solutions of Cylindrical Systems

The simplest and best known "cylindrical system" is "probably" the system (1.5.6) describing the motion of the mathematical pendulum. The right-hand side is *periodic in one of the phase coordinates* $x_1 = \theta$. This is called a *cyclical coordinate*. In points of the phase plane whose x_1 coordinate differs in an integer multiple of 2π, the system behaves in the same way since in such points the vector field assumes equal values. Therefore, points with x_1 and with $x_1 + 2k\pi$ ($k = \pm 1, \pm 2, ...$) can be identified; thus, the phase plane can be considered isomorphic to a circular cylinder of radius one—hence the name.

A system of differential equations is called *cylindrical* if its vector field is periodic in some (at least one) phase coordinate. Apart from this, if the system is not autonomous, it may be periodic in the time variable too. Though this chapter is about autonomous systems, in this section we shall treat *non-autonomous periodic systems as well* because it would not be reasonable to treat cylindrical systems in two sections separating the autonomous and non-autonomous cases. If a vector function that is *linear in the time variable* is substituted into a cylindrical system for a cyclical coordinate, a periodic function is obtained, i.e. solutions that are linear in the cyclical coordinates behave, in a sense, like periodic solutions. Such solutions, *"periodic solutions of the second kind"*, have already been considered by Poincaré [1890,1899]. Since then, besides the mathematical pendulum, several important applications occurred (see e.g. Cook-Louisell-Yocom [1958], Stanisic-Mlakar [1969], Barbashin-Tabueva [1969] and Blehman [1971]), and several authors have studied the existence and stability of such solutions (see Diliberto-Hufford [1956], Akulenko [1967], Akulenko-Volosov [1967a,1967b], Barbashin [1967a,1967b] Nazarov [1970]). Here we shall adopt a probably more straightforward definition for "periodic solutions of the second kind" grasping one of their characteristic properties and shall apply also a somewhat more general definition of "cylindrical systems" (see Farkas [1968,1970,1971,1975b]).

Definition 5.3.1. We say that the function $\varphi \in C^1(\mathbf{R}, \mathbf{R}^n)$ is *a D-*

periodic solution (derivo-periodic in earlier formulations) with period $T > 0$ if its derivative $\dot{\varphi}$ is periodic with period T.

Lemma 5.3.1. *The function* $\varphi \in C^1(\mathbf{R}, \mathbf{R}^n)$ *is D-periodic with period* $T > 0$ *if and only if there exist a constant vector* $a \in \mathbf{R}^n$ *and a periodic function* $\nu \in C^1(\mathbf{R}, \mathbf{R}^n)$ *with period* T *such that*

$$\varphi(t) \equiv at + \nu(t). \tag{5.3.1}$$

Proof. The condition is, obviously, sufficient. Suppose that φ is D-periodic with period $T > 0$. Then

$$\dot{\varphi}(t + T) \equiv \dot{\varphi}(t).$$

Integrating both sides from 0 to t we get

$$\varphi(t + T) \equiv \varphi(t) + \varphi(T) - \varphi(0).$$

Introducing the notation

$$a := (\varphi(T) - \varphi(0))/T = (1/T) \int_0^T \dot{\varphi}(t)dt \tag{5.3.2}$$

and defining a function ν by $\nu(t) := \varphi(t) - at$, we have that $\nu \in C^1$ and

$$\begin{aligned}
\nu(t + T) &= \varphi(t + T) - at - aT \\
&= \varphi(t) + \varphi(T) - \varphi(0) - at - \varphi(T) + \varphi(0) \\
&= \nu(t);
\end{aligned}$$

thus, ν is periodic with period T. $\square$

It is easy to see that a D-periodic function φ determines uniquely the vector $a \in \mathbf{R}^n$ and the periodic function ν occurring in the expression (5.3.1); we call $a \in \mathbf{R}^n$ the *coefficient vector* of the D-periodic function φ. It is the integral mean of the derivative $\dot{\varphi}$. Obviously, the periodic functions (in the ordinary sense) are D-periodic functions with zero coefficient vector. We may start off from the derivative $\dot{\varphi}$, i.e. we may consider a continuous T-periodic function. According to Lemma 4.1.4 its integral function is T-periodic if and only if its mean value (over a period) is zero, i.e. if and only if the coefficient vector defined by (5.3.2) is zero. Otherwise the integral function is D-periodic with period T.

We shall give now a slight generalization of the concept of a cylindrical function.

Definition 5.3.2. Let $a \in \mathbf{R}^n$ be a constant vector and $T > 0$ a constant; we say that the function $F : \mathbf{R}^n \times \mathbf{R}^m \to \mathbf{R}^n$ is *periodic in the vector variable* x *with vector period* aT if for every $x \in \mathbf{R}^n$, $z \in \mathbf{R}^m$ we have

$$F(x + aT, z) \equiv F(x, z).$$

Introducing the vector period as the product of a vector and a positive scalar is rather artificial; however, it will be convenient when we shall study D-periodic solutions of systems that are periodic in a scalar and in a vector variable both.

We have to clarify the relation of this Definition to the concept of a "cylindrical function". *If F is cylindrical*, i.e. periodic in one or more variables *then it is periodic in the vector variable*. Indeed, suppose, say, that there are non-zero constants $a_1, a_2 \in \mathbf{R}$, $T > 0$ such that

$$F(x_1 + a_1 T, x_2, ...) \equiv F(x_1, x_2, ...),$$
$$F(x_1, x_2 + a_2 T, ...) \equiv F(x_1, x_2, ...),$$

then, clearly, also

$$F(x_1 + a_1 T, x_2 + a_2 T, x_3, ...) \equiv F(x_1, x_2, x_3, ...)$$

and F is periodic in the vector variable. In this case F has more than one vector periods, namely, $(a_1 T, a_2 T, 0, ..., 0)$, $(a_1 T, 0, ..., 0)$, $(0, a_2 T, 0, ..., 0)$. The converse implication is not valid. A function may be periodic in the vector variable without having a single cylindrical or rather cyclical variable. As an example consider the two dimensional vector field $F = (F_1, F_2)$ where

$$F_i(x_1, x_2) := \sin(c_1 x_1 + c_2 x_2) + \sin(d_1 x_2 + d_2 x_2) \qquad (i = 1, 2),$$

$c_j \neq 0$, $d_j \neq 0$ are real constants and the ratios c_j / d_j are irrational $(j = 1, 2)$. Then F is *not* periodic in either variable; at the same time, one may always find integers k, l, a positive T and a non-zero $a = (a_1, a_2) \in \mathbf{R}^2$ in such a way that

$$c_1 a_1 T + c_2 a_2 T = 2\pi k, \qquad d_1 a_1 T + d_2 a_2 T = 2\pi l,$$

and this means that F is periodic in its vector variable with vector period aT. Another, probably more striking, example is $(F_1(x, y), F_2(x, y)) = ((x_1 + x_2)^k, (x_1 + x_2)^l)$ where $k, l \in \mathbf{N}$. This function is periodic in the vector variable; for any $\alpha \in \mathbf{R}$ the vector $(\alpha, -\alpha)$ is a vector period.

However, if F is periodic in the vector variable with vector period $aT \neq 0$, then by an orthogonal coordinate transformation we may achieve that $aT/|aT|$ be the first vector of the base of the coordinate system. In this transformed system, F will be periodic in its first variable.

First we shall sketch how results about the existence, isolation and stability of periodic solutions can be extended to D-periodic solutions of periodic non-autonomous systems that are also periodic in the vector variable. Secondly, we shall treat autonomous systems periodic in the vector variable.

Let $a \in \mathbf{R}^n$ and $X \subset \mathbf{R}^n$ be an open and connected set that has the following property: if $x \in X$, then for all $t \in \mathbf{R}$ also $(x + at) \in X$. Let $f \in C^0(\mathbf{R} \times X, \mathbf{R}^n)$, $f'_x \in C^0(\mathbf{R} \times X, \mathbf{R}^{n^2})$, assume that f is periodic in the variable $t \in \mathbf{R}$ with period $T > 0$ and it is periodic in the vector variable $x \in X$ with vector period aT, i.e.

$$f(t + T, x) \equiv f(t, x), \qquad f(t, x + aT) \equiv f(t, x),$$

and consider the system of differential equations

$$\dot{x} = f(t, x). \tag{5.3.3}$$

We are interested in those D-periodic solutions of (5.3.3) whose period is T and coefficient vector a.

The following Lemma is an analogue of Lemma 2.2.1.

Lemma 5.3.2. *The solution $\varphi : \mathbf{R} \to X$ of (5.3.3) is D-periodic with period T and coefficient vector a if and only if there exists a $t_0 \in \mathbf{R}$ such that*

$$\varphi(t_0 + T) = \varphi(t_0) + aT. \tag{5.3.4}$$

Proof. Clearly, (5.3.4) is a necessary condition (for arbitrary $t_0 \in \mathbf{R}$). We have to prove that it is sufficient as well. First, it is to be observed that if φ is a solution, then the function $\varphi(t) + aT$ is a solution too because

$$(\varphi(t) + aT)^\bullet \equiv \dot{\varphi}(t) \equiv f(t, \varphi(t)) \equiv f(t, \varphi(t) + aT)$$

where we have used the fact that f is periodic in the vector variable x with vector period aT. We know from the proof of Lemma 2.2.1 that also the function $\varphi(t + T)$ is a solution, but by (5.3.4)

$$
\begin{aligned}
[\varphi(t + T)]_{t=t_0} &= \varphi(t_0 + T) = \varphi(t_0) + aT \\
&= [\varphi(t) + aT]_{t=t_0};
\end{aligned}
$$

hence, because of the uniqueness of solutions, $\varphi(t + T) \equiv \varphi(t) + aT$. Differentiating this last identity we see that φ is D-periodic indeed with period T, and then from the identity itself and from (5.3.2) it follows that a is its coefficient vector. $\square$

Fixed point methods can be used to prove the existence of D-periodic solutions. Without going into details we show how the Poincaré-Andronov operator (4.1.3) can be generalized to this case. Suppose that a subset $X_0 \subset X$ exists such that every solution $\varphi(t, 0, x^0)$ of (5.3.3) with initial condition in X_0, i.e. $x^0 \in X_0$, is defined over the closed interval $[0, T]$ at least. Then the operator $\varphi_T : X_0 \to X$ defined by

$$\varphi_T(x^0) := \varphi(T, 0, x^0) - aT \tag{5.3.5}$$

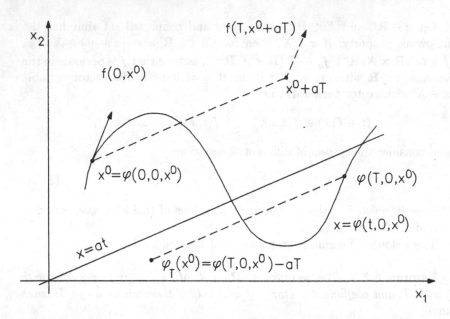

FIGURE 5.3.1. The Poincaré-Andronov map (5.3.5) of a periodic system which is also periodic in the vector variable. The field at the points x^0 and $x^0 + aT$ is equal provided that the time difference is the period T; thus, if $\varphi(T, 0, x^0)$ was equal to $x^0 + aT$, we could expect D-periodic behaviour.

is called the *Poincaré-Andronov operator* of system (5.3.3) (see Figure 5.3.1).

The previous Lemma and the definition of the Poincaré-Andronov operator yield the analogue of Theorem 4.1.1:

Theorem 5.3.3. *System (5.3.3) has a D-periodic solution with period T and coefficient vector a if and only if the operator (5.3.5) has a fixed point $x^0 \in X^0$; if $x^0 = \varphi_T(x^0)$, then the solution $\varphi(t, 0, x^0)$ is D-periodic with period T and coefficient vector a.*

We note that existence results for D-periodic solutions of scalar n-th order differential equations have been achieved recently by Andrès [1990].

Conditions for the isolation and the stability of periodic solutions were relying heavily on the properties of the variational system. Now, if p is a D-periodic solution of system (5.3.3) with period T and coefficient vector a, then the functions $f(t, p(t))$ and $f_x'(t, p(0))$ are periodic with period T, as this can be seen easily using the periodicity of f in t and in the vector variable x and (5.3.1). Thus, the variational system of (5.3.3) with respect to the D-periodic solution p,

$$\dot{y} = f_x'(t, p(t))y, \tag{5.3.6}$$

is a periodic system with period T. The only problem that arises in proving theorems analogous to those of Section 4.2 about stability and isolation is that if p is a genuine D-periodic solution, then its path is no longer a compact set and the function p is not bounded.

There holds the analogue of Theorem 4.2.1.

Theorem 5.3.4. *If all the characteristic multipliers of the variational system (5.3.6) are in modulus less than 1, then the D-periodic solution p of system (5.3.3) is asymptotically stable.*

Proof. The proof is similar to that of Theorem 4.2.1. The only point that is to be shown is that when writing system (5.3.3) in the form analogous to (4.2.2) the remainder term will be $o(|x|)$ uniformly in t.

Proceeding as at the beginning of Section 4.2, system (5.3.3) is written in the form

$$\dot{z} = f(t, p(t) + z) - f(t, p(t))$$

or

$$\dot{z} = f'_x(t, p(t))z + r(t, z)$$

where, using the notation $r = (r_1, ..., r_n)$,

$$r_k(t, z) = < f'_{kx}(t, p(t) + \eta_k z) - f'_{kx}(t, p(t)), z > \qquad (5.3.7)$$

$$0 < \eta_k < 1 \qquad (k = 1, 2, ..., n).$$

We have to show that $r(t, z) = o(|z|)$ uniformly in $t \in \mathbf{R}$. Denote the path of p by γ and the path of the restriction of p to $[0, T]$ by $\tilde{\gamma}$, i.e.

$$\gamma := \{x = p(t) : t \in \mathbf{R}\}, \quad \tilde{\gamma} := \{x = p(t) : 0 \le t \le T\}.$$

The compact set $\tilde{\gamma}$ has a neighbourhood $U_{\tilde{\gamma}}$ of some radius $\rho_0 > 0$ such that its closure $\overline{U}_{\tilde{\gamma}}$ is a subset of X, i.e.

$$U_{\tilde{\gamma}} = \{x \in \mathbf{R}^n : \text{dist}\,(x, \tilde{\gamma}) < \rho_0\}, \quad \overline{U}_{\tilde{\gamma}} \subset X.$$

Let us denote the ρ_0 neighbourhood of the complete path γ by U_γ:

$$U_\gamma = \{x \in \mathbf{R}^n : \text{dist}\,(x, \gamma) < \rho_0\}.$$

We show that also $\overline{U}_\gamma \subset X$. Indeed, if $x \in \overline{U}_\gamma$, then there exists a time $t \in \mathbf{R}$ such that $|x - p(t)| \le \rho_0$ and there is an $m \in \mathbf{N}$ such that $mT \le t < (m+1)T$. Since p is D-periodic with period T and coefficient vector a, we have $p(t + T) = p(t) + aT$, and

$$\begin{aligned} |x - maT - p(t - mT)| &= |x - maT - (p(t) - maT)| \\ &= |x - p(t)| \le \rho_0; \end{aligned}$$

hence, because of $0 \leq t - mT < T$ we have $(x - maT) \in \overline{U}_{\tilde{\gamma}} \subset X$. But then the assumption on the set X yields $x \in X$. We know that f'_x is uniformly continuous on the compact set $[0, T] \times \overline{U}_{\tilde{\gamma}} \subset \mathbf{R} \times \mathbf{R}^n$. Since f'_x is periodic in t with period T and in the vector variable x with vector period aT, and we have just seen that for $x \in \overline{U}_\gamma$ there exists an $m \in \mathbf{N}$ such that $(x - maT) \in \overline{U}_{\tilde{\gamma}}$, f'_x is uniformly continuous on the whole set $\mathbf{R} \times \overline{U}_{\tilde{\gamma}}$. Thus, (5.3.7) implies that $r(t, z) = o(|z|)$ uniformly in $t \in \mathbf{R}$. $\square$

Isolation can be treated analogously as it has been done in connection with periodic solutions of periodic systems. A definition similar to Definition 4.2.1 can be given and the following Theorem can be proved exactly as it has been done in the case of Theorem 4.2.7.

Theorem 5.3.5. *If the number 1 is not a characteristic multiplier of the variational system (5.3.6), then the D-periodic solution p with period T and vector coefficient a of system (5.3.3) is isolated.*

We turn now to the study of D-periodic solutions of *autonomous systems* that are periodic in the vector variable. Let $X \subset \mathbf{R}^n$ be an open and connected set, $a \in \mathbf{R}^n$ a constant vector, and suppose that $x \in X$ implies $(x + at) \in X$ for every $t \in \mathbf{R}$. Let the function $f \in C^1(X, \mathbf{R}^n)$ be periodic in the vector variable x with vector period a, i.e. $f(x + a) \equiv f(x)$, and consider the autonomous system of differential equations

$$\dot{x} = f(x). \tag{5.3.8}$$

We shall be interested in D-periodic solutions of (5.3.8) that can be written in the form

$$\psi(t) = a_\psi t + \nu(t), \qquad \nu(t + T) = \nu(t), \qquad a_\psi = a/T$$

for some $T > 0$. The difference compared to the case of a non-autonomous periodic system is that (5.3.8) does not determine the period of a possible periodic or D-periodic solution. If a D-periodic solution is found with period $T > 0$, then it will be "admissible", i.e. substituted into the right-hand side of the system it will yield a T-periodic function if and only if its coefficient vector a_ψ is parallel to the vector period a of the system, more exactly if and only if $a_\psi = a/T$ or an integer multiple of this value. The analogue of Lemma 5.3.2 is in this case:

Lemma 5.3.6. *The solution $\psi : \mathbf{R} \to X$ of (5.3.8) is D-periodic with period $T > 0$ and coefficient vector $a_\psi = a/T$ if and only if there exists a $t_0 \in \mathbf{R}$ such that*

$$\psi(t_0 + T) = \psi(t_0) + a. \tag{5.3.9}$$

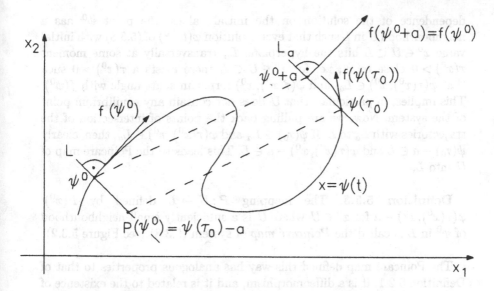

FIGURE 5.3.2. The Poincaré map attached to system (5.3.8) which is periodic in the vector variable with vector period a.

Proof. It is identical to the proof of Lemma 5.3.2. $\square$

In the case of system (5.3.8), which is periodic in the vector variable, the Poincaré map is to be defined in a modified form if we want to use it for the detection of D-periodic solutions. We follow the construction preceding Definition 5.2.1 with the necessary modifications. Assume that (5.3.8) has a non-constant solution $\psi : \mathbf{R}_+ \to X$ and set $\psi^0 = \psi(0)$. Consider an $n-1$ dimensional hyperplane L passing through the point ψ^0 and transversal to the path of ψ. (It may be, for instance, orthogonal to the tangent vector $\dot\psi(0) = f(\psi^0)$.) We displace the hyperplane parallel to the point $\psi^0 + a$, denote the displaced hyperplane by L_a, and assume that the path of ψ hits L_a at some moment $\tau_0 > 0$ (see Figure 5.3.2) and that $\dot\psi(\tau_0)$ forms an acute angle with $\dot\psi(0) = f(\psi^0)$. This latter requirement means that the path of ψ crosses the plane

$$L_a = \{x \in \mathbf{R}^n : < x - (\psi^0 + a), \ f(\psi^0) > = 0\}$$

in the same direction at $\psi(\tau_0)$ as it does the plane

$$L = \{x \in \mathbf{R}^n : < x - \psi^0, \ f(\psi^0) > = 0\}$$

at ψ^0. (This requirement is important in case the vector period a of the system is orthogonal to $f(\psi^0)$ or forms an obtuse angle with it. It is instructive to draw the picture in these cases.) Because of the continuous dependence of the solution on the initial values, the point ψ^0 has a neighbourhood U in L such that every solution $\varphi(t, x^0)$ of (5.3.8) with initial value $x^0 \in U \subset L$ hits the hyperplane L_a transversally at some moment $\tau(x^0) > 0$ close to τ_0, i.e. for $x^0 \in U \subset L$ there exists a $\tau(x^0) > 0$ such that $\varphi(\tau(x^0), x^0) \in L_a$, and $\dot{\varphi}(\tau(x^0), x^0)$ forms an acute angle with $f(\psi^0)$. This implies, in particular, that U does not contain any equilibrium point of the system. Now, we are pulling back the points of intersection of the trajectories with L_a to L. If $\psi(\tau_0) \in L_a$ and $\varphi(\tau(x^0), x^0) \in L_a$, then, clearly $\psi(\tau_0) - a \in L$ and $\varphi(\tau(x^0), x^0) - a \in L$. This leads to the Poincaré map of U into L.

Definition 5.3.3. The mapping $P : U \to L$ defined by $P(x^0)$: $\varphi(\tau(x^0), x^0) - a$ for $x^0 \in U$ where U is a sufficiently small neighbourhood of ψ^0 in L is called the *Poincaré map* of system (5.3.8) (see Figure 5.3.2).

The Poincaré map defined this way has analogous properties to that of Definition 5.2.1. It is a diffeomorphism, and it is related to the existence of a D-periodic solution in the following way. System (5.3.8) has a D-periodic solution with period $T > 0$ and coefficient vector a/T if and only if the Poincaré map has a fixed point: $P(x^0) = x^0$ and $\tau(x^0) = T$.

We note, however, that system (5.3.8) may be considered as a system with vector period ma where m is an arbitrary (say, positive) integer. Now we may define the iterates of the Poincaré map by

$$P^m(x^0) := P(P^{m-1}(x^0)) \qquad (m = 2, 3, \ldots) \qquad (5.3.10)$$

provided that $P^{m-1}(x^0) \in U$, the domain of the map. In the case $m = 2$ we have

$$P^2(x^0) = P(\varphi(\tau(x^0), x^0)) - a) = P(x^1 - a)$$

where $x^1 := \varphi(\tau(x^0), x^0) \in L_a$, or

$$P^2(x^0) = \varphi(\tau(x^1 - a), x^1 - a) - a$$

where it is assumed that $x^1 - a \in U$. Now, (5.3.8) has the following basic property: if $\varphi(t, x^1)$ is the solution that assumes x^1 at $t = 0$, then $\varphi(t, x^1) - a$ is a solution too and it assumes the initial value $x^1 - a$ at $t = 0$, i.e.

$$\varphi(t, x^1 - a) \equiv \varphi(t, x^1) - a. \qquad (5.3.11)$$

Indeed,

$$(\varphi(t, x^1) - a)^\bullet \equiv \dot{\varphi}(t, x^1) \equiv f(\varphi(t, x^1)) \equiv f(\varphi(t, x^1) - a)$$

because f is periodic in x with vector period a. Applying (5.3.11) we obtain

$$P^2(x^0) = \varphi(\tau(x^1 - a), x^1) - 2a,$$

or applying Definition 1.3.1 (iii)

$$P^2(x^0) = \varphi(\tau(x^1 - a) + \tau(x^0), x^0) - 2a.$$

This means that the motion described by the solution $\varphi(t, x^0)$ reaches the plane

$$L_{2a} = \{x \in \mathbf{R}^n : < x - (\psi^0 + 2a), \ f(\psi^0) > = 0\}$$

at the moment $\tau(x^0) + \tau(x^1 - a)$ and then it is "pulled back" by $2a$ into the plane L. Similarly, by mathematical induction

$$P^m(x^0) = \varphi\left(\tau(x^0) + \sum_{k=1}^{m-1} \tau(x^k - a), x^0\right) - ma \qquad (m = 2, 3, ...)$$

where $x^k = \varphi(\tau(x^{k-1} - a), x^{k-1} - a)$, and it is assumed that $x^k - a \in U$ ($k = 2, 3, ... m - 1$). Clearly, $\varphi(\tau(x^0) + \sum_{k=1}^{m-1} \tau(x^k - a), x^0) \in L_{ma}$ where $L_{ma} = \{x \in \mathbf{R}^n : < x - (\psi^0 + ma), \ f(\psi^0) >= 0\}$. Now, if x^0 is a fixed point of P^m, i.e. $P^m(x^0) = x^0$, then this means that $\varphi(t, x^0)$ is a D-periodic solution with period

$$T = \tau(x^0) + \sum_{k=1}^{m-1} \tau(x^k - a)$$

and coefficient vector ma/T. This follows from $\varphi(T, x^0) = x^0 + ma$ which is a version of (5.3.9) with a replaced by ma; cf. Problem 5.7.3.

Assume now that the role of ψ in the construction of the Poincaré map is played by a non-constant D-periodic solution p of (5.3.8) with period $T > 0$ and coefficient vector $a_p = a/T$. Then the Poincaré map can be linearized at $p^0 = p(0)$ exactly as this has been done around formulae (5.2.1) and (5.2.2). We may also consider the variational system of (5.3.8) with respect to the solution p:

$$\dot{y} = f'(p(t))y; \qquad (5.3.12)$$

it is, clearly, a T-periodic homogeneous linear system of differential equations. Since p is non-constant, $\dot{p}(t) \not\equiv 0$ and it is a periodic solution of (5.3.12) with period T. Hence, the number 1 is a characteristic multiplier of system (5.3.12). There holds the analogue of Theorem 5.2.2.

Theorem 5.3.7. *The spectrum of the linearization of the Poincaré map attached to the D-periodic solution p with period $T > 0$ and coefficient vector a/T is equal to the set of the characteristic multipliers of the variational system (5.3.12) minus the number 1 which is to be deleted once from this set.*

Proof. The proof is similar to the proof of Theorem 5.2.2. $\square$

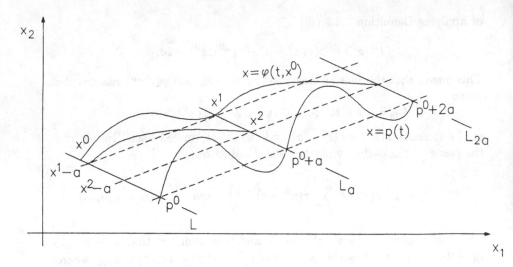

FIGURE 5.3.3. A D-periodic solution p with a contractive Poincaré map: $x^1 - a = P(x^0)$, $x^2 - a = P^2(x^0) = P(x^1 - a)$.

Just as in the proof of Theorem 5.2.2, it is easy to see that if $n - 1$ characteristic multipliers are in modulus less than one, then the Poincaré map is contractive, and this means that the orbit of p attracts the solutions from a neighbourhood (see Figure 5.3.3). And indeed the analogue of the Andronov-Witt Theorem 5.1.2 holds for D-periodic solutions of systems periodic in the vector variable:

Theorem 5.3.8. *If 1 is a simple characteristic multiplier of the variational system (5.3.12), and the remaining $n - 1$ characteristic multipliers are in modulus less than one, then the D-periodic solution p of (5.3.8) is asymptotically orbitally stable having the asymptotic phase property.*

Proof. The proof is analogous to the proof of Theorem 5.1.2 with the exception of proving that $r(t, z)$ of (5.1.14) is $o(|z|)$ uniformly in $t \in \mathbf{R}_+$. This presents a problem because $p(t)$ is no longer a bounded (periodic) function. However, this can be proved as has been done in the proof of Theorem 5.3.4. □

We note that for cylindrical systems the analogue of the Andronov-Witt Theorem was proved first by Vejvoda [1959].

Definition 5.3.4. We say that the non-constant D-periodic solution

p with period $T > 0$ and coefficient vector a/T of system (5.3.8) is *non-isolated* if to each $\varepsilon > 0$ one may find an $x^0 \in \mathbf{R}^n$ such that

(i) $0 < \operatorname{dist}(x^0, \gamma) < \varepsilon$ where γ is the path of p;

(ii) $\varphi(t, x^0)$ is a D-periodic solution with period $\tau(x^0)$ and coefficient vector $a/\tau(x^0)$;

(iii) $|\tau(x^0) - T| < \varepsilon$.

If these conditions do not hold, then the solution p is said to be *isolated*.

The following Theorem can be proved similarly to the proof of Theorem 5.2.3.

Theorem 5.3.9. *If the number 1 is a simple characteristic multiplier of the variational system (5.3.12), then the D-periodic solution p of system (5.3.8) is isolated.*

Example 5.3.1. *The mathematical pendulum without damping* (cf. Example 1.5.4). If it is assumed that there is no damping, the differential equation of the mathematical pendulum of length $l > 0$ is

$$\ddot{\theta} = -(g/l)\sin\theta \qquad (5.3.13)$$

where g is the gravitational acceleration. With the notations $x = \theta$, $y = \dot{x} = \dot{\theta}$, the system equivalent to (5.3.13)

$$\dot{x} = y, \qquad \dot{y} = -(g/l)\sin x \qquad (5.3.14)$$

admits the first integral (the energy integral)

$$V(x, y) := y^2/2 + (g/l)\int_0^x \sin s\, ds = y^2/2 + (g/l)(1 - \cos x).$$

System (5.3.14) is periodic in the vector variable with vector period $a = [2\pi, 0]$. It is sufficient to construct the phase portrait of the system in a vertical strip of width 2π in the phase plane x, y and then shift the picture to the right and to the left by integer multiples of 2π. The equations of the trajectories are obtained by making the energy integral equal to a constant $c \geq 0$, called the *energy level*:

$$(1/2)y^2 + (g/l)(1 - \cos x) = c$$

or

$$y = \pm\sqrt{2}(c - g/l + (g/l)\cos x)^{1/2}. \qquad (5.3.15)$$

The equilibria are $(m\pi, 0)$ $(m = 0, \pm1, \pm2, ...)$. It is easy to see that in the strip $-\pi \leq x \leq \pi$:

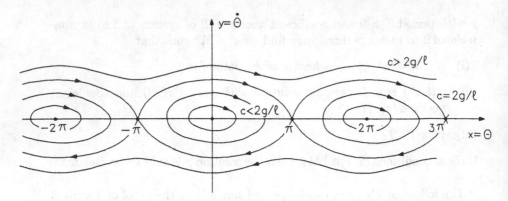

FIGURE 5.3.4. Phase portrait of the mathematical pendulum without damping. If the energy level c is below $2g/l$ the solutions are periodic, if it is above this value they are D-periodic.

(i)   If $0 < c < 2g/l$, then the path is closed it surrounds the origin and has a central symmetry.

(ii)   If $c = 2g/l$, then two heteroclinic trajectories $y = \pm\sqrt{2g/l}(1 + \cos x)^{1/2}$, $(-\pi < x < \pi)$ and the two equilibria $(\pm\pi, 0)$ joined by them are obtained.

(iii)   If $c > 2g/l$, then we have two trajectories (one in the upper, the other one in the lower half plane according to whether the sign $+$ or $-$ is chosen) which are graphs of functions periodic with period 2π (in the variable x).

See Figure 5.3.4.

Now, (5.3.15) is, in fact, a first order separable differential equation for the angular displacement function $\theta(t)$:

$$\dot\theta = \pm\sqrt{2}((c - g/l) + (g/l)\cos\theta)^{1/2}.$$

Integrating under the assumption that $\theta(0) = 0$ we obtain (taking only the plus sign into consideration)

$$
\begin{aligned}
t &= (1/\sqrt{2})\int_0^\theta (c - g/l + (g/l)\cos s)^{-1/2}ds \\
&= (l/2g)^{1/2}\int_0^\theta (cl/g - 1 + \cos s)^{-1/2}ds \\
&= (1/2)(l/g)^{1/2}\int_0^\theta (k^2 - \sin^2(s/2))^{-1/2}ds \qquad (5.3.16)
\end{aligned}
$$

where $k^2 = cl/(2g)$. Substituting $\sin(s/2) = k \sin z$ we obtain

$$t = (l/g)^{1/2} \int_0^\psi (1 - k^2 \sin^2 z)^{-1/2} dz \qquad (5.3.17)$$

where $\sin(\theta/2) = k \sin \psi$. The integral on the right-hand side is the so-called *elliptic integral of the first kind* (see Farkas [1964]) and is, usually, denoted by

$$F(k, \psi) := \int_0^\psi (1 - k^2 \sin^2 z)^{-1/2} dz.$$

In case (i) where the energy level c is small we have $0 < k^2 < 1$. It is easy to see that $\dot\theta = 0$ when $\sin(\theta/2) = \pm k$, i.e. when $\psi = \pm\pi/2$. Because of symmetry the integration up to this value yields the quarter of the period T, i.e. the period of the periodic solution corresponding to the energy level $0 < c = 2gk^2/l < 2g/l$ is

$$T(k) = 4(l/g)^{1/2} \int_0^{\pi/2} (1 - k^2 \sin^2 z)^{-1/2} dz = 4(l/g)^{1/2} F(k)$$

where $F(k)$ is the *complete elliptic integral of the first kind*. Obviously, T is an *increasing function of* k, i.e. of the energy level in $0 < k < 1$. Hence, by Problem 5.7.3 the function $\theta(t)$ is D-periodic with period

$$T(k) = 2\pi b = (l/g)^{1/2} \int_0^\pi (k^2 - \sin^2(\theta/2))^{-1/2} d\theta$$

and coefficient $1/b$. Here, obviously, the *period is a decreasing function of the energy level*, i.e. of k in the interval $1 < k < \infty$. This corresponds to intuition: the higher the energy level, the faster the pendulum is rotating. It is easy to compute the maximal and the minimal angular velocities: $\max \dot\theta = \sqrt{2c}$, $\min \dot\theta = \sqrt{2(c - 2g/l)}$. The difference between the maximal and the minimal angular velocity is

$$\max \dot\theta - \min \dot\theta = (\sqrt{2g/l})/(\sqrt{c} + \sqrt{c - 2g/l});$$

hence, while the maximal angular velocity is increasing with the energy, the difference between the maximal and the minimal angular velocity is decreasing.

Example 5.3.2. *The stability of closed geodesics* (see Farkas [1968]). Let M denote an n dimensional differentiable manifold with a positive definite Riemannian metric (see Appendix 3), V a simple coordinate neighbourhood of M with metric tensor g_{ik} and Christoffel symbols Γ^i_{jk} in the coordinate system $y = (y^1, ..., y^n)$. Let $y = p(s)$, $s \in [0, \infty]$, be the equation of a closed geodesic in natural (arc length) parametrization. The natural parametrization is characterized by the condition that the norm of the tangent vector $\dot{p}(s)$ is identically equal to 1: $g_{ik}(p(s))\dot{p}^i(s)\dot{p}^k(s) \equiv 1$ where the Einstein convention has been applied. An index appearing in a term in a lower and in an upper position means summation for that index from 1 to n. We assume that the geodesic $\gamma = \{y = p(s) : 0 \leq s < \infty\}$ stays inside the coordinate neighbourhood V, i.e. $\gamma \subset V$. Let $y = \varphi(s, y_0, \dot{y}_0)$ be the equation of any other geodesic in V *given also in natural parametrization* where $\varphi(0, y_0, \dot{y}_0) = y_0$, $\dot{\varphi}(0, y_0, \dot{y}_0) = \dot{y}_0$. We say that the geodesic γ is *stable* (in the Liapunov sense) (i) if a $\rho > 0$ can be found such that $|y_0 - p(0)| < \rho$, $|\dot{y}_0 - \dot{p}(0)| < \rho$ implies that φ is defined for $s \in [0, \infty)$ and $\varphi(s, y_0, \dot{y}_0) \in V$; (ii) given an arbitrary $\varepsilon > 0$, a $0 < \delta < \rho$ can be found such that $|y_0 - p(0)| < \delta$, $|\dot{y}_0 - \dot{p}(0)| < \delta$ implies $|\varphi(s, y_0, \dot{y}_0) - p(s)| < \varepsilon$ and $|\dot{\varphi}(s, y_0, \dot{y}_0) - \dot{p}(s)| < \varepsilon$ for $s \in [0, \infty)$. It is easy to see that if p is stable in one coordinate system, then it is stable in any admissible coordinate system too.

Now, as we know (see Appendix 3), the geodesics in "canonical parametrization" satisfy the system of differential equations

$$d^2 y^i/dt^2 + \Gamma_{jl}^i \, dy^j/dt \, dy^l/dt = 0 \qquad (i = 1, 2, ..., n) \qquad (5.3.18)$$

If the $2n$ dimensional vector $x = (x_1, ..., x_{2n})$ is introduced by $x_k = y^k$, $x_{n+k} = \dot{y}^k$ $(k = 1, 2, ..., n)$, then the normal form of system (5.3.18) will be

$$\dot{x}_k = x_{n+k}, \qquad \dot{x}_{n+k} = -\sum_{j,l=1}^{n} \Gamma_{jl}^k x_{n+j} x_{n+l} \qquad (k = 1, ..., n),$$

$$(5.3.19)$$

and one could easily expect that the stability of a geodesic is equivalent to the Liapunov stability of the corresponding solution of system (5.3.19). However, this is not the case; a solution of (5.3.18) or (5.3.19) is a geodesic but it is not necessarily given in natural parametrization. The canonical parameter t in (5.3.18) is the arc length if and only if $g_{ik}(\varphi(t))\dot{\varphi}^i(t)\dot{\varphi}^k(t) \equiv 1$ along the solution φ of (5.3.18), i.e. if and only if

$$\sum_{i,k=1}^{n} g_{ik}(x_1, ..., x_n)x_{n+i}x_{n+k} = 1$$

along the corresponding solution of (5.3.19). If a geodesic is given in canonical parametrization, i.e. it satisfies (5.3.18), then its tangent vector is of constant length $g_{ik}\dot{y}^i\dot{y}^k \equiv$ constant, but it is not necessarily a unit vector. This means that

$$U(x) := \sum_{i,k=1}^{n} g_{ik}(x_1, ..., x_n)x_{n+i}x_{n+k} - 1$$

is a first integral of system (5.3.19), and $U(x) = 0$, i.e.

$$\sum_{i,k=1}^{n} g_{ik}(x_1, ..., x_n)x_{n+i}x_{n+k} - 1 = 0 \qquad (5.3.20)$$

is an invariant manifold (see Appendix 3). If our geodesic p is given in natural parametrization, then its path belongs to the invariant manifold (5.3.20), and its stability is equivalent to the *stability with respect to solutions whose path belongs to the same manifold*. Thus, system (5.3.19) is to be restricted to the manifold (5.3.20) and the stability of the geodesic p given in natural parametrization is equivalent to the Liapunov stability of p considered as a solution of the system restricted to (5.3.20).

As an illustration of these ideas consider the unit sphere S^2 with centre in the origin whose equation in the Cartesian orthogonal coordinate system x, y, z is $x^2 + y^2 + z^2 = 1$. This is a two dimensional compact Riemannian manifold. We need a parametrization that provides a coordinate neighbourhood containing a complete closed geodesic, i.e. a great circle in its interior. Such a parametrization is

$$x = -2y^1/(1 + {y^1}^2 + {y^2}^2), \qquad y = 2y^2/(1 + {y^1}^2 + {y^2}^2),$$

$$z = (1 - ({y^1}^2 + {y^2}^2))/(1 + {y^1}^2 + {y^2}^2)$$

(the parameter curves are circles passing through the "south pole" $(0, 0, -1)$ and have tangents parallel to the x- and y-axes, respectively. A simple calculation yields the components g_{ik} of the metric tensor, the Christoffel symbols Γ^i_{jl} and the differential equation of the geodesics given in canonical parametrization. Omitting the details, applying the notations $x_1 = y^1$, $x_2 = y^2$, $x_3 = \dot{y}^1$, $x_4 = \dot{y}^2$ we write down the normal form of the system of differential equations

$$\dot{x}_1 = x_3, \qquad \dot{x}_2 = x_4,$$
$$\dot{x}_3 = 2(x_1 x_3^2 + 2x_2 x_3 x_4 - x_1 x_4^2)/(1 + x_1^2 + x_2^2),$$
$$\dot{x}_4 = 2(-x_2 x_3^2 + 2x_1 x_3 x_4 + x_2 x_4^2)/(1 + x_1^2 + x_2^2). \qquad (5.3.21)$$

The equation of the invariant manifold (5.3.20) is now

$$4(x_3^2 + x_4^2)/(1 + x_1^2 + x_2^2)^2 - 1 = 0. \qquad (5.3.22)$$

A periodic solution of (5.3.21) having a path in the manifold (5.3.22) is

$$p(t) = [\cos t, \sin t, -\sin t, \cos t] \qquad (5.3.23)$$

(this corresponds, actually, to the "equator" of the sphere given in natural parametrization). A parametric representation of the three dimensional hypersurface (5.3.22) of the four dimensional space of the x's is

$$x_1 = u_1, \qquad x_2 = u_2,$$

$$x_3 = (1/2)(1 + u_1^2 + u_2^2)\cos u_3, \qquad x_4 = (1/2)(1 + u_1^2 + u_2^2)\sin u_3. \qquad (5.3.24)$$

The restriction of system (5.3.21) to the manifold (5.3.24) is easily computable. It is

$$\dot{u}_1 = (1/2)(1 + u_1^2 + u_2^2)\cos u_3, \qquad \dot{u}_2 = (1/2)(1 + u_1^2 + u_2^2)\sin u_3,$$

$$\dot{u}_3 = u_1 \sin u_3 - u_2 \cos u_3. \tag{5.3.25}$$

The latter system, which is periodic in the vector variable with vector period $(0, 0, 2\pi)$, admits the D-periodic solution

$$q(t) = [\cos t, \sin t, t + \pi/2]$$

with period 2π and coefficient vector $[0, 0, 1]$. Obviously, $p(t) = x(u)|_{u=q(t)}$ where $x(u)$ represents the parametrization (5.3.24). Thus, the problem of the stability of the geodesic p is reduced to the problem of settling the Liapunov stability of the solution q with respect to the three-dimensional system (5.3.25). The variational system (5.3.25) with respect to the D-periodic solution q is the 2π-periodic system

$$
\begin{aligned}
\dot{z}_1 &= -\sin t \cos t \cdot z_1 - \sin^2 t \cdot z_2 - \cos t \cdot z_3, \\
\dot{z}_2 &= \cos^2 t \cdot z_1 + \sin t \cos t z_2 - \sin t \cdot z_3, \\
\dot{z}_3 &= \cos t \cdot z_1 + \sin t \cdot z_2.
\end{aligned}
$$

As we know, $\dot{q}(t) = [-\sin t, \cos t, 1]$ is a periodic solution of the latter system. With some trouble one may find more independent solutions. A fundamental matrix is

$$Z(t) = \begin{pmatrix} -\sin t & \cos t & \sin t \cos t \\ \cos t & \sin t \cos t & \sin^2 t \\ 1 & \sin t & -\cos t \end{pmatrix}.$$

Hence, it is seen that all solutions are periodic with period 2π; thus, the number 1 is a triple characteristic multiplier. This does not settle the problem of stability; however, it is perfectly clear that a great circle of the sphere is stable with respect to the great circles given in natural parametrization: if a point is given near to the fixed great circle on the sphere and a direction in the tangent plane at this point near to the direction of the tangent of the fixed great circle at the nearest point, then the great circle through the given point in the given direction will stay close to the fixed great circle in the sense that after having covered the same arc length on both circles, the points reached and the respective directions of the tangents at these points will be close to each other.

Further examples of systems that are periodic in the vector variable can be found in Barbashin-Tabueva [1969]. We note also that the concept of a D-periodic solution has been generalized by Andrès [1987] who introduced solutions with periodic *second* derivatives.

5.4 Existence of Periodic Solutions

In this Section some methods are presented that enable one to settle the existence problem of periodic solutions for systems of dimension higher than two. Generalizations of the Poincaré-Bendixson theory (see Section 3.1), fixed point theorems and degree theory (see Appendix 2) are applied, and in case of three dimensional systems geometry, the drawing of figures helps in some cases. However, it is to be understood that no general method, nothing like the planar Poincaré-Bendixson theory, exists, and all the known results make use of special properties of the particular system or type of systems concerned (like conservativeness, monotonicity, etc.) in establishing the existence of a periodic solution.

We start with the generalization of the Poincaré-Bendixson theory for higher dimensional systems due to Russel A. Smith [1979,1980$_a$,1981]. The autonomous system

$$\dot{x} = f(x) \tag{5.4.1}$$

is considered where $f \in C^1(X, \mathbf{R}^n)$, $X \subset \mathbf{R}^n$ an open and connected set. Let $\varphi : [-\infty, \infty] \to X$ be a solution of (5.4.1) and $\gamma = \{x \in \mathbf{R}^n : x = \varphi(t), -\infty < t < \infty\}$ its path. An *omega limit point* or an *alpha limit point* of γ is defined exactly as this has been done in case of $n = 2$ preceding Lemma 3.1.3. We define the *omega limit set* $\Omega(\gamma)$ and the *alpha limit set* $A(\gamma)$ of γ similarly. There holds the following:

Theorem 5.4.1 *(R.A. Smith [1980$_a$]). Let $K \subset X$ be a compact set and suppose that there exists a real symmetric $n \times n$ matrix P having 2 negative and $n - 2$ positive eigenvalues such that for any pair of solutions $\varphi^1(t)$ and $\varphi^2(t)$ of (5.4.1) staying in K for every $t \in \mathbf{R}$, i.e. $\varphi^1(t), \varphi^2(t) \in K$, $t \in \mathbf{R}$, we have $U(\varphi^1(t) - \varphi^2(t)) \leq 0$ for $t \in \mathbf{R}$ where U is the quadratic form $U(x) := x'Px$; if for a positive semitrajectory γ^+ of (5.4.1), we have $\gamma^+ \subset K$, and the omega limit set $\Omega(\gamma^+)$ of γ^+ does not contain any equilibria then $\Omega(\gamma^+)$ contains at least one periodic orbit.*

Compare the Poincaré-Bendixson Theorem 3.1.8, and note that even with the additional assumptions on (5.4.1) this is not a full generalization of Theorem 3.1.8 since now Theorem 5.4.1 allows more than one periodic orbit in $\Omega(\gamma^+)$.

Proof. The essence of the proof is that the conditions imposed make it possible to project the omega limit set into a (two dimensional) plane in a homeomorphic way and draw the conclusion from the projected phase picture.

The assumptions on the quadratic form $U(x)$ imply that there is a regular linear coordinate transformation $x = Qy$ where x and y are the column vectors of the old and new coordinates, respectively, and Q is a regular real $n \times n$ matrix such that U assumes its canonical form

$$U(Qy) = y_1^2 + \cdots + y_{n-2}^2 - y_{n-1}^2 - y_n^2$$

in the new coordinates. Let $\Pi : \mathbf{R}^n \to \mathbf{R}^2$ be the linear mapping defined by

$\Pi x = \sqrt{2}[y_{n-1}, y_n]$. Since $|y|^2 = |Q^{-1}x|^2 = \sum_{i=1}^n y_i^2$, we have

$$2|Q^{-1}x|^2 \geq 2(y_{n-1}^2 + y_n^2) = |\Pi x|^2 = |Q^{-1}x|^2 - U(x), \quad x \in \mathbf{R}^n. \quad (5.4.2)$$

Let φ^1 and φ^2 be two solutions of (5.4.1) staying in K for $t \in \mathbf{R}$. Substituting $x = \varphi^1(t) - \varphi^2(t)$ into (5.4.2) we obtain

$$\sqrt{2}|Q^{-1}||\varphi^1(t) - \varphi^2(t)| \geq |\Pi\varphi^1(t) - \Pi\varphi^2(t)| \geq |Q|^{-1}|\varphi^1(t) - \varphi^2(t)|,$$

$$t \in \mathbf{R}, \quad (5.4.3)$$

where matrix norms (A1.22) attached to the Euclidean vector norm (A1.21) are used and the conditions imposed upon U are taken into account. Since the mapping Π is linear, $\Pi\varphi^1(t)$ is differentiable and $(\Pi\varphi^1(t))^{\cdot} = \Pi\dot{\varphi}^1(t)$. We replace $\varphi^2(t)$ in (5.4.3) by $\varphi^1(t+h)$ where h is an arbitrary constant; this gives

$$|\Pi\varphi^1(t+h) - \Pi\varphi^1(t)| \geq |Q|^{-1}|\varphi^1(t+h) - \varphi^1(t)|.$$

Dividing by h and making h tend to zero we get

$$|(\Pi\varphi^1(t))^{\cdot}| \geq |Q|^{-1}|\dot{\varphi}^1(t)|. \quad (5.4.4)$$

Since K is compact and $\gamma^+ \subset K$, the omega limit set $\Omega(\gamma^+)$ is bounded, non-empty and $\Omega(\gamma^+) \subset K$. Now, also $\Omega(\gamma^+)$ is a compact set which consists of complete orbits. This is because Lemmata 3.1.3 and 3.1.6 can be proved in this n dimensional case exactly as this has been done there. Let $\omega^1, \omega^2 \in \Omega(\gamma^+)$ and set $\varphi^i(t) = \varphi(t, \omega^i)$ $(i = 1, 2)$ where $\varphi(t, x)$ is the solution with initial value $x = \varphi(0, x)$. Thus, (5.4.3) implies that for every $\omega^1, \omega^2 \in \Omega(\gamma^+)$

$$\sqrt{2}|Q^{-1}||\omega^1 - \omega^2| \geq |\Pi\omega^1 - \Pi\omega^2| \geq |Q|^{-1}|\omega^1 - \omega^2|,$$

and this means that Π is a homeomorphism between $\Omega(\gamma^+)$ and $\Pi\Omega(\gamma^+)$, i.e. it is one-to-one and continuous along with its inverse Π^{-1}. Thus, disjoint complete orbits in $\Omega(\gamma^+)$ map onto disjoint plane curves and periodic orbits in $\Omega(\gamma^+)$ map onto simple closed plane curves. If $\omega(\gamma^+)$ contains a complete orbit γ_1 that is not a periodic orbit, then the image $\Pi\gamma_1$ does not intersect itself. If the path of the solution $\varphi^1(t) = \varphi(t, \omega^1)$, $\omega^1 \in \Omega(\gamma^+)$ is denoted by $\gamma_1 \subset \Omega(\gamma^+)$, then the tangent vector to the plane curve $\Pi\gamma_1$ at the point $\Pi\varphi^1(t)$, the vector $(\Pi\varphi^1(t))^{\cdot}$, is non-zero by (5.4.4). Also the omega limit set $\Omega(\gamma^1)$ of γ_1 is contained in $\omega(\gamma^+)$ because the latter is compact.

Now choose solutions of (5.4.1), $\varphi^1(t)$, $\varphi^2(t)$, $\varphi^3(t)$, with respective trajectories $\gamma_1, \gamma_2, \gamma_3$ such that $\gamma_1 \subset \Omega(\gamma^+)$, $\gamma_2 \subset \Omega(\gamma_1)$, $\gamma_3 \subset \Omega(\gamma_2)$. Then $\Omega(\gamma_2) \subset \Omega(\gamma_1) \subset \Omega(\gamma^+)$. Set $\omega^3 = \varphi^3(0) \in \gamma_3$ and let $v = \Pi\dot{\varphi}^3(0)$ be the non-zero vector tangential to the plane curve $\Pi\gamma_3$ at the point $\Pi\omega^3$. Consider the open disk $B(\Pi\omega^3, \delta)$ in the plane y_{n-1}, y_n with centre in $\Pi\omega^3$

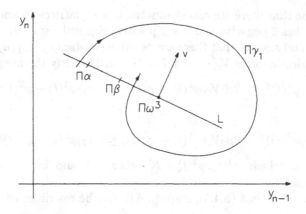

FIGURE 5.4.1. The Bendixson sack constructed for the projected phase picture in the proof of Theorem 5.4.1.

and radius $\delta > 0$. Since ω^3 is an omega limit point of both γ_1 and γ_2, the plane curves $\Pi\gamma_1$ and $\Pi\gamma_2$ must pass through $B(\Pi\omega^3, \delta)$ an infinite number of times as $t \to \infty$. When $\Pi\varphi^1(t)$ is close to $\Pi\omega^3$, by (5.4.3) $\varphi^1(t)$ is close to $\omega^3 = \varphi^3(0)$ but then $\dot{\varphi}^1(t)$ is close to $\dot{\varphi}^3(0)$ and, further, $\Pi\dot{\varphi}^1(t)$ is close to $v = \Pi\dot{\varphi}^3(0)$. It follows that if the radius δ of the neighbourhood of $\Pi\omega^3$ is sufficiently small, then the tangent vectors of the arcs of $\Pi\gamma_1$ contained in $B(\Pi\omega^3, \delta)$ are close to v. The same argument applies to the arcs of $\Pi\gamma_2$ that lie in $B(\Pi\omega^3, \delta)$. Consider the "transversal" straight line segment L in the plane y_{n-1}, y_n that passes through $\Pi\omega^3$ and is perpendicular to v. Then each single arc of $\Pi\gamma_1$ and $\Pi\gamma_2$ in $B(\Pi\omega^3, \delta)$ may intersect L at most once.

We prove now that either γ_1 or γ_2 is a periodic orbit. Suppose that this is false. Then neither of the plane curves $\Pi\gamma_1$, $\Pi\gamma_2$ can intersect itself. Every time $\Pi\gamma_2$ intersects the transversal, L does so at a different point. Let $\alpha, \beta \in \gamma_2$ be such that $\Pi\alpha$ and $\Pi\beta$ are successive intersections of $\Pi\gamma_2$ with L. Then $\Pi\alpha \neq \Pi\beta$, and α, β are omega limit points of γ_1 because $\gamma_2 \subset \langle \gamma_1 \rangle$. Hence $\Pi\gamma_1$ intersects L infinitely many times arbitrarily close to both $\Pi\alpha$ and $\Pi\beta$. However, after $\Pi\gamma_1$ has cut L near $\Pi\alpha$ and then near $\Pi\beta$ it cannot cut L again near $\Pi\alpha$ because we have here a Bendixson sack (cf. what has been written following Corollary 3.1.9, and also Figure 5.4.1). This contradiction proves the Theorem. $\square$

In order to have a full generalization of the Poincaré-Bendixson Theorem, further conditions are to be imposed upon system (5.4.1). While treating this topic, R.A.Smith [1980a] is followed further. The following hypothesis will be used:

(H) Assume that there are real symmetric $n \times n$ matrices P_1 and P_2 such that $P_1 - P_2$ has 2 negative and $n - 2$ positive eigenvalues; let $K \subset X$ be a compact set and assume that there are positive constants $\mu_1, \varepsilon_1, \mu_2, \varepsilon_2$ such that the quadratic forms $V_i(x) := x'P_i x$ $(i = 1, 2)$ satisfy the inequalities

$$(V_1(\varphi^1(t) - \varphi^2(t)))\dot{} + 2\mu_1 V_1(\varphi^1(t) - \varphi^2(t)) \le -\varepsilon_1|\varphi^1(t) - \varphi^2(t)|^2 \quad (5.4.5)$$

and

$$(V_2(\varphi^1(t) - \varphi^2(t)))\dot{} - 2\mu_2 V_2(\varphi^1(t) - \varphi^2(t)) \le -\varepsilon_2|\varphi^1(t) - \varphi^2(t)|^2 \quad (5.4.6)$$

for every t at which $\varphi^1(t), \varphi^2(t) \in K$ where φ^1 and φ^2 are arbitrary solutions of (5.4.1).

It is to be noted that (5.4.5), resp. (5.4.6) can be rewritten as

$$(V_1(\varphi^1(t) - \varphi^2(t))\exp(2\mu_1 t))\dot{} \le -\varepsilon_1|\varphi^1(t) - \varphi^2(t)|^2 \exp(2\mu_1 t) \quad (5.4.7)$$

and

$$(V_2(\varphi^1(t) - \varphi^2(t))\exp(-2\mu_2 t))\dot{} \le -\varepsilon_2|\varphi^1(t) - \varphi^2(t)|^2 \exp(-2\mu_2 t). \quad (5.4.8)$$

Lemma 5.4.2. *Suppose that (H) holds, and (5.4.1) has a positive semi-trajectory $\gamma^+ \subset K$, Let $U(x) := V_1(x) - V_2(x)$, and the matrix Q and the linear mapping $\Pi : \mathbf{R}^n \to \mathbf{R}^2$ be defined for this U as in the proof of Theorem 5.4.1; if $\varphi^1(t)$ and $\varphi^2(t)$ are solutions of (5.4.1) that are defined and satisfy $\varphi^1(t), \varphi^2(t) \in K$ for $t \in [t_0, \infty]$ with some $t_0 \in \mathbf{R}$, then*

(i) *either there are constants $h \ge 0$, $c > 0$ such that*

$$|\varphi^1(t) - \varphi^2(t + h)| \le ce^{-\mu_1 t}, \qquad t \ge t_0, \qquad (5.4.9)$$

(ii) *or for every $k > 0$ there exists a $\tau(k) \ge t_0$ such that*

$$|\varphi^1(t) - \varphi^2(t + h)| \le |Q||\Pi\varphi^1(t) - \Pi\varphi^2(t + h)| \qquad (5.4.10)$$

for all $t \ge \tau(k)$ and all h with $0 \le h \le k$.

Proof. Since $\varphi^1(t), \varphi^2(t) \in K$ for $t \in [t_0, \infty]$, the solutions φ^1 and φ^2 are bounded on this interval. Therefore, $V_2(\varphi^1(t) - \varphi^2(t))\exp(-2\mu_2 t) \to 0$ as $t \to \infty$. But by (5.4.8), $V_2(\varphi^1(t) - \varphi^2(t))\exp(-2\mu_2 t)$ is decreasing; hence, it is non-negative for $t \ge t_0$. The C^1 property of f and the compactness of K implies that a $\rho > 0$ exists such that

$$|f(\varphi^1(t)) - f(\varphi^2(t))| \le \rho|\varphi^1(t) - \varphi^2(t)|, \qquad t \ge t_0.$$

Applying this inequality we have

$$exp(-2\mu_1 t)((\varphi^1(t) - \varphi^2(t))^2 \exp(2\mu_1 t))\dot{}$$
$$= 2\mu_1(\varphi^1(t) - \varphi^2(t))^2 + 2(\varphi^1(t) - \varphi^2(t))'(\dot{\varphi}^1(t) - \dot{\varphi}^2(t))$$
$$= 2\mu_1(\varphi^1(t) - \varphi^2(t))^2 + 2(\varphi^1(t) - \varphi^2(t))'(f(\varphi^1(t)) - f(\varphi^2(t)))$$
$$\le 2(\mu_1 + \rho)(\varphi^1(t) - \varphi^2(t))^2, \qquad t \ge t_0$$

where the prime denotes transposition of a column vector into a row vector. The last inequality and (5.4.7) imply

$$(2(\mu_1 + \rho)V_1(\varphi^1(t) - \varphi^2(t)) \exp(2\mu_1 t) + \varepsilon_1(\varphi^1(t) - \varphi^2(t))^2 \exp(2\mu_1 t))^. \leq 0;$$

thus, according to (5.4.7) and the last inequality both functions

$$\begin{aligned} &V_1(\varphi^1(t) - \varphi^2(t)) \exp(2\mu_1 t), \\ &(2(\mu_1 + \rho)V_1(\varphi^1(t) - \varphi^2(t)) + \varepsilon_1(\varphi^1(t) - \varphi^2(t))^2) \exp(2\mu_1 t) \end{aligned} \qquad (5.4.11)$$

are decreasing in $[t_0, \infty)$. In particular, for $t_0 \leq t_1 \leq t$ we have

$$V_1(\varphi^1(t_1) - \varphi^2(t_1)) \exp(2\mu_1 t_1) \geq V_1(\varphi^1(t) - \varphi^2(t)) \exp(2\mu_1 t). \quad (5.4.12)$$

Now, there are two possibilities: (i) If $V_1(\varphi^1(t) - \varphi^2(t)) \exp(2\mu_1 t)$ is bounded below in $[t_0, \infty)$, then both monotonous functions in (5.4.11) tend to finite limit as $t \to \infty$, and, as a consequence, $(\varphi^1(t) - \varphi^2(t))^2 \exp(2\mu_1 t)$ tends to a finite limit. Hence, a constant $c > 0$ exists such that

$$c^2 \geq (\varphi^1(t) - \varphi^2(t))^2 \exp(2\mu_1 t), \qquad t \in [t_0, \infty). \quad (5.4.13)$$

In what has been done up to this point $\varphi^2(t)$ can be replaced by $\varphi^2(t + h)$ with arbitrary $h \geq 0$. The choice of $c > 0$, naturally, depends on h. Thus, *if there exists an $h \geq 0$ such that $V_1(\varphi^1(t) - \varphi^2(t + h)) \exp(2\mu_1 t)$ is bounded below*, then we arrive at (5.4.13) with $\varphi^2(t)$ replaced by $\varphi^2(t + h)$, and statement (i) of the Theorem is valid. (ii) *If no such h exists*, i.e. for arbitrary $h \geq 0$ the function $V_1(\varphi^1(t) - \varphi^2(t + h)) \exp(2\mu_1 t)$ is *not* bounded below, then there is a $t_1(h) \geq t_0$ such that $V_1(\varphi^1(t_1(h)) - \varphi^2(h + t_1(h))) < 0$, but then (5.4.12) implies that $V_1(\varphi^1(t) - \varphi^2(h + t)) < 0$ for $t \geq t_1(h)$. We have shown at the start that $V_2(\varphi^1(t) - \varphi^2(t + h)) \geq 0$ for $t \in [t_0, \infty]$; therefore, $U(\varphi^1(t) - \varphi^2(t + h)) < 0$ for $t \in [t_1(h), \infty]$. Then (5.4.2) implies

$$\begin{aligned} |\Pi\varphi^1(t) - \Pi\varphi^2(t + h)| &\geq |Q^{-1}(\varphi^1(t) - \varphi^2(t + h))| \\ &\geq |Q|^{-1}|\varphi^1(t) - \varphi^2(t + h)| \end{aligned} \quad (5.4.14)$$

for $t \geq t_1(h)$. Now we prove that having fixed $k > 0$, the function $t_1(h)$ can be chosen bounded above for $h \in [0, k]$. For each integer $j \geq 1$ define the set $E_j = \{h \in [0, k] : \text{there exists } t \in [t_0, t_0 + j] \text{ such that } V_1(\varphi^1(t) - \varphi^2(t + h)) < 0\}$. In the present case, E_j is, clearly, non-empty, relatively open, $E_{j+1} \supset E_j$ $(j = 1, 2, ...)$, and $[0, k] \subset \cup E_j$. By the Heine-Borel Theorem there exists a positive integer p such that $[0, k] \subset E_p$. This means that if $h \in [0, k]$, then $t_1(h) \in [t_0, t_0 + p]$, and (5.4.14) holds for $t \geq t_0 + p$, i.e. statement (ii) of the Theorem is true with $\tau(k) = t_0 + p$. $\square$

Now a full generalization of the Poincaré-Bendixson Theorem is already possible.

Theorem 5.4.3 *(R.A. Smith [1980$_a$]). Suppose that (H) holds and (5.4.1) has a positive semitrajectory $\gamma^+ \subset K$; if the omega limit set $\Omega(\gamma^+)$ does not contain any equilibrium point of (5.4.1), then it consists of a single periodic orbit.*

Proof. If φ^1 and φ^2 are solutions of (5.4.1) such that $\varphi^1(t), \varphi^2(t) \in K$ for $t \in \mathbf{R}$ then the boundedness of $\varphi^1(t) - \varphi^2(t)$ implies that

$$\lim_{t \to -\infty} V_1(\varphi^1(t) - \varphi^2(t)) \exp(2\mu_1 t) = 0,$$

$$\lim_{t \to \infty} V_2(\varphi^1(t) - \varphi^2(t)) \exp(-2\mu_2 t) = 0.$$

By (5.4.7) and (5.4.8) the functions in these limit relations are decreasing in $\mathbf{R}$ and, therefore,

$$V_1(\varphi^1(t) - \varphi^2(t)) \exp(2\mu_1 t) \le 0 \le V_2(\varphi^1(t) - \varphi^2(t)) \exp(-2\mu_2 t), \qquad t \in \mathbf{R}.$$

Hence, $U(\varphi^1(t) - \varphi^2(t)) \le 0$ for $t \in \mathbf{R}$ where $U = V_1 - V_2$. Thus, U satisfies the requirements of Theorem 5.4.1, so that by that Theorem $\Omega(\gamma^+)$ contains a periodic orbit γ_p. We have to show that now $\gamma_p = \Omega(\gamma^+)$.

Let $\varphi^1(t)$ and $\varphi^2(t)$ be those solutions whose trajectories are γ^+ and γ_p, respectively, and apply Lemma 5.4.2. If (5.4.9) holds, then, obviously, $\Omega(\gamma^+) = \gamma_p$. Suppose that Lemma 5.4.2 (i) is false. Then (ii) holds, i.e. inequality (5.4.10) holds for $t \ge \tau(k)$, $0 \le h \le k$, where $k > 0$ can be chosen arbitrarily. Choose the period T of φ^2 for k. The inequality (5.4.10) implies that the arc of $\Pi\gamma^+$ corresponding to $t \ge \tau(T)$ could intersect $\Pi\gamma_p$ only if γ^+ intersects γ_p. But in this case $\gamma^+ = \gamma_p$ and $\Omega(\gamma^+) = \gamma_p = \gamma^+$. Let us delete the arc corresponding to $0 \le t < \tau(T)$ from γ^+ and suppose that $\Pi\gamma^+$ does not intersect $\Pi\gamma_p$. In this case, γ^+ is not a periodic orbit since otherwise it could not have a disjoint periodic orbit in its omega limit set. Also (5.4.10) shows that $\Pi\varphi^1(t)$ is close to $\Pi\gamma_p$ only if $\varphi^1(t)$ is close to γ_p. Now apply Lemma 5.4.2 again with the same $\varphi^1(t)$ and with $\varphi^2(t) := \varphi^1(t + T/2)$. If (5.4.9) holds, then φ^1 satisfies the conditions in Problem 5.7.5, therefore; $\Omega(\gamma^+) = \gamma_p$. If (i) of the Lemma is false, then we have (5.4.10), i.e.

$$|\varphi^1(t) - \varphi^1(t + h + T/2)| \le |Q| \|\Pi\varphi^1(t) - \Pi\varphi^1(t + h + T/2)|$$

for all $t \ge \tau(T)$ and $0 \le h \le T$. Since γ^+ is not a periodic orbit, this shows that

$$\Pi\varphi^1(t) \ne \Pi\varphi^1(t + h + T/2) \text{ for } t \ge \tau(T), \ 0 \le h \le T. \qquad (5.4.15)$$

Now we prove that $\Pi\gamma^+$ approaches the simple closed curve $\Pi\gamma_p$ spirally as $t \to \infty$. Choose an arbitrary point x^0 on γ_p and let L be a transversal straight line segment passing through Πx^0 and perpendicular to the non-zero tangent vector of $\Pi\gamma_p$ at Πx^0. Since $x^0 \in \Omega(\gamma^+)$, there exist points of

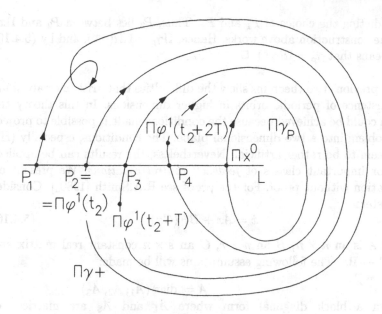

FIGURE 5.4.2. Proving that $\Omega(\gamma^+) = \gamma_p$ in Theorem 5.4.3.

intersection of L and $\Pi\gamma^+$ arbitrarily close to Πx^0. Let $P_1 = \Pi\varphi^1(t_1)$ and $P_2 = \Pi\varphi^1(t_2)$ be two points of intersection with L such that $\tau(T) < t_1 < t_2$ and P_2 lies between P_1 and Πx^0. Because of continuous dependence on initial values we may suppose that P_2 is so close to Πx^0 that $\Pi\varphi^1(t)$ remains close to $\Pi\gamma_p$ in $t_2 \leq t \leq t_2 + 3T$. Then there are at least two more points of intersection of $\Pi\gamma^+$ with L for $t_2 \leq t \leq t_2 + 3T$ because three complete revolutions are made on $\Pi\gamma_p$ in this time interval. Denote by P_3 and P_4 these points of intersection, the first being close to $\Pi\varphi^1(t_2 + T)$, the second to $\Pi\varphi^1(t_2 + 2T)$ (see Figure 5.4.2). We state that P_3 lies between P_2 and P_4. If this was not the case, then the arc P_2P_3 of $\Pi\gamma^+$ would intersect the arc P_3P_4 at a point where $\Pi\varphi^1(t) = \Pi\varphi^1(t + \tilde{T})$ where $0 < \tilde{T} < 3T/2$, say, but this contradicts (5.4.15) and asserts the statement. Assume now that P_3 lies between P_2 and Πx^0, then P_4 lies between P_3 and Πx^0. Continuing this construction, successive arcs P_3P_4, P_4P_5, ... of $\Pi\gamma^+$ are obtained not intersecting each other and meeting L in the monotonous sequence of points P_3, P_4, P_5, ... which tends to Πx^0. This proves that $\Pi\gamma^+$ tends to $\Pi\gamma_p$ in a spiralling way as $t \to \infty$. We have to prove that P_3 lies between P_2 and Πx^0 indeed. If this was not the case, then P_2 would lie between P_3 and Πx^0. In this case we may follow $\Pi\gamma^+$ in the backward direction from P_2 obtaining arcs of $\Pi\gamma^+$ that do not intersect each other because of (5.4.15) and intersect L at points that are closer to Πx^0 each time until the point P_1 is reached. This would mean that P_1 lies between P_2 and Πx^0, contradicting the choice of P_1 and P_2. Thus, P_3 lies between P_2 and Πx^0 and the construction above works. Hence, $\Pi \gamma_p = \Pi \Omega(\gamma^+)$, and by (5.4.10) this means that $\gamma_p = \Omega(\gamma^+)$. $\square$

The previous two theorems show the difficulties that arise in establishing the existence of periodic orbits in higher dimensions. In this theory the results could be achieved because the conditions made it possible to project the problem into a two dimensional plane. The conditions, especially (H), may seem to be rather artificial. Nevertheless, the results can be applied, e.g., for important classes of *feedback control equations*. We present an application without proof. For the proof see R.A.Smith [1980$_a$]. Consider the system

$$\dot{x} = Ax + BF(Cx) \tag{5.4.16}$$

where A is an $n \times n$, B an $n \times r$, C an $s \times n$ constant real matrix and $F : \mathbf{R}^s \to \mathbf{R}^r$. The following assumptions will be made:

(i) $$A = \operatorname{diag}(A_1, A_2, A_3)$$
in a block diagonal form where A_1 and A_3 are matrices of type $n_1 \times n_1$, $n_3 \times n_3$, respectively, and A_2 is a 2×2 matrix ($n_1 + n_3 + 2 = n$); there exist constants $\mu_1, \mu_2 > 0$ such that

$$\operatorname{Re} \lambda^1 < -\mu_1 \qquad \text{for all eigenvalues } \lambda^1 \text{ of } A_1,$$
$$-\mu_1 < \operatorname{Re} \lambda^2 < \mu_2 \qquad \text{for both eigenvalues } \lambda^2 \text{ of } A_2,$$
$$\mu_2 < \operatorname{Re} \lambda^3 \qquad \text{for all eigenvalues } \lambda^3 \text{ of } A_3.$$

(ii) $$F \in C^1(\mathbf{R}^s, \mathbf{R}^r), \qquad |F'_{iy_k}(y)| \le \gamma$$
for $y \in \mathbf{R}^s$ ($i = 1, 2, ..., r; k = 1, 2, ..., s$), and either γ or $|C|$ and $|B|$ are sufficiently small.

(iii)   There exist a constant $\beta > 0$ and an $r \times s$ matrix K such that

$$\lim_{|y| \to \infty} |y|^{-\beta}(F(y) - Ky) = 0.$$

(iv) $Ax + BF(Cx) = 0$ if and only if $x = 0$.

(v)   Either $A + BF'_y(0)C$ has exactly $n_3 + 2$ eigenvalues with positive real part and $A + BKC$ has exactly n_3 eigenvalues with non-negative real part, or $A + BF'_y(0)C$ has exactly n_3 eigenvalues with non-negative real part and $A + BKC$ has exactly $n_3 + 2$ eigenvalues with positive real part.

Theorem 5.4.4. *If conditions (i)-(v) hold, then (5.4.16) has a periodic orbit.*

For systems satisfying somewhat similar conditions to those of Theorems 5.4.1 - 5.4.3, not only the Poincaré-Bendixson Theorem but also Dulac's Criterion for the non-existence of periodic orbits (Theorem 3.5.16) and the theory connected with the Poincaré index (Section 3.5) can be generalized (see R.A.Smith [1981]). The difficulties that may arise are shown by the simple three dimensional system of Problem 5.7.6.

We turn now to another method using topological degree (see Appendix 2) in establishing the existence of a periodic orbit. In this part of the Section, Grasman's [1977] results are presented. System (5.4.1) will be transformed by the "polar transformation"

$$P: \qquad x_1 = y_2 \cos y_1, \qquad x_2 = y_2 \sin y_1,$$
$$x_i = y_i \qquad (i = 3, 4, ..., n) \qquad (5.4.17)$$

into

$$\dot{y} = g(y) \qquad (5.4.18)$$

the latter being, by the way, a cylindrical system in the variable y_1. By transformation P, the relations between the derivatives are

$$y_2 \dot{y}_1 = -\dot{x}_1 \sin y_1 + \dot{x}_2 \cos y_1, \qquad (5.4.19)$$
$$\dot{y}_2 = \dot{x}_1 \cos y_1 + \dot{x}_2 \sin y_1,$$

and this means that the conditions imposed upon the system must ensure a reasonable definition for $g_1(y_1, ..., y_n)$ at $y_2 = 0$.

Theorem 5.4.5 (*Grasman [1977]*). *Suppose that*

(i) *there is a compact neighbourhood $0 \in M \subset X$ that is star shaped from 0 and is positively invariant for (5.4.1);*

(ii) *the origin 0 is the only equilibrium of (5.4.1) in M, and it is hyperbolic: two eigenvalues of $f'_x(0)$ have positive real parts and $n - 2$ eigenvalues negative real parts;*

(iii) *the stable manifold of 0 contains the set $\{x \in \mathbf{R}^n : x_1 = x_2 = 0\} \cap M$;*

(iv) *system (5.4.1) can be transformed by (5.4.17) into (5.4.18) where $g \in C^1(P^{-1}M, \mathbf{R}^n)$ and $g_1(y) \neq 0$ for $y \in P^{-1}M$.*

Then (5.4.1) has a (non-constant) periodic orbit.

A few explanations are appropriate. The requirement that M be a compact neighbourhood of the origin means that M is compact, but Int M is not empty and $0 \in$ Int M. That M is star shaped means that each ray from 0 intersects ∂M in exactly one point. Condition (iii) is fairly strong: it means that the stable manifold (see Appendix 3) contains that part of the $n - 2$ dimensional subspace orthogonal to the x_1, x_2 plane that is contained in M. This obviously implies that the eigenspace corresponding to those eigenvalues of $f'(0)$ that have negative real parts is orthogonal to the x_1, x_2 plane. (Normally this eigenspace is tangent to the stable manifold but now it is supposed to belong to it, at least, in M.) Since the stable manifold is invariant, if $x_1 = x_2 = 0$, then $\dot{x}_1 = \dot{x}_2 = 0$ must hold, i.e.

$$f_i(0, 0, x_3, ..., x_n) \equiv 0 \qquad (i = 1, 2). \qquad (5.4.20)$$

This implicit requirement ensures that g_1 can be defined at least continuously at $y_2 = 0$. Indeed, if $y_2 = 0$, then by (5.4.17) both $x_1 = x_2 = 0$ hence, the right-hand side of (5.4.19) is

$$\dot{x}_2 \cos y_1 - \dot{x}_1 \sin y_1 = f_2(0, 0, x_3, ..., x_n) \cos y_1 - f_1(0, 0, x_3, ..., x_n) \sin y_1 \equiv 0,$$

so that

$$g_1(y) = \begin{cases} (1/y_2)(f_2(y_2 \cos y_1, y_2 \sin y_1, y_3, ...) \cos y_1 \\ \quad - f_1(y_2 \cos y_1, y_2 \sin y_1, y_3...) \sin y_1), & y_2 \neq 0 \\ \\ (f'_{2x_1}(0, 0, y_3, ...) \cos y_1 + f'_{2x_2}(0, 0, y_3, ...) \sin y_1) \cos y_1 \\ \quad - (f'_{1x_1}(0, 0, y_3, ...) \cos y_1 + f'_{1x_2}(0, 0, y_3, ...) \sin y_1) \sin y_1 & y_2 = 0 \end{cases}$$
$$(5.4.21)$$

where also (5.4.20) was taken into account, the latter identity implying that $f'_{ix_k}(0, 0, x_3, ...) \equiv 0$ $(i = 1, 2; k = 3, 4, ..., n)$. Thus, (iii) and $f \in C^1$ implies that also $g_1 \in C^0(P^{-1}M, \mathbf{R})$. Condition (iv) strengthens this requirement still further. (However, if $f \in C^2$, then $g_1 \in C^1$ automatically, and the only additional condition of (iv) is that $g_1(y) \neq 0$.) Since the transformation P is periodic in y_1 and its Jacobian is zero at $y_2 = 0$, it is not one-to-one and, therefore, it is not invertible in the strict sense of the word. Therefore, $P^{-1}M$, simply, means the set

$$P^{-1}M := \{y \in \mathbf{R}^n : Py \in M\}.$$

The compactness of M implies that $P^{-1}M \cap \{y \in \mathbf{R}^n : 0 \leq y_1 \leq 2\pi\}$ is compact. Since the functions g_i are by assumption continuous and periodic in y_1 with period 2π $(i = 1, 2, ..., n)$, they assume their respective suprema and infima in $P^{-1}M$.

Proof. The outline of the proof is that an $n - 1$ dimensional section of M will be constructed that is mapped into itself by the flow generated by system (5.4.1). Then by a degree-theoretic argument it will be shown that this mapping has a fixed point apart from the origin.

We may assume without loss of generality that $g_1(y) > 0$ for $y \in P^{-1}M$ and then there exists a $k > 0$ such that $g_1(y) \geq k$ for $y \in P^{-1}M$. Let $\psi_t : P^{-1}M \to P^{-1}M$, $t \geq 0$, be the flow generated by (5.4.18). Clearly, $P^{-1}M$ is positively invariant for this flow; $\psi_t(y) = \psi(t, y)$ is the solution of (5.4.18) with initial value $y \in P^{-1}M$. Since $\dot{y}_1 = g_1(y) \geq k > 0$ for every $y \in P^{-1}M$, there exists a unique $\tau(y) > 0$ such that

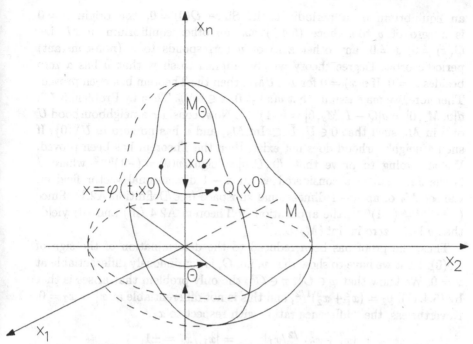

FIGURE 5.4.3. Mapping of the set M_θ in the plane $x_2 = x_1 \tan \theta$ into itself by the "Poincaré map" Q. Here the x_n-axis is the stable manifold: points on x_n are moved towards the origin along this axis, i.e. they are staying inside M_θ.

$(\psi_{\tau(y)}(y))_1 = \psi_1(\tau(y), y) = y_1 + 2\pi$. A simple application of the implicit function theorem yields that $\tau \in C^1(P^{-1}M, \mathbf{R}_+)$. For an arbitrarily fixed $\theta \in \mathbf{R}$, set

$$M_\theta := \{x \in M : y_1 = \theta, \ x = P(y)\},$$

i.e. by (5.4.17), M_θ is that part of the $n-1$ dimensional hyperplane $x_2 = x_1 \tan \theta$ that is contained in M (see Figure 5.4.3). Since M is star shaped, M_θ is also star shaped from 0. Let $\varphi(t, x)$ be the solution of (5.4.1) for which $\varphi(0, x) = x$, and consider the "Poincaré map" (cf. Definition 5.2.1) $Q : M_\theta \to M_\theta$ defined by $Q(x) := \varphi(\tau(y), x)$ where $x = P(y) \in M_\theta$. The expression "Poincaré map" stands between quotation marks because M_θ is not transversal to the flow at points of intersection of the set M_θ with the stable manifold (at points where $y_2 = 0$, i.e. $x_1 = x_2 = 0$) and at the origin 0. Still τ is well defined above for every $y \in P^{-1}M$, and Q maps points of M_θ where there is no transversality also into M_θ. Also, clearly, $Q \in C^0(M_\theta)$ and $Q \in C^1(M_\theta \backslash \{0\})$, however, we shall see that there is no problem with the continuous differentiability even at $x = 0$. Now, consider the vector field $v : M_\theta \to \mathbf{R}^{n-1}$ defined by $v(x) := Q(x) - x$ for $x \in M_\theta$, i.e. at each point of M_θ the value of v is a vector of the $n-1$ dimensional plane $x_2 = x_1 \tan \theta$. Obviously, each zero of v corresponds to a fixed point of Q, i.e. a closed trajectory of (5.4.1) that may be an equilibrium or a periodic orbit. Since $Q(0) = 0$, the origin $x = 0$ is a zero of v too. Since (5.4.1) has no other equilibrium in M, i.e. $Q(x) \neq 0$, $x \neq 0$, any other zero of v corresponds to a (non-constant) periodic orbit. Degree theory will be applied to show that v has a zero besides $x = 0$. If $v(x) = 0$ for $x \in \partial M_\theta$, then the Theorem has been proved. Therefore, we may assume that $v(x) \neq 0$, $x \in \partial M_\theta$. Then by Problem 5.7.7, $d[v, M_\theta, 0] = d[Q - I, M_\theta, 0] = (-1)^{n-1}$. Now, consider a neighbourhood U of 0 in M_θ such that $0 \in U$, $\overline{U} \subset \text{Int } M_\theta$, and v has no zero in $\overline{U} \backslash \{0\}$. If such a neighbourhood does not exist, then the Theorem has been proved. We are going to prove that $d[v, U, 0] = \text{sign } Jv(0) = (-1)^{n-2}$ where J is the Jacobian of v considered as an $n - 1$ dimensional vector field on the set M_θ of an $n - 1$ dimensional subspace (see Definition A2.1). Since $(-1)^{n-2} \neq (-1)^{n-1}$, the application of Theorem A2.4 (iii) and (ii) yields that v has a zero in $\text{Int } (M_\theta \backslash U)$.

Thus, the proof has been reduced to the determination of the sign of $Jv(0)$. First we have to show that v, i.e. Q, is continuously differentiable at $x = 0$. We know that $\varphi \in C^1$, $\tau \in C^1$; the only problem that arises is that by (5.4.17), $y_2 = (x_1^2 + x_2^2)^{1/2}$, and this is not differentiable at $x_1 = x_2 = 0$. Nevertheless, the "difference ratio" with respect to x_1,

$$(x_1^2 + x_2^2)^{1/2}/x_1 \big|_{x_2=0} = |x_1|/x_1 = \pm 1,$$

is bounded along with the partial derivative $\partial y_2/\partial x_1 = x_1/(x_1^2 + x_2^2)^{1/2}$ (and similarly for x_2). Let us introduce the notations

$$y^0 = (\theta, 0, ..., 0), \quad \tau_0 = \tau(y^0), \quad \Delta\tau_j = \tau(\theta, 0, ..., \Delta y_j, ..., 0) - \tau_0$$

$$(j = 2, 3, ..., n).$$

Then

$$Q'_{x_i}(0) = \lim_{\Delta x_i \to 0} \sum_{j=2}^{n} ((\varphi(\tau_0 + \Delta\tau_j, 0) - \varphi(\tau_0, 0))/\Delta\tau_j)(\Delta\tau_j/\Delta y_j)\Delta y_j/\Delta x_i$$

$$+\varphi'_{x_i}(\tau_0, 0) \qquad (i = 1, 2).$$

Since $\varphi(t, 0) \equiv 0$, $\dot\varphi(t, 0) \equiv 0$. Hence, the first factor in each term of the sum tends to zero, the second one tends to some finite limit and the third one is bounded even when $j = 2$. Thus, $Q'_{x_i}(0) = \varphi'_{x_i}(\tau_0, 0)$ $(i = 1, 2)$, and it is easy to see that these derivatives are continuous also at $x_1 = x_2 = 0$. As a consequence, $Q \in C^1(M_\theta)$.

The problem is now that though v is an $n - 1$ dimensional vector field with values in the $n - 1$ dimensional hyperplane $x_2 = x_1 \tan \theta$, formally it is defined through the flow φ as an n dimensional field on $\mathbf{R}^n$. First we shall allow $x \in \mathbf{R}^n$ and also $y \in P^{-1}M$ in $\tau(y)$ in the definition of v, and in the second step we shall restrict v to M_θ. Thus, using again the fact that $\dot\varphi(t, x) \equiv 0$ we have for the $n \times n$ matrix

$$v'_x = Q'_x(0) - I = \varphi'_x(\tau_0, 0) - I,$$

clearly, $y^0 = (\theta, 0, ..., 0)$ corresponds to the origin $x = 0$. As we know (see (1.1.5) et seq.) the matrix $\varphi'_x(t, 0)$ is that fundamental matrix of the variational system $\dot{y} = f'_x(0)y$ that assumes the unit matrix I at $t = 0$. This means that $\varphi'_x(t, 0) = \exp(f'_x(0)t)$ (see (1.2.14). Hence,

$$v'_x(0) = \exp(f'_x(0)\tau_0) - I.$$

Denote the eigenvalues of $f'_x(0)$ by λ_k $(k = 1, 2, ..., n)$. We may assume according to condition (ii) of the Theorem that $\operatorname{Re} \lambda_1, \operatorname{Re} \lambda_2 > 0$ and $\operatorname{Re} \lambda_j < 0$ $(j = 3, 4, ..., n)$. The eigenvalues of $\exp(f'_x(0)\tau_0) - I$ are then $\exp(\lambda_k\tau_0) - 1$ $(k = 1, 2, ..., n)$ (see Corollary A1.8 and also Theorem A1.12). The $n - 2$ dimensional eigensubspace of $f'_x(0)$ that corresponds to the eigenvalues λ_j $(j = 3, ..., n)$ with negative real part is the tangent space of the stable manifold of 0 that, in turn, by (iii) of the Theorem contains the set $\{x \in \mathbf{R}^n : x_1 = x_2 = 0\} \cap M$. Thus, this eigensubspace is the $n - 2$ dimensional plane $x_1 = x_2 = 0$ which is contained in the $n - 1$ dimensional plane $x_2 = x_1 \tan\theta$ of M_θ. The eigenvectors and eigensubspaces of $f'_x(0)$ coincide with those of $\exp(f'_x(0)\tau_0) - I = v'_x(0)$, so that the linearization of v at $x = 0$ has $x_1 = x_2 = 0$ as an $n - 2$ dimensional eigensubspace corresponding to its eigenvalues $\exp(\lambda_j\tau_0) - 1$ $(j = 3, ..., n)$. Thus, these $n - 2$ numbers among which there might be equal ones form the set of $n - 2$ eigenvalues (counted with multiplicities) of the linearization of the restriction of v to the set M_θ. Since $\operatorname{Re} \lambda_j < 0$,

$$\operatorname{Re}(\exp(\lambda_j\tau_0) - 1) < 0 \qquad (j = 3, ..., n). \qquad (5.4.22)$$

We have to find the $(n - 1)$-st eigenvalue of the restriction of v to M_θ. First we show that the two eigenvalues of $f'_x(0)$ with positive real parts are conjugate complex numbers. Because of (5.4.20)

$$f'_x(0) = \begin{pmatrix} f'_{1x_1}(0) & f'_{1x_2}(0) & 0 & \cdots & 0 \\ f'_{2x_1}(0) & f'_{2x_2}(0) & 0 & \cdots & 0 \\ f'_{3x_1}(0) & f'_{3x_2}(0) & f'_{3x_3}(0) & \cdots & f'_{3x_4}(0) \\ \cdot & \cdot & \cdot & \cdots & \cdot \\ f'_{nx_1}(0) & f'_{nx_2}(0) & f'_{nx_3}(0) & \cdots & f'_{nx_n}(0) \end{pmatrix}.$$

Applying the notations $f_{ik} = f'_{ix_k}(0)$ $(i, k = 1, 2, ..., n)$,

$$D_2 f = \begin{pmatrix} f_{11} & f_{12} \\ f_{21} & f_{22} \end{pmatrix}, \qquad D_{n-2}f = \begin{pmatrix} f_{33} & \cdots & f_{3n} \\ \cdot & \cdots & \cdot \\ f_{n3} & \cdots & f_{nn} \end{pmatrix},$$

we see that the eigenvalues of the matrix $f'_x(0)$ are those of $D_2 f$ and $D_{n-2}f$. The eigenvalues λ_1, λ_2 with positive real parts are those of $D_2 f$ since if λ_j is an eigenvalue of $D_{n-2}f$, then a corresponding eigenvector of $f'_x(0)$ has the first two coordinates equal to zero, i.e. it is in the stable eigenspace implying that $\operatorname{Re} \lambda_j < 0$. By (5.4.21)

$$
\begin{aligned}
g_1(y^0) &= g_1(\theta, 0, ..., 0) \\
&= f_{21} \cos^2 \theta + (f_{22} - f_{11}) \cos \theta \sin \theta - f_{12} \sin^2 \theta \\
&= -\cos^2 \theta (f_{12} \tan^2 \theta + (f_{11} - f_{22}) \tan \theta - f_{21}).
\end{aligned}
$$

Since by (iv) of the Theorem $g_1(y^0) \neq 0$ for any $\theta \in \mathbf{R}$, the quadratic expression in brackets cannot have a root, so that its discriminant must be negative:

$$(f_{11} - f_{22})^2 + 4f_{12}f_{21} < 0. \tag{5.4.23}$$

But the eigenvalues of $D_2 f$ are

$$
\begin{aligned}
\lambda_{1,2} &= (f_{11} + f_{22})/2 \pm (1/2)((f_{11} + f_{22})^2 - 4(f_{11}f_{22} - f_{12}f_{21}))^{1/2} \\
&= (f_{11} + f_{22})/2 \pm i\omega \tag{5.4.24}
\end{aligned}
$$

where

$$\omega = (-4f_{12}f_{21} - (f_{11} - f_{22})^2)^{1/2}/2 > 0,$$

the expression under the square root being positive in view of (5.4.23). Thus, $\lambda_2 = \bar{\lambda}_1 \neq \lambda_1$. Next we show that $\tau_0 = 2\pi/\omega$. By (5.4.21) $g_1(y_1, 0, ..., 0) = f_{21} \cos^2 y_1 + (f_{22} - f_{11})(\cos y_1 \sin y_1) - f_{12} \sin^2 y_1$, and $g_i(y_1, 0, ..., 0) = f_i(0, 0, ..., 0) = 0$ $(i = 2, 3, ..., n)$. Thus, for the solution $\psi(t) := \psi(t, y_0)$ of (5.4.18) we have $\psi_i(t) \equiv 0$ $(i = 2, ..., n)$, and $\dot{\psi}_1(t) = f_{21} \cos^2 \psi_1(t) + (f_{22} - f_{11}) \cos \psi_1(t) \sin \psi_1(t) - f_{12} \sin^2 \psi_1(t)$. The last equation can be integrated, and we obtain $\psi_1(t) = \tan^{-1}(k_1 \tan(\omega t + k_2) + k_3)$ where k_1, k_2, k_3 are to be determined appropriately, and by an abuse of notation, $\tan^{-1}$ is to be understood such a way that $\psi_1((\pi/2 - k_2)/\omega - 0) = \pi/2 = \psi_1((\pi/2 - k_2)/\omega + 0)$, $\psi_1((3\pi/2 - k_2)/\omega - 0) = 3\pi/2 = \psi_1((3\pi/2)/\omega + 0)$, so that $\psi_1((2\pi - k_2)/\omega) = 2\pi$, and $\psi_1(2\pi/\omega) = \theta + 2\pi$ since $\psi_1(0) = \tan^{-1}(k_1 \tan k_2 + k_3) = \theta$ by assumption, i.e. $\tau_0 = 2\pi/\omega$, indeed. This result and (5.4.24) yield $\exp(\lambda_1 \tau_0) - 1 = \exp(\tau_0(f_{11} + f_{22})/2 + 2\pi i) - 1 = \exp(\tau_0(f_{11} + f_{22})/2) - 1 > 0$ since the real part of λ_1 is positive. Thus, the $(n-1)$-st eigenvalue of $v'_x(0)$ is real and positive. As a consequence, the corresponding eigenvector is real, and since it is also an eigenvector of $f'_x(0)$ belonging to an eigenvalue of the latter matrix with a positive real part, it cannot be orthogonal to the x_1, x_2 plane. Now, we may choose θ so that this eigenvector be parallel to the plane of M_θ. Thus, we have established that $\exp(\lambda_1 \tau_0) - 1 > 0$ is the missing eigenvalue of the linearization of the restriction of v to M_θ. The Jacobian of v considered as an $n - 1$ dimensional vector field on M_θ is the product of the $n - 1$ eigenvalues

$$Jv(0) = (\exp(\tau_0(f_{11} + f_{22})/2) - 1) \prod_{j=3}^{n} (\exp(\lambda_j \tau_0) - 1).$$

Here the first factor is positive. If there are $0 \leq l \leq (n-2)/2$ conjugate complex pairs and $n - 2 - 2l$ negative numbers among $\exp(\lambda_j \tau_0) - 1$

$(j = 3, ..., n)$, then taking into account (5.4.22)

$$\text{sign } Jv(0) = (-1)^{n-2-2l} = (-1)^{n-2} \neq (-1)^{n-1}. \qquad \square$$

The following example which illustrates the use of this Theorem concerns the famous Leonard-May model of population dynamics having a large literature; see May-Leonard [1975].

Example 5.4.1 *(The May-Leonard model with immigration, Grasman [1977])*. The competitive system

$$\begin{aligned}
\dot{N}_1 &= N_1(1 - N_1 - \alpha N_2 - \beta N_3) + \varepsilon, \\
\dot{N}_2 &= N_2(1 - \beta N_1 - N_2 - \alpha N_3) + \varepsilon, \\
\dot{N}_3 &= N_3(1 - \alpha N_1 - \beta N_2 - N_3) + \varepsilon
\end{aligned} \qquad (5.4.25)$$

will be studied where $0 < \beta < 1 < \alpha$, $\alpha + \beta > 2$ and $0 < \varepsilon$. Here N_1, N_2, N_3 are the quantities of three competing species; the constants α, β which measure the "interspecific competition" are chosen so that species 2 outcompetes species 1, species 3 outcompetes species 2 and species 1 outcompetes species 3 (cf. Sections 4.4 and 4.5). In the case $\varepsilon = 0$, i.e. there is no immigration then $(\overline{N}(0), \overline{N}(0), \overline{N}(0))$ where $\overline{N}(0) = 1/(1 + \alpha + \beta)$ is a positive equilibrium, and $\alpha + \beta > 2$ is exactly the condition of its instability. In this case the system has an equilibrium on each positive axis N_i and a closed heteroclinic orbit contained in the boundary of the positive octant joining the three equilibria on the axes. All the trajectories starting in the interior of the positive octant except a pair of trajectories (contained, by the way, in the ray $N_1 = N_2 = N_3 > 0$) forming the stable manifold of the interior equilibrium spiral outward towards this heteroclinic orbit.

The situation is different if $\varepsilon > 0$, i.e. there is a constant immigration of the same rate for each species participating in the competition. In this case the positive octant of the N_1, N_2, N_3 space is positively invariant since $\dot{N}_i = \varepsilon > 0$ if $N_i = 0$ $(i = 1, 2, 3)$. Straightforward calculations yield the following results. The interior equilibrium point is now $(\nu, \nu, \nu) = (\overline{N}(\varepsilon), \overline{N}(\varepsilon), \overline{N}(\varepsilon))$ where $\nu = (1 + (1 + 4\varepsilon\rho)^{1/2})/(2\rho)$, $\rho := 1 + \alpha + \beta > 3$. At this equilibrium the eigenvalues are

$$\begin{aligned}
\lambda_1 &= -(1 + 4\varepsilon\rho)^{1/2} < 0, \\
\lambda_{2,3} &= 1 - \nu(\rho + 3)/2 \pm i(\beta - \alpha)\nu\sqrt{3}/2.
\end{aligned}$$

The straight line $N_1 = N_2 = N_3 > 0$ is invariant, and it is the stable manifold of the equilibrium (ν, ν, ν) corresponding to the negative eigenvalue λ_1. For the conjugate complex pair of eigenvalues λ_2, λ_3 we have $\text{Re } \lambda_2 = \text{Re } \lambda_3 > 0$ if and only if $\rho_1(\varepsilon) < \rho < \rho_2(\varepsilon)$ and $\varepsilon < 1/12$ where

$$0 < \rho_1(\varepsilon) = (1 - 3\varepsilon - (1 - 12\varepsilon)^{1/2})/\varepsilon < (1 - 3\varepsilon + (1 - 12\varepsilon)^{1/2})/\varepsilon =: \rho_2(\varepsilon).$$

Clearly, $\rho_1(\varepsilon) \to 3$, $\rho_2(\varepsilon) \to \infty$ as $\varepsilon \to 0$, so that for a given value of $\rho > 3$ one may determine the exact values of ε for which the required inequality holds. It turns out that if

$$\varepsilon < 2(\rho - 3)/(\rho + 3)^2, \qquad (5.4.26)$$

then $\operatorname{Re} \lambda_2 = \operatorname{Re} \lambda_3 > 0$. It is also clear that if $\varepsilon > 2(\rho - 3)/(\rho + 3)^2$ for a fixed value of $\rho > 3$, then the immigration stabilizes the internal equilibrium, all three eigenvalues have a negative real part. What happens if there is immigration but it is not too large, more exactly, if it satisfies (5.4.26)? We shall show that the conditions of Theorem 5.4.5 hold, i.e. there is a periodic orbit in the interior of the positive octant of the N_1, N_2, N_3 space.

First, it is easy to see that if $N_i = b := (1 + (1 + 4\varepsilon)^{1/2})/2$, then $\dot{N}_i \le 0$, and that if $0 < N_i = \delta \le (1/2)(1 - b(\alpha + \beta) + (1 - b(\alpha + \beta))^2 + 4\varepsilon)^{1/2}) > 0$, and $0 \le N_j \le b$, $(j \ne i)$ $(i = 1, 2, 3)$, then $\dot{N}_i \ge 0$. Hence, the cube

$$M_N := \{(N_1, N_2, N_3) \in \mathbf{R}^3 : \delta \le N_i \le b, \quad i = 1, 2, 3\}$$

is a positively invariant convex set, and $(\nu, \nu, \nu) \in \operatorname{Int} M_N$ if $\delta > 0$ is sufficiently small. The linear coordinate transformation

$$
\begin{aligned}
x_1 &= (-N_1 + N_3)/\sqrt{2}, & x_2 &= (-N_1 + 2N_2 - N_3)/\sqrt{6}, \\
x_3 &= (N_1 + N_2 + N_3)/\sqrt{3} - \nu\sqrt{3} & & (5.4.27)
\end{aligned}
$$

moves the origin into the point (ν, ν, ν) and the stable manifold $N_1 = N_2 = N_3$ into the x_3-axis ($x_1 = x_2 = 0$). By this transformation, M_N is transformed into a convex (hence star shaped) compact set M whose interior contains the origin of the x_1, x_2, x_3 space. We perform now the transformation P given by (5.4.17). Only the first differential equation is needed. The calculations yield

$$\dot{y}_1 = (\sqrt{3}/2)p(N_1, N_2 N_3)/q(N_1, N_2, N_3) \qquad (5.4.28)$$

where

$$
\begin{aligned}
p(N_1, N_2, N_3) &= (\beta - 1)(N_1^2 N_2 + N_2^2 N_3 + N_3^2 N_1) \\
&\quad + (1 - \alpha)(N_1 N_2^2 + N_2 N_3^2 + N_3 N_1^2) \\
&\quad + 3(\alpha - \beta)N_1 N_2 N_3, \\
q(N_1, N_2, N_3) &= N_1^2 + N_2^2 + N_3^2 - (N_1 N_2 + N_2 N_3 + N_3 N_1), \\
(N_1, N_2, N_3) &\notin \{(N_1, N_2, N_3) \in \mathbf{R}^3 : N_1 = N_2 = N_3\},
\end{aligned}
$$

and, naturally, the substitution of the x's and the y's is to be performed. After having performed these, this differential equation assumes the form

$$\dot{y}_1 = 2(\beta - \alpha)\nu + (2/\sqrt{3})(\beta - \alpha)y_3$$

$$+\sqrt{6}(\beta - \alpha)y_2 \sin y_1 \cos^2 y_1$$
$$+\sqrt{2}(\alpha + \beta - 2)y_2 \sin^2 y_1 \cos y_1$$
$$+\sqrt{2/3}(\alpha - \beta)y_2 \sin^3 y_1$$
$$-(\sqrt{2}/3)(\alpha + \beta - 2)y_2 \cos^3 y_1 \qquad (5.4.29)$$

which is already valid in the whole $P^{-1}M$. The denominator of the right-hand side of (5.4.28) is

$$q(N_1, N_2, N_3) = (3/2)(x_1^2 + x_2^2) > 0, \qquad x_1 \neq 0 \neq x_2,$$

the numerator p is non-positive on the boundary of the positive octant, and if we start from any point of this boundary in the direction $[1, 1, 1]$, the directional derivative in this direction is

$$(1/\sqrt{3})(\partial/\partial N_1 + \partial/\partial N_2 + \partial/\partial N_3)p(N_1, N_2, N_3)$$
$$= (\beta - \alpha)q(N_1, N_2, N_3) < 0 \quad \text{unless} \quad N_1 = N_2 = N_3.$$

Thus, $\dot{y}_1 = g_1(y) < 0$ unless $N_1 = N_2 = N_3$ which is the same as $x_1 = x_2 = 0$ or $y_2 = 0$. On the stable manifold $y_2 = 0$ we use (5.4.29). Then $\dot{y}_1 = g_1(y_1, 0, y_3) = 2(\beta - \alpha)(\nu + y_3/\sqrt{3})$ but $y_3 = x_3 > -\nu\sqrt{3}$ by (5.4.27); hence, $g_1(y_1, 0, y_3) < 0$. Thus, $g_1(y) \neq 0$, $y \in P^{-1}M$. This result implies also that there is no equilibrium in M_N apart from (ν, ν, ν) outside the stable manifold, but clearly there can be no equilibrium apart from (ν, ν, ν) on the stable manifold $N_1 = N_2 = N_3$ either. This means that all the conditions of Theorem 5.4.5 hold, *i.e. if $0 < \beta < 1 < \alpha$, $\alpha + \beta > 2$ and $\varepsilon < 2(\rho - 3)/(\rho + 3)^2$ where $\rho = 1 + \alpha + \beta$, then system (5.4.25) has a periodic orbit in the interior of the positive octant.*

The main idea in R.A. Smith's (Theorem 5.4.1) and Grasman's (Theorem 5.4.5) results, namely that under certain conditions an n dimensional system can be projected onto a two dimensional one so that the Poincaré-Bendixson Theorem can be applied, has recently been made to work by Mallet-Paret and H.L. Smith [1990] for "monotone cyclic feedback systems". We describe this result without going into details. Let $X \in \mathbf{R}^n$ be a connected open set such that its projection $X^i \subset \mathbf{R}^2$ to the x_{i-1}, x_i coordinate plane is convex ($i = 1, 2, ..., n$; $x_0 := x_n$), let $f : X \to \mathbf{R}^n$ be such that its i-th coordinate f_i depends on x_{i-1} and x_i only, $f_i \in C^1(X^i, \mathbf{R})$ ($i = 1, 2, ..., n$; $x_0 := x_n$), and

$$\text{sign } f'_{ix_{i-1}}(x_{i-1}, x_i) = 1 \quad \text{or} \quad -1 \quad \text{for} \quad \text{all} \quad (x_{i-1}, x_i) \in X_i \quad (5.4.30)$$

$$(i = 1, 2, ..., n).$$

Then

$$\dot{x}_i = f_i(x_{i-1}, x_i) \qquad (i = 1, 2, ..., n; \ x_0 : x_n) \qquad (5.4.31)$$

is said to be a *monotone cyclic feedback system*. The explanation of the name is straightforward. The growth of the coordinate x_i is affected by itself and by x_{i-1} only. Because of (5.4.30) the growth of x_{i-1} increases the growth rate of x_i if $f'_{ix_{i-1}} > 0$ and inhibits it if $f'_{ix_{i-1}} < 0$ monotonously. The identification of x_0 with x_n produces a cycle and a feedback. The system is said to be a *negative or a positive feedback system* according to whether

$$\prod_{i=1}^{n} f'_{ix_{i-1}}(x_{i-1}, x_i)$$

is negative or positive. Such systems arise in mathematical models of cellular control systems in which the coordinates x_i represent the concentration of certain molecules in the cell. Let $\Pi^i : \mathbf{R}^n \to \mathbf{R}^2$ be the projection defined by $\Pi^i(x) = (x_{i-1}, x_i)$. Mallet-Paret and H.L. Smith prove among other things that if $x^0 \in X$, the positive semitrajectory γ^+ of x^0 is bounded and its closure $\overline{\gamma}^+$ is contained in X; then denoting its omega limit set by $\Omega(\gamma^+)$, the projection $\Pi^i : \Omega(\gamma^+) \to \mathbf{R}^2$ is one-to-one, and $\Omega(\gamma^+)$ is either an equilibrium point or a non-trivial periodic orbit or a set consisting of equilibria and heteroclinic and homoclinic orbits joining them (cf. Poincaré-Bendixson Theorem 3.1.8 and Theorem 3.1.11).

This result has been applied by Hofbauer, Mallet-Paret and H.L. Smith [1991] to the *hypercycle system*

$$\dot{x}_i = x_i(F_i(x_{i-1}, x_i) - \sum_{j=1}^{n} x_j F_j(x_{j-1}, x_j))$$

$$(i = 1, 2, ..., n)$$

introduced in order to model the evolution of macromolecules and, perhaps, life (see Hofbauer-Sigmund [1988]). Hofbauer, Mallet-Paret and H.L. Smith have proved the existence of a periodic orbit for the hypercycle system under fairly general conditions.

In the rest of this Section we shall study a system of harmonic oscillators coupled by odd non-linear terms without damping. Under certain conditions the existence of a family of odd periodic solutions will be proved. The character of the problem is completely different from that of the problems treated in this Section earlier, still degree theory will be applied. The results presented are due to Lazer [1975]. Related results can be found, e.g., in Pliss [1965]. The system

$$\ddot{x}_k + \lambda_k x_k = g_k(x) \qquad (k = 1, 2, ..., n) \qquad (5.4.32)$$

will be considered where $\lambda_k \in \mathbf{R}$, $g_k \in C^2(U, \mathbf{R})$, $U \subset \mathbf{R}^n$ a neighbourhood of the origin $x = 0$ $(k = 1, 2, ..., n)$ and

$$g_k(0) = g'_{kx_j}(0) = 0 \qquad (k, j = 1, 2, ..., n). \qquad (5.4.33)$$

The existence of a family of "*small amplitude*" periodic solutions will be established around the origin $x = 0$ which is an equilibrium point of the system. In this particular case "small amplitude" will mean that the first derivative $\dot{x}$ of the solutions will be small. The "*blowing up*" procedure will be applied. This consists of the application of the coordinate transformation $x = \varepsilon y$ where $\varepsilon > 0$ is small, which reduces the problem of finding a family of "small amplitude" periodic solutions of the original system to the existence of ("large amplitude") periodic solutions for a one parameter family of differential equations. The following Lemmata will be needed.

Lemma 5.4.6 *Let $U \subset \mathbf{R}^n$ be a convex set that contains the origin in its interior, $g \in C^m(U, \mathbf{R})$ $(m \geq 1)$ and $g(0) = 0$; then there are functions $g_i \in C^{m-1}(U, \mathbf{R})$ $(i = 1, 2, ..., n)$ such that*

$$g(x) = \sum_{i=1}^{n} x_i g_i(x), \qquad g_i(0) = g'_{x_i}(0) \qquad (i = 1, 2, ..., n).$$

Proof. In U there holds the identity

$$g(x) \;\; = \;\; \int_0^1 (d/dt) g(tx_1, ..., tx_n) dt = \int_0^1 \sum_{i=1}^{n} x_i g'_{x_i}(tx_1, ..., tx_n) dt$$

$$= \;\; \sum_{i=1}^{n} x_i \int_0^1 g'_{x_i}(tx_1, ..., tx_n) dt.$$

The functions $g_i(x) := \int_0^1 g'_{x_i}(tx_1, ..., tx_n) dt$ belong, clearly, to the C^{m-1} class and $g_i(0) = g'_{x_i}(0)$. $\square$

Lemma 5.4.7. *If $g_k \in C^2$ and conditions (5.4.33) hold, then there exists an $\varepsilon_0 > 0$ such that for the function*

$$h_k(y, \varepsilon) := \begin{cases} g_k(\varepsilon y)/\varepsilon, & \varepsilon > 0 \\ 0, & \varepsilon = 0 \end{cases}$$

there holds $h_k \in C^1(B(0, 2) \times [0, \varepsilon_0], \mathbf{R})$ and $h_k(y, 0) = 0$ $(k = 1, 2, ..., n)$ (here $B(0, 2) = \{y \in \mathbf{R}^n : |y| < 2\}$).

Proof. The statement is a straightforward consequence of the previous Lemma applied to g_k. $\square$

Theorem 5.4.8 *(Lazer [1975]). Assume that an odd number of the λ_k's of system (5.4.32) are positive and equal:*

$$\omega^2 = \lambda_k > 0 \quad (k = 1, 2, ..., 2l + 1), \qquad \omega > 0 \tag{5.4.34}$$

$$\lambda_k/\omega^2 \neq m^2 \quad (m = 1, 2, ...; k = 2l + 2, ..., n); \tag{5.4.35}$$

and the functions g_k are odd:

$$g_k(-x) = -g_k(x) \qquad (k = 1, 2, ..., n).$$

Then there exists an $\varepsilon_0 > 0$ such that for $0 < \varepsilon \le \varepsilon_0$ there exists an odd periodic solution $p(t, \varepsilon)$ of (5.4.32) of least period $T(\varepsilon) > 0$ satisfying the conditions $|\dot{p}(0, \varepsilon)| = \varepsilon$ and $T(\varepsilon) \to 2\pi/\omega$ as $\varepsilon \to 0$.

Before proving the Theorem we show first that if an even number of the λ_k's are positive and equal, then the system may not have any non-constant periodic solution; and secondly that if the non-resonance condition (5.4.35) is violated, there may be no periodic solution satisfying the conditions. Consider first the system

$$\ddot{x}_1 + x_1 = x_2(x_1^2 + x_2^2), \qquad \ddot{x}_2 + x_2 = -x_1(x_1^2 + x_2^2).$$

Here $\lambda_1 = \lambda_2 = 1$. If $p(t) = (p_1(t), p_2(t))$ is a T-periodic solution, then substituting it into the system, multiplying the first equation by $p_2(t)$ and the second by $-p_1(t)$ and adding them up we obtain

$$p_2(t)\ddot{p}_1(t) - \ddot{p}_2(t)p_1(t) = (p_1^2(t) + p_2^2(t))^2.$$

Integrating from 0 to T we get

$$\begin{aligned} 0 &= \int_0^T (p_2(t)\dot{p}_1(t) - p_1(t)\dot{p}_2(t))\dot{\,} dt = \int_0^T (p_2(t)\ddot{p}_1(t) - \ddot{p}_2(t)p_1(t))dt \\ &= \int_0^T (p_1^2(t) + p_2^2(t))^2 dt. \end{aligned}$$

Hence, $p_1(t) \equiv p_2(t) \equiv 0$, i.e. there is no non-constant periodic solution. Secondly, consider the system

$$\ddot{x}_1 + x_1 = 0, \qquad \ddot{x}_2 + 9x_2 = 4x_1^3, \qquad \ddot{x}_3 + 9x_3 = 0.$$

Here $\lambda_1 = \omega^2 = 1$ and $\lambda_2 = \lambda_3 = 9\lambda_1$ so that (5.4.35) is violated. If $p(t) = (p_1(t), p_2(t), p_3(t))$ is a periodic solution of this system, then $p_1(t) = A\cos t + B\sin t$, and so if $p_1(t) \not\equiv 0$, then the period must be an integer multiple of 2π. Applying Theorem 2.1.3, $p_2(t)$ can be 2π-periodic if and only if

$$\int_0^{2\pi} 4p_1^3(t)\sin 3t\, dt = \int_0^{2\pi} 4p_1^3(t)\cos 3t\, dt = 0.$$

This leads to the conditions $\pi A(A^2 - 3B^2) = \pi B(-B^2 + 3A^2) = 0$, i.e. $A = B = 0$. Hence, $p_1(t) \equiv 0$, $p_2(t) = C\cos 3t + D\sin 3t$, $p_3(t) = E\cos 3t + F\sin 3t$. This means that the system does not have a non-constant periodic solution with least period close to $2\pi = 2\pi/\omega$.

Proof. Choose $\varepsilon_0 > 0$ according to Lemma 5.4.7 and perform the blowing up transformation $x = \varepsilon y$ for $0 \leq \varepsilon \leq \varepsilon_0$. Then system (5.4.32) becomes

$$\ddot{y}_k + \lambda_k y_k = h_k(y, \varepsilon) \qquad (k = 1, ..., n), \qquad (5.4.36)$$

i.e. a family of systems depending on the parameter ε. For $a \in \mathbf{R}^n$ denote by $\psi(t, a, \varepsilon)$ the solution of (5.4.36) that satisfies

$$\psi(0, a, \varepsilon) = 0, \qquad \dot{\psi}(0, a, \varepsilon) = a. \qquad (5.4.37)$$

The Lemma implies that a $0 < \varepsilon_1 \leq \varepsilon_0$ exists such that for $0 \leq \varepsilon < \varepsilon_1$ and for $|a| < 2$, say, $\psi(t, a, \varepsilon)$ is defined on the interval $t \in [-2\pi/\omega, 2\pi/\omega]$ and it is continuously differentiable with respect to a and ε. For $\varepsilon = 0$ we obtain the uncoupled system

$$\ddot{y}_k + \lambda_k y_k = 0 \qquad (k = 1, 2, ..., n),$$

so that $\psi(t, a, 0)$ is obviously given by

$$\psi_k(t, a, 0) = (a_k/\omega) \sin \omega t \qquad (k = 1, 2, ..., 2l + 1),$$

$$\psi_k(t, a, 0) = \begin{cases} (a_k/(-\lambda_k)^{1/2}) \sinh(-\lambda_k)^{1/2} t & \text{if } \lambda_k < 0 \\ a_k t & \text{if } \lambda_k = 0 \\ (a_k/\lambda_k^{1/2}) \sin \lambda_k^{1/2} t & \text{if } \lambda_k > 0 \end{cases}$$

$$(k = 2l + 2, ..., n). \qquad (5.4.38)$$

We shall need the Jacobian of ψ with respect to a

$$\begin{aligned} J_a \psi(t, a, 0) &= \det \psi_a'(t, a, 0) \\ &= \left(\frac{\sin \omega t}{\omega}\right)^{2l+1} \Delta(t) \end{aligned} \qquad (5.4.39)$$

where

$$\Delta(t) = \prod_{k=2l+2}^{n} \psi_{k a_k}'(t, a, 0).$$

$\Delta(\pi/\omega) \neq 0$ in view of (5.4.35). Therefore, we can choose a $0 < \delta < \pi/\omega$ such that

$$\Delta(t) \neq 0, \qquad t \in [\pi/\omega - \delta, \pi/\omega + \delta].$$

Hence, by (5.4.39) taking into account that $0 < \delta\omega < \pi$

$$\begin{aligned} 0 &\neq \text{sign } J_a \psi(\pi/\omega - \delta, a, 0) = \text{sign } (\sin^{2l+1}(\pi - \delta\omega) \Delta(\pi/\omega - \delta)) \\ &= \text{sign } \Delta(\pi/\omega - \delta) = \text{sign } \Delta(\pi/\omega + \delta) \\ &\neq \text{sign } J_a \psi(\pi/\omega + \delta, a, 0) = \text{sign } (\sin^{2l+1}(\pi + \delta\omega) \Delta(\pi/\omega + \delta)) \\ &= (-1)^{2l+1} \text{sign } \Delta(\pi/\omega + \delta) = -\text{sign } \Delta(\pi/\omega + \delta) \neq 0. \qquad (5.4.40) \end{aligned}$$

(Note that this was the step where the oddness of $2l + 1$ was used.)

Now, set $D = \{a \in \mathbf{R}^n : |a| \leq 1\}$ and define the mapping $F_{(t,\varepsilon)} : \overline{D} \to \mathbf{R}^n$ by $F_{(t,\varepsilon)}(a) := \psi(t, a, \varepsilon)$. We shall be concerned with the topological degree $d[F, D, 0]$ of F with respect to the set D and to $0 \in \mathbf{R}^n$ (see Definition A2.1 et seq.). We see from (5.4.38) that the mapping $F_{(t,0)}$ is linear, and from (5.4.40) that the maps $F_{(\pi/\omega - \delta, 0)}$, $F_{(\pi/\omega + \delta, 0)}$ are non-singular, so that by Definition A2.1 (see also Theorem A2.11)

$$d[F_{(\pi/\omega \pm \delta, 0)}, D, 0] = \text{sign } J_a\psi(\pi/\omega \pm \delta, 0, 0) \neq 0.$$

The solutions $\psi(t, a, \varepsilon)$ are continuous functions of the parameter ε so that a $0 < \varepsilon(\delta) \leq \varepsilon_1$ exists such that for $0 \leq \varepsilon \leq \varepsilon(\delta)$ and $|a| = 1$, i.e. $a \in \partial D$, we have

$$|\psi(\pi/\omega \pm \delta, a, \varepsilon) - \psi(\pi/\omega \pm \delta, a, 0)| < |\psi(\pi/\omega \pm \delta, a, 0)|;$$

thus $\psi(\pi/\omega \pm \delta, a, \varepsilon)$ is never of opposite direction to $\psi(\pi/\omega \pm \delta, a, 0)$, so that by the Poincaré-Bohl Theorem (Corollary A2.6) $d[F_{(\pi/\omega \pm \delta, \varepsilon)}, D, 0]$ is defined for $0 \leq \varepsilon \leq \varepsilon(\delta)$, and

$$\begin{aligned} d[F_{(\pi/\omega \pm \delta, \varepsilon)}, D, 0] &= d[F_{(\pi/\omega \pm \delta, 0)}, D, 0] \\ &= \text{sign } J_a\psi(\pi/\omega \pm \delta, 0, 0). \end{aligned}$$

But this means by (5.4.40) that

$$d[F_{(\pi/\omega - \delta, \varepsilon)}, D, 0] \neq d[F_{(\pi/\omega + \delta, \varepsilon)}, D, 0].$$

As a consequence, the conclusion of Theorem A2.5 (invariance by homotopy) is not true. This implies that $F_{(\pi/\omega - \delta, \varepsilon)}$ *cannot* be homotopic to $F_{(\pi/\omega + \delta, \varepsilon)}$, i.e. there exist a number $T(\varepsilon)$ and a point $a(\varepsilon) \in \partial D$ such that

$$\pi/\omega - \delta < T(\varepsilon) < \pi/\omega + \delta, \tag{5.4.41}$$

$$|a(\varepsilon)| = 1, \qquad \psi(T(\varepsilon), a(\varepsilon), \varepsilon) = 0. \tag{5.4.42}$$

Consider the solution of (5.4.36) defined by $q(t, \varepsilon) := \psi(t, a(\varepsilon), \varepsilon)$. This solution is defined on the interval $[-2\pi/\omega, 2\pi/\omega]$. We show that it can be extended to $(-\infty, \infty)$ as a $2T(\varepsilon)$-periodic odd solution. Indeed, since the functions $h_k(y, \varepsilon)$ are odd in y, i.e. $h_k(-y, \varepsilon) = -h_k(y, \varepsilon)$ $(k = 1, 2, ..., n)$, the function $\tilde{q}(t, \varepsilon) := -q(-t, \varepsilon)$ is also a solution of system (5.4.34). But $\tilde{q}(0, \varepsilon) = -q(0, \varepsilon) = 0 = q(0, \varepsilon)$ and $\dot{\tilde{q}}(0, \varepsilon) = \dot{q}(0, \varepsilon) = a(\varepsilon)$ by (5.4.37); therefore, because of the uniqueness of solutions of initial value problems, $-q(-t, \varepsilon) \equiv \tilde{q}(t, \varepsilon) \equiv q(t, \varepsilon)$ for $t \in [-2\pi/\omega, 2\pi/\omega]$. This implies that $q(-T(\varepsilon), \varepsilon) = -q(T(\varepsilon), \varepsilon) = q(T(\varepsilon), \varepsilon) = 0$, and $\dot{q}(-T(\varepsilon), \varepsilon) = \dot{q}(T(\varepsilon), \varepsilon)$. By Lemma 2.2.1 this implies that $q(t, \varepsilon)$ is defined over $(-\infty, \infty)$ and $q(t + 2T(\varepsilon), \varepsilon) = q(t, \varepsilon)$. The implication of the result for the original system (5.4.32) is that $p(t, \varepsilon) := \varepsilon q(t, \varepsilon)$ is a periodic solution of the latter system with period $2T(\varepsilon)$ for $0 < \varepsilon \leq \varepsilon(\delta)$ and $|\dot{p}(0,\varepsilon)| = \varepsilon|\dot{q}(0,\varepsilon)| = \varepsilon|a(\varepsilon)| = \varepsilon$, i.e. p is not constant.

It remains to show that the least period of p, i.e. of q, tends to $2\pi/\omega$ as $\varepsilon \to 0$. Suppose that this is not the case. Then there exist a number $0 < \alpha < \pi/(3\omega)$, a sequence $\varepsilon_m > 0$ $(m = 1, 2, ...)$ and a sequence t_m $(m = 1, 2, ...)$ such that $\varepsilon_m \to 0$, $m \to \infty$, $0 < t_m \leq \pi/\omega - \alpha$ and $q(t_m, \varepsilon_m) = 0$ $(m = 1, 2, ...)$. (In this case, clearly, $2t_m$ is a period of $q(t, \varepsilon_m)$.) We may assume that $0 < \alpha \leq t_m \leq \pi/\omega - \alpha$ because if we had $0 < t_m \leq \alpha$, then for some positive integer j the inequality $\alpha \leq jt_m \leq \pi/\omega - \alpha$ would hold and also $q(jt_m, \varepsilon_m) = 0$ since $q(t, \varepsilon_m)$ is odd and $2t_m$-periodic. We may also assume without loss of generality that $t_m \to t^* \in [\alpha, \pi/\omega - \alpha]$, and since $|\dot{q}(0,\varepsilon_m)| = 1$ that $\dot{q}(0, \varepsilon_m) \to a^*$ with $|a^*| = 1$, and further that $T(\varepsilon_m) \to T^* \in [\pi/\omega - \delta, \pi/\omega + \delta]$, as $m \to \infty$. The continuous dependence of solutions on the initial data and on the parameter implies that $\psi(t^*, a^*, 0) = \psi(T^*, a^*, 0) = 0$. We recall that δ was chosen in such a way that $\Delta(t) \neq 0$, i.e. $\psi_k(t, a, 0) \neq 0$ for $k = 2l + 2, ..., n$ on $t \in [\pi/\omega - \delta, \pi/\omega + \delta]$ unless $a_k = 0$ $(k = 2l + 2, ..., n)$. Thus, $\psi(T^*, a^*, 0) = 0$ implies $\psi_k(t, a^*, 0) \equiv 0$ $(k = 2l + 2, ..., n)$. Now, by the first formula of (5.4.38), $\psi_k(t, a^*, 0) \neq 0$ for $t \in [\alpha, \pi/\omega - \alpha]$ $(k = 1, 2, ..., 2l + 1)$ unless $a_k = 0$ $(k = 1, 2, ..., 2l + 1)$. Thus, $\psi(t^*, a^*, 0) = 0$ implies $\psi_k(t, a^*, 0) \equiv 0$ $(k = 1, 2, ..., 2l + 1)$. This means that the solution $\psi(t, a^*, 0)$ is identically zero: $\psi(t, a^*, 0) \equiv 0$ contradicting $|\dot{\psi}(0, a^*, 0)| = |a^*| = 1$. This contradiction implies that $2T(\varepsilon) \to 2\pi/\omega$ as $\varepsilon \to 0$. $\square$

The proof shows that the "blowing up transformation" leads to the one parameter family of systems (5.4.36) which at $\varepsilon = 0$ is an uncoupled system of harmonic oscillators. This system, i.e. (5.4.36) at $\varepsilon = 0$, has a $2\pi/\omega$ periodic solution

$$q(t, 0) = [(a_1/\omega) \sin \omega t, ..., (a_{2l+1}/\omega) \sin \omega t, 0, ..., 0]$$

which satisfies $|\dot{q}(0, 0)| = 1$ if $\sum_{k=1}^{2l+1} a_k^2 = 1$. In the case $\varepsilon > 0$ the periodic solution of (5.4.36) emanates from this solution and generates the solution $p(t, \varepsilon) = \varepsilon q(t, \varepsilon)$ of the original system. It is to be observed that the substitution of $\varepsilon = 0$ now leads us to the trivial solution $0 \equiv p(t, 0)$ of system (5.4.32) which does not carry any information.

The following Theorem can be proved analogously. Its proof is left to the reader as a non-trivial exercise.

Theorem 5.4.9 *(Lazer [1975]). Assume that an odd number of the λ_k's of system (5.4.32) are positive and equal:*

$$\omega^2 = \lambda_k > 0 \qquad (k = 1, 2, ..., 2l + 1), \qquad \omega > 0,$$

$$\lambda_k/\omega^2 \neq m^2 \qquad (m = 1, 2, ...; k = 2l + 2, ..., n)$$

and that $\lambda_k \neq 0$ $(k = 1, 2, ..., n)$; then there exists an $\varepsilon_0 > 0$ such that for $0 < \varepsilon \leq \varepsilon_0$ there exists an even periodic solution $p(t, \varepsilon)$ of (5.4.32) of least period $T(\varepsilon) > 0$ satisfying the conditions $|p(0, \varepsilon)| = \varepsilon$ and $T(\varepsilon) \to 2\pi/\omega$ as $\varepsilon \to 0$.

Observe that to prove the existence of odd periodic solutions we had to assume that the functions g_k are odd, we do not need some similar assumption in order to prove the existence of even periodic solutions.

5.5 Competitive and Cooperative Systems, Existence in Dimension Three

Competitive and cooperative systems play an important role in biological and economic applications. They have now a fairly developed theory due mainly to Hirsch [1982, 1985] and also to H.L. Smith [1986] and others. We give here a short treatment of those results only that are needed in establishing the existence of periodic orbits for such systems. The implication of these results is particularly instructive in case the system is three dimensional. In the second part of this Section we shall present a result of Hastings and Murray [1975] concerning the existence of a periodic orbit of the Field-Körös-Noyes model for the Belousov-Zhabotinskij reaction. Apart from the great interest in this phenomenon which has been growing all the time in the past twenty years, the method applied by Hastings and Murray is characteristic and fairly often used for three dimensional systems.

Let $X \subset \mathbf{R}^n$ be an open and connected set, $f \in C^1(X, \mathbf{R}^n)$, and consider the system

$$\dot{x} = f(x). \tag{5.5.1}$$

Definition 5.5.1. We say that system (5.5.1) is *cooperative*, resp. *competitive* if

$$f'_{ix_k}(x) \geq 0, \quad \text{resp}. \quad f'_{ix_k}(x) \leq 0, \quad \text{for} \quad x \in X$$

$$(i, k = 1, 2, ..., n; i \neq k) \tag{5.5.2}$$

(cf. Section 4.5).

In particular, if (5.5.1) is a *Kolmogorov system*, i.e. $f_i(x) = x_i F_i(x)$, $F_i \in C^1(\mathbf{R}_+^n, \mathbf{R}^n)$ $(i = 1, 2, ..., n)$, then conditions (5.5.2) are equivalent to

$$F'_{ix_k}(x) \geq 0, \quad \text{resp}. \quad F'_{ix_k}(x) \leq 0, \quad \text{for} \quad x \in \mathbf{R}_+^n$$

$$(i, k = 1, 2, ..., n; i \neq k). \tag{5.5.3}$$

If (5.5.1) is an *autonomous Lotka-Volterra system*, i.e. $f_i(x) = x_i F_i(x)$ where $F_i(x) = b_i - \sum_{j=1}^{n} a_{ij} x_j$, then conditions (5.5.3) are equivalent to $a_{ik} \leq 0$, resp. $a_{ik} \geq 0$ $(i, k = 1, 2, ..., n; i \neq k)$. The intuitive meaning of the conditions is, in case x_i is the quantity of species i that is sharing an ecological "habitat" with the rest of the species, that in the cooperative case the increase of the quantity of species k is beneficial to the increase of the quantity of species i for any $i \neq k$; in the competitive case, the same has an adverse affect on species i.

Before embarking on the study of the basic properties of cooperative and competitive systems, some preparation is needed. The following notations reflecting the partial ordering of vectors $x, y \in \mathbf{R}^n$ will be used: $x < y$ if $x_i < y_i$ for all $i = 1, 2, ..., n$; $x \leq y$ if $x_i \leq y_i$ for all $i = 1, 2, ..., n$. Similar notations will be used in case of real matrices of the same type.

Definition 5.5.2. Let $\varphi(t, x)$ denote the solution of (5.5.1) with initial value x, i.e. $\varphi(0, x) = x$; we say that the flow generated by (5.5.1) is *monotone*, resp. *strongly monotone* if for any $x, y \in X$ such that $x \leq y$, $x \neq y$ we have

$$\varphi(t, x) \leq \varphi(t, y), \qquad \text{resp.} \qquad \varphi(t, x) < \varphi(t, y) \qquad \text{for} \qquad t > 0$$

in their common domain of definition.

Cooperativity with a slight restriction on the domain X implies monotonicity as we shall see after having proved *Kamke's Theorem*:

Theorem 5.5.1 (*Kamke [1932]*). *Let $X \subset \mathbf{R}^n$ be a convex, open and connected set, $g \in C^1(\mathbf{R} \times X, \mathbf{R}^n)$, the functions g_i non-decreasing in x_k $(k = 1, ..., n; i = 1, 2, ..., n; k \neq i)$ and consider the system*

$$\dot{x} = g(t, x); \tag{5.5.4}$$

let $\varphi : [t_0, t_1] \to X$ and $\psi : [t_0, t_1] \to X$ be two solutions, $t_0 < t_1$; if $\varphi(t_0) < \psi(t_0)$, resp. $\varphi(t_0) \leq \psi(t_0)$, then $\varphi(t) < \psi(t)$, resp. $\varphi(t) \leq \psi(t)$ for all $t \in [t_0, t_1]$.

Proof. First the statement with the strict inequalities will be proved. By continuity there is a $\delta > 0$ such that $\varphi(t) < \psi(t)$, $t \in [t_0, t_0 + \delta)$. Suppose that there is a "first" moment $t_2 \in [t_0, t_1]$ such that the strict inequality holds in $[t_0, t_2)$ and there is a coordinate, the first say, of the two solutions that assumes the same value at t_2, i.e. $\varphi(t) < \psi(t)$, $t \in [t_0, t_2)$ and $\varphi_1(t_2) = \psi_1(t_2)$. Then we may write

$$
\begin{aligned}
(\psi_1(t) - \varphi_1(t))^{\cdot} &\equiv g_1(t, \psi(t)) - g_1(t, \varphi(t)) \\
&\equiv g_1(t, \psi(t)) - g_1(t, \psi_1(t), \varphi_2(t), ..., \varphi_n(t)) \\
&\quad + g_1(t, \psi_1(t), \varphi_2(t), ..., \varphi_n(t)) - g_1(t, \varphi(t)).
\end{aligned}
$$

Introducing the notations $u(t) := \psi_1(t) - \varphi_1(t)$,

$$
\begin{aligned}
u^0 : &= u(t_0) = \psi_1(t_0) - \varphi_1(t_0) > 0, \\
p(t) : &= g_1(t, \psi(t)) - g_1(t, \psi_1(t), \varphi_2(t), ..., \varphi_n(t)), \\
h(t, u(t)) : &= g_1(t, \varphi_1(t) + u(t), \varphi_2(t), ..., \varphi_n(t)) - g_1(t, \varphi(t))
\end{aligned}
$$

we may write

$$
\dot{u}(t) \equiv p(t) + h(t, u(t)).
$$

By the assumptions $p(t) \geq 0$ for $t \in [t_0, t_2]$, and applying Lemma 5.4.6, the last identity can be written in the form

$$
\dot{u}(t) \equiv p(t) + a(t, u(t))u(t)
$$

where a is a continuous function. Thus, the function u satisfies the scalar linear differential equation $\dot{y} = a(t, u(t))y + p(t)$ where u has already been substituted into a. Applying (1.2.11) we obtain

$$
u(t_2) = \exp \int_{t_0}^{t_2} a(t, u(t))dt \left(u^0 + \int_{t_0}^{t_2} p(t) \exp \left(- \int_{t_0}^{t} a(s, u(s))ds \right) dt \right),
$$

i.e. $u(t_2) = \psi_1(t_2) - \varphi_1(t_2) > 0$, a contradiction that proves the statement involving the strict inequalities.

In order to prove the statement concerning the inequalities that allow equalities of some coordinates we introduce the one (positive) parameter family of differential equations

$$
\dot{x} = g(t, x) + \mu c, \qquad \mu \geq 0, \qquad c = [1, 1, ..., 1]
$$

with the initial condition $x(t_0) = \psi(t_0) + \mu c$. We denote the solution of this system satisfying the initial condition by $\sigma(t, \mu)$. Since $g(t, x) + \mu c > g(t, x)$, and $\psi(t_0) + \mu c > \varphi(t_0)$ for $\mu > 0$ by Problem 5.7.8, $\sigma(t, \mu) > \varphi(t)$ for $t \in [t_0, t_1]$, $\mu > 0$. But because of the continuous dependence on initial values and parameters (Theorem 1.1.3), $\sigma(t, \mu) \to \psi(t)$ as $\mu \to +0$. Hence, $\psi(t) \geq \varphi(t)$ for $t \in [t_0, t_1]$. □

Corollary 5.5.2. *If the domain X of the right-hand side of (5.5.1) is convex and the system is cooperative, then its flow is monotone.*

A stronger result is valid for systems that are "irreducible". We say that the $n \times n$ real *matrix* $A = [a_{ik}]$ is *irreducible* if the linear mapping $A : \mathbf{R}^n \to \mathbf{R}^n$ does not map any proper linear subspace of dimension at least one spanned by a subset of the standard basis $[1, 0, ..., 0], [0, 1, ...,], ..., [0, ..., 0, 1]$ into itself. Alternatively, this can be expressed the following way: A is an irreducible matrix if whenever the set $\{1, 2, ..., n\}$ is split up as the union of two disjoint proper subsets H and H', then for every $i \in H$ there exists $j, k \in H'$ such that $a_{ij} \neq 0$, $a_{ki} \neq 0$.

We say that the C^1 function $f : X \to \mathbf{R}^n$ is irreducible if its derivative matrix $f_x'(x)$ is irreducible for $x \in X$.

Lemma 5.5.3 (Hirsch [1985]). Denote the flow generated by system (5.5.1) by $\varphi(t, x)$, $t \in \mathbf{R}$, $x \in X$, assume that X is convex and the system is cooperative; then

(i) $\varphi_x'(t, x)$ is a non-negative matrix for $t \geq 0$;

(ii) if f is irreducible, then $\varphi_x'(t, x)$ is a positive matrix for $t \geq 0$.

Proof. Fix $x \in X$ and consider the variational system with respect to the solution $\varphi(t, x)$:

$$\dot{y} = A(t)y \qquad (5.5.5)$$

where $A(t) := f_x'(\varphi(t, x))$. By (1.1.5) et seq. $M(t) := \varphi_x'(t, x)$ is the fundamental matrix of (5.5.5) that assumes the unit matrix at $t = 0$, i.e. $M(0) = I$. Considering (5.5.5) as a matrix differential equation and denoting the elements of A and M by a_{ik} and m_{ik}, respectively, we may say that the n^2 dimensional vector $[m_{11}, ..., m_{nn}]$ satisfies the n^2 dimensional system of differential equations

$$\dot{m}_{ik} = G_{ik}(t, m_{11}, ..., m_{nn}) \qquad (i, k = 1, 2, ..., n) \qquad (5.5.6)$$

where $G_{ik}(t, m_{11}, ..., m_{nn}) := \sum_{j=1}^n a_{ij}(t)m_{jk}$. Clearly, $\partial G_{ik}/\partial m_{rs} \geq 0$ if $(i, k) \neq (r, s)$ since $a_{ij}(t) = f_{ix_j}'(\varphi(t, x)) \geq 0$ $(i \neq j)$, because of the cooperativity of (5.5.1). Since $m_{ik}(0) = \delta_{ik} \geq 0$ (where δ_{ik} is the Kronecker delta) by Kamke's Theorem 5.5.1, $m_{ik}(t) \geq 0$, the trivial solution of (5.5.6) for $t \geq 0$ $(i, k = 1, 2, ..., n)$. Thus, $M(t) = \varphi_x'(t, x) \geq 0$ for $t \geq 0$, $x \in X$ and (i) has been proved.

Now assume that f is irreducible. If for $t_0 > 0$ we have $M(t_0) > 0$, then by Kamke's Theorem, $M(t) > 0$ for $t > t_0$. Suppose that $M(t) > 0$ does not hold for all $t > 0$. Then there exists a $t_1 > 0$ such that $m_{ij}(t_1) = 0$ for some (i, j). But in this case according to what has been said in the second sentence of this paragraph, for every $t \in [0, t_1]$ there exists a pair of indices $(k(t), l(t))$ such that $m_{kl}(t) = 0$. Denote the subset of the interval $[0, t_1]$ where m_{kl} is zero by T_{kl}. Since $\bigcup_{k,l=1}^n T_{kl} = [0, t_1]$, at least one T_{kl} must have a non-empty interior, i.e. there exists a pair of indices (k, l) such that $m_{kl}(t) \equiv 0$, $t \in (t_2, t_3)$ where $0 \leq t_2 < t_3 \leq t_1$. But this implies that also $\dot{m}_{kl}(t) \equiv 0$, $t \in (t_2, t_3)$. However, we shall show that if $m_{kl}(t_0) = 0$ for some $t_0 > 0$, then $\dot{m}_{kl}(t_0) > 0$ must hold. Let S denote the set of indices r such that $m_{rl}(t_0) = 0$, i.e.

$$S = \{r \in \mathbf{N} : 1 \leq r \leq n, \quad m_{rl}(t_0) = 0\}.$$

S is non-empty since $k \in S$, and $S \neq \{1, 2, ..., n\}$ since the matrix $M(t_0)$ is regular. The matrix $A(t_0)$ is irreducible. Therefore, there exists a

$j \in \{1, ..., n\} \backslash S$ such that $a_{kj}(t_0) \neq 0$. Since k belongs to S, $j \neq k$, so that the cooperativity implies that $a_{kj}(t_0) > 0$. Thus,

$$\dot{m}_{kl}(t_0) = \sum_{r=1}^{n} a_{kr}(t_0) m_{rl}(t_0) = \sum_{\substack{r=1 \\ r \neq k}}^{n} a_{kr}(t_0) m_{rl}(t_0),$$

since $m_{kl}(t_0) = 0$. In the last sum, each term is the product of non-negative numbers, and $a_{kj}(t_0) m_{jl}(t_0) > 0$; thus, $\dot{m}_{kl}(t_0) > 0$. This contradiction implies that $M(t) > 0$ for all $t > 0$. $\square$

Corollary 5.5.4. *If the domain X of the right-hand side of (5.5.1) is convex, the system is cooperative, and f is irreducible, then the flow $\varphi(t, x)$ generated by the system is strongly monotone.*

Proof. For $a, b \in X$, $a \leq b$, $a \neq b$ we may write

$$\varphi(t, b) - \varphi(t, a) = \int_0^1 (d\varphi(t, a + (b - a)s)/ds) ds$$

$$= \int_0^1 \varphi'_x(t, a + (b - a)s)(b - a) ds > 0$$

since the column vector $b - a$ has non-negative elements and at least one of them is positive, and all the elements of the matrix φ'_x are positive everywhere by the Lemma. $\square$

The previous results based on Kamke's Theorem concerned cooperative systems. Now, cooperative systems are not too interesting from the point of view of periodic solutions since, as we shall soon see, they cannot have attracting closed orbits. However, Kamke's Theorem has the following implication for *competitive* systems.

Corollary 5.5.5. *Suppose that X is convex, and system (5.5.1) is competitive, let $\varphi : [t_0, t_1] \to X$ and $\psi : [t_0, t_1] \to X$ be two solutions, $t_0 < t_1$; if $\varphi(t_0)$ and $\psi(t_0)$ are not related by $<$ nor by $>$, then $\varphi(t_1)$ and $\psi(t_1)$ are not related either.*

Proof. Reversing the time by $\tau = -t$, system (5.5.1) goes into

$$\frac{dx}{d\tau} = -f(x). \tag{5.5.7}$$

If (5.5.1) is competitive, then (5.5.7) is cooperative, and $\tau_1 := -t_1 < -t_0 =: \tau_0$; further, $\tilde{\varphi}(\tau) := \varphi(-\tau)$ and $\tilde{\psi}(\tau) := \psi(-\tau)$ are solutions of (5.5.7). If $\tilde{\varphi}(\tau_1) = \varphi(-\tau_1) = \varphi(t_1) < \psi(t_1) = \psi(-\tau_1) = \tilde{\psi}(\tau_1)$ hold, then by Theorem 5.5.1, $\varphi(t_0) = \tilde{\varphi}(\tau_0) < \tilde{\psi}(\tau_0) = \psi(t_0)$ would hold and similarly for $>$. $\square$

After having gone through this preparatory study of some basic properties of cooperative, resp. competitive systems, we continue with the study of their periodic orbits. Two Lemmata will be proved first.

Lemma 5.5.6. *Assume that X is convex, system (5.5.1) is cooperative and irreducible, and denote the subset of X where f is a non-negative or non-positive vector by S:*

$$S := \{x \in X : f(x) \geq 0 \quad or \quad f(x) \leq 0\};$$

if γ is a periodic orbit of (5.5.1), then $\gamma \subset X \backslash S$, and no pair of distinct points on γ can be related by $\leq$.

Proof. If $a \in \gamma$ and $f(a) \geq 0$, then $f_i(a) > 0$ for some i since γ is a not an equilibrium. The solution $\varphi(t, a)$ of (5.5.1) is a periodic solution whose path is γ. The derivative $\dot\varphi(t, a)$ is a periodic solution of the variational system of (5.5.1) with respect to the solution $\varphi(t, a)$:

$$\ddot\varphi(t, a) = f'_x(\varphi(t, a))\dot\varphi(t, a)$$

and at $t = 0$ it assumes the value $\dot\varphi(0, a) = f(a)$. Hence, $\dot\varphi(t, a) = f(\varphi(t, a)) = \varphi'_x(t, a)f(a)$ since $\varphi'_x(t, a)$ is the fundamental matrix of the variational system that assumes the matrix I at $t = 0$. By Lemma 5.5.3 (ii), $\varphi'_x(t, a) > 0$ for $t > 0$; as a consequence, $f(\varphi(t, a)) > 0$ for $t > 0$. But this is impossible because γ is a periodic orbit. Let a and $\varphi(t, a)$ be two distinct points on γ where $0 < t < T$, the least positive period on γ. If $a \leq \varphi(t, a)$, then because of the strong monotonicity of the flow (Corollary 5.5.4) $\varphi(t, a) < \varphi(t, \varphi(t, a)) = \varphi(2t, a) < \varphi(3t, a) < \cdots < \varphi(mt, a) < \cdots (m = 1, 2, \dots)$. Since the sequences of all the coordinates are, clearly, bounded, $\varphi(mt, a) \to b \in \gamma$ as $m \to \infty$. But then because of the continuous dependence of the solutions on the initial values, $\varphi(t, b) = b$, contradicting the assumption that $t > 0$ is less than the least positive period. $\square$

The following Lemma due to H.L. Smith [1986] shows that *an irreducible, cooperative system cannot have an attracting periodic orbit.* We note that this is true even without the assumption of the irreducibility (see Hirsch [1985]).

Lemma 5.5.7. *Assume that X is convex and that system (5.5.1) is cooperative and irreducible; if γ is a periodic orbit of the system, then it has a characteristic multiplier λ^* that is simple, real, greater than 1 and exceeds in modulus all other characteristic multipliers.*

Proof. Let $a \in \gamma$, $\varphi(t, a)$ a periodic solution with path γ and $T > 0$, the least positive period. Then as in the proof of the previous Lemma

$\dot{\varphi}(t,a) = \varphi'_x(t,a)f(a)$, and

$$f(a) = \dot{\varphi}(0,a) = \dot{\varphi}(T,a) = \varphi'_x(T,a)f(a),$$

expressing the fact that 1 is a characteristic multiplier and $f(a)$ a corresponding eigenvector of the principal matrix $\varphi'_x(T,a)$ (see (2.2.5) and (1.3.7)). By the previous Lemma, $f(a) \not\geq 0$, $f(a) \not\leq 0$ and by Corollary 5.5.4, $\varphi'_x(T,a) > 0$. By Perron's Theorem A1.13 the matrix $\varphi'_x(T,a)$ has a simple positive eigenvalue λ^* which is greater than the modulus of any other eigenvalue and λ^* has a positive eigenvector:

$$\varphi'_x(T,a)v = \lambda^* v, \qquad v > 0.$$

Now, if 1 is a multiple eigenvalue, then it cannot be λ^* which is simple. If 1 is a simple eigenvalue, then it has no eigenvector linearly independent of $f(a)$ which is neither positive nor negative; hence, $\lambda^* \neq 1$ and as a consequence $\lambda^* > 1$. $\square$

The previous two Lemmata have straightforward implications for competitive systems. Since reversing the time a competitive system transforms into a cooperative one, *Lemma 5.5.6 remains valid if the word cooperative is replaced by competitive in its text.* If the time reversal $t = -\tau$ is performed, then system (5.5.1) goes into $dx/d\tau = -f(x)$. Applying the notations in the proof of Lemma 5.5.7, $\varphi(-\tau,a)$ is a periodic solution of period $T > 0$ of the time reversed system, and $\varphi'_x(-T,a)$ is the principal matrix of the corresponding variational system. But by (2.2.6)

$$\varphi'_x(-T,a)\varphi'_x(Ta) = \varphi'_x(0,a) = I,$$

i.e. $\varphi'_x(-T,a) = (\varphi'_x(T,a))^{-1}$. Thus, the characteristic multipliers of the time reversed system are the reciprocals of those of the original one. This means that Lemma 5.5.7 is replaced now by

Lemma 5.5.8. *Assume that X is convex and that system (5.5.1) is competitive and irreducible; if γ is a periodic orbit of the system, then it has a characteristic multiplier λ_* that is simple, satisfies $0 < \lambda_* < 1$ and is less than the modulus of any other characteristic multiplier.*

According to this Lemma an irreducible competitive system may have attracting periodic orbits. However, the case $n = 2$ is an exception. If there is a closed Jordan curve in the plane, then, obviously, it contains necessarily points related by $\leq$ or, alternatively, its tangent vector performing a rotation of 360 degrees will be positive (and negative) at some parts. By Lemma 5.5.6 and the remark following the proof of Lemma 5.5.7 this implies that *an irreducible two dimensional cooperative or competitive system cannot have a periodic orbit.* We note that this is true even without the assumption on irreducibility (see Hirsch [1982]).

The previous theory has important implications concerning the $n = 3$ case. We are going to present two Theorems concerning periodic solutions of three dimensional systems. Unfortunately, we have to omit the rigorous proofs of these Theorems because they would require a long preparation (the proofs can be found in the respective references). However, we shall give somewhat intuitive sketches of the proofs because these are instructive and show some important characteristics of cooperative and competitive systems.

Theorem 5.5.9 (Hirsch [1982]). *Assume that $n = 3$, the connected open set $X \subset \mathbf{R}^3$ is convex, system (5.5.1) is cooperative or competitive, and let $M \subset X$ be a compact alpha or omega limit set which contains no equilibrium point. Then:*

(i) *M is either a periodic orbit or a two dimensional topological cylinder of periodic orbits;*

(ii) *if the system is cooperative and M is an omega limit set, then M is a periodic orbit;*

(iii) *if M contains a hyperbolic periodic orbit, then it is a periodic orbit.*

Sketch of Proof. We note first that a periodic orbit is called hyperbolic if 1 is a simple characteristic multiplier and the modulus of the other characteristic multipliers is different from 1.

The proof is based on the fact that a limit set M of an n dimensional cooperative or competitive system is "compressible" along any positive vector. *Compressible along a vector $v \in \mathbf{R}^n$* means that the set M can be projected to an $n - 1$ dimensional hyperplane orthogonal to v in such a way that the projection is a homeomorphism, its inverse satisfies some smoothness condition, and the trajectories in M are projected to the trajectories of some "fairly smooth" flow in the hyperplane preserving the orientation on the curves. The compressibility of a limit set of a cooperative or competitive system can be proved relatively easily using the fact that no two distinct points of such a limit set can be related by $\leq$, and similarly to Lemma 5.5.6, if a is a limit point, then $a \in X \backslash S$. These are properties that can be proved fairly easily too. In the case $n = 3$, compressibility means that the limit set can be projected into a two dimensional plane, and the Poincaré-Bendixson theory can be applied. The possibility that the limit set may consist of isolated periodic orbits joined by trajectories spiralling to them as $t \to \pm\infty$ can be ruled out by applying some results of topological dynamics. $\square$

Recently Hirsch [1990] proved that the case of a two dimensional topological cylinder of periodic orbits can be ruled out, i.e. the conditions of Theorem 5.5.9 imply that M is a periodic orbit. This can be proved relatively easily if the system is irreducible. There holds the following:

Theorem 5.5.10 *(H.L.Smith [1986]). Assume that $n = 3$, the connected open set $X \subset \mathbf{R}^3$ is convex, system (5.5.1) is irreducible and cooperative or competitive, and let $M \subset X$ be a compact alpha or omega limit set which contains no equilibrium point; then M is a periodic orbit.*

Sketch of Proof. Since alpha, resp. omega limit sets of competitive systems are omega, resp. alpha limit sets of cooperative systems, we may consider cooperative systems only. By virtue of Theorem 5.5.9 (ii) we have to prove the statement for alpha limit sets only. Thus, suppose that $M = A(z)$ for some $z \in X$ where $A(z)$ is the alpha limit set of the point z. Suppose that $M = A(z)$ is a cylinder of periodic orbits. Thus, each point on this cylinder must be an alpha limit point of the solution $\varphi(t, z)$ with initial value z. Because of Lemma 5.5.7 (cf. also Section 5.6) each periodic orbit has a two dimensional "unstable manifold", a topological cylinder such that the solutions with trajectories in such a cylinder tend away from the periodic orbit, or, rather, these solutions tend towards the periodic orbit as $t \to -\infty$. Taking the union of these "unstable cylinders" corresponding to a continuum of periodic orbits we obtain a "three dimensional set" which contains an open neighbourhood of at least one periodic orbit in its interior. But since each point of each periodic orbit is an alpha limit point, $\varphi(t, z)$ must lie in this neighbourhood for some large modulus negative value of t. As a consequence, $\varphi(t, z)$ lies on one of the "unstable cylinders" of a periodic orbit, and this implies that $\varphi(t, z)$ tends to this particular periodic orbit as $t \to -\infty$. This contradicts to the assumption that the *whole* cylinder M is the alpha limit set. □

We note that the previous Theorem does not rule out the possibility that a cylinder of periodic orbits occurs in the case of a three dimensional irreducible cooperative or competitive system but says that such a cylinder cannot be a limit set.

We have seen that the monotonicity of a flow generated by a competitive or a cooperative system leads directly to existence results of periodic orbits in dimension three. Graphic intuition still applicable in dimension three may help in establishing the existence of periodic orbits even in the case when the flow is not monotone. The *"geometric method"* to be described in the sequel makes use of the fact that the three dimensional space is divided into two by a two dimensional plane. This makes the application of the *"torus principle"* relatively easy. The torus principle says that if a toroidal set of $\mathbf{R}^3$ bounded by a two dimensional torus surface is positively invariant, a cross section of the torus is transversal to all the trajectories, and every trajectory that starts in this cross section at time zero meets it again at some positive time, then the toroidal set contains a periodic orbit. This is a straightforward consequence of Brouwer's Fixed Point Theorem (Corollary A2.9). One of the first applications showing how this principle is to be used in order to establish the existence of a periodic orbit for a three dimensional system is due to Pliss [1966]. Since then several authors have applied the method. We shall present here a result due to Hastings and Murray [1975] concerning the Field-Körös-Noyes model of the *Belousov-Zhabotinskij reaction* which has been in the focus of research ever since Belousov published his observation of temporal oscillations in the concentrations of intermediaries when citric acid was oxidized by acid bromate in the presence of a cerium ion catalyst in 1959. To cut a long story short the Field-Körös-Noyes description of the phenomenon leads to the system

$$\begin{aligned}
\dot{x} &= s(y - xy + x - qx^2), \\
\dot{y} &= (pz - y - xy)/s, \\
\dot{z} &= w(x - z)
\end{aligned} \tag{5.5.8}$$

where s, q, p, w are positive constants determined by the reaction velocities and other chemical characteristics of the reagents, and x, y and z are proportional to the concentrations of H Br O_2, Br$^-$ and Ce(IV), respectively. The chemical data determine the values of q and s; they are

$$q = 8.375 \times 10^{-6}, \qquad s = 77.27, \tag{5.5.9}$$

and we assume that

$$1 \leq p < 2.412. \tag{5.5.10}$$

(This assumption is somewhat more restrictive than that of Hastings and Murray but yields the result in a more apparent way.) In the sequel we shall describe the procedure leading to the existence of a periodic orbit without performing all the calculations in detail.

The system has two equilibria with non-negative coordinates: the origin $(0, 0, 0)$ and the point $C = (x_0, y_0, z_0)$ where

$$z_0 = x_0, \qquad y_0 = px_0/(1 + x_0) = (1 + p - qx_0)/2,$$

$$x_0 = ((1 - p - q) + ((1 - p - q)^2 + 4q(1 + p))^{1/2})/(2q).$$

It is easy to see that the origin is always unstable. At the equilibrium point $C = (x_0, y_0, z_0)$ the linearized system is

$$\begin{pmatrix} x \\ y \\ z \end{pmatrix}^{\bullet} = \begin{pmatrix} s(-y_0 + 1 - 2qx_0) & s(1 - x_0) & 0 \\ -y_0/s & -(1 + x_0)/s & p/s \\ w & 0 & -w \end{pmatrix} \begin{pmatrix} x \\ y \\ z \end{pmatrix}. \tag{5.5.11}$$

The characteristic polynomial is

$$\lambda^3 + a_2\lambda^2 + a_1\lambda + a_0$$

where
$$a_2 = w + x_0(1/s + 3qs/2) + s(p-1)/2 + 1/s. \qquad (5.5.12)$$

It can be shown that $a_2 > 0$, $a_0 > 0$. The Hurwitz condition of stability $a_1 a_2 > a_0$, see (1.4.12), is violated, i.e. the polynomial has at least one root with positive real part if

$$\begin{aligned} 0 \; < \; & w < -(E^2 + p(1-x_0))/(2E) \\ & + \;((E^2 + p(1-x_0))^2 - 4E^2(2qx_0^2 + x_0(q-1) + p))^{1/2}/(2E) \quad (5.5.13) \end{aligned}$$

where $E = sy_0 + (1/s + 2qs)x_0 + 1/s - s$. Condition (5.5.10) ensures that

$$2qx_0^2 + x_0(q-1) + p < 0; \qquad (5.5.14)$$

hence, the set of w's satisfying (5.5.13) is not empty. We choose such a w. Denoting the roots of the characteristic polynomial by $\lambda_1, \lambda_2, \lambda_3$ and taking into account that $\lambda_1 \lambda_2 \lambda_3 = -a_0 < 0$, we obtain that one root is negative: $\lambda_1 < 0$, say, and the remaining two roots have positive real parts: $\lambda_2 = \alpha_2 + i\beta$, $\lambda_3 = \alpha_3 - i\beta$ with $\alpha_2 > 0$, $\alpha_3 > 0$ (if $\beta \neq 0$, then, naturally, $\alpha_2 = \alpha_3$). The expressions for the coordinates of the equilibrium point and inequality (5.5.14) imply that for small positive q (like that in (5.5.9))

$$q < 1 < 2 < x_0 < 1/q, \qquad 1 < z_0 < 1/q,$$

and

$$y_1 < y_0 < y_2 \quad \text{where} \quad y_1 := pq/(1+q), \quad y_2 := p/(2q).$$

It can be shown that the closed rectangular box

$$B := \{(x,y,z) \in \mathbf{R}^3 : 1 \le x \le 1/q, \; y_1 \le y \le y_2, \; 1 \le z \le 1/q\}$$

is positively invariant for system (5.5.8). We divide B into eight boxes B_i ($i = 1, 2, ..., 8$) in the following way

$$\begin{aligned} B_1 : &= \{(x,y,z) : 1 \le x \le x_0, \; y_1 \le y \le y_0, \; 1 \le z \le z_0\}, \\ B_2 : &= \{(x,y,z) : x_0 \le x \le 1/q, \; y_1 \le y \le y_0, \; 1 \le z \le z_0\}, \\ B_3 : &= \{(x,y,z) : 1 \le x \le x_0, \; y_0 \le y \le y_2, \; 1 \le z \le z_0\}, \\ B_4 : &= \{(x,y,z) : x_0 \le x \le 1/q, \; y_0 \le y \le y_2, \; 1 \le z \le z_0\}, \\ B_5 : &= \{(x,y,z) : 1 \le x \le x_0, \; y_1 \le y \le y_0, \; z_0 \le z \le 1/q\}, \\ B_6 : &= \{(x,y,z) : x_0 \le x \le 1/q, \; y_1 \le y \le y_0, \; z_0 \le z \le 1/q\}, \\ B_7 : &= \{(x,y,z) : 1 \le x \le x_0, \; y_0 \le y \le y_2, \; z_0 \le z \le 1/q\}, \\ B_8 : &= \{(x,y,z) : x_0 \le x \le 1/q, \; y_0 \le y \le y_2, \; z_0 \le z \le 1/q\}. \end{aligned}$$

We shall show that with exactly two exceptions trajectories that have a point in B_4 or in B_5 must leave these boxes, and that no trajectory can

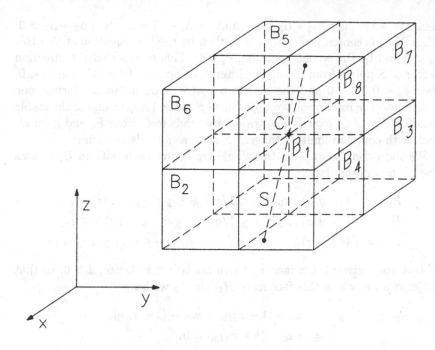

FIGURE 5.5.1. The positively invariant box $B = \bigcup_{i=1}^{8} B_i$ of system (5.5.8) around the equilibrium point $C = (x_0, y_0, z_0)$. The tangent line of the stable manifold of C passes through boxes B_4 and B_5.

enter B_4 and B_5 from B. Further we shall show that any solution whose path has a point in $B \backslash (B_4 \cup B_5)$ has to proceed from box to box in the sequential order

$$B_3 \to B_1 \to B_2 \to B_6 \to B_8 \to B_7 \to B_3 \to \cdots \qquad (5.5.15)$$

(see Figure 5.5.1).

First we note that the stable manifold W_s of the equilibrium C is one dimensional, and its "tangent space" is the straight line S passing through C in the direction of the eigenvector $e = [e_x, e_y, e_z]$ corresponding to the negative eigenvalue λ_1 (see Definition A3.6 et seq.). The coordinates of e satisfy

$$
\begin{aligned}
s(-y_0 + 1 - 2qx_0)e_x &&+ s(1 - x_0)e_y && &= \lambda_1 e_x, \\
-(y_0/s)e_x && -((1 + x_0)/s)e_y && +(p/s)e_z &= \lambda_1 e_y, \quad (5.5.16) \\
we_x && && -we_z &= \lambda_1 e_z.
\end{aligned}
$$

But by (5.5.12) the condition (5.5.10) implies that

$$a_2 = -\lambda_1 - \lambda_2 - \lambda_3 = -\lambda_1 - \alpha_2 - \alpha_3 > w$$

and also

$$a_2 = \lambda_1 - \alpha_2 - \alpha_3 > (1 + x_0)/s.$$

Hence, $-\lambda_1 - w > \alpha_2 + \alpha_3 > 0$ and $-\lambda_1 - (1 + x_0)/s > \alpha_2 + \alpha_3 > 0$. Thus, if e_x is chosen positive, $e_x > 0$, then by the last equation of (5.5.16), $e_z < 0$ and by the second equation, $e_y > 0$. This means that the direction vector of S points from C into the interior of box B_4. (If we choose $e_x < 0$, then $e_y < 0$, $e_z > 0$ and the direction vector opposite to the former one points into the interior of box B_5.) Since S is the tangent line of the stable manifold W_s of C, exactly one trajectory tends to C from B_4 and from B_5 each with common limiting tangent S at C as t tends to infinity.

We show that apart from this single trajectory each path in B_4 crosses one of the internal faces

$$
\begin{aligned}
F_1 &= \{(x, y, z) : x_0 < x \le 1/q, \quad y_0 \le y \le y_2, \quad z = z_0\}, \\
F_2 &= \{(x, y, z) : x_0 \le x \le 1/q, \quad \quad y = y_0, \quad 1 \le z < z_0\}, \\
F_3 &= \{(x, y, z) : \quad \quad x = x_0, \quad \quad y_0 < y \le y_2, 1 \le z \le z_0\}
\end{aligned}
$$

of this box outward. On face F_1 there holds $z < x$; hence, $\dot{z} > 0$, so that trajectories reaching this face leave B_4. On F_2 we have

$$
\begin{aligned}
\dot{y} &= pz - (1 + x)y_0 < pz_0 - (1 + x_0)y_0 \\
&= px_0 - (1 + x_0)y_0 = 0,
\end{aligned}
$$

so that trajectories reaching this face leave B_4. In order to handle face F_3 we show that the right-hand side (divided by s) of the first differential equation in system (5.5.8), $g(x, y) := y - xy + x - qx^2$, is negative if $x_0 \le x \le 1/q$, $y_0 < y \le y_2$ and positive if $1 < x \le x_0$, $y_1 \le y < y_0$. This clearly implies that $\dot{x} < 0$ on F_3. To show this we observe that $g_y'(x, y) = 1 - x < 0$ for $x > 1$. Since $g(x_0, y_0) = 0$, it follows that $g(x_0, y) > 0$ if $y < y_0$ and $g(x_0, y) < 0$ if $y > y_0$. By this we have shown that $\dot{x} < 0$ on F_3 and trajectories that reach this face leave B_4. However, for future use, we prove the rest of the statement about g. We have $g_x'(x_0, y_0) = -y_0 + 1 - 2qx_0$. But $y_0 = px_0/(1 + x_0)$ and $qx_0^2 + qx_0 - x_0 + px_0 = 1 + p$; thus,

$$
-y_0 + 1 - 2qx_0 = -(p + qx_0 + qx_0^2)/(1 + x_0) < 0,
$$

and for xs near enough to x_0 we have $g(x, y_0) \gtrless 0$ according to whether $x \lessgtr x_0$. If one of these inequalities failed for some $x \in (1, 1/q]$, then $g(x, y_0) = 0$ would hold for some $x \in (1, 1/q]$. But the roots of the quadratic equation $g(x, y_0) = 0$ are of opposite sign, and one of them is $x = x_0$. Thus, $g(x, y_0) > 0$ for $x \in (1, x_0)$, and $g(x, y_0) < 0$ for $x \in (x_0, 1/q]$. But for fixed $x > 1$ the function $g(x, \bullet)$ is decreasing, hence,

$$
g(x, y) < 0 \quad \text{for} \quad x_0 < x \le 1/q, \quad y_0 \le y \le y_2,
$$

$$
g(x, y) > 0 \quad \text{for} \quad 1 < x \le x_0, \quad y_1 \le y < y_0.
$$

We have proved that no solution can enter B_4 from B. A similar argument yields the analogous conclusion for B_5.

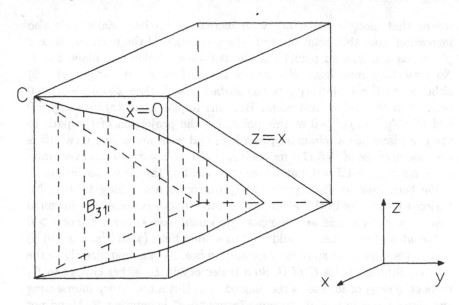

FIGURE 5.5.2. Box B_3 divided up by $\dot{z} = 0$ and $\dot{x} = 0$.

Next consider box B_2 where $z < x$, so $\dot{z} > 0$. Along the face $x = x_0$, $y_1 \leq y \leq y_0$, $1 \leq z \leq z_0$ we see that $\dot{x} = sg(x_0, y) > 0$. On the edge $x = x_0$, $y = y_0$, $1 \leq z \leq z_0$ we have $\dot{y} < 0$. Hence, any trajectory in B_2 tends to the face $z = z_0$ and cannot cross into B_4 or B_1, nor can it tend to C. Thus, a trajectory in B_2 must intersect the plane $z = z_0$ and cross into B_6. Similarly, trajectories in B_7 pass into B_3.

Now we turn to box B_3. The equation $\dot{z} = 0$ or $x = z$ determines a "diagonal plane" that divides B_3 into two halves, an "upper" and a "lower" one. Consider the lower half where $z < x$ or $\dot{z} > 0$. The surface $\dot{x} = 0$ or $g(x, y) = 0$ divides this lower half of box B_3 into two parts, the "left-hand" one B_{31} where $\dot{x} > 0$, $\dot{z} > 0$ and the "right-hand" one B_{32} where $\dot{x} < 0$, $\dot{z} > 0$ (see Figure 5.5.2). The region B_{31} is bounded from the left by the triangle $1 \leq x \leq x_0$, $y = y_0$, $1 \leq z < x$. We shall show that any path in B_{31} must leave it through this triangle and enter box B_1. In B_{31} and on the plane $z = x$, $1 \leq x < x_0$, $y \geq y_0$ we find that

$$\begin{aligned}
\dot{y} &= (1/s)(pz - y - xy) \leq (1/s)(px - y_0 - xy_0) \\
&= (1/s)(x(p - y_0) - y_0) < (1/s)(x_0(p - y_0) - y_0) \\
&= (1/s)(pz_0 - y_0 - x_0 y_0) = 0 \quad\quad (5.5.17)
\end{aligned}$$

where $y_0 = px_0/(1 + x_0) < p$ was used. B_{31} is bounded from above by that part of the plane $z = x$ where $\dot{x} > 0$. Thus, if $z = x$, i.e. $\dot{z} = 0$ and $\dot{x} > 0$, then $\ddot{z} = w\dot{x} > 0$, and as a consequence at a point of this upper boundary to B_{31} we have $dz/dx = 0$ $d^2z/dx^2 = (\ddot{z}\dot{x} - \dot{z}\ddot{x})/\dot{x}^3 > 0$. This means that along a trajectory with increasing x, the variable z is also increasing from this point onward (the projection of the path to the x, z plane has a minimum point) but the path stays below the plane $z = x$. No trajectory may leave B_{31} across the surface $\dot{x} = 0$ (or $g(x, y) = 0$) either since if a trajectory hits the surface $\dot{x} = 0$, then $\ddot{x} = \dot{y}(1 - x) > 0$ because of (5.5.17) at this point. But this implies that $dx/dy = \dot{x}/\dot{y} = 0$ and $d^2x/dy^2 = \ddot{x}/\dot{y}^2 > 0$ at this point, i.e. the projection of the path to the y, x plane has a minimum point here and with *decreasing* y (which is the case because of (5.5.17)) moves inside B_{31}. If $\dot{z} = \dot{x} = 0$ (and naturally $\dot{y} < 0$ again), then $\ddot{z} = 0$, but $\dddot{z} = w\ddot{x} > 0$ yielding the same conclusion.

We turn now to those parts of B_3 which do not belong to B_{31}. No trajectory can stay in the whole future either above or below the diagonal plane $z = x$, for if this were the case, we would have either $\dot{z} < 0$ or $\dot{z} > 0$ in the whole future, i.e. z would tend to a finite limit (with it also x and y) and this is impossible since no trajectory in box B_3 can tend directly to the only equilibrium point C of B. So a trajectory in B_3 either crosses to B_1 through $y = y_0$ or intersects the plane $z = x$. But a trajectory intersecting the plane $z = x$ in box B_3 "coming from above" is entering B_{31} (and not B_{32}). This is so because at the point of intersection of the path with the plane $z = x$ we have $\dot{z} = 0$, and $\dot{z}$ is locally increasing (above the plane it is negative, below positive), i.e. $\ddot{z} = w\dot{x} \geq 0$. If $\dot{x} > 0$, then the path has entered B_{31}. If $\dot{x} = 0$ at the point of intersection, $\ddot{z} = w\ddot{x} = w\dot{y}(1 - x) > 0$ because of (5.5.17). But this means that $\ddot{z}$, i.e. $\dot{x}$ is locally increasing; hence, $\dot{x} > 0$, and we are again in B_{31}. We have proved that every trajectory in B_3 leaves this box crossing into B_1. One may show similarly that every path in B_6 must cross into B_8.

Finally, consider box B_1. If a path gets to B_1 from B_5, then it might cross into B_3, but we have seen that in this case it crosses back to B_1 across the face $y = y_0$ but below the line $\dot{y} = 0$ (whose equation, by the way, is $z = (x + 1)y_0/p$). We may assume, therefore, that at the point of entry into B_1 we have $\dot{y} > 0$, and in B_1 there holds $\dot{x} > 0$. So if the path ever enters the region $\dot{z} \geq 0$ which is characterized in this box also by $z \geq x$, then it remains there until it crosses into B_2. If $\dot{z} < 0$ held on the path and $\dot{y}$ became zero, $\dot{y} = 0$, then $\ddot{y} = (1/s)(p\dot{z} - \dot{x}y) < 0$ would hold, i.e. with increasing t the value of y would decrease, and this means that the path cannot reenter B_3. Since $\dot{x} > 0$ and the path cannot get into B_5 it must cross the plane $z = x$ and then enter B_2. Similarly, trajectories in B_8 must cross into B_7. Thus, we have proved that solutions whose path has a point in $B \setminus (B_4 \cup B_5)$ oscillate in the box B in the sequential order (5.5.15).

Now, we choose the face F that separates box B_6 from B_2, i.e.

$$F = \{(x, y, z) : x_0 \leq x \leq 1/q, \ y_1 \leq y \leq y_0, \ z = z_0\}$$

and consider solutions $\varphi(t, (x, y, z))$ of system (5.5.8) with initial values in this closed bounded set. We have proved that for any $(x, y, z) \in F$ there exists a smallest positive $\tau(x, y, z)$ such that $\varphi(\tau(x, y, z), (x, y, z)) \in F$.

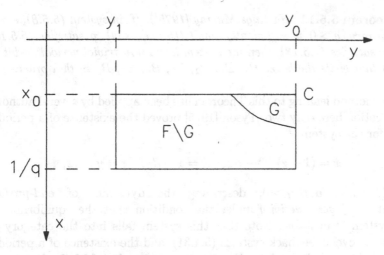

FIGURE 5.5.3. Face F of boxes B_2 and B_6 (projected onto the coordinate plane x, y).

This mapping $\varphi : F \to F$ can be called the *Poincaré map* of the system (cf. Definition 5.2.1), and it is a diffeomorphism by Theorem 5.2.1 (it can be checked, easily, that each trajectory except the point trajectory of the equilibrium intersects F transversally); hence, it is continuous. As a consequence, Brouwer's Fixed Point Theorem (Corollary A2.9) can be applied. The problem is that the equilibrium point $C = (x_0, y_0, z_0)$ belongs to F. If we can prove that the mapping φ has a fixed point apart from C then we have proved the existence of a non-trivial periodic orbit in B. We do this in the following way. We consider the tangent line S of the one-dimensional stable manifold belonging to C and surround it by an elliptical cylinder of small diameter. In a sufficiently small neighbourhood of C all trajectories that cross the surface of this cylinder do it in the outward direction. We have shown that the straight line S, i.e. the axis of the cylinder, lies in the interior of boxes B_4 and B_5 (apart from the point C). We choose the diameter of the cylinder so small that the neighbourhood of C in which trajectories may cross the cylinder only outward should contain the intersection of the solid cylinder with boxes B_3, B_1, B_2, B_6, B_8, B_7. Thus, a trajectory may enter the cylinder only from boxes B_4 and B_5. Denote the intersection of the interior of this solid cylinder with the face F by G (see Figure 5.5.3). Then φ maps $F \backslash G$ into itself, since a path with initial values in $F \backslash G$ cannot enter boxes B_4 and B_5 and, therefore, cannot enter the cylinder. Hence, φ has a fixed point in $F \backslash G$ that is different from C, the latter belonging to G.

Thus, leaving some details of the calculations to the reader we have proved the following:

Theorem 5.5.11 *(Hastings-Murray [1975]). If, in system (5.5.8), $s > 0$ is arbitrary, $q > 0$ is sufficiently small ((5.5.9), say), p satisfies (5.5.10) and w satisfies (5.5.13), then the system has a non-trivial periodic orbit in box B that visits the boxes B_3, B_1, B_2, B_6, B_8, B_7, B_3 in that order.*

The method leading to this Theorem has been applied by several authors. We mention here only two. Tyson [1975] proved the existence of a periodic orbit for the system

$$\dot{x} = (1+z)^{-m} - \alpha x, \quad \dot{y} = x - \beta y, \quad \dot{z} = y - \gamma z,$$

$m \in \{1, 2, ...\}$, $\alpha, \beta, \gamma > 0$, describing the dynamics of *end-product inhibition of gene activity* under the condition that the equilibrium of the system is unstable. Note that this system falls into the category of monotone cyclic feedback systems (5.4.31), and the existence of a periodic orbit could be deduced from the more recent results of Mallet-Paret and H.L. Smith [1990]. Dai [1981] has studied the *predator-prey system with memory*

$$
\begin{aligned}
\dot{N}(t) &= \varepsilon N(t)(1 - N(t)/K - P(t)\alpha/\varepsilon) \\
\dot{P}(t) &= -\gamma P(t) + \beta P(t) \int_{-\infty}^{t} N(\tau)G(t-\tau)d\tau
\end{aligned}
\qquad (5.5.18)
$$

where $\varepsilon, \alpha, \gamma, \beta, K > 0$ and $G(t) = a\exp(-at)$ with $a > 0$. He proved that if $\gamma < K\beta$ and $0 < a < K\beta - \gamma\varepsilon/(K\beta)$, then system (5.5.18) has a non-constant periodic solution. We shall return to system (5.5.18) in Chapter 7.

5.6   Invariant and Integral Manifolds of Periodic Solutions

We give here an informal treatment to the problem of invariant and integral manifolds that can be attached to periodic solutions. The term "invariant manifold" is used in the sense of Appendix 3 and of Section 1.3, i.e. it is a differentiable manifold in the phase space $\mathbf{R}^n$ invariant under the flow considered. We use the term *"integral manifold"* for manifolds in the $n+1$ dimensional space $\mathbf{R} \times \mathbf{R}^{n+1}$ of the "time" $t \in \mathbf{R}$ and $x \in \mathbf{R}^n$, i.e. where the *integral curves*, the graphs of the solutions are drawn; this is in accordance with Definition 1.1.2, but a new formal definition will be given below. First we shall treat invariant and integral manifolds that can be attached to an isolated periodic solution. Then manifolds consisting of orbits, resp. integral curves of periodic solutions will be considered. The most important theorems are presented but, unfortunately, because of lack of space, proofs will not be provided. The proofs of these theorems about

manifolds are lengthy and need a lot of preparation; however, the references
will, naturally, be given.

Consider the autonomous system

$$\dot{x} = f(x) \tag{5.6.1}$$

where $f \in C^k(\mathbf{R}^n, \mathbf{R}^n)$, $k \geq 1$, assume that $p : \mathbf{R} \to \mathbf{R}^n$ is a non-constant
periodic solution with (least positive) period $T > 0$, denote the orbit of p
by $\gamma = \{x \in \mathbf{R}^n : x = p(t), \quad t \in \mathbf{R}\}$ and consider the variational system of
(5.6.1) with respect to p:

$$\dot{y} = f'_x(p(t))y. \tag{5.6.2}$$

Suppose, further, that $n - 1$ characteristic multipliers of (5.6.2) have
modulus different from 1. By Theorem 5.2.3 this ensures that the periodic
solution p is isolated. Now, consider an $n - 1$ dimensional hyperplane L
passing through a point, $p(0)$ say, of γ and transversal to it. If $\Phi(t)$ denotes
the fundamental matrix of (5.6.2) that satisfies $\Phi(0) = I$, the unit matrix,
then $\dot{p}(0) \neq 0$ is the eigenvector corresponding to the simple eigenvalue 1 of
the principal matrix $\Phi(T)$ (see (2.2.5) and Theorem 2.2.2). For L we may
choose the $n - 1$ dimensional eigenspace of $\Phi(T)$ corresponding to the $n - 1$
eigenvalues of $\Phi(T)$ which have modulus different from 1. The Poincaré
map (see Definition 5.2.1) is defined in a neighbourhood of $p(0)$ in L. The
linearization of this map at $p(0)$ has the $n - 1$ multipliers of (5.6.2) different
from 1 as eigenvalues (see Theorem 5.2.2). A decomposition theorem,
analogous to Theorem A3.1, according to which the one parameter family
of maps (A3.14) can be transformed into (A3.23) is valid to this single
map too (see Hartman [1964] p. 239). Assuming that $s \geq 0$ characteristic
multipliers are in modulus less than 1 and $u \geq 0$ are in modulus greater
than 1, obviously $s + u = n - 1$ and $c = 0$ (cf. Assumption (A) in Appendix
3), this implies that the fixed point $p(0)$ of this Poincaré map has an s
dimensional stable and an $u = n - 1 - s$ dimensional unstable manifold.
These stable and unstable manifolds (of the same dimension s, resp. u) can
be considered in every point of the periodic orbit γ, and one may expect
that they can be joined into an $s + 1$ dimensional and an $u + 1$-dimensional
differentiable manifold of $\mathbf{R}^n$, the first containing the orbits that tend to γ
exponentially as $t \to \infty$ and the second containing those orbits that tend
to γ exponentially as $t \to -\infty$. See Figure 5.6.1 where $n = 3$, $s = 1$, $u = 1$.
This is, in fact, the case as has been shown by Kelley [1967] for $k \geq 3$ and
by Hirsch-Pugh-Shub [1977] for $k \geq 1$. Actually Kelley's Theorem is more
general; he shows the exsistence of a *stable*, a *"centre-stable"* a *centre*, a
"centre-unstable" and an *unstable manifold* and the local uniqueness of the
stable and the unstable manifolds under the general assumption that 1 may
be a multiple characteristic multiplier and, in addition, some multipliers
may have modulus 1. We omit here the definitions and refer to Definition
A3.6 which provides the definitions of the stable and unstable manifolds in
the case of an equilibrium point or, rather, to the explanations following
this Definition.

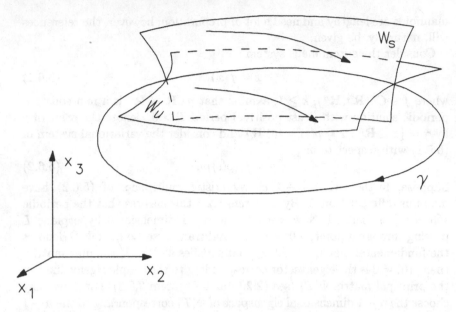

FIGURE 5.6.1. Stable W_s and unstable W_u manifolds of an isolated periodic orbit γ in the phase space $\mathbf{R}^3$; $s = 1$, $u = 1$, dim $W_s = 2$, dim $W_u = 2$, $W_s \cap W_u = \gamma$.

In spite of the fact that the construction described above is a nice generalization for periodic orbits of the corresponding notions related to equilibria it does not yield all the information that could be expected. For instance, if we restrict the system to the stable manifold, then γ is an orbitally asymptotically stable periodic orbit with *asymptotic phase* (see Andronov-Witt Theorem 5.1.2 and Definition 5.1.3). This is the case with the original system if $s = n - 1$ and $u = 0$. The stable manifold which is then simply a neighbourhood of γ contains all the trajectories that tend to γ as $t \to \infty$ but provides no information about the asymptotic phase property.

In order to obtain this information from the representation of the system we consider it in the space $\mathbf{R} \times \mathbf{R}^n$ of t and x and represent the solutions with their integral curves. As we know, the periodic orbit γ is the trajectory of all the solution $p(t - \alpha)$ where $\alpha \in \mathbf{R}$ can be arbitrarily given. The *integral curves* of these solutions form the set

$$M = \{(t, x) \in \mathbf{R} \times \mathbf{R}^n : x = p(t - \alpha), \ \alpha \in \mathbf{R}\}. \qquad (5.6.3)$$

M is a cylinder with axis t, obviously, a two dimensional differentiable manifold that is said to be an "integral manifold" of system (5.6.1). If $n - 1$ characteristic multipliers of the variational system (5.6.2) are in modulus less than 1, then p is orbitally asymptotically stable with asymptotic phase. This means first of all that if (t_0, x^0) is sufficiently near M, then the solution $\varphi(t, t_0, x^0)$ of system (5.6.1) that satisfies $\varphi(t_0, t_0, x^0) = x^0$ stays near M

for $t > t_0$ and $\text{dist}\,((t, \varphi(t, t_0, x^0)), M) \to 0$ as $t \to \infty$. But this is not all. The asymptotic phase property ensures that there is an integral curve in M such that φ tends to this curve as $t \to \infty$, i.e. there exists an $\alpha(t_0, x^0)$ such that

$$\lim_{t \to \infty} |\varphi(t, t_0, x^0) - p(t - \alpha(t_0, x^0))| = 0.$$

We have to use now the initial time t_0 in the identification of the solutions of an autonomous system because we deal with their integral curves.

The integral curves of system (5.6.1) could be considered the trajectories of an autonomous system of dimension $n + 1$ obtained from (5.6.1) by "attaching the time t as a new coordinate", i.e. by extending the system by the differential equation $\dot{x}_{n+1} = 1$. Then the system

$$\dot{x}_i = f_i(x_1, ..., x_n) \quad (i = 1, 2, ..., n), \quad \dot{x}_{n+1} = 1 \qquad (5.6.4)$$

is an autonomous system in $\mathbf{R}^{n+1}$. Clearly, $(p_1(t), ..., p_n(t), t)$ is a D-periodic solution of (5.6.4) (see Section 5.3) that is periodic in the vector variable $(x_1, ..., x_n, x_{n+1})$ with vector period $(0, ..., 0, a_{n+1}T)$ where $a_{n+1} \in \mathbf{R}$ is arbitrary, since the right-hand side of (5.6.4) does not depend on x_{n+1}. However, in this case our D-periodic solution cannot be considered as isolated anymore since it is a member of the one parameter family $(p_1(t), ..., p_n(t), t + b)$, $b \in \mathbf{R}$ arbitrary. The trajectories of these solutions form the cylinder M given by (5.6.3) with the difference that now $t + b$ is not the first but the $n + 1$-st coordinate. See Definition 5.3.4. Theorem 5.3.9 implies then that 1 cannot be a simple characteristic multiplier of the variational system of (5.6.4) with respect to the D-periodic solution in question. Indeed, this variational system is

$$\dot{\tilde{y}} = \begin{pmatrix} f'_x(p(t)) & 0 \\ 0 & 0 \end{pmatrix} \tilde{y}$$

where $\tilde{y} \in \mathbf{R}^{n+1}$. If $\Phi(T)$ is the principal matrix of (5.6.2), then the $(n + 1) \times (n + 1)$ matrix

$$\begin{pmatrix} \Phi(T) & 0 \\ 0 & 1 \end{pmatrix}$$

is the principal matrix of the extended variational system. Since 1 is, by assumption, a simple characteristic multiplier of (5.6.2) it is a double eigenvalue of the latter matrix. One could continue the study of M along these lines, but we turn now to the general study of "integral manifolds" consisting of integral curves belonging to periodic solutions.

Definition 5.6.1. We say that the differentiable manifold $M \subset \mathbf{R} \times \mathbf{R}^n$ is an *integral manifold* of system (5.6.1) with respect to a solution $\psi : \mathbf{R} \to \mathbf{R}^n$ if $(t, \psi(t)) \in M$ for all $t \in \mathbf{R}$, and if the integral curve $\Gamma := \{(t, x) \in \mathbf{R} \times \mathbf{R}^{n+1} : x = \psi(t), \ t \in \mathbf{R}\}$ has a neighbourhood U_Γ such that if $(t_0, x^0) \in U_\Gamma \cap M$, then for the integral curve of the solution with initial values (t_0, x^0) there holds $(t, \varphi(t, t_0, x^0)) \in M$ as long as $(t, \varphi(t, t_0, x^0)) \in U_\Gamma$.

The manifold defined by (5.6.3), clearly, satisfies all the requirements. We are going to describe how the solutions of a system behave near a manifold consisting of integral curves belonging to periodic solutions. We shall present here two theorems due to Aulbach [1981]. Other important contributions to the theory are Hale-Stokes [1960], Hale [1969], Mitropolskij-Lykova [1973] and Pliss [1977]. The results to be presented are analogous to those concerning manifolds of equilibria (see Theorem A3.5) but now the situation is more complicated.

We consider system (5.6.1) with the assumption that $k = 2$, i.e. $f \in C^2(\mathbf{R}^n, \mathbf{R}^n)$ and denote, as before, the solution that assumes x^0 at t_0 by $\varphi(t, t_0, x^0)$. Suppose that

(a)   system (5.6.1) admits an m parameter family of periodic solutions

$$\psi(t, b, \alpha) := p(\omega(b)t + \alpha, \beta) \qquad (5.6.5)$$

where $p(\tau + 2\pi, b) = p(\tau, b)$ for all $\tau \in \mathbf{R}$ and $b \in V$ an open set or a single point in $\mathbf{R}^{m-1}$, $\alpha \in \mathbf{R}$, $\omega(b) > 0$ for $b \in V$, ω and p are in C^2 and p is three times continuously differentiable with respect to τ (if V is a single point, then $m = 1$ and (5.6.5) represents a single closed orbit in $\mathbf{R}^n$ and a two dimensional cylinder (actually, M of (5.6.3)) in $\mathbf{R} \times \mathbf{R}^n$);

(b)   for all $\tau \in \mathbf{R}$ and $b \in V$

$$\text{rank } [p'_\tau(\tau, b), p'_b(\tau, b)] = m;$$

(c)   for all $b \in V$ the variational system

$$\dot{y} = f'_x(\psi(t, b, \alpha))y \qquad (5.6.6)$$

has s characteristic multipliers in modulus less than 1, u characteristic multipliers in modulus greater than 1, and $s + u + m = n$.

We note that conditions (a) and (b) ensure that

$$M := \{(t, x) \in \mathbf{R} \times \mathbf{R}^n : x = \psi(t, b, \alpha) \ t \in \mathbf{R}, \ b \in V, \ \alpha \in \mathbf{R}\} \qquad (5.6.7)$$

is an $m + 1$ dimensional C^2 manifold of the space $\mathbf{R} \times \mathbf{R}^n$, and that it is an integral manifold of (5.6.1). The vector functions ψ'_{b_i} $(i = 1, 2, ..., m-1)$ along with $\dot{\psi}$ are linearly independent solutions of the variational system (5.6.6). Denote the fundamental matrix of (5.6.6) that assumes I at $t = 0$ by $\Phi(t)$ and the period of ψ by $T(b) = 2\pi/\omega(b)$. The derivative of ψ

with respect to b_i is not necessarily periodic. Differentiating the identity $\psi(t + T(b), b, \alpha) \equiv \psi(t, b)$ by b_i we obtain

$$\dot{\psi}(t + T(b), b, \alpha)T'_{b_i}(b) + \psi'_{b_i}(t + T(b), b, \alpha) \equiv \psi'_{b_i}(t, b, \alpha),$$

hence, substituting $t = 0$

$$\dot{\psi}(T(b), b, \alpha)T'_{b_i}(b) + \psi'_{b_i}(T(b), b, \alpha) = \psi'_{b_i}(0, b, \alpha) \quad (i = 1, 2, ..., m - 1).$$

Applying the formulae $\dot{\psi}(T(b), b, \alpha) = \Phi(T(b))\dot{\psi}(0, b, \alpha)$, $\psi'_{b_i}(T(b), b) = \Phi(T(b))\psi'_b(0, b)$ we get

$$\Phi(T(b))(\dot{\psi}(0, b, \alpha)T'_{b_i}(b) + \psi'_{b_i}(0, b, \alpha)) = \psi'_{b_i}(0, b, \alpha)$$

or, taking into account that $\dot{\psi}(T(b), b, \alpha) = \dot{\psi}(0, b, \alpha)$,

$$\Phi(T(b))\psi'_{b_i}(0, b, \alpha) = \psi'_{b_i}(0, b, \alpha) - \dot{\psi}(0, b, \alpha)T'_{b_i}(b)$$

$$(i = 1, 2, ..., m - 1),$$

and this along with the identity $\Phi(T(b))\dot{\psi}(0, b, \psi) = \dot{\psi}(0, b, \alpha)$ means that *1 is an m-tuple characteristic multiplier.* (Cf. Corollary 2.2.4 but observe that now system (5.6.6) does not necessarily have m independent periodic solutions. In the special case when the m parameter family of periodic solutions is isochronous, i.e. $T'_{b_i} = 0$ $(i = 1, 2, ..., m - 1)$, then the variational system has, indeed, m linearly independent periodic solutions.)

Besides the integral manifold $M \subset \mathbf{R} \times \mathbf{R}^n$ given by (5.6.7) we shall consider also the corresponding invariant manifold $M_0 \subset \mathbf{R}^n$ in the phase space consisting of the *orbits* of the periodic solutions belonging to the family

$$M_0 := \{x \in \mathbf{R}^n : x = \psi(t, b, \alpha), \ t \in \mathbf{R}, \ b \in V, \ \alpha \in \mathbf{R}\}. \tag{5.6.8}$$

We note that because of the particular dependence of the family on the parameter α (the "initial phase"), we do not need t in the definition of M_0, i.e.

$$M_0 = \{x \in \mathbf{R}^n : x = \psi(0, b, \alpha), \ b \in V, \ \alpha \in \mathbf{R}\}. \tag{5.6.9}$$

Fusing Aulbach's two theorems into one, the properties of the solutions in the neighbourhood of M, resp. M_0 are characterised by the following:

Theorem 5.6.1 *(Aulbach [1981]). Suppose that in (5.6.1) $f \in C^2$ and conditions (a), (b) and (c) hold; then*

(i)   *there exist two functions*

$$w_s : \mathbf{R} \times \mathbf{R}^n \times V \times \mathbf{R} \to \mathbf{R}^{u+m}$$

and

$$w_u : \mathbf{R} \times \mathbf{R}^n \times V \times \mathbf{R} \to \mathbf{R}^{s+m}$$

such that $w_s, w_u \in C^0$, are continuously differentiable with respect to $t \in \mathbf{R}$; for all $b \in V$ and $\alpha \in \mathbf{R}$

$$w_s(t, \psi(t, b, \alpha), b, \alpha) \equiv 0, \qquad w_u(t, \psi(t, b, \alpha), b, \alpha) \equiv 0;$$

for any $b^0 \in V$, $\alpha_0 \in \mathbf{R}$ there exist $\beta, \gamma, \delta > 0$ such that if for $t_0 \in \mathbf{R}$, $x^0 \in \mathbf{R}^n$ there hold $w_s(t_0, x^0, b^0, \alpha_0) = 0$, resp. $w_u(t_0, x^0, b^0, \alpha_0) = 0$, and $|x^0 - \psi(t_0, b^0, \psi_0)| < \delta$, then the solution $\varphi(t, t_0, x^0)$ is defined for all $t \geq t_0$, resp. $t \leq t_0$, and the relations

$$w_s(t, \varphi(t, t_0, x^0) b^0, \alpha_0) = 0, \tag{5.6.10}$$

$$|\varphi(t, t_0, x^0) - \psi(t, b^0, \alpha_0)| \leq \beta \exp(-\gamma(t - t_0)) \tag{5.6.11}$$

hold for all $t \geq t_0$, resp.

$$w_u(t, \varphi(t, t_0, x^0), b^0, \alpha_0) = 0,$$

$$|\varphi(t, t_0, x^0) - \psi(t, b^0, \alpha_0)| \leq \beta \exp(\gamma(t - t_0))$$

hold for all $t \leq t_0$;

(ii) if $\varphi(t)$ is a solution whose omega limit set, resp. alpha limit set is contained in the subset $M_0^\star$ of M_0 defined by

$$M_0^\star = \{x \in M_0 : b \in V^\star \text{ a compact subset of } V\},$$

then there exist $b^0 \in V$, $\alpha_0 \in \mathbf{R}$ such that

$$|\varphi(t) - \psi(t, b^0, \alpha_0)| \to 0 \text{ exponentially}$$

as $t \to \infty$, resp. $t \to -\infty$.

In the interpretation of this Theorem, in order to avoid confusion we shall treat only the statements concerning w_s, i.e. the behaviour as t tends to *plus* infinity. The interpretation of the respective statements is analogous. Clearly, for each periodic solution $\psi(t, b^0, \alpha_0)$ of the family, the set $W_s := \{(t, x) \in \mathbf{R} \times \mathbf{R}^n : w_s(t, x, b^0, \alpha_0) = 0\}$ plays the role of the "stable integral manifold", and the set $W_{s0} := \{x \in \mathbf{R}^n : w_s(t, x, b^0, \alpha_0) = 0\}$ plays the role of the "stable invariant manifold". The problem is that they are not necessarily *differentiable* manifolds. Nevertheless, we may imagine

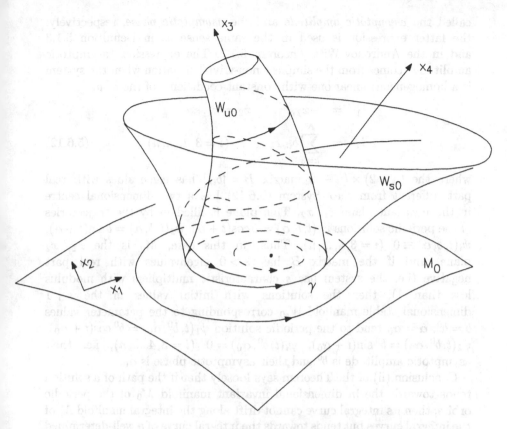

FIGURE 5.6.2. Phase picture of a four dimensional autonomous system which has a two parameter family of periodic solutions (the one parameter family of the corresponding periodic orbits forms the two dimensional manifold M_0 imagined as having the x_1, x_2 plane as tangent plane); it is assumed that the system has a multiplier less than 1 and a multiplier greater than 1; W_{s0} and W_{u0} are the stable and the unstable manifolds, respectively, of a periodic solution of the family whose path is γ.

them, normally, as $s + 2$, resp. $s + 1$ dimensional surfaces containing the integral curve, resp. the path of the periodic solution and containing the integral curves, resp. paths of those solutions that tend to our periodic solution $\psi(t, b^0, \alpha_0)$ exponentially as $t \to \infty$, i.e. for which (5.6.11) holds (see Figure 5.6.2). For such a solution $\varphi(t, t_0, x^0)$, the values b^0 and α_0 are called the *asymptotic amplitude* and the *asymptotic phase*, respectively; the latter expression is used in the same sense as in Definition 5.1.3 and in the Andronov-Witt Theorem 5.1.2. The expression "asymptotic amplitude" comes from the simplest non-trivial situation when the system is a homogeneous linear one with constant coefficients of the form

$$\dot{x}_1 = -x_2, \qquad \dot{x}_2 = x_1,$$

$$\dot{x}_i = \sum_{k=3}^{N} b_{ik} x_k \qquad (i = 3, 4, ..., n) \qquad (5.6.12)$$

where the $(n-2) \times (n-2)$ matrix $B = [b_{ik}]$ has eigenvalues with real part different from zero. System (5.6.12) has a two dimensional centre in the invariant plane x_1, x_2. This plane is filled in by the trajectories of the periodic solutions $\psi_1(t, b, \alpha) = b \cos(t + \alpha)$, $\psi_2(t, b, \alpha) = b \sin(t + \alpha)$, $\psi_i(t, b, \alpha) \equiv 0$ $(i = 3, 4, ..., n)$. Thus, in this case, M_0 is the x_1, x_2 plane, and if the matrix B has $s > 0$ eigenvalues with real part negative (i.e. the system has s characteristic multipliers with modulus less than 1), then the solutions with initial values in the $s+1$ dimensional stable manifold W_{s0} corresponding to the parameter values $b = b^0$, $\alpha = \alpha_0$ tend to the periodic solution $\psi_1(t, b^0, \alpha_0) = b^0 \cos(t + \alpha_0)$, $\psi_2(t, b^0, \alpha_0) = b^0 \sin(t + \alpha_0)$, $\psi_i(t, b^0, \alpha_0) = 0$ $(i = 3, 4, ..., n)$, i.e. their asymptotic amplitude is b^0 and their asymptotic phase is α_0.

Conclusion (ii) of the Theorem says loosely that if the path of a solution tends towards the m dimensional invariant manifold M_0 of the periodic orbits, then its integral curve cannot drift along the integral manifold M of the integral curves but tends towards the integral curve of a well-determined periodic solution (cf. the corresponding property of an invariant manifold consisting of equilibria, Theorem A3.4-A3.5).

Putting things together concisely we see that if there is an m parameter family of periodic solutions, each periodic solution is "as hyperbolic as it can be", i.e. the variational system with respect to such a periodic solution has 1 as an m-tuple multiplier, s multipliers have modulus less than 1, u multipliers have modulus greater than 1 and $m + s + u = n$ (no further multiplier exists with modulus 1), then each periodic solution has an $s+1$ dimensional "stable manifold" W_{s0} a $u+1$ dimensional unstable manifold W_{u0} and a "centre manifold" of dimension m which is the manifold M_0 of the periodic orbits.

We conclude this Section by the implication of the previous Theorem in case $u = 0$, i.e. in the case when the manifold M_0 is attracting.

Theorem 5.6.2 *(Aulbach [1981]). Under the assumptions of Theorem 5.6.1, if $u = 0$, i.e. $s = n - m$ characteristic multipliers of system (5.6.6) are in modulus less than 1 for all $b \in V$, then*

(i)   *there exists an open neighbourhood $U \subset \mathbf{R}^n$ of the invariant manifold M_0 such that for every $x^0 \in U$ there exists a $b^0 = b(x^0) \in V$,*

an $\alpha_0 = \alpha(x^0) \in \mathbf{R}$, *constants* $\beta = \beta(x^0) > 0$ *and* $\gamma = \gamma(x^0) > 0$
satisfying

$$|\varphi(t, 0, x^0) - \psi(t, b^0, \alpha_0)| \le \beta(x^0) \exp(-\gamma(x^0)t)$$

for all $t \ge 0$;

(ii) *for any compact* $V^* \subset V$ *there exist an open neighbourhood* U^* *of*
$M_0^* = \{x \in M_0 : b \in V^*\}$ *and constants* $\beta^* > 0$, $\gamma^* > 0$ *such that for*
all $x^0 \in U^*$ *and* $t \ge 0$ *we have*

$$|\varphi(t, 0, x^0) - \psi(t, b^0, \alpha_0)| < \beta^* \exp(-\gamma^* t).$$

This Theorem says, clearly, that the invariant manifold is attracting, and that if we restrict the amplitudes of the periodic solutions forming the manifold M_0 to a compact set, then the attraction is uniform in the initial value x^0 in a neighbourhood.

5.7 Problems

Problem 5.7.1. Show that for the isolated periodic solution of Example 5.2.1 all three characteristic multipliers are equal to 1.

Problem 5.7.2. Show that if the dimension of system (5.3.8) periodic in the vector variable with *least* vector period $a \ne 0$ (meaning that no non-zero vector parallel to a and in modulus smaller is a vector period) is two, then it cannot have a D-periodic solution with coefficient vector ma ($m \ge 2$ an integer), unless the solution is a linear function of t. (Hint: use the fact that trajectories of an autonomous system cannot intersect each other.)

Problem 5.7.3. Show that if $\varphi : \mathbf{R} \to \mathbf{R}$ is a scalar strictly monotonous continuous D-periodic function with period $T > 0$ and coefficient $a \ne 0$, then its inverse function φ^{-1} is also a D-periodic function with period $|aT|$ and coefficient $1/a$.

Problem 5.7.4. Consider the cylinder whose equation in the Cartesian orthogonal coordinate system x, y, z is $x^2 + y^2 = 1$. Investigate the stability of a closed geodesic, i.e. a parallel circle on this surface on the lines of Example 5.3.2. A parametrization of the cylinder that contains a parallel circle in the interior of a coordinate neighbourhood is

$$x = y^2/(y^{1^2} + y^{2^2})^{1/2}, \quad y = y^1/(y^{1^2} + y^{2^2})^{1/2}, \quad z = \ln(y^{1^2} + y^{2^2})^{1/2}.$$

Show that the normal form of the differential equation of the geodesics is

$$\begin{aligned}
\dot{x}_1 &= x_3, \qquad \dot{x}_2 = x_4, \\
\dot{x}_3 &= (x_1 x_3^2 + 2x_2 x_3 x_4 - x_1 x_4^2)/(x_1^2 + x_2^2), \\
\dot{x}_4 &= (-x_2 x_3^2 + 2x_1 x_3 x_4 + x_2 x_4^2)/(x_1^2 + x_2^2)
\end{aligned} \qquad (5.7.1)$$

where $x_1 = y^1$, $x_2 = y^2$, $x_3 = \dot{y}^1$, $x_4 = \dot{y}^2$, and that the equation of the invariant manifold corresponding to (5.3.20) is $x_3^2 + x_4^2 - x_1^2 - x_2^2 = 0$. Show that $p(t) = [\cos t, \sin t, \sin t, \cos t]$ is a solution of system (5.7.1). Determine the restriction of system (5.7.1) to the given invariant manifold and the solution of the restricted system corresponding to p. Determine the characteristic multipliers of the variational system with respect to this solution. (Hint: a possible parametrization of the invariant manifold is $x_1 = u_1 \cosh u_3$, $x_2 = u_2 \cosh u_3$, $x_3 = u_1 \sinh u_3 + u_2$, $x_4 = u_2 \sinh u_3 - u_1$.)

Problem 5.7.5. Let $\varphi : [0, \infty) \to \mathbf{R}^n$ be a bounded solution of system $\dot{x} = f(x)$ where $f \in C^1(\mathbf{R}^n, \mathbf{R}^n)$. Prove that if $|\varphi(t+h) - \varphi(t)| \leq c \exp(-\lambda t)$ for some positive constants h, c, λ and for all $t \in [0, \infty)$, then the omega limit set of φ is either a single periodic orbit or a single equilibrium point. (Hint: show that $\varphi(t+kh)$ $(k = 1, 2, ...)$ is a uniformly convergent functional sequence in $t \in [0, \infty)$ whose limit is a periodic solution with period $h > 0$.)

Problem 5.7.6. Show that the system $\dot{x} = xz - y$, $\dot{y} = yz + x$, $\dot{z} = 1 - x^2 - y^2$ has no equilibrium but has a (non-trivial) periodic orbit.

Problem 5.7.7. Let $D \subset \mathbf{R}^n$ be an open bounded set "star shaped" from the origin (the last condition meaning that each ray from the origin intersects the boundary ∂D in exactly one point), and consider the continuous mapping $f : \overline{D} \to \overline{D}$; prove that if $f(x) \neq x$ for $x \in \partial D$, then the topological degree of $f - I$ where I is the identity is $(-1)^n$, i.e. $d[f - I, D, 0] = (-1)^n$. (Hint: since D is star shaped for $x \in \partial D$, the vector $f(x) - x$ is never of opposite direction to $-x$; apply the Poincaré-Bohl Theorem (Corollary A2.6) and calculate the degree of $-I$.)

Problem 5.7.8. Let $f, g \in C^1(\mathbf{R} \times X, \mathbf{R}^n)$ where $X \subset \mathbf{R}^n$ is a convex, open and connected set, $f(t, x) \geq g(t, x)$ for $(t, x) \in \mathbf{R} \times X$, the functions g_i non-decreasing in x_k $(k = 1, 2, ..., n;$ $i = 1, 2, ..., n;$ $k \neq i)$, and consider the systems $\dot{x} = f(t, x)$, $\dot{x} = g(t, x)$. Prove that if $t_0 < t_1$, and $\psi : [t_0, t_1] \to X$ and $\varphi : [t_0, t_1] \to X$ are solutions of the first and the second system, respectively, and $\psi(t_0) > \varphi(t_0)$, then $\psi(t) > \varphi(t)$ for $t \in [t_0, t_1]$. (Hint: cf. the first part of the proof of Theorem 5.5.1.)

Problem 5.7.9. Consider the four dimensional linear system transformed to canonical form:

$$\dot{x}_1 = -x_2, \quad \dot{x}_2 = x_1, \quad \dot{x}_3 = -x_3, \quad \dot{x}_4 = x_4.$$

Determine the integral manifold M and the invariant manifold M_0 of the periodic solutions, the functions w_s and w_u provided by Theorem 5.6.1 and the stable and the unstable invariant manifolds W_{s0} and W_{u0} of each periodic solution.

6
Perturbations

The "perturbation method" is, probably, the most powerful and most widely used method for the determination of the existence and stability of periodic solutions. It goes back to Poincaré [1899] and is often called *"Poincaré's method"*. The essence of the method can be explained easily. A one parameter family of systems of differential equations is considered and it is assumed that for a certain value of the parameter μ, say, for $\mu = 0$ the system has a periodic solution. Under some non-criticality assumptions it can be proved that for sufficiently small $|\mu|$ the system has a periodic solution too, close to the periodic solution of the $\mu = 0$ case. One may show also that under some non-criticality conditions the stability of the "unperturbed ($\mu = 0$) periodic solution" is inherited by the perturbed one. If the family depends analytically on the parameter μ, then for sufficiently small $|\mu|$ the "perturbed periodic solution" can be developed into a power series with respect to μ. The coefficients of this series can be determined successively by solving systems of linear differential equations. It is to be noted that the perturbation method is applied in many situations concerning solutions of algebraic equations, difference equations, etc.; however, in this book its application to the existence and stability problem of periodic solutions of ODEs is treated only.

In Section 1 periodically perturbed periodic systems will be treated. In Section 2 autonomous systems having an isolated periodic solution will be considered and the existence of periodic solutions will be proved under "controllably periodic perturbations". The existence of perturbed periodic solutions of autonomous systems under autonomous perturbations will follow as a special case. The stability problem of perturbed periodic solutions of autonomous systems will be treated in Section 3. As an application of the previous results, Van der Pol's equation under controllably periodic perturbations will be considered in Section 4. In Sections 5 and 6 the important "method of averaging" and the basic theory of "singular perturbations" (when the *derivative* of the unknown function is multiplied by the "small parameter") will be treated, respectively. Finally, in Section 7 it will be shown how attractive sets arise as a result of perturbations of periodic solutions.

6.1   Periodic Perturbations of Periodic Systems

In this Section we shall treat the one parameter family of differential equations

$$\dot{x} = f(t,x) + \mu g(t,x,\mu) \tag{6.1.1}$$

where $f, f_x' \in C^0(\mathbf{R} \times X)$, $X \subset \mathbf{R}^n$ an open and connected domain, $g, g_x', g_\mu' \in C^0(\mathbf{R} \times X \times I_\mu)$ where I_μ is an open interval containing $\mu = 0$, and f and g are periodic in the variable t with (not necessarily the least positive) period $T > 0$. The system

$$\dot{x} = f(t,x) \tag{6.1.2}$$

is called the *unperturbed equation*, whereas (6.1.1) is the perturbed one, and it will be assumed that (6.1.2) has a periodic solution $p : \mathbf{R} \to X$ with period T; this solution will be called the *unperturbed periodic solution*.

System (6.1.1) is a special case of system

$$\dot{x} = F(t,x,\mu) \tag{6.1.3}$$

where $F, F_x' F_\mu' \in C^0(\mathbf{R} \times X \times I_\mu)$ and $F(t+T,x,\mu) = F(t,x,\mu)$ for all $(t,x,\mu) \in \mathbf{R} \times X \times I_\mu$; however, if also $F_{\mu\mu}'' \in C^0$, then (6.1.3) can be written in the form (6.1.1) in an obvious way.

It is to be observed that results concerning (6.1.1), resp. (6.1.3) cover also *the case when the period of the system depends on the parameter*. Indeed, suppose that the conditions imposed upon (6.1.3) hold with the difference that

$$F(t+T(\mu),x,\mu) = F(t,x,\mu)$$

for all $(t,x,\mu) \in \mathbf{R} \times X \times I_\mu$ where the function $T : I_\mu \to \mathbf{R}_+$ is strictly positive and $T \in C^2$. If time is rescaled by the transformation $t = sT(\mu)$, then (6.1.3) assumes the form

$$dx/ds = G(s,x,\mu),$$

and here the right-hand side $G(s,x,\mu) := T(\mu)F(sT(\mu),x,\mu)$ is periodic with period 1 for all $\mu \in I_\mu$:

$$G(s+1,x,\mu) = G(s,x,\mu).$$

In order to formulate the classical Theorem concerning the existence of perturbed periodic solutions, we need also the variational system of the unperturbed equation (6.1.2) with respect to its periodic solution p:

$$\dot{y} = f'_x(t, p(t))y, \tag{6.1.4}$$

(cf. Coddington-Levinson [1955] p. 348 or Hartman [1964] p. 415).

Theorem 6.1.1. *If the variational system (6.1.4) does not have non-trivial T-periodic solutions or, equivalently, 1 is not a characteristic multiplier, then $\mu = 0$ has an open neighbourhood $U_\mu \subset I_\mu$ such that for each $\mu \in U_\mu$ the perturbed system (6.1.1) has one and only one periodic solution $q(t, \mu)$ with period T with the properties that $q \in C^1(\mathbf{R} \times U_\mu)$ and $q(t, 0) = p(t)$.*

Note that by Theorem 4.2.7 the assumption of this Theorem implies that the periodic solution p is isolated. Note also that the conditions do not rule out the case when $p(t) \equiv$ constant. In this case the perturbed periodic solution $q(t, \mu)$ may be (but is not necessarily) an equilibrium too. Observe, further, that if p is a non-constant periodic solution, then the condition of the Theorem *cannot hold in case the unperturbed system (6.1.2) is autonomous*, since in this case $\dot{p}$ is a non-trivial periodic solution of the variational system (6.1.4) (see (1.3.7)).

Proof. The essence of the proof is the application of the Implicit Function Theorem to the condition of periodicity. Denote the solution of (6.1.1) that assumes the initial value x^0 at $t = 0$ by $\varphi(t, x^0, \mu)$. The solution p is defined on $\mathbf{R}$; because of the continuous dependence of the solutions on the initial values and on the parameters (see Theorem 1.1.3), if x^0 is close enough to $p(0)$ and $|\mu|$ is sufficiently small, then the solution $\varphi(t, x^0, \mu)$ is defined on the interval $[0, T]$, say. φ is T-periodic if and only if

$$\varphi(T, x^0, \mu) - x^0 = 0 \tag{6.1.5}$$

(see Lemma 2.2.1). Since $\varphi(t, p(0), 0) \equiv p(t)$, (6.1.5) is satisfied by $x^0 = p(0)$, $\mu = 0$. By the Implicit Function Theorem if the Jacobian of the right-hand side of (6.1.5) is non-zero at $x^0 = p(0)$, $\mu = 0$, then there is one and only one function $\xi : U_\mu \to X$, $U_\mu \subset I_\mu$ a neighbourhood of $\mu = 0$, such that $\xi \in C^1$, $\xi(0) = p(0)$ and $\varphi(T, \xi(\mu), \mu) - \xi(\mu) \equiv 0$. But this means that to each $\mu \in U_\mu$ there belongs a unique initial value $\xi(\mu)$ such that the solution $q(t, \mu) := \varphi(t, \xi(\mu), \mu)$ of system (6.1.1) is T-periodic. Now, the Jacobian of the right-hand side of (6.1.5) with respect to x^0 at $x^0 = p(0)$, $\mu = 0$ is $\det[\varphi'_{x^0}(T, p(0), 0) - I]$ where I is the unit matrix. But $\varphi'_{x^0}(T, p(0), 0)$ is the principal matrix of the variational system (6.1.4) (see (2.2.5), (1.1.5) et seq.), and by assumption 1 is *not* one of its eigenvalues; thus, the Jacobian is not zero. $\square$

Corollary 6.1.2. *If, besides the conditions imposed above, in the variables x, μ, the functions f and g are in the C^k class, $k \geq 2$ (resp. analytic), and the conditions of the Theorem hold, then the function q representing the family of T-periodic solutions of the family of systems (6.1.1) is also in the C^k class (resp. analytic) in $\mu \in U_\mu$ where U_μ is an open neighbourhood of $\mu = 0$.*

In the analytic case *Poincaré's method* can be applied for the determination of the power series expansion (with respect to the parameter μ) of the function q. Poincaré's method has a vast literature; see Poincaré [1899], Coddington-Levinson [1955], p. 350 and Proskuryakov [1977]. The essence of the method is that the coefficients of the power series expansion can be determined by successively solving systems of inhomogeneous linear differential equations with periodic coefficients whose coefficient matrices are the same at each step. Assume that f and g are analytic and the conditions of Theorem 6.1.1 hold; then by the Corollary, the family of periodic solutions can be developed into a power series

$$q(t, \mu) = \sum_{k=0}^{\infty} q^k(t)\mu^k \tag{6.1.6}$$

convergent for $t \in \mathbf{R}$ and $\mu \in U_\mu$ with T-periodic coefficients. This power series is substituted into system (6.1.1) and the right hand-side is developed into a power series with respect to μ and the coefficients of equal powers of μ are made equal:

$$
\begin{aligned}
\dot{q}(t, \mu) &= \sum_{k=0}^{\infty} \dot{q}^k(t)\mu^k = f(t, q(t, \mu)) + \mu g(t, q(t, \mu), \mu) \\
&= f(t, p(t)) + (f'_x(t, p(t))q'_\mu(t, 0) + g(t, p(t), 0))\mu \\
&\quad + (1/2!)(f''_{xx}(t, p(t))q'_\mu(t, 0)q'_\mu(t, 0) + f'_x(t, p(t))q''_{\mu\mu}(t, 0) \\
&\quad + g'_x(t, p(t), 0)q'_\mu(t, 0) + g'_\mu(t, p(t), 0))\mu^2 + \cdots \\
&= f(t, p(t)) + (f'_x(t, p(t))q^1(t) + g(t, p(t), 0))\mu \\
&\quad + (1/2!)(f''_{xx}(t, p(t))q^1(t)q^1(t) + 2f'_x(t, p(t))q^2(t) \\
&\quad + g'_x(t, p(t), 0)q^1(t) + g'_\mu(t, p(t), 0))\mu^2 + \cdots
\end{aligned}
$$

where $f''_{xx}(t, p(t))q^1(t)q^1(t)$ is a shorthand notation for the vector whose i-th coordinate is $\sum_{j,l=1}^{n} f''_{ix_j x_l}(t, p(t))q_j^1(t)q_l^1(t)$ $(i = 1, 2, ..., n)$. Since $q^0(t) = q(t, 0) = p(t)$, equating equal powers of μ we obtain the following sequence of systems:

$$
\begin{aligned}
\dot{p}(t) &= f(t, p(t)), \\
\dot{q}^1(t) &= f'_x(t, p(t))q^1(t) + g(t, p(t), 0), \\
\dot{q}^2(t) &= f'_x(t, p(t))q^2(t) + G^2(t, p(t), q^1(t))
\end{aligned}
$$

where

$$G^2(t, p(t), q^1(t)) = (1/2)(f''_{xx}(t, p(t))q^1(t)q^1(t)$$
$$+ g'_x(t, p(t), 0)q^1(t) + g'_\mu(t, p(t), 0)),$$

and so on; clearly,

$$\dot{q}^k(t) = f'_x(t, p(t))q^k(t) + G^k(t, p(t), q^1(t), ..., q^{k-1}(t)), \qquad (k = 1, 2, ...).$$

where G^k is a known function $(G^1(t, p(t)) := g(t, p(t), 0))$. The "zero-th" system is a known identity. If the variational system (6.1.4) has been solved, then these systems can be solved successively; in the k-th equation the forcing term G^k depends on $p, q^1, ..., q^{k-1}$ which have been obtained by solving the previous systems. Actually, the k-th system is

$$\dot{y} = f'_x(t, p(t))y + G^k(t, p(t), q^1(t), ..., q^{k-1}(t)) \qquad (k = 1, 2, ...), \quad (6.1.7)$$

and we are interested in its T-periodic solution. The coefficient q^k in the power series expansion of q is a T-periodic solution of this system. Applying the notations in the proof of Theorem 6.1.1, $q(t, \mu) = \varphi(t, \xi(\mu), \mu)$; hence, $q(0, \mu) = \xi(\mu)$ where ξ is the initial value (at $t = 0$) of the periodic solution: an analytic function of μ. Thus,

$$\xi(\mu) = \sum_{k=0}^{\infty} \xi^k \mu^k = \dot{q}(0, \mu) = \sum_{k=0}^{\infty} q^k(0)\mu^k;$$

hence, $q^k(0) = \xi^k$ $(k = 1, 2, ...)$, $q^0(0) = p(0) = \xi^0$. This initial value, i.e. the k-th coefficient in the power series expansion of the initial value function ξ determines the solution q^k of (6.1.7). However, system (6.1.7) has only one T-periodic solution by Theorem 2.3.1. In case the fundamental matrix solution $\Phi(t)$ of (6.1.4) that satisfies $\Phi(0) = I$ is known, then applying formula (1.2.11)

$$q^k(t) = \Phi(t)(\xi^k + \int_0^t \Phi^{-1}(s)G^k(s, p(s), q^1(s), ..., q^{k-1}(s))ds). \quad (6.1.8)$$

One does not have to expand $\xi(\mu)$ into power series; the periodicity condition $q^k(T) = q^k(0) = \xi^k$ yields

$$\xi^k = \Phi(T)(\xi^k + \int_0^T \Phi^{-1}(s)G^k(s, p(s), q^1(s), ..., q^{k-1}(s))ds) \quad (6.1.9)$$

which determines ξ^k uniquely $(\det(\Phi(T) - I) \neq 0$, since 1 is not a characteristic multiplier $(k = 1, 2, ...)$.

If the basic condition of Theorem 6.1.1 is fulfilled, then it is easy to show that, loosely speaking, the asymptotic stability of the unperturbed periodic solution is inherited by the perturbed ones. More exactly there holds:

Theorem 6.1.3. *If all the characteristic multipliers of the variational system (6.1.4) are in modulus less than 1, then for sufficiently small $|\mu|$ the perturbed periodic solution $q(t, \mu)$ of system (6.1.1) is asymptotically stable.*

Proof. The assumption implies that the condition of Theorem 6.1.1 is satisfied. Consider the variational system of the perturbed system (6.1.1) with respect to its periodic solution $q(t, \mu)$:

$$\dot{y} = (f'_x(t, q(t, \mu)) + \mu g'_x(t, q(t, \mu), \mu))y \qquad (6.1.10)$$

and denote its fundamental matrix that assumes the unit matrix I at $t = 0$ by $\Psi(t, \mu)$. Since (6.1.10) depends continuously on μ, Ψ is a continuous function of μ for small $|\mu|$, and, clearly, $\Psi(t, 0) = \Phi(t)$, the fundamental matrix of (6.1.4) assuming I at $t = 0$. In particular, $\Psi(T, 0) = \Phi(T, 0)$, and $\Psi(T, \mu)$ is a continuous function of μ. Since its eigenvalues are also continuous functions of μ and the eigenvalues of $\Phi(T)$ are in modulus less than 1, for sufficiently small $|\mu|$ the eigenvalues of $\Psi(T, \mu)$ are also less than 1. By Theorem 4.2.1 this implies that the periodic solution q is asymptotically stable. $\square$

Example 6.1.1. The simplest non-trivial case in which the previous theory can be applied is a periodically forced linear system with constant coefficient subject to small periodic perturbations, that is, the perturbed system

$$\dot{x} = Ax + b(t) + \mu g(t, x, \mu) \qquad (6.1.11)$$

where A is a constant $n \times n$ matrix, $b, g, g'_x, g'_\mu \in C^0$ and for some $T > 0 : b(t + T) = b(t)$, $g(t + T, x, \mu) = g(t, x, \mu)$. If we suppose that A *is a stable matrix*, i.e. all its eigenvalues have negative real parts then by Theorem 2.1.1 the unperturbed system

$$\dot{x} = Ax + b(t)$$

has exactly one T-periodic solution. The variational system with respect to this (and any other) solution is

$$\dot{y} = Ay$$

which, clearly, has no periodic solution apart from the trivial one. Thus, by Theorem 6.1.1 the perturbed system (6.1.11) has a unique periodic solution with period T for each μ whose modulus is sufficiently small. By Theorem 6.1.3 this periodic solution is asymptotically stable.

In order to be more specific and to illustrate Poincaré's method, consider the system

$$\begin{aligned}
\dot{x}_1 &= \alpha x_1 - x_2 + \cos t + \mu g_1(t, x_1, x_2, \mu), \\
\dot{x}_2 &= x_1 + \alpha x_2 + \sin t + \mu g_2(t, x_1, x_2, \mu)
\end{aligned} \qquad (6.1.12)$$

where
$g_i, g'_{ix_k}, g'_{i\mu} \in C^0$, $g_i(t + 2\pi, x_1, x_2, \mu) = g_i(t, x_1, x_2, \mu)$ $(i = 1, 2; \ k = 1, 2)$, and $\alpha < 0$. The unperturbed system is

$$\begin{aligned}
\dot{x}_1 &= \alpha x_1 - x_2 + \cos t, \\
\dot{x}_2 &= x_1 + \alpha x_2 + \sin t.
\end{aligned}$$

Elementary calculations yield that $p(t) := -(1/\alpha)[\cos t, \sin t]$ is the only 2π periodic solution of the latter system. The variational system is $\dot{y}_1 = \alpha y_1 - y_2$, $\dot{y}_2 = y_1 + \alpha y_2$. The eigenvalues of its coefficient matrix are $\lambda_{1,2} = \alpha \pm i$, and the fundamental matrix that assumes the unit matrix at $t = 0$ is

$$\Phi(t) = e^{\alpha t} \begin{pmatrix} \cos t & -\sin t \\ \sin t & \cos t \end{pmatrix};$$

its inverse is

$$\Phi^{-1}(t) = e^{-\alpha t} \begin{pmatrix} \cos t & \sin t \\ -\sin t & \cos t \end{pmatrix}.$$

Thus, for $|\mu|$ sufficiently small, (6.1.12) has a unique 2π-periodic solution

$$q(t, \mu) = p(t) + \mu q^1(t) + \cdots$$

where $p(t) = -(1/\alpha)[\cos t, \sin t]$, and according to (6.1.8) $q^1(t)$ is given by

$$\begin{aligned}
q_1^1(t) = {}& e^{\alpha t}(\xi_1^1 \cos t - \xi_2^1 \sin t) \\
& + e^{\alpha t} \cos t \int_0^t e^{-\alpha s}(\cos s \, g_1(s, -(1/\alpha)\cos s, -(1/\alpha)\sin s, 0) \\
& + \sin s \, g_2(s, -(1/\alpha)\cos s, -(1/\alpha)\sin s, 0))ds \\
& - e^{\alpha t} \sin t \int_0^t e^{-\alpha s}(-\sin s \, g_1(s, -(1/\alpha)\cos s, -(1/\alpha)\sin s, 0) \\
& + \cos s \, g_2(s, -(1/\alpha)\cos s, -(1/\alpha)\sin s, 0))ds, \\
q_2^1(t) = {}& e^{\alpha t}(\xi_1^1 \sin t + \xi_2^1 \cos t) \\
& + e^{\alpha t} \sin t \int_0^t e^{-\alpha s}(\cos s \, g_1(s, -(1/\alpha)\cos s, -(1/\alpha)\sin s, 0) \\
& + \sin s \, g_2(s, -(1/\alpha)\cos s, -(1/\alpha)\sin s, 0))ds \\
& + e^{\alpha t} \cos t \int_0^t e^{-\alpha s}(-\sin s \, g_1(s, -(1/\alpha)\cos s, -(1/\alpha)\sin s, 0) \\
& + \cos s \, g_2(s, -(1/\alpha)\cos s, -(1/\alpha)\sin s, 0))ds.
\end{aligned}$$

Here $\xi^1 = [\xi_1^1, \xi_2^1] = q^1(0)$, and according to (6.1.9)

$$\xi^1 = (I - \Phi(T))^{-1}\Phi(T) \int_0^T \Phi^{-1}(s)g(s, p(s), 0)ds.$$

In our case

$$(I - \Phi(2\pi))^{-1}\Phi(2\pi) = \begin{pmatrix} \frac{\exp(2\alpha\pi)}{1-\exp(2\alpha\pi)} & 0 \\ 0 & \frac{\exp(2\alpha\pi)}{1-\exp(2\alpha\pi)} \end{pmatrix}$$

thus

$$
\begin{aligned}
\xi_1^1 &= \frac{\exp(2\alpha\pi)}{1 - \exp(2\alpha\pi)} \int_0^{2\pi} e^{-\alpha s}(\cos s \, g_1(s, -(1/\alpha)\cos s, -(1/\alpha)\sin s, 0) \\
&\quad + \sin s \, g_2(s, -(1/\alpha)\cos s, -(1/\alpha)\sin s, 0))ds, \\
\xi_2^1 &= \frac{\exp(2\alpha\pi)}{1 - \exp(2\alpha\pi)} \int_0^{2\pi} e^{-\alpha s}(-\sin s \, g_1(s, -(1/\alpha)\cos s, -(1/\alpha)\sin s, 0) \\
&\quad + \cos s \, g_2(s, -(1/\alpha)\cos s, -(1/\alpha)\sin s, 0))ds.
\end{aligned}
$$

The characteristic exponents, i.e. the eigenvalues of the (constant) coefficient matrix of the unperturbed variational system have negative real parts: $\mathrm{Re}\,\lambda_i = \alpha < 0$ $(i = 1, 2)$; equivalently, the characteristic multipliers, i.e. the eigenvalues of the principal matrix $\Phi(2\pi)$ are $\exp(2\alpha\pi)$, double, and it is in modulus less than 1. Hence, for $|\mu|$ sufficiently small the perturbed periodic solution $q(t, \mu)$ is asymptotically stable.

The situation is much more intricate in the case when , unlike as in the previous theory, the variational system of the unperturbed equation has non-trivial periodic solutions. This is the case, for instance, when f in (6.1.1) does not depend on t, i.e. when the unperturbed system is autonomous, and its periodic solution p is non-constant. We shall treat this case in the next Section. For the general non-autonomous case, see, e.g., a series of papers by Lewis [1955$_a$, 1955$_b$, 1956].

In the rest of this Section we give a survey of the most important results concerning the case when the unperturbed system is linear. The *periodic quasilinear system*

$$\dot{x} = A(t)x + \mu g(t, x, \mu) \tag{6.1.13}$$

will be considered where $A \in C^0$ is an $n \times n$ matrix, $g, g_x', g_\mu' \in C^0$ and for some $T > 0$, $A(t + T) = A(t), g(t + T, x, \mu) = g(t, x, \mu)$. The theory of system (6.1.13) has an enormous literature and has been treated in several monographs and textbooks (see e.g. Coddington-Levinson [1955], Cronin [1964] and Rouche-Mawhin [1973]). Therefore, we shall present the following theorems without proofs.

In case the unperturbed variational system which is now identical to the unperturbed system

$$\dot{x} = A(t)x \tag{6.1.14}$$

has no non-trivial T-periodic solution, the previous theory is applicable. In this case the unperturbed periodic solution is the trivial solution $p(t) \equiv 0$. For small $|\mu|$ the perturbed system (6.1.13) has a unique one parameter family of periodic solutions $q(t, \mu)$ such that $q(t, 0) \equiv 0$.

Assume now that (6.1.14) *has at least one non-trivial periodic solution with period T*. The difficulties that arise in this case can be conceived recalling Theorem 2.3.4 which establishes the conditions for the existence of periodic solutions of inhomogeneous linear periodic systems. If we suppose that (6.1.13) has a periodic solution $q(t, \mu)$ with period T and substitute this solution into g, only then we are confronted with an inhomogeneous linear T-periodic system

$$\dot{x} = A(t)x + \mu g(t, q(t, \mu), \mu),$$

and we know that this system has a T-periodic solution, namely, q, but as we know from Theorem 2.3.4 the condition for this is that $g(t, q(t, \mu), \mu)$ be orthogonal to the periodic solutions of the system adjoint to (6.1.14):

$$\int_0^T \eta^{i\prime}(t)g(t, q(t, \mu), \mu)dt = 0 \qquad (i = 1, 2, ..., k) \qquad (6.1.15)$$

where $\eta^{i\prime}$ is the transpose of the column vector η^i, the latter functions $(i = 1, 2, ..., k)$ forming a maximal set of linearly independent T-periodic solutions of the adjoint system $\dot{\eta} = -A'(t)\eta$ (cf. (2.3.5) and (2.3.10); now the functions are supposed to be real). The trouble is that, naturally, we do not know q, so we cannot expect simple conditions for the existence of a periodic solution for (6.1.13).

First, the unperturbed system is *reducible*; hence, by a regular linear T-periodic coordinate transformation $x = P(t)z$ where $P \in C^1$, $P(t + T) = P(t)$ and $\det P(t) \neq 0$, it can be transformed into a linear system with constant coefficients (see Theorem 2.2.6). This coordinate transformation carries the perturbed system (6.1.13) into

$$\dot{z} = Bz + \mu P^{-1}(t)g(t, P(t)z, \mu)$$

or, denoting the coefficient of μ by $G(t, z, \mu)$, into

$$\dot{z} = Bz + \mu G(t, z, \mu) \qquad (6.1.16)$$

where B is a constant (possibly complex) $n \times n$ matrix, $G, G_z', G_\mu' \in C^0$ and $G(t + T, z, \mu) = G(t, z, \mu)$. Since, by assumption, (6.1.14) has a non-trivial T-periodic solution, so does $\dot{z} = Bz$. As a consequence, B must have at least one eigenvalue of the form $\lambda = ik2\pi/T$, $k \in \mathbf{Z}$.

In the sequel we shall deal with system (6.1.16) assuming, for the sake of simplicity, that B is a real matrix (as we know from Theorem 2.2.6 if we had considered the original system as a $2T$-periodic one, then it could have been transformed into (6.1.16) with a real constant coefficient matrix). We also suppose that B has u pairs of eigenvalues (counted with multiplicities) of the form $\lambda_j = \pm i k_j 2\pi/T$ where k_j is a non-zero integer ($j = 1, 2, .., u$), 0 is a v-tuple eigenvalue, and the rest of the eigenvalues whose number is w are not of the form $ik2\pi/T$, $k \in \mathbf{Z}$, $2u + v + w = n$, $u + v \geq 1$. The last inequality ensures that the unperturbed system

$$\dot{y} = By \qquad (6.1.17)$$

has at least one (independent) non-trivial T-periodic solution. Then B can be brought to the real canonical form

$$B = \text{diag} \left[B_1, ..., B_\beta, C_1, ..., C_\gamma, D \right]$$

where B_j is a $2\beta_j$ order quadratic matrix of the form

$$B_j = \begin{pmatrix} S_j & 0 & \cdots & 0 & 0 \\ I_2 & S_j & \cdots & 0 & 0 \\ 0 & I_2 & \cdots & 0 & 0 \\ \cdot & \cdot & \cdots & \cdot & \cdot \\ 0 & 0 & \cdots & I_2 & S_j \end{pmatrix}, \qquad S_j = \begin{pmatrix} 0 & -k_j 2\pi/T \\ k_j 2\pi/T & 0 \end{pmatrix}$$

$$(j = 1, 2, ..., \beta \leq u);$$

I_2 is the 2×2 unit matrix and the zeroes stand for the 2×2 zero matrix, C_j is a γ_j order quadratic matrix of the form

$$C_j = \begin{pmatrix} 0 & 0 & \cdots & 0 & 0 \\ 1 & 0 & \cdots & 0 & 0 \\ 0 & 1 & \cdots & 0 & 0 \\ \cdot & \cdot & \cdots & \cdot & \cdot \\ 0 & 0 & \cdots & 1 & 0 \end{pmatrix}, \qquad (j = 1, 2, ..., \gamma)$$

and D is a w order quadratic matrix not having an eigenvalue of the form $ik2\pi/T$, $k \in \mathbf{Z}$, $\sum_{j=1}^{\beta} 2\beta_j + \sum_{j=1}^{\gamma} \gamma_j + w = n$. We assume that B has already this form.

Denote the solution of (6.1.16) that assumes the initial value z^0 at $t = 0$ by $\varphi(t, z^0, \mu)$. Substituting this solution into G first and then applying formula (2.1.4) we obtain

$$\varphi(t, z^0, \mu) = \exp(Bt)z^0 + \mu \int_0^t \exp((t - s)B)G(s, \varphi(s, z^0, \mu)\mu)ds.$$

Applying Lemma 2.2.1, φ is periodic with period T iff

$$(\exp(TB) - I)z^0 + \mu \int_0^T \exp((T - s)B)G(s, \varphi(s, z^0, \mu)\mu)ds = 0. \quad (6.1.18)$$

For $\mu = 0$, (6.1.16) is reduced to the unperturbed system (6.1.17), and the periodicity condition becomes

$$(\exp(TB) - I)z^0 = 0. \qquad (6.1.19)$$

Now, it can be shown that the special structure of B implies that if $z(0)$ is a solution of (6.1.19), then all its components $z_i(0)$ are zero except, possibly, those with subscript

$$i = 2 \sum_{l=1}^{h} \beta_l - 1, 2 \sum_{l=1}^{h} \beta_l, 2 \sum_{l=1}^{\beta} \beta_l + \sum_{l=1}^{m} \gamma_l,$$

$$(h = 1, 2, ..., \beta; \quad m = 1, 2, ..., \gamma).$$

There are all together $2\beta + \gamma$ indices like this (at least one of these coordinates must be different from zero in order to have a non-trivial T-periodic solution). Denote the set of these $2\beta + \gamma$ indices by E. It can be seen that for any vector z the coordinates of the vector $(\exp(TB) - I)z$ which belong to the indices

$$j = 1, 2, 2 \sum_{l=1}^{h} \beta_l + 1, 2 \sum_{l=1}^{h} \beta_l + 2, 2 \sum_{l=1}^{\beta} \beta_l + 1, 2 \sum_{l=1}^{\beta} \beta_l + \sum_{l=1}^{m} \gamma_l + 1$$

$(h = 1, 2, ..., \beta - 1; m = 1, 2, ..., \gamma - 1)$ are zero. There are all together $2\beta + \gamma$ indices like this and their set will be denoted by S. If $z(\mu)$ satisfies (6.1.18), it follows then that the coordinates of the vector defined by the integral corresponding to these indices must be zero:

$$\left[\int_0^T \exp((T - s)B)G(s, \varphi(s, z(\mu), \mu), \mu)ds\right]_j = 0, \qquad j \in S, \qquad \mu \neq 0$$

where $[\quad]_j$ represents the j-th coordinate of the vector in brackets. In particular, if $\mu = 0$, then $\varphi(s, z(0), 0) = \exp(Bs)z(0)$ where $z(0)$ is a solution of (6.1.19); thus, this solution of the unperturbed system (6.1.17) is T-periodic. Taking the limit $\mu \to 0$ for $j \in S$ we get

$$H_j(z(0)) := \left[\int_0^T \exp((T - s)B)G(s, \exp(Bs)z(0), 0)ds\right]_j = 0.$$

Performing the transformation $t = T - s$ and taking into account that G and $\exp(Bs)z(0)$ are T-periodic functions we obtain

$$H_j(z(0)) = \left[\int_0^T \exp(Bt)G(-t, \exp(-Bt)z(0), 0)dt\right]_j = 0, \qquad j \in S.$$

$$(6.1.20)$$

We know that $z_i(0) = 0$ for $i \notin E$. Thus, (6.1.20) can be considered as $2\beta + \gamma$ equations for the $2\beta + \gamma$ unknowns that are $z_i(0)$, $i \in E$. More exactly in order to have a periodic solution with period T for the perturbed system, the system of equations

$$H_j(z) = 0, \qquad j \in S, \qquad (6.1.21)$$

must have a solution z_i, $i \in E$, the rest of the coordinates of the solution vector z being equal to zero. This motivates the conditions of the following Theorem which we present without the proof:

Theorem 6.1.4 *(Coddington-Levinson [1955] p.362). If the conditions imposed upon B and G hold, (6.1.19) has a solution $z(0)$, this $z(0)$ satisfies also (6.1.21), the corresponding periodic solution of (6.1.17) is denoted by $p(t) = \exp(Bt)z(0)$, and the Jacobian of the system of equations (6.1.21) with respect to the unknowns z_i, $i \in E$, does not vanish at $z = z(0)$; then for sufficiently small $|\mu|$, the perturbed system (6.1.16) has one and only one T-periodic solution $q(t, \mu)$ such that $q \in C^0$ and $q(t, 0) = p(t)$.*

We note that, though, in general, checking the conditions of this Theorem may not be easy, they do not require any information about the solutions of the perturbed system; in particular, equations (6.1.21) can be written down without knowing anything but the fundamental matrix of the unperturbed linear system. We note also that, naturally, in the general case the unperturbed system (6.1.17) has more than one non-trivial periodic solution, and if the conditions of the Theorem are fulfilled, then a one parameter family of periodic solutions of the perturbed system emanates from each.

Often before treating the existence problem of periodic solutions, the perturbed system is brought into "*standard form*". We describe this procedure here in an important special case and shall return to this point in Section 5 when dealing with the method of "*averaging*". Consider again system (6.1.16) assuming that

$$G, G'_z, G'_\mu \in C^0, \quad G(t + T, z, \mu) = G(t, z, \mu), \quad T > 0, \quad (6.1.22)$$

and B is a constant $n \times n$ matrix having $u > 0$ eigenvalues (counted with multiplicities) of the form $\lambda_j = ik_j 2\pi/T$, $k_j \in \mathbf{Z}$ $(j = 1, 2, ..., u)$ all simple in the minimal polynomial and $v = n - u$ eigenvalues that are not of the form $ik2\pi/T$, $k \in \mathbf{Z}$. Then the matrix B can be transformed into a block diagonal matrix of the form $B = \text{diag}\,[C, D]$ where C is an $u \times u$ matrix with the eigenvalues $\lambda_j = ik_j 2\pi/T$ $(j = 1, 2, ..., u)$ and D is a $v \times v$ matrix that has no eigenvalue of the form $ik2\pi/T$, $k \in \mathbf{Z}$. Assuming that B is already in this form, system (6.1.16) is, in fact,

$$\begin{aligned}
\dot{z}^1 &= Cz^1 + \mu G^1(t, z, \mu), \\
\dot{z}^2 &= Dz^2 + \mu G^2(t, z, \mu)
\end{aligned} \quad (6.1.23)$$

where $z = z^1 \oplus z^2$, $z^1 \in \mathbf{R}^u$, $z^2 \in \mathbf{R}^v$, $G^1 : \mathbf{R} \times \mathbf{R}^n \times I_\mu \to \mathbf{R}^u$, $G^2 : \mathbf{R} \times \mathbf{R}^n \times I_\mu \to \mathbf{R}^v$, I_μ being an open interval containing $\mu = 0$, and G^1, G^2 satisfying conditions (6.1.22). The conditions imply that every solution of the system $\dot{z}^1 = Cz^1$ is T-periodic, i.e. its fundamental matrix $\exp(Ct)$ is T-

periodic and, naturally, regular. The T-periodic coordinate transformation

$$z^1 = \exp(Ct)x^1, \qquad z^2 = x^2$$

where $x^1 \in \mathbf{R}^u$, $x^2 = \mathbf{R}^v$ carries system (6.1.23) into

$$
\begin{aligned}
\dot{x}_1 &= \mu \exp(-Ct)G^1(t, \exp(Ct)x^1 \oplus x^2, \mu), \\
\dot{x}_2 &= Dx^2 + \mu G^2(t, \exp(Ct)x^1 \oplus x^2, \mu)
\end{aligned}
\qquad (6.1.24)
$$

where the right-hand side is, clearly, again T-periodic in t. Equations (6.1.24) is a quasi-linear system given in *standard form*. The unperturbed system $\dot{x}^1 = 0$, $\dot{x}^2 = Dx^2$ has only constant T-periodic solutions; they are $x = x^1 \oplus x^2 = c \oplus 0$ where $c \in \mathbf{R}^u$ is a constant and 0 denotes the zero vector in $\mathbf{R}^v$.

We shall present a Theorem without proof concerning the important special case when $v = 0$, i.e. when *all the solutions of the original unperturbed system $\dot{y} = By$ are T-periodic*. In this case C in (6.1.23) is an $n \times n$ matrix, every eigenvalue of which is of the form $ik2\pi/T$, $k \in \mathbf{Z}$ and is simple in the minimal polynomial, and the second equation of (6.1.23) disappears. The standard form (6.1.24) is then

$$\dot{x} = \mu g(t, x, \mu) \qquad (6.1.25)$$

where $g, g'_x, g'_\mu \in C^0$, $g(t+T, x, \mu) = g(t, x, \mu)$. Here the unperturbed system is $\dot{x} = 0$, every constant $x = c$ being a T-periodic solution of the latter. The following holds:

Theorem 6.1.5 *(Rouche-Mawhin [1973] p.127). Suppose that $c \in \mathbf{R}^n$ satisfies the equation*

$$g^0(x) := (1/T) \int_0^T g(t, x, 0)dt = 0, \qquad (6.1.26)$$

i.e. $g^0(c) = 0$ and the Jacobian of g^0 at c is non-zero:

$$\det g_x^{0'}(c) \neq 0;$$

then to every sufficiently small $|\mu|$ there belongs a unique T-periodic solution of (6.1.25), $q(t, \mu)$, such that $q \in C^0$ and $q(t, 0) \equiv c$.

Observe that since now the coefficient matrix of the unperturbed system is the n dimensional zero matrix and, as a consequence, its fundamental matrix is the unit matrix I, the last Theorem is a special case of Theorem 6.1.4, equation (6.1.26) being, in fact, (6.1.21) with $S = \{1, 2, ..., n\}$.

There are results concerning systems that are seemingly quasi-linear; however, the existence of periodic solutions is established by them on a prescribed "large" interval of the perturbation parameter μ. These are not really perturbation results since they concern "large" perturbations. For instance the following Theorem due to Jane Cronin [1974] is proved applying Leray-Schauder degree (see Definition A2.9). Consider the system

$$\dot{x} = A(t)x + \mu g(t, x, \mu) \qquad (6.1.27)$$

where the matrix $A \in C^1(\mathbf{R})$, $g \in C^1(\mathbf{R} \times \mathbf{R}^n \times [0,1])$, $A(t+T) = A(t)$, $g(t+T, x, \mu) = g(t, x, \mu)$ for some $T > 0$.

Theorem 6.1.6 *(Jane Cronin [1974]). If the unperturbed equation $\dot{x} = A(t)x$ has no non-trivial T-periodic solution, g'_x is bounded over $[0, T] \times \mathbf{R}^n \times [0,1]$, and there exist positive constants $r, \varepsilon \in (0,1)$ and k such that for $|x| \geq r$, $\mu \in [0,1]$ and $t \in [0, T]$ we have $|g(t, x, \mu)| < k|x|^{1-\varepsilon}$, then (6.1.27) has a T-periodic solution for each $\mu \in [0, 1]$.*

Here the existence of a periodic solution is established not only in a small neighbourhood of $\mu = 0$ but also on the μ interval $[0, 1]$. The price to be paid is the assumption that g'_x is bounded over the whole space and that $|g|$ is increasing "sublinearly" for large $|x|$.

6.2   Controllably Periodic Perturbations of Autonomous Systems

In astronomical, mechanical, physical, chemical, biological, economical, etc., applications one is often confronted with models in which a non-linear autonomous system is subject to small periodic perturbations. If the unperturbed autonomous system has an equilibrium point, and we are interested in the problem of what happens to this equilibrium under perturbation, then the conditions of Theorem 6.1.1, Corollary 6.1.2 and Theorem 6.1.3 may hold. In this case the unperturbed variational system is a homogeneous system with a constant coefficient matrix that may be, for instance, a stable matrix. However, if the unperturbed autonomous system has a non-constant periodic solution, then the conditions of these theorems cannot hold, since in this case the derivative of this solution is a non-trivial periodic solution of the unperturbed variational system. In order to overcome this difficulty, the concept of a *"controllably periodic perturbation"* has been introduced in Farkas [1970, 1971, 1972], the idea being that the period of the perturbation can be chosen appropriately.

In this Section, first the existence problem of perturbed periodic solutions will be treated under controllably periodic perturbations, Poincaré's method will be applied for the approximate determination of the periodic solution and estimates will be given for its domain of existence in terms of the perturbation parameter. Then the implications of this theory will be considered for the case when the perturbation is of fixed period. The latter case is related to the "*entrainment of frequency*" or "*synchronization*" problem that will also be dealt with.

The perturbed system

$$\dot{x} = f(x) + \mu g(t/T, x, \mu, T) \tag{6.2.1}$$

will be considered where $f \in C^1(X)$, $X \subset \mathbf{R}^n$ an open and connected set, $g, g'_x, g'_\mu, g'_T \in C^0(\mathbf{R} \times X \times I_\mu \times I_T)$, I_μ being an open interval containing $\mu = 0$, I_T being an open interval containing $T_0 > 0$ and *not* containing $T = 0$, and it will be assumed that

$$g(s+1, x, \mu, T) \equiv g(s, x, \mu, T),$$

i.e. g is periodic in the variable t with period $T > 0$. Thus, the period of the perturbation occurs in the perturbation as a parameter and it is assumed that it can be chosen appropriately. Therefore, (6.2.1) is called a *controllably periodically perturbed system*. It will be assumed that the unperturbed system, i.e. (6.2.1) at $\mu = 0$,

$$\dot{x} = f(x), \tag{6.2.2}$$

has a non-constant periodic solution $p : \mathbf{R} \to X$ with period $T_0 > 0$, i.e. $p(t + T_0) \equiv p(t) \not\equiv \text{const}$. The notations $p^0 = p(0) = (p_1^0, ..., p_n^0)$, $\dot{p}^0 = \dot{p}(0) = (\dot{p}_1^0, ..., \dot{p}_n^0)$ are introduced. The conditions imply that $\dot{p}^0 \neq 0$. Without loss of generality we may assume that $\dot{p}^0$ is parallel to the first coordinate axis, implying that $\dot{p}_1^0 \neq 0$ $\dot{p}_i^0 = 0$, $(i = 2, ..., n)$, or $\dot{p}^0 = (\dot{p}_1^0, 0, ..., 0)$. We denote the solution of (6.2.1) that assumes the value $x^0 \in X$ at $t = t_0$ by $\varphi(t, t_0, x^0, \mu, T)$. Consider the hyperplane passing through the point p^0 and orthogonal to the tangent vector $\dot{p}^0$ (the equation of it is $x_1 = p_1^0$). Because of the continuous dependence of the solutions on the initial values and the parameters if $|t_0|$, $|x^0 - p^0|$, $|\mu|$ and $|T - T_0|$ are sufficiently small, the path of the solution $\varphi(t, t_0, x^0, \mu, T)$ intersects this hyperplane at some moment near to $t = 0$ and at some point near to p^0. We denote the moment at which this intersection takes place and whose modulus is minimal by θ and the point of intersection by $p^0 + h$ where h is orthogonal to $\dot{p}^0 = (\dot{p}_1^0, 0, ..., 0)$, i.e. $h = (0, h_2, ..., h_n)$. Thus, solutions with initial values near to $t = 0$, $x^0 = p^0$ can be characterized (and parametrized) by these θ and h; therefore, in the sequel we shall consider solutions in the form $\varphi(t, \theta, p^0 + h, \mu, T)$. We have $\varphi(\theta, \theta, p^0 + h, \mu, T) = p^0 + h$ where h is an $n - 1$ dimensional vector $h = (0, h_2, ..., h_n)$, and there is no moment $t \in (-|\theta|, |\theta|)$ such that $\varphi(t, \theta, p^0 + h, \mu, T)$ belongs to the hyperplane $x_1 = p_1^0$. We shall need the variational system of the unperturbed system (6.2.2) with respect to the periodic solution p:

$$\dot{y} = f'_x(p(t))y. \tag{6.2.3}$$

We know that $\dot{p}$ is a non-trivial T_0-periodic solution of this T_0-periodic system; hence, by Corollary 2.2.3 the number 1 is its characteristic multiplier.

Now, we are able to prove

Theorem 6.2.1 *(Farkas [1971]). If* 1 *is a simple characteristic multiplier of (6.2.3), then there are* $\mu_0 > 0$ *and* $\theta_0 > 0$ *such that to each pair* (μ, θ) *satisfying* $|\mu| < \mu_0$, $|\theta| < \theta_0$ *there belongs a unique period* $\tau(\mu, \theta)$ *and a unique vector* $h(\mu, \theta) = (0, h_2(\mu, \theta), ..., h_n(\mu, \theta))$ *such that*

$$q(t, \mu, \theta) := \varphi(t, \theta, p^0 + h(\mu, \theta), \mu, \tau(\mu, \theta))$$

is a periodic solution with period $\tau(\mu, \theta)$ *of the perturbed system (6.2.1) with* $T = \tau(\mu, \theta)$, *the functions* τ *and* h *are in the* C^1 *class,* $\tau(0,0) = T_0$, $h(0,0) = 0$, $q(t, 0, 0) = p(t)$.

Proof. If $|\theta|$, $|h|$, $|\mu|$ and $|T - T_0|$ are sufficiently small, then the corresponding solution φ is defined on the closed interval $[\theta, \theta + 2T]$, say. Then the periodicity condition is

$$z(\theta, h, \mu, T) := \varphi(\theta + T, \theta, p^0 + h, \mu, T) - (p^0 + h) = 0. \qquad (6.2.4)$$

This system of equations has a solution: $\theta = 0$, $h = 0$, $\mu = 0$, $T = T_0$. We shall show that the conditions of the Implicit Function Theorem hold; thus, $h = (0, h_2, ..., h_n)$ and T can be expressed from (6.2.4) uniquely as C^1 functions of μ and θ in a neighbourhood of $(\mu, \theta) = (0, 0)$. The function z is, clearly, continuously differentiable in a neighbourhood of the point $P = (\theta, h, \mu, T) = (0, 0, 0, T_0)$. We show that its Jacobian is non-zero at P.

$$z'_{iT}(P) = \dot{\varphi}_i(T, 0, p^0, 0, T_0) = \dot{p}_i(T_0) = \dot{p}_i^0 = \delta_{i1}\dot{p}_i^0,$$

$$(i = 1, 2, ..., n)$$

where δ_{i1} is the Kronecker symbol and it was taken into account that at $\mu = 0$ the function z_i does not depend on its last argument,

$$z'_{ih_k}(P) = \varphi'_{ih_k}(P) - \delta_{ik} \qquad (i = 1, 2, ..., n; k = 2, ..., n)$$

where δ_{ik} is again the Kronecker symbol. Thus,

$$J = \frac{\partial(z_1, z_2, ..., z_n)}{\partial(T, h_2, ..., h_n)}\bigg|_P = \begin{vmatrix} \dot{p}_1^0 & \varphi'_{1h_2} & \cdots & \varphi'_{1h_n} \\ 0 & \varphi'_{2h_2} - 1 & \cdots & \varphi'_{2h_n} \\ \cdot & \cdot & \cdots & \cdot \\ 0 & \varphi'_{nh_2} & \cdots & \varphi'_{nh_n} - 1 \end{vmatrix}_P$$

or

$$J = \dot{p}_1^0 \begin{vmatrix} \varphi'_{2h_2} - 1 & \cdots & \varphi'_{2h_n} \\ \cdot & \cdots & \cdot \\ \varphi'_{nh_2} & \cdots & \varphi'_{nh_n} - 1 \end{vmatrix}_P.$$

Now, consider the solution $\varphi(t, 0, x^0, 0, T)$. Its derivative matrix $\varphi'_{x^0}(t, 0, p^0, 0, T_0)$ is the fundamental matrix of the variational system

(6.2.3) that assumes I, the unit matrix at $t = 0$ (see (1.1.5) et seq.). Denote the first column of the matrix φ'_{x^0} by ψ, i.e.

$$\psi(t) := \varphi'_{x^0_1}(t, 0, p^0, 0, T_0).$$

This is a solution of (6.2.3) and $\psi(0) = [1, 0, ..., 0] = (1/\dot{p}^0_1)[\dot{p}^0_1, 0, ..., 0] = (1/\dot{p}^0_1)\dot{p}^0$; hence, because of the uniqueness of solutions, $\psi(t) = (1/\dot{p}^0_1)\dot{p}(t)$. As a consequence, ψ is periodic with period T_0, $\psi(t + T_0) = \psi(t)$, and in particular $\psi(T_0) = \psi(0)$. The characteristic multipliers of (6.2.3) are the roots of the characteristic polynomial which can be written in the following form because of the properties of the first column:

$$\det[\varphi'_{x^0}(T_0, 0, p^0, 0, T_0) - \lambda I]$$

$$= \begin{vmatrix} 1 - \lambda & \varphi'_{1x^0_2} & \cdots & \varphi'_{1x^0_n} \\ 0 & \varphi'_{2x^0_2} - \lambda & \cdots & \varphi'_{2x^0_n} \\ \vdots & \vdots & \cdots & \\ 0 & \varphi'_{nx^0_2} & \cdots & \varphi'_{nx^0_n} - \lambda \end{vmatrix}_{t=T_0}$$

$$= (1 - \lambda) \begin{vmatrix} \varphi'_{2x^0_2} - \lambda & \cdots & \varphi'_{2x^0_n} \\ \vdots & \cdots & \\ \varphi'_{nx^0_2} & \cdots & \varphi'_{nx^0_n} - \lambda \end{vmatrix}_{t=T_0}.$$

Since, by assumption, 1 is a simple characteristic multiplier, the last determinant is non-zero at $\lambda = 1$; hence,

$$J = \dot{p}^0_1 \begin{vmatrix} \varphi'_{2h_2} - 1 & \cdots & \varphi'_{2h_n} \\ \vdots & \cdots & \\ \varphi'_{nh_2} & \cdots & \varphi'_{nh_n} - 1 \end{vmatrix}_P = \dot{p}^0_1 \begin{vmatrix} \varphi'_{2x^0_2} - 1 & \cdots & \varphi'_{2x^0_n} \\ \vdots & \cdots & \\ \varphi'_{nx^0_2} & \cdots & \varphi'_{nx^0_n} - 1 \end{vmatrix}_{t=T_0} \neq 0.$$

Thus, by the Implicit Function Theorem there are $\mu_0 > 0$ and $\theta_0 > 0$ and uniquely determined functions $\tau : U \to \mathbf{R}_+, h : U \to \mathbf{R}^{n-1}$ where $U = \{(\mu, \theta) \in \mathbf{R}^2 : |\mu| < \mu_0, |\theta| < \theta_0\}$ such that $\tau, h \in C^1$, $\tau(0, 0) = T_0$, $h(0, 0) = 0$ and $z(\theta, h(\mu, \theta), \mu, \tau(\mu, \theta)) \equiv 0$ for $(\mu, \theta) \in U$ (by an abuse of notation we consider $h = (0, h_2, ..., h_n)$). Since (6.2.4) is the periodicity condition for each $(\mu, \theta) \in U$, the unique periodic solution of (6.2.1) with period $\tau(\mu, \theta)$ near T_0 and with path near the path of p is $q(t, \mu, \theta) := \varphi(t, \theta, p^0 + h(\mu, \theta), \mu, \tau(\mu, \theta))$ provided that in (6.2.1) also $\tau(\mu, \theta)$ is substituted for T (this is why we say that the period of the perturbation is controllable). The properties of the functions τ and h imply that $g \in C^1$ too and that $q(t, 0, 0) \equiv p(t)$. $\square$

A few important remarks are to be added. First, for $\mu = 0$ and $|\theta| < \theta_0$ the unique periodic solution with period "near to T_0" and path "near the path of p" is $p(t - \theta)$ which has the same period and the same path. Thus,

$$\tau(0, \theta) \equiv T_0, \qquad h(0, \theta) \equiv 0, \qquad q(t, 0, \theta) \equiv p(t - \theta). \tag{6.2.5}$$

If a non-zero μ is fixed, $0 < |\mu| < \mu_0$, then "in the generic case" the function $\tau(\mu, \bullet)$ is already not constant. This means that to each $\mu \in (-\mu_0, \mu_0)$, $\mu \neq 0$ there belongs a *one parameter family of periods* and a *one parameter family of corresponding periodic systems* that have a periodic solution near p. We shall use this result when we shall apply the results to perturbations with a fixed period.

Secondly, suppose that T_0 is the least positive period of the solution p. Naturally, p and the perturbed system (6.2.1) can be considered periodic with period mT_0 and mT, respectively, with $m = 1, 2, 3, \ldots$. *A periodic solution of (6.2.1) with least positive period mT ($m \geq 2$) is called a subharmonic solution.* (The *harmonics* of a T-periodic solution are the terms in its Fourier series; these have periods T/m ($m = 1, 2, 3, \ldots$).) The uniqueness part of the previous Theorem does not exclude the possibility of the existence of periodic solutions near p with least positive period mT near mT_0 ($m \geq 2$). If p is considered periodic with period mT_0 ($m \geq 2$), then also the variational system (6.2.3) is considered as such. Denoting its fundamental matrix that assumes the unit matrix I at $t = 0$, by $\Phi(t)$ *the characteristic multipliers of system (6.2.3) considered as an mT_0-periodic system* are the eigenvalues of the principal matrix $\Phi(mT_0)$. But by (2.2.6), $\Phi(mT_0) = \Phi((m-1)T_0)\Phi(T_0)$, i.e. $\Phi(mT_0) = (\Phi(T_0))^m$; thus, the characteristic multipliers of system (6.2.3) considered as an mT_0-periodic system are exactly the m-th powers of its characteristic multipliers considered as a T_0-periodic system. If (6.2.3) considered as a T_0-periodic system has (besides 1) an m-th root of unity $\exp(i2k\pi/m)$, k an integer satisfying $1 \leq k \leq m - 1$, as a characteristic multiplier, then (6.2.3) considered as an mT_0-periodic system has 1 as a multiple (not simple) characteristic multiplier. In this case, if (6.2.1) is considered a system with' period mT, then the conditions of Theorem 6.2.1 do not hold, p may be non-isolated (as an mT_0-periodic solution of the unperturbed system, see Definition 5.2.2 and Example 5.2.1), and (6.2.1) with $T = \tau(\mu, \theta)$ of the Theorem may have periodic solutions with least positive period $m\tau(\mu, \theta)$, i.e. subharmonic solutions besides $q(t, \mu, \theta)$. (For non-existence of subharmonic solutions in case of *autonomous* perturbations, see Urabe [1967]).

Thirdly, an analogous Theorem is valid in case (6.2.1) is periodic in the vector variable x and p is D-periodic (see Definitions 5.3.1 and 5.3.2). The same condition guarantees the existence of a D-periodic solution of the perturbed system near p with period near T_0 and coefficient vector near that of p.

Theorem 6.2.1 contains a classical perturbation theorem about *autonomous perturbations* of autonomous systems as a special case. Suppose that in (6.2.1) g does not depend on t (and on T, of course), i.e. consider the perturbed system

$$\dot{x} = f(x) + \mu g(x, \mu) \tag{6.2.6}$$

where $f \in C^1(X)$, $X \subset \mathbf{R}^n$ an open and connected set, $g, g_x', g_\mu' \in C^0(X \times I_\mu)$, I_μ being an open interval containing $\mu = 0$, and assume as before that the unperturbed system (6.2.2) has a non-constant periodic solution $p : \mathbf{R} \to X$ with period $T_0 > 0$. The following Theorem (see e.g. Hartman [1964] p. 416 or Coddington-Levinson [1955] p.352) is a simple consequence of Theorem 6.2.1.

Theorem 6.2.2. *If 1 is a simple characteristic multiplier of (6.2.3), then there is a $\mu_0 > 0$ such that to each $\mu \in (-\mu_0, \mu_0)$ there belongs one and only one periodic solution $r(t, \mu)$ of (6.2.6) with period $T(\mu)$ such that $T, r \in C^1$, $T(0) = T_0$, $r(t, 0) \equiv p(t)$.*

Proof. Since (6.2.6) is a special case of (6.2.1) and the conditions of Theorem 6.2.1 hold, the functions $\tau(\mu, \theta)$, $h(\mu, \theta)$ and $q(t, \mu, \theta)$ provided by that Theorem exist with the properties described there. Set $T(\mu) := \tau(\mu, 0)$ and $r(t, \mu) = q(t, \mu, 0)$. This period T as a function of μ and periodic solution r satisfy the requirements of the present Theorem. The uniqueness of r follows from the uniqueness of q, taking into account that now, since the perturbed system is also autonomous, $q(t, \mu, \theta) = r(t - \theta, \mu)$, and as a consequence, $\tau(\mu, \theta) \equiv T(\mu)$. (For a fixed $\mu \in (-\mu_0, \mu_0)$ we have again a one parameter family of periodic solutions, but they belong now to the same system and have the same orbit and the same period. Clearly, $h(\mu, \theta)$ does not depend on θ either.) $\square$

We turn now to the problem of the determination of the perturbed periodic solution and its period. An appropriate version of Poincaré's method (see the text following Corollary 6.1.2) will be applied; however, first a Corollary analogous to Corollary 6.1.2 will be announced.

Corollary 6.2.3. *If, besides the conditions imposed above, f and g are in the C^k class, $k \geq 2$, resp. analytic in x, μ and T and the conditions of Theorem 6.2.1 (and Theorem 6.2.2) hold, then the functions τ, h, q, T and r of these Theorems are also in the C^k class, resp. analytic in μ.*

Proof. This is an immediate consequence of Theorem 1.1.4, resp. the note following (1.1.14), and of the Implicit Function Theorem concerning C^k, resp. analytic functions. $\square$

In the sequel we assume that in (6.2.1) f and g are analytic and that the conditions of Theorem 6.2.1 hold. The assumptions and notations applied in the preparation and the proof of this Theorem will be used. In particular, $\dot{p}^0 = \dot{p}(0) = [\dot{p}_1^0, 0, ..., 0] \neq 0$ and h denotes a vector orthogonal to $\dot{p}^0$, i.e.

$h = [0, h_2, ..., h_n]$. We know that for $\mu \in (-\mu_0, \mu_0)$, $\theta \in (-\theta_0, \theta_0)$ we have a uniquely determined $\tau(\mu, \theta)$ and $h(\mu, \theta)$ such that the system

$$\dot{x} = f(x) + \mu g(t/\tau(\mu, \theta), x, \mu, \tau(\mu, \theta)) \qquad (6.2.7)$$

has a uniquely determined $\tau(\mu, \theta)$-periodic solution

$$q(t, \mu, \theta) = \varphi(t, \theta, p^0 + h(\mu, \theta), \mu, \tau(\mu, \theta)); \qquad (6.2.8)$$

the functions τ, h and q are analytic in μ (for fixed t and θ) and $\tau(0, 0) = T_0$, $h(0, 0) = 0$, $q(t, 0, 0) \equiv p(t)$. In order to apply Poincaré's method, first a new independent variable s is introduced for fixed μ and θ by the transformation

$$t = \theta + s\tau(\mu, \theta). \qquad (6.2.9)$$

In this variable system, (6.2.7) assumes the form

$$dx/ds = \tau(\mu, \theta)\big(f(x) + \mu g(s + \theta/\tau(\mu, \theta)x, \mu, \tau(\mu, \theta))\big) \qquad (6.2.10)$$

whose solution corresponding to (6.2.8) is

$$\psi(s, \mu, \theta) := q(\theta + s\tau(\mu, \theta), \mu, \theta).$$

System (6.2.10) is, clearly, periodic in s with period 1, and ψ is its 1-periodic solution. We expand the solution ψ and the period τ into a power series by powers of μ for fixed $\theta \in (-\theta_0, \theta_0)$ and t:

$$\psi(s, \mu, \theta) = \sum_{k=0}^{\infty} \mu^k \psi^k(s, \theta),$$

$$\tau(\mu, \theta) = \sum_{k=0}^{\infty} \mu^k \tau_k(\theta).$$

By (6.2.5)

$$\psi^0(s, \theta) = \psi(s, 0, \theta) = q(\theta + sT_0, 0, \theta) = p(sT_0). \qquad (6.2.11)$$

Since the period of ψ, the number 1, is independent of μ, the derivatives of ψ with respect to μ and, hence, the coefficients of the power series of ψ are also periodic in s with period 1. Further, we have

$$\psi(0, \mu, \theta) = q(\theta, \mu, \theta) = p^0 + h(\mu, \theta),$$

and since the first coordinate of the vector h is zero, the first coordinate of the vector $\psi(0, \mu, \theta)$ is equal to the first coordinate of p^0:

$$\psi_1(0, \mu, \theta) \equiv p_1^0. \qquad (6.2.12)$$

The last identity and (6.2.11) imply that the first coordinates of the derivatives of the vector ψ with respect to μ at $s = 0$ and, hence, the first coordinates of the vectors ψ^k $(k = 1, 2, 3, ...,)$ at $s = 0$ are zero:

$$\psi_1^k(0, \theta) \equiv 0 \qquad (k = 1, 2, 3, ...). \tag{6.2.13}$$

Substituting the power series of ψ and τ into system (6.2.10) we obtain the identity

$$\sum_{k=0}^{\infty} \mu^k d\psi^k(s, \theta)/ds \equiv \sum_{k=0}^{\infty} \mu^k \tau_k(\theta) \big(f(\psi, \mu, \theta)$$
$$+ \mu g(s + \theta/\tau(\mu, \theta), \psi(s, \mu, \theta), \mu, \tau(\mu, \theta))\big).$$

Expanding the second factor on the right-hand side by powers of μ and equating the coefficients of equal powers of μ on both sides yields, the identities

$$dp(sT_0)/ds \equiv T_0 f(p(sT_0)), \tag{6.2.14}$$
$$d\psi^1(s, \theta)/ds \equiv T_0 f_x'(p(sT_0))\psi^1(s, \theta)$$
$$+ T_0 g(s + \theta/T_0, p(sT_0), 0, T_0)$$
$$+ \tau_1(\theta) f(p(sT_0)), \tag{6.2.15}$$

and similar but increasingly complicated identities for ψ^k $(k = 2, 3, ...)$. The last identity means that the function $\psi^1(\bullet, \theta)$ satisfies the differential system

$$dz/ds = T_0 f_x'(p(sT_0))z + T_0 g(s + \theta/T_0, p(sT_0), 0, T_0)$$
$$+ \kappa f(p(sT_0)) \tag{6.2.16}$$

if the value of the parameter κ is chosen as $\tau_1(\theta)$.

We show that the 1-periodicity of ψ^1, the condition (6.2.13), i.e. that its first coordinate is zero, and (6.2.15) characterize ψ^1 and τ_1 uniquely. The proof of the following Theorem follows the outlines of the proof of an analogous theorem concerning autonomous perturbations (see Coddington-Levinson [1955] p. 355).

Theorem 6.2.4 (*Farkas [1972, 1975$_a$]*). *If for fixed $\theta \in (-\theta_0, \theta_0)$ the function $\zeta \in C^1(\mathbf{R}, \mathbf{R}^n)$ and the real number κ satisfy (6.2.16), ζ is periodic with period $1, \zeta(s + 1) \equiv \zeta(s)$, and its first coordinate at $s = 0$ is zero, $\zeta_1(0) = 0$, then $\zeta(s) \equiv \psi^1(s, \theta)$ and $\kappa = \tau_1(\theta)$.*

Proof. Introducing the notations

$$w(s) := \psi^1(s, \theta) - \zeta(s), \qquad \gamma = (\tau_1(\theta) - \kappa)/T_0$$

substituting $z = \zeta(s)$ and κ into (6.2.16) and subtracting (6.2.16) form (6.2.15) we obtain that the function w is a solution of the inhomogeneous linear system

$$dw/ds = T_0 f'_x(p(sT_0))w + \gamma T_0 f(p(sT_0))$$

which by (6.2.14) can be written in the form

$$dw/ds = T_0 f'_x(p(sT_0))w + \gamma dp(sT_0)/ds. \qquad (6.2.17)$$

The corresponding homogeneous linear system is

$$dv/ds = T_0 f'_x(p(sT_0))v \qquad (6.2.18)$$

which is the variational system of the system $dx/ds = T_0 f(x)$ with respect to its 1-periodic solution $p(sT_0)$ (see (6.2.14)). The derivative of $p(sT_0)$, i.e. $dp(sT_0)/ds$ and, hence,

$$v^1(s) := (1/(T_0 \dot{p}_1^0)) dp(sT_0)/ds$$

is a 1-periodic solution of (6.2.18). Because of the conditions imposed upon $\dot{p}(0)$ we have

$$v^1(0) = [1, 0, ..., 0].$$

It is easy to see by direct substitution that the function $\gamma T_0 \dot{p}_1^0 s v^1(s)$ is a solution of the inhomogeneous system (6.2.17). Hence, the difference

$$v^2(s) := w(s) - \gamma T_0 \dot{p}_1^0 s v^1(s)$$

is also a solution of the homogeneous system (6.2.18). Denote the fundamental matrix of (6.2.18) that assumes the unit matrix I at $s = 0$ by $V(s)$. Then

$$\begin{aligned} w(s) &= v^2(s) + \gamma T_0 \dot{p}_1^0 s v^1(s) \\ &= V(s)v^2(0) + \gamma \dot{p}_1^0 s v^1(s). \end{aligned}$$

Since $w(s)$ is periodic with period 1, we have

$$0 = w(1) - w(0) = V(1)v^2(0) - I v^2(0) + \gamma T_0 \dot{p}_1^0 v^1(0)$$

where also the periodicity of v^1 with period 1 has been used. Hence

$$(V(1) - I)v^2(0) = -\gamma T_0 \dot{p}_1^0 v^1(0). \qquad (6.2.19)$$

The first column vector of the matrix $V(0) = I$ is $v^1(0)$, so that the solution $v^1(s)$ is the first column vector of the fundamental matrix $V(s)$. Denote the entries of the matrix $V(1)$ by c_{ik} ($i, k = 1, 2, ..., n$) and the coordinates of the vector $v^2(0)$ by v_k^2 ($k = 1, 2, .., n$). Since $v^1(s)$ is 1-periodic, we have for the first column of $V(1)$ that $v^1(1) = v^1(0)$, i.e. $c_{i1} = \delta_{i1}$, the Kronecker symbol $(i = 1, 2, ..., n)$. Thus, (6.2.19) can be written in the form

$$\sum_{k=2}^{n} c_{ik}v_k^2 = -\gamma T_0 \dot{p}_1^0,$$

$$\sum_{k=2}^{n}(c_{ik} - \delta_{ik})v_k^2 = 0 \qquad (i = 2, 3, ..., n). \qquad (6.2.20)$$

The eigenvalues of $V(1)$ are the characteristic multipliers of the system (6.2.18), but these are identical (with their multiplicities) to the characteristic multipliers of system (6.2.3) since if $Y(t)$ is a fundamental matrix of (6.2.3), then $Y(sT_0)$ is a fundamental matrix of (6.2.18). By the assumptions of the Theorem, the number 1 is a simple characteristic multiplier, i.e. a simple eigenvalue of $V(1)$; this implies that the coefficient matrix of the system consisting of the last $n - 1$ equations of system (6.2.20) is regular; hence, $v_2^2 = v_3^2 = \cdots = v_n^2 = 0$. Substituting this into the first equation of (6.2.20) yields $\gamma = 0$, i.e. $\kappa = \tau_1(\theta)$. Further, since the vector $v^2(0)$ is a scalar multiple of the vector $v^1(0)$ and because of $\gamma = 0$, we have $w(s) \equiv v^2(s)$; therefore, $w(s) \equiv cv^1(s)$ where c is some scalar number. But the first coordinate of ψ^1 and ζ hence that of w is zero at $s = 0$; at the same time the first coordinate of $v^1(0)$ is 1. From this $c = 0$ and, finally, $w(s) \equiv 0$ follows. $\square$

We note that *all we need in order to determine the series expansion of the perturbed periodic solution and that of the period is a fundamental matrix of the unperturbed variational system (6.2.3)*. If we know such a fundamental matrix, then we know a fundamental matrix of (6.2.18) and the general solution of the inhomogeneous system (6.2.16) can be written out explicitly. From this, $\psi^1(s, \theta)$ and $\tau_1(\theta)$ can be determined easily. The further coefficients ψ^k and τ_k $(k = 2, 3, ...)$ in the series expansions of the periodic solution ψ and the period τ satisfy systems similar to (6.2.16). The similarity means that the homogeneous linear part is the same and only the inhomogeneous term (the forcing term) varies and becomes more and more complicated (cf. (6.1.7)), but still it contains only functions which have already been determined.

Note also that as we have shown in Theorem 6.2.2, the autonomous perturbation case (6.2.6) is a special case of a controllably periodic perturbation, so that the variant of Poincaré's method having been presented here is applicable to (6.2.6) too, provided that f and g are analytic there.

Note finally that the method for the approximate determination of the perturbed periodic solution and its period can be applied also when f and g are not analytic but are in the C^k class $(k \geq 1)$. If we determine k terms in the formal series expansions of ψ and τ, they will approximate the exact functions with an error of $o(\mu^k)$ as $\mu \to 0$.

Though the previous theory is *local* in the sense that it guarantees the existence of a perturbed periodic solution only for "small values of $|\mu|$ and $|\theta|$", still there is a possibility to determine explicitly the radius of the neighbourhood of $(0,0)$ in the μ, θ plane in which the periodic solution exists. Results in this direction have been achieved by Lewis [1955$_b$], Freedman, [1968, 1971], Farkas [1978$_a$] and others. We shall present here the results based on Farkas [1978$_a$] which concern controllably periodic perturbations. Some preparation is needed. First some notations will be introduced, then an Implicit Function Theorem providing estimates for the domain of the implicitly defined function will be proved; finally in several lemmata and a theorem estimates will be given for the domain in the parameter plane where the perturbed periodic solutions exist.

In this study, the vector norm $|u| = \max_i |u_i|$ will be used where $u = [u_1, ..., u_m]$. We introduce a *norm for non-square matrices*. Let $D = [d_{ik}]$ be an $n \times m$ matrix $(i = 1, 2, ..., n; k = 1, 2, ..., m)$. Its *norm* is defined by

$$|D| := m \max_{i,k} |d_{ik}|.$$

It is easy to see that besides some other basic properties $|Du| \le |D||u|$ where u is an m dimensional column vector (and Du is, naturally, an n dimensional column vector). If, in particular, the matrices A and B are quadratic matrices of order n, then $|AB| \le |A||B|$. We shall apply "tensors with three subscripts". If $E = [e_{ikl}]$ $(i, k, l = 1, 2, ..., n)$, its *norm* is defined by

$$|E| := n^2 \max_{ikl} |e_{ikl}|.$$

The inner product of E and a vector $x \in \mathbf{R}^n$ is the matrix

$$Ex = \left[\sum_{l=1}^{n} e_{ikl}x_l \right] \qquad (i, k = 1, 2, ..., n).$$

If $y \in \mathbf{R}^n$ is a further vector, then

$$Exy = \left[\sum_{k,l=1}^{n} e_{1kl}x_ly_k, ..., \sum_{k,l=1}^{n} e_{nkl}x_ly_k \right] \in \mathbf{R}^n.$$

It is easy to see that $|Ex| \le |E||x|$ and $|Exy| \le |E||x||y|$. Finally, if $f \in C^2(\mathbf{R}^n, \mathbf{R}^n)$, then we shall adopt the notation $f''_{xx} = [f''_{ix_kx_l}(x)]$ $(i, k, l = 1, 2, ..., n)$.

The following Implicit Function Theorem is due to Lewis [1938]. We present it here with the proof given in Farkas [1978$_a$]. Let $u \in \mathbf{R}^m$ and $v \in \mathbf{R}^n$ denote m and n dimensional vectors, respectively, $a > 0$, $\beta > 0$,

$$W := \{(u,v) \in \mathbf{R}^m \times \mathbf{R}^n : |u| \le a, |v| \le \beta\},$$

$z \in C^2(W, \mathbf{R}^n)$, and consider the system of n equations with $n + m$ unknowns,

$$z(u, v) = 0. \tag{6.2.21}$$

If $\det z'_v(u, v) \neq 0$ for $(u, v) \in W$, then the linear system of equations

$$z'_v(u, v)d^h + z'_{v_h}(u, v) = 0 \tag{6.2.22}$$

defines the functions $d^h : W \to \mathbf{R}^n$ uniquely for each $h = 1, 2, ..., m$ and $d^h = \text{col}\,[d_1^h, d_2^h, ..., d_n^h] \in C_W^1$. Moreover, since $|\det z'_v|$ is continuous and non-zero on the compact set W, it is larger than a positive number, and consequently there exists a $\Delta > 0$ such that the norm of the $n \times n$ matrix $D = [d^1, d^2, ..., d^n]$ is less than Δ:

$$|D(u, v)| \leq \Delta, \qquad (u, v) \in W. \tag{6.2.23}$$

Theorem 6.2.5 *(Lewis [1938], Farkas [1978a]). Suppose that $z(0, 0) = 0$ and $\det z'_v(u, v) \neq 0$ for $(u, v) \in W$, and let*

$$0 < c < \min(a, \beta/\Delta) \tag{6.2.24}$$

and $U := \{u \in \mathbf{R}^m : |u| < c\}$. Then there exists one and only one function $w : U \to \mathbf{R}^n$ such that $w \in C^1$, $|w(u)| < \beta$ for $u \in U$, $w(0) = 0$, and

$$z(u, w(u)) \equiv 0, \qquad for \qquad u \in U. \tag{6.2.25}$$

Proof. If a function $w : U \to \mathbf{R}^n$ is in the C^1 class, $|w(u)| < \beta$ for $u \in U$, and w satisfies the condition (6.2.25), then differentiating the latter identity with respect to u_h we get

$$z'_v(u, w(u))w'_{u_h}(u) + z'_{u_h}(u, w(u)) \equiv 0 \qquad (h = 1, 2, ..., m),$$

i.e. $w'_{u_h}(u) \equiv d^h(u, w(u))$. Conversely, if a function w is in the C^1 class, $|w(u)| < \beta$ in U, $w(0) = 0$, and w satisfies the system of partial differential equations

$$v'_{u_h} = d^h(u, v) \qquad (h = 1, 2, ..., m), \tag{6.2.26}$$

then, clearly, it satisfies (6.2.25) as well.

For system (6.2.26) the "condition of complete integrability"

$$d_{u_l}^{h\prime} + d_v^{h\prime}d^l - d_{u_h}^{l\prime} - d_v^{l\prime}d^h = 0 \qquad (h, l = 1, 2, ..., m) \tag{6.2.27}$$

holds in W. To prove (6.2.27) assume that the solution of (6.2.22) has been substituted into (6.2.22) and differentiate this identity first with respect to u_l, then with respect to v. In the latter case we multiply the derived identity by the column vector d^l from the right. In the first case we get

$$z''_{vu_l}d^h + z'_v d_{u_l}^{h\prime} + z''_{u_h u_l} = 0;$$

in the second

$$z''_{vv}d^h d^l + z'_v d^{h'}_v d^l + z''_{u_h v}d^l = 0.$$

(Here, as it was pointed out

$$z''_{vv}d^h d^l = \sum_{i,j=1}^{n} z''_{v_i v_j}d^h_j d^l_i,$$

which is considered as a column vector. The writing out of the arguments has been supressed everywhere.) Adding the last two identities, we obtain

$$z''_{vu_l}d^h + z'_v d^h_{u_l} + z''_{u_h u_l} + z''_{vv}d^h d^l + z'_v d^{h'}_v d^l + z''_{u_h v}d^l = 0.$$

Interchanging the roles of h and l and subtracting the two identities yields

$$z'_v(d^{h'}_{u_l} + d^{h'}_v d^l - d^{l'}_{u_h} - d^{l'}_v d^h) = 0.$$

Since the matrix z'_v is regular in W, equation (6.2.27) follows.

In what follows we are going to show that (6.2.26) has a solution $w \in C^1$ for which the conditions $|w(u)| < \beta$ in U and $w(0) = 0$ hold.

Let u^* be an arbitrary point in U and consider the system of ordinary differential equations

$$\frac{dx}{dt} = D(u^*t, x)u^* \qquad (6.2.28)$$

with the initial condition $x(0, u^*) = 0$. The right-hand side of (6.2.28) where the vector u^* occurs as a parameter is continuously differentiable for $u^* \in U$ (i.e. $|u^*| < c$), $|x| < \beta$ and $|t| < a/c$ (recall that by (6.2.24) $a/c > 1$). Let us denote the supremum of the right-hand side of (6.2.28) by M. According to (6.2.24)

$$M = \sup_{\substack{|u^*|<c \\ |t|<a/c \\ |x|<\beta}} |D(u^*t, x)u^*| \le \sup_{\substack{|u^*|<c \\ |t|<a/c \\ |x|<\beta}} |D(u^*t, x)||u^*| \le \Delta c < \beta.$$

By the local existence and uniqueness Theorem 1.1.1, system (6.2.28) has a unique solution $x(t, u^*)$ satisfying the initial condition $x(0, u^*) = 0$ defined in the interval $(-\alpha, \alpha)$ where

$$\alpha = \min(a/c, \beta/M) > 1$$

and $|x(t, u^*)| < \beta$ for $t \in (-\alpha, \alpha)$. As is well known, this solution is a continuously differentiable function of the parameter u^* for all $t \in (-\alpha, \alpha)$ and $u^* \in U$.

Now we are in the position to construct the required solution of (6.2.26). Let us define a function $w : U \to \mathbf{R}^n$ in the following way:

$$w(u^*) := x(1, u^*), \qquad u^* \in U. \qquad (6.2.29)$$

Clearly $w \in C^1$, $|w(u^*)| < \beta$ in U and $w(0) = x(1,0) = 0$. We have to show that the function defined by (6.2.29) is a solution of system (6.2.26).

For this purpose let us differentiate the identity

$$\dot{x}(t, u^*) \equiv D(u^*t, x(t, u^*))u^*$$

with respect to the coordinates u_h^* ($h = 1, 2, ..., m$) of the vector parameter u^*:

$$\dot{x}'_{u_h^*}(t, u^*) \equiv \left(D'_{u_h}(u^*t, x(t, u^*))t \sum_{i=1}^{n} D'_{v_i}(u^*t, x(t, u^*))x'_{iu_h^*}(t, u^*) \right)u^*$$
$$+ d^h(u^*t, x(t, u^*)).$$

If we take into consideration also that $x(0, u^*) \equiv 0$ and, hence, $x'_{u_h^*}(0, u^*) \equiv 0$, we see that the function $x'_{u_h^*}(t, u^*)$ is a solution of the inhomogeneous linear system

$$\dot{y} = \sum_{i=1}^{n} D'_{v_i}(u^*t, x(t, u^*))u^* y_i$$
$$+ D'_{u_h}(u^*t, x(t, u^*))u^*t + d^h(u^*t, x(t, u^*)) \qquad (6.2.30)$$

satisfying also the initial condition $y(0) = 0$. At the same time, by substituting the function $y = td^h(u^*t, x(t, u^*))$ into (6.2.30) and applying (6.2.28) and (6.2.27) we can see that the latter function satisfies the same system and, clearly, the same initial condition. Thus,

$$x'_{u_h^*}(t, u^*) \equiv td^h(u^*t, x(t, u^*)).$$

Applying the last identity at $t = 1$ we get

$$w'_{u_h^*}(u^*) = x'_{u_h^*}(1, u^*) = d^h(u^*, x(1, u^*)) = d^h(u^*, w(u^*))$$

$$(h = 1, 2, ..., m).$$

With the star dropped in the last identity, this shows that the function defined in (6.2.29) (which is to be read also without the star) satisfies the system (6.2.26) of partial differential equations. Thus, the existence of a function w satisfying all the requirements of the theorem in the whole region U has been proved.

The uniqueness of the function $w : U \to \mathbf{R}^n$ can easily be proved by some standard method. It is, clearly, sufficient to prove that (6.2.26) has only one solution satisfying the initial condition $v(0) = 0$ and $|v(u)| < \beta$ in U. Assume that $w : U \to \mathbf{R}^n$ is such a solution and define a function of a single variable by

$$w^1(u_1) := w(u_1, 0, ..., 0), \qquad |u_1| < c.$$

This function, obviously, satisfies the system of ordinary differential equations

$$\frac{dw^1(u_1)}{du_1} = d^1(u_1, 0, ..., 0, w^1(u_1)), \qquad |u_1| < c,$$

and $w^1(0) = 0$. Thus, w^1 is unique and, hence, the values of w are also uniquely determined for $|u_1| < c$, $u_2 = \cdots = u_m = 0$. Let $|u_1^*| < c$ and consider the function of a single variable

$$w^2(u_2) = w(u_1^*, u_2, 0, ..., 0), \qquad |u_2| < c.$$

This function, obviously, satisfies the system

$$\frac{dw^2(u_2)}{du_2} = d^2(u_1^*, u_2, 0, ..., 0, w^2(u_2)), \qquad |u_2| < c,$$

and the initial condition $w^2(0) = w^1(u_1^*)$. Thus, w^2 is unique and, hence, the values of w are also uniquely determined for $|u_1| < c$, $|u_2| < c$, $u_3 = \cdots = u_m = 0$. If we proceed further in a similar manner, the uniqueness of the solution in U follows. $\square$

In what follows we return to the existence problem of the periodic solution of the perturbed system (6.2.1). The aim is to put the problem into a form where the Implicit Function Theorem 6.2.5 can be applied to the periodicity condition (6.2.4). The main purpose and the main difficulty of this study is to give explicit estimates to the constants a, β, Δ and c that determine the neighbourhood U: the existence domain of the perturbed periodic solution in the parameter plane. We simplify system (6.2.1) a little; the problem could be treated similarly if we worked with the original system, but the formulae would be still more burdensome. Besides that, we need now a C^2 right-hand side. Therefore, we introduce the problem again, though, keep to the notations applied in connection with (6.2.1) as far as possible.

Let $B \subset \mathbf{R}^n$ be an open bounded ball with centre in the origin, $f \in C^2(\overline{B}, \mathbf{R}^n)$, $g \in C^2(\mathbf{R} \times \overline{B}, \mathbf{R}^n)$ where $\overline{B}$ is the closure of B and consider the system

$$\dot{x} = f(x) + \mu g(t/T, x) \tag{6.2.31}$$

where $\mu \in \mathbf{R}$, $T > 0$, and g is periodic in the variable t with period T, i.e. $g(s+1, x) \equiv g(s, x)$, $s \in \mathbf{R}$, $x \in \overline{B}$. Suppose that the unperturbed system

$$\dot{x} = f(x) \tag{6.2.32}$$

has a non-constant periodic solution $p: \mathbf{R} \to B$ with period $T_0 > 0$, and that the number 1 is a simple characteristic multiplier of the variational system

$$\dot{y} = f_x'(p(t))y. \tag{6.2.33}$$

We use again the notations $p^0 = p(0) = [p_1^0, ..., p_n^0]$ and $\dot{p}^0 = \dot{p}(0) = [\dot{p}_1^0, ..., \dot{p}_n^0]$ and assume without loss of generality that $\dot{p}_1^0 \neq 0$, $\dot{p}_i^0 = 0$ ($i = 2, 3, ..., n$). Vectors orthogonal to $\dot{p}^0$ will be denoted by h, i.e. $h = [0, h_2, ..., h_n]$, and the solution of (6.2.31) that crosses the hyperplane passing through p^0 and orthogonal to $\dot{p}^0$ at the point $p^0 + h$ at time $t = \theta$ will be denoted by $\varphi(t, \theta, p^0 + h, \mu, T)$. This solution is periodic with period T if and only if (6.2.4) holds. The derivative matrix of z with respect to the variables $T, h_2, ..., h_n$ will be denoted by ∂z:

$$\partial z(\theta, h, \mu, T) = [\dot{\varphi}(\theta + T, \theta, p^0 + h, \mu, T) + \varphi'_T(\theta + T, \theta, p^0 + h, \mu, T),$$
$$\varphi'_{h_2}(\theta + T, \theta, p^0 + h, \mu, T) - e^2, ..., \varphi'_{h_n}(\theta + T, \theta, p^0 + h, \mu, T) - e^n]$$

where $e^k = \text{col}\,[0, 0, ..., 1, ..., 0]$ (the 1 at the k-th place) ($k = 2, ..., n$). We have proved in the proof of Theorem 6.2.1 that the determinant of this matrix at $(\theta, h, \mu, T) = (0, 0, 0, T_0)$ is non-zero. The determinant, the Jacobian of the system of equations (6.2.4), will be denoted by $J(\theta, h, \mu, T) := \det(\partial z(\theta, h, \mu, T))$, so that

$$J(0, 0, 0, T_0) \neq 0. \tag{6.2.34}$$

By Theorem 6.2.1 the conditions above guarantee the existence of the functions $\tau(\mu, \theta)$, $h(\mu, \theta)$ satisfying certain properties such that $\varphi(t, \theta, p^0 + h(\mu, \theta), \mu, \tau(\mu, \theta))$ is a $\tau(\mu, \theta)$-periodic solution of the system

$$\dot{x} = f(x) + \mu g(t/\tau(\mu, \theta), x).$$

Our aim is to give an estimate for the domain of τ and h. For this purpose estimates are needed for the norm of the difference $J(\theta, h, \mu, T) - J(0, 0, 0, T)$ and for Δ ocurring in (6.2.23).

The following quantities are considered to be known:

$$F_0 = \max_{x \in B} |f(x)|, \qquad G_0 = \max_{\substack{x \in B \\ s \in \mathbf{R}}} |g(s, x)|,$$

$$F_1 = \max_{x \in B} |f'_x(x)|, \qquad G_1 = \max_{\substack{x \in B \\ s \in \mathbf{R}}} |g'_x(s, x)|,$$

$$F_2 = \max_{x \in B} |f''_{xx}(x)|, \qquad G_s = \max_{\substack{x \in B \\ s \in \mathbf{R}}} |g'_s(s, x)|. \tag{6.2.35}$$

$\Phi(t)$ will denote the fundamental matrix solution of the unperturbed variational system (6.2.33) that assumes the unit matrix at $t = 0$, i.e. $I = \Phi(0)$. This matrix is considered to be known along with its inverse $\Phi^{-1}(t)$ and

$$K := \max_{t \in [-T_0/2, T_0]} |\Phi(t)|, \qquad K_{-1} := \max_{t \in [-T_0/2, T_0]} |\Phi^{-1}(t)|. \tag{6.2.36}$$

It is to be noted that the first column of $\Phi(t)$ is $(1/\dot{p}_1^0)\dot{p}(t)$ and the i-th one $(i = 2, 3, ..., n)$ is $\varphi'_{h_i}(t; 0, p^0, 0, T_0)$. Thus,

$$P := \max |\dot{p}(t)| \le (K/n)|\dot{p}_1^0|, \qquad |\varphi'_{h_i}(t; 0, p^0, 0, T_0)| \le K/n, \qquad (6.2.37)$$
$$t \in [-T_0/2, T_0] \ (i = 2, 3, ..., n).$$

In the following estimates, of course, each F, G and K can be replaced by an upper bound.

The path of the periodic solution p will be denoted by $\gamma = \{x \in \mathbf{R}^n : x = p(t), t \in [0, T_0]\}$ and its distance from the boundary of the ball B by

$$\sigma = \text{dist}(\gamma, \partial B) > 0.$$

The domain in which the "initial phase" θ and the period T varies will be restricted from now on to

$$|\theta| < T_0/2, \qquad |T - T_0| < T_0/2. \qquad (6.2.38)$$

Since T_0 need not be the least positive period of the function p, conditions (6.2.38) do not restrict generality, at least, as long as periods greater than T_0 are concerned.

First of all we need bounds for $|\mu|$ and $|h|$ such that for arbitrary θ and T satisfying (6.2.38) the solution $\varphi(t; \theta, p^0 + h, \mu, T)$ of (6.2.31) should be defined and its path contained in B for $t \in [\theta, \theta + T]$.

Lemma 6.2.6. *If μ and h are such that*

$$(3/2)G_0 T_0|\mu| + |h| < \sigma \exp(-(3/2)F_1 T_0), \qquad (6.2.39)$$

then the solution $\varphi(t; \theta, p^0 + h, \mu, T)$ is defined and $\varphi(t; \theta, p^0 + h, \mu, T) \in B$ for all θ and T satisfying (6.2.38) and $t \in [\theta, \theta + T]$.

Proof. Consider the identities

$$\dot{\varphi}(t; \theta, p^0 + h, \mu, T) \equiv f(\varphi(t; \theta, p^0 + h, \mu, T)) + \mu g(t/T, \varphi(t; \theta, p^0 + h, \mu, T)),$$
$$(6.2.40)$$
$$\dot{p}(t - \theta) \equiv f(p(t - \theta)). \qquad (6.2.41)$$

Integrate the difference of these identities from θ to $t > \theta$:

$$\varphi(t; \theta, p^0 + h, \mu, T) - p^0 - h - p(t - \theta) + p^0$$
$$\equiv \int_\theta^t \left(f(\varphi(u; \theta, p^0 + h, \mu, T)) - f(p(u - \theta)) \right.$$
$$\left. + \mu g(u/T, \varphi(u; \theta, p^0 + h, \mu, T)) \right) du.$$

Provided that the path of the solution φ is in B in the interval $[\theta, t)$ we get from here

$$|\varphi(t; \theta, p^0 + h, \mu, T) - p(t - \theta)|$$
$$\le |h| + \int_\theta^t \left(F_1|\varphi(u; \theta, p^0 + h, \mu, T) - p(u - \theta)| + G_0|u| \right) du.$$

Applying Gronwall's lemma (see Problem 1.6.6) we obtain

$$|\varphi(t;\theta,p^0+h,\mu,T)-p(t-\theta)| \le (G_0|u|(t-\theta)+|h|)\exp(F_1(t-\theta)).$$

The solution $\varphi(t;\theta,p^0+h,\mu,T)$ is defined and its path is contained in B as long as

$$\text{dist}\,(\varphi(t;\theta,p^0+h,\mu,T),\gamma) < \sigma. \tag{6.2.42}$$

The left-hand side of the last inequality is less than the left-hand side of the previous one. Since $0 \le t-\theta \le T < (3/2)T_0$, inequality (6.2.42) follows from (6.2.39) for all $t \in [\theta,\theta+T]$. $\square$

From now on it will be assumed that μ and h satisfy (6.2.39). The following three Lemmata yield estimates for the variations of the columns of the matrix $\partial z(\theta,h,\mu,T)$. The abbreviated notations

$$\delta\varphi'_{h_i}(t) = \varphi'_{h_i}(t;\theta,p^0+h,\mu,T) - \varphi'_{h_i}(t+T_0-T-\theta;0,p^0,0,T_0)$$
$$(i=2,3,...,n),$$
$$\delta f'_x(t) = f'_x(\varphi(t;\theta,p^0+h,\mu,T)) - f'_x(p(t+T_0-T-\theta))$$

will be used. Obviously, $\delta\varphi'_{h_i}(\theta+T)$ is the difference between the i-th columns of the matrices $\partial z(\theta,h,\mu,T)$ and $\partial z(0,0,0,T_0)$.

Lemma 6.2.7. *If conditions (6.2.38) and (6.2.39) are satisfied, then*

$$|\delta\varphi'_{h_i}(\theta+T)|$$
$$\le \ ((3/2)((3/2)G_0T_0|\mu|+|h|+F_0|T-T_0|)F_2T_0\exp[(3/2)F_1T_0]$$
$$+(3/2)G_1T_0|\mu|+F_1|T-T_0|)(K/n)\exp[(3/2)T_0(F_1+G_1|\mu|)]$$
$$(i=2,3,...,n). \tag{6.2.43}$$

Proof. Differentiating the identity (6.2.40) with respect to h_i we get

$$\dot{\varphi}'_{h_i}(t;\theta,p^0+h,\mu,T)$$
$$\equiv \ \big(f'_x(\varphi(t;\theta,p^0+h,\mu,T))$$
$$+\mu g'_x(t/T,\varphi(t;\theta,p^0+h,\mu,T))\big)\varphi'_{h_i}(t;\theta,p^0+h,\mu,T).$$
$$\tag{6.2.44}$$

Substituting $\theta=0$, $h=0$, $\mu=0$, $T=T_0$ and writing $t+T_0-T-\theta$ for t, we obtain the identity

$$\dot{\varphi}'_{h_i}(t+T_0-T-\theta;0,p^0,0,T_0)\equiv f'_x(p(t+T_0-T-\theta))\varphi'_{h_i}(t+T_0-T-\theta;0,p^0,0,T_0).$$

Subtracting the last identity from the previous one and integrating the difference from θ to $t > \theta$, we get

$$\varphi'_{h_i}(t;\theta;p^0+h,\mu,T) - \varphi'_{h_i}(t+T_0-T-\theta;0,p^0,0,T_0)$$

$$\equiv \; \varphi'_{h_i}(\theta; \theta, p^0 + h, \mu, T) - \varphi'_{h_i}(T_0 - T; 0, p^0, 0, T_0)$$

$$+ \int_\theta^t \Big(\big(f'_x(\varphi(u; \theta, p^0 + h, \mu, T)) \big.$$

$$+ \mu g'_x(u/T, \varphi(u; \theta, p^0 + h, \mu, T)) \big) \big(\varphi'_{h_i}(u; \theta, p^0 + h, \mu, T)$$

$$- \varphi'_{h_i}(u + T_0 - T - \theta; 0, p^0, 0, T_0) \big)$$

$$+ \big(f'_x(\varphi(u; \theta, p^0 + h, \mu, T)) + \mu g'_x(u/T, \varphi(u; \theta, p^0 + h, \mu, T))$$

$$- f'_x(p(u + T_0 - T - \theta)) \big) \varphi'_{h_i}(u + T_0 - T - \theta; 0, p^0, 0, T_0) \Big) du,$$

or applying the notations previously introduced, we get

$$\delta\varphi'_{h_i}(t) \; \equiv \; \delta\varphi'_{h_i}(\theta) + \int_\theta^t \Big(\big(f'_x(\varphi(u; \theta, p^0 + h, \mu, T)) \big.$$

$$+ \mu g'_x(u/T, \varphi(u; \theta, p^0 + h, \mu, T)) \big) \delta\varphi'_{h_i}(u)$$

$$+ \big(\delta f'_x(u) + \mu g'_x(u/T, \varphi(u; \theta, p^0 + h, \mu, T)) \big)$$

$$\bullet \varphi'_{h_i}(u + T_0 - T - \theta; 0, p^0, 0, T_0) \Big) du. \qquad (6.2.45)$$

In order to continue, an estimate is needed for $|\delta\varphi'_{h_i}(\theta)|$ and $|\delta f'_x(t)|$. Clearly,

$$\delta\varphi'_{h_i}(\theta) \; = \; \varphi'_{h_i}(\theta; \theta, p^0 + h, \mu, T) - \varphi'_{h_i}(T_0 - T; 0, p^0, 0, T_0)$$

$$= \; e^i - \varphi'_{h_i}(T_0 - T; 0, p^0, 0, T_0)$$

$$= \; \varphi'_{h_i}(0; 0, p^0, 0, T_0) - \varphi'_{h_i}(T_0 - T; 0, p^0, 0, T_0).$$

As it is well known, $\varphi'_{h_i}(t; 0, p^0, 0, T_0)$ satisfies system (6.2.33); thus,

$$|\delta\varphi'_{h_i}(\theta)| = \Big| \int_0^{T_0 - T} f'_x(p(t)) \varphi'_{h_i}(t; 0, p^0, 0, T_0) dt \Big|.$$

Hence, applying (6.2.35) and (6.2.37), we have

$$|\delta\varphi'_{h_i}(\theta)| \le F_1(K/n)|T - T_0|. \qquad (6.2.46)$$

Applying mean value theorems for the elements of the matrix $\delta f'_x(t)$ we easily obtain

$$|\delta f'_x(t)| \le F_2 |\varphi(t; \theta, p^0 + h, \mu, T) - p(t + T_0 - T - \theta)|. \qquad (6.2.47)$$

To estimate the right-hand side, one may proceed as in the proof of Lemma 6.2.6. Besides (6.2.4) one has to consider, instead of (6.2.41), the identity

$$\dot{p}(t + T_0 - T - \theta) \equiv f(p(t + T_0 - T - \theta)).$$

The procedure yields

$$\begin{aligned}
|\varphi(t;\theta,p^0+h,\mu,T) &- p(t+T_0-T-\theta)| \\
&\leq (G_0|\mu|(t-\theta)+|h|+F_0|T-T_0|)\exp(F_1(t-\theta)) \\
&\leq ((3/2)G_0T_0|\mu|+|h|+F_0|T-T_0|)\exp((3/2)F_1T_0). \quad (6.2.48)
\end{aligned}$$

Applying the last inequality in (6.2.47), we get

$$|\delta f'_x(t)| \leq F_2((3/2)G_0T_0|\mu|+|h|+F_0|T-T_0|)\exp((3/2)F_1T_0), \ t\in[\theta;\theta+T].$$
$$(6.2.49)$$

Making use of (6.2.46), (6.2.49), (6.2.35) and (6.2.37) we get from (6.2.45) the following inequality:

$$\begin{aligned}
|\delta\varphi'_{h_i}(t)| \leq \ & F_1(K/n)|T-T_0| + \int_\theta^t ((F_1+|\mu|G_1)|\delta\varphi'_{h_i}(u)| \\
& +(F_2((3/2)G_0T_0|\mu|+|h|+F_0|T-T_0|)\exp((3/2)F_1T_0) \\
& +|\mu|G_1)(K/n))du, \qquad T\in[\theta,\theta+T].
\end{aligned}$$

Applying Gronwall's Lemma (see Problem 1.6.6) we get

$$\begin{aligned}
|\delta\varphi'_{h_i}(t)| \leq \ & ((F_2((3/2)G_0T_0|\mu|+|h|+F_0|T-T_0|)\exp((3/2)F_1T_0) \\
& +G_1|\mu|)(K/n)(t-\theta) \\
& +F_1(K/n)|T-T_0|)\exp[(F_1+|\mu|G_1)(t-\theta)].
\end{aligned}$$

Substituting $t=\theta+T$ and taking (6.2.38) into account, we conclude (6.2.43) readily. $\square$

Lemma 6.2.8. *If conditions (6.2.38) and (6.2.39) are satisfied, then*

$$\begin{aligned}
|\dot\varphi(\theta+T;\theta,p^0+h,\mu,T) &- \dot\varphi(T_0;0,p^0,0,T_0)| \\
= \ & |\dot\varphi(\theta+T;\theta,p^0+h,\mu,T)-\dot p(T_0)| \\
\leq \ & F_1((3/2)G_0T_0|\mu|+|h|+F_0|T-T_0|)\exp((3/2)F_1T_0)+G_0|\mu|.
\end{aligned}$$
$$(6.2.50)$$

Proof. Substituting $t=\theta+T$ into (6.2.40), $t=\theta+T_0$ into (6.2.41) and subtracting the two identities, we obtain

$$\begin{aligned}
|\dot\varphi(\theta+T;\theta,p^0&+h,\mu,T)\dot p(T_0)| \\
= \ & |f(\varphi(\theta+T;\theta,p^0+h,\mu,T))-f(p(T_0)) \\
& +\mu g((\theta+T)/T,\varphi(\theta+T;\theta,p^0+h,\mu,T))| \\
\leq \ & F_1|\varphi(\theta+T;\theta,p^0+h,\mu,T)-p(T_0)|+|\mu|G_0.
\end{aligned}$$

Taking into account (6.2.48) at $t=\theta+T$, we can conclude (6.2.50). $\square$

Lemma 6.2.9. *If conditions (6.2.38) and (6.2.39) are satisfied, then*

$$|\varphi'_T(\theta + T; \theta, p^0 + h, \mu, T)| \le |\mu|((2\theta + 3T_0)/T_0)G_s \exp[(3/2)T_0(F_1 + G_1|\mu|)].$$
$$(6.2.51)$$

Proof. Differentiating (6.2.40) with respect to T, we get

$$\dot{\varphi}'_T(t; \theta, p^0 + h, \mu, T)$$
$$\equiv (f'_x(\varphi(t; \theta, p^0 + h, \mu, T))$$
$$+\mu g'_x(t/T, \varphi(t; \theta, p^0 + h, \mu, T)))\varphi'_T(t; \theta, p^0 + h, \mu, T)$$
$$-\mu(t/T^2)g'_s(t/T, \varphi(t; \theta, p^0 + h, \mu, T)).$$

Integrating from θ to $t > \theta$, we obtain

$$\varphi'_T(t; \theta, p^0 + h, \mu, T) - \varphi'_T(\theta; \theta; p^0 + h, \mu, T)$$
$$\equiv \int_\theta^t \Big((f'_x(\varphi(u; \theta, p^0 + h, \mu, T))$$
$$+\mu g'_x(u/T, x(u; \theta, p^0 + h, \mu, T)))\varphi'_T(u; \theta, p^0 + h, \mu, T)$$
$$-\mu(u/T^2)g'_s(u/T, \varphi(u, \theta, p^0 + h, \mu, T)) \Big) du.$$

The second term on the left-hand side is zero since $x(\theta; \theta, p^0 + h, \mu, T) \equiv p^0 + h$ does not depend on T. Taking into account (6.2.35) and (6.2.38) for $t \in [\theta, \theta + T]$, we have the inequality

$$|\varphi'_T(t; \theta, p^0 + h, \mu, T)|$$
$$\le \int_\theta^t ((F_1 + |\mu|G_1)|\varphi'_T(u; \theta, p^0 + h, \mu, T)| + |\mu|(\theta + T)G_s/T^2)du.$$

Applying Gronwall's lemma (see Problem 1.6.6) we have

$$|\varphi'_T(u; \theta, p^0 + h, \mu, T)| \le |\mu|((\theta + T)/T^2)G_s T \exp[(F_1 + |\mu|G_1)T],$$
$$t \in [\theta, \theta + T].$$

With (6.2.38) taken into account again and with the substitution of $t = \theta + T$ into the last inequality (6.2.51) follows. $\square$

The following two Lemmata are needed for giving estimates for Δ of (6.2.22) in our case.

Lemma 6.2.10. *If conditions (6.2.38) and (6.2.39) are satisfied, then*

$$|\varphi'_\mu(\theta + T; \theta, p^0 + h, \mu, T) - \varphi'_\mu(T_0; 0, p^0, 0, T_0)|$$
$$\le (3/2)G_s(|T - T_0| + |\theta|)$$
$$+(3/2)(G_1T_0 + F_2KK_{-1}G_0T_0^2)((3/2)G_0T_0|\mu| + |h| + F_0|T - T_0|)$$
$$\cdot \exp((3/2)F_1T_0) + (3/2)G_1KK_{-1}G_0T_0^2|\mu| + KK_{-1}G_0|T - T_0|)$$
$$\cdot \exp[(3/2)T_0(F_1 + G_1|\mu|)].$$
$$(6.2.52)$$

Proof. The proof is similar to the proof of Lemma 6.2.7 and will not be given in detail. However, attention is drawn to the fact that $\varphi'_\mu(t; 0, p^0, 0, T_0)$ satisfies the inhomogeneous linear system

$$\dot{y} = f'_x(p(t))y + g(t/T_0, p(t))$$

and $\varphi'_\mu(0; 0, p^0, 0, T_0) = 0$; thus,

$$\varphi'_\mu(t; 0, p^0, 0, T_0) = \Phi(t) \int_0^t \Phi^{-1}(u)g(u/T_0, p(u))du. \quad \square$$

Lemma 6.2.11. *If conditions (6.2.38) and (6.2.39) are satisfied, then*

$$|\varphi'_\theta(\theta + T; \theta, p^0 + h, \mu, T) - \varphi'_\theta(T_0; 0, p^0, 0, T_0)|$$
$$\leq ((3/2)((3/2)G_0T_0|\mu| + |h| + F_0|T - T_0|)F_2T_0 \exp((3/2)F_1T_0)$$
$$+(3/2)G_1T_0|\mu| + F_1|T - T_0|)P \exp[(3/2)T_0(F_1 + G_1|\mu|)]. \quad (6.2.53)$$

Proof. The proof is similar to that of Lemma 6.2.7 and will not be given in detail. It is to be noted that $\varphi(t; \theta, p^0, 0, T_0) \equiv p(t - \theta)$ and hence $\varphi'_\theta(t; \theta, p^0, 0, T_0) \equiv -p(t - \theta)$. Thus $|\varphi'_\theta(t; \theta, p^0, 0, T_0)| \leq P$ (see (6.2.37)). $\square$

Applying the Implicit Function Theorem 6.2.5 to the periodicity condition (6.2.4) we arrive at (6.2.22) where $(u_1, u_2) = (\mu, \theta)$ and $(v_1, v_2, ..., v_n) = (T, h_2, ..., h_n)$, and these Lemmata make it possible to give an estimate for Δ of (6.2.23). If the derivative matrix of z with respect to $T, h_2, ..., h_n$ is calculated, we obtain, taking into account the remark preceding (6.2.37), that at $(\theta, h, \mu, T) = (0, 0, 0, T_0)$

$$\partial z(0, 0, 0, T_0) = \begin{pmatrix} p_1^0 & 0 & \cdots & 0 \\ 0 & 0 & \cdots & 0 \\ \cdot & \cdot & \cdots & \cdot \\ 0 & 0 & \cdots & 0 \end{pmatrix} + \Phi(T_0) - I.$$

We consider this matrix to be known along with its inverse which will be denoted by

$$\partial z^{-1} := (\partial z(0, 0, 0, T_0))^{-1}$$

(without writing out the arguments) for short.

The following formulae lead us to explicit values of the constants a, β and Δ occurring in Theorem 6.2.5.

Let $a_1, a_2, a_3, \beta_1, \beta_2, \beta_3$ be arbitrary positive numbers satisfying the following inequalities, respectively:

$$(3/2)G_0T_0a_1 + \beta_1 < \sigma \exp(-3F_1T_0/2), \quad (6.2.54)$$

$$a_2((3/2)F_1G_0T_0 \exp(3F_1T_0/2) + G_0 + 4G_s \exp[3T_0(F_1 + G_1a_2)/2])$$
$$+\beta_2((1 + F_0)F_1 \exp(3F_1T_0/2) < \frac{1}{n|\partial z^{-1}|}, \quad (6.2.55)$$

$$((3a_3/2)((3/2)F_2G_0T_0^2 \exp(3F_1T_0/2) + G_1T_0)$$
$$+\beta_3((3/2)(1 + F_0)F_2T_0 \exp(3F_1T_0/2) + F_1)) \exp(3T_0G_1a_3/2)$$
$$< \frac{1}{|\partial z^{-1}|K} \exp(-3F_1T_0/2). \qquad (6.2.56)$$

Further, let us define

$$a := \min(a_1, a_2, a_3, T_0/2)$$
$$\beta := \min(\beta_1, \beta_2, \beta_3, T_0/2),$$
$$H_1 := n(a((3/2)F_1G_0T_0 \exp(3F_1T_0/2) + G_0$$
$$+4G_s \exp[3T_0(F_1 + G_1a)/2])$$
$$+\beta(1 + F_0)F_1 \exp(3F_1T_0/2)),$$
$$H_2 := (3a/2)((3/2)F_2G_0T_0^2 \exp(3F_1T_0/2) + G_1T_0)$$
$$+\beta((3/2)(1 + F_0)F_2T_0 \exp(3F_1T_0/2) + F_1))$$
$$\cdot K \exp[3T_0(F_1 + G_1a)/2],$$
$$H := \max(H_1, H_2),$$
$$\Delta_1 := |\partial z^{-1}|KK_{-1}G_0T_0 + \frac{|\partial z^{-1}|}{1 - |\partial z^{-1}|H}\Big(((3/2)G_s(a + \beta)$$
$$+(3/2)(G_1T_0 + F_2KK_{-1}G_0T_0^2)$$
$$\cdot((3/2)G_0T_0a + \beta + F_0\beta) \exp(3F_1T_0/2)$$
$$+(3/2)G_1KK_{-1}G_0T_0^2a + KK_{-1}G_0\beta) \exp[3T_0(F_1 + G_1a)/2]$$
$$+|\partial z^{-1}|HKK_{-1}G_0T_0\Big),$$
$$\Delta_2 := \frac{|\partial z^{-1}|}{1 - |\partial z^{-1}|H}\Big(F_1((3/2)G_0T_0a + \beta + F_0\beta) \exp(3F_1T_0/2)$$
$$+G_0a + ((3/2)((3/2)G_0T_0a + \beta + F_0\beta)F_2T_0 \exp(3F_1T_0/2)$$
$$+(3/2)G_1T_0a + F_1\beta)P \exp[3T_0(F_1 + G_1a)/2]\Big),$$
$$\Delta := 2\max(\Delta_1, \Delta_2). \qquad (6.2.57)$$

Finally we introduce the notation

$$U = \{(\mu, \theta) \in \mathbf{R}^2 : |\mu| < \min(a, \beta/\Delta), |\theta| < \min(a, \beta/\Delta)\}.$$

Now, it is already possible to prove the following:

Theorem 6.2.12 *(Farkas [1978a]). Under the conditions following (6.2.31), to each $(\mu, \theta) \in U$ there belongs one and only one $\tau(\mu, \theta) > 0$ and one and only one $h(\mu, \theta) = (0, h_2(\mu, \theta), ..., h_n(\mu, \theta))$ such that the solution of system*

$$\dot{x} = f(x) + \mu g(t/\tau(\mu, \theta), x)$$

that assumes the value $p^0 + h(\mu, \theta)$ at $t = \theta$ is periodic with period $\tau(\mu, \theta)$,
$\tau \in C^1(U, \mathbf{R})$, $h \in C^1(U, \mathbf{R}^n)$, $|\tau(\mu, \theta) - T_0| < \beta \le T_0/2$, $|h(\mu, \theta)| < \beta$, for
$(\mu, \theta) \in U$, and $\tau(0,0) = T_0$, $h(0,0) = 0$.

Proof. The proof consists of checking the validity of Theorem 6.2.5 for
the system of equations (6.2.4). We have for the variation of the derivative
matrix of the system

$$|\partial z(\theta, h, \mu, T) - \partial z(0,0,0,T_0)|$$
$$= n \max\{|\dot\varphi(\theta + T, \theta, p^0 + h, \mu, T) - \dot p(T_0)| + |\varphi'_T(\theta + T, \theta, p^0 + h, \mu, T)|,$$
$$|\varphi'_{h_i}(\theta + T, \theta, p^0 + h, \mu, T) - \varphi'_{h_i}(T_0, 0, p^0, 0, T_0)|(i = 2, 3, ..., n)\}.$$

Applying Lemmata 6.2.8, 6.2.9 and 6.2.7 and taking into consideration
(6.2.55), (6.2.56) we see that for $|\mu| \le a$, $|\theta| \le a$, $|T - T_0| \le \beta$ and $|h| \le \beta$:

$$|\partial z(\theta, h, \mu, T) - \partial z(0,0,0,T_0)| \le H < 1/|\partial z^{-1}|.$$

Thus, by (6.2.34) and Ostrowski's Theorem A1.19 the matrix $\partial z(\theta, h, \mu, T)$
is regular. (We note that the matrix norm applied in this study:
$|A| := n \max_{i,k} |a_{ik}|$ for the $n \times n$ matrix $A = [a_{ik}]$ is greater than the
natural norm (A1.24), and it is easy to see that (A1.31) remains valid,
as a consequence.)
We have two systems for (6.2.22), namely

$$\partial z(\theta, h, \mu, T)d^1 = -\varphi'_\mu(\theta + T, \theta, p^0 + h, \mu, T) \qquad (6.2.58)$$

and

$$\partial z(\theta, h, \mu, T)d^2 = -(\dot\varphi(\theta + T, \theta, p^0 + h, \mu, T) + \varphi'_\theta(\theta + T, \theta, p^0 + h, \mu, T)). \qquad (6.2.59)$$

As it can be seen from the proof of Lemma 6.2.10

$$\varphi'_\mu(T_0, 0, p^0, 0, T_0) = \Phi(T_0) \int_0^{T_0} \Phi^{-1}(t)g(t/T_0, p(t))dt.$$

Clearly, $\dot\varphi(T_0, 0, p^0, 0, T_0) = \dot p(T_0) = \dot p(0) = \dot p^0$, and since
$\varphi(t, \theta, p^0, 0, T_0) \equiv p(t - \theta)$, we have $\varphi'_\theta(T_0, 0, p^0, 0, T_0) = -\dot p^0$. Therefore, at
$(\theta, h, \mu, T) = (0,0,0,T_0)$, systems (6.2.58) and (6.2.59) assume the forms

$$\partial z(0,0,0,T_0)d^1 = -\Phi(T_0) \int_0^{T_0} \Phi^{-1}(t)g(t/T_0, p(t))dt,$$

$$\partial z(0,0,0,T_0)d^2 = 0,$$

respectively. Hence,

$$d^1 = -\partial z^{-1}\Phi(T_0) \int_0^{T_0} \Phi^{-1}(t)g(t/T_0, p(t))dt, \qquad d^2 = 0,$$

at $(\theta, h, \mu, T) = (0, 0, 0, T_0)$. Estimates can be given for $d^1(\theta, h, \mu, T)$ and $d^2(\theta, h, \mu, T)$ and, thus, for the $n \times 2$ matrix, $D(\theta, h, \mu, T) = [d^1, d^2]$ by applying Lemmata 6.2.10, 6.2.8 and 6.2.11 and Ostrowski's Theorem A1.19. It turns out that for $|\mu| \leq a$, $|\theta| \leq a$, $|T - T_0| \leq \beta$, $|h| \leq \beta$ we have $|d^1(\theta, h, \mu, T)| \leq \Delta_1$ $|d^2(\theta, h, \mu, T)| \leq \Delta_2$, and as a consequence

$$|D(\theta, h, \mu, T)| \leq \Delta,$$

where Δ is defined by (6.2.57). Thus, condition (6.2.23) is fulfilled with this Δ, and this completes the proof. □

The theory of controllably periodic perturbations expounded earlier may yield answers to existence problems of periodic solutions in case the perturbation of the autonomous system is of fixed period. The theory of periodic perturbations (forcing) of autonomous systems has a vast literature. We mention here only some titles that have a closer relationship to the material dealt with here. These are the now classical paper of Loud [1959] and papers due to Duistermaat [1970], Amel'kin-Gaishun-Ladis [1975], Rosenblat-Cohen [1980, 1981], Wiggins-Holmes [1987], and Hausrath-Manasevich [1988]. Most books on non-linear oscillations like Minorsky [1947], Hale [1963], Stoker [1950], Roseau [1966], Kryloff-Bogoliuboff [1949], Hagedorn [1988], etc., treat the problem too. The results about controllably periodic perturbations provide an answer to some problems related to "*the entrainment of frequency*" or "*synchronisation*". This problem is put forward the best, probably, in Minorsky's [1947] classical text (p.341): "If a periodic electromotive force of frequency ω is applied to an oscillator tuned to a frequency ω_0, one observes the well known effect of beats, or heterodyning, which can be heard through a headphone in a circuit inductively coupled to the oscillator. As the difference between the two frequencies decreases, the pitch of the sound decreases, and from linear theory one may expect that the beat frequency should decrease indefinitely as $|\omega - \omega_0| \to 0$. In reality, the sound in the headphone disappears suddenly at a certain value of the difference $(\omega - \omega_0)$, and it is found that the oscillator frequency ω_0 falls in synchronism with, or is *entrained* by, the external frequency ω within a certain band of frequencies. This phenomenon is called *entrainment of frequency*, and the band of frequency in which the entrainment occurs is called the band or the *zone of entrainment*... in which both frequencies coalesce and there exists only one frequency ω. On the basis of linear theory, the difference $|\omega - \omega_0|$ should be zero for only one value of $\omega = \omega_0$ ... The phenomenon of entrainment of frequency is a manifestation of the non-linearity of the system and cannot be accounted for by linear theory." Besides the texts already quoted, this phenomenon is extensively treated by Blehman [1971] where also many applications can be found.

Thus, let $X \subset \mathbf{R}^n$ be an open and connected set, I_μ an open interval containing $\mu = 0$, the functions $f : X \to \mathbf{R}^n$ and $\tilde{g} : \mathbf{R} \times X \times I_\mu \to \mathbf{R}^n$

continuous and for fixed $t \in \mathbf{R}$ analytic in the variables $(x, \mu) \in X \times I_\mu$; suppose also that $\tilde{g}$ is periodic in the variable t with period $\tau_0 > 0$, and consider the periodically perturbed autonomous system

$$\dot{x} = f(x) + \mu \tilde{g}(t, x, \mu). \tag{6.2.60}$$

Suppose, further, that the unperturbed system

$$\dot{x} = f(x) \tag{6.2.61}$$

has a non-constant periodic solution $p : \mathbf{R} \to X$ with period $T_0 > 0$, and that the variational system with respect to p

$$\dot{y} = f'_x(p(t))y \tag{6.2.62}$$

has 1 as a simple characteristic multiplier. We draw the attention to the important fact that in treating controllably periodic perturbations we did not assume that T_0 is the least positive period of p.

System (6.2.60) will be embedded into a controllably periodically perturbed system. We define the function $g : \mathbf{R} \times X \times I_\mu \times I_T \to \mathbf{R}^n$ where $I_T = \{T \in \mathbf{R} : |T - T_0| < \beta\}$ for some $0 < \beta < T_0$ by

$$g(t/T, x, \mu) := \tilde{g}(t\tau_0/T, x, \mu).$$

This function is, clearly, periodic in the variable t with period T (for fixed T). Suppose that $\tau_0 \in I_T$. Then for $T = \tau_0$ the function g is periodic in t with period τ_0, and

$$g(t/\tau_0, x, \mu) \equiv \tilde{g}(t, x, \mu).$$

For the system

$$\dot{x} = f(x) + \mu g(t/T, x, \mu)$$

the conditions of Theorem 6.2.1 are valid (if g did not depend on μ, then also Theorem 6.2.12 could be applied, and actually a theorem analogous to this last one could be worked out for g's depending on μ). Thus, $(\mu, \theta) = (0, 0)$ has a neighbourhood $U \subset \mathbf{R}^2$ and an analytic function $\tau : U \to \mathbf{R}$ exists such that for $(\mu, \theta) \in U$ the system

$$\dot{x} = f(x) + \mu g(t/\tau(\mu, \theta), x, \mu)$$

has a periodic solution $q(t, \mu, \theta)$ with period $\tau(\mu, \theta)$ satisfying the properties announced in Theorem 6.2.1. This, obviously, implies that *if for a given* $\mu \in I_\mu$ *there exists a* $\theta \in \mathbf{R}$ *such that* $(\mu, \theta) \in U$ *and* $\tau(\mu, \theta) = \tau_0$ *then* (6.2.60) *has a* τ_0-*periodic solution.* This means that the existence of a periodic solution of the perturbed system (6.2.60) with the period τ_0 of the perturbation for every sufficiently small $|\mu|$ depends on whether *the range of the function* $\tau(\mu, \bullet)$ *for each* μ *with sufficiently small modulus contains* τ_0. If the difference between τ_0 and T_0 is large compared to the value of

T_0, we may replace T_0 with an integer multiple of it which may already be close to τ_0 (since T_0 of Theorem 6.2.1 need not be the least positive period of the unperturbed periodic solution).

The range of the function $\tau : U \to \mathbf{R}$ may be determined with the help of its power series expansion

$$\tau(\mu,\theta) = \sum_{k=0}^{\infty} \mu^k \tau_k(\theta)$$

which, in turn, can be calculated, in principle, as it has been done following (6.2.10). As this has been pointed out right after the proof of Theorem 6.2.4 for the determination of the coefficients of this power series, nothing else but a fundamental matrix of (6.2.62) is needed. By (6.2.5), $\tau_0(\theta) = \tau(0,\theta) \equiv T_0$. If $|\mu|$ is sufficiently small and $\theta = O(\mu)$, then one may use the approximation

$$\tau(\mu,\theta) \approx T_0 + \tau_1(\theta)\mu \qquad (6.2.63)$$

for practical purposes. The value of $\tau_1(\theta)$ can be determined applying Theorem 6.2.4.

We deduce now the system of equations that determine $\tau_1(\theta)$ explicitly. The fundamental matrix of (6.2.62) that assumes I at $t = 0$ will be denoted by $\Phi(t)$. As has been pointed out in the proof of Theorem 6.2.4, $\Phi(sT_0)$ is the fundamental matrix of (6.2.18). Applying formula (1.2.11) to the inhomogeneous system (6.2.16), the solution $\zeta(s)$ of the latter that satisfies the initial condition $\zeta(0) = c \in \mathbf{R}^n$ is

$$\zeta(s) = \Phi(sT_0)c$$
$$+ \Phi(sT_0) \int_0^s \Phi^{-1}(\sigma T_0)(T_0 g(\sigma + \theta/T_0, p(\sigma T_0), 0, T_0) + \kappa f(p(\sigma T_0)))d\sigma.$$

According to Theorem 6.2.4, if $\zeta(s+1) \equiv \zeta(s)$ and $\zeta_1(0) = c_1 = 0$, then $\kappa = \tau_1(0)$. In our case

$$g(\sigma + \theta/T_0, \ p(\sigma T_0), 0, T_0) \equiv \tilde{g}(\sigma\tau_0 + O\tau_0, p(\sigma T_0), 0).$$

By Lemma 2.2.1 ζ is periodic with period 1 if and only if

$$0 = \zeta(1) - \zeta(0) = (\Phi(T_0) - I)c$$
$$+ \Phi(T_0) \int_0^1 \phi^{-1}(sT_0)(T_0\tilde{g}(s\tau_0 + \theta\tau_0/T_0, p(sT_0), 0) + \kappa f(p(sT_0)))ds,$$

that is, iff

$$(\Phi(T_0) - I)c + \kappa\Phi(T_0) \int_0^1 \Phi^{-1}(sT_0)fp(sT_0))ds$$
$$= -T_0\Phi(T_0) \int_0^1 \Phi^{-1}(sT_0)\tilde{g}(s\tau_0 + \theta\tau_0/T_0, p(sT_0), 0)ds \quad (6.2.64)$$

where according to the second condition upon ζ the vector $c = \text{col}\,[0, c_2, ..., c_n]$. Equation (6.2.64) is an inhomogeneous linear system of equations for the unknowns $c_2, ..., c_n, \kappa$ which has a unique solution by Theorem 6.2.4, and in this solution $\kappa = \tau_1(\theta)$. In (6.2.64) only the right-hand side depends on θ. If this function of θ is determined, then we may determine the values of the function $\tau_1(\theta)$ in the domain of this function by solving the linear system.

By Theorem 6.2.4, θ may vary in some interval $(-\theta_0, \theta_0)$, $\theta_0 > 0$. We may imagine that the *range* of τ_1 is an open, bounded interval containing zero (naturally this need not be the case). Then for "small" fixed μ the range of the right-hand side of (6.2.63) is an interval containing T_0. The smaller $|\mu|$ the "amplitude of the external excitation" is, the smaller will be this interval, i.e. the "zone of entrainment" (as Minorsky put it; this time we use periods instead of frequencies). So this theory settles the problem of synchronization in principle for small amplitude periodic excitations at least.

6.3   The Stability of Perturbed Periodic Solutions

The problem of the stability of the perturbed periodic solution can be solved easily if the system is either a periodically perturbed periodic one or an autonomous one under autonomous perturbation. In the first case this problem was settled by Theorem 6.1.3 according to which if the unperturbed periodic solution is asymptotically stable and this stability is determined by the linearization, then the same is true for the perturbed periodic solution for sufficiently small perturbations. Similarly in the second case, as we shall see, if the unperturbed periodic solution is orbitally asymptotically stable and this is determined by the variational system (or in other words the conditions of the Andronov-Witt Theorem 5.1.2 hold), then the perturbed periodic solution is also orbitally asymptotically stable for sufficiently small perturbations. There is no wonder in this. In the first case all the characteristic multipliers of the unperturbed variational system are in the interior of the unit circle, and since these multipliers are continuous functions of the "perturbation parameter" μ, for $|\mu|$ sufficiently small they stay in the interior. In the second case, one multiplier is 1, and $n - 1$ multipliers are in the interior of the unit circle. If the perturbation is autonomous, then the perturbed variational system has 1 also as a characteristic multiplier (since it has a non-trivial periodic solution, cf. Theorem 6.2.2), and because of continuity, the remaining $n - 1$ multipliers stay inside the unit circle. Thus, in the first case Theorem 6.1.3, in the second the Andronov-Witt Theorem 5.1.2 applies to the perturbed system.

The situation is different if an autonomous system is subject to a periodic (non-autonomous) perturbation. If the *unperturbed autonomous system* has a non-constant periodic solution to which the Andronov-Witt

Theorem 5.1.2 applies, i.e. the unperturbed variational system has besides the simple multiplier 1, all the $n - 1$ characteristic multipliers in the interior of the unit circle, we cannot tell what will be the case with the *perturbed non-autonomous system*. To be sure, if the perturbation is sufficiently small, then the $n - 1$ multipliers stay in the interior of the unit circle, but the characteristic multiplier 1 may move either into the interior of the unit circle or out of it (for the comparison of three cases, see Figure 6.3.1).

Thus, we return now to system (6.2.1) with all the assumptions made there and suppose that the conditions of Theorem 6.2.1 hold. Then $(\mu, \theta) = (0, 0)$ has an open neighbourhood $U \subset \mathbf{R}^2$ such that there exists a unique "period function" $\tau \in C^1(U, \mathbf{R})$ such that the system

$$\dot{x} = f(x) = \mu g(t/\tau(\mu, \theta), x, \mu, \tau(\mu, \theta)) \tag{6.3.1}$$

has a unique periodic solution $q(t, \mu, \theta)$ with period $\tau(\mu, \theta)$ satisfying $\tau(0, 0) = T_0$ (the period of the non-constant periodic solution $p(t)$ of the unperturbed system $\dot{x} = f(x)$), and $q(t, 0, 0) = p(t)$. We now establish conditions for the asymptotic stability of the perturbed solution q. Consider the variational system of (6.3.1) with respect to the solution q,

$$\dot{y} = \left(f'_x(q(t, \mu, \theta)) + \mu g'_x(t/\tau(\mu, \theta), q(t, \mu, \theta), \mu, \tau(\mu, \theta))\right)y, \tag{6.3.2}$$

and denote the fundamental matrix of the latter system that assumes the unit matrix I at $t = 0$ by $\Phi(t, \mu, \theta)$. The corresponding principal matrix is

$$C(\mu, \theta) := \Phi(\tau(\mu, \theta), \mu, \theta).$$

If $(\mu, \theta) = (0, 0)$, then (6.3.2) is reduced to (6.2.3):

$$\dot{y} = f'_x(p(t))y \tag{6.3.3}$$

and

$$C(0, 0) = \Phi(T_0, 0, 0). \tag{6.3.4}$$

The matrix function $C : U \to \mathbf{R}^{n^2}$ is, obviously, continuous, and the roots of a polynomial are continuous functions of its coefficients (see, e.g., Fricke [1924], vol. I. p. 92). Since, by the assumption of Theorem 6.2.1, the number 1 is a simple characteristic multiplier of (6.3.3), i.e. a simple eigenvalue of $C(0, 0)$; therefore, $(\mu, \theta) = (0, 0)$ has an open neighbourhood $U_1 \subset U$, 1 has an open neighbourhood $W \subset \mathbf{R}$, and there exists one and only one function

$$\lambda : U_1 \to W \tag{6.3.5}$$

such that $\lambda \in C^0$, $\lambda(0, 0) = 1$ and for each $(\mu, \theta) \in U_1$ the real number $\lambda(\mu, \theta)$ is the unique (simple) eigenvalue of the matrix $C(\mu, \theta)$ in W. We shall need the following:

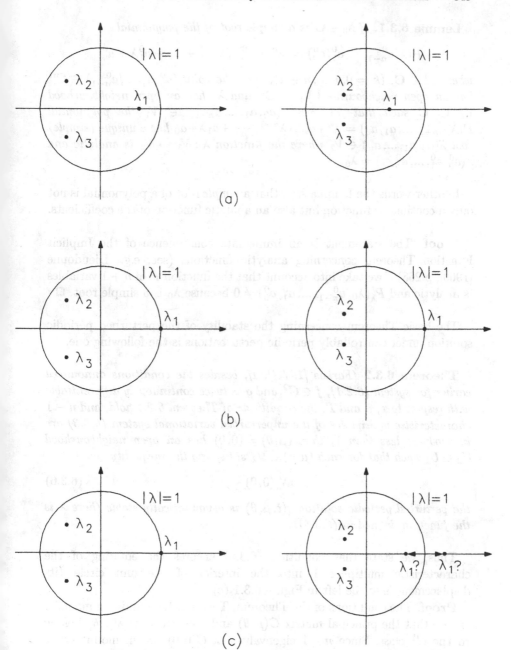

FIGURE 6.3.1. Characteristic multipliers of (a) an asymptotically stable periodic solution of a periodically perturbed periodic system, (b) an orbitally asymptotically stable periodic solution of an autonomously perturbed autonomous system, (c) an orbitally asymptotically stable periodic solution of a periodically perturbed autonomous system; the first column is the unperturbed case and the second column is the perturbed case.

Lemma 6.3.1. *If $\lambda_0 \in \mathbf{C}$ is a simple root of the polynomial*

$$P(\lambda, a_{n-1}^0, ..., a_1^0, a_0^0) := \lambda^n + a_{n-1}^0 \lambda^{n-1} + \cdots + a_1^0 \lambda + a_0^0$$

where $a_k^0 \in \mathbf{C}$, $(k = 0, 1, ..., n-1)$, then the point $(a_0^0, a_1^0, ..., a_{n-1}^0) \in \mathbf{C}^n$ has an open neighbourhood $V_1 \subset \mathbf{C}^n$ and λ_0 has an open neighbourhood $V_2 \subset \mathbf{C}$ such that for every $(a_0, a_1, ..., a_{n-1}) \in V_1$ the polynomial $P(\lambda, a_{n-1}, ..., a_1, a_0) = \lambda^n + a_{n-1}\lambda^{n-1} + \cdots + a_1\lambda + a_0$ has a unique (simple) root $\lambda(a_0, a_1, ..., a_n) \in V_2$ where the function $\lambda : V_1 \to V_2$ is analytic and $\lambda(a_0^0, a_1^0, ..., a_{n-1}^0) = \lambda_0$.

In other words the Lemma says that a *simple* root of a polynomial is not only a continuous function but also an analytic function of the coefficients.

Proof. The statement is an immediate consequence of the Implicit Function Theorem concerning analytic functions (see, e.g., Dieudonné [1960], p.268) if we take into account that the function P of $n + 1$ variables is analytic and $P'_\lambda(\lambda_0, a_{n-1}^0, ..., a_1^0, a_0^0) \neq 0$ because λ_0 is a simple root. $\square$

The basic Theorem concerning the stability of the perturbed periodic solution under controllably periodic perturbations is the following one.

Theorem 6.3.2 *(Farkas [1971]). If, besides the conditions announced earlier for system (6.2.1), $f \in C^2$ and g is twice continuously differentiable with respect to x, μ and T, the conditions of Theorem 6.2.1 hold, and $n-1$ characteristic multipliers of the unperturbed variational system (6.3.3) are in modulus less than 1, then $(\mu, \theta) = (0,0)$ has an open neighbourhood $U_2 \subset U_1$ such that for each $(\mu, \theta) \in U_2$ satisfying the inequality*

$$\mu\lambda'_\mu(0,0) < 0 \tag{6.3.6}$$

the perturbed periodic solution $q(t, \mu, \theta)$ is asymptotically stable (here λ is the function defined in (6.3.5)).

Thus, we state that condition (6.3.6) ensures the entering of the characteristic multiplier 1 into the interior of the unit circle (its displacement is to the left in Figure 6.3.1 (c)).

Proof. The conditions of the Theorem, Theorem 1.1.4 and Lemma 6.3.1 ensure that the principal matrix $C(\mu, \theta)$ and its eigenvalue $\lambda(\mu, \theta)$ belong to the C^1 class. Since $n-1$ eigenvalues of $C(0,0)$ are in modulus less than 1, the point $(\mu, \theta) = (0,0)$ has a neighbourhood $U_3 \subset U_1$ such that for $(\mu, \theta) \in U_3$, $n-1$ eigenvalues of $C(\mu, \theta)$ are in modulus less than 1 and stay outside a neighbourhood of the number 1 on the complex plane. We know from (6.2.5) that $q(t, 0, \theta) \equiv p(t - \theta)$ and $\tau(0, \theta) \equiv T_0$, so that for $\mu = 0$ system (6.3.2) is reduced to

$$\dot{y} = f'_x(p(t - \theta))y. \tag{6.3.7}$$

It is easy to see that the matrix

$$\Phi_\theta(t) := \Phi(t - \theta, 0, 0)$$

is a fundamental matrix of (6.3.7) (in general, different from $\Phi(t, 0, \theta)$). The corresponding principal matrix C_θ is by (2.2.4)

$$
\begin{aligned}
C_\theta &\equiv \Phi_\theta^{-1}(t)\Phi_\theta(t + T_0) \equiv \Phi^{-1}(t - \theta, 0, 0,)\Phi(t + T_0 - \theta, 0, 0) \\
&\equiv \Phi^{-1}(0, 0, 0)\Phi(T_0, 0, 0) \equiv C(0, 0).
\end{aligned}
$$

We see that for $\mu = 0$ the characteristic multipliers of system (6.3.2), i.e. (6.3.7), do not depend on θ; in particular,

$$\lambda(0, \theta) \equiv 1 \tag{6.3.8}$$

for sufficiently small $|\theta|$. Let $\lambda_\mu'(0, 0) > 0$ (resp. < 0); then because of the continuity of λ_μ', the point $(\mu, \theta) = (0, 0)$ has a neighbourhood $U_4 \subset U_3$ in which $\lambda_\mu'(\mu, \theta) > 0$ (resp. < 0), $(\mu, \theta) \in U_4$. Thus, for each fixed θ_0 for which $(0, \theta_0) \in U_4$, the function $\lambda(\mu, \theta_0)$ is strictly increasing (resp. decreasing), $(\mu, \theta_0) \in U_4$. Since $\lambda(0, \theta_0) = 1$, it follows that $(\mu, \theta) = (0, 0)$ has a neighbourhood $U_2 \subset U_4$ such that for every $(\mu, \theta) \in U_2$ for which $\mu < 0$ (resp. $\mu > 0$) holds, we have $-1 < \lambda(\mu, \theta) < 1$. Thus, for $(\mu, \theta) \in U_2$, satisfying (6.3.6) every characteristic multiplier of the perturbed variational system (6.3.2) is in modulus less than 1. By Theorem 4.2.1 this implies the asymptotic stability of the periodic solution q. □

This Theorem says that if the periodic solution of the unperturbed system is orbitally asymptotically stable and this can be decided by linearization (by the Andronov-Witt Theorem), then if $\lambda_\mu'(0, 0)$ is positive (negative), for negative (positive) μ's with sufficiently small modulus the perturbed periodic solution is asymptotically stable. Clearly, μ's satisfying (6.3.6) exist if and only if $\lambda_\mu'(0, 0) \neq 0$. In case this derivative is zero, this criterion cannot be applied. If this is the case, $\lambda_\mu'(0, 0) = 0$, then it still may happen that $(\mu, \theta) = (0, 0)$ is a conditional local strict maximum point of the function $\lambda(\mu, \theta)$ under the condition $\mu \neq 0$. This, clearly, implies that $(\mu, \theta) = (0, 0)$ has an open neighbourhood in which the corresponding perturbed periodic solution is asymptotically stable for *both positive and negative values of $\mu \neq 0$*. To check whether this is the case may not be easy since, because of (6.3.8), standard sufficient criteria of local maxima cannot be applied.

The proof of the previous Theorem also yields:

Theorem 6.3.3 *(Farkas [1971]). If the conditions of the previous Theorem hold, then $(\mu, \theta) = (0, 0)$ has a neighbourhood U_2 such that for each $(\mu, \theta) \in U_2$ satisfying $\mu\lambda_\mu'(0, 0) > 0$ the perturbed periodic solution $q(t, \mu, \theta)$ is unstable.*

As in the case of the existence problem, the existence theorem for controllably periodic perturbations (Theorem 6.2.1) implied the existence theorem for autonomous perturbations (Theorem 6.2.2); now also in case the perturbation is autonomous, the stability condition of the perturbed periodic solution is a simple consequence of Theorem 6.3.2. We consider system (6.2.6),

$$\dot{x} = f(x) + \mu g(x, \mu) \tag{6.3.9}$$

with the conditions imposed upon it, i.e. $f \in C^1(X)$, $X \subset \mathbf{R}^n$ open and connected, $g, g'_x, g'_\mu \in C^0(X, I_\mu)$, I_μ being an open interval containing $\mu = 0$, and the unperturbed system

$$\dot{x} = f(x)$$

has a non-constant periodic solution p with period $T_0 > 0$. There holds the following:

Theorem 6.3.4. *If* 1 *is a simple characteristic multiplier of the variational system (6.3.3) and the other $n - 1$ multipliers are in modulus less than* 1, *then there is a $\mu_1 > 0$ such that for each $\mu \in (-\mu_1, \mu_1)$ the periodic solution $r(t, \mu)$ of (6.3.9) whose existence is guaranteed by Theorem 6.2.2 is orbitally asymptotically stable with asymptotic phase.*

Proof. This follows easily from Theorem 6.3.2. Now the variational system (6.3.2) has the form

$$\dot{y} = f'_x(r(t, \mu)) + \mu g'_x(r(t, \mu), \mu))y. \tag{6.3.10}$$

As we know from Theorem 6.2.2, $r(t, 0) = p(t)$, and its period $T(\mu)$ is such that $T(0) = T_0$. The fundamental matrix of (6.3.10) that assumes I at $t = 0$ is denoted by $\Phi(t, \mu)$. The corresponding principal matrix is

$$C(\mu) := \Phi(T(\mu), \mu).$$

It is a continuous function of μ and so are its eigenvalues, the characteristic multipliers of (6.3.10). Denote the unique eigenvalue of $C(\mu)$ that corresponds to (6.3.5) by $\lambda(\mu)$; this is characterized by the condition that $\lambda(0) = 1$. However, since now the perturbed system is also autonomous and $r(t, \mu)$ is a non-constant periodic solution of it for sufficiently small $|\mu|$, it follows that 1 is an eigenvalue of $C(\mu)$; hence, $\lambda(\mu) \equiv 1$. The remaining $n - 1$ eigenvalues of $C(0)$ are in modulus less than 1; so because of the continuity a $\mu_1 > 0$ exists such that for $\mu \in (-\mu_1, \mu_1)$ the $n - 1$ corresponding eigenvalues of $C(\mu)$ are also in modulus less than 1. The conclusion follows then from the Andronov-Witt Theorem 5.1.2. $\square$

Note that now, unlike in Theorem 6.3.2, we did not have to assume that f and g are twice continuously differentiable because we needed only the continuous dependence of the characteristic multipliers on the perturbation parameter and not their differentiability.

One may imagine that the condition (6.3.6) of the stability of the perturbed periodic solution is only of theoretical nature since, seemingly, the determination of the quantity $\lambda'_\mu(0,0)$ requires the perturbed periodic solution, the principal matrix of the perturbed variational system, its eigenvalues, etc. We shall show that this is *not* the case. An explicit formula will be given for $\lambda'_\mu(0,0)$ which requires only the unperturbed periodic solution and the fundamental matrix of the unperturbed variational system.

In the sequel we assume that the conditions imposed upon system (6.2.1) and upon the unperturbed system (6.2.2), the conditions of Theorem 6.2.1 hold, and also (like in Theorem 6.3.2) that $f \in C^2$ and g is twice continuously differentiable with respect to x, μ and T. We shall make use of the transformation (6.2.9) and the notations introduced subsequently, i.e. substituting the perturbed period $\tau(\mu,\theta)$ into the system, in the transformed independent variable s where $t = \theta + s\tau(\mu,\theta)$ it assumes the form (6.2.10)

$$dx/ds = \tau(\mu,\theta)(f(x) + \mu g(s + \theta/\tau(\mu,\theta), x, \mu, \tau(\mu,\theta))), \qquad (6.3.11)$$

the latter system having the 1-periodic solution

$$\psi(s,\mu,\theta) := q(\theta + s\tau(\mu,\theta), \mu, \theta). \qquad (6.3.12)$$

Since now analyticity is not assumed, instead of the power series expansions of ψ and τ with respect to μ we have only the truncated formulae

$$\begin{aligned}\psi(s,\mu,\theta) &= \psi^0(s,\theta) + \mu\psi^1(s,\theta) + o(\mu) \\ &= p(sT_0) + \mu\psi^1(s,\theta) + o(\mu)\end{aligned} \qquad (6.3.13)$$

by (6.2.11) and

$$\tau(\mu,\theta) = \tau_0(\theta) + \mu\tau_1(\theta) + o(\mu) = T_0 + \mu\tau_1(\theta) + o(\mu) \qquad (6.3.14)$$

by (6.2.5). As before, the fundamental matrix of the perturbed variational system (6.3.2) that assumes the value I at $t = 0$ is denoted by $\Phi(t,\mu,\theta)$. Thus, the identity

$$\dot{\Phi}(t,\mu,\theta) \equiv \left(f'_x(q(t,\mu,\theta)) + \mu g'_x(t/\tau(\mu,\theta), q(t,\mu,\theta), \mu, \tau(\mu\theta))\right)\Phi(t,\mu,\theta) \qquad (6.3.15)$$

holds. As we have seen, $\Phi_\theta(t) := \Phi(t - \theta, 0, 0)$ is a fundamental matrix of the variational system (6.3.7). In particular $\Phi_0(t) = \Phi(t,0,0)$ is the fundamental matrix of the unperturbed variational system (6.3.3) that assumes I at $t = 0$. Clearly,

$$\Phi(t,0,\theta) \equiv \Phi_0(t - \theta)\Phi_0^{-1}(-\theta) = \Phi_\theta(t)\Phi_\theta^{-1}(0) \qquad (6.3.16)$$

since both are fundamental matrices of (6.3.7) and both assume I at $t = 0$. The principal matrix

$$C(\mu, \theta) = \Phi(\tau(\mu, \theta), \mu, \theta)$$

of the perturbed variational system (6.3.2) is continuously differentiable; hence,

$$C(\mu, \theta) = C_0(\theta) + \mu C_1(\theta) + \mu R(\mu, \theta) \qquad (6.3.17)$$

where the matrices C_0, C_1 and R are continuous and $R(0, \theta) \equiv 0$ in a neighbourhood of $(\mu, \theta) = (0, 0)$. Applying the identity $\tau(0, \theta) \equiv T_0$ and (6.3.16)

$$C_0(\theta) = C(0, \theta) = \Phi(T_0, 0, \theta) = \Phi_0(T_0 - \theta)\Phi_0^{-1}(-\theta).$$

In particular, at $\theta = 0$ we obtain

$$C_0(0) = \Phi_0(T_0),$$

the principal matrix of the unperturbed variational system (6.3.3).

We continue with the determination of the coefficient $C_1(\theta)$ of the expansion (6.3.17) which is a more difficult task:

$$C_1(\theta) \;=\; C_\mu'(0, \theta) = \dot{\Phi}(\tau(0, \theta), 0, \theta)\tau_\mu'(0, \theta) + \Phi_\mu'(\tau(0, \theta), 0, \theta)$$

$$\;=\; \dot{\Phi}(T_0, 0, \theta)\tau_1(\theta) + \Phi_\mu'(T_0, 0, \theta) \qquad (6.3.18)$$

where we used the expansion (6.3.14). The determination of the first term of the right-hand side is easy since (6.3.16) is a fundamental matrix of (6.3.7):

$$\dot{\Phi}(T_0, 0, \theta) \;=\; f_x'(p(T_0 - \theta))\Phi(T_0, 0, \theta)$$

$$\;=\; f_x'(p(T_0 - \theta))\Phi_0(T_0 - \theta)\Phi_0^{-1}(-\theta). \qquad (6.3.19)$$

In order to determine the second term on the right-hand side of (6.3.18) differentiate the identity (6.3.15) with respect to μ at $\mu = 0$:

$$\dot{\Phi}_\mu'(t, 0, \theta) \;=\; \left(f_{xx}''(p(t - \theta))q_\mu'(t, 0, \theta) + g_x'(t/T_0, p(t - \theta), 0, T_0) \right)\Phi(t, 0, \theta)$$

$$+ f_x'(p(t - \theta))\Phi_\mu'(t, 0, \theta) \qquad (6.3.20)$$

where notations introduced before (6.2.21) are used. From the expansion (6.3.13) returning from s to the independent variable t we may write

$$q(t, \mu, \theta) \;\equiv\; \psi((t - \theta)/\tau(\mu, \theta), \mu, \theta)$$

$$\;\equiv\; p((t - \theta)T_0/\tau(\mu, \theta)) + \mu\psi^1((t - \theta)/\tau(\mu, \theta), \theta) + o(\mu),$$

and differentiating with respect to μ at $\mu = 0$

$$q_\mu'(t, 0, \theta) \;=\; (\partial/\partial\mu)p((t - \theta)T_0/\tau(\mu, \theta))|_{\mu=0} + \psi^1((t - \theta)/T_0, \theta)$$

$$\;=\; \psi^1((t - \theta)/T_0, \theta) - \dot{p}(t - \theta)(t - \theta)\tau_1(\theta)/T_0$$

where (6.3.14) has been used. Substituting this expression and also (6.3.16) into (6.3.20) and denoting the first term on the right-hand side of this formula by $B(t, \theta)$ we obtain

$$\begin{aligned} B(t, \theta) &= f''_{xx}\big(p(t - \theta)\big)(\psi^1((t - \theta)/T_0, \theta) - \dot{p}(t - \theta)(t - \theta)\tau_1(\theta)/T_0) \\ &\quad + g'_x(t/T_0, p(t - \theta), \theta, T_0)\Phi_0(t - \theta)\Phi_0^{-1}(-\theta). \end{aligned}$$

Thus, we see that for fixed θ the matrix function $\Phi'_\mu(t, 0, \theta)$ satisfies the inhomogeneous matrix differential equation

$$\dot{Y} = f'_x(p(t - \theta))Y + B(t, \theta),$$

and since $\Phi(0, \mu, \theta) \equiv I$, it satisfies $\Phi'_\mu(0, 0, \theta) = 0$ also. Since a fundamental matrix of the corresponding homogeneous system (6.3.7) is $\Phi_\theta(t) = \Phi_0(t - \theta)$, the "method of the variation of constants" (cf. (1.2.11)) yields

$$\Phi'_\mu(t, 0, \theta) = \Phi_0(t - \theta)\int_0^t \Phi_0^{-1}(u - \theta)B(u, \theta)du.$$

Substituting this expression and (6.3.19) into (6.3.18) we get finally that

$$\begin{aligned} C_1(\theta) &= \tau_1(\theta)f'_x(p(T_0 - \theta))\Phi_0(T_0 - \theta)\Phi_0^{-1}(-\theta) \\ &\quad + \Phi_0(T_0 - \theta)\int_0^{T_0} \Phi_0^{-1}(t - \theta)B(t, \theta)dt. \end{aligned}$$

In particular,

$$C_1(0) = \tau_1(0)f'_x(p(T_0))\Phi_0(T_0) + \Phi_0(T_0)\int_0^{T_0} \Phi_0^{-1}(t)B(t, 0)dt \qquad (6.3.21)$$

where

$$\begin{aligned} B(t, 0) &= \big(f''_{xx}(p(t))(\psi^1(t/T_0, 0) - \dot{p}(t)t\tau_1(0)/T_0) \\ &\quad + g'_x(t/T_0, p(t), 0, T_0)\big)\Phi_0(t). \end{aligned}$$

After having determined the first two terms in the expansion (6.3.17) we give an explicit expression for the derivative (with respect to μ) of the eigenvalue $\lambda(\mu, \theta)$ of the matrix $C(\mu, \theta)$ that assumes the value 1 at $(\mu, \theta) = (0, 0)$, i.e. $\lambda(0, 0) = 1$ (cf. Theorem 6.3.2). The notations

$$(-1)^n d(\lambda, \mu, \theta) = \det(C(\mu, \theta) - \lambda I), \qquad (6.3.22)$$

$$d(\lambda, \mu, \theta) = \lambda^n + \alpha_{n-1}(\mu, \theta)\lambda^{n-1} + \cdots + \alpha_1(\mu, \theta)\lambda + \alpha_0(\mu, \theta)$$

will be introduced for the characteristic polynomial of $C(\mu, \theta)$. Clearly, the point $(\mu, \theta) = (0, 0)$ has a neighbourhood where the coefficients α_k are continuously differentiable ($k = 0, 1, ..., n - 1$), and, as a consequence, $d \in C^1$ as a function of a complex and two real variables. Further, the

i-th row of the matrices $C_0(\theta)$, $C_1(\theta)$, $R(\mu,\theta)$ occurring in (6.3.17) and that of the unit matrix I will be denoted by $c_i^0(\theta)$, $c_i^1(\theta)$, $r_i(\mu,\theta)$ and e_i, respectively $(i = 1, 2, ..., n)$. Thus, these matrices are

$$C_0(\theta) = \begin{pmatrix} c_1^0(\theta) \\ \vdots \\ c_n^0(\theta) \end{pmatrix}, \qquad C_1(\theta) = \begin{pmatrix} c_1^1(\theta) \\ \vdots \\ c_n^1(\theta) \end{pmatrix},$$

$$R(\mu,\theta) = \begin{pmatrix} r_1(\mu,\theta) \\ \vdots \\ r_n(\mu,\theta) \end{pmatrix}, \qquad I = \begin{pmatrix} e_1 \\ \vdots \\ e_n \end{pmatrix}.$$

Theorem 6.3.5 *(Farkas [1972]). If the conditions of Theorem 6.3.2 hold, then for the eigenvalue function $\lambda(\mu,\theta)$ occurring there we have*

$$\lambda_\mu'(0,0) = ((-1)^{n+1}/d_\lambda'(1,0,0)) \sum_{i=1}^{n} \det \begin{pmatrix} c_1^0(0) - e_1 \\ \vdots \\ c_{i-1}^0(0) - e_{i-1} \\ c_i^1(0) \\ c_{i+1}^0(0) - e_{i+1} \\ \vdots \\ c_n^0(0) - e_n \end{pmatrix} \qquad (6.3.23)$$

where the i-th term of the sum is obtained from $\det((C_0(0) - I)$ by replacing the i-th row by $c_i^1(0)$ $(i = 1, 2, ..., n)$.

Observe that $d(\lambda, 0, 0)$ is the characteristic polynomial of $\Phi_0(T_0) = C_0(0)$. This formula, (6.3.21) and Theorem 6.2.4 imply that for the determination of $\lambda_\mu'(0,0)$ only the unperturbed periodic solution p and the unperturbed fundamental matrix $\Phi_0(t)$ is needed.

Proof. Applying (6.3.17) the determinant (6.3.22) will be written in the form of a sum of determinants by the sums occurring in its last row:

$$(-1)^n d(\lambda, \mu, \theta) = \det(C(\mu,\theta) - \lambda I)$$

$$= \det \begin{pmatrix} c_1^0(\theta) + \mu c_1^1(\theta) + \mu r_1(\mu,\theta) - \lambda e_1 \\ \vdots \\ c_n^0(\theta) + \mu c_n^1(\theta) + \mu r_n(\mu,\theta) - \lambda e_n \end{pmatrix}$$

$$= \det \begin{pmatrix} c_1^0(\theta) - \lambda e_1 + \mu c_1^1(\theta) + \mu r_1(\mu,\theta) \\ \vdots \\ c_{n-1}^0(\theta) - \lambda e_{n-1} + \mu c_{n-1}^1(\theta) + \mu r_{n-1}(\mu,\theta) \\ c_n^0(\theta) - \lambda e_n \end{pmatrix}$$

$$+ \quad \mu \det \begin{pmatrix} c_1^0(\theta) - \lambda e_1 + \mu c_1^1(\theta) + \mu r_1(\mu,\theta) \\ \vdots \\ c_{n-1}^0(\theta) - \lambda e_{n-1} + \mu c_{n-1}^1(\theta) + \mu r_{n-1}(\mu,\theta) \\ c_n^1(\theta) \end{pmatrix}$$

$$+ \quad \mu \det \begin{pmatrix} c_1^0(\theta) - \lambda e_1 + \mu c_1^1(\theta) + \mu r_1(\mu,\theta) \\ \vdots \\ c_{n-1}^0(\theta) - \lambda e_{n-1} + \mu c_{n-1}^1(\theta) + \mu r_{n-1}(\mu,\theta) \\ r_n(\mu,\theta) \end{pmatrix}.$$

The determinant in the third term is zero at $\mu = 0$ because $r_n(0,\theta) = 0$. The determinant in the second term is written in the form of the sum of two determinants by the sum occurring in its $(n-1)$-st row: the first determinant will have $c_{n-1}^0 - \lambda e_{n-1}$ in its $(n-1)$-st row, and from the second, the factor μ can be taken out from the $(n-1)$-st row so that this term will be multiplied already by μ^2. Proceeding on like this also in the first term we get finally that

$$(-1)^n d(\lambda,\mu,\theta) = \det(C_0(\theta) - \lambda I)$$

$$+ \mu \sum_{i=1}^n \det \begin{pmatrix} c_1^0(\theta) - \lambda e_1 \\ \vdots \\ c_i^1(\theta) \\ \vdots \\ c_n^0(\theta) - \lambda e_n \end{pmatrix} + \mu r(\mu,\theta) \quad (6.3.24)$$

where $r \in C^0$ is a real valued function such that $r(0,\theta) = 0$. By our assumptions the number 1 is a simple root of the polynomial $d(\lambda,0,0)$, i.e. $d(1,0,0) = 0$ and $d_\lambda'(1,0,0,) \neq 0$. Therefore, as we have already proved (cf. Lemma 6.3.1), the unique function $\lambda(\mu,\theta)$ that makes $d(\lambda,\mu,\theta) = 0$ an identity and assumes 1 at $(\mu,\theta) = (0,0)$ belongs to the C^1 class, and by the Implicit Function Theorem

$$\lambda_\mu'(0,0) = -d_\mu'(1,0,0)/d_\lambda'(1,0,0).$$

Differentiating (6.3.24) with respect to μ at $(\lambda, \mu, \theta) = (1, 0, 0)$ we obtain

$$d'_\mu(1,0,0) = (-1)^n \sum_{i=1}^{n} \det \begin{pmatrix} c_1^0(0) - e_1 \\ \vdots \\ c_i^1 \\ \vdots \\ c_n^0(0) - e_n \end{pmatrix} \tag{6.3.25}$$

yielding (6.3.23). $\square$

Note that in the proof of this Theorem the assumption about the location of the remaining $n - 1$ characteristic multipliers is not needed; we need only that 1 be a simple multiplier.

The derivative $\lambda'_\mu(0,0)$ can also be expressed in terms of the row vectors of the matrices $C_0(0)$ and $C_1(0)$ only. Taking into account that

$$d(\lambda,0,0) = (-1)^n \det(\Phi_0(T_0) - \lambda I) = (-1)^n \det(C_0(0) - \lambda I)$$

and differentiating according to the differentiation rule of determinants we get

$$d'_\lambda(1,0,0) = (-1)^n \sum_{i=1}^{n} \det \begin{pmatrix} c_1^0(0) - e_1 \\ \vdots \\ c_{i-1}^0(0) - e_{i-1} \\ -e_i \\ c_{i+1}^0(0) - e_{i+1} \\ \vdots \\ c_n^0(0) - e_n \end{pmatrix} \tag{6.3.26}$$

where the i-th term of the sum is obtained from $\det((C_0(0) - I)$ by replacing the i-th row by $-e_i$ $(i = 1, 2, ..., n)$. Thus, combining (6.3.23) and (6.3.26),

$$\lambda'_\mu(0,0) = -\sum_{i=1}^{n} \det \begin{pmatrix} c_1^0(0) - e_1 \\ \vdots \\ c_i^1(0) \\ \vdots \\ c_n^0(0) - e_n \end{pmatrix} \Bigg/ \sum_{i=1}^{n} \det \begin{pmatrix} c_1^0(0) - e_1 \\ \vdots \\ -e_i \\ \vdots \\ c_n^0(0) - e_n \end{pmatrix}. \tag{6.3.27}$$

Hence, we obtain the following stability criterion.

Theorem 6.3.6 *(Farkas [1975b]). Suppose that the conditions of*

Theorem 6.3.2 hold and that

$$\text{sign} \sum_{i=1}^{n} \det \begin{pmatrix} c_1^0(0) - e_1 \\ \vdots \\ c_i^1(0) \\ \vdots \\ c_n^0(0) - e_n \end{pmatrix} = (-1)^k \text{ sign} \sum_{i=1}^{n} \det \begin{pmatrix} c_1^0(0) - e_1 \\ \vdots \\ -e_i \\ \vdots \\ c_n^0(0) - e_n \end{pmatrix};$$

then $(\mu, \theta) = (0,0)$ has a neighbourhood W such that for $(\mu, \theta) \in W$, if $k = 0$, then for $\mu > 0$; if $k = 1$, then for $\mu < 0$ the perturbed periodic solution $q(t, \mu, \theta)$ of system (6.2.1) is asymptotically stable.

Proof. This is an immediate consequence of Theorem 6.3.2 and formula (6.3.27). □

In the case $\lambda'_{\mu}(0,0) = 0$, Theorems 6.3.2 and 6.3.6 cannot be used for the determination of stability. In this case it may happen that $(\mu, \theta) = (0,0)$ is a maximum point of the function $\lambda(\mu, \theta)$, i.e. that $\lambda(\mu, \theta) \leq 1$ in a neighbourhood. The maximum cannot be strict since $\lambda(0, \theta) \equiv 1$. Therefore, the second differential of the function $\lambda(\mu, \theta)$, provided that it exists, cannot be negative definite, and well known sufficient criteria of maxima cannot be applied. However, if we fix $\theta = 0$, the problem can be handled. In this special case El-Owaidy [1972, 1975$_a$] managed to expand $C(\mu, 0)$ up to second order terms,

$$C(\mu, 0) = C_0(0) + \mu C_1(0) + \mu^2 C_2(0) + \mu^2 R(\mu, 0)$$

where R is continuous and $R(0,0) = 0$ provided that, naturally, f and g are three times continuously differentiable. He managed to give an explicit expression for the matrix $C_2(0)$ and for the second derivative $\lambda''_{\mu\mu}(0,0)$. Clearly, if $\lambda'_{\mu}(0,0) = 0$ and $\lambda''_{\mu\mu}(0,0) < 0$ and the rest of the conditions of Theorem 6.3.2 hold, then we have asymptotic stability for sufficiently small $|\mu| > 0$ and $\theta = 0$.

6.4 Controllably Periodic Perturbations of Van der Pol's Equation

The theory developed in the previous two Sections will be applied now to Van der Pol's equation. We rely on Farkas,I.-Farkas [1972] and on Tran Van Nhung [1983].

Consider Van der Pol's differential equation (cf. (3.2.5))

$$\ddot{u} + m(u^2 - 1)\dot{u} + u = 0 \tag{6.4.1}$$

where $m > 0$. In Section 3.2 we have shown that this equation has a unique non-constant periodic solution which will be denoted here by $u_m(t)$. Its least positive period will be denoted by $T_m > 0$. We may assume without loss of generality that

$$a := u_m(0) > 0, \qquad \dot{u}_m(0) = 0. \qquad (6.4.2)$$

We shall write equation (6.4.1) in the form of a system of first order equations following Loud [1959] by introducing the notations

$$
\begin{aligned}
x_1 &= -\dot{u} + m(a^3/3 - a) - m(u^3/3 - u), \\
x_2 &= u - a.
\end{aligned}
\qquad (6.4.3)
$$

Then (6.4.1) is equivalent to the system

$$
\begin{aligned}
\dot{x}_1 &= x_2 + a, \\
\dot{x}_2 &= -x_1 + m(a^3/3 - a) - m((x_2 + a)^3/3 - (x_2 + a)), \quad (6.4.4)
\end{aligned}
$$

or, with the notations $x := (x_1, x_2)$,

$$f(x) := (x_2 + a, -x_1 + m(a^3/3 - a) - m((x_2 + a)^3/3 - (x_2 + a)),$$

to the system

$$\dot{x} = f(x). \qquad (6.4.5)$$

The periodic solution of (6.4.5) corresponding to $u_m(t)$ is

$$p(t) = (-\dot{u}_m(t) + m(a^3/3 - a) - m(u_m^3(t)/3 - u_m(t)), u_m(t) - a).$$

The variational system of (6.4.5) with respect to the periodic solution p is

$$\dot{y} = f'_x(p(t))y \qquad (6.4.6)$$

where

$$f'_x(p(t)) = \begin{pmatrix} 0 & 1 \\ -1 & -m(u_m^2(t) - 1) \end{pmatrix}.$$

We know that $\dot{p}(t) = (\dot{u}_m(t), \ddot{u}_m(t))$ is a solution of (6.4.6). If the solution of the differential equation

$$\ddot{v} + m(u_m^2(t) - 1)\dot{v} + v = 0$$

that satisfies the initial conditions

$$v(0) = 0, \qquad \dot{v}(0) = 1/a \qquad (6.4.7)$$

is denoted by $v_m(t)$, then $(v_m(t), \dot{v}_m(t))$ is a second independent solution of (6.4.6). Denote the Wronskian of these two solutions by

$$W(t) = \begin{vmatrix} u_m(t) & v_m(t) \\ \dot{u}_m(t) & \dot{v}_m(t) \end{vmatrix}.$$

Conditions (6.4.2) and (6.4.7) imply that $W(0) = 1$; hence, by Liouville's formula (Theorem 1.2.4)

$$W(t) = \exp\left(-m \int_0^t (u_m^2(r) - 1)dr\right).$$

The fundamental matrix of (6.4.6) that assumes the unit matrix I at $t = 0$ is

$$\Phi(t) = \begin{pmatrix} u_m(t)/a & av_m(t) \\ \dot{u}_m(t)/a & a\dot{v}_m(t) \end{pmatrix}. \tag{6.4.8}$$

Clearly, $W(t) = \det \Phi(t)$; thus, the determinant of the principal matrix is

$$\det \Phi(T_m) = W(T_m) = \exp\left(-m \int_0^{T_m} (u_m^2(t) - 1)dt\right). \tag{6.4.9}$$

The product of the characteristic multipliers is equal to this determinant, and one of the characteristic multipliers of system (6.4.6) is 1 since the system has $\dot{p}(t)$ as a non-trivial T_m-periodic solution. Thus, the second characteristic multiplier is equal to (6.4.9). As we have pointed out following (5.1.29), the integral in (6.4.9) is positive, and as a consequence, the second characteristic multiplier is a positive number less than 1, i.e. the periodic solution $p(t)$ is orbitally asymptotically stable with asymptotic phase (see Theorem 5.1.2).

We turn now to controllably periodic perturbations of Van der Pol's equation. Denote the phase plane of $(u, \dot{u})$ by $\mathbf{R}^2$, let I_μ be an open interval containing $\mu = 0$, I_T an open interval of positive numbers containing $T = T_m$, $\gamma \in C^2(\mathbf{R} \times \mathbf{R}^2 \times I_\mu \times I_T, \mathbf{R})$ a periodic function in the first variable with period 1, i.e.

$$\gamma(s + 1, u, \dot{u}, \mu, T) \equiv \gamma(s, u, \dot{u}, \mu, T),$$

and consider the perturbed Van der Pol's equation

$$\ddot{u} + m(u^2 - 1)\dot{u} + u = \mu\gamma(t/T, u, \dot{u}, \mu, T), \qquad m > 0. \tag{6.4.10}$$

Applying the transformation (6.4.3) this equation is equivalent to the system

$$\dot{x} = f(x) + \mu g(t/T, x, \mu, T) \tag{6.4.11}$$

where f is the same as in (5.4.5) and

$g(s, x, \mu, T)$
$$:= (-\gamma(t/T, x_2 + a, -x_1 + m(a^3/3 - a) - m((x_2 + a)^3/3 - (x_2 + a)), \mu, T), 0).$$

Condition (6.4.2) implies that $p(0) = (0, 0)$ and $\dot{p}(0) = (a, 0)$. We denote, as in the general theory following (6.2.2), the solution of (6.4.11) that assumes the value $x = (0, h)$ at $t = \theta$ by $\varphi(t, \theta, h, \mu, T)$ and the

corresponding solution of the scalar second order differential equation
(6.4.10) by $u(t, \theta, a + h, \mu, T)$ (an abuse of notation). As seen from
(6.4.3), $u(\theta, \theta, a + h, \mu, T) = a + \varphi_2(\theta, \theta, h, \mu, T) = a + h$. There holds the
following:

Theorem 6.4.1 *(Farkas, I.-Farkas [1972]). For $m > 0$ the point
$(\mu, \theta) = (0, 0)$ has a neighbourhood such that to each (μ, θ) in this
neighbourhood there corresponds unique period $\tau(\mu, \theta)$ and a real number
$h(\mu, \theta)$ such that the solution*

$$w(t, \mu, \theta) := u(t, \theta, a + h(\mu, \theta), \mu, \tau(\mu, \theta))$$

*of the perturbed Van der Pol equation (6.4.10), where also $T = \tau(\mu, \theta)$ is to
be substituted, is periodic with period $\tau(\mu, \theta)$, the solution w as a function
of (μ, θ) and the functions τ and h belong to the C^2 class, $\tau(0, 0) = T_m$,
$h(0, 0) = 0$ and $w(t, 0, 0) = u_m(t)$.*

Proof. This is an immediate consequence of Theorem 6.2.1 if we take
into account also that f is analytic, $g \in C^2$, and that for $m > 0$, (6.4.9) is
less than 1. $\square$

We apply now Poincaré's method, the notations and the procedure
preceding Theorem 6.2.4 for the determination of the first approximations
to the periodic solution and its period. Note that if the periodic solution
$u_m(t)$ of the unperturbed Van der Pol equation is known, then $v_m(t)$ can be
determined easily since both are solutions of the same homogeneous linear
second order differential equation, and then by (6.4.8) the fundamental
matrix $\Phi(t)$ is explicitly given. We introduce the following notation:

$$\gamma^*(t) := \gamma((t + \theta)/T_m, u_m(t), \dot{u}_m(t), 0, T_m).$$

Theorem 6.4.2 *(Farkas,I.-Farkas [1972]). Under the same conditions
as in Theorem 6.4.1 in a sufficiently small neighbourhood of $(0, 0)$ the period
and the periodic solution of (6.4.10) corresponding to (μ, θ) are given by*

$$\tau(\mu, \theta) = T_m + \mu \tau_1(\theta) + o(\mu) \tag{6.4.12}$$

where

$$\tau_1(\theta) = \int_0^{T_m} (\gamma^*(r)/W(r))(\dot{v}_m(r) - \dot{u}_m(r)v_m(T_m)/(a(1 - W(T_m)))) dr \tag{6.4.13}$$

and

$$w(t, \mu, \theta) = u_m(t - \theta) + \mu\left(\dot{v}_m(t - \theta)\int_0^{t-\theta} (\gamma^*(r)\dot{u}_m(r)/W(r)) dr\right.$$

$$\left. -\dot{u}_m(t - \theta)\int_0^{t-\theta} (\gamma^*(r)v_m(r)/W(r)) dr\right.$$

$$+(\dot{v}_m(t-\theta)W(T_m)/(1-W(T_m)))\int_0^{T_m}(\gamma^*(r)\dot{u}_m(r)/W(r))dr\Big)$$

$$+o(\mu),\tag{6.4.14}$$

respectively.

Proof. Poincaré's method as developed preceding Theorem 6.2.4 will be applied. We denote the $\tau(\mu,\theta)$-periodic solution of system (6.4.11) with $T=\tau(\mu,\theta)$ corresponding to $w(t,\mu,\theta)$ by $q(t,\mu,\theta)$, apply transformation (6.2.9), $t=\theta+s\tau(\mu,\theta)$, and write

$$\psi(s,\mu,\theta) = q(t+s\tau(\mu,\theta),\mu,\theta)=p(sT_m)+\mu\psi^1(s,\theta)+o(\mu),$$
$$\tau(\mu,\theta) = T_m+\mu\tau_1(\theta)+o(\mu).$$

By Theorem 6.2.4, $z=\psi^1$ is a solution of system (6.2.16) with $\kappa=\tau_1(\theta)$. The identity expressing this has the form

$$d\psi^1(s,\theta)/ds \equiv T_m\begin{pmatrix}0 & 1\\ -1 & -m(u_m^2(sT_m)-1)\end{pmatrix}\psi^1(s,\theta)$$
$$+\ \tau_1(\theta)\begin{pmatrix}u_m(sT_m)\\ \dot{u}_m(sT_m)\end{pmatrix}+T_mg(s+\theta/T_m,p(sT_m),0,T_m).$$

The corresponding homogeneous system (cf (6.2.8)) has $\Phi(sT_m)$ as its fundamental matrix assuming I at $s=0$. Introducing the notations

$$c=(c_1,c_2)=\psi^1(0,\theta),$$
$$b(r)=\tau_1(\theta)(u_m(rT_m),\dot{u}_m(rT_m))+T_mg(r+\theta/T_m,p(rT_m),0,T_m),$$

and applying the method of the variation of the constants (Theorem 1.2.5) we have

$$\psi^1(s,\theta)=\Phi(sT_m)c+\Phi(sT_m)\int_0^s\Phi^{-1}(rT_m)b(r)dr.\tag{6.4.15}$$

By Theorem 6.2.4 the first coordinate of ψ^1 at $s=0$ is zero, i.e. $\psi_1^1(0,\theta)=c_1=0$, and ψ^1 is periodic with period 1, i.e. $\psi^1(1,\theta)=\psi^1(0,\theta)$. The last equation, the periodicity condition, represents a linear algebraic system of two equations for the two unknowns $\tau_1(\theta)$ and c_2. By solving this system we obtain the expression (6.4.13) for $\tau_1(\theta)$, and substituting the value of c_2 (and $c_1=0$) into (6.4.15), we get

$$\psi_1^1(s,\theta) = su_m(sT_m)\tau_1(\theta)$$
$$+\Big(W(T_m)v_m(sT_m)/(1-W(T_m))\Big)\int_0^{T_m}(\gamma^*(r)\dot{u}_m(r)/W(r))dr$$
$$-u_m(sT_m)\int_0^{sT_m}(\gamma^*(r)v_m(r)/W(r))dr$$

$$+v_m(sT_m)\int_0^{sT_m}(\gamma^*(r)\dot{u}_m(r)/W(r))dr,$$

$$\psi_2^1(s,\theta) = \dot{v}_m(sT_m)\int_0^{sT_m}(\gamma^*(r)\dot{u}_m(r)/W(r))dr$$

$$-\dot{u}_m(sT_m)\int_0^{sT_m}(\gamma^*(r)\dot{v}_m(r)/W(r))dr$$

$$+(\dot{v}_m(sT_m)W(T_m)/(1-W(T_m)))\int_0^{T_m}(\gamma^*(r)\dot{u}_m(r)/W(r))dr$$

$$+s\dot{u}_m(sT_m)\tau_1(\theta). \tag{6.4.16}$$

Taking into account (6.4.3) we get

$$w(\theta+s\tau(\mu,\theta),\mu,\theta) = a+\psi_2(s,\mu,\theta)=a+p_2(sT_m)+\mu\psi_2^1(s,\theta)+o(\mu)$$
$$= u_m(sT_m)+\mu\psi_2^1(s,\theta)+o(\mu)$$

or, returning to the original variable t,

$$w(t,\mu,\theta)=u_m((t-\theta)T_m/\tau(\mu,\theta))+\mu\psi_2^1((t-\theta)/\tau(\mu,\theta),\theta)+o(\mu).$$

Substituting (6.4.16) into the last expression and expanding everything again by μ up to order one, (6.4.14) is obtained. $\square$

We give now explicit conditions for the stability of the perturbed periodic solution. It is to be observed that the transformation (6.4.3) is such that the stability properties (Liapunov stability, asymptotic stability, instability) of a bounded solution of (6.4.10) that has a bounded derivative (this is the case, e.g., if a periodic solution is treated) are equivalent to the stability properties of the corresponding solution of system (6.4.11).

Under the conditions imposed upon the perturbed Van der Pol equation (6.4.10) let (μ,θ) belong to the neighbourhood of the point $(\mu,\theta)=(0,0)$ whose existence is guaranteed by Theorem 6.4.1, and let $\tau(\mu,\theta)$ be the corresponding period and $w(t,\mu,\theta)$ the periodic solution with period $\tau(\mu,\theta)$ of the differential equation

$$\ddot{u}+m(u^2-1)\dot{u}+u=\mu\gamma(t/\tau(\mu,\theta),u,\dot{u},\mu,\tau(\mu,\theta)). \tag{6.4.17}$$

There holds the following:

Theorem 6.4.3 (Farkas,I.-Farkas [1972]). *Under the conditions imposed upon (6.4.10) for sufficiently small $|\mu|$ and $|\theta|$ the periodic solution $w(t,\mu,\theta)$ of (6.4.17) is asymptotically stable, resp. unstable if*

$$\mu\lambda_\mu'(0,0)<0, \qquad resp. \qquad \mu\lambda_\mu'(0,0)>0$$

where

$$\lambda_\mu'(0,0) = \int_0^{T_m}\Big(\gamma_t'(t/T_m,u_m(t),\dot{u}_m(t),0,T_m)(\dot{v}_m(t)$$

$$-v_m(T_m)\dot{u}_m(t)/(a(1-W(T_m))))/W(t)\Big)dt. \tag{6.4.18}$$

(Here γ_t' denotes the partial derivative of the function γ occurring in (6.4.10) with respect to the variable t.)

Proof. In the proof instead of (6.4.17) we deal with system (6.4.11) substituting $T = \tau(\mu, \theta)$ and with its periodic solution corresponding to $w(t, \mu, \theta)$ that has been denoted by $q(t, \mu, \theta)$. As in the general theory $\lambda(\mu, \theta)$ denotes the characteristic multiplier of the variational system of (6.4.11) with $T = \tau(\mu, \theta)$ with respect to the periodic solution $q(t, \mu, \theta)$ that satisfies $\lambda(0, 0) = 1$.

The first part of the Theorem is an immediate consequence of Theorems 6.3.2-6.3.3. Only formula (6.4.18) is left to be proved.

In the proof we apply Theorem 6.3.5 and use the notations introduced preceding it with the obvious modifications. For the matrices occurring in the expansion (6.3.17) we have now by (6.4.8)

$$C_0(0) = \Phi(T_m) = \begin{pmatrix} 1 & av_m(T_m) \\ 0 & a\dot{v}_m(T_m) \end{pmatrix},$$

$$
\begin{aligned}
W(T_m) &= \det \Phi(T_m) = a\dot{v}_m(T_m), & (6.4.19) \\
d(\lambda, 0, 0) &= \lambda^2 - \lambda(1 + a\dot{v}_m(T_m)) + a\dot{v}_m(T_m), \\
d_\lambda'(1, 0, 0) &= 1 - a\dot{v}_m(T_m) = 1 - W(T_m).
\end{aligned}
$$

For the determination of the matrix $C_1(0)$ we use formula (6.3.21). A long but simple calculation follows from which we omit the details, writing down for future use only that

$$
\begin{aligned}
B(t, 0) &= \left(f_{xx}''(p(t))q_\mu'(t, 0, 0) + g_x'(t/T_m, p(t), 0, T_m)\right)\Phi(t), \\
q_\mu'(t, 0, 0) &= \psi^1(t/T_m, 0) - \dot{p}(t)t\tau_1(0)/T_m
\end{aligned}
$$

$$f_{xx}''(p(t))q_\mu'(t, 0, 0) = \begin{pmatrix} 0 & 0 \\ 0 & -2mu_m(t)q_{2\mu}'(t, 0, 0) \end{pmatrix} \qquad (6.4.20)$$

and introduce the row vector

$$
\begin{aligned}
w(t): \quad = \quad & [\dot{v}_m(t) - v_m(T_m)\dot{u}_m(t)/(a(1 - W(T_m))), \\
& -v_m(t) + v_m(T_m)u_m(t)/(a(1 - W(T_m)))]/W(t).
\end{aligned}
$$

The calculation yields by formula (6.3.23)

$$
\begin{aligned}
\lambda_\mu'(0, 0) \quad = \quad & -av_m(T_m)\tau_1(0)/(1 - W(T_m)) \\
& + \int_0^{T_m} w(t)\left(f_{xx}''(p(t))q_\mu'(t, 0, 0) + g_x'(t/T_m, p(t), 0, T_m)\right)\dot{p}(t)dt
\end{aligned}
$$

$$(6.4.21)$$

where the integrand is the product of the row vector w, a 2×2 matrix and the column vector $\dot{p}$.

It is easy to verify that the row vector $w(t)$ is periodic with period T_m and that its transpose, the column vector $w'(t)$, satisfies

$$\dot{w}'(t) \equiv -\Big(f'_x(p(t))\Big)' w'(t),$$

or taking the transpose of both sides, the row vector w satisfies

$$\dot{w}(t) \equiv -w(t)f'_x(p(t)). \tag{6.4.22}$$

Further, it is easy to see that $q'_\mu(t,0,0)$ satisfies the inhomogeneous system

$$\dot{q}'_\mu(t,0,0) \equiv f'_x(p(t))q'_\mu(t,0,0) + g(t/T_m,p(t),0,T_m). \tag{6.4.23}$$

Now consider the following derivative

$$\Big(w(t)(f'_x(p(t))q'_\mu(t,0,0) + g(t/T_m,p(t),0,T_m))\Big)^{\bullet}$$

$$= \dot{w}(t)\big(f'_x(p(t))q'_\mu(t,0,0) + g(t/T_m,p(t),0,T_m)\big)$$

$$+ w(t)\Big(f''_{xx}(p(t))q'_\mu(t,0,0)\dot{p}(t) + f'_x(p(t))\dot{q}'(t,0,0)$$

$$+ g'_t(t/T_m,p(t),0,T_m) + g'_x(t/T_m,p(t),0,T_m)\Big).$$

(Here g'_t and g'_x are the partial derivative with respect to t and the derivative matrix with respect to x of the vector function $g(t/T_m,x,0,T_m)$, respectively.) Substituting expressions (6.4.22) and (6.4.23) into the last identity for the integrand of (6.4.21) we obtain

$$w(t)\big(f''_{xx}(p(t))q'_\mu(t,0,0) + g'_x(t/T_m,p(t),0,T_m)\big)\dot{p}(t)$$

$$= \big(w(t)(f'_x(p(t))q'_\mu(t,0,0) + g(t/T_m,p(t),0,T_m))\big)^{\bullet}$$

$$- w(t)g'_t(t/T_m,p(t),0,T_m).$$

Substituting the last expression into (6.4.21), taking into account the periodicity of $w(t)$, $f'_x(p(t))$ and $g(t/T_m,p(t),0,T_m)$, expression (6.4.20) and the relation between g and γ, we arrive at (6.4.18). □

The unique non-constant periodic solution $u_m(t)$ of Van der Pol's equation (6.4.1) can be determined by a computer or, alternatively, in the next Section, methods will be shown how to determine the solution approximately developing it by powers of m. We saw that having determined this solution even numerically or approximately we may determine everything that is needed to find the perturbed periodic solution and its period (the period that is to be substituted for T into (6.4.10)) approximately (see (6.4.12)-(6.4.13)). Having determined the range of the period function $\tau(\mu, \theta)$ for fixed μ we may apply these results in case Van der Pol's equation is forced by a small amplitude periodic forcing term of *fixed period*. Formula (6.4.18) shows also that having determined $u_m(t)$ we may find everything needed to tell whether the perturbed periodic solution is asymptotically stable for positive or for negative μ's provided, of course, that $\lambda'_\mu(0,0) \neq 0$. In case $\lambda'_\mu(0,0) = 0$, then El Owaidy's [1972, 1975a] results may be applied to settle the stability problem.

Tran Van Nhung [1983] has performed the calculations preceding Theorem 6.2.12 and applied this Theorem to the perturbed Van der Pol equation determining the domain (or rather a lower estimate of the domain) in the μ, θ parameter plane of the perturbed periodic solution and its period. Because of lack of space we omit the details.

Liénard's equation (3.2.7) under controllably periodic perturbations was treated analogously by El Owaidy [1972,1975b], Farkas-Abdel Karim [1972] and Nguyen Van Su-Tran Van Nhung [1987].

6.5   Averaging

The method of averaging is one of the most effective perturbation techniques which may provide existence and stability results for periodic and "quasi-periodic" solutions of non-linear systems and may also yield the asymptotic determination of the periodic and "quasi-periodic" solutions. It has a vast literature; some of the most important references are Sanders-Verhulst [1985], Murdock [1988], Mitropolskij-Homa [1983], Samoilenko-Ronto [1979], and Volosov-Morgunov [1971]. Most of the classical texts like Minorsky [1947], Hale [1969], and Urabe [1967] also treat the method. In this Section we shall give a concise introductory survey of this method as well as of some related techniques, primarily the method of Krylov and Bogoliubov [1934]. Since the basic results have already been published in several books we shall omit most of the proofs.

Definition 6.5.1. We say that the function $f : \mathbf{R} \to \mathbf{R}^n$ is *quasi-periodic with (angular) frequencies* $\omega_1, ..., \omega_m$ $(\omega_k > 0, k = 1, 2, ..., m)$ if $f(t) = F(\omega_1 t, ..., \omega_m t)$ where $F : \mathbf{R}^m \to \mathbf{R}^n$ is a periodic function with period 2π in each variable, i.e. if

$$F(t_1, ..., t_k + 2\pi, ..., t_m) \equiv F(t_1, ..., t_k, ..., t_m) \quad (k = 1, 2, ..., m).$$

A quasi-periodic function usually is not periodic in the ordinary sense since the composite function of F and of $\omega_1 t, ..., \omega_m t$ is periodic "*with respect to the variable t at the k-th place*" with period $2\pi/\omega_k$, and the ratio $2\pi/\omega_k : 2\pi/\omega_l$, i.e. ω_l/ω_k is not necessarily rational for $k \neq l$. In the special case when all the *basic frequencies* $\omega_1, ..., \omega_m$ are in rational ratio to each other, the quasi-periodic function is periodic in the ordinary sense.

There is a powerful generalization of the concept of a quasi-periodic function; this is the property of "almost periodicity", see Levitan [1953] or Besicovitch [1954]. Informally a function is said to be almost periodic if its basic frequencies form an infinite set. Because of lack of space our studies will not be extended in this direction.

Most texts on the method of averaging start with results concerning systems given in "*standard form*", i.e. systems of the form

$$\dot{x} = \mu F(t, x)$$

where $t \in \mathbf{R}$, $x \in \mathbf{R}^n$, F satisfies certain conditions and $\mu \in \mathbf{R}$ is a small parameter. How we arrive at the standard form from a typical perturbation problem (cf. (6.1.1))

$$\dot{x} = f(t, x) + \mu g(t, x, \mu) \tag{6.5.1}$$

is seldom explained. Sanders-Verhulst [1985] is an exception. The process that reduces the last perturbed system to standard form is a generalization of the method of the variation of constants (cf. (1.2.11)). We assume that (6.5.1) satisfies the conditions of the existence and uniqueness theorem and that the solutions of the unperturbed system

$$\dot{y} = f(t, y) \tag{6.5.2}$$

are known. Denote the solution of (6.5.2) that assumes the value y at $t = 0$ by $\varphi(t, y)$. Now, varying the constant, i.e. the initial value y, we assume that the perturbed system (6.5.1) has a solution of the form $x(t) = \varphi(t, \eta(t))$ where $\eta(t)$ is some differentiable function. The question is that if $x(t)$ satisfies (6.5.1), then what differential system is satisfied by $\eta(t)$? Substituting into (6.5.1) we obtain

$$\begin{aligned} \dot{x}(t) &= \partial\varphi(t, \eta(t))/\partial t + (\partial\varphi(t, \eta(t))/\partial y)\dot{\eta}(t) \\ &= f(t, \varphi(t, \eta(t))) + \mu g(t, \varphi(t, \eta(t), \mu)). \end{aligned}$$

But $\partial\varphi(t, \eta(t))/\partial t = \dot{\varphi}(t, \eta(t)) = f(t, \varphi(t, \eta(t)))$, so that $\dot{\eta}(t) = \mu\varphi_y'^{-1}(t, \eta(t))g(t, \varphi(t, \eta(t)), \mu)$; note that $\varphi_y'(t, y)$ is a fundamental matrix of the variational system of (6.5.2) with respect to the solution $\varphi(t, y)$, hence it is regular. We conclude that if $\eta(t)$ is a solution of the system in standard form

$$\dot{z} = \mu\varphi_y'^{-1}(t, z)g(t, \varphi(t, z), \mu), \tag{6.5.3}$$

then $\varphi(t, \eta(t))$ is a solution of (6.5.1). Moreover, obviously, if $\eta(0) = \eta^0$, then also $\varphi(0, \eta(0)) = \eta^0$. Observe that if the function g does not depend on μ, then the right-hand side of the standard form depends on μ linearly.

In the important particular case when the unperturbed system is linear, i.e. if (6.5.1) is a "*quasi-linear system*"

$$\dot{x} = A(t)x + \mu g(t, x, \mu) \tag{6.5.4}$$

with a continuous matrix function $A(t)$, then $\varphi(t, y) = \Phi(t)y$ where $\Phi(t)$ is the fundamental matrix of

$$\dot{y} = A(t)y \tag{6.5.5}$$

that satisfies $\Phi(0) = I$, the unit matrix. In this case (6.5.3) assumes the form

$$\dot{z} = \mu\Phi^{-1}(t)g(t, \Phi(t)z, \mu). \tag{6.5.6}$$

If $\eta(t)$ is a solution of the last system, then $\varphi(t, \eta(t)) = \Phi(t)\eta(t)$ satisfies (6.5.4), and $\varphi(0, \eta(0)) = \Phi(0)\eta(0) = I\eta(0)$. In the case $A(t) \equiv A$, a constant matrix, $\Phi(t) = \exp(At)$, and (6.5.6) becomes

$$\dot{z} = \mu\exp(-At)g(t, \exp(At)z, \mu). \tag{6.5.7}$$

If A has eigenvalues with negative real parts, for instance, if A is a stable matrix, then even if g is a bounded function, the right-hand side of (6.5.7) may tend to infinity as $t \to \infty$, and this may cause problems about the existence of solutions over $\mathbf{R}_+$.

If the unperturbed system (6.5.2) (in particular (6.5.5)) is periodic with period $T_0 > 0$ and *all its solutions* $\varphi(t, y)$ *are periodic with period* T_0 and if also $g(t, x, \mu)$ is periodic in t with period $T > 0$, the right-hand side of (6.5.3) (in particular (6.5.6) or (6.5.7)) may be periodic or quasi-periodic depending on whether $T = T_0$ or not (or rather if the ratio T/T_0 is rational or not). In the first case if $\eta(t)$ is a T_0-periodic solution of (6.5.3), then $\varphi(t, \eta(t))$ is a T_0-periodic solution of the perturbed system (6.5.1). In the second case, if $\eta(t)$ is a quasi-periodic solution of (6.5.3), then $\varphi(t, \eta(t))$ is also quasi-periodic.

There is another way of reaching the standard form of a perturbation problem that is at the same time related to and a starting point of several classical perturbation methods. Though the method has been generalized to higher dimensional systems (see, e.g., Urabe [1967] and Hale [1969]) we shall present it in its classical form concerning second order scalar differential equations, i.e. "systems with one degree of freedom". Consider the scalar differential equation

$$\ddot{x} + \omega_0^2 x = \mu g(t, x, \dot{x}, \mu) \tag{6.5.8}$$

where μ is a real small parameter and $\omega_0 > 0$. The unperturbed system

$$\ddot{y} + \omega_0^2 y = 0$$

is the harmonic oscillator; all its solutions are periodic with period $2\pi/\omega_0$, i.e. it is *"fully oscillatory"*. Its general solution is $y(t) = a\cos(\omega_0 t + \theta)$ where the amplitude $a \geq 0$ and the phase $\theta \in \mathbf{R}$ are arbitrary. We assume now that for small $|\mu|$ the perturbation, i.e. the right-hand side of (6.5.8), causes the amplitudes and the phases of the solutions to vary slowly in time without affecting the oscillatory behaviour of the solutions. In other words we assume that (6.5.8) has solutions of the form

$$x(t) = a(t)\cos(\omega_0 t + \theta(t)). \tag{6.5.9}$$

We have to determine the differential equations to be satisfied by the amplitude function a and the phase function θ. Differentiating x we obtain (suppressing the writing out of the argument t where possible)

$$\dot{x} = \dot{a}\cos(\omega_0 t + \theta) - a\dot{\theta}\sin(\omega_0 t + \theta) - a\omega_0\sin(\omega_0 t + \theta).$$

The additional condition

$$\dot{a}\cos(\omega_0 t + \theta) - a\dot{\theta}\sin(\omega_0 t + \theta) = 0 \qquad (6.5.10)$$

is imposed. Then

$$\ddot{x} = -\dot{a}\omega_0\sin(\omega_0 t + \theta) - a\dot{\theta}\omega_0\cos(\omega_0 t + \theta) - a\omega_0^2\cos(\omega_0 t + \theta).$$

Substituting into (6.5.8) we obtain

$$\dot{a}\omega_0\sin(\omega_0 t + \theta) + a\dot{\theta}\omega_0\cos(\omega_0 t + \theta)$$
$$= -\mu g(t, a\cos(\omega_0 t + \theta), -a\omega_0\sin(\omega_0 t + \theta), \mu). \quad (6.5.11)$$

From (6.5.10) and (6.5.11), $\dot{a}$ and $\dot{\theta}$ can be expressed easily:

$$\dot{a} = -(\mu/\omega_0)g(t, a\cos(\omega_0 t + \theta), -a\omega_0\sin(\omega_0 t + \theta), \mu)\sin(\omega_0 t + \theta),$$
$$\dot{\theta} = -(\mu/(\omega_0 a))g(t, a\cos(\omega_0 t + \theta), -a\omega_0\sin(\omega_0 t + \theta), \mu)\cos(\omega_0 t + \theta).$$
$$(6.5.12)$$

We see that (6.5.12) is already in standard form and it is equivalent to (6.5.8) provided, naturally, that $a(t) \neq 0$.

We return now to the general perturbation problem in standard form. Let $X \subset \mathbf{R}^n$ be an open and connected set, $F \in C^0(\mathbf{R}_+ \times X, \mathbf{R}^n)$, $F'_x \in C^0(\mathbf{R}_+ \times X, \mathbf{R}^{n^2})$, $|F|$ and the matrix norm $|F'_x|$ bounded, and consider the system

$$\dot{x} = \mu F(t, x) \qquad (6.5.13)$$

where $\mu \in \mathbf{R}$ is a "small parameter". We construct "*the averaged system*" of (6.5.13) by replacing the right hand side by its "time average" over the interval $[0, \infty)$. Set

$$G(x) := \lim_{T \to \infty} (1/T) \int_0^T F(t, x)\,dt \qquad (6.5.14)$$

provided that the limit exists for $x \in X$; the *averaged system* corresponding to (6.1.13) is

$$\dot{y} = \mu G(y). \qquad (6.5.15)$$

The essence of the *method of averaging* consists of gaining information about the solutions of (6.5.13) in case solutions of the averaged system (6.5.15) are known.

Two general theorems will be announced first. For the proofs see the references quoted. The first one is the *Bogoliubov-Demidovich Theorem* establishing the nearness of solutions of (6.5.13) to solutions of (6.5.15) *on finite intervals.*

Theorem 6.5.1 (*see Demidovich [1967ₐ] and Mitropolskij-Homa [1983]*). *Besides the conditions imposed upon F, let $M \geq |F(t,x)|$ for $(t,x) \in \mathbf{R}_+ \times X$ where the Euclidean norm is used, suppose that the limit (6.5.14) exists uniformly in $x \in X$, that for any given $x^0 \in D$ the averaged system (6.5.15) has a unique solution $\psi(t, \mu)$ satisfying $\psi(0, \mu) = x^0$ and that at $\mu = 1$ this solution is defined on the interval $[0, T_1]$, $T_1 > 0$, and denote the solution of (6.5.13) that assumes the value x^0 at $t = 0$ by $\varphi(t, \mu)$ and the distance of x^0 from the boundary ∂X by $\rho > 0$. Then for each $\eta > 0$ there exists a $\mu_0 > 0$ such that for $0 \leq \mu \leq \mu_0$*

$$|\varphi(t, \mu) - \psi(t, \mu)| < \eta$$

on the interval $t \in [0, L/\mu]$ where $L = \min(T_1, \rho/(M\sqrt{n}))$.

It is said sometimes that *"for small μ the difference between $\varphi(t, \mu)$ and $\psi(t, \mu)$ is small on the time scale $1/\mu$."*

The following Theorem due to Banfi and Filatov establishes the nearness of the solutions of the original and the averaged system on infinite intervals provided that the participating solution of the averaged system is uniformly asymptotically stable (see Definition 1.4.6).

Theorem 6.5.2 (*Mitropolskij-Homa [1983]*). *Besides the conditions imposed upon F, suppose that for any $x \in X$ the limit*

$$G(x) = \lim_{T \to \infty} (1/T) \int_t^{t+T} F(t, x) dt$$

exists uniformly in $t \in \mathbf{R}_+$ and that the function G is bounded, assume further that for some $x^0 \in X$ the solution $\psi(t, \mu)$ of the averaged system

$$\dot{y} = \mu G(y)$$

satisfying $\psi(0, \mu) = x^0$ is defined for $t \in \mathbf{R}_+$, its path has a δ-neighbourhood that is contained in X and that $\psi(\tau/\mu, \mu)$ is a uniformly asymptotically stable solution of $dy/d\tau = G(y)$. Then for each $\eta > 0$ there exists a $\mu_0 > 0$ such that for $0 \leq \mu < \mu_0$

$$|\varphi(t, \mu) - \psi(t, \mu)| < \eta, \qquad t \in \mathbf{R}_+$$

where φ is the solution of (6.5.13) satisfying $\varphi(0, \mu) = x^0$.

We note that there are estimates available for μ_0 if the permissible error η is prescribed (see Schapiro-Sethna [1976]).

We turn now to the application of the method of averaging to periodic systems. Consider the system in standard form

$$\dot{x} = \mu F(t, x, \mu) \tag{6.5.16}$$

where $F, F'_x, F'_\mu \in C^0(\mathbf{R} \times X \times I_\mu)$, $X \subset \mathbf{R}^n$ an open and connected set, I_μ an open interval containing $\mu = 0$ and suppose that F is periodic in the variable t with period $T_0 > 0$:

$$F(t + T_0, x, \mu) \equiv F(t, x, \mu).$$

It is easy to see that in this periodic case

$$
\begin{aligned}
G(x): \quad &= \lim_{T \to \infty} (1/T) \int_0^T F(t, x, 0) dt \\
&= (1/T_0) \int_0^{T_0} F(t, x, 0) dt.
\end{aligned}
\tag{6.5.17}
$$

Consider the averaged system

$$\dot{y} = \mu G(y). \tag{6.5.18}$$

There holds the following:

Theorem 6.5.3 *(Urabe [1967], Samoilenko-Ronto [1979]). (i) System (6.5.16) has a periodic solution of period T_0 for each sufficiently small $|\mu|$ only if the averaged system (6.5.18) has an equilibrium point; (ii) if (6.5.18) has an equilibrium point $y^0 \in X$ and*

$$\det G'_y(y_0) \neq 0, \tag{6.5.19}$$

then system (6.5.16) has a unique periodic solution of period T_0 in a neighbourhood of y^0 for each sufficiently small $|\mu|$; (iii) in the latter case if the real parts of all the eigenvalues of the matrix $\mu G'_y(y_0)$ are negative, then (the equilibrium y^0 of (6.5.18) and) the corresponding T_0-periodic solution of (6.5.16) is asymptotically stable.

Proof. Denote the solution of (6.5.16) that assumes $x^0 \in X$ at $t = 0$ by $\varphi(t, x^0, \mu)$, i.e.

$$\varphi(0, x^0, \mu) = x^0. \tag{6.5.20}$$

φ is continuously differentiable with respect to x^0 and μ and, clearly, $\varphi(t, x^0, 0) \equiv x^0$. Fix μ for a moment and consider the function $\varphi(t, x^0, s\mu)$. The partial derivative of the latter function with respect to s is $\mu \varphi'_\mu(t, x^0, s\mu)$; hence,

$$\varphi(t, x^0, \mu) - x^0 = \mu \int_0^1 \varphi'_\mu(t, x^0, s\mu) ds. \tag{6.5.21}$$

Introducing the notation

$$\tilde{\varphi}(t, x^0, \mu) := \int_0^1 \varphi'_\mu(t, x^0, s\mu)ds, \tag{6.5.22}$$

obviously, $\tilde{\varphi}(0, x^0, \mu) = 0$ and by (6.5.21) and (6.5.16)

$$\begin{aligned}
\dot{\tilde{\varphi}}(t, x^0, \mu) &\equiv F(t, \varphi(t, x^0, \mu), \mu) \\
&\equiv F(t, x^0 + \mu\tilde{\varphi}(t, x^0, \mu), \mu).
\end{aligned} \tag{6.5.23}$$

Thus, the function $\tilde{\varphi}$ defined by (6.5.22) is a solution of the system

$$\dot{z} = F(t, x^0 + \mu z, \mu)$$

satisfying the initial condition $z(0) = 0$. As a consequence, $\tilde{\varphi}$ is continuously differentiable with respect to x^0 and μ. As we know (Lemma 2.2.1) the solution φ of (6.5.16) is T_0-periodic if and only if

$$\varphi(T_0, x^0, \mu) - x^0 = 0.$$

By (6.5.21) this condition is equivalent to

$$\tilde{\varphi}(T_0, x^0, \mu) = 0, \tag{6.5.24}$$

provided that $\mu \neq 0$. If (6.5.24) has a solution in x^0 for each sufficiently small $|\mu|$, then by continuity $\tilde{\varphi}(T_0, x^0, 0) = 0$ must have a solution too. But by (6.5.23) and (6.5.17)

$$\tilde{\varphi}(T_0, x^0, 0) = \int_0^{T_0} F(t, x^0, 0)dt = T_0 G(x^0);$$

hence, condition (6.5.24) is equivalent to

$$G(x^0) = 0. \tag{6.5.25}$$

Thus, (6.5.16) has a T_0-periodic solution for each $|\mu|$ only if (6.5.25) has a solution, and this proves (i).

Assume now that (6.5.25) has a solution $y^0 \in X$, and (6.5.19) holds. This implies then that $\tilde{\varphi}(T_0, y^0, 0) = 0$ and that

$$\det \tilde{\varphi}'_{x^0}(T_0, y^0, 0) \neq 0.$$

According to the Implicit Function Theorem, equation (6.5.24) has a unique solution $x^0(\mu)$ for each μ with sufficiently small modulus such that $x^0(\mu)$ is continuously differentiable and $x^0(0) = y^0$. This means that

$$p(t, \mu) := \varphi(t, x^0(\mu), \mu) = x^0(\mu) + \mu\tilde{\varphi}(t, x^0(\mu), \mu) \tag{6.5.26}$$

is a uniquely determined T_0-periodic solution of (6.5.16) for $|\mu|$ sufficiently small; $\mu \neq 0$ proving (ii).

In order to prove (iii) consider the variational system of (6.5.16) with respect to the periodic solution (6.5.26),

$$\dot{z} = \mu F_x'(t, p(t, \mu), \mu)z, \qquad (6.5.27)$$

and denote the fundamental matrix of this system that assumes I at $t = 0$ by $\Phi(t, \mu)$. This fundamental matrix will be written in the form

$$\Phi(t, \mu) = I + \mu \Phi_1(t, \mu)$$

since the variational system at $\mu = 0$ is $\dot{z} = 0$, and, therefore, $\Phi(t, 0) \equiv I$. Substituting the expression for $\Phi(t, \mu)$ into (6.5.27) we obtain

$$\mu \dot{\Phi}_1(t, \mu) \equiv \mu F_x'(t, p(t, \mu), \mu)(I + \mu \Phi_1(t, \mu)),$$

i.e.

$$\dot{\Phi}_1(t, \mu) = F_x'(t, p(t, \mu), \mu). \qquad (6.5.28)$$

Here the right-hand side is a continuous function of μ; hence, $\Phi_1(t, \mu)$, being its integral function, is also a continuous function. $\Phi(0, \mu) = I$ implies that $\Phi_1(0, \mu) = 0$, $\mu \neq 0$, and by continuity also $\Phi_1(0, 0) = 0$. Integrating (6.5.28) from 0 to T_0 yields

$$\Phi_1(T_0, \mu) = \int_0^{T_0} F_x'(t, p(t, \mu), \mu)dt.$$

At $\mu = 0$ we have $p(t, 0) \equiv x^0(0) = y^0$, and

$$\Phi_1(T_0, 0) = \int_0^{T_0} F_x'(t, y^0, 0)dt = T_0 G_x'(y^0)$$

by (6.5.17). If all the eigenvalues of $\mu G_x'(y^0)$ have negative real parts, then the same is true for $\mu T_0 G_x'(y^0)$ (since T_0 is a positive factor), and by continuity, for sufficiently small $|\mu| \neq 0$ also all the eigenvalues of $\mu \Phi_1(T_0, \mu)$ have negative real parts. (To be sure, depending on the sign of the real parts of the eigenvalues belonging to the matrix $G_x'(y^0)$ this is true either for $\mu > 0$ or for $\mu < 0$, the two cases excluding each other.) Denote an arbitrary eigenvalue of $\Phi_1(T_0, \mu)$ by $\lambda(\mu) = \alpha(\mu) + i\beta(\mu)$ where α and β are bounded continuous real functions in a neighbourhood of $\mu = 0$. Then the corresponding eigenvalue of $\mu \Phi_1(T_0, \mu)$ is $\mu \lambda(\mu) = \mu \alpha(\mu) + i\mu\beta(\mu)$, and that of the principal matrix

$$\Phi(T_0, \mu) = I + \mu \Phi_1(T_0, \mu),$$

is $1 + \mu\lambda(\mu) = 1 + \mu\alpha(\mu) + i\mu\beta(\mu)$. Thus,

$$|1 + \mu\lambda(\mu)|^2 = 1 + 2\mu\alpha(\mu) + \mu^2(\alpha^2(\mu) + \beta^2(\mu)).$$

If, by assumption, $\mu\alpha(\mu) < 0$, then for sufficiently small $|\mu| \neq 0$ this identity implies that $|1 + \mu\lambda(\mu)| < 1$, i.e. all the characteristic multipliers are in modulus less than 1; hence, by Theorem 4.2.1 the periodic solution $p(t, \mu)$ is asymptotically stable. $\square$

The method of averaging will be illustrated by

Example 6.5.1 *(Van der Pol's equation considered as a perturbation of the harmonic oscillator).* Van der Pol's equation (3.2.5) has been treated already by several methods. It is assumed now in the form

$$\ddot{x} + x = \mu(1 - x^2)\dot{x}, \tag{6.5.29}$$

x is a scalar function and μ a small parameter. The unperturbed equation $\ddot{y} + y = 0$ is the linear harmonic oscillator whose general solution is $y(t) = a\cos(t + \theta)$ with $a, \theta \in \mathbf{R}$ arbitrary. The method applied to (6.5.8) will be used first, i.e. we assume the solutions of (6.5.29) in the form

$$x(t) = a(t)\cos(t + \theta(t)).$$

Then according to (6.5.12) the amplitude and phase function satisfy

$$\begin{aligned} \dot{a} &= \mu a(1 - a^2\cos^2(t + \theta))\sin^2(t + \theta), \\ \dot{\theta} &= \mu(1 - a^2\cos^2(t + \theta))\sin(t + \theta)\cos(t + \theta), \end{aligned} \tag{6.5.30}$$

and this is already a system in standard form, though unlike the original equation it is no longer autonomous. We are interested in the 2π-periodic solutions of the periodic system (6.5.30). Performing the averaging, (6.5.30) is replaced by the averaged system

$$\dot{\alpha} = \mu G_1(\alpha, \psi), \qquad \dot{\psi} = \mu G_2(\alpha, \psi)$$

where

$$G_1(\alpha, \psi) = (1/2\pi)\int_0^{2\pi} \alpha(1 - \alpha^2\cos^2(t + \psi))\sin^2(t + \psi)dt,$$

$$G_2(\alpha, \psi) = (1/2\pi)\int_0^{2\pi} (1 - \alpha^2\cos^2(t + \psi))\sin(t + \psi)\cos(t + \psi)dt.$$

Integrating these expressions,

$$G_1(\alpha, \psi) = (1 - \alpha^2/4)\alpha/2, \qquad G_2(\alpha, \psi) = 0$$

are obtained, i.e. the averaged system is

$$\dot{\alpha} = \mu(1 - \alpha^2/4)\alpha/2, \qquad \dot{\psi} = 0. \tag{6.5.31}$$

If we try to apply the previous Theorem we see that the *necessary condition* for the existence of a 2π-periodic solution for system (6.5.30) is fulfilled since the averaged system has equilibria, namely, $(0, \psi_0)$, $(2, \psi_0)$, $(-2, \psi_0)$

where ψ_0 is arbitrary. In the neighbourhood of $(0, \psi_0)$, system (6.5.30) has a periodic solution indeed, namely, $a \equiv 0$ and $\theta(t)$, a solution of $\dot{\theta} = \mu \sin(t + \theta) \cos(t + \theta)$. (This first order differential equation can be integrated by quadratures, see Kamke [1959].) Irrespective of the form of $\theta(t)$ the corresponding periodic solution of Van der Pol's equation (6.5.29) is the trivial one, $x \equiv 0$. The form of (6.5.31) and (6.5.30) implies that if an $(\alpha(t), \psi(t))$, resp. an $(a(t), \theta(t))$ is a solution, then $(-\alpha(t), \psi(t))$, resp. $(-a(t), \theta(t))$ is a solution too; therefore, $(-2, \psi_0)$ need not be treated separately.

If we use $(2, \psi_0)$ and calculate the derivative matrix of the right-hand side of (6.5.31) at this point, we obtain

$$\begin{pmatrix} (1 - \alpha^2/2)\mu/2 & 0 \\ 0 & 0 \end{pmatrix}_{\alpha=2} = \begin{pmatrix} -\mu/2 & 0 \\ 0 & 0 \end{pmatrix}.$$

Thus, one of the eigenvalues is $-\mu/2$, the other one is zero, i.e. the sufficient conditions of the previous Theorem are not satisfied. A critical situation like this was to be expected since Van der Pol's equation is autonomous, so its limit cycle cannot be asymptotically but orbitally asymptotically stable. The Theorems of Section 3.2 imply that for $\mu > 0$, (6.5.29) has an orbitally asymptotically stable limit cycle. The present results show that for small positive μ it is near to $x(t) = 2\cos(t + \psi_0)$. Note that at $\alpha = 0$ the eigenvalues of the derivative matrix are $\mu/2$ and zero, and we know already that for $\mu > 0$ the trivial solution $x = 0$ is unstable.

We have no place here to treat the classical methods of nonlinear oscillations in detail. A recent elementary introduction to these can be found in Hagedorn [1988] which deals with methods of "harmonic balance", "equivalent linearization", etc. Most of these methods are related to the method of Krylov and Bogoliubov, which in turn is closely related to averaging (see Minorsky [1947], Krylov-Bogoliubov [1934], Bogoliubov-Mitropolskij [1965]). The method of Krylov and Bogoliubov is a generalization of how (6.5.12) was obtained from (6.5.8) combined with averaging. In the case of the autonomous perturbed linear oscillator

$$\ddot{x} + \omega_0^2 x = \mu g(x, \dot{x}) \tag{6.5.32}$$

the method consists of assuming the solution in the form of an "asymptotic series" (which may be divergent):

$$x(t) = a(t) \sin s(t) + \sum_{k=1}^{\infty} \mu^k u_k(a(t), s(t))$$

where it is supposed that a and s satisfy differential equations of the form

$$\dot{a} = \sum_{k=1}^{\infty} \mu^k P_k(a), \qquad \dot{s} = \omega_0 + \sum_{k=1}^{\infty} \mu^k Q_k(a),$$

the functions $u_k(a, s)$ are periodic in s with period 2π, and

$$\int_0^{2\pi} u_k(a, s) \sin s \, ds = \int_0^{2\pi} u_k(a, s) \cos s \, ds = 0$$

$$(k = 1, 2, ...).$$

Substituting into (6.5.32) one can, successively, determine u_k, P_k, and Q_k $(k = 1, 2, ...)$.

We note that there is a relatively new powerful method, the so-called *numerical-analytic method* due to Samoilenko and Ronto [1979, 1990] providing existence theorems of periodic solutions and also means for the approximate determination of the solution.

6.6   Singular Perturbations and Relaxation Oscillations

The singular perturbation technique that considers the differential equation embedded into a one parameter family of differential equations, where the "small parameter" multiplies the derivative of the unknown function in some of the equations, is widely used for solving different problems related to both ordinary and partial differential equations. While the solution of a regular perturbation problem is, mostly, a convergent power series of the "small parameter", a singular perturbation problem leads, usually, to a divergent series or to a series that is not a power series. For general reading cf. e.g. the monographs Lomov [1981], Chang-Howes [1984] and the papers Nipp [1988], Hoppensteadt, F. [1974], for the handling of divergent series see e.g. Bender-Orszag [1978]. Here we shall be concerned only with the applications of the method to the study of periodic solutions. The basic references in this direction are Mishchenko-Rozov [1980] and Grasman, J. [1987] (cf. also the references in these monographs). We shall present here an informal introduction to the method with the basic theorems whose proofs will be omitted.

We start with a fairly simple case when the perturbed system is assumed in the form

$$\begin{aligned}
\dot{x}_i &= f_i(x) \qquad (i = 1, 2, ..., n-1), \\
\varepsilon \dot{x}_n &= f_n(x)
\end{aligned}$$

(6.6.1)

where $f_i \in C^1(X, \mathbf{R})$, X being an open and connected set in $\mathbf{R}^n$ $(i = 1, 2, ..., n)$ and $\varepsilon \in \mathbf{R}$ is a small parameter. The unperturbed system is considered to be the one obtained from (6.6.1) by substituting $\varepsilon = 0$, i.e.

$$\begin{aligned}
\dot{x}_i &= f_i(x) \qquad (i = 1, 2, ..., n-1), \\
0 &= f_n(x).
\end{aligned}$$

(6.6.2)

As we see, the character of the system has changed. In the unperturbed system we have $n - 1$ differential equations for the n unknown functions $x = [x_1, x_2, ..., x_n]$ and a "finite" scalar equation that is to be satisfied by them.

We may assume that the unperturbed system (6.6.2) has a periodic solution $p : \mathbf{R} \to X$ with period $T_0 > 0$, i.e. $p \in C^1$, $p(t + T_0) \equiv p(t)$, and

$$\dot{p}_i(t) \equiv f_i(p(t)) \qquad (i = 1, 2, ..., n - 1),$$

$$f_n(p(t)) \equiv 0.$$

The question arises of whether in this case the perturbed system (6.6.1) has also a periodic solution near to $p(t)$ for sufficiently small values of $|\varepsilon|$. Denote the path of p by

$$\gamma = \{x \in \mathbf{R}^n : x = p(t), \ t \in \mathbf{R}\}.$$

The awkwardness of (6.6.2) can be overcome if x_n can be expressed uniquely from the last equation in terms of $x_1, ..., x_{n-1}$ in a neighbourhood of the path γ. By the Implicit Function Theorem, if $f'_{nx_n}(x) \neq 0$ for $x \in \gamma$, then a function $F \in C^1$ of $n - 1$ variables exists such that

$$f_n(x_1, ..., x_{n-1}, F(x_1, ..., x_{n-1})) \equiv 0$$

and

$$p_n(t) = F(p_1(t), ..., p_{n-1}(t)), \qquad t \in [0, T_0]. \tag{6.6.3}$$

In this case the $n - 1$ dimensional periodic vector function $\tilde{p}(t) := [p_1(t), ..., p_{n-1}(t)]$ is a solution of the $n - 1$ dimensional system of differential equations

$$\dot{x}_i = f_i(x_1, ..., x_{n-1}, F(x_1, ..., x_{n-1})) \qquad (i = 1, 2, ..., n - 1), \tag{6.6.4}$$

and $p_n(t)$ is defined by the right-hand side of (6.6.3). An existence and uniqueness result concerning the perturbed system (6.6.1) can be expected if $\tilde{p}(t)$ is an isolated periodic solution of (6.6.4) (cf. Definition 5.2.2 and Theorem 5.2.3). In order to give a sufficient condition for this we have to set up the variational system of (6.6.4) with respect to the periodic solution $\tilde{p}(t)$. It is

$$\dot{y}_i = \sum_{k=1}^{n-1} (f'_{ix_k}(p(t)) + f'_{ix_n}(p(t)) F'_{x_k}(\tilde{p}(t))) y_k$$

$$(i = 1, 2, ..., n - 1),$$

or applying implicit differentiation

$$\dot{y}_i = \sum_{k=1}^{n-1} (f'_{ix_k}(p(t)) - f'_{ix_n}(p(t)) f'_{nx_k}(p(t)) / f'_{nx_n}(p(t))) y_k$$

$$(i = 1, 2, ..., n - 1). \tag{6.6.5}$$

According to Theorem 5.2.3, if 1 is a simple characteristic multiplier of the $n - 1$ dimensional variational system (6.6.5), then $\tilde{p}(t)$ is isolated.

Theorem 6.6.1 *(Friedrichs-Wasow [1946]). If system (6.6.2) has a periodic solution $p(t)$ with period $T_0 > 0$ and path γ,*

$$f'_{n x_n}(x) \neq 0, \qquad x \in \gamma, \tag{6.6.6}$$

and 1 is a simple characteristic multiplier of the variational system (6.6.5), then an $\varepsilon_0 > 0$ exists such that for every $|\varepsilon| < \varepsilon_0$, system (6.6.1) has a unique periodic solution $\psi(t, \varepsilon)$ with period $T(\varepsilon)$ and with the properties ψ depends continuously on ε, $\psi(t, \varepsilon) \to p(t)$ as $\varepsilon \to 0$ uniformly in $t \in \mathbf{R}$, $T(\varepsilon) \to T_0$ as $\varepsilon \to 0$.

The essence of the proof (see Friedrichs-Wasow [1946]) is an application of the Implicit Function Theorem. Though this paper does not contain explicitly a stability theorem, its results imply the following Theorem in a straightforward way.

Theorem 6.6.2. *If the conditions of the previous Theorem hold, and $n - 2$ characteristic multipliers of the unperturbed variational system (6.6.5) are in modulus less than 1, then (the solution $\tilde{p}(t)$ of (6.6.4) and) the periodic solution $\psi(t, \varepsilon)$ of (6.6.1) are orbitally asymptotically stable with the asymptotic phase property for sufficiently small positive $\varepsilon > 0$.*

The proof of this Theorem goes by showing that as $\varepsilon \to +0$ the principal matrix of the variational system of (6.6.1) with respect to $\psi(t, \varepsilon)$ tends to a matrix whose submatrix of the first $n - 1$ rows and columns is the principal matrix of (6.6.5) and whose last column is zero. Thus, the characteristic multipliers of the perturbed variational system tend to those of (6.6.5) and to 0 as the missing n-th one. The Andronov-Witt Theorem 5.1.2 then yields the result.

The two theorems above illustrate the "smooth version" of the singular perturbation method. The unperturbed system is, in fact, a restriction of the original system to the $n - 1$ dimensional hypersurface $f_n(x) = 0$. It has a smooth periodic solution with path in this hypersurface, and the perturbed system for small values of the parameter ε has a periodic solution near the unperturbed one, i.e. near the hypersurface. While this situation may well occur, the "non-smooth or discontinuous version" of the method is more interesting and important. A singular perturbation problem arises, usually and naturally, when very rapid and slow variations of the state variables interchange with each other during a "cycle". If such a behaviour is observed in some real life phenomenon, then the mathematical modelling of the situation leads often to a singular perturbation problem in which some of the state variables are "slow," and others are "fast". The general form of a *singular perturbation system of ordinary differential equations* is

$$\varepsilon\dot{x} = f(x,y),$$
$$\dot{y} = g(x,y) \qquad (6.6.7)$$

where $x = [x_1, ..., x_n]$ is an n-vector, $y = [y_1, ..., y_m]$ is an m-vector, $f \in C^1(\mathbf{R}^n \times \mathbf{R}^m, \mathbf{R}^n)$, $g \in C^1(\mathbf{R}^n \times \mathbf{R}^m, \mathbf{R}^m)$ and $\varepsilon > 0$ is a small positive parameter. If $(x,y) \in \mathbf{R}^n \times \mathbf{R}^m$ is away from the manifold defined by the system of equations

$$f(x,y) = 0 \qquad (6.6.8)$$

and $\varepsilon > 0$ is small, then $|\dot{x}|$ is large, i.e. the $x = [x_1, ..., x_n]$ variables are changing rapidly compared to the changes in the $y = [y_1, ..., y_m]$ variables. Therefore, x represents the *fast variables* and y the *slow ones*. The unperturbed system belonging to (6.6.7) is

$$0 = f(x,y), \qquad \dot{y} = g(x,y). \qquad (6.6.9)$$

If the Jacobian of f with respect to x is non-zero,

$$J_x f(x,y) = \det f'_x(x,y) \neq 0, \qquad (6.6.10)$$

at the points x, y where $f(x,y) = 0$, then this equation determines an m dimensional manifold in the $n + m$ dimensional space. The set of points satisfying the first equation of (6.6.9) will be denoted by

$$M := \{(x,y) \in \mathbf{R}^n \times \mathbf{R}^m : f(x,y) = 0\}.$$

Besides systems (6.6.7) and (6.6.9), the n dimensional system

$$\varepsilon\dot{x} = f(x,y) \qquad (6.6.11)$$

will be considered for small values of ε where $y \in \mathbf{R}^m$ will be considered as parameters in the system. Suppose that for some $y^0 \in \mathbf{R}^m$ there exists an $x^0 \in \mathbf{R}^n$ such that $f(x^0, y^0) = 0$, i.e. $(x^0, y^0) \in M$. Then $x = x^0$ is an equilibrium point of system (6.6.11) taken at the parameter $y = y^0$. If $f'_x(x^0, y^0)$ is a stable matrix, then this equilibrium is asymptotically stable, and at the same time (6.6.10) holds at (x^0, y^0).

Now combining informations obtainable from systems (6.6.7), (6.6.9) and (6.6.11) we are to construct *"non-smooth trajectories"* for the unperturbed system (6.6.9). Suppose that $(x^0, y^0) \in M$ and that $\det f'_x(x^0, y^0) \neq 0$. Then by the Implicit Function Theorem the first equation of (6.6.9) determines uniquely a function $F \in C^1(V, \mathbf{R}^n)$ where $V \subset \mathbf{R}^m$ is an open neighbourhood of y^0 such that $F(y^0) = x^0$, $f(F(y), y) \equiv 0$. This way, i.e.

expressing x as a function of y from the first equation of (6.6.9), it is made equivalent to

$$\dot{y} = g(F(y), y). \tag{6.6.12}$$

Having solved this last equation, the corresponding values of x are determined by $x = F(y)$.

Denote by M_s and M_0 those parts of the set M where f'_x is a stable matrix and $\det f'_x = 0$, respectively, i.e.

$$M_s: \quad = \quad \{(x, y) \in M : f'_x(x, y) \text{ is stable}\},$$
$$M_0: \quad = \quad \{(x, y) \in M : \det f'_x(x, y) = 0\}.$$

Suppose that $(x^0, y^0) \in M_s$, denote the solution of (6.6.12) that assumes y^0 at $t = 0$ by $\eta(t)$ and let $\xi(t) := F(\eta(t))$. Assume that $(\xi(t), \eta(t)) \in M_s$ for $t \in [0, t_1)$ and that $(\xi(t_1), \eta(t_1)) \in M_0$. For an arbitrarily fixed value of $\bar{t} \in [0, t_1)$ the point $\xi(\bar{t})$ is an asymptotically stable equilibrium of (6.6.11) with $y = \eta(\bar{t})$. If $\varepsilon > 0$ is small, then y in (6.6.7) is varying slowly relative to the variation of x; therefore, it can be considered as approximately equal to the constant $\eta(\bar{t})$, whereas the x coordinates of the solutions of (6.6.7) with initial values near $(\xi(\bar{t}), \eta(\bar{t}))$ tend rapidly to $\xi(\bar{t})$. Hence, we may consider the arc $(\xi(t), \eta(t))$, $t \in [0, t_1)$, attracting with respect to the perturbed system (6.6.7).

At $(\xi(t_1), \eta(t_1))$ the function F and the system (6.6.12) is no longer uniquely defined, and one cannot tell what a solution of (6.6.9) with initial values $(\xi(t_1), \eta(t_1))$ will do. However, it may happen that $(\xi(t_1), \eta(t_1))$ is in (or on the boundary of) the domain of attraction of an other equilibrium of the system

$$\varepsilon \dot{x} = f(x, \eta(t_1)). \tag{6.6.13}$$

Denote this equilibrium by ξ^2; the point $(\xi^2, \eta(t_1)) \in M_s$ is then, normally, on an other leaf of the manifold or rather on another component of M_s (see Figure 6.6.1). In this case we may imagine that the state of the system jumps from $(\xi(t_1), \eta(t_1))$ to $(\xi^2, \eta(t_1))$ in a very short time. The line connecting $(\xi(t_1), \eta(t_1))$ with $(\xi^2, \eta(t_1))$ cannot be considered as part of a path belonging to system (6.6.9) since trajectories of this system must lie on M. Nevertheless, this line or rather curve that stays in the n dimensional plane $\{(x, y) \in \mathbf{R}^n \times \mathbf{R}^m : y = \eta(t_1)\}$ can be considered as the limiting position of the solution of (6.6.13) that tends to $\xi(t_1)$ as t tends to $-\infty$ and to ξ^2 as $t \to \infty$, provided that such a solution exists. Remember that $\xi(t_1)$ and ξ^2 are equilibria of (6.6.13). Taking into account again that in (6.6.7), away from M, the variable y is almost constant while x is varying rapidly this curve can be near a path of (6.6.7) too. Having arrived at $(\xi^2, \eta(t_1))$ the motion continues according to system (6.6.12), i.e. we imagine that the solution of (6.6.12) has *jumped* instantaneously from

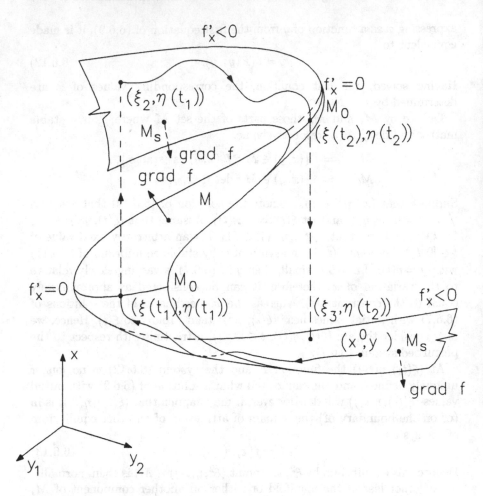

FIGURE 6.6.1. A non-smooth trajectory of system (6.6.9), $n = 1$, $m = 2$ with the "slow manifold" M and two jumps.

$(\xi(t_1), \eta(t_1))$ to $(\xi^2, \eta(t_1))$, and this last point serves now as an initial value. However, the system (6.6.12) has to be changed too, since now we have to replace the function F with the solution F_2 of the first equation of (6.6.9) that assumes ξ^2 at $y = \eta(t_1)$, i.e. $F_2(\eta(t_1)) = \xi^2$. So the next part of our "non-smooth trajectory" will be the path of the solution of system

$$\dot{y} = g(F_2(y), y) \qquad (6.6.14)$$

that assumes the initial value $\eta(t_1)$ at $t = t_1$. More exactly, for $t \geq t_1$ we denote again by $\eta(t)$ that solution of (6.6.14) that assumes the previously reached value $\eta(t_1)$ at $t = t_1$ and set $\xi(t) := F_2(\eta(t))$; $(\xi(t), \eta(t))$ is our solution for $t > t_1$. We may assume that $(\xi(t), \eta(t)) \in M_s$ for $t \in [t_1, t_2)$ but $(\xi(t_2), \eta(t_2)) \in M_0$. At $t = t_2$ we may have another jump like before, and this jump may take us back to the original leaf of M_s that contained (x^0, y^0), etc.; see Figure 6.6.1. In particular, it may happen that at this jump we hit the first branch of our trajectory, i.e. if the jump goes from $(\xi(t_2), \eta(t_2))$ to $(\xi^3, \eta(t_2)) \in M_s$, it may happen that $(\xi^3, \eta(t_2)) \in \{(x, y) \in M_s : x = \xi(t), y = \eta(t), \ 0 \le t < t_1\}$. In this case a closed path has been obtained.

The construction above led to the concept of a *non-smooth trajectory* that is, in fact, a *discontinuous* trajectory consisting of several pieces of trajectories belonging to systems (6.6.12), (6.6.14), etc., joined by instantaneous jumps along limiting positions of heteroclinic trajectories belonging to systems (6.6.13), $\varepsilon \dot{x} = f(x, \eta(t_2))$, etc. While those parts of the non-smooth trajectories that belong to (6.6.12), etc., belong to the stable leaves of M, the branches along which the jumps take place are contained in the planes $y = \eta(t_1)$, etc.

If a closed non-smooth trajectory has been found, the question arises of whether the perturbed system (6.6.7) has a closed orbit in some neighbourhood. The construction performed above can be found in greater detail in Mishchenko-Rozov [1980] which contains also the exact conditions and some theorems providing the answer to this question. The proofs of these fairly complicated theorems can be found in Vasil'eva-Butuzov [1973].

We shall give the results following Mishchenko-Rozov only in the two dimensional case: $n = 1$, $m = 1$, i.e. the system

$$\varepsilon \dot{x} = f(x, y), \qquad \dot{y} = g(x, y), \qquad \varepsilon > 0 \qquad (6.6.15)$$

where $f \in C^2(\mathbf{R}^2, \mathbf{R})$, $g \in C^1(\mathbf{R}^2, \mathbf{R})$ and the corresponding unperturbed system

$$f(x, y) = 0, \qquad \dot{y} = g(x, y) \qquad (6.6.16)$$

is considered. We suppose that the curve

$$M = \{(x, y) \in \mathbf{R}^2 : f(x, y) = 0\}$$

consists of regular points only, i.e.

$$|\operatorname{grad} f(x, y)|^2 = (f_x'(x, y))^2 + (f_y'(x, y))^2 > 0, \qquad (x, y) \in M,$$

the points where $f_x'(x, y) = 0$ form a discrete (though not necessarily finite) set on M, i.e. they are isolated, and these points are non-degenerate, i.e. if at $(x, y) \in M$ we have $f_x'(x, y) = 0$, then $f_{xx}''(x, y) \ne 0$. As in the higher dimensional case the notations

$$M_s = \{(x, y) \in M : f_x'(x, y) < 0\},$$
$$M_0 = \{(x, y) \in M : f_x'(x, y) = 0\}$$

are introduced. M_s consists of the *stable branches* of M, and M_0 is the set of *critical points*. It is assumed, finally, that $M_s \cup M_0$ does not contain equilibria of the perturbed system (6.6.15), i.e. that $g(x, y) \neq 0$, $(x, y) \in M_s \cup M_0$. There holds the following:

Theorem 6.6.3 *(Mishchenko-Rozov [1980]). If the conditions imposed above hold, and the unperturbed system (6.6.16) has a closed non-smooth trajectory, then an $\varepsilon_0 > 0$ exists such that for each $0 < \varepsilon < \varepsilon_0$ there exists a unique orbitally asymptotically stable periodic orbit of the perturbed system (6.6.15); this periodic orbit tends to the closed non-smooth trajectory of (6.6.16) uniformly as $\varepsilon \to 0$.*

If the conditions of the Theorem hold, then the motion along the limit cycle of the perturbed system is called a *relaxation oscillation* for small positive values of the parameter ε. Such a periodic motion consists of slow portions near the stable branches of the curve $f(x, y) = 0$ and fast portions near the jumps of the unperturbed system. Along the latter y is nearly constant and x is varying rapidly. The approximation of such a relaxation oscillation is, usually, not an easy task: different approximations are to be used for the slow and for the fast branches. There are methods for the approximate determination of the period too where the times needed to cover the fast portions are more difficult to obtain; see Mishchenko-Rozov [1980]. We shall illustrate the method on Van der Pol's equation.

Example 6.6.1 *(Van der Pol's equation with large damping coefficient).* Consider Van der Pol's differential equation (3.2.5) in the form

$$\ddot{x} + \mu(x^2 - 1)\dot{x} + x = 0, \qquad (6.6.17)$$

and assume now that $\mu > 0$ is large (cf. Example 6.5.1 where μ was considered to be small.) We have seen there that for small positive values of μ the limit cycle is near $x(t) = 2\cos(t + \psi_0)$, $\dot{x}(t) = -2\sin(t + \psi_0)$, i.e. in the x, $\dot{x}$ plane it is approximately circular. Now that μ is supposed to be large, a different phase plane will be introduced and also the time will be transformed. We write the equation in the form

$$-x/\mu = \ddot{x}/\mu + (x^2 - 1)\dot{x}$$

and denote the integral of the right-hand side by

$$y := \dot{x}/\mu + x^3/3 - x.$$

Then

$$\dot{x}/\mu = x - x^3/3 + y.$$

Introducing the new "slow time" by $\tau = t/\mu$ (for large $\mu > 0$) we obtain

$$\begin{aligned} (1/\mu^2)dx/d\tau &= x - x^3/3 + y, \\ dy/d\tau &= \mu\dot{y} = \ddot{x} + \mu(x^2 - 1)\dot{x} = -x. \end{aligned}$$

Finally, denoting $\varepsilon := 1/\mu^2$ we arrive at the system

$$\varepsilon dx/d\tau = x - x^3/3 + y,$$
$$dy/d\tau = -x \qquad (6.6.18)$$

where $\varepsilon > 0$ is a small parameter; if $\mu \to \infty$, then $\varepsilon \to +0$.

Now, the curve M of the theory is the cubic parabola $f(x,y) := x - x^3/3 + y = 0$, i.e. $y = x^3/3 - x$. The stable branches and the critical points are obtained from $f'_x(x,y) = 1 - x^2 < 0$ and from $1 - x^2 = 0$, respectively. Thus,

$$M_s = \{(x,y) \in M : |x| > 1\}, \quad M_0 = \{(-1, 2/3), (1, -2/3)\},$$

and $g(x,y) := -x \neq 0$ for $(x,y) \in M_s \cup M_0$. Looking at the phase picture in (Figure 6.6.2) and assuming a transversal at some point of the non-smooth closed trajectory of the unperturbed system $x - x^3/3 + y = 0$, $dy/d\tau = -x$, it is fairly easy to establish the existence of a limit cycle for the perturbed system applying a Fixed Point Theorem (see Appendix 2); however, all the conditions of Theorem 6.6.3 hold so that this Theorem guarantees the existence and asymptotic orbital stability of a periodic orbit for sufficiently small $\varepsilon > 0$, i.e. sufficiently large μ.

The time required to complete a cycle on the closed non-smooth trajectory will be denoted by T_0. Because of the symmetry this is twice the time required to complete one branch of M_s along the cubic parabola $y = x^3/3 - x$, say, the one between the points $(-2, -2/3)$ and $(-1, 2/3)$ denoted by M_{s1}. We may determine this from the second equation of the system written in the form $d\tau/dy = -1/x$:

$$T_0/2 = \int_0^{T_0/2} d\tau = -\int_{M_{s1}} dy/x = -\int_{-2/3}^{2/3} dy/x$$

where x is to be substituted from $y = x^3/3 - x$. Performing this substitution instead we obtain

$$T_0/2 = -\int_{-2}^{-1} ((x^2 - 1)/x)dx = \int_{-2}^{-1} (1/x - x)dx = 3/2 - \ln 2.$$

The period $\tilde{T}_\varepsilon$ of the periodic solution of the perturbed system (6.6.18) is approximately equal to T_0, i.e. $\tilde{T}_\varepsilon \approx 3 - 2\ln 2$ or returning to the original time t the period is $T_\mu \approx \mu(3 - 2\ln 2)$ for large μ. In the case $\mu = 100$, say, this approximate expression yields the period with an error less than 1%; cf. the numerical results in Grasman [1987].

More about relaxation oscillations can be found also in Levi [1981]. The singular perturbation technique was applied by Grasman-Jansen [1979] to the synchronization phenomenon of weakly coupled biological oscillators. More recently Rinaldi and Muratori [1992] have set up a three dimensional singular perturbation model to describe observed relaxation oscillations in forest-pest outbreaks.

(a)

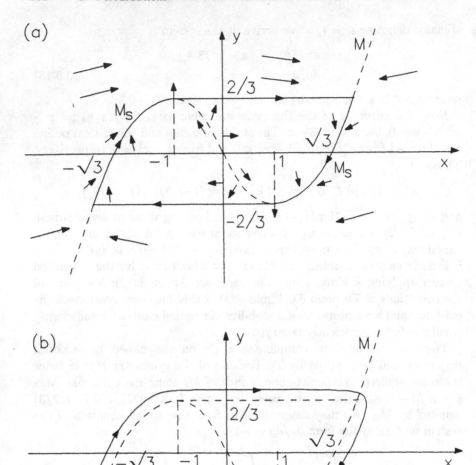

(b)

FIGURE 6.6.2. (a) The slow manifold M of system (6.6.18) with the phase picture of the perturbed system and the closed non-smooth trajectory of the unperturbed system. (b) The limit cycle representing the relaxation oscillation of system (6.6.18).

6.7 Aperiodic Perturbations

In this closing section of the chapter on perturbations the problem dealt with will be what happens to a periodic solution if the system is subject to not too large aperiodic perturbations. The problem is important from the practical point of view in mathematical modelling since real world systems are always subject to some extent to aperiodic perturbations. It is also related to the theory of stochastically perturbed differential equations in which direction we shall not stray (see Has'minskij [1969], Bunke [1972] and Tran Van Nhung [1982]).

First, following Yoshizawa [1966] the most important facts will be presented about the *stability of sets*. Then we shall present results of Farkas [1980, 1981] and Kertész [1987, 1988] that show that a uniform asymptotically stable periodic solution of a periodic system under aperiodic perturbations carries an attractive solid spiralling set, a "snake" around its integral curve whose thickness and whose region of attractivity can be estimated. Finally, since this construction does not work for orbitally asymptotically stable periodic solutions of autonomous systems, Nguyen Van Minh's and Tran Van Nhung's [1991] result will be presented showing that if such a system is perturbed, an attractive tube arises, the thickness of whose wall can be estimated.

Assume that $f \in C^0(\mathbf{R}_+ \times \mathbf{R}^n, \mathbf{R}^n)$, $f'_x \in C^0(\mathbf{R}_+ \times \mathbf{R}^n, \mathbf{R}^{n^2})$ and consider the system

$$\dot{x} = f(t, x). \tag{6.7.1}$$

Let $M \subset \mathbf{R}_+ \times \mathbf{R}^n$ and for $\tau \geq 0$ denote by M_τ, the intersection of the subset M with the hyperplane $t = \tau$ of $\mathbf{R} \times \mathbf{R}^n$. Suppose that $M_\tau \neq \emptyset$ (the empty set) for $\tau \geq 0$, and denote the solution of (6.7.1) that assumes the value $x^0 \in \mathbf{R}^n$ at $t = t_0 \geq 0$ by $\varphi(t, t_0, x^0)$.

Definition 6.7.1. We say that the set M is *positively invariant* if $(t_0, x^0) \in M$ implies $(t, \varphi(t, t_0, x^0)) \in M$ for all $t \geq t_0$.

Definition 6.7.2. If there is a compact subset $Q \subset \mathbf{R}^n$ such that $M_t \subset Q$ for all $t \geq 0$, then we say that the set M_t is *bounded*.

Definition 6.7.3. We say that the bounded set M is a *stable set* of (6.7.1) if for any $\varepsilon > 0$ and $t_0 \geq 0$ there exists a $\delta(t_0, \varepsilon) > 0$ such that dist $(x^0, M_{t_0}) < \delta$ implies dist $(\varphi(t, t_0, x^0), M_t) < \varepsilon$ for all $t \geq t_0$; if δ does not depend on t_0, then we say that M is a *uniformly stable set*.

Definition 6.7.4. We say that the bounded set M is *uniformly asymptotically stable* if it is uniformly stable and if there exists a $\delta_0 > 0$ and for any $\varepsilon > 0$, there exists a $T(\varepsilon) > 0$ such that for any $t_0 \geq 0$ the inequality dist$(x^0, M_{t_0}) < \delta_0$ implies dist $(\varphi(t, t_0, x^0), M_t) < \varepsilon$ for all $t \geq t_0 + T(\varepsilon)$.

Besides the conditions imposed upon the set M following formula (6.7.1) it will be assumed that M_t is varying in a "Lipschitz continuous way", i.e. for any compact set $Q \subset \mathbf{R}_+ \times \mathbf{R}^n$ there exists a number $K > 0$ such that for every $(t_1, x), (t_2, x) \in Q$ there holds

$$| \operatorname{dist} (x, M_{t_1}) - \operatorname{dist} (x, M_{t_2})| \leq K|t_1 - t_2|. \qquad (6.7.2)$$

We present now two Theorems concerning the stability of sets without proofs. The proofs, which would require much preparation, can be found in the monograph to be quoted. The first Theorem is an "inverse theorem" of Liapunov's direct method: it states the existence of a Liapunov function for a set which is stable.

The system (6.7.1) is considered and also a set M satisfying the conditions above; the notation

$$D_\rho = \{(t, x) \in \mathbf{R}_+ \times \mathbf{R}^n : 0 \leq t < \infty, \ \operatorname{dist} (x, M_t) \leq \rho\} \qquad (6.7.3)$$

will be introduced where $\rho > 0$ is fixed. If a scalar function $V : \mathbf{R}_+ \times D_\rho \to \mathbf{R}$ is continuous and satisfies the *Lipschitz condition*

$$|V(t, x^1) - V(t, x^2)| \leq L|x^1 - x^2|, \qquad (t, x^1), (t, x^2) \in D_\rho$$

with some constant $L > 0$, then the notation

$$\dot{V}_{(6.7.1)}(t, x) := \lim_{h \to +0} \sup \ (V(t + h, x + hf(t, x)) - V(t, x))/h$$

will be applied for $(t, x) \in D_\rho$. If V is differentiable, this, obviously, coincides with the derivative of V with respect to the system (6.7.1) (see Definition 1.5.3).

Theorem 6.7.1 *(Yoshizawa [1966] p.114). Suppose that $|f'_x|$ is bounded on D_ρ; if M is a uniformly asymptotically stable set for (6.7.1), then there exist a $0 < \rho' < \rho$, a Liapunov function $V \in C^0(\mathbf{R}_+ \times D_{\rho'}, \mathbf{R})$ and a constant $c > 0$ such that*

(i) *there exist functions $h_1, h_2 \in \mathcal{H}$ (for the function class $\mathcal{H}$ see Definition 1.5.2) for which*

$$h_1(\operatorname{dist} (x, M_t)) \leq V(t, x) \leq h_2(\operatorname{dist} (x, M_t))$$

holds;

(ii) *there holds the Lipschitz condition*

$$|V(t, x^1) - V(t, x^2)| \leq L|x^1 - x^2|, \qquad (t, x^1), (t, x^2) \in D_{\rho'}$$

with some $L > 0$;

(iii)
$$\dot{V}_{(6.7.1)}(t,x) \leq -cV(t,x), \qquad (t,x) \in D_{\rho'}.$$

The proof of this Theorem proceeds by constructing directly the Liapunov function with the desired properties.

We turn now to perturbations of system (6.7.1),

$$\dot{x} = f(t,x) + \mu g(t,x,\mu), \tag{6.7.4}$$

where $g, g'_x, g'_\mu \in C^0(\mathbf{R}_+ \times \mathbf{R}^n \times I_\mu)$, I_μ being an open interval containing $\mu = 0$. It will be assumed that

(a) (6.7.1) has a bounded, uniformly asymptotically stable set M satisfying the conditions following (6.7.1), including (6.7.2);

(b) $|f'_x|$ is bounded in D_ρ;

(c) $|g(t,x,\mu)|$ and $|g'_x(t,x,\mu)|$ are bounded in $(t,x,\mu) \in (D_\rho \setminus M) \times I_\mu$.

Theorem 6.7.2 *(Yoshizawa [1966] p.134). If conditions (a), (b) and (c) hold, and V is a Liapunov function for the unperturbed system (6.7.1) with the properties guaranteed by Theorem 6.7.1, then for sufficiently small $\alpha > 0$ the set*

$$M_\alpha := \{(t,x) \in D : V(t,x) \leq \alpha\}$$

is closed, satisfies $M \subset M_\alpha \subset D_{\rho'}$, for sufficiently small $|\mu(\alpha)|$ it is a uniformly asymptotically stable positively invariant set of the perturbed system (6.7.4) where $\mu = \mu(\alpha)$ is to be substituted, and the function $\mu(\alpha)$ can be chosen so that as $\mu \to 0$ also $\alpha \to 0$ and $M_\alpha \to M$.

Thus, this Theorem says that under some natural conditions a bounded uniformly asymptotically stable set may become "thicker" if the system is subject to small perturbations but it persists. We note that here somewhat stronger assumptions were applied than in the original Yoshizawa theorem.

We turn now to the problem of small aperiodic perturbations of periodic systems. The periodic system

$$\dot{x} = f(t,x) \tag{6.7.5}$$

is considered where $f, f'_x, f''_{xx} \in C^0(\mathbf{R}_+ \times \mathbf{R}^n)$ and $f(t+T,x) \equiv f(t,x)$ for some $T > 0$. It is assumed that (6.7.5) has a periodic solution $p : [0,\infty) \to \mathbf{R}^n$ with period $T : p(t+T) \equiv p(t)$, and that all the characteristic multipliers of the variational system

$$\dot{y} = f'_x(t,p(t))y \tag{6.7.6}$$

are in modulus less than 1. This means (see Theorem 4.2.1) that p is a uniformly asymptotically stable solution. This, obviously, implies that its integral curve

$$\Gamma := \{(t,x) \in \mathbf{R}_+ \times \mathbf{R}^n : x = p(t), t \in \mathbf{R}_+\}$$

is a uniformly asymptotically stable set. Now, suppose that system (6.7.5) is perturbed by an aperiodic function $g : \mathbf{R}_+ \times \mathbf{R}^n \to \mathbf{R}^n$ for which we have $g, g'_x \in C^0(\mathbf{R}_+ \times \mathbf{R}^n)$, and for any compact set $Q \subset \mathbf{R}^n$ there exists an $\eta > 0$ such that

$$|g(t,x)| < \eta, \qquad (t,x) \in \mathbf{R}_+ \times Q \qquad (6.7.7)$$

and also g'_x is bounded on $\mathbf{R}_+ \times Q$, i.e. along with (6.7.5) we consider the perturbed system

$$\dot{x} = f(t,x) + g(t,x). \qquad (6.7.8)$$

If the function $x(t)$ satisfies (6.7.8), then $z(t) := x(t) - p(t)$ satisfies

$$
\begin{aligned}
\dot{z}(t) &= \dot{x}(t) - \dot{p}(t) = f(t,p(t)+z(t)) + g(t,p(t)+z(t)) - f(t,p(t)) \\
&= f'_x(t,p(t))z(t) + h(t,z(t)) + g(t,p(t)+z(t))
\end{aligned}
$$

where the function $h(t,z)$ is periodic in t with period T and $|h(t,z)| = o(|z|)$. Denote the path of the periodic solution p by

$$\gamma := \{x \in \mathbf{R}^n : x = p(t), \ t \in \mathbf{R}_+\},$$

and choose a closed $0 < \rho_1$-neighbourhood

$$\overline{U}(\gamma, \rho_1) := \{x \in \mathbf{R}^n : \text{dist}(x, \gamma) \le \rho_1\}$$

of this path. The periodicity and the C^2 property of f implies that a $\rho_2 > 0$ exists such that

$$|h(t,z)| \le \rho_2 |z|^2, \qquad t \in \mathbf{R}_+, \qquad |z| < \rho_1. \qquad (6.7.9)$$

In particular cases, ρ_2 can be determined from the second partial derivatives of f. Thus, if x satisfies (6.7.8), then $z = x - p$ satisfies the system

$$\dot{z} = f'_x(t,p(t))z + h(t,z) + g(t,p(t)+z). \qquad (6.7.10)$$

This is equivalent to (6.7.8), whereas the system

$$\dot{z} = f'_x(t,p(t))z + h(t,z) \qquad (6.7.11)$$

is the unperturbed system (6.7.5) transformed for the variation with respect to its solution p.

We shall give now explicit estimates for an attractive set of the perturbed system (6.7.8) around the integral curve Γ of the unperturbed periodic solution p and for its domain of attractivity. First a Liapunov function will be constructed. As we know (see Theorem 2.2.6) there exists a regular, T- or $2T$-periodic C^1 matrix $P(t)$ such that $P(0) = I$ and the transformation $y = P(t)u$ carries the system (6.7.6) into a system with a real constant coefficient matrix

$$\dot{u} = Bu. \qquad (6.7.12)$$

The assumption that all the characteristic multipliers of (6.7.6) are in modulus less than 1 implies that all the eigenvalues of the $n \times n$ matrix B have negative real parts. The transformation matrix P satisfies

$$\dot{P}(t) \equiv f'_x(t, p(t))P(t) - P(t)B. \tag{6.7.13}$$

We also know (see Theorem 1.5.6) that a positive definite symmetric matrix W can be found such a way that the derivative of the quadratic form $w(u) = u'Wu$ with respect to system (6.7.12) is negative definite. Moreover, it is easy to see that positive constants $\rho'_3 > 0$ and $\rho_3 > 0$ exist such that the inequalities

$$\dot{w}_{(6.7.12)}(u) \leq -\rho'_3 w(u) \leq -\rho_3 |u|^2, \qquad u \in \mathbf{R}^n, \tag{6.7.14}$$

hold. Let us substitute $u = P^{-1}(t)y$ into w. What we get

$$v(t, y) := w(P^{-1}(t)y) = y'P^{-1'}(t)WP^{-1}(t)y$$

is a positive definite quadratic form for each $t \in \mathbf{R}_+$. Differentiating the identity $P(t)P^{-1}(t) \equiv I$ and taking into account (6.7.13) we get

$$\dot{P}^{-1}(t) \equiv P^{-1}(t)f'_x(t, p(t)) + BP^{-1}(t).$$

Differentiating v with respect to system (6.7.6) and applying the last identity we obtain

$$\begin{aligned}
\dot{v}_{(6.7.6)}(t, y) &= y'P^{-1'}(t)(B'W + WB)P^{-1}(t)y \\
&= \dot{w}_{(6.7.12)}(P^{-1}(t)y).
\end{aligned}$$

In view of (6.7.14) there holds

$$\dot{v}_{(6.7.6)}(t, y) \leq -\rho_3 |P^{-1}(t)y|^2, \qquad t \in \mathbf{R}_+, \qquad y \in \mathbf{R}^n.$$

However,

$$\begin{aligned}
|P^{-1}(t)y|^2 &= \; < P^{-1}(t)y, P^{-1}(t)y > \\
&= \; < y, P^{-1'}(t)P^{-1}(t)y >
\end{aligned}$$

where $< ., . >$ denotes scalar product. The last equation implies that $P^{-1'}(t)P^{-1}(t)$ is a positive definite matrix for all $t \geq 0$. Denote its least eigenvalue by $\lambda_p(t)$. Since $P^{-1} \in C^1$ and it is periodic, $\lambda_p \in C^0(\mathbf{R}_+, (0, \infty))$ and it is also periodic; hence,

$$\lambda := \min_{t \in \mathbf{R}_+} \lambda_p(t) > 0. \tag{6.7.15}$$

Thus

$$|P^{-1}(t)y|^2 \geq \lambda_p(t)|y|^2 \geq \lambda|y|^2,$$

so that

$$\dot{v}_{(6.7.6)}(t, y) \leq -\rho_3 \lambda |y|^2, \qquad t \in \mathbf{R}_+, \qquad y \in \mathbf{R}^n. \qquad (6.7.16)$$

We continue by establishing the negativity of the derivative of v with respect to the system (6.7.10)

$$\dot{v}_{(6.7.10)}(t, z) = \dot{v}_{(6.7.6)}(t, z) + < \mathrm{grad}_z v(t, z), h(t, z) >$$
$$+ < \mathrm{grad}_z v(t, z), g(t, p(t) + z) > .$$

Since v is quadratic in z and the coefficients are continuous and periodic in t, its gradient is linear in z, and there exists a constant $\beta > 0$ such that

$$|\mathrm{grad}_z v(t, z)| \leq \beta |z|, \qquad t \in \mathbf{R}_+, \qquad z \in \mathbf{R}^n. \qquad (6.7.17)$$

Applying (6.7.16), (6.7.17), (6.7.9) and (6.7.7) where η is the positive constant corresponding to the set $Q = \overline{U}(\gamma, \rho_1)$ and also the Cauchy-Buniakovskij inequality we obtain the inequality

$$\dot{v}_{(6.7.10)}(t, z) \leq -\rho_3 \lambda |z|^2 + \beta |z| \rho_2 |z|^2 + \beta |z| \eta, \qquad t \in \mathbf{R}_+, \qquad |z| < \rho_1.$$

If z is restricted by $|z| < \rho_3 \lambda / (2\beta \rho_2)$, say, then the inequality

$$\dot{v}_{(6.7.10)}(t, z) \leq -(\rho_3 \lambda / 2) |z|^2 + \beta \eta |z|, \qquad t \in \mathbf{R}_+,$$

$$|z| < \min(\rho_1, \rho_3 \lambda / (2\beta \rho_2))$$

follows. Here the right-hand side is negative if $2\beta \eta / (\rho_3 \lambda) < |z|$, i.e.

$$\dot{v}_{(6.7.10)}(t, z) \leq -(\rho_3 \lambda / 2) |z|^2 + \beta \eta |z| < 0 \qquad (6.7.18)$$

if

$$t \in \mathbf{R}_+ \quad \text{and} \quad 2\beta \eta / (\rho_3 \lambda) < |z| < \min(\rho_1, \rho_3 \lambda / (2\beta \rho_2)). \qquad (6.7.19)$$

The set defined by the last inequality is *not* the empty set provided that

$$0 < \eta < \min\left(\rho_1 \rho_3 \lambda (2\beta), \rho_3^2 \lambda^2 / (4\beta^2 \rho_2)\right). \qquad (6.7.20)$$

Having constructed the Liapunov function v and established its most important properties some notations will be introduced. Set

$$\alpha_1 : = \min v(t, z), \quad t \in [0, T], \quad |z| = \min(\rho_1, \rho_3 \lambda / (2\beta \rho_2)),$$
$$\alpha_2 : = \max v(t, z), \quad t \in [0, T], \quad |z| = 2\beta \eta / (\rho_3 \lambda).$$

Clearly $\alpha_i > 0$ $(i = 1, 2)$. Denote the least and the largest eigenvalue of the positive definite coefficient matrix $P^{-1'}(t) W P^{-1}(t)$ of v by $\lambda_{v1}(t)$ and $\lambda_{v2}(t)$, respectively. It is easy to see that

$$0 < \lambda_{v1}(t)|z|^2 \leq v(t, z) \leq \lambda_{v2}(t)|z|^2, \qquad t \in \mathbf{R}_+, \qquad z \neq 0,$$

and that there are directions in the space of $z \in \mathbf{R}^n$ in which the respective equalities hold. Hence,

$$\alpha_1 = \lambda_1 (\min(\rho_1, \rho_3 \lambda/(2\beta\rho_2)))^2, \qquad \alpha_2 = \lambda_2 (2\beta\eta/(\rho_3\lambda))^2$$

where $\lambda_1 = \min \lambda_{v1}(t) > 0$, $\lambda_2 = \max \lambda_{v2}(t) > 0$, since $\lambda_{vi} \in C^0(\mathbf{R}_+, (0, \infty))$ and it is periodic $(i = 1, 2)$.

Defining two sets by

$$\begin{aligned} M_\eta : &= \{(t, x) \in \mathbf{R}_+ \times \mathbf{R}^n : v(t, x - p(t)) \le \alpha_2, t \in \mathbf{R}_+\}, \\ A : &= \{(t, x) \in \mathbf{R}_+ \times \mathbf{R}^n : v(t, x - p(t)) < \alpha_1, t \in \mathbf{R}_+\} \end{aligned}$$

there holds:

Theorem 6.7.3 *(Farkas [1981]). If the conditions imposed upon systems (6.7.5), (6.7.6) and (6.7.8) hold, and η in (6.7.7) satisfies the inequality*

$$0 < \eta < (1/2)(\lambda_1/\lambda_2)^{1/2} \min(\rho_1 \rho_3 \lambda/\beta, \rho_3^2 \lambda^2/(2\beta^2 \rho_2)), \qquad (6.7.21)$$

then M_η is a uniformly asymptotically stable invariant set of system (6.7.8) whose region of attractivity (in $\mathbf{R}_+ \times \mathbf{R}^n$) contains the set A.

Note that the positive constants ρ_1 and ρ_2 can be determined from the unperturbed system and its periodic solution p, the positive constants ρ_3, λ, β, λ_1 and λ_2 are known provided that the unperturbed variational system (6.7.6) has been solved and the quadratic Liapunov function w has been constructed; these data determine the function v and the constants α_1 and α_2.

Proof. Inequality (6.7.21) implies $\alpha_2 < \alpha_1$; thus, $M_\eta \subset A$. The positive numbers α_1 and α_2 had been defined so that the set $A \setminus M_\eta$ is contained in the set defined by (6.7.19) where $z = x - p(t)$ is to be understood. Hence, (6.7.18) holds in $A \setminus M_\eta$; also (6.7.21) implies (6.7.20). Thus, the proof of Yoshizawa's Theorem 6.7.2 can be applied to establish the uniform asymptotic stability of the set M_η. Obviously, $\Gamma \subset M_\eta$ and $M_\eta \to \Gamma$ as $\eta \to 0$. The validity of (6.7.18) for $(t, x - p(t)) \in A \setminus M_\eta$ implies that A is part of the region of attractivity. $\square$

Observe that both M_η and its basin A can be imagined as the union of solid ellipsoidal sets that cover the graph Γ of the periodic solution p in $\mathbf{R}_+ \times \mathbf{R}_n$: "a snake that has swallowed plenty of mice" (see Figure 6.7.1). Both M_η and A are "periodic sets" in the following sense. If the intersection of M_η and A with the hyperplane $t = s \ge 0$ is denoted by $M_\eta(s)$ and $A(s)$, respectively, then the latter sets considered as subsets of $\mathbf{R}^n$ have the following property:

$$M_\eta(s + T) = M_\eta(s), \qquad A(s + T) = A(s), \qquad s \in \mathbf{R}_+.$$

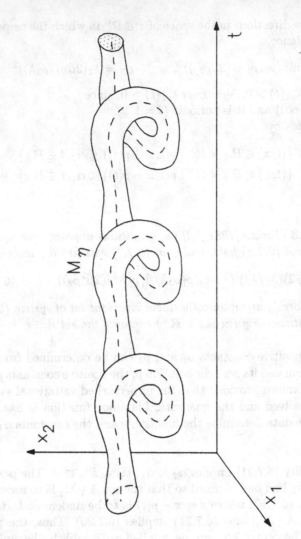

FIGURE 6.7.1. "The snake that has swallowed plenty of mice": the uniformly asymptotically stable set M_η of system (6.7.8) around the integral curve Γ of the periodic solution of system (6.7.5).

As an application consider the following:

Example 6.7.1 *(Duffing's equation under aperiodic perturbations, Farkas [1980]).* We are treating here the damped periodically excited Duffing equation (cf. (4.3.15), the notations are changed here)

$$\ddot{x}_1 = -k^2 x_1 + m(-b\dot{x}_1 + cx_1^3) + a\sin t \qquad (6.7.22)$$

where $k > 0$ is *not* an integer, $b > 0$, $m > 0$, $a, c \in \mathbf{R}$, $c \neq 0$. We assume that m is such that (6.7.22) has an asymptotically stable periodic solution p_1 with period 2π. We know that this is the case if m is sufficiently small. The perturbed equation will be assumed in the form

$$\ddot{x}_1 = -k^2 x_1 + m(-b\dot{x}_1 + cx_1^3) + a\sin t + G(t, x_1, \dot{x}_1) \qquad (6.7.23)$$

where $G, G'_{x_1}, G'_{\dot{x}_1} \in C^0(\mathbf{R}_+ \times \mathbf{R}^2, \mathbf{R})$; for any compact set $Q \subset \mathbf{R}^2$ there exists an $\eta > 0$ such that

$$|G(t, x_1, \dot{x}_1)| < \eta, \qquad (t, x_1, \dot{x}_1) \in \mathbf{R}_+ \times Q,$$

and G'_{x_1} and $G'_{\dot{x}_1}$ are also bounded on $\mathbf{R}_+ \times Q$.

We are giving explicit conditions for η in order to ensure the existence of a uniformly asymptotically stable set for (6.7.23) around the graph of the function $(p_1, \dot{p}_1)$ in the $t, x_1, \dot{x}_1$ space. This set will be given explicitly and also its domain of attractivity will be estimated.

Denote $x_2 = \dot{x}_1$, $p_2 = \dot{p}_1$, $x = (x_1, x_2)$, $p = (p_1, p_2)$, $z = (z_1, z_2)$ and $P = \max |p_1(t)|$. The variational system with respect to the periodic solution p of the system corresponding to (6.7.22) is

$$\dot{z}_1 = z_2, \qquad \dot{z}_2 = (-k^2 + 3mcp_1^2(t))z_1 - mbz_2. \qquad (6.7.24)$$

The coefficient matrix of this system can be written in the form

$$\begin{pmatrix} 0 & 1 \\ -k^2 & -mb \end{pmatrix} + \begin{pmatrix} 0 & 0 \\ 3mcp_1^2(t) & 0 \end{pmatrix}.$$

In constructing a Liapunov function to the homogeneous linear periodic system (6.7.24) Chetaev's method (see (2.4.6) et seq.) will be followed. First a positive definite quadratic Liapunov function will be constructed whose derivative with respect to the asymptotically stable system

$$\dot{z}_1 = z_2, \qquad \dot{z}_2 = -k^2 z_1 - mbz_2 \qquad (6.7.25)$$

is $-(\kappa_1 z_1^2 + \kappa_2 z_2^2)$ where κ_1, κ_2 are positive constants to be chosen later. The coefficients of this quadratic form

$$v(z) = v(z_1, z_2) = v_{11}z_1^2 + 2v_{12}z_1 z_2 + v_{22}z_2^2$$

are determined from Liapunov's matrix equation (1.5.9). This yields

$$v_{11} = (\kappa_1(m^2b^2 + k^2) + \kappa_2k^4)/(2mbk^2),$$
$$v_{12} = \kappa_1/(2k^2), \qquad v_{22} = (\kappa_1 + \kappa_2k^2)/(2mbk^2).$$

It is easy to see that the quadratic form

$$v(z) = z_1^2(\kappa_1(m^2b^2 + k^2) + \kappa_2k^4)/(2mbk^2)$$
$$+ z_1z_2\kappa_1/k^2 + z_2^2(\kappa_1 + \kappa_2k^2)/(2mbk^2)$$

is positive definite, indeed, for arbitrary positive values of κ_1 and κ_2, and the derivative of v with respect to system (6.7.25) is

$$\dot{v}_{(6.7.25)}(z) = -(\kappa_1z_1^2 + \kappa_2z_2^2).$$

The derivative of the same v with respect to system (6.7.24) is

$$\dot{v}_{(6.7.24)}(t,z) = -\kappa_1(1 - 3mcp_1^2(t)/k^2)z_1^2$$
$$+ z_1z_23cp_1^2(t)(\kappa_1 + \kappa_2k^2)/(bk^2) - \kappa_2z_2^2.$$

We want to establish conditions under which this quadratic form is negative definite; the details of the fairly straightforward but cumbersome calculations will be omitted. Choose $\kappa_1 = k^4b^2$, $\kappa_2 = 9c^2P^2$, and suppose that

(M1) $3|c|P^2 < kb$,
(M2) either $c < 0$ or $0 < m < (k^2b^2 - 9c^2P^4)/(4b^2cP^2)$;
then the derivative of the positive definite quadratic form

$$v(z) = z_1^2k^2(m^2b^4 + k^2b^2 + 9c^2P^4)/(2mb)$$
$$+ k^2b^2z_1z_2 + z_2^2(k^2b^2 + 9c^2P^4)/(2mb) \qquad (6.7.26)$$

with respect to system (6.7.24),

$$\dot{v}_{(6.7.24)}(t,z) = -(z_1^2(k^4b^2 - 3mk^2b^2cp_1^2(t))$$
$$- z_1z_23cp_1^2(t)(k^2b^2 + 9c^2P^4)/b$$
$$+ 9c^2P^4z_2^2)$$

is negative definite. In the calculation it is to be taken into account that (M2) implies

(M2') either $c < 0$ or $0 < m < k^2/(3cP^2)$.
This result implies that a $\rho_3 > 0$ exists such that

$$\dot{v}_{(6.7.24)}(t,z) \leq -\rho_3|z|^2, \qquad t \in \mathbf{R}_+, \qquad z \in \mathbf{R}^2; \qquad (6.7.27)$$

more exactly, if (M1) and (M2) hold and

(M3) either $c < 0$ and

$$0 < \rho_3 \leq \min\left(k^4 b^2, \frac{9c^2 P^4 (3k^4 b^4 - 18k^2 b^2 c^2 P^4 - 81c^4 P^8)}{4b^2(k^4 b^2 + 3mk^2 b^2 |c| P^2 + 9c^2 P^4)} \right)$$

or $c > 0$ and

$$0 < \rho_3 \leq \min\left(k^2 b^2 (k^2 - 3mc P^2), \frac{(9c^2 P^4)^2 (k^2 b^2 - 9c^2 P^4)}{4b^2(k^4 b^2 + 9c^2 P^4)} \right),$$

then (6.7.27) holds.

We turn now to the unperturbed non-linear equation (6.7.22). Introducing the new variables $z_1 = x_1 - p_1(t)$, $z_2 = x_2 - p_2(t)$ the system equivalent to (6.7.22) is

$$\dot{z}_1 = z_2, \quad \dot{z}_2 = (-k^2 + 3mc p_1^2(t))z_1 - mb z_2$$
$$+ 3mc(p_1(t) + z_1/3)z_1^2$$

(cf. (6.7.11)). Choosing an arbitrary $\rho_1 > 0$ and restricting ourselves to $|z| < \rho_1$ the nonlinearity of the last system can be estimated by

$$|3mc(p_1(t) + z_1/3)z_1^2| \leq 3m|c|(P + \rho_1/3)z_1^2 = \rho_2 z_1^2,$$

$$t \in \mathbf{R}_+, \qquad |z| < \rho_1$$

where

$$\rho_2 = 3m|c|(P + \rho_1/3) \qquad (6.7.28)$$

(cf (6.7.9)). We have determined ρ_1, ρ_2, ρ_3. In order to apply Theorem 6.7.3 we have to find β, λ_1, and λ_2 (λ does not occur, or rather $\lambda = 1$ because we applied Chetaev's method).

For the determination of β an estimate of the gradient of v is needed (cf. (6.7.17)). This can be obtained in a fairly straightforward way. The result is that

$$|\operatorname{grad} v(z)| \leq \beta |z|, \qquad |z| \in \mathbf{R}^2$$

if β is chosen the following way:

$$\beta = \sqrt{2}k^2(b^4 + (m^2 b^4 + k^2 b^2 + 9c^2 P^4)^2/(\kappa^2 m^2 b^2))^{1/2} \qquad (6.7.29)$$

where

$$\kappa^2 = \begin{cases} k^2 & \text{if } 0 < k^2 < 1 \\ 1 & \text{if } k^2 > 1. \end{cases} \qquad (6.7.30)$$

The missing λ_1 and λ_2 are the less and the larger eigenvalues of the (constant) coefficient matrix of the positive definite quadratic form (6.7.26). A simple but lengthy calculation yields

$$\lambda_{1,2} = \frac{1}{4mb}(m^2 k^2 b^4 + k^4 b^2 + k^2 9c^2 P^4 + k^2 b^2 + 9c^2 P^4 \pm A^{\frac{1}{2}})$$

where

$$A = (m^2k^2b^4)^2 + (k^4b^2)^2 + (k^29c^2P^4)^2 + (k^2b^2)^2 + (9c^2P^4)^2$$
$$+ 2(k^4b^2k^29c^2P^4 + k^2b^29c^2P^4 + k^4b^2m^2k^2b^4 + k^29c^2P^4m^2k^2b^4$$
$$- m^2k^2b^49c^2P^4 - k^4b^2k^2b^2 - k^2(9c^2P^4)^2 - 2k^4b^29c^2P^4 + k^2b^2m^2k^2b^4).$$

Now an analysis of the terms in the last expression shows that

$$A < (m^2k^2b^4 + k^4b^2 - k^29c^2P^4 + k^2b^2 + 9c^2P^4)^2 \quad \text{if} \quad k^2 < 1,$$
$$A < (m^2k^2b^4 + k^4b^2 + k^29c^2P^4 + k^2b^2 - 9c^2P^4)^2 \quad \text{if} \quad k^2 > 1$$

and in both cases the corresponding $\lambda_{1,2}$ are positive. Using this result we can give a reasonable positive lower estimate of λ_1 and an upper estimate of λ_2:

$$0 < \kappa^2\frac{9c^2P^4}{2mb} < \lambda_1 \leq \lambda_2 < \frac{m^2k^2b^4 + k^4b^2 + k^2b^2 + \kappa^29c^2P^4}{2mb}$$

where (6.7.30) was used. Thus, with

$$\lambda_1^* = \kappa^2\frac{9c^2P^4}{2mb}, \qquad \lambda_2^* = \frac{m^2k^2b^4 + k^4b^2 + k^2b^2 + \kappa^29c^2P^4}{2mb}, \qquad (6.7.31)$$

we have

$$\lambda_1^*|z|^2 \leq v(z) \leq \lambda_2^*|z|^2.$$

Now everything has been determined for the application of Theorem 6.7.3.

Proposition. *Assume that (M1), (M2), (M3) hold, β is given by (6.7.29) and let*

$$0 < \eta < \frac{3\kappa|c|P^2\rho_3\left(-3P + (9P^2 + 2\rho_3/(\beta m|c|))^{1/2}\right)}{4\beta(m^2k^2b^4 + k^4b^2 + k^2b^2 + 9\kappa^2c^2P^4)^{1/2}},$$

$$\alpha_1 : = (-3P + (9P^2 + 2\rho_3/(\beta m|c|))^{1/2})^29c^2P^4\kappa^2/(4mb)$$
$$\alpha_2 : = (m^2k^2b^4 + k^4b^2 + k^2b^2 + 9\kappa^2c^2P^4)2\beta^2\eta^2/(mb\rho_3^2);$$

then

$$M_\eta := \{(t,x_1,x_2) \in \mathbf{R}_+ \times \mathbf{R}^2 : v(x - p(t)) \leq \alpha_2, \quad t \in \mathbf{R}_+\}$$

is a uniformly asymptotically stable set of the system corresponding to (6.7.23), and the set

$$B := \{(t,x_1,x_2) \in \mathbf{R}_+ \times \mathbf{R}^2 : v(x - p(t)) < \alpha_1, \quad t \in \mathbf{R}_+\}$$

is part of its region of attractivity.

Proof. Substituting the values λ_1^*, λ_2^* and (6.7.28) into the formulae preceding Theorem 6.7.3 for α_2 we obtain the expression stated in the proposition. For α_1 we get

$$\alpha_1 = \frac{9\kappa^2 c^2 P^4}{2mb}\left(\min\left(\rho_1, \frac{\rho_3}{6m|c|\beta(P + \rho_1/3)}\right)\right)^2.$$

Here ρ_1 is an arbitrary positive number. Clearly, ρ_1 is to be fixed such a way that α_1 becomes maximal, i.e. we set

$$\alpha_1 = \frac{9\kappa^2 c^2 P^4}{2mb}\max_{0 < \rho_1 < \infty}\left(\min\left(\rho_1, \frac{\rho_3}{6m|c|\beta(P + \rho_1/3)}\right)\right)^2.$$

It is easy to see that $\min\left(\rho_1, \rho_3/(6m|c|\beta(P + \rho_1/3))\right)$ is maximal where $\rho_1 = \rho_3/(6m|c|\beta(P + \rho_1/3))$, i.e. at

$$\rho_1 = (1/2)\left(-3P + (9P^2 + 2\rho_3/(\beta m|c|))^{1/2}\right).$$

Substituting this into the expression for α_1, the formula of the Proposition is obtained. The same substitution into (6.7.21) yields the best (largest) upper estimate for η given in the Proposition. $\square$

Observe that in all the formulae defining the attractor and its region of attractivity besides the parameters of the equation (6.7.22) and the upper band η of the perturbing term G, only the amplitude P of the unperturbed periodic solution of (6.7.22) occurs. Detailed calculations leading to the determination of this periodic solution can be found in Malkin [1956]. If we set $m = 0$ in (6.7.22), then the 2π-periodic solution is $(a/(k^2 - 1))\sin t$. This shows that for small values of m the amplitude a of the periodic forcing term comes into the picture through P.

Kertész [1988] has improved the estimates provided by Theorem 6.7.3 considerably and also extended the range of the applicability of this Theorem (see also Kertész [1987]). The main result will be presented here without proof. Consider the system

$$\dot{z} = A(t)z + h(t, z) + k(t, z) \tag{6.7.32}$$

(cf. (6.7.10) where the $n \times n$ matrix A is continuous and periodic, $h, h'_z, k, k'_z \in C^0(\mathbf{R}_+ \times \mathbf{R}^n)$, and assume that a $\rho_1 > 0$ and a $\rho_2 > 0$ exist such that for $t \in \mathbf{R}_+$, $|z| < \rho_1$ there holds $|h(t, z)| \leq \rho_2|z|^2$ and $|k(t, z)| < \eta(|z|)$ where η is some positive function. Suppose further that for the system

$$\dot{z} = A(t)z \tag{6.7.33}$$

there exists a continuous, periodic matrix function $V(t)$ such that the quadratic form

$$v(t, z) = z'V(t)z$$

is positive definite and that a constant $\rho_3 > 0$ exists such that

$$\dot{v}_{(6.7.33)}(t, z) < -\rho_3|z|^2, \qquad t \in \mathbf{R}_+, \qquad z \in \mathbf{R}^n.$$

Finally, denote by

$$\lambda_1 : = \min_i \min_{t \in \mathbf{R}_+} (\text{ the } i\text{-th eigenvalue of } V(t)) > 0,$$

$$\lambda_2 : = \max_i \max_{t \in \mathbf{R}_+} (\text{ the } i\text{-th eigenvalue of } V(t)) > 0.$$

Theorem 6.7.4 *(Kertész [1988]). Let a and b be positive constants such that*

$$0 < a < b \le \rho_1, \qquad \lambda_2 a^2 < \lambda_1 b^2,$$

and suppose that for $a < |z| < b$ there holds

$$0 < \eta(|z|) < -\rho_2|z|^2 + |z|\rho_3/(2\lambda_2);$$

then for any $\alpha_1, \alpha_2 > 0$ such that

$$\lambda_2 a^2 < \alpha_2 < \alpha_1 < \lambda_1 b^2$$

the set

$$M_\eta := \{(t, z) \in \mathbf{R}_+ \times \mathbf{R}^n : v(t, z) \le \alpha_2, \ t \in \mathbf{R}_+\}$$

is a uniformly asymptotically stable set of (6.7.32), and its region of attractivity contains the set

$$A_\eta := \{(t, z) \in \mathbf{R}_+ \times \mathbf{R}^n : v(t, z) < \alpha_1, \ t \in \mathbf{R}_+\}.$$

This Theorem makes it possible to determine a thinner M_η (smaller α_2) and a larger A_η (larger α_1) than those provided by Theorem 6.7.3 in case η is a constant.

The previous theory (Theorems 6.7.3 and 6.7.4) works also when the unperturbed system (6.7.5) is autonomous with an asymptotically stable *equilibrium point*, and analogous estimates can be given for the uniformly asymptotically stable set of the perturbed system (6.7.8) that contains the integral curve of the equilibrium, i.e. a straight line parallel to the t-axis in its interior and for its region of attractivity (see Farkas [1982]).

On the other hand, if the unperturbed autonomous system has a non-constant periodic solution, one cannot proceed on as above. We know (see the beginning of Section 5.1) that a non-constant periodic solution of an autonomous system cannot be asymptotically stable and its integral curve cannot be isolated in the sense of Definition 4.2.1. As a consequence no "snake that has swallowed plenty of mice" exists for a corresponding perturbed system. One may handle the situation, instead, in the following way. Consider the unperturbed autonomous system

$$\dot{x} = f(x) \tag{6.7.34}$$

where $f \in C^1(\mathbf{R}^n, \mathbf{R}^n)$, and suppose that it has a non-constant periodic solution $p : \mathbf{R} \to \mathbf{R}^n$ with period $T > 0$, and that the conditions of the Andronov-Witt Theorem 5.1.2 hold, i.e. the variational system

$$\dot{y} = f_x'(p(t))y \tag{6.7.35}$$

has $n - 1$ characteristic multipliers with modulus less than 1. In this case, p is orbitally asymptotically stable with asymptotic phase. If the orbit of p is denoted by

$$\gamma := \{x \in \mathbf{R}^n : x = p(t), \ t \in \mathbf{R}\},$$

then the integral curve of p is imbedded into the cylinder $\mathbf{R} \times \gamma \subset \mathbf{R} \times \mathbf{R}^n$ that consists of the integral curves of the solutions $p(t + \alpha)$, $\alpha \in \mathbf{R}$. The cylinder $\mathbf{R} \times \gamma$ is a uniformly asymptotically stable set of (6.7.34). Suppose that (6.7.34) is subject to a perturbation

$$\dot{x} = f(x) + g(t, x) \tag{6.7.36}$$

where $g, g_x' \in C^0(\mathbf{R} \times \mathbf{R}^n)$; for any compact set $Q \in \mathbf{R}^n$ there exists an $\eta > 0$ such that

$$|g(t, x)| < \eta, \qquad (t, x) \in \mathbf{R} \times Q,$$

and g_x' is also bounded in $\mathbf{R} \times Q$. By Yoshizawa's Theorem 6.7.2 for sufficiently small $\eta > 0$, system (6.7.36) has a uniformly asymptotically stable positively invariant set M_η which contains the cylinder $\mathbf{R} \times \gamma$ in its interior. M_η is to be imagined as a cylinder with a thick wall of periodically varying cross section. Nguyen Van Minh and Tran Van Nhung [1991] have given explicit estimates of the perturbed uniformly asymptotically stable set M_η and of its region of attractivity. Because of lack of space we are unable to give the results here in detail. These results give, in fact, an explicit estimate to a stable two dimensional integral manifold (see Definition 5.6.1) whose existence has been established by Jack Hale in the following Theorem which is presented here without the proof.

Theorem 6.7.5 *(Hale [1969] p.244). Under the hypotheses announced above, the periodic orbit γ has a neighbourhood $U \subset \mathbf{R}^n$ and there exists an $\eta_1 > 0$ such that for $0 < \eta \leq \eta_1$ the perturbed system (6.7.36) has a two dimensional integral manifold*

$$P_\eta := \{(t, x) \in \mathbf{R} \times \mathbf{R}^n : x = p(\theta) + v(t, \theta, \eta), \quad (t, \theta) \in \mathbf{R}^2\}$$

where $v(t, \theta, 0) = 0$, $v(t, \theta, \eta) = v(t, \theta + T, \eta)$, v *is almost periodic (T-periodic) in* t *if* g *is almost periodic (T-periodic) in* t, *and* P_η *is asymptotically stable.*

6.8 Problems

Problem 6.8.1. Find as general as possible particular cases of perturbed systems (6.2.31) such that inequalities (6.2.55) and (6.2.56) or just the first of them be linear in the unknowns a_i, β_i ($i = 2, 3$).

Problem 6.8.2. Suppose that Liénard's differential equation $\ddot{x} + f(x)\dot{x} + g(x) = 0$ satisfies the conditions of Theorem 3.2.2 and that $\xi : \mathbf{R} \to \mathbf{R}$ is a non-constant periodic solution with period $T_0 > 0$. Under what condition can it be guaranteed that Liénard's equation with controllably periodic perturbation

$$\ddot{x} + f(x)\dot{x} + g(x) = \mu\gamma(t/T, x, \dot{x}, \mu, T)$$

($\gamma \in C^1$ and periodic in t with period T) has a periodic solution with period τ near to T_0 provided that $T = \tau$ is substituted into the equation for small $|\mu|$?

Problem 6.8.3. Suppose that the generalized Liénard differential equation $\ddot{x} + f(x, \dot{x})\dot{x} + g(x) = 0$ satisfies the conditions of Theorem 3.2.3 and that $\xi : \mathbf{R} \to \mathbf{R}$ is a non-constant periodic solution with period $T_0 > 0$. Answer the question analogous to that of the previous problem concerning the perturbed equation

$$\ddot{x} + f(x, \dot{x})\dot{x} + g(x) = \mu\gamma(t/T, x, \dot{x}, \mu, T).$$

Problem 6.8.4. Suppose that system $\dot{x} = f(x) + \mu g(t/T, x, \mu, T)$ satisfies the conditions imposed upon (6.2.1) and that it is periodic in the vector argument x with vector period aT, $a \in \mathbf{R}^n$, and suppose further that the unperturbed system $\dot{x} = f(x)$ has a D-periodic solution $p(t)$ with period T_0, coefficient vector a^0 (see Definitions 5.3.1-5.3.2) and that $aT = a^0 T_0$. Prove the analogues of Theorems 6.2.1 and 6.3.2 (see Farkas [1971]).

Problem 6.8.5. Consider the singularly perturbed system $\varepsilon\dot{x} = x^2 + y^2 - 1$, $\dot{y} = x$. Determine the solutions and the trajectories of the unperturbed system ($\varepsilon = 0$) and the equation of the trajectories of the perturbed one. Determine a periodic solution of the perturbed system whose path tends to the circle $x^2 + y^2 = 1$ as $\varepsilon \to 0$. (Hint: the differential equation of the trajectories of the perturbed system can be solved by the method of integrating factors.)

Problem 6.8.6. (*Farkas* [1981], *Kertész* [1988]). Consider the aperiodically perturbed linear differential equation $\ddot{x} + 3\dot{x} + 2x = \sin t + \eta \sin(t^2)$, $\eta > 0$. Estimate its asymptotically stable set by Theorem 6.7.3 and by Theorem 6.7.4. (Remark: the sharpness of these results can be judged if the asymptotically stable solution of the perturbed equation is determined that has the same initial values as the periodic solution of the unperturbed equation.)

7
Bifurcations

In this chapter the emphasis will be on the *"Andronov-Hopf bifurcation"*, the generic mathematical model of the phenomenon how a real world system depending on a parameter is losing the stability of an equilibrium state as the parameter is varied, giving rise to small stable or unstable oscillations. This will be treated in Section 2; applications in population dynamics will be presented in Sections 3 and 4. In Section 1 the underlying theory of structural stability will be dealt with in a concise form. According to the general structure of this book this Section ought to go into the Appendix; however, in this last chapter it yields, probably, an easier reference standing at the start.

The last two Sections go beyond the scope of this book. Though this book is about ordinary differential equations, it would not be complete if it did not mention, at least, some implications and generalizations of the theory to more general situations. In Section 5 differential equations with delay, i.e. functional differential equations, will be treated briefly with emphasis on the occurrence and stability of periodic solutions. In Section 6 we glimpse discrete dynamical systems and period doubling bifurcations.

The applications of bifurcation theory (and also of perturbation theory of the previous chapter) involves enormous algebraic computations in concrete situations, as it will be seen. Nowadays, more and more computer algebra is used in treating such problems. Unfortunately, there is no space here to go in that direction; the interested reader may refer, e.g., to Rand-Armbruster [1987].

7.1   Structural Stability and Bifurcations

When a real world phenomenon, a natural law of mechanics, physics, chemistry, biology or an economic, a historical, a linguistic, a psychological, etc., situation is modelled mathematically, some factors are necessarily neglected, and if the model is deterministic like a differential equation, then stochastic effects are not taken into consideration at all. If one is fortunate, then the neglected factors and the stochastic effects influence the phenomenon only weakly. This means that the (non-existent) "exact unknown mathematical model" of the situation differs only little from the model on hand. Because of this, a differential equation considered as a mathematical model of a real world phenomenon can be relied upon only if a small change of the right-hand side does not change the qualitative properties of the solutions. If small perturbations of the right-hand side cause big qualitative changes in the phase picture of the differential equation, then the latter can be used as a reliable model only with reservation.

We are dealing here with dynamical systems only i.e. with flows generated by autonomous systems of differential equations (see Section 1.3). It is assumed that all solutions are defined on the whole real line (see (1.3.5) et seq.). A dynamical system will be considered "*structurally stable*" or "*robust*" if its phase picture is qualitatively similar, "*equivalent*" to the phase picture of every "*neighbouring*" dynamical system. Naturally, we must give mathematically meaningful definitions to all these expressions: "neighbouring", "equivalent", "structurally stable".

The original somewhat more restrictive definition of robustness is due to Andronov and Pontriagin [1937]. Here we are relying, first of all, upon Peixoto [1959, 1962], Arnold [1983], Guckenheimer-Holmes [1983] and Wiggins [1988, 1990]. The theory works on flows defined on compact manifolds because otherwise we are unable to introduce a reasonable metric or topology in the space of flows. Alternatively, we may consider flows defined on compact subsets of $\mathbf{R}^n$ that are homeomorphic to the compact unit ball provided that the boundary of the subset is everywhere transversal to the flow, i.e. all trajectories having a common point with the boundary cross it transversally inward, say.

Thus, in the sequel, M denotes an n dimensional orientable (see Definition A3.2) compact C^2-manifold or a compact subset of $\mathbf{R}^n$ C^2-diffeomorphic to the compact unit ball and with boundary transversal to the flow.

Consider two flows on M generated by

$$\dot{x} = f(x) \qquad \text{and} \qquad \dot{x} = g(x) \qquad (7.1.1)$$

where $f, g \in C^1(M, \mathbf{R}^n)$.

Definition 7.1.1. The C^1-*distance* of the two flows (7.1.1) is by definition

$$\text{dist}\,(f,g) := \max_{x \in M} |f(x) - g(x)| + \max_{x \in M} |f'_x(x) - g'_x(x)|$$

where any vector and matrix norm may be used.

Because of the compactness of M, clearly, the C^1-distance is well defined for any $f, g \in C^1(M, \mathbf{R}^n)$. The use of the C^1-distance (C^1-norm) is intuitively necessary; it implies that if two flows are close, then their linearizations are also near each other. Since the eigenvalues of the linearization at an equilibrium point, say, carry important information about the flow, this is a must. If the "C^0-norm" were used, i.e. the distance were the maximum of the difference of the vector fields then arbitrary close systems might have completely different eigenvalues.

Equivalence of two flows is defined the following way.

Definition 7.1.2. We say that the two flows generated by systems (7.1.1) are (*orbitally topologically*) *equivalent* if there is a homeomorphism of M onto M that maps the trajectories of one flow onto the trajectories of the other, preserving the direction of the motion along the paths.

Thus, two equivalent flows have topologically identical (or similar) phase pictures. Since both flows are in the C^1 class, it is natural to ask: Why don't we use a diffeomorphism in the Definition instead of a homeomorphism? The answer is that a definition based on a diffeomorphism would be too "fine" and would leave inequivalent too many flows. We show this presently. Suppose that we require that there should be a diffeomorphism carrying the directed trajectories of flow one into the directed trajectories of flow two, and assume that $h : M \to M$ is such a diffeomorphism. Denoting the "time" in the first equation with t, in the second by s and the phase variable in the second equation by y instead of x, the systems are

$$\frac{dx}{dt} = f(x), \qquad \frac{dy}{ds} = g(y), \qquad x, y \in M. \qquad (7.1.2)$$

Since h is a diffeomorphism, it induces a linear mapping of the vectors in the tangent space T_y of M at $y \in M$ onto the vectors of the tangent space T_x of M at the corresponding point $x = h(y) \in M$ (cf. the paragraph preceding (A3.3)). Thus, by the mapping h, the tangent vector $g(y) \in T_y$ of the path of the second flow passing through y goes into the vector $h'(y)g(y) = h'(h^{-1}(x))g(h^{-1}(x)) \in T_x$. The equivalence of the flows means that this vector must be a positive scalar multiple of the tangent vector $f(x)$ of the path of the first flow passing through x:

$$h'(h^{-1}(x))g(h^{-1}(x)) = \mu(x)f(x) \qquad (7.1.3)$$

where $\mu \in C^0(M, \mathbf{R})$, $\mu(x) > 0$, $x \in M$. Assume without loss of generality that $y = x = 0$ is an equilibrium point of both systems. Expanding f, g, h and h^{-1} at this point we may write

$$\begin{aligned}
f(x) &= Ax + F(x), & g(y) &= By + G(y), \\
h(y) &= Cy + H(y), & h^{-1}(x) &= C^{-1}x + K(x)
\end{aligned}$$

where A, B and C are $n \times n$ matrices, C is regular, $F(x) = o(|x|)$, $G(y) = o(|y|)$, $H(y) = o(|y|)$, $K(x) = o(|x|)$. Substituting these expressions into (7.1.3) we obtain

$$\left[(C + H'(y))(By + G(y))\right]_{y=h^{-1}(x)} = \mu(x)(Ax + F(x))$$

or

$$(C + H'(h^{-1}(x)))(BC^{-1}x + BK(x) + G(h^{-1}(x))) = \mu(x)(Ax + F(x)),$$

i.e.

$$CBC^{-1}x + L(x) = \mu(x)(Ax + F(x)), \qquad L(x) = o(|x|).$$

Dividing the last identity by $|x| \neq 0$ and tending with x to zero from an arbitrarily fixed direction this implies that $CBC^{-1}e = \mu(0)Ae$ for an arbitrary unit vector e. Hence,

$$CBC^{-1} = \mu(0)A.$$

This means that if a *diffeomorphism* established the equivalence of the two flows, then for two equivalent flows in corresponding points of equilibria the eigenvalues of either flow must be positive scalar multiples of the other with the same positive factor. Thus, if, e.g., the eigenvalues of the matrix B were -2 and -5 and those of A were -4 and -10, then the two systems could be equivalent, but if the eigenvalues of A were -4 and -11, they could not. This, obviously, would not serve the purpose, and therefore a coarser definition is used, leaving less different classes of inequivalent flows. If as in the Definition a *homeomorphism* establishes the equivalence, then the eigenvalues of equivalent flows in corresponding equilibria are not in such a tight correspondence to each other, though clearly, if, say, the equilibrium of one of the flows is attractive, then the corresponding equilibrium of the other flow must be attractive too.

Now we may define the structural stability of a flow or of the corresponding system of differential equations. The space $C^1(M, \mathbf{R}^n)$ of the vector fields on M will be equipped with the *metric* generated by the C^1-distance of Definition 7.1.1.

Definition 7.1.3. We say that the system $\dot{x} = f(x)$ (or the corresponding flow) is *structurally stable (robust)* if f has a neighbourhood in $C^1(M, \mathbf{R}^n)$ such that our flow is equivalent to every flow generated by vector fields (i.e. systems) in this neighbourhood.

If a system is structurally stable in this sense and it is used as a mathematical model of a natural law or a real life phenomenon, then one can rely on the results gained by it since small effects that have been neglected at the model building will not change the qualitative properties of the solutions. There are two basic questions to be answered in connection with this concept. First, how can structurally stable systems be characterized, i.e. how can one determine whether a system is structurally stable or not? Secondly, do structurally stable systems form a sufficiently large subset of $C^1(M, \mathbf{R}^n)$? Both problems could be solved positively *in dimension 2*. The characterization was given by Andronov and Pontriagin [1937] and then by Peixoto [1959, 1962] applying an improved definition. We state the basic theorems in this Section without proof.

Theorem 7.1.1 *(Andronov-Pontriagin [1937], Peixoto [1959, 1962]). Let the system $\dot{x} = f(x)$ be such that $f \in C^1(M, \mathbf{R}^2)$ where M is a two dimensional orientable compact C^2-manifold (or alternatively $M \subset \mathbf{R}^2$ is C^2-diffeomorphic to the compact unit disk and has boundary ∂M transversal to f); the system is structurally stable if and only if*

(i) it has only a finite number of equilibrium points and all these are hyperbolic, i.e. no eigenvalue has a zero real part;

(ii) all the alpha and omega limit sets consist of equilibria and closed orbits only;

(iii) a path does not connect a saddle point with another saddle point, i.e. there is no path whose alpha limit set and omega limit set would both be a saddle point;

(iv) it has only a finite number of closed orbits and all these are "hyperbolic", i.e. the number 1 is a simple characteristic multiplier for each.

Conditions (i) and (iv) are intuitively clear; condition (ii) excludes cases when homoclinic and heteroclinic trajectories (see the note following Theorem 3.1.11) are parts of limit sets or when the whole M is the limit set that may happen to a dynamical system on the two dimensional torus when the "rotation number" is irrational (see Arnold [1983]). That a system having a "saddle connection" contrary to (iii) cannot be structurally stable, i.e. arbitrary small perturbations may change qualitatively the phase picture, can be seen from Figure 7.1.1

The answer to the second question is given in the following two Theorems where we have to separate the case when M is a compact differentiable

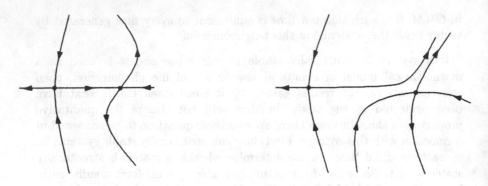

FIGURE 7.1.1. Small perturbation of a saddle connection.

manifold from the case when it is a compact subset of $\mathbf{R}^2$.

Theorem 7.1.2 *(Peixoto [1959, 1962]). Let M be a two dimensional orientable compact C^2-manifold without boundary; the structurally stable systems defined on M form an open and dense subset of $C^1(M, \mathbf{R}^2)$.*

Let $M \subset \mathbf{R}^2$ be C^2-diffeomorphic to the unit disk, and denote the C^1 vector fields defined on M and transversal to ∂M by $C_T^1(M, \mathbf{R}^2)$.

Theorem 7.1.3 *(Peixoto [1959]). The structurally stable systems form an open and dense subset of $C_T^1(M, \mathbf{R}^2)$.*

Unfortunately, in dimensions higher than two no characterization of structurally stable systems similar to the Andronov-Pontriagin-Peixoto Theorem 7.1.1 exists. Still for a long time it was a strong belief that structurally stable systems form an open and dense subset of the space of dynamical systems. While it is obvious that the set of structurally stable systems is open, density could not be proved and Smale's discovery of the "horseshoe" has shown that it is not true (Smale [1966, 1967] see also Guckenheimer-Holmes [1983] and Wiggins [1988, 1990]). Smale's horseshoe is a structurally unstable dynamical system that has a neighbourhood of similar structurally unstable systems. We say, vaguely, that a property is *generic* if it is present in the "vast majority of the systems", for instance, if an open and dense subset of the systems has this property. (There are more exact definitions of genericity that are weaker, but we shall not need them; see e.g. Wiggins [1988, 1990]). Thus, loosely speaking, structurally stable systems are generic in dimension two but are not in higher dimensions.

We see that structural stability is a too strong requirement, and one has to retreat from the position that only structurally stable systems could be accepted as models of real life phenomena. There are several ways by which one may attempt to overcome this difficulty. It is possible to treat the problem *locally* in the neighbourhood of an equilibrium point or a periodic orbit, say. One may define the *"local distance"*, the *"local equivalence"* and the *"local structural stability"* of flows.

Secondly, there is a robustness definition introduced by Zeeman [1988] that reduces the problem to the "structural stability" of real valued functions of n real variables. Now, the latter problem is well in hand. A function $f \in C^2(\mathbf{R}^n, \mathbf{R})$ is said to be a *Morse function* (i) if in any compact subset of $\mathbf{R}^n$ it has only a finite number of *critical points*, i.e. points where grad $f(x) = 0$; if (ii) all critical points are *non-degenerate*, i.e. $\det[f''_{x_i x_k}(x)] \neq 0$ where grad $f(x) = 0$. Being a Morse function is a *structurally stable property*, i.e. every function in a C^2-neighbourhood of a Morse function is a Morse function. Moreover, Morse functions are generic in $C^2(\mathbf{R}^n, \mathbf{R})$.

Thirdly, we may consider systems and accept them as realistic models of real world phenomena if they have *"structurally stable properties"* that are relevant from the physical point of view (thus agreeing with the philosophy expounded by Guckenheimer and Holmes [1983] p. 259). We say that a *property is relatively structurally stable* if small perturbations of the system have that property. Naturally, we have to tell what kind of perturbations we do allow. The structural stability of the property at hand is then *relative* to the allowable perturbations. For instance, if only homogeneous linear systems with constant coefficients are considered, then the property that the origin is a hyperbolic equilibrium point (all the eigenvalues have real parts different from zero) is relatively structurally stable. In this case if the stable and the unstable subspaces have dimension s and u, respectively (see (A3.16) et seq.), $s + u = n$, the dimension, then, clearly, this is true for every linear system with constant coefficient matrix sufficiently close to the given one. To be a (stable, say) node is not a structurally stable property since if the coefficient matrix has a double negative (real) eigenvalue, then arbitrarily close matrices may have, instead of this, a pair of complex conjugate eigenvalues with negative real and small non-zero imaginary parts, i.e. the node turns into a (stable) spiral point. More generally if a flow has a hyperbolic equilibrium point, then this property is structurally stable relative to flows in a sufficiently small C^1-neighbourhood *that have an equilibrium point in the neighbourhood of the given one*. Now, if the n dimensional system $\dot{x} = f(x)$ is considered, then its equilibrium is the point of intersection of the $n - 1$-dimensional hypersurfaces $f_i(x) = 0$ $(i = 1, 2, ..., n)$. We say that differentiable manifolds of $\mathbf{R}^n$ intersect *transversally* at a point if the direct sum of their tangent spaces at the point of intersection spans the space $\mathbf{R}^n$. It can be proved that transversal intersection is preserved by small peturbations (see Arnold [1983]). On the other hand, if the intersection is not transversal, i.e. the tangent spaces of $f_i(x) = 0$ $(i = 1, 2..., n)$ do not span $\mathbf{R}^n$ at the equilibrium, then it is easy to see that the equilibrium cannot be hyperbolic ($\det [f'_{i x_k}(x)] = 0$), so we may say that having a hyperbolic equilibrium point is a structurally stable property.

Instead of studying structurally stable properties relative to *arbitrary* C^1 perturbations, the more rewarding approach is to restrict the study to a parametrized family of systems

$$\dot{x} = f(x, \mu) \tag{7.1.4}$$

where $\mu \in \mathbf{R}^m$, $x \in \mathbf{R}^n$, $f \in C^r(\mathbf{R}^n \times \mathbf{R}^m, \mathbf{R}^n)$, $r \geq 1$. There are two approaches. The more ambitious one is to ask which are the parameter values μ where (7.1.4) is structurally stable (or locally structurally stable), and at which *critical values* of μ does structural stability break down. The less ambitious approach is to try to find those values of the parameters at which the system has a certain property that prevails when the parameters are varied a little, i.e. to look for the structural stability of certain properties *relative to the family of systems (7.1.4)*, and to find those critical values of μ at which the property is lost giving way to another property. The first approach is that of René Thom's *catastrophe theory* (see Thom [1972] and, e.g., Poston-Stewart [1978]). Thom's famous classification theorem gives a complete classification of cases of how structural stability can be lost for *gradient systems*:

$$\dot{x} = -\operatorname{grad} V(x, \mu) \tag{7.1.5}$$

where $V \in C^2(\mathbf{R}^n \times \mathbf{R}^m, \mathbf{R})$ provided that the dimension m of the parameter space is less than or equal to 5. Here we cannot stray in this direction, but there are two facts about gradient systems that have to be mentioned. First we quote, without proof, Smale's Theorem about the characterization of structurally stable gradient systems.

Theorem 7.1.4 *(Smale [1961]). Gradient systems for which all equilibria are hyperbolic and all intersections of stable and unstable manifolds transversal are structurally stable.*

The properties of the Morse functions and of transversal intersection imply that structurally stable gradient systems are generic in the space of gradient systems.

The second fact to be mentioned especially in a book about periodic solutions is the following negative statement.

Theorem 7.1.5. *A gradient system can have neither a (non-constant) periodic orbit nor a homoclinic orbit.*

Proof. Suppose that $p(t)$ is a non-constant periodic solution of (7.1.5) with period $T > 0$. (Forget about the μ in the system or rather fix it). Consider V along the solution p and differentiate the composite function $V \circ p$ with respect to t:

$$\frac{d}{dt}(V \circ p)(t) = \langle \operatorname{grad} V(p(t)), \dot{p}(t) \rangle = -(\operatorname{grad} V(p(t)))^2 \leq 0$$

and equality cannot occur, otherwise the path of p would contain an equilibrium point. Thus, the composite function $V \circ p$ is strictly monotonous decreasing, in particular, $V(p(0)) > V(p(T)) = V(p(0))$, a contradiction. The non-existence of a homoclinic orbit is proved analogously. $\square$

The second approach to (7.1.4) mentioned above is what is usually called *bifurcation theory*. The properties studied are, first of all, the existence, *the number* and the stability of equilibria and/or periodic solutions. One expects that there are open sets of the parameter space in which the number and the stability of the existing equilibria or periodic orbits do not change upon the variation of the parameters, and at the boundary of such domains a change occurs, for instance, the existing single stable equilibrium loses its stability and gives rise to two stable equilibria (hence the expression bifurcation which was, probably, introduced by Henri Poincaré), or the system had no equilibrium point and at a certain value of μ equilibria arise (usually in pairs), or at a certain value of the parameters periodic orbits appear around the equilibrium point. In this book the last phenomenon is the most important, and this will be the topic of the following Section. Here we are showing some bifurcation phenomena in which only equilibrium points are involved.

Example 7.1.1. *The saddle-node bifurcation.* Consider the one parameter family of systems

$$\dot{x} = -x^2 + \mu, \qquad \dot{y} = -y. \qquad (7.1.6)$$

For $\mu < 0$ the system has no equilibrium point. As μ is increased at $\mu = 0$, the origin $(x, y) = (0, 0)$ becomes a unique equilibrium point of the system. The eigenvalues of the linearization at the origin are 0 and -1. The phase picture can easily be drawn (see Figure 7.1.2 (b)): in the left-hand half plane ($x < 0$) it looks like one around a saddle, in the right-hand half plane it resembles a node (cf. (1.4.13) et seq. and Figure 1.4.3). This type of equilibrium is called a *saddle-node*. As μ is increased further the equilibrium "*bifurcates*" into two equilibria: for $\mu > 0$ the equilibrium point $E_1 = (-\sqrt{\mu}, 0)$ is a saddle, the eigenvalues of the linearization at E_1 are $\lambda_1^{(1)}(\mu) = 2\sqrt{\mu}$, $\lambda_2^{(1)}(\mu) = -1$; the equilibrium point $E_2 = (\sqrt{\mu}, 0)$ is a node, the eigenvalues at E_2 are $\lambda_1^{(2)}(\mu) = -2\sqrt{\mu}$, $\lambda_2^{(2)}(\mu) = -1$ (it is an improper node for $0 < \mu \neq 1/4$, a proper node for $\mu = 1/4$). As $\mu \to +0$ both equilibria tend to $(0, 0)$ and the eigenvalues tend to 0 and 1.

The second equation of system (7.1.6) is attached to the first one only in order to show the saddle-like and the node-like characters which would not come out in one dimension; however, it is worthwhile to draw the picture of this bifurcation in case we have just a (one dimensional) scalar equation

$$\dot{x} = -x^2 + \mu. \qquad (7.1.7)$$

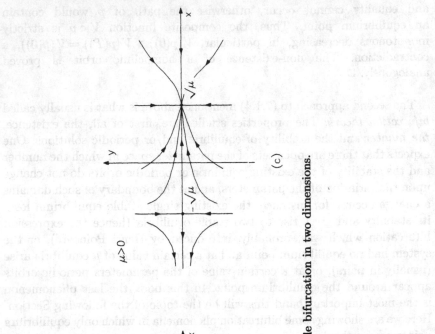

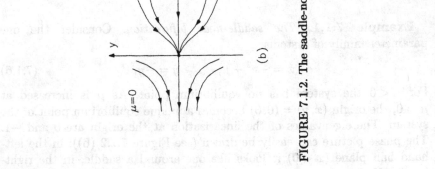

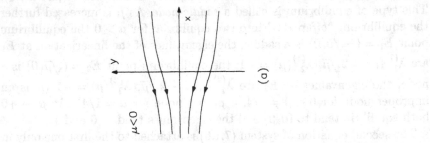

FIGURE 7.1.2. The saddle-node bifurcation in two dimensions.

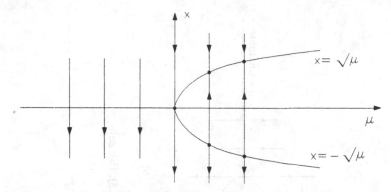

FIGURE 7.1.3. The saddle-node bifurcation in one dimension. The equilibria on the upper branch of the parabola are the stable nodes, those on the lower branch are the saddles, the origin is a "saddle-node".

In this case one can show what is happening in the μ, x plane (see Figure 7.1.3).

The saddle-node bifurcation is not something unusual or peculiar, it turns up generically in systems of arbitrary dimension: it is a typical phenomenon of the appearance or disappearance of an equilibrium (see Sotomayor's Theorem in Guckenheimer-Holmes [1983]).

Example 7.1.2. *The pitchfork bifurcation.* Consider the one parameter family of scalar differential equations

$$\dot{x} = -x^3 + \mu x. \tag{7.1.8}$$

The right-hand side of the equation is an odd function of x for arbitrary $\mu \in \mathbf{R}$; therefore, $x = 0$ is an equilibrium point always. Denoting the right-hand side by $f(x, \mu) := -x^3 + \mu x$ the derivative with respect to x is $f'_x(x, \mu) = -3x^2 + \mu$. Thus, $x = 0$ is asymptotically stable for $\mu < 0$ and unstable for $\mu > 0$. It is easy to see by integration that at $\mu = 0$ the equilibrium $x = 0$ is still asymptotically stable but it is no longer "linearly stable": the eigenvalue is zero. However, for $\mu > 0$ two other equilibria appear: $x = \pm\sqrt{\mu}$. Clearly $f'_x(\pm\sqrt{\mu}, \mu) = -2\mu < 0$; hence, both are asymptotically stable. Thus, as μ is increased through $\mu = 0$ the zero equilibrium loses its stability and two new stable equilibria emerge. This is called a *pitchfork bifurcation*. In this particular example it is a *supercritical* bifurcation meaning that the new equilibria arise above the critical parameter value $\mu = 0$ where the zero equilibrium is already unstable (see Figure 7.1.4 (a)).

Changing the minus sign to plus at x^3 in (7.1.8) we get the differential equation

$$\dot{x} = x^3 + \mu x. \tag{7.1.9}$$

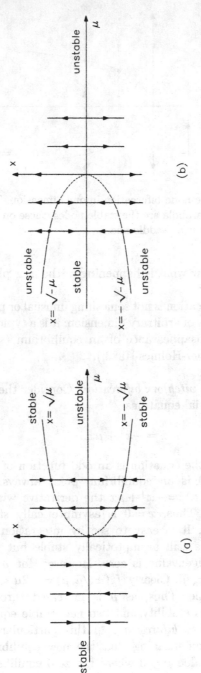

FIGURE 7.1.4. The pitchfork bifurcation. (a): supercritical case (7.1.8), (b): subcritical case (7.1.9).

The situation is analogous to the previous one with the difference that now at $\mu = 0$ the zero equilibrium is already unstable and that the non-zero equilibria ("the pitchfork") appear below the critical value $\mu = 0$ and they are unstable. This is called a *subcritical* pitchfork bifurcation (see Figure 7.1.4 (b)).

The pitchfork bifurcation occurs generically in higher order systems that have certain symmetry such as the oddness of the right-hand side as shown here (cf. Guckenheimer-Holmes [1983]).

7.2   The Andronov-Hopf Bifurcation

It is often observed, one may even say it is a common experience, that real world systems that have a stable steady state (an asymptotically stable equilibrium) may lose the stability of this steady state and may begin to oscillate first with small amplitudes and then wilder and wilder as a parameter of the system is varied in a certain direction (its value is increased or decreased). Formerly stable civil engineering structures, electrical networks, ecological communities and economic situations may begin to oscillate and then, as the amplitudes get larger, collapse, break down or just remain oscillatory as some relevant parameter is varied. The mathematical modelling of such phenomena leads to systems of differential equations depending on a parameter and having an isolated equilibrium point that is stable if the parameter value belongs to some interval but loses its stability as the parameter crosses the boundary of this interval. At the same time, in the neighbourhood of the critical value of the parameter where the stability of the equilibrium is lost, small amplitude periodic solutions occur. The appearance of small amplitude periodic solutions simultaneously with the loss of stability of an equilibrium is a generic phenomenon in systems depending on a parameter and *having an isolated equilibrium for all values of the parameter*. The important thing is that the linearization of the system at the equilibrium point must have a pair of complex conjugate eigenvalues (which naturally depend on the parameter), and at the critical value of the parameter this pair has to cross from the left-hand half plane to the right-hand one in the complex plane. This bifurcation of small amplitude periodic solutions has already been studied by H. Poincaré. The systematic study of the conditions and the proof of the corresponding bifurcation theorem was done by Andronov and Leontovich [1937] in the two dimensional case (to which the n dimensional case can also be reduced). The proof of the bifurcation theorem in the n dimensional case is due to Hopf [1942]. The "*Andronov-Hopf bifurcation*" has a vast literature. We are relying mainly on Arnold [1983], Marsden-McCracken [1976], Guckenheimer-Holmes [1983], Hassard-Kazarinoff-Wan [1981], and Negrini-Salvadori [1979].

Before presenting the theory we are showing what is to be expected on a simple non-trivial example.

Example 7.2.1. Consider the one parameter family of systems

$$\dot{x} = y - x(x^2 + y^2 - \mu), \qquad \dot{y} = -x - y(x^2 + y^2 - \mu). \qquad (7.2.1)$$

It is easy to see that the unique equilibrium is the origin $(0,0)$ for arbitrary $\mu \in \mathbf{R}$. The linearized system at $(0,0)$ is

$$\dot{x} = \mu x + y, \qquad \dot{y} = -x + \mu y.$$

The eigenvalues are $\lambda_{1,2}(\mu) = \mu \pm i$; hence, the origin is a stable spiral point for $\mu < 0$ and an unstable one for $\mu > 0$. The polar transformation $x = r\cos\theta$, $y = r\sin\theta$, $r \geq 0$ carries the system into

$$\dot{r} = -r(r^2 - \mu), \qquad \dot{\theta} = -1. \qquad (7.2.2)$$

This shows that all trajectories cross circles with sufficiently large radius and centre in the origin inward, i.e. all solutions are bounded on $t \in [0, \infty)$. Thus, by Theorem 3.1.11 for $\mu > 0$ when the only equilibrium $(0,0)$ cannot belong to the omega limit set of the solutions, the solutions must have a periodic orbit as an omega limit set. Indeed, for $\mu > 0$ the circle $x^2 + y^2 = \mu$ is a periodic orbit, and from (7.2.2) it is clear that all the solutions except the equilibrium tend to it as $t \to \infty$. It is easy to see that a periodic solution having this circle of radius $\sqrt{\mu}$ for its path is $p_1(t) = -\sqrt{\mu}\cos t$, $p_2(t) = \sqrt{\mu}\sin t$, and the variational system with respect to this periodic solution is

$$\begin{aligned} \dot{z}_1 &= -2z_1\mu\cos^2 t + z_2(1 + 2\mu\cos t \sin t), \\ \dot{z}_2 &= z_1(-1 + 2\mu\cos t \sin t) - 2z_2\mu\sin^2 t. \end{aligned}$$

As we know, one of the characteristic multipliers of this system is 1, and the other λ by Liouville's Theorem 1.2.4 and by the remark preceding (5.1.27) is

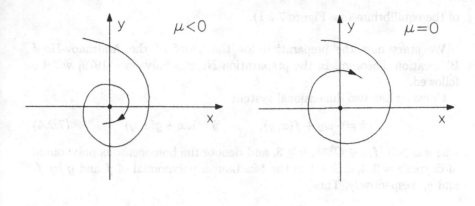

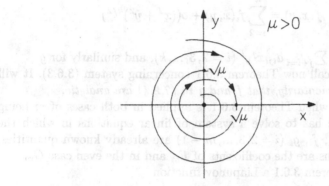

FIGURE 7.2.1. Phase portrait of (7.2.1).

$$0 < \lambda = 1 \cdot \lambda \;\; = \;\; \exp \int_0^{2\pi} (-2\mu \cos^2 t - 2\mu \sin^2 t)\,dt$$
$$= \;\; \exp(-4\mu\pi) < 1. \qquad (7.2.3)$$

Thus, also by the Andronov-Witt Theorem 5.1.2 the periodic solution is asymptotically orbitally stable with asymptotic phase.

We see that as μ is increased from negative values to positive ones at $\mu = 0$ the equilibrium loses its stability (to be sure at $\mu = 0$ the system in polar coordinates is $\dot{r} = -r^3$, $\dot{\theta} = -1$, thus the origin is still asymptotically stable but not "linearly stable"), and small amplitude ($\sqrt{\mu}$) periodic solutions bifurcate from it. These periodic solutions "inherit" the stability of the equilibrium (see Figure 7.2.1).

We start now the preparation for the proof of the Andronov-Hopf Bifurcation Theorem. In the preparation Negrini-Salvadori [1979] will be followed.

Consider the two dimensional system

$$\dot{x} = -\omega y + f(x, y), \qquad \dot{y} = \omega x + g(x, y) \tag{7.2.4}$$

where $\omega > 0$, $f, g \in C^{k+1}$, $k \geq 3$, and denote the homogeneous polynomial of degree $i = 2, 3, ..., k+1$ in the MacLaurin polynomial of f and g by f_i and g_i, respectively. Thus,

$$f(x, y) = \sum_{i=2}^{k} f_i(x, y) + o((x^2 + y^2)^{k/2})$$

where $f_i(x, y) = \sum_{j+l=i} a_{ijl} x^j y^l$ $(i = 2, 3, ..., k)$, and similarly for g.

We have to recall now Theorem 3.6.1 concerning system (3.6.3). It will be *assumed, momentarily, that f and g in (7.2.4) are analytic.*

We note that when Theorem 3.6.1 is applied in both cases of m being even or odd one has to solve a system of linear equations in which the coefficients of F_i, f_i, g_i $(i = 2, 3, ..., m-1)$ are already known quantities, and the unknowns are the coefficients of F_m and in the even case G_m.

Thus, by Theorem 3.6.1 a Liapunov function

$$F(x, y) = x^2 + y^2 + \sum_{i=3}^{m} F_i(x, y) \tag{7.2.5}$$

can be determined where F_i is a homogeneous polynomial of degree i $(i = 3, ..., m)$ such that

$$\dot{F}_{(7.2.4)}(x, y) = G_m(x^2 + y^2)^{m/2} + o((x^2 + y^2)^{m/2})$$

where $m \geq 4$ is an even number and $G_m \in \mathbf{R}$ is uniquely determined. Extending Definition 3.6.1 we set:

Definition 7.2.1. Let

$$M := \max\{m \geq 3 : \text{there exist } F_3, ..., F_{m-1} \text{ such that}$$

$$\tilde{F}_j = 0 \quad (j = 3, ..., m-1)\}$$

(where $\tilde{F}_j$ is the homogeneous polynomial of degree j in the expansion of $\dot{F}_{(7.2.4)}$) if this maximum exists, and $M := \infty$ if such a maximum does not exist; M is called the *index* of system (7.2.4); the number

$$G := \begin{cases} 0 & \text{if } M = \infty \\ G_M & \text{if } M \in \mathbf{N} \end{cases}$$

is called the *Poincaré-Liapunov constant* of (7.2.4).

If the index M is a positive integer, then it must be even and G_M is the first non-zero Poincaré-Liapunov *coefficient* (the $M/2$-th one).

Theorem 7.2.1. *If the index $M = \infty$, then system (7.2.4) admits an analytic first integral of the form*

$$F(x,y) = x^2 + y^2 + \sum_{i=3}^{\infty} F_i(x,y),$$

the origin $(0,0)$ is stable in the Liapunov sense but not asymptotically, and all solutions in a sufficiently small neighbourhood of the origin are periodic; if the index $M \in \mathbf{N}$, then there exists a polynomial Liapunov function of the form (7.2.5) with $m = M$ such that

$$\dot{F}_{(7.2.4)}(x,y) = G(x^2 + y^2)^{M/2} + \psi(x,y)$$

where $G \neq 0$, and ψ is analytic whose expansion starts with terms of degree greater than or equal to $M + 1$. In this case if $G < 0$, then the origin is asymptotically stable; if $G > 0$, then it is repelling, i.e. solutions in a neighbourhood tend to it as $t \to -\infty$.

We return now to the general (non-analytic) case, that is, *we assume again that $f, g \in C^{k+1}$, $k \geq 3$ in (7.2.4).*

Definition 7.2.2. Let h be an integer between 2 and k, i.e. $2 \leq h \leq k$; we say that the origin is *h-asymptotically stable* (resp. *h-repelling*) if for arbitrary $\xi, \eta \in C^0(\mathbf{R}^2, \mathbf{R})$ such that both ξ and η are $o((x^2 + y^2)^{h/2})$ for the system

$$\dot{x} = -\omega y + \sum_{i=2}^{h} f_i(x,y) + \xi(x,y),$$

$$\dot{y} = \omega x + \sum_{i=2}^{h} g_i(x,y) + \eta(x,y); \qquad (7.2.6)$$

the origin is asymptotically stable (resp. repelling), and h is the least integer between 2 and k with this property.

This Definition of h-asymptotic stability is an obvious generalization of "linear stability". The equilibrium point of an autonomous system is linearly stable if its linearization determines its stability (i.e. if all the eigenvalues have negative real parts; this is also called exponential asymptotic stability since in this case the solutions tend to the equilibrium exponentially as $t \to \infty$). If this is not the case as in (7.2.4), then it may happen that one may go into the expansion of the right-hand side up to the h-th degree terms, and the system with right-hand side truncated up to this degree already determines the stability, irrespective of what follows in the expansion. There holds:

Theorem 7.2.2 *(Negrini-Salvadori [1979]). Let $2 \le h \le k$; the equilibrium $(x, y) = (0, 0)$ of system (7.2.4) is h-asymptotically stable (resp. h-repelling) if and only if the index of the system*

$$\dot{x} = -\omega y + \sum_{i=2}^{h} f_i(x, y),$$

$$\dot{y} = \omega x + \sum_{i=2}^{h} g_i(x, y) \qquad (7.2.7)$$

is equal to $h + 1$, and the Poincaré-Liapunov constant of (7.2.7) is negative (resp. positive).

Observe that if the origin is h-asymptotically stable or h-repelling, then it is a weak spiral point of order $(h - 1)/2$ by Definition 3.6.1.

Proof. Suppose, first, that $(0, 0)$ is an h-asymptotically stable solution of (7.2.4). Putting $\xi = 0$, $\eta = 0$ in (7.2.6) this means that $(0, 0)$ is an asymptotically stable solution of the truncated system (7.2.7). By Theorem 7.2.1 this implies that the index M of (7.2.7) is an even number and the Poincaré-Liapunov constant G of (7.2.7) is negative. Hence, there exists a polynomial (7.2.5) with $m = M$ such that

$$\dot{F}_{(7.2.7)}(x, y) = G(x^2 + y^2)^{M/2} + o((x^2 + y^2)^{M/2}).$$

The index M must be greater than or equal to $h + 1$ because if we had $M \le h$, then $(0, 0)$ of (7.2.7) and (7.2.4) would be $(h - 1)$-asymptotically stable at most. We show that M cannot be greater than $h + 1$. Suppose that $M > h + 1$ and choose

$$\xi(x, y) = ax(x^2 + y^2)^{M/2 - 1}, \qquad \eta(x, y) = ay(x^2 + y^2)^{M/2 - 1}$$

with $a > -G/2$ in (7.2.6). Then

$$\dot{F}_{(7.2.6)}(x, y) = (2a + G)(x^2 + y^2)^{M/2} + o((x^2 + y^2)^{M/2})$$

where $2a + G > 0$, i.e. $(0, 0)$ would be repelling, contrary to the assumption.

Secondly, suppose that the index of system (7.2.7) is $h + 1$ and its Poincaré-Liapunov constant is negative. Then there exists a polynomial (7.2.5) with $m = h + 1$ whose derivative with respect to (7.2.6) is

$$\dot{F}_{(7.2.6)}(x, y) = G(x^2 + y^2)^{(h+1)/2} + o((x^2 + y^2)^{(h+1)/2})$$

with $G < 0$. This implies that for some $l \leq h$ the equilibrium $(0,0)$ of (7.2.4) is l-asymptotically stable. By the first part of the proof, if $l < h$ held, then the index would be less than $h + 1$, contrary to the assumption.

The statement about the h-repelling property can be proved similarly. □

The concept of h-asymptotic stability (resp. h-repelling property) can be extended to higher dimensional systems. Consider the system

$$\dot{x} = f(x) \tag{7.2.8}$$

where $f \in C^{k+1}(\mathbf{R}^n, \mathbf{R}^n)$, $k \geq 3$, $f(0) = 0$, $f'_x(0)$ has a pair of pure imaginary eigenvalues $\pm i\omega$, $\omega > 0$ and the other $n - 2$ eigenvalues have negative real parts.

Definition 7.2.3. Let $2 \leq h \leq k$; we say that $x = 0$ is an h-asymptotically stable (resp. h-unstable) equilibrium of (7.2.8) if it is an h-asymptotically stable (resp. h-repelling) equilibrium of the restriction of the system to the two dimensional centre manifold corresponding to the eigenvalues $\pm i\omega$ (see Theorems A3.1-A3.2, Definition A3.7 and Lemma A1.20).

We now state and prove the *Andronov-Hopf Bifurcation Theorem* in a relatively simple form. More sophisticated versions can be found in Hassard-Kazarinoff-Wan [1981].

Theorem 7.2.3 *(Hopf [1942], see also Marsden-McCracken [1976]). Consider the system*

$$\dot{x} = f(x, \mu) \tag{7.2.9}$$

where $f \in C^{k+1}(\mathbf{R}^n \times \mathbf{R})$, $k \geq 4$ and $f(0, \mu) \equiv 0$; suppose that for small $|\mu|$ the matrix $f'_x(0, \mu)$ has a pair of complex conjugate eigenvalues $\alpha(\mu) \pm i\omega(\mu)$, $\omega(\mu) > 0$, $\alpha(0) = 0$, $\alpha'(0) > 0$ (the derivative of the real part with respect to the parameter μ is positive) and the other $n - 2$ eigenvalues have negative real part; then

(i) *there exist a $\delta > 0$ and a function $\mu \in C^{k-2}((-\delta, \delta), \mathbf{R})$ such that for $\varepsilon \in (-\delta, \delta)$ the system $\dot{x} = f(x, \mu(\varepsilon))$ has a periodic solution $p(t, \varepsilon)$ with period $T(\varepsilon) > 0$, also $T \in C^{k-2}$, $\mu(0) = 0$, $T(0) = 2\pi/\omega(0)$, $p(t, 0) \equiv 0$, and the amplitude of this periodic solution (the approximate distance of the corresponding periodic orbit from the origin) is proportional to $\sqrt{|\mu(\varepsilon)|}$;*

(ii) *the origin $(x, \mu) = (0,0)$ of the space $\mathbf{R}^n \times \mathbf{R}$ has a neighbourhood $U \subset \mathbf{R}^n \times \mathbf{R}$ that does not contain any periodic orbit of (7.2.9) but those of the family $p(t, \varepsilon)$, $\varepsilon \in (-\delta, \delta)$;*

(iii) *if the origin $x = 0$ is a 3-asymptotically stable (resp. 3-unstable) 3-unstable equilibrium of the system $\dot{x} = f(x, 0)$, then $\mu(\varepsilon) > 0$ (resp.*

$\mu(\varepsilon) < 0$ *for* $\varepsilon \neq 0$, *and the periodic solution* $p(t, \varepsilon)$ *is asymptotically orbitally stable (resp. unstable).*

The conditions of the Theorem are natural and are satisfied typically. The condition that $x = 0$ is an equilibrium for all μ is not an essential restriction of generality since typically the system of equations $f(x, \mu) = 0$ can be expected to have a solution $\overline{x}(\mu)$ for each μ, and if the origin is displaced to $\overline{x}(\mu)$, then the transformation $y = x - \overline{x}(\mu)$ carries the system into one that already satisfies this condition. If the equilibrium $x = 0$ is linearly asymptotically stable for μ's in an interval, i.e. all the eigenvalues of $f_x'(0, \mu)$ have negative real parts, then as μ is increased (or decreased) one may expect that at a certain value of μ either a negative eigenvalue crosses the imaginary axis or a pair of complex conjugate eigenvalues crosses into the right-hand half plane. It is "unlikely" and generically does not happen in a *one parameter* family of systems that two pairs of complex eigenvalues or a pair and a real eigenvalue cross simultaneously into the positive half of the complex plane resulting in the destabilization of the equilibrium point. (In the case the family depends on two or more parameters, such a situation may generically occur, giving rise to so-called "codimension two or higher bifurcations"; see e.g., Langford [1979] and Golubitsky-Schaeffer [1985]). Now, the first case, a simple real eigenvalue changing sign has been treated briefly in the previous Section: a typical example being the pitchfork bifurcation. The second case, a pair of complex eigenvalues crossing the imaginary axis, is covered by the Andronov-Hopf Theorem. The condition $\alpha'(0) > 0$ is expressed by saying that the pair of complex conjugate eigenvalues crosses the imaginary axis with non-zero speed. This is also a generic requirement, though it is not absolutely necessary: the existence part of the Theorem remains valid even in the degenerate case when this derivative is zero but $\alpha(\mu) \lessgtr 0$ according to whether $\mu \lessgtr 0$.

The statements (i), (ii) and (iii) state the existence, the uniqueness and the stability of the bifurcating periodic solution, respectively. The uniqueness is to be understood as it is expressed in (ii) and it is to be observed that it does *not* mean that to each sufficiently small μ there belongs exactly one periodic orbit. The point is that the function $\mu(\varepsilon)$ may assume the same value at different values of ε arbitrary near $\varepsilon = 0$ (cf. Negrini-Salvadori [1979]) resulting in several periodic orbits corresponding to the same system (to the same μ) but to different ε's, or even it may happen that $\mu(\varepsilon) \equiv 0$.

More is true than what is expressed in (iii): *if the origin* $x = 0$ *is an* h-*asymptotically stable (resp. h-unstable) equilibrium of* $\dot{x} = f(x, 0)$ $(3 \leq h \leq k)$, *then* $p(t, \varepsilon)$ *is asymptotically orbitally stable (resp. repelling).* For the proof of this stronger statement, see Negrini-Salvadori [1979]. Under the assumptions of the Theorem the origin is not linearly asymptotically stable since it has a pair of pure imaginary eigenvalues. If it is a centre

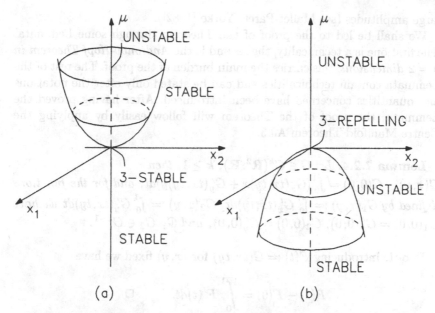

FIGURE 7.2.2. Supercritical (a) and subcritical (b) Andronov-Hopf bifurcation at $\mu = 0$; the paraboloid-like surfaces are the unions of the periodic orbits belonging to the different values of μ; these orbits are orbitally asymptotically stable in case (a) and unstable in case (b).

not only of the linearized system but also of $\dot{x} = f(x, 0)$, then $\mu(\varepsilon) \equiv 0$, the first two statements can be proved fairly easily and, clearly, the bifurcating periodic solutions that appear all for $\mu = 0$ cannot be orbitally asymptotically stable (neither completely repelling, to be sure). However, if $x = 0$ is *not* a centre for $\dot{x} = f(x, 0)$, then generically it must be either 3-asymptotically stable or 3-unstable. If this is not the case, i.e. $x = 0$ is 5-, 7-, ... k-asymptotically stable (or unstable), then the application of the Theorem is extremely difficult because of technical reasons, though, now there are already algorithms and symbolic manipulation computer programs available for these applications (see e.g. Kertész-Kooij [1991]).

Apart from the different degeneracies described in the previous two paragraphs, the two typical situations that arise if the condition in (iii) is satisfied are shown on Figure 7.2.2. The first case when $x = 0$ is 3-asymptotically stable is called a *supercritical bifurcation* since then the periodic orbits appear for $\mu > 0$, i.e. above the critical value $\mu = 0$ of the bifurcation parameter. The second case when $x = 0$ is 3-unstable is called a *subcritical bifurcation*: in this case the periodic orbits appear below the critical value $\mu = 0$, i.e. for negative values of μ. The respective terms of *soft* and *hard bifurcation* are also used: "soft" because then the system begins to oscillate stably with small amplitudes; "hard" because then the behaviour of the system is unforeseeable, usually it starts to oscillate with large amplitudes (see Mallet-Paret, Yorke [1982]).

We shall be led to the proof of the Theorem through some Lemmata. The first one is a technicality, the second is the *Andronov-Hopf Theorem in $n = 2$ dimensions*; this carries the main burden of the proof. The rest of the Lemmata contain technicalities and can be stated only after the notations and quantities concerned have been introduced. After having proved the Lemmata, the proof of the Theorem will follow easily by applying the Centre Manifold Theorem A3.3.

Lemma 7.2.4. *Let $G \in C^k(\mathbf{R}^2, \mathbf{R})$, $k \geq 1$; then*
$G(x, y) - G(0, 0) = \int_0^1 (G'_x(tx, ty)x + G'_y(tx, ty)y)dt$, *and for the functions defined by* $G_1(x, y) := \int_0^1 G'_x(tx, ty)dt$, $G_2(x, y) := \int_0^1 G'_y(tx, ty)dt$ *we have* $G_1(0, 0) = G'_x(0, 0)$, $G_2(0, 0) = G'_y(0, 0)$, *and* $G_1, G_2 \in C^{k-1}$.

Proof. Introducing $F(t) := G(tx, ty)$ for (x, y) fixed we have

$$F(1) - F(0) = \int_0^1 F'(t)dt. \qquad \square$$

Now follows the *Andronov-Hopf Theorem in two dimensions*.

Lemma 7.2.5 *(Andronov-Leontovich [1937], see also Marsden-McCracken [1976]). Consider the system*

$$\dot{x} = f(x, \mu) \qquad (7.2.10)$$

where $f \in C^{k+1}(\mathbf{R}^2 \times \mathbf{R}, \mathbf{R}^2)$, $k \geq 4$ and $f(0, \mu) = 0$; suppose that for small $|\mu|$ the 2×2 matrix $f'_x(0, \mu)$ has a pair of complex conjugate eigenvalues $\alpha(\mu) \pm i\omega(\mu)$, $\omega(\mu) > 0$, $\alpha(0) = 0$, $\alpha'(0) > 0$; then

(i) *there exist a $\delta > 0$ and a function $\mu \in C^{k-1}((-\delta, \delta), \mathbf{R})$ such that for $x_1^0 \in (-\delta, \delta)$ the solution $p(t, x_1^0) := \varphi(t, (x_1^0, 0), \mu(x_1^0))$ of the system $\dot{x} = f(x, \mu(x_1^0))$ is periodic with period $T(x_1^0) > 0$; also $T \in C^{k-1}$, $T(0) = 2\pi/\omega(0)$, and $\mu(0) = \mu'(0) = 0$;*

(ii) *the origin $(x_1, x_2, \mu) = (0, 0, 0)$ of the space $\mathbf{R}^2 \times \mathbf{R}$ has a neighbourhood $U \subset \mathbf{R}^2 \times \mathbf{R}$ that does not contain any periodic orbit of (7.2.10) but those of the family $p(t, x_1^0)$, $x_1^0 \in (-\delta, \delta)$;*

(iii) *if the origin $(x_1, x_2) = (0, 0)$ is a 3-asymptotically stable (resp. 3-repelling) equilibrium of the system $\dot{x} = f(x, 0)$, then $\mu(x_1^0) > 0$ (resp. $\mu(x_1^0) < 0$) for $x_1^0 \neq 0$, and the periodic solutions $p(t, x_1^0)$ are asymptotically orbitally stable (resp. repelling).*

Here, naturally, $\varphi(t, (x_1^0, x_2^0), \mu)$ denotes the solution of (7.2.10) that satisfies the initial condition $\varphi(0, (x_1^0, x_2^0), \mu) = x^0 = (x_1^0, x_2^0)$.

Proof. We may assume without loss of generality that system (7.2.10) has already been transformed to the normal form

$$\dot{x} = f_x'(0,\mu)x + g(x,\mu) \tag{7.2.11}$$

where

$$f_x'(0,\mu) = \begin{pmatrix} \alpha(\mu) & -\omega(\mu) \\ \omega(\mu) & \alpha(\mu) \end{pmatrix} \tag{7.2.12}$$

and $g \in C^k(\mathbf{R}^2 \times \mathbf{R}, \mathbf{R}^2)$, $|g(x,\mu)| = o(|x|)$ (see Lemma A1.20). Introducing polar coordinates by the transformation $x_1 = r\cos\theta$, $x_2 = r\sin\theta$, system (7.2.10) goes into the equivalent system

$$\dot{r} = f_r(r,\theta,\mu), \qquad \dot{\theta} = f_\theta(r,\theta,\mu) \tag{7.2.13}$$

where

$$f_r(r,\theta,\mu) := f_1(r\cos\theta, r\sin\theta, \mu)\cos\theta + f_2(r\cos\theta, r\sin\theta, \mu)\sin\theta,$$

$$f_\theta(r,\theta,\mu) := \begin{cases} -f_1(r\cos\theta, r\sin\theta, \mu)\sin\theta/r \\ \quad +f_2(r\cos\theta, r\sin\theta, \mu)\cos\theta/r, & r \neq 0 \\ \omega(\mu), & r = 0. \end{cases}$$

Taking into account (7.2.12) and that

$$(1/r)f_i(r\cos\theta, r\sin\theta, \mu) = \cos\theta \int_0^1 f_{ix_1}'(rt\cos\theta, rt\sin\theta, \mu)dt$$

$$+ \sin\theta \int_0^1 f_{ix_2}'(rt\cos\theta, rt\sin\theta, \mu)dt$$

by Lemma 7.2.4 we obtain that $f_\theta \in C^k$. Clearly, $f_r \in C^{k+1}$.

We are going to construct a Poincaré map (see Definition 5.2.1) for system (7.2.10) whose fixed point will yield the periodic solution. In order to bring the bifurcation parameter into play in a natural way we consider it as a third state variable and attach the differential equation $\dot{\mu} = 0$ to both systems (7.2.10) and (7.2.13). Thus, the equivalent systems

$$\dot{x}_1 = f_1(x_1, x_2, \mu), \qquad \dot{x}_2 = f_2(x_1, x_2, \mu), \qquad \dot{\mu} = 0 \tag{7.2.14}$$

and

$$\dot{r} = f_r(r,\theta,\mu), \qquad \dot{\theta} = f_\theta(r,\theta,\mu), \qquad \dot{\mu} = 0 \tag{7.2.15}$$

will be considered. We need a Poincaré map for the flow generated by (7.2.14); however, we cannot find a transversal "hyperplane", i.e. a transversal straight line in the origin since $f_1(0,0,\mu) = f_2(0,0,\mu) = 0$. At the same time while $f_r(0,\theta,\mu) = 0$, $f_\theta(0,\theta,\mu) = \omega(\mu) \neq 0$, so that the Poincaré map will be constructed first for system (7.2.15). The solution of (7.2.14) that assumes the initial values x_1, x_2, μ at $t = 0$ will be denoted by $(\varphi(t, x_1, x_2, \mu), \mu)$; the corresponding solution of (7.2.15) that assumes the initial values r, θ, μ at $t = 0$ will be $(\tilde{\varphi}(t, r, \theta, \mu), \mu)$. System (7.2.15) is cylindrical (see Definition 5.3.2 et seq.): it is periodic in the state variable θ with period 2π. It is easy to see that

$$(\tilde{\varphi}(t, 0, \theta, \mu), \mu) = (0, \theta + \omega(\mu)t, \mu) \tag{7.2.16}$$

so that

$$(\tilde{\varphi}(2\pi/\omega(\mu), 0, 0, \mu), \mu) = (0, 2\pi, \mu), \tag{7.2.17}$$

$$(\tilde{\varphi}(2\pi/\omega(0), 0, 0, 0), 0) = (0, 2\pi, 0). \tag{7.2.18}$$

We see that the solution of (7.2.15) with initial values $r = 0$, $\theta = 0$, $\mu = 0$ reaches the plane $\theta = 2\pi$ in the space $\mathbf{R}^3$ of r, θ, μ at time $t = 2\pi/\omega(0)$. Because of the continuous dependence of the solution on the initial values and because the plane $\theta = 0$ is transversal to the flow at $(r, \theta, \mu) = (0, 0, 0)$ (since $f_\theta(0, 0, 0) = \omega(0) > 0$) there exists a neighbourhood

$$\tilde{U}_0 = \{(r, 0, \mu) \in \mathbf{R}^3 : (r^2 + \mu^2)^{1/2} < \delta_0\}$$

of $(r, \theta, \mu) = (0, 0, 0)$ in this plane such that every solution with initial values in $\tilde{U}_0$ reaches the plane $\theta = 2\pi$ in finite time $T(r, \mu)$ near $T(0, 0) := 2\pi/\omega(0)$ in some neighbourhood of the point $(r, \theta, \mu) = (0, 2\pi, 0)$, denoted by

$$\tilde{U}_{2\pi} = \{(r, 2\pi, \mu) \in \mathbf{R}^3 : (r^2 + \mu^2)^{1/2} < \delta_{2\pi}\}, \quad \delta_{2\pi} > 0.$$

See Figure 7.2.3. This way a mapping $\tilde{P}$ of $\tilde{U}_0$ into $\tilde{U}_{2\pi}$ has been defined: $\tilde{P} : \tilde{U}_0 \to \tilde{U}_{2\pi}$ which acts so that for every $(r, 0, \mu) \in \tilde{U}_0$

$$\tilde{P}(r, 0, \mu) = (\tilde{\varphi}(T(r, \mu), r, 0, \mu), \mu).$$

If the first and the second coordinates of $\tilde{\varphi}$ are denoted by $\tilde{r}$ and $\tilde{\theta}$, respectively, then $\tilde{r}(T(r, \mu), r, 0, \mu)$ is some value such that $(\tilde{r}(T(r, \mu), r, 0, \mu), 2\pi, \mu) \in \tilde{U}_{2\pi}$ and $\tilde{\theta}(T(r, \mu), r, 0, \mu) = 2\pi$. The time $T(r, \mu)$ needed to get from $\tilde{U}_0$ to $\tilde{U}_{2\pi}$ is obtained from

$$\tilde{\theta}(t, r, 0, \mu) = 2\pi.$$

Since $\tilde{\theta}(2\pi/\omega(0), 0, 0) = 2\pi$ by (7.2.18), and $\dot{\tilde{\theta}}(2\pi/\omega(0), 0, 0, 0) = \omega(0) > 0$, the Implicit Function Theorem can be applied, and as a consequence, the function $T : \tilde{U}_0 \to \mathbf{R}$ is in the C^k class. As a consequence also $\tilde{P} \in C^k$. By the way, (7.2.17) implies that $T(0, \mu) = 2\pi/\omega(\mu)$. The mappping $\tilde{P}$ generates a Poincaré map P for system (7.2.14) in the following way. By the polar coordinate transformation the neighbourhood $\tilde{U}_0$ goes into

$$U_0 = \{(x_1, 0, \mu) \in \mathbf{R}^3 : (x_1^2 + \mu^2)^{1/2} < \delta_0\}$$

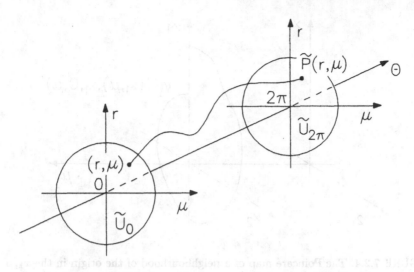

FIGURE 7.2.3. The mapping $\widetilde{P}$ of $\widetilde{U}_0$ into $\widetilde{U}_{2\pi}$ by the flow generated by (7.2.15).

since $x_1 = r\cos 0 = r$, $x_2 = r\sin 0 = 0$. For $(x_1, 0, \mu) \in U_0$ we define $P : U_0 \to \mathbf{R}^2$ (where $\mathbf{R}^2$ is the x_1, μ plane defined by $x_2 = 0$) by

$$
\begin{aligned}
P(x_1, 0, \mu) &= \big(\tilde{r}(T(x_1, \mu), x_1, 0, \mu)\cos(2\pi), \\
&\qquad \tilde{r}(T(x_1, \mu), x_1, 0, \mu)\sin(2\pi), \mu\big) \\
&= \big(\varphi_1(T(x_1, \mu), x_1, 0, \mu), 0, \mu\big). \qquad (7.2.19)
\end{aligned}
$$

Indeed, the solutions of (7.2.14) are obtained from the corresponding solution of (7.2.15) by

$$
\begin{aligned}
\varphi_1(t, x_1, x_2, \mu) &= \tilde{r}(t, r, \theta, \mu)\cos\tilde{\theta}(t, r, \theta, \mu), \\
\varphi_2(t, x_1, x_2, \mu) &= \tilde{r}(t, r, \theta, \mu)\sin\tilde{\theta}(t, r, \theta, \mu), \qquad \mu = \mu,
\end{aligned}
$$

provided that $x_1 = r\cos\theta$, $x_2 = r\sin\theta$. The point $P(x_1, 0, \mu)$ is the first point on the positive semitrajectory of the point $(x_1, 0, \mu)$ in which this trajectory cuts the plane x_1, μ on the same side where x_1 is situated (see Figure 7.2.4).

Subtracting x_1 from the first coordinate of the point $P(x_1, 0, \mu)$ we obtain the "*displacement function*" V which shows how much x_1 is displaced by travelling once around the μ-axis along the trajectory which starts at x_1. Thus,

$$
V(x_1, \mu) := \varphi_1(T(x_1, \mu), x_1, 0, \mu) - x_1, \qquad (x_1, \mu) \in U_0.
$$

We show that to each sufficiently small x_1 there belongs a μ such that V is zero at this point. If for some $x_1 \neq 0$ the displacement function is zero,

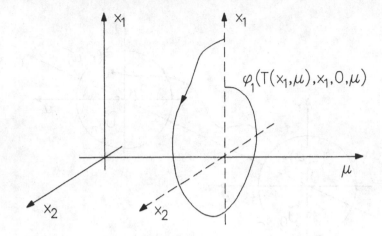

FIGURE 7.2.4. The Poincaré map of a neighbourhood of the origin in the x_1, μ plane into this plane.

then the path through this x_1 is, obviously, closed. The partial derivatives of V will be needed. Since $V(0, \mu) \equiv 0$,

$$V_\mu'(0, \mu) \equiv V_\mu'(0, 0) = 0, \tag{7.2.20}$$

$$\begin{aligned} V_{x_1}'(x_1, \mu) &= \dot{\varphi}_1(T(x_1, \mu), x_1, 0, \mu) T_{x_1}'(x_1, \mu) \\ &\quad + \varphi_{1x_1}'(T(x_1, \mu), x_1, 0, \mu) - 1. \end{aligned}$$

If $x_1 = 0$, then $\varphi_1(t, 0, 0, \mu) \equiv 0 \ (\equiv \varphi_2(t, 0, 0, \mu))$. Therefore, the first term on the right-hand side of the last formula is zero at $x_1 = 0$; thus,

$$V_{x_1}'(0, \mu) = \varphi_{1x_1}'(T(0, \mu), 0, 0, \mu) - 1.$$

But we know (see (1.1.5) et seq.) that $\varphi_x'(t, 0, 0, \mu)$ is the fundamental matrix of the linearized system

$$\dot{y} = f_x'(0, \mu)y$$

that satisfies $\varphi_x'(0, 0, 0, \mu) = I$, the unit matrix. Taking into account the form (7.2.12) of $f_x'(0, \mu)$ this fundamental matrix is

$$\varphi_x'(t, 0, 0, \mu) = \begin{pmatrix} \exp(\alpha(\mu)t)\cos\omega(\mu)t & -\exp(\alpha(\mu)t)\sin\omega(\mu)t \\ \exp(\alpha(\mu)t)\sin\omega(\mu)t & \exp(\alpha(\mu)t)\cos\omega(\mu)t \end{pmatrix}.$$

Since $T(0, \mu) = 2\pi/\omega(\mu)$ we obtain

$$\varphi_{1x_1}'(T(0, \mu), 0, 0, \mu) = \exp(2\pi\alpha(\mu)/\omega(\mu)) \tag{7.2.21}$$

and

$$V'_{x_1}(0,\mu) = \exp(2\pi\alpha(\mu)/\omega(\mu)) - 1. \qquad (7.2.22)$$

Now $V(0,0) = 0$, but since $\alpha(0) = 0$, besides (7.2.20) we have $V'_{x_1}(0,0) = 0$. Thus, the Implicit Function Theorem cannot be applied directly. A function $\tilde{V}$ will be introduced that is zero for $x_1 \neq 0$ exactly when V is zero and to which the Implicit Function Theorem can be applied.

$$\tilde{V}(x_1,\mu) = \begin{cases} V(x_1,\mu)/x_1, & x_1 \neq 0, \\ V'_{x_1}(0,\mu) & , & x_1 = 0, \end{cases} \qquad (x_1,\mu) \in U_0.$$

We show that $\tilde{V} \in C^{k-1}$, $\tilde{V}(0,0) = 0$ and $\tilde{V}'_\mu(0,0) \neq 0$. By Lemma 7.2.4

$$V(x_1,\mu) = \int_0^1 V'_{x_1}(tx_1,\mu)x_1 dt;$$

hence,

$$V(x_1,\mu)/x_1 = \int_0^1 V'_{x_1}(tx_1,\mu)dt,$$

and as a consequence

$$\lim_{x_1 \to 0} V(x_1,\mu)/x_1 = V'_{x_1}(0,\mu).$$

Since $V \in C^k$, $\tilde{V} \in C^{k-1}$, and by (7.2.22) $\tilde{V}(0,0) = V'_{x_1}(0,0) = 0$; further

$$\begin{aligned} \tilde{V}'_\mu(0,0) &= V''_{x_1\mu}(0,0) = 2\pi((\alpha'(0)\omega(0) - \alpha(0)\omega'(0))/\omega^2(0))\exp 0 \\ &= 2\pi\alpha'(0)/\omega(0) > 0. \end{aligned}$$

Thus, the Implicit Function Theorem can be applied to $\tilde{V}$, i.e. there exist a $\delta > 0$ and a function $\mu : (-\delta,\delta) \to \mathbf{R}$ such that $\mu \in C^{k-1}$, $\mu(0) = 0$ and $\tilde{V}(x_1,\mu(x_1)) \equiv 0$, i.e. $V(x_1,\mu(x_1)) \equiv 0$. This means that the solution of the system $\dot{x} = f(x,\mu(x_1))$ with initial values $(x_1,0)$ is periodic with period $T(x_1,\mu(x_1))$; the period T is a C^{k-1} function of x_1 and $T(0,0) = 2\pi/\omega(0)$. This periodic solution will be denoted by

$$p(t,x_1) := \varphi(t,(x_1,0),\mu(x_1)).$$

The orbit of this solution cuts the "other half" of the x_1-axis at some point $\tilde{x}_1(x_1)$ at some moment t between 0 and $T(x_1,\mu(x_1))$, $\tilde{x}_1(x_1) \to 0$ as $x_1 \to 0$, sign $\tilde{x}_1(x_1) = -$ sign x_1, $\mu(\tilde{x}_1(x_1)) = \mu(x_1)$ and the orbit of the solution

$$p(t,\tilde{x}_1(x_1)) = \varphi(t,(\tilde{x}_1(x_1),0),\mu(\tilde{x}_1(x_1)))$$

is the same as that of $p(t,x_1)$. Hence,

$$\text{sign}\,(\mu(\tilde{x}_1(x_1)) - \mu(0))/\tilde{x}_1(x_1) = -\text{sign}\,(\mu(x_1) - \mu(0))/x_1,$$

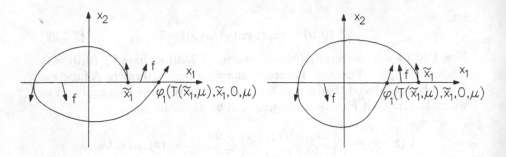

FIGURE 7.2.5. The Bendixson sack in proving the uniqueness of the periodic family in the two dimensional Andronov-Hopf Theorem; the two possible cases are shown in the plane parallel to the x_1, x_2 coordinate plane at some $\mu \neq 0$.

and this implies that

$$\mu'(0) = \lim_{x_1 \to 0} \mu(\tilde{x}_1(x_1))/\tilde{x}_1(x_1) = -\lim_{x_1 \to 0} \mu(x_1)/x_1 = -\mu'(0),$$

i.e. $\mu'(0) = 0$. This way (i) of the Lemma has been proved completely.

Observe that this result implies that

$$\mu(x_1) = (1/2)\mu''(0)x_1^2 + o(x_1^2). \tag{7.2.23}$$

We prove now the uniqueness of the family of periodic orbits. The previously established properties of the flow generated by (7.2.15) imply that the origin $(x_1, x_2, \mu) = (0, 0, 0)$ of $\mathbf{R}^3$ has a small neighbourhood such that the path γ of each solution with initial values in this neighbourhood crosses the plane x_1, μ at some point $(\tilde{x}_1, \mu)$. If this neighbourhood is chosen sufficiently small, then $(\tilde{x}_1, \mu)$ belongs to the domain of definition U_0 of the Poincaré map P (7.2.19). Now, $f_2(0, 0, \mu) = 0$ and $f'_{2x_1}(0, 0, \mu) = \omega(\mu) > 0$; therefore, the origin $(x_1, \mu) = (0, 0)$ of the x_1, μ plane has a neighbourhood such that for (x_1, μ) in this neighbourhood

$$f_2(x_1, 0, \mu) \gtrless 0 \quad \text{according to whether} \quad x_1 \gtrless 0. \tag{7.2.24}$$

If the original neighbourhood of $(0, 0, 0) \in \mathbf{R}^3$ is chosen sufficiently small, then the point $(\tilde{x}_1, \mu)$ at which the path γ belonging to any initial point in this neighbourhood crosses the plane x_1, μ belongs to that neighbourhood of $(0, 0) \in \mathbf{R}^2$ in which also (7.2.24) holds. Suppose that γ is closed, i.e. it is a periodic orbit. If $P(\tilde{x}_1, 0, \mu) \neq (\tilde{x}_1, 0, \mu)$, then this path cannot be a periodic orbit since we have a Bendixson sack (see text following Corollary 3.1.9 and Figure 7.2.5.) into which or out of which our path can get no more. Therefore, $P(\tilde{x}_1, 0, \mu) = (\tilde{x}_1, 0, \mu)$ must hold, i.e. we must have

$$V(\tilde{x}_1, \mu) = \varphi_1(T(\tilde{x}_1, \mu), \tilde{x}_1, 0, \mu) - \tilde{x}_1 = 0$$

proving that in a sufficiently small neighbourhood of the origin no periodic orbit may lie but those of the family $p(t, x_1)$.

Having proved (ii) of the Lemma we turn now to (iii). The statement concerning the stability will be proved only; the corresponding statement about the instability of the bifurcating periodic orbits can be proved similarly. We prove the asymptotic orbital stability of the bifurcating periodic orbits by showing that the derivative with respect to x_1 of the first coordinate of the Poincaré map P defined by (7.2.19) is in modulus less than 1 for $|x_1|$ and $|\mu|$ small. By Theorem 5.2.2 this implies that for small $|x_1|$ the second characteristic multiplier of the variational system

$$\dot{y} = f'_x(p(t, x_1), \mu(x_1))y$$

of (7.2.10) with respect to the periodic solution $p(t, x_1)$ is in modulus less than 1, so that the conditions of the Andronov-Witt Theorem 5.1.2 are satisfied.

The first coordinate P_1 of the Poincaré map (7.2.19) is, by the definition of the displacement function V,

$$P_1(x_1, \mu) := \varphi_1(T(x_1, \mu), x_1, 0, \mu) = V(x_1, \mu) + x_1,$$

so that $P'_{1x_1}(x_1, \mu) = V'_{x_1}(x_1, \mu) + 1$. We already know by (7.2.22) that $V'_{x_1}(0, 0) = 0$, implying by continuity that for small $|x_1|$, $|\mu|$ also $V'_{x_1}(x_1, \mu) > -2$. Thus, if $V'_{x_1}(x_1, \mu)$ is negative, then statement (iii) is proved. But by Lemma 7.2.6

$$
\begin{aligned}
V(x_1, 0) &= V(0, 0) + V'_{x_1}(0, 0)x_1 + (1/2)V''_{x_1 x_1}(0, 0)x_1^2 \\
&\quad + (1/6)V'''_{x_1 x_1 x_1}(0, 0)x_1^3 + o(|x_1|^3) \\
&= (1/6)V'''_{x_1 x_1 x_1}(0, 0)x_1^3 + o(|x_1|^3), \\
V'_{x_1}(x_1, 0) &= (1/2)V'''_{x_1 x_1 x_1}(0, 0)x_1^2 + o(|x_1|^2),
\end{aligned}
$$

and by Lemma 7.2.7 the 3-asymptotic stability of the origin, i.e. $G = G_4 < 0$ implies that $V'''_{x_1 x_1 x_1}(0, 0) < 0$. Hence, $V'_{x_1}(x_1, 0) < 0$, i.e. by continuity $V'_{x_1}(x_1, \mu) < 0$ for small $|\mu|$, and as a consequence $|P'_1(x_1, \mu)| < 1$ for $x_1 \neq 0$, $|x_1|$, $|\mu|$ small. Formulae (7.2.28) and (7.2.31) show that $\mu''(0) > 0$, and since $\mu(0) = \mu'(0) = 0$, this means by (7.2.23) that $\mu(x_1) > 0$ for $|x_1|$ small, $x_1 \neq 0$. $\square$

Now the two technical Lemmata applied in the previous proof will be proved. System (7.2.10), i.e. (7.2.11) for $\mu = 0$, will be written in the following form:

$$
\begin{aligned}
\dot{x}_1 &= -\omega x_2 + g_{12}(x_1, x_2, 0) + g_{13}(x_1, x_2, 0) + o((x_1^2 + x_2^2)^{3/2}), \\
\dot{x}_2 &= \omega x_1 + g_{22}(x_1, x_2, 0) + g_{23}(x_1, x_2, 0) + o((x_1^2 + x_2^2)^{3/2}) \quad (7.2.25)
\end{aligned}
$$

where $\omega > 0$ is shorthand for $\omega(0)$, g_{12} and g_{22} are homogeneous polynomials of degree 2, and g_{13}, g_{23} are homogeneous polynomials of degree 3. The coefficients of these polynomials are needed, so they are written out in detail:

$$
\begin{aligned}
g_{12}(x_1, x_2, 0) &= a_1 x_1^2 + a_2 x_1 x_2 + a_3 x_2^2, \\
g_{13}(x_1, x_2, 0) &= b_1 x_1^3 + b_2 x_1^2 x_2 + b_3 x_1 x_2^2 + b_4 x_2^3, \\
g_{22}(x_1, x_2, 0) &= c_1 x_1^2 + c_2 x_1 x_2 + c_3 x_2^2, \\
g_{23}(x_1, x_2, 0) &= d_1 x_1^3 + d_2 x_1^2 x_2 + d_3 x_1 x_2^2 + d_4 x_2^3. \quad (7.2.26)
\end{aligned}
$$

The calculations of the following Lemma can be found more or less in Liapunov [1892].

Lemma 7.2.6. *For the displacement function V defined in the proof of the previous Lemma we have*

$$
V(0,0) = V'_{x_1}(0,0) = V'_\mu(0,0) = 0, \quad (7.2.27)
$$

$$
V''_{x_1 \mu}(0,0) = 2\pi \alpha'(0)/\omega(0) > 0, \quad (7.2.28)
$$

$$
V''_{x_1 x_1}(0,0) = 0, \quad (7.2.29)
$$

and

$$
\begin{aligned}
V'''_{x_1 x_1 x_1}(0,0) &= (3\pi/2\omega^2)(\omega(3b_1 + b_3 + d_2 + 3d_4) \\
&\quad + a_1 a_2 + a_2 a_3 - c_1 c_2 - c_2 c_3 + 2(a_3 c_3 - a_1 c_1)); \quad (7.2.30)
\end{aligned}
$$

also

$$
V'''_{x_1 x_1 x_1}(0,0) = -3V''_{x_1 \mu}(0,0)\mu''(0). \quad (7.2.31)
$$

Proof. Formulae (7.2.27) and (7.2.28) have been proved in the proof of the previous Lemma where we have also shown that $\mu(0) = \mu'(0) = 0$. In order to prove (7.2.29), differentiate the identity $V(x_1, \mu(x_1)) \equiv 0$ twice:

$$
\begin{aligned}
&V'_{x_1}(x_1, \mu(x_1)) + V'_\mu(x_1, \mu(x_1))\mu'(x_1) \equiv 0, \\
&V''_{x_1 x_1}(x_1, \mu(x_1)) + 2V''_{x_1, \mu}(x_1, \mu(x_1))\mu'(x_1) + V''_{\mu\mu}(x_1, \mu(x_1))(\mu'(x_1))^2 \\
&\quad + V'_\mu(x_1, \mu(x_1))\mu''(x_1) \equiv 0. \quad (7.2.32)
\end{aligned}
$$

From here, substituting $x_1 = 0$, (7.2.29) follows.

Differentiating (7.2.32) once more (we note that $\mu \in C^{k-1}$, $k \geq 4$) we obtain

$$
\begin{aligned}
&V'''_{x_1 x_1 x_1}(x_1, \mu(x_1)) + 3V'''_{x_1 x_1 \mu}(x_1, \mu(x_1))\mu'(x_1) \\
&\quad + 3V'''_{x_1 \mu\mu}(x_1, \mu(x_1))(\mu'(x_1))^2 + 2V''_{x_1 \mu}(x_1, \mu(x_1))\mu''(x_1) \\
&\quad + V'''_{\mu\mu\mu}(x_1, \mu(x_1))(\mu'(x_1))^3 + 3V''_{\mu\mu}(x_1, \mu(x_1))\mu'(x_1)\mu''(x_1) \\
&\quad + V''_{x_1 \mu}(x_1, \mu(x_1))\mu''(x_1) + V'_\mu(x_1, \mu(x_1))\mu'''(x_1) \equiv 0.
\end{aligned}
$$

Substituting $x_1 = 0$ and taking into account the formulae that had been proved already, we obtain

$$V'''_{x_1 x_1 x_1}(0,0) + 3V''_{x_1 \mu}(0,0)\mu''(0) = 0$$

from which (7.2.31) follows.

Formula (7.2.30) requires more work, naturally. The polar transformation applied in the proof of the previous Lemma is used also here, this time in the $\mu = 0$ case. Then the right-hand sides of (7.2.25) become

$$f_1(r\cos\theta, r\sin\theta, 0) = -\omega r\sin\theta + r^2 g_{12}(\theta) + r^3 g_{13}(\theta) + o(r^3),$$
$$f_2(r\cos\theta, r\sin\theta, 0) = \omega r\cos\theta + r^2 g_{22}(\theta) + r^3 g_{23}(\theta) + o(r^3)$$

where g_{12}, g_{22} and g_{13}, g_{23} are quadratic and cubic forms, respectively, of $\cos\theta$ and $\sin\theta$. (There is an abuse of notation here which, perhaps, helps the identification of the terms with those of (7.2.25).) The right-hand sides of the equivalent system (7.2.13) in polar coordinates at $\mu = 0$ are now

$$f_r(r,\theta,0) = r^2(g_{12}(\theta) + rg_{13}(\theta) + o(r))\cos\theta$$
$$+ r^2(g_{22}(\theta) + rg_{23}(\theta) + o(r))\sin\theta,$$
$$f_\theta(r,\theta,0) = \omega - r(g_{12}(\theta) + rg_{13}(\theta) + o(r))\sin\theta$$
$$+ r(g_{22}(\theta) + rg_{23}(\theta))\cos\theta.$$

By what has been said following (7.2.18), the solutions of (7.2.13) that belong to sufficiently small initial values $r = r_0$, $\theta = 0$ and now $\mu = 0$ stay near (7.2.16) and reach the value $\theta = 2\pi$ in finite time, so that if the differential equation of the trajectories

$$dr/d\theta = f_r(r,\theta,0)/f_\theta(r,\theta,0)$$

$$= \frac{r^2((g_{12}(\theta) + rg_{13}(\theta) + o(r))\cos\theta + (g_{22}(\theta) + rg_{23}(\theta) + o(r))\sin\theta)}{\omega + r(-(g_{12}(\theta) + rg_{13}(\theta) + o(r))\sin\theta + (g_{22}(\theta) + rg_{23}(\theta) + o(r))\cos\theta)}$$
(7.2.33)

is considered, and its solution that assumes the value $r = r_0$ at $\theta = 0$ is denoted by $\hat{r}(\theta, r_0)$, then for $|r_0|$ sufficiently small, $\hat{r}(\theta, r_0)$ is defined for $\theta \in [0, 2\pi]$. Since the right-hand side of the differential equation (7.2.33) is in the C^k class, the solution $\hat{r}$ is also k times continuously differentiable with respect to r_0, and can be expanded in the following way:

$$\hat{r}(\theta, r_0) = u_1(\theta)r_0 + u_2(\theta)r_0^2 + \cdots + u_k(\theta)r_0^k + o(r_0^k) \qquad (7.2.34)$$

where $u_i \in C^1$ ($i = 1, 2, ..., k$), and in particular,

$$\hat{r}(2\pi, r_0) = u_1(2\pi)r_0 + u_2(2\pi)r_0^2 + \cdots + u_k(2\pi)r_0^k + o(r_0^k).$$

By the relation between the solutions of systems (7.2.13) and (7.2.11) (cf. (7.2.19)), clearly,

$$\varphi_1(T(x_1, 0), x_1, 0, 0) = \hat{r}(2\pi, x_1)\cos(2\pi)$$
$$= u_1(2\pi)x_1 + u_2(2\pi)x_1^2 + \cdots + u_k(2\pi)x_1^k + o(x_1^k),$$

and for $\mu = 0$ the displacement function V is given by

$$V(x_1, 0) = (u_1(2\pi) - 1)x_1 + u_2(2\pi)x_1^2 + \cdots + u_k(2\pi)x_1^k + o(x_1^k). \quad (7.2.35)$$

This means that $V'_{x_1}(0,0) = u_1(2\pi) - 1$ and

$$V^{(l)}_{x_1 \ldots x_1}(0,0) = l! u_l(2\pi) \qquad (l = 2, 3, \ldots, k). \qquad (7.2.36)$$

Expression (7.2.30) will be obtained through the determination of $u_3(2\pi)$. Now

$$r_0 = \hat{r}(0, r_0) = u_1(0)r_0 + u_2(0)r_0^2 + \cdots + u_k(0)r_0^k + o(r_0^k),$$

so that

$$u_1(0) = 1, \qquad u_i(0) = 0 \qquad (i = 2, \ldots, k). \qquad (7.2.37)$$

Substituting (7.2.34) into the differential equation (7.2.33) and equating equal powers of r_0 we obtain successive differential equations for the determination of the functions $u_i(\theta)$ $(i = 1, 2, \ldots, k)$:

$$u'_1(\theta)r_0 + u'_2(\theta)r_0^2 + u'_3(\theta)r_0^3 + \cdots$$

$$\equiv \frac{(u_1(\theta)r_0 + u_2(\theta)r_0^2 + \cdots)^2((g_{12}(\theta) + \cdots)\cos\theta + (g_{22}(\theta) + \cdots)\sin\theta)}{\omega + (u_1(\theta)r_0 + u_2(\theta)r_0^2 + \cdots)(-(g_{12}(\theta) + \cdots)\sin\theta + (g_{22}(\theta) + \cdots)\cos\theta)}$$

$$\equiv (u_1^2(\theta)r_0^2 + 2u_1(\theta)u_2(\theta)r_0^3 + \cdots)((g_{12}(\theta)\cos\theta + g_{22}(\theta)\sin\theta + \cdots)(1/\omega)$$

$$\cdot(1 - (1/\omega)(u_1(\theta)r_0 + u_2(\theta)r_0^2 + \cdots)(-g_{12}(\theta)\sin\theta + g_{22}(\theta)\cos\theta + \cdots) + \ldots).$$

Thus, $u'_1(\theta) = 0$, i.e. taking into account (7.2.37), $u_1(\theta) \equiv 1$.

$$\begin{aligned} u'_2(\theta) &= (1/\omega)u_1^2(\theta)(g_{12}(\theta)\cos\theta + g_{22}(\theta)\sin\theta) \\ &= (1/\omega)(g_{12}(\theta)\cos\theta + g_{22}(\theta)\sin\theta). \end{aligned} \qquad (7.2.38)$$

Hence,

$$u_2(2\pi) = u_2(2\pi) - u_2(0) = (1/\omega)\int_0^{2\pi} (g_{12}(\theta)\cos\theta + g_{22}(\theta)\sin\theta)d\theta = 0$$

since the integrand is a cubic (i.e. odd) polynomial of $\cos\theta$ and $\sin\theta$. (By the way, this proves (7.2.29) once again since $V''_{x_1 x_1}(0,0) = 2u_2(2\pi)$ by (7.2.36).)

$$\begin{aligned} u'_3(\theta) &= (1/\omega)\Big(g_{13}(\theta)\cos\theta + g_{23}(\theta)\sin\theta \\ &\quad -(1/\omega)(g_{12}(\theta)\cos\theta + g_{22}(\theta)\sin\theta)(-g_{12}(\theta)\sin\theta + g_{22}(\theta)\cos\theta) \\ &\quad +2u_2(\theta)(g_{12}(\theta)\cos\theta + g_{22}(\theta)\sin\theta)\Big), \end{aligned}$$

or taking into account (7.2.38),

$$u'_3(\theta) = (1/\omega)(g_{13}(\theta)\cos\theta + g_{23}(\theta)\sin\theta)$$
$$-(1/\omega^2)(g_{12}(\theta)\cos\theta + g_{22}(\theta)\sin\theta)(-g_{12}(\theta)\sin\theta + g_{22}(\theta)\cos\theta)$$
$$+(u_2(\theta))^{2\prime}.$$

Substituting $x_1 = \cos\theta$, $x_2 = \sin\theta$ into (7.2.26), writing out the expressions for $g_{ij}(\theta)$ in detail, integrating from 0 to 2π and taking into account that $u_2(2\pi) = u_2(0) = 0$, we obtain

$$u_3(2\pi) = (1/\omega)((b_1 + d_4)\int_0^{2\pi} \sin^4\theta d\theta$$
$$+(1/4)(b_3 + d_2)\int_0^{2\pi} \sin^2 2\theta d\theta)$$
$$-(1/\omega^2)\Big(2(c_1c_2 - a_1a_2)\int_0^{2\pi} \cos^4\theta\sin^2\theta d\theta$$
$$+2(c_2c_3 - a_2a_3)\int_0^{2\pi} \sin^4\theta\cos^2\theta d\theta$$
$$+a_1c_1\int_0^{2\pi} (\cos^6\theta - \cos^4\theta\sin^2\theta)d\theta$$
$$+a_3c_3\int_0^{2\pi} (\sin^4\theta\cos^2\theta - \sin^6\theta)d\theta$$
$$+(a_3c_1 + a_2c_2 + a_1c_3)\int_0^{2\pi} (\cos^4\theta\sin^2\theta - \cos^2\theta\sin^4\theta)d\theta\Big).$$

Performing the integration, the expression

$$u_3(2\pi) = (\pi/(4\omega^2))(\omega(3b_1 + b_3 + d_2 + 3d_4)$$
$$-c_1c_2 - c_2c_3 + a_1a_2 + a_2a_3 - 2(a_1c_1 - a_3c_3))$$

follows. By (7.2.36) this yields (7.2.30). $\square$

We deduce now *Bautin's Formula* for the second Poincaré-Liapunov coefficient G_4 of system (7.2.25) and establish the relation between the latter and $V'''_{x_1x_1x_1}(0,0)$.

Lemma 7.2.7 *(Bautin [1949], Andronov-Leontovich-Gordon-Mayer [1967]). The second Poincaré-Liapunov coefficient of system (7.2.25) is given by*

$$G_4 = (1/4)(3b_1 + b_3 + d_2 + 3d_4)$$
$$+(1/(4\omega))(a_1a_2 + a_2a_3 - c_1c_2 - c_2c_3 + 2(a_3c_3 - a_1c_1))$$
$$= (\omega/(6\pi))V'''_{x_1x_1x_1}(0,0). \tag{7.2.39}$$

Proof. This will prove, at the same time, Theorem 3.6.1 for $m = 4$. Assume a Liapunov function for system (7.2.25) in the form (7.2.5)

$$F(x_1, x_2) = x_1^2 + x_2^2 + F_3(x_1, x_2) + F_4(x_1, x_2) \qquad (7.2.40)$$

where

$$
\begin{aligned}
F_3(x_1, x_2) &= A_1 x_1^3 + A_2 x_1^2 x_2 + A_3 x_1 x_2^2 + A_4 x_2^3, \\
F_4(x_1, x_2) &= B_1 x_1^4 + B_2 x_1^3 x_2 + B_3 x_1^2 x_2^2 + B_4 x_1 x_2^3 + B_5 x_2^4
\end{aligned}
$$

with undetermined coefficients A_i and B_j ($i = 1, ..., 4; \quad j = 1, ..., 5$). These coefficients will be determined in such a way that

$$\dot{F}_{(7.2.25)}(x_1, x_2) = G_4(x_1^2 + x_2^2)^2 + o((x_1^2 + x_2^2)^2). \qquad (7.2.41)$$

Differentiating (7.2.40) with respect to system (7.2.25) we obtain

$$
\begin{aligned}
\dot{F}_{(7.2.25)}(x_1, x_2) = \; & (2x_1 + 3A_1 x_1^2 + 2A_2 x_1 x_2 + A_3 x_2^2 \\
& + 4B_1 x_1^3 + 3B_2 x_1^2 x_2 + 2B_3 x_1 x_2^2 + B_4 x_2^3)\dot{x}_1 \\
& + (2x_2 + A_2 x_1^2 + 2A_3 x_1 x_2 + 3A_4 x_2^2 \\
& + B_2 x_1^3 + 2B_3 x_1^2 x_2 + 3B_4 x_1 x_2^2 + 4B_5 x_2^3)\dot{x}_2.
\end{aligned}
$$

Here for $\dot{x}_1$ and $\dot{x}_2$, the right-hand sides of (7.2.25) are substituted and then the coefficients of the corresponding terms on the left- and on the right-hand side of (7.2.41), respectively, are equated. First, the third degree terms are written down on the left-hand side of (7.2.41) and their sum is made to be equal to zero since no such term occurs on the right-hand side:

$$
\begin{aligned}
& 2x_1 g_{12}(x_1, x_2, 0) + 2x_2 g_{22}(x_1, x_2, 0) \\
& + (3A_1 x_1^2 + 2A_2 x_1 x_2 + A_3 x_2^2)(-\omega x_2) \\
& \quad + (A_2 x_1^2 + 2A_3 x_1 x_2 + 3A_4 x_2^2)\omega x_1 \equiv 0,
\end{aligned}
$$

i.e.

$$
\begin{aligned}
& 2x_1(a_1 x_1^2 + a_2 x_1 x_2 + a_3 x_2^2) + 2x_2(c_1 x_1^2 + c_2 x_1 x_2 + c_3 x_2^2) \\
& -\omega x_2(3A_1 x_1^2 + 2A_2 x_1 x_2 + A_3 x_2^2) + \omega x_1(A_2 x_1^2 + 2A_3 x_1 x_2 + 3A_4 x_2^2) \equiv 0.
\end{aligned}
$$

For the coefficients this yields the system of equations

$$
\begin{aligned}
2a_1 + \omega A_2 &= 0, & 2c_3 - \omega A_3 &= 0, \\
2a_2 + 2c_1 - 3\omega A_1 + 2\omega A_3 &= 0, & \\
2a_3 + 2c_2 - 2\omega A_2 + 3\omega A_4 &= 0, &
\end{aligned}
$$

which determines the A_i's uniquely:

$$A_1 = (2a_2 + 2c_1 + 4c_3)/(3\omega), \qquad A_2 = -2a_1/\omega,$$
$$A_3 = 2c_3/\omega, \qquad A_4 = -(4a_1 + 2a_3 + 2c_2)/(3\omega). \quad (7.2.42)$$

Equating the terms of degree 4 on both sides of (7.2.41) we obtain

$$2x_1(b_1x_1^3 + b_2x_1^2x_2 + b_3x_1x_2^2 + b_4x_2^3)$$
$$+ 2x_2(d_1x_1^3 + d_2x_1^2x_2 + d_3x_1x_2^2 + d_4x_2^3)$$
$$+ (3A_1x_1^2 + 2A_2x_1x_2 + A_3x_2^2)(a_1x_1^2 + a_2x_1x_2 + a_3x_2^2)$$
$$+ (A_2x_1^2 + 2A_3x_1x_2 + 3A_4x_2^2)(c_1x_1^2 + c_2x_1x_2 + c_3x_2^2)$$
$$- \omega x_2(4B_1x_1^3 + 3B_2x_1^2x_2 + 2B_3x_1x_2^2 + B_4x_2^3)$$
$$+ \omega x_1(B_2x_1^3 + 2B_3x_1^2x_2 + 3B_4x_1x_2^2 + 4B_5x_2^3) \equiv G_4(x_1^4 + 2x_1^2x_2^2 + x_2^4).$$

Equating the corresponding coefficients:

$$3a_1A_1 + c_1A_2 + \omega B_2 + 2b_1 = G_4,$$
$$3a_2A_1 + 2a_1A_2 + c_2A_2 + 2c_1A_3 - 4\omega B_1 + 2\omega B_3 + 2b_2 + 2d_1 = 0,$$
$$3a_3A_1 + 2a_2A_2 + a_1A_3 + c_3A_2 + 2c_2A_3 + 3c_1A_4$$
$$- 3\omega B_2 + 3\omega B_4 + 2b_3 + 2d_3 = 2G_4,$$
$$2a_3A_2 + a_2A_3 + 2c_3A_3 + 3c_2A_4 - 2\omega B_3 + 4\omega B_5 + 2b_4 + 2d_3 = 0,$$
$$a_3A_3 + 3c_3A_4 - \omega B_4 + 2d_4 = G_4.$$

Substituting the A_i's from (7.2.42) we see that the second and the fourth equation can be solved independently of the others for B_1, B_3, B_5 having an infinite number of solutions. This does not interest us. The first, third and fifth equations assume the forms

$$\omega B_2 - G_4 = -(2/\omega)(a_1a_2 + 2a_1c_3) - 2b_1,$$
$$3\omega B_2 - 3\omega B_4 - 2G_4 = -(2/\omega)(2a_1a_2 + 2a_1c_1 - a_2a_3 - 2a_3c_3$$
$$+ c_1c_2 - 2c_2c_3) + 2b_3 + 2d_2,$$
$$\omega B_4 + G_4 = -(2/\omega)(2a_1c_3 + c_2c_3) + 2d_4.$$

This system has a unique solution for the three unknowns B_2, B_4 and G_4. For G_4 we get (7.2.39), the second equation resulting from the comparison with (7.2.30). $\square$

Having proved the last two Lemmata we have completed the proof of the two dimensional Andronov-Hopf Theorem (Lemma 7.2.5). Thus, if G_4 of (7.2.39) is negative, then it is the Poincaré-Liapunov constant of the system ensuring the asymptotic orbital stability of the bifurcating periodic solutions. If it is positive, then the bifurcating periodic solutions are unstable. If $G_4 = 0$, then one has to try to determine the stability of the periodic solutions by calculating G_6 (if this is zero, then G_8, etc.); however, as it was mentioned earlier, G_6, etc., cannot reasonably be calculated by hand, and some symbolic manipulation software is to be used.

We have proved that if $G_4 < 0$, then $\mu''(0) > 0$, and if x_1 is restricted to small positive values, $x_1 \in [0, \delta]$ for some $\delta > 0$, then the function $\mu(x)$ is strictly increasing in this interval. This means that we can take its inverse, substitute it for the parameter x_1 in the family of periodic solutions $p(t, x_1)$, and have the bifurcating family of periodic solutions be parametrized by μ. In this case to each sufficiently small positive μ there belongs a unique periodic orbit $p(t, \mu)$ (by an abuse of notation) that is orbitally asymptotically stable. There holds, obviously, the following:

Corollary 7.2.8. *If the conditions of Lemma 7.2.5 hold including the conditions in (iii), then there exists a $\overline{\mu} > 0$ such that for each $\mu \in (0, \overline{\mu})$ (resp. $(-\overline{\mu}, 0)$) system (7.2.10) has a periodic solution $p(t, \mu)$ of period $T(\mu) > 0$ such that $p \in C^{k-1}$, $T \in C^{k-1}$, $\lim_{\mu \to 0} p(t, \mu) = 0$, $\lim_{\mu \to 0} T(\mu) = 2\pi/\omega(0)$; in a sufficiently small neighbourhood of $(x_1, x_2, \mu) = (0, 0, 0)$ the system has no periodic orbit apart from those of the family $p(t, \mu)$; the periodic solution $p(t, \mu)$ is asymptotically orbitally stable (resp. repelling).*

Expression (7.2.23) gives the approximate functional relation between the bifurcation parameter μ and the initial value x_1 of the corresponding periodic solution. The latter can be considered as an "*approximation of the amplitude*" of the bifurcating periodic solution (since the initial value x_2 is zero). From (7.2.23), $x_1 \approx (2\mu/\mu''(0))^{1/2}$. Applying (7.2.28), (7.2.31) and (7.2.39) the approximate amplitude is given, for small $\mu > 0$, by

$$x_1 \approx (-2\alpha'(0)/G_4)^{1/2}\mu^{1/2} \tag{7.2.43}$$

in the supercritical case $G_4 < 0$. This formula shows that the amplitude is increasing proportionally to the square root of μ, justifying Figure 7.2.2. In practical estimates the constant of proportionality may serve one well.

Proof *of the n dimensional Andronov-Hopf Theorem 7.2.3.* The essence of the proof is the reduction of the problem to the two dimensional theorem (Lemma 7.2.5) by an application of the Centre Manifold Theorem (Theorems A3.2 and A3.3). System (7.2.9) will be written in the form

$$\dot{x} = f_x'(0, \mu)x + g(x, \mu) \tag{7.2.44}$$

where $g \in C^k$, $g(0, \mu) \equiv 0$, $g_x'(0, \mu) \equiv 0$, $g_\mu'(0, \mu) \equiv 0$. By our assumptions, $f_x'(0, \mu)$ has a pair of conjugate complex eigenvalues $\alpha(\mu) \pm i\omega(\mu)$, $\alpha(0) = 0$, $\alpha'(0) > 0$, $\omega(\mu) > 0$ and $n - 2$ eigenvalues with negative real parts. If (7.2.44) is taken at $\mu = 0$, we have

$$\dot{x} = f_x'(0, 0)x + g(x, 0) \tag{7.2.45}$$

where the coefficient matrix of the linear part has a pair of pure imaginary eigenvalues $\pm i\omega(0)$ and $n-2$ eigenvalues with negative real parts. The (two dimensional) eigenplane corresponding to the imaginary eigenvalues will be denoted by $\mathbf{R}^c = \mathbf{R}^2$, the $n-2$ dimensional eigenspace corresponding to the rest of the eigenvalues by $\mathbf{R}^s = \mathbf{R}^{n-2}$. By Theorem A3.2, system (7.2.45) has a two dimensional centre manifold M_2 whose tangent plane at $x = 0$ is $\mathbf{R}^c$. This is the two dimensional manifold to which the problem ought to be reduced. However, M_2 is *not* a centre manifold of system (7.2.44) for $\mu \neq 0$ since for such μ the latter system does not have a centre manifold having no eigenvalue with zero real part. We overcome this difficulty by introducing the bifurcation parameter μ as an $n+1$-st state variable attaching the differential equation $\dot{\mu} = 0$ to (7.2.44), i.e. the following $n+1$ dimensional system of differential equations is considered:

$$\dot{\mu} = 0,$$
$$\dot{x} = f'_x(0,\mu)x + g(x,\mu), \tag{7.2.46}$$

or since $f(0,\mu) \equiv 0$ implies $f'_\mu(0,0) = 0$, equivalently,

$$\dot{\mu} = 0,$$
$$\dot{x} = f'_x(0,0)x + \hat{g}(x,\mu) \tag{7.2.47}$$

where $\hat{g}(0,\mu) \equiv 0$, $\hat{g}(x,\mu) = o(|x| + |\mu|)$, $\hat{g} \in C^k$. The linearization of this system has three eigenvalues with zero real part, $0, \pm i\omega(0)$, and the rest of the eigenvalues have negative real parts, so that this system has a *three dimensional centre manifold $M_3 \subset \mathbf{R}^{n+1}$* that is, obviously, the direct product of the two dimensional centre manifold M_2 of system (7.2.45) and of the μ-axis denoted by $\mathbf{R}_\mu$, i.e. $M_3 = M_2 \times \mathbf{R}_\mu$ (the eigenvector corresponding to the eigenvalue 0 is $[1,0,...,0] \in \mathbf{R}^{n+1}$; for every $\mu \in \mathbf{R}$ the point $(\mu,0,...,0) \in M_3$, since every point of the μ-axis is an equilibrium of system (7.2.47)). Introducing the vector $[1,0,...,0] \in \mathbf{R}^{n+1}$, the real and the imaginary parts of the eigenvector belonging to the eigenvalue $i\omega(0)$ and $n-2$ linearly independent vectors from $\mathbf{R}^s$ as a new base, the corresponding linear transformation carries (7.2.47) into

$$\dot{\mu} = 0, \qquad \begin{pmatrix} y_1 \\ y_2 \end{pmatrix}^{\bullet} = C \begin{pmatrix} y_1 \\ y_2 \end{pmatrix} + \begin{pmatrix} \tilde{g}_1(y_1,...,y_n,\mu) \\ \tilde{g}_2(y_1,...,y_n,\mu) \end{pmatrix},$$

$$\begin{pmatrix} y_3 \\ \vdots \\ y_n \end{pmatrix}^{\bullet} = S \begin{pmatrix} y_3 \\ \vdots \\ y_n \end{pmatrix} + \begin{pmatrix} \tilde{g}_3(y_1,...,y_n,\mu) \\ \cdots \\ \tilde{g}_n(y_1,...,y_n,\mu) \end{pmatrix} \tag{7.2.48}$$

where $[y_1,y_2] \in \mathbf{R}^c$, $[y_3,...,y_n] \in \mathbf{R}^s$, $\tilde{g}_i \in C^k$, $\tilde{g}(y,\mu) = o(|y| + |\mu|)$, $\tilde{g}(0,\mu) \equiv 0$,

$$C = \begin{pmatrix} 0 & -\omega(0) \\ \omega(0) & 0 \end{pmatrix}$$

and the eigenvalues of the $(n-2) \times (n-2)$ matrix S are the eigenvalues of $f'_x(0,0)$ with negative real parts. Let $h : \mathbf{R}^c \times \mathbf{R}_\mu \to \mathbf{R}^s$ be the C^k mapping whose graph is the centre manifold M_3 (cf. Theorem A3.2). By this Theorem, $h(0) = 0$, $h'_{y_1}(0,0,0) = h'_{y_2}(0,0,0) = 0$, $h'_\mu(0,0,0) = 0$, so that the tangent space of M_3 at $(y_1, y_2, \mu) = (0,0,0)$ is $\mathbf{R}^c \times \mathbf{R}_\mu$. The coordinates of the function h will be denoted by $h = (h_3, h_4, ..., h_n)$, and the coordinate transformation

$$z_1 = y_1, \quad z_2 = y_2, \quad z_i = y_i - h_i(y_1, y_2, \mu) \quad (i = 3, 4, ..., n) \quad (7.2.49)$$

will be applied. System (7.2.48) goes into

$$\dot\mu = 0, \quad \begin{pmatrix} z_1 \\ z_2 \end{pmatrix}^{\bullet} = C \begin{pmatrix} z_1 \\ z_2 \end{pmatrix} + \begin{pmatrix} \tilde{g}_1(z_1, z_2, z_3 + h_3(z_1, z_2, \mu), ..., \mu) \\ \tilde{g}_2(z_1, z_2, z_3 + h_3(z_1, z_2, \mu), ..., \mu) \end{pmatrix},$$

$$\begin{pmatrix} z_3 \\ \vdots \\ z_n \end{pmatrix}^{\bullet} = S \begin{pmatrix} z_3 \\ \vdots \\ z_n \end{pmatrix} + \begin{pmatrix} G_3(z_1, ..., z_n, \mu) \\ \vdots \\ G_n(z_1, ..., z_n, \mu) \end{pmatrix} \quad (7.2.50)$$

where $\tilde{g}_1, \tilde{g}_2$, $G_i \in C^k$, $G_i(z_1, ..., z_n, \mu) = o(|z| + |\mu|)$, $G_i(z_1, z_2, 0, ..., 0, \mu) \equiv 0$ $(i = 3, ..., n)$, the last condition being a consequence of the invariance of M_3 that is characterized by $z_3 = \cdots = z_n = 0$.

The restriction of the system to the centre manifold M_3 is achieved by substituting $z_3 = z_4 = \cdots = z_n = 0$, so that the restricted system is

$$\dot\mu = 0, \quad \begin{pmatrix} z_1 \\ z_2 \end{pmatrix}^{\bullet} = C \begin{pmatrix} z_1 \\ z_2 \end{pmatrix} + \begin{pmatrix} \tilde{g}_1(z_1, z_2, h_3(z_1, z_2, \mu), ..., \mu) \\ \tilde{g}_2(z_1, z_2, h_3(z_1, z_2, \mu), ..., \mu) \end{pmatrix}. \quad (7.2.51)$$

System (7.2.50) is equivalent to (7.2.47), i.e. to (7.2.46) in a neighbourhood of the centre manifold. The coordinate transformation (7.2.49) is such that the z_i coordinate $(i = 3, 4, ..., n)$ of a point shows how far the y_i coordinate of the same point is from the corresponding point on the centre manifold above the "coordinate plane" $(y_1, y_2) = (z_1, z_2) \in \mathbf{R}^c$. The linearization of (7.2.50) at the equilibrium $(z, \mu) = (y, \mu) = (x, \mu) = (0,0)$ is

$$\dot\mu = 0, \quad \begin{pmatrix} z_1 \\ z_2 \end{pmatrix}^{\bullet} = C \begin{pmatrix} z_1 \\ z_2 \end{pmatrix}, \quad \begin{pmatrix} z_3 \\ \vdots \\ z^n \end{pmatrix}^{\bullet} = S \begin{pmatrix} z_3 \\ \vdots \\ z_n \end{pmatrix}. \quad (7.2.52)$$

However, one may also linearize system (7.2.50) at some other equilibrium point $(z, \mu) = (0, \mu)$, $\mu \neq 0$. The point $(0, \mu) \in M_3$, the centre manifold belonging to the equilibrium $(z, \mu) = (0,0)$. One may introduce a μ-dependent linear coordinate transformation to maintain the block diagonal form of the linearized system. At $(0, \mu)$ the eigenvalues are $\alpha(\mu) \pm i\omega(\mu)$, and since these are simple roots of the characteristic polynomial, they are C^k functions of μ implying that the corresponding, say, unit eigenvectors are also C^k functions of μ. The rest of the eigenvalues at $(0, \mu)$ have negative real parts and among them there may be multiple ones. Though a multiple root of a polynomial is also a continuous function of the coefficients, it is no longer in the C^1 class. Nevertheless, if one does not want to bring the matrix corresponding to the eigenvalues with negative real parts to canonical form, one may choose base vectors in the $n - 2$ dimensional eigenspace corresponding to those eigenvalues that depend in a C^k way on μ. This way after a suitable linear transformation the linearization of (7.2.50) at $(0, \mu)$ is

$$\dot{\mu} = 0, \qquad \begin{pmatrix} z_1 \\ z_2 \end{pmatrix}^{\bullet} = C_\mu \begin{pmatrix} z_1 \\ z_2 \end{pmatrix}, \qquad \begin{pmatrix} z_3 \\ \vdots \\ z_n \end{pmatrix}^{\bullet} = S_\mu \begin{pmatrix} z_3 \\ \vdots \\ z_n \end{pmatrix}$$

where $C, S \in C^k$,

$$C_\mu = \begin{pmatrix} \alpha(\mu) & -\omega(\mu) \\ \omega(\mu) & \alpha(\mu) \end{pmatrix}, \qquad S_\mu \to S, \quad \mu \to 0.$$

The restriction of the system to the invariant manifold M_3 is then

$$\dot{\mu} = 0, \qquad \begin{pmatrix} z_1 \\ z_2 \end{pmatrix}^{\bullet} = C_\mu \begin{pmatrix} z_1 \\ z_2 \end{pmatrix} + \begin{pmatrix} G_1(z_1, z_2, \mu) \\ G_2(z_1, z_2, \mu) \end{pmatrix} \qquad (7.2.53)$$

where, by an abuse of notation, the coordinates after the μ-dependent linear transformation have also been denoted by z_1, z_2 (for $\mu = 0$ they are the "old" z_1, z_2 coordinates, indeed), $G_i \in C^k$, $G_i(0, 0, \mu) \equiv 0$, $G'_{iz_k}(0, 0, \mu) \equiv 0$ $(i = 1, 2; k = 1, 2)$.

Now, we drop the differential equation $\dot{\mu} = 0$ again and look at μ as a bifurcation parameter. Clearly, (7.2.53) satisfies all the conditions of Lemma 7.2.5 with the difference that the right-hand side is not C^{k+1} but only C^k. Therefore, (i), (ii) and (iii) of this Lemma hold true with the difference that the function $\mu \in C^{k-2}$ and also the period T as a function of the parameter (denoted this time by ε) is also in the C^{k-2} class. If the family of the $T(\varepsilon)$-periodic solutions is denoted by $p(t, \varepsilon)$, then this family is unique in the z_1, z_2 plane in the sense of (ii). If (7.2.50) is also considered as an n dimensional system depending on the parameter μ, the first equation $\dot{\mu} = 0$ being dropped, then we see that $[z_1, z_2, z_3, ..., z_n] = [p_1(t, \varepsilon), p_2(t, \varepsilon), 0, ..., 0]$ is a family of periodic solutions for this system with $\mu = \mu(\varepsilon)$, and all the properties expressed in (i) of Theorem 7.2.3 hold. The uniqueness in $\mathbf{R}^n$ in the sense of (ii) of this Theorem follows from Theorem A3.3, i.e. from the attractivity of the invariant set M_3: if an initial point $(\mu = \mu)$, $(z_1^0, z_2^0, z_3^0, ..., z_n^0)$ is given outside M_3 in a neighbourhood of $(z, \mu) = (0, 0)$, then the corresponding solution cannot be periodic since it must tend to M_3 as $t \to \infty$.

Turning now to the proof of (iii) note that if the bifurcating periodic solutions are unstable in the invariant manifold M_3, then *a fortiori* they are unstable in $\mathbf{R}^n$ too, so that only the statement concerning the stability is to be proved. The assumption about the 3-asymptotic stability of the origin for $\mu = 0$ implies that the bifurcating periodic solution $p(t, \varepsilon)$ is orbitally asymptotically stable with respect to the two dimensional system (7.2.52) with $\mu = \mu(\varepsilon)$ (and that $\mu(\varepsilon) > 0$). In the last part of the proof of Lemma 7.2.5 it was also shown that the derivative of the Poincaré map generated by the two dimensional flow is in modulus less than 1 in a sufficiently small neighbourhood of the origin $z_1 = 0$, $\mu = 0$. Let $|\varepsilon|$ be so small that the path of $p(t, \varepsilon)$ lies in this neighbourhood (clearly $p(t, \varepsilon) \to 0$ as $\varepsilon \to 0$). Consider now the variational system of (7.2.9) with respect to $[p_1(t, \varepsilon), p_2(t, \varepsilon), 0, ..., 0]$:

$$\dot{y} = f'_x((p_1(t, \varepsilon), p_2(t, \varepsilon), 0, .., 0), \mu(\varepsilon))y \qquad (7.2.54)$$

and denote its fundamental matrix that assumes the unit matrix at $t = 0$ by $\Phi(t, \varepsilon)$. The characteristic multipliers are the eigenvalues of $\Phi(T(\varepsilon), \varepsilon)$. But as $\varepsilon \to 0$ also $\mu(\varepsilon) \to 0$, $p(t, \varepsilon) \to 0$ and $T(\varepsilon) \to 2\pi/\omega(0)$, so that system (7.2.54) tends to the system with constant coefficients

$$\dot{y} = f'_x(0, 0)y,$$

and $\Phi(t, \varepsilon)$ tends to the fundamental matrix of the latter system that assumes the unit matrix at $t = 0$:

$$\lim_{\varepsilon \to 0} \Phi(t, \varepsilon) = \exp(f'_x(0, 0)t),$$

in particular,

$$\lim_{\varepsilon \to 0} \Phi(T(\varepsilon), \varepsilon) = \exp(f'_x(0, 0)2\pi/\omega(0)).$$

One characteristic multiplier of $\Phi(T(\varepsilon), \varepsilon)$ is 1 because (7.2.54) has a non-trivial periodic solution. Since $n - 2$ eigenvalues of $f'_x(0, 0)$ have negative real parts, the matrix on the right-hand side has $n - 2$ eigenvalues with modulus less than 1. If $|\varepsilon|$ is sufficiently small, those $n - 2$ eigenvalues of $\Phi(T(\varepsilon), \varepsilon)$ that tend to eigenvalues of the right-hand side in modulus less than 1 are also in modulus less than 1. There is a difficulty only with the single eigenvalue of $\Phi(T(\varepsilon), \varepsilon)$ that tends to the eigenvalue 1 of the exp matrix. However, by Theorem 5.2.2 the characteristic multipliers different from 1 are the eigenvalues of the linearization of the Poincaré map generated by the flow. The "missing characteristic" multiplier must be the eigenvalue of the restriction of the Poincaré map to the centre manifold, i.e. the derivative of the restricted Poincaré map that is in modulus less than 1 by the last part of the proof of Lemma 7.2.5. Thus, for $|\varepsilon|$ sufficiently small, $n - 1$ characteristic multipliers are in modulus less than 1; therefore, by the

Andronov-Witt Theorem 5.1.2, $(p_1(t, \varepsilon), p_2(t, \varepsilon), 0, ..., 0)$ is asymptotically orbitally stable. □

We note that Theorem 7.2.3 remains valid, *mutatis mutandis*, if the condition $\alpha'(0) > 0$ is replaced by $\alpha'(0) < 0$. In this case, clearly, the equilibrium is asymptotically stable for $\mu > 0$, and it is unstable for $\mu < 0$. The Andronov-Hopf bifurcation occurs then as the bifurcation parameter μ is *decreased* through $\mu = 0$. If the periodic orbits occur for $\mu < 0$ and are orbitally asymptotically stable, we call the bifurcation *supercritical*; if they occur for $\mu > 0$ and are unstable, then we say that the bifurcation is *subcritical*.

Non-trivial applications of the theory will follow in the next two Sections.

There is no space here to go into more intricate problems about the Andronov-Hopf bifurcation. Two problems will be mentioned only with some references. The first one is: How far the family of the bifurcating periodic solutions can be continued as the bifurcation parameter μ is increased in the supercritical case and decreased in the subcritical case (under the assumption $\alpha'(0) > 0$) and how this family ceases to exist? This problem is related to the phenomenon of how two or more families of periodic orbits may bifurcate from a "degenerate" periodic orbit. Some important references concerning these problems are Mallet-Paret, Yorke [1982] and Alexander, Yorke [1983]. The other problem is: what happens when a system which undergoes an Andronov-Hopf bifurcation is periodically perturbed at the same time? Some references concerning this problem are: H.L.Smith [1981], Rosenblat, Cohen [1981], Bajaj [1986] and Namachchivaya, Ariaratnam [1987]. We have to observe that none of the authors having attacked this problem tried to apply yet the theory of controllably periodic perturbations (see Sections 6.2 and 6.3).

7.3   A Predator-Prey Model with Memory

In this Section a predator-prey model is treated which is an improved one compared to the classical Lotka-Volterra model (see Section 3.4) in two respects. First, the *intraspecific competition* in the prey species is taken into account: this has a saturation effect, and as a consequence the prey is *not* growing exponentially in the absence of predation but tends to a finite limit. Secondly, while it is acceptable that the predator quantity has an instantaneous effect on the growth rate of prey, obviously, the present growth rate of a predator depends not only on the present quantity of food but also on past quantities (in the period of gestation, say). Therefore, a *"delay term"* is built into the differential equation concerning the predator. Conditions are established for the existence of a positive equilibrium of the ecological system made up by the two species, for its stability, its bifurcation and for the appearance of stable periodic behaviour. The presentation is based on Farkas [1984$_a$], Farkas, A.-Farkas [1988], Farkas, A.-Farkas-Szabó

[1987], Farkas, A.-Farkas-Szabó [1988], and Szabó [1987].

Denote the specific growth rate of prey, the predation rate, the mortality of predator and the conversion rate by $\varepsilon > 0$, $\alpha > 0$, $\gamma > 0$, $\beta > 0$, respectively (cf. Section 3.4), and the carrying capacity of the environment (for the prey) by $K > 0$ (cf. Example 3.1.1, (3.4.10) and Example 4.1.1). Let $G : [0, \infty) \to \mathbf{R}_+$ be a C^1 density function satisfying

$$\int_0^\infty G(s)ds = 1, \tag{7.3.1}$$

and denote the quantity of prey and predator at time t by $N(t)$ and $P(t)$, respectively. The system governing the dynamics of the community is taken up in the form

$$\dot{N}(t) = \varepsilon N(t)(1 - N(t)/K - P(t)\alpha/\varepsilon),$$
$$\dot{P}(t) = -\gamma P(t) + \beta P(t) \int_{-\infty}^t N(\tau)G(t - \tau)d\tau. \tag{7.3.2}$$

Several authors have studied this system mainly under the assumption that the density function is exponentially decaying, $G(s) = a \exp(-as)$, $a > 0$, and have established conditions for the existence of periodic solutions; see Volterra [1931], Cushing [1977], Dai [1981], MacDonald [1977, 1978], Wörz-Busekros [1978]. Here two cases will be treated; first the case when

$$G(s) = a \exp(-as), \qquad a > 0 \tag{7.3.3}$$

("*exponentially fading memory*") and, secondly, the case when

$$G(s) = a^2 s \exp(-as), \qquad a > 0 \tag{7.3.4}$$

("*memory with a hump*"). System (7.3.2) is an integro-differential system, or in other words, a system with *continuous infinite delay*. We shall treat systems with delay briefly in Section 5; however, as Fargue [1973] has shown, it, in the case that the density function G is a solution of a homogeneous linear differential equation with constant coefficients, i.e. it is a polynomial multiplied by an exponential function (7.3.2), is equivalent to a system of ordinary differential equations of higher dimension.

Exponentially fading memory. Assume that G is given by (7.3.3) and introduce the notation

$$Q(t) : = \int_{-\infty}^t N(\tau)G(t - \tau)d\tau$$
$$= a \int_{-\infty}^t N(\tau) \exp(-a(t - \tau))d\tau, \qquad t \geq 0. \tag{7.3.5}$$

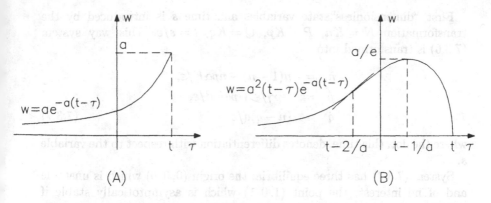

FIGURE 7.3.1. (A) The density function of the exponentially fading memory. (B) The density function of the memory with a hump.

Then (7.3.2) goes into

$$\dot{N} = \varepsilon N(1 - N/K) - \alpha NP,$$
$$\dot{P} = -\gamma P + \beta PQ,$$
$$\dot{Q} = a(N - Q) \qquad (7.3.6)$$

where the last equation is obtained by straightforward differentiation. The equivalence of (7.3.2) and (7.3.6) on $\mathbf{R}_+$ is to be understood in the following sense. Let $\overline{C}^0(-\infty, 0]$ denote the set of real valued functions that are continuous and bounded on $(-\infty, 0]$. Let $(N, P) : [0, \infty) \to \mathbf{R}^2$ be the solution of (7.3.2) that corresponds to the *"initial function"* $\tilde{N} \in \overline{C}^0(-\infty, 0]$ and the initial value $P_0 = P(0)$ (i.e. on $(-\infty, 0)$, N is considered to be equal to $\tilde{N}$); then $(N, P, Q) : [0, \infty) \to \mathbf{R}^3$ where Q is given by (7.3.5) is the solution of (7.3.6) that satisfies the initial conditions $N(0) = \tilde{N}(0)$, $P(0) = P_0$ and

$$Q(0) = Q_0 = a \int_{-\infty}^0 \tilde{N}(\tau) \exp(a\tau)d\tau,$$

and vice versa. (Clearly, if the initial values $N(0)$, $P(0)$ and $Q(0)$ for (7.3.6) are prescribed, then these do not determine the initial function $\tilde{N}$ of the corresponding solution of (7.3.2) uniquely.) Thus, instead of the delay equation (7.3.2) with exponentially fading memory (Figure 7.3.1 (A) shows how past moments are weighted when the effect of the quantity of prey is calculated), the system (7.3.6) of ordinary differential equations will be studied on $t \in [0, \infty)$.

First "dimensionless" state variables and time s is introduced by the transformation $N = Kn$, $P = Kp$, $Q = Kq$, $t = s/\varepsilon$. This way system (7.3.6) is transformed into

$$
\begin{aligned}
\dot{n} &= n(1-n) - np\alpha K/\varepsilon, \\
\dot{p} &= -p\gamma/\varepsilon + pqK\beta/\varepsilon, \\
\dot{q} &= (n-q)a/\varepsilon
\end{aligned}
\tag{7.3.7}
$$

where the dot, this time, denotes differentiation with respect to the variable s.

System (7.3.7) has three equilibria: the origin $(0,0,0)$ which is unstable and of no interest; the point $(1,0,1)$ which is asymptotically stable if $\gamma/(K\beta) > 1$ and unstable if $\gamma/(K\beta) < 1$, the former condition meaning that the limiting value of prey density K and the rate of conversion of prey into predator β are not large enough to support the predator population which, eventually, dies out. The third equilibrium is

$$
(n_0, p_0, q_0) = (\gamma/(K\beta), (1 - \gamma/(K\beta))\varepsilon/(\alpha K), \gamma/(K\beta))
$$

which is in the positive octant of n, p, q space iff

$$
\gamma/(K\beta) < 1.
\tag{7.3.8}
$$

This condition means that the time-dependent specific growth rate of the predator is positive, at least, when Q (in the second equation of (7.3.6)) assumes the value K. We have seen that this implies that (7.3.7) has no asymptotically stable equilibrium except, possibly, (n_0, p_0, q_0).

In order to check the stability of (n_0, p_0, q_0) we linearize system (7.3.7) at this point. The coefficient matrix is

$$
\begin{pmatrix}
-\gamma/(K\beta) & -\alpha\gamma/(\varepsilon\beta) & 0 \\
0 & 0 & (1 - \gamma/(K\beta))\beta/\alpha \\
a/\varepsilon & 0 & -a/\varepsilon
\end{pmatrix}
$$

and the characteristic equation assumes the form

$$
\lambda^3 + (\gamma/(K\beta) + a/\varepsilon)\lambda^2 + \lambda\gamma a/(K\beta\varepsilon) + (1 - \gamma/(K\beta))a\gamma/\varepsilon^2 = 0. \tag{7.3.9}
$$

By the Routh-Hurwitz criteria (1.4.12) this is a stable polynomial iff (7.3.8) holds and

$$
(\gamma/(K\beta) + a/\varepsilon)\gamma a/(K\beta\varepsilon) > (1 - \gamma/(K\beta))a\gamma/\varepsilon^2,
$$

i.e. iff besides (7.3.8) we have

$$
a > K\beta - \gamma - \gamma\varepsilon/(K\beta).
$$

If the right-hand side of this inequality is negative or zero which taking into account (7.3.8) roughly means that the specific growth rate ε of prey is large enough, then (n_0, p_0, q_0) is asymptotically stable for all positive "a".

A more interesting situation arises if

$$K\beta - \gamma - \gamma\varepsilon/(K\beta) > 0. \tag{7.3.10}$$

In this case the equilibrium is losing its stability at the positive value

$$a_0 = K\beta - \gamma - \gamma\varepsilon/(K\beta) \tag{7.3.11}$$

as a decreases, i.e. the influence of the past increases (see Figure 7.3.1 (A)). Note that (7.3.10) implies (7.3.8). This is an instance of the rule of thumb according to which the increase of time lag has a destabilizing effect.

Introducing the notation $b = 1/(K\beta)$, (7.3.10) assumes the form

$$1 - \gamma b - \gamma\varepsilon b^2 > 0 \tag{7.3.12}$$

and the value (7.3.11) is expressed by

$$a_0 = (1 - \gamma b - \gamma\varepsilon b^2)/b. \tag{7.3.13}$$

In what follows, (7.3.12) will be assumed and this, as we have mentioned earlier, implies (7.3.8), i.e. $1 - \gamma b > 0$. At the value $a = a_0$ the characteristic polynomial (7.3.9) assumes the form

$$\begin{aligned}
\lambda^3 &+ (1/(b\varepsilon) - \gamma/\varepsilon)\lambda^2 + \lambda(1 - \gamma b - \gamma\varepsilon b^2)\gamma/\varepsilon \\
&+ (1 - \gamma b)(1 - \gamma b - \gamma\varepsilon b^2)\gamma/(b\varepsilon^2) \\
&= (\lambda^2 + (1 - \gamma b - \gamma\varepsilon b^2)\gamma/\varepsilon)(\lambda + (1 - \gamma b)/(b\varepsilon)).
\end{aligned}$$

The eigenvalues are

$$\lambda_0(a_0) = -(1 - \gamma b)/b\varepsilon$$

which is negative and $\lambda_{1,2}(a_0) = \pm i\omega$, where

$$\omega = ((1 - \gamma b - \gamma\varepsilon b^2)\gamma/\varepsilon)^{1/2} > 0. \tag{7.3.14}$$

In order to check the validity of Theorem 7.2.3 we have to determine the derivative with respect to "a" of the real part of the smooth extension of the root $\lambda_1(a_0)$. Let us denote by $\lambda_1(a)$ the root of (7.3.9) that assumes the value $i\omega$ at a_0 and by

$$F(\lambda, a) = \lambda^3 + (\gamma b + a/\varepsilon)\lambda^2 + \lambda a\gamma b/\varepsilon + (1 - \gamma b)\gamma a/\varepsilon^2$$

the characteristic polynomial in (7.3.9) as a function of the parameter "a". Since $F(\lambda_1(a_0), a_0) = F(i\omega, a_0) = 0$ and $i\omega$ is a simple root of the polynomial $F(\lambda, a_0)$, the smooth function λ_1 is uniquely determined by

$F(\lambda_1(a), a) \equiv 0$, $\lambda_1(a_0) = i\omega$. We are going to determine the derivative of the implicit function λ_1 at a_0:

$$\lambda_1'(a_0) = -F_a'(i\omega, a_0)/F_\lambda'(i\omega, a_0)$$
$$= \gamma b^2 \frac{(\gamma b + i\omega)(2\gamma b(1 - \gamma b - \gamma\varepsilon b^2) + i2\omega(1 - \gamma b))}{4\gamma^2 b^2(1 - \gamma b - \gamma\varepsilon b)^2 + 4\omega^2(1 - \gamma b)^2}.$$

Hence, we have

$$\frac{d\,\mathrm{Re}\,\lambda_1(a_0)}{da} = \mathrm{Re}\,\frac{d\lambda_1(a_0)}{da} = -\frac{\gamma b^2}{2}\frac{1 - \gamma b - \gamma\varepsilon b^2}{\varepsilon\gamma b^2(1 - \gamma b - \gamma\varepsilon b^2) + (1 - \gamma b)^2} < 0.$$

If the bifurcation parameter

$$\mu = 1/a - 1/a_0, \qquad a(\mu) = a_0/(1 + a_0\mu) \qquad (7.3.15)$$

is introduced in system (7.3.7), we see that the equilibrium (n_0, p_0, q_0) is asymptotically stable for negative values of μ and is losing its stability at $\mu = 0$. Obviously,

$$\left.\frac{d\,\mathrm{Re}\,\lambda_1(a(\mu))}{d\mu}\right|_{\mu=0} = \frac{\gamma}{2}\frac{(1 - \gamma b - \gamma\varepsilon b^2)^3}{\varepsilon\gamma b^2(1 - \gamma b - \gamma\varepsilon b^2) + (1 - \gamma b)^2} > 0.$$

The facts established about the behaviour of the eigenvalues mean that the conditions of the Andronov-Hopf Bifurcation Theorem (7.2.3) hold (apart, for the time being, from the conditions of stability of the bifurcating closed orbits).

Having proved (i) and (ii) of Theorem 7.2.3 we shall deduce the condition for the supercriticality, resp. subcriticality of the bifurcation. First, the system will be restricted to the centre manifold (cf. (7.2.49)), then Bautin's Formula (Lemma 7.2.7) will be written out; finally, this formula will be replaced by a simpler one equivalent to it from the point of view of sign.

First, the origin is moved to the equilibrium point (n_0, p_0, q_0) by the transformation

$$x = n - n_0 = n - \gamma b, \qquad y = p - p_0 = p - \varepsilon(1 - \gamma b)/(\alpha K),$$

$$z = q - q_0 = q - \gamma b.$$

At $\mu = 0$, i.e. $a = a_0$, the following system is obtained:

$$\dot{x} = -x\gamma b - y\alpha\gamma Kb/\varepsilon - x^2 - xy\alpha K/\varepsilon,$$
$$\dot{y} = z(1 - \gamma b)/(\alpha Kb) + yz/(\varepsilon b),$$
$$\dot{z} = x(1 - \gamma b - \varepsilon\gamma b^2)/(\varepsilon b) - z(1 - \gamma - \varepsilon\gamma b^2)/(\varepsilon b). \qquad (7.3.16)$$

We have to determine the $\mu = 0$ section of the three dimensional centre manifold of the "suspended system" obtained from (7.3.16) by attaching

$\mu = 0$. The eigenvectors corresponding to the eigenvalues $\lambda_0(a_0)$ and $i\omega$ are $s^0 = \mathrm{col}\,[\gamma^2 b^2/\omega^2, \varepsilon/(\alpha K), -1]$ and

$$s^1 = \mathrm{col}\,[1, 0, 1] + i\,\mathrm{col}\,[\gamma b/\omega, -1(1 - \gamma b)/(\omega \alpha K b), 0],$$

respectively. Introducing s^0 and the real and the imaginary parts of s^1 as a new base, we perform the coordinate transformation

$$\mathrm{col}\,[x, y, z] = T\,\mathrm{col}\,[x_1, x_2, x_3]$$

where the matrix of the coordinate transformation is

$$T = \begin{pmatrix} 1 & \gamma b/\omega & \gamma^2 b^2/\omega^2 \\ 0 & -(1 - \gamma b)/(\omega \alpha K b) & \varepsilon/(\alpha K) \\ 1 & 0 & -1 \end{pmatrix}.$$

In the new coordinates, (7.3.16) assumes the form

$$\begin{aligned}
\dot{x}_1 &= \omega x_2 + W(-\varepsilon(1 - \gamma b)U_1(x_1, x_2, x_3) + \gamma b U_2(x_1, x_2, x_3)), \\
\dot{x}_2 &= -\omega x_1 + W(-\omega \varepsilon^2 b U_1(x_1, x_2, x_3) - \omega(1 + (\gamma b/\omega)^2)U_2(x_1, x_2, x_3)), \\
\dot{x}_3 &= -x_3(1 - \gamma b)/(\varepsilon b) + W(-\varepsilon(1 - \gamma b)U_1(x_1, x_2, x_3) + \gamma b U_2(x_1, x_2, x_3))
\end{aligned}$$

$$\tag{7.3.17}$$

where

$$\begin{aligned}
W &= \omega^2/(\gamma((1 - \gamma b)^2 + \varepsilon^2 \omega^2 b^2)), \\
U_1(x_1, x_2, x_3) &= x_1^2 + x_2^2((\gamma b/\omega)^2 - \gamma b(1 - \gamma b)/(\omega^2 \varepsilon b)) \\
&\quad + x_3^2((\gamma b/\omega)^4 + (\gamma b/\omega)^2) + x_1 x_2(2\gamma b/\omega - (1 - \gamma b)/(\omega \varepsilon b)) \\
&\quad + x_2 x_3(2(\gamma b/\omega)^3 + \gamma b/\omega - (\gamma b/\omega)^2(1 - \gamma b)/(\omega \varepsilon b)) \\
&\quad + x_1 x_3(2(\gamma b/\omega)^2 + 1), \\
U_2(x_1, x_2, x_3) &= -x_1 x_2(1 - \gamma b)/(\omega b) + \varepsilon x_1 x_3 + x_2 x_3(1 - \gamma b)/(\omega b) - \varepsilon x_3^2.
\end{aligned}$$

Now we determine approximately the $\mu = 0$ section of the center manifold M which is characterized by the conditions that it is tangent to the x_1, x_2 plane at the origin (the eigenspace corresponding to the eigenvalues $\pm i\omega$) and is locally invariant with respect to the flow of system (7.3.17). Thus, we may assume its equation in the form $x_3 = h(x_1, x_2)$ with the conditions $h(0, 0) = h'_{x_i}(0, 0) = 0$ $(i = 1, 2)$, $h \in C^k$, for arbitrary fixed positive integer k which we choose to be larger than 3.

Set $h'_{x_i}(x_1, x_2) = h_{i1}x_1 + h_{i2}x_2 + 0(|x|^2)$, where $i = 1, 2$, $x = [x_1, x_2]$, $h_{12} = h_{21}$. Then

$$h(x_1, x_2) = \frac{1}{2}(h_{11}x_1^2 + 2h_{12}x_1 x_2 + h_{22}x_2^2) + o(|x|^3).$$

If $(x_1(s), x_2(s), x_3(s))$ is a solution of (7.3.17) near the origin with a value on M, then it stays locally in M, i.e.,

$$x_3(s) \equiv h(x_1(s), x_2(s)).$$

As a consequence,

$$\dot{x}_3(s) - h'_{x_1}(x_1(s), x_2(s))\dot{x}_1(s) - h'_{x_2}(x_1(s), x_2(s))\dot{x}_2(s) \equiv 0, \qquad (7.3.18)$$

i.e. using (7.3.17) we get the identity

$$-h(x_1, x_2)(1 - \gamma b)/(\varepsilon b) + W(-\varepsilon(1 - \gamma b)U_1(x_1, x_2, h(x_1, x_2))$$
$$+\gamma b U_2(x_1, x_2, h(x_1, x_2))) - (h_{11}x_1 + h_{12}x_2 + O(|x|^2))(\omega x_2 + O(|x|^2))$$
$$-(h_{12}x_1 + h_{22}x_2 + O(|x|^2))(-\omega x_1 + O(|x|^2)) \equiv 0.$$

Omitting terms of order at least three,

$$(h_{11}x_1^2 + 2h_{12}x_1x_2 + h_{22}x_2^2)(1 - \gamma b)/(2\varepsilon b)$$
$$+W\varepsilon(1 - \gamma b)(x_1^2 + x_2^2((\gamma b/\omega)^2 - \gamma(1 - \gamma b)/(\omega^2\varepsilon))$$
$$+x_1x_2(2\gamma b/\omega - (1 - \gamma b)/(\omega\varepsilon b)))$$
$$+x_1x_2W\gamma(1 - \gamma b)/\omega + \omega h_{11}x_1x_2 + \omega h_{12}x_2^2 - \omega h_{12}x_1^2 - \omega h_{22}x_1x_2 \equiv 0.$$

Equating the coefficients with zero we get

$$(1 - \gamma b)h_{11} - 2\omega\varepsilon b h_{12} = -2\varepsilon^2 bW(1 - \gamma b),$$
$$\omega^2\varepsilon b h_{11} + \omega(1 - \gamma b)h_{12} - \omega^2\varepsilon b h_{22} = -W(1 - \gamma b)(2\varepsilon^2\gamma b^2 - \varepsilon(1 - \gamma b) + \varepsilon\gamma b),$$
$$2\omega^3\varepsilon b h_{12} + \omega^2(1 - \gamma b)h_{22} = -2W\varepsilon(1 - \gamma b)(\varepsilon\gamma^2 b^3 - \gamma b(1 - \gamma b)).$$

This system can be solved uniquely for h_{ik}. One gets

$$h_{11} = -\frac{2W\varepsilon^3 b^2(\omega^2 + \gamma b(2 - \gamma b))}{4\omega^2\varepsilon^2 b^2 + (1 - \gamma b)^2},$$

$$h_{12} = \frac{W(1 - \gamma b)\varepsilon(-4\varepsilon^2\gamma^2 b^4 + (1 - \gamma b)(2\varepsilon\gamma b^2 + 1 - 2\gamma b))}{\omega(4\omega^2\varepsilon^2 b^2 + (1 - \gamma b)^2)},$$

$$h_{22} = \frac{2W\varepsilon^2\gamma b^2(1 - \gamma b)(1 + 2\varepsilon b)}{4\omega^2\varepsilon^2 b^2 + (1 - \gamma b)^2}.$$

In order to restrict system (7.3.17) to the $\mu = 0$ section M of the center manifold, the new coordinates $y_1 = x_1$, $y_2 = x_2$, $y_3 = x_3 - h(x_1, x_2)$ are to be introduced. In this coordinate system the equation of M is $y_3 = 0$, i.e., M is the "y_1, y_2 plane". By this coordinate transformation, system (7.3.17) goes into

$$\dot{y}_1 = \omega y_2 + W[-\varepsilon(1 - \gamma b)U_1(y_1, y_2, y_3 + h(y_1, y_2))$$
$$+\gamma b U_2(y_1, y_2, y_3 + h(y_1, y_2))],$$
$$\dot{y}_2 = -\omega y_1 + W[-\omega\varepsilon^2 b U_1(y_1, y_2, y_3 + h(y_1, y_2))$$
$$-\omega(1 + (\gamma b/\omega)^2)U_2(y_1, y_2, y_3 + h(y_1, y_2))],$$
$$\dot{y}_3 = \ldots .$$

Restricting the system to M we have to write $y_3 = 0$ everywhere and take into account that $\dot{y}_3 = 0$ (see (7.3.18)). Thus, omitting the last insignificant equation and writing in the expressions for U_i we get the two dimensional system

$$
\begin{aligned}
\dot{y}_1 \;=\;& \omega y_2 + W[-\varepsilon(1-\gamma b)y_1^2 - \varepsilon(1-\gamma b)((\gamma b/\omega)^2 - \gamma b(1-\gamma b)/\omega^2 \varepsilon b)y_2^2 \\
&+ y_1 y_2 (1 - 2\gamma b - 2\varepsilon\gamma b^2)(1 - \gamma b)/\omega b] \\
&+ W h(y_1, y_2)[(-2\varepsilon(\gamma b/\omega)^2 - \varepsilon + 2\varepsilon\gamma b(\cdot, b/\omega)^2 + 2\varepsilon\gamma b)y_1 \\
&+ y_2(1 - 2\varepsilon b(\gamma b/\omega)^2 - \varepsilon b + \gamma b(1-\gamma b)/\omega^2)(1-\gamma b)\gamma/\omega] + O(|y|^4) \\
\dot{y}_2 \;=\;& -\omega y_1 + W[-\omega\varepsilon^2 b y_1^2 - \omega\varepsilon^2 b((\gamma b/\omega)^2 - \gamma b(1-\gamma b)/(\omega^2\varepsilon b))y_2^2 \\
&+ ((1-\gamma b)(1+(\gamma b/\omega)^2 + \varepsilon b)/b - 2\varepsilon^2\gamma b^2)y_1 y_2] \\
&+ W h(y_1, y_2)[-(2(\gamma b/\omega)^2\varepsilon b + \varepsilon b + 1 + (\gamma b/\omega)^2)\omega\varepsilon y_1 \\
&- \omega\varepsilon^2 b(2(\gamma b/\omega)^3 + \gamma b/\omega - (\gamma b/\omega)^2(1-\gamma b)/(\omega\varepsilon b) \\
&+ (1-\gamma b)/(\omega\varepsilon^2 b^2) + (\gamma b/\omega)^2(1-\gamma b)/(\omega\varepsilon^2 b^2))y_2] + O(|y|^4).
\end{aligned}
$$

$$(7.3.19)$$

Applying Bautin's Formula (Lemma 7.2.7, in Farkas [1984a] this was done via the method of Negrini-Salvadori [1979]); we get that

$$\operatorname{sign} G_4 = \operatorname{sign} \Phi$$

where

$$
\begin{aligned}
\Phi \;=\;& (1-\gamma b)((1+2\varepsilon b)(-2\gamma^2 b(1-\gamma b)^2 \\
&- 2\gamma^2\varepsilon b^2(-(1-2\gamma b)(1-\gamma b - \gamma\varepsilon b^2) - 2\gamma\varepsilon b^2(1-\gamma b) \\
&+ \gamma^2\varepsilon^2 b^4)) - (\gamma^3\varepsilon b^3(1-\gamma b)/\omega^2 + \gamma\varepsilon b(1-\gamma b) + \gamma^2\varepsilon^2 b^3) \\
&\times ((1-\gamma b)(1-2\gamma b) + 2\gamma\varepsilon b^2(1-\gamma b - 2\gamma\varepsilon b^2))).
\end{aligned}
$$

Szabó [1987] has shown that

$$
\begin{aligned}
\operatorname{sign} G_4 = \operatorname{sign} \Phi \;=\;& -\operatorname{sign}(K\beta - \gamma - \gamma\varepsilon/(K\beta) - \varepsilon^2\gamma/(K\beta(2\gamma+\varepsilon))) \\
\;=\;& -\operatorname{sign}(a_0 - \varepsilon^2\gamma/(K\beta(2\gamma+\varepsilon))).
\end{aligned}
$$

These calculations yield the following:

Theorem 7.3.1 (*Farkas [1984a], Szabó [1987]*). *If* $a_0 = K\beta - \gamma - \gamma\varepsilon/(K\beta) > 0$ *and*

$$a_0 - \varepsilon^2\gamma/(K\beta(2\gamma+\varepsilon)) > 0 \qquad (resp. < 0), \qquad\qquad (7.3.20)$$

then there exists a $\delta > 0$ *such that for each* $a \in (a_0 - \delta, a_0)$ *(resp.*

$a \in (a_0, a_0 + \delta))$, *system (7.3.6) has a unique periodic orbit in a neighbourhood of the equilibrium point*

$$(N_0, P_0, Q_0) = (\gamma/\beta, (1 - \gamma/(K\beta))\varepsilon/\alpha, \gamma/\beta); \qquad (7.3.21)$$

and the corresponding periodic solution is asymptotically orbitally stable (resp. unstable).

As we have already observed the influence of the past, the delay is increasing with the decrease of the parameter a. The result expressed by this Theorem is an instance of the thumb rule according to which *delay destabilizes*. However, this "rule" is to be applied with caution because *sometimes delay has a stabilizing effect* (see Stépán [1986]).

Note also that by the Andronov-Hopf Bifurcation Theorem the period of the bifurcating periodic solutions (for a near to a_0) is approximately given by

$$T_0 = 2\pi/\omega = 2\pi(\varepsilon/\gamma)^{1/2}(1 - \gamma/(K\beta) - \gamma\varepsilon/(K\beta)^2)^{-1/2}. \qquad (7.3.22)$$

The previous bifurcation result can be put into a more general context. It is easy to see that the parameter α can be transformed out of the system (by introducing αP as the second state variable), and the bifurcation depends on only three independent parameters. If new parameters are introduced by

$$u = \varepsilon/\gamma, \qquad v = K\beta/\gamma, \qquad w = \gamma/a, \qquad (7.3.23)$$

and the situation is considered in the three dimensional u, v, w parameter space, then we have:

Corollary 7.3.2 *(Farkas, A. - Farkas - Szabó [1988]). The equation of the bifurcation surface in the u, v, w space is*

$$w(v^2 - v - u) - v = 0; \qquad (7.3.24)$$

if this surface is crossed, the equilibrium (7.3.1) of system (7.3.6) undergoes an Andronov-Hopf bifurcation; the bifurcation is supercritical, resp. subcritical according as

$$2v - 1 - ((8u^2 + 9u + 2)/(u + 2))^{1/2} > 0, \qquad resp. \quad < 0 \qquad (7.3.25)$$

at the crossing.

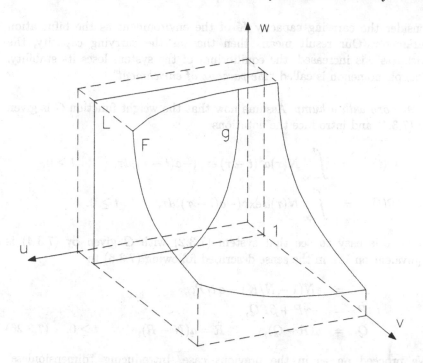

FIGURE 7.3.2. Bifurcation diagram for system (7.3.6) in the three dimensional parameter space u, v, w. Between the vertical plane $L : v = 1$ and the surface F given by (7.3.24), the equilibrium (7.3.21) is asymptotically stable. As F is crossed, an Andronov-Hopf bifurcation occurs that is supercritical, resp. subcritical according to whether the crossing takes place below or above the curve g (corresponding to (7.3.25)). (From Farkas, A.-Farkas-Szabó, *J. Math. Biol.* 26 (1988) 93-103.)

Proof. Equation (7.3.24) is in fact $a = a_0$ in the new parameters; (7.3.25) is (7.3.20). □

The situation is shown on Figure 7.3.2. By condition (7.3.8) only the $v > 1$ part of the positive octant of u, v, w space is interesting. The shape of the bifurcation surface F shows that the formerly stable equilibrium point (7.3.21) is destabilized and the system begins to oscillate if either ε is decreased or K is increased or the delay $1/a$ is increased (while the rest of the parameters are kept constant in each case). Since $\gamma, \varepsilon, \beta, 1/a$ may be considered as genetically determined parameters, the most reasonable is to consider the carrying capacity K of the environment as the bifurcation parameter. Our result means then that as the carrying capacity, the "richness", is increased, the equilibrium of the system loses its stability. This phenomenon is called "*the paradox of enrichment*".

Memory with a hump. Assume now that the weight function G is given by (7.3.4) and introduce the notations

$$Q(t): = \int_{-\infty}^{t} N(\tau)a^2(t-\tau)\exp(-a(t-\tau))d\tau, \qquad t \geq 0,$$

$$R(t): = \int_{-\infty}^{t} N(\tau)a\exp(-a(t-\tau))d\tau, \qquad t \geq 0.$$

Then it is easy to see that system (7.3.2) with G given by (7.3.4) is equivalent on $\mathbf{R}_+$ in the sense described following (7.3.6) to

$$\begin{aligned}
\dot{N} &= \varepsilon N(1 - N/K) - \alpha NP, \\
\dot{P} &= -\gamma P + \beta PQ, \\
\dot{Q} &= a(R - Q), \qquad \dot{R} = a(N - R), \qquad t \geq 0. \quad (7.3.26)
\end{aligned}$$

We proceed on as in the previous case. Introducing "dimensionless" coordinates by

$$N = Kn, \quad P = Kp, \quad Q = Kq, \quad R = Kr, \quad t = s/\varepsilon,$$

system (7.3.26) goes into

$$\begin{aligned}
\dot{n} &= n(1 - n) - np\alpha K/\varepsilon, \\
\dot{p} &= -p\gamma/\varepsilon + pqK\beta/\varepsilon, \\
\dot{q} &= (r - q)a/\varepsilon, \\
\dot{r} &= (n - r)a/\varepsilon \quad (7.3.27)
\end{aligned}$$

where the dot denotes differentiation with respect to s this time.

System (7.3.27) has three equilibria: $(0,0,0,0)$ is uninteresting and unstable for arbitrary positive values of the parameters; $(1,0,1,1)$ represents the absence of predators. It is easy to see that this equilibrium is asymptotically stable if $\gamma/(K\beta) > 1$ and unstable if $\gamma/(K\beta) < 1$. We concentrate on the third equilibrium

$$(n_0, p_0, q_0, r_0) = (\gamma/(K\beta), (1 - \gamma/(K\beta))\varepsilon/(K\alpha), \gamma/(K\beta), \gamma/(K\beta)) \quad (7.3.28)$$

which represents the coexistence of the two species provided that it is in the positive orthant of n, p, q, r space, i.e. if (cf. (7.3.8))

$$\gamma/(K\beta) < 1. \quad (7.3.29)$$

In what follows this will be assumed, implying that (7.3.28) is the only possibly stable equilibrium of the system.

The characteristic polynomial of system (7.3.27) linearized at (7.3.28) is

$$\lambda^4 + \left(\frac{\gamma}{K\beta} + \frac{2a}{\varepsilon}\right)\lambda^3 + \left(\frac{a^2}{\varepsilon^2} + \frac{2a\gamma}{K\beta\varepsilon}\right)\lambda^2 + \frac{a^2\gamma}{K\beta\varepsilon^2}\lambda + \left(1 - \frac{\gamma}{K\beta}\right)\frac{\gamma a^2}{\varepsilon^3}. \quad (7.3.30)$$

The Routh-Hurwitz criterion boils down to

$$\frac{\gamma^2}{(K\beta)^3}(2a - K\beta + \gamma)\varepsilon^2 + \frac{4a\gamma}{(K\beta)^2}(a - K\beta + \gamma)\varepsilon$$

$$+ \frac{2a^2}{K\beta}(a - 2(K\beta - \gamma)) > 0. \quad (7.3.31)$$

Now, the left-hand side is a quadratic polynomial in ε. There are three possible cases:

(i) If $a < (1/2)(K\beta - \gamma)$, then, clearly, the left-hand side of (7.3.31) is negative for all positive $\varepsilon > 0$, i.e. the equilibrium (7.3.28) is unstable. This is the case of the "large delay" $1/a$. (Note that this case does not occur in the previous case of exponentially fading memory.)

(ii) If $a > 2(K\beta - \gamma)$, then (7.3.31) holds for all positive ε, i.e. the equilibrium (7.3.28) is asymptotically stable. This is the case of the "small delay" $1/a$;

(iii) If

$$(1/2)(K\beta - \gamma) < a < 2(K\beta - \gamma), \quad (7.3.32)$$

then, clearly, (7.3.31) holds for large ε and it does not hold for small positive ε. In this case the left-hand side of (7.3.31) has a single positive root

$$\varepsilon_0 = \frac{2K\beta a}{\gamma} \frac{\sqrt{2(K\beta - \gamma)} - \sqrt{a}}{2\sqrt{a} - \sqrt{2(K\beta - \gamma)}}. \quad (7.3.33)$$

The equilibrium (7.3.28) is asymptotically stable for $\varepsilon > \varepsilon_0$ and unstable for $0 < \varepsilon < \varepsilon_0$.

We consider ε (the maximal growth rate of prey) as a bifurcation parameter and try to find out what happens if ε is decreased across the critical value ε_0. The characteristic polynomial (7.3.30) has the following decomposition at $\varepsilon = \varepsilon_0 : (\lambda^2 + \omega^2)(\lambda^2 + A_1\lambda + A_0)$ where

$$\omega^2 = \frac{\gamma^2}{(K\beta)^2} \frac{(\sqrt{K\beta - \gamma} - \sqrt{2a})^2}{2\sqrt{a}(\sqrt{2(K\beta - \gamma)} - \sqrt{a})} > 0, \quad (7.3.34)$$

$$A_1 = \frac{\gamma}{K\beta} \frac{\sqrt{a}}{\sqrt{2(K\beta - \gamma)} - \sqrt{a}} > 0,$$

$$A_0 = \frac{\gamma^2(K\beta - \gamma)}{2(K\beta)^2 a} \frac{\sqrt{a}(2\sqrt{a} - \sqrt{2(K\beta - \gamma)})}{(\sqrt{a} - \sqrt{2(K\beta - \gamma)})^2} > 0.$$

Thus, at $\varepsilon = \varepsilon_0$ the characteristic polynomial has a pair of imaginary eigenvalues

$$\lambda(\varepsilon_0) = i\omega, \qquad \overline{\lambda}(\varepsilon_0) = -i\omega,$$

and the remaining two eigenvalues have negative real parts. Let us denote the root of (7.3.30) that assumes the value $i\omega$ at $\varepsilon = \varepsilon_0$ by $\lambda(\varepsilon)$. Routine calculation shows that

$$\frac{d\,\mathrm{Re}\,\lambda(\varepsilon_0)}{d\varepsilon} < 0.$$

The conditions of the Andronov-Hopf Theorem 7.2.3 hold and we have established the following:

Theorem 7.3.3 (*Farkas, A. - Farkas [1988]*). *If conditions (7.3.29) and (7.3.32) hold, then system (7.3.27) undergoes an Andronov-Hopf bifurcation at $\varepsilon = \varepsilon_0$, i.e. the unique equilibrium (7.3.28) in the interior of the positive orthant loses its stability as ε is decreased through ε_0 and the system has small amplitude periodic solutions in a neighbourhood of (n_0, p_0, q_0, r_0) for some ε whose angular frequency is near ω given by (7.3.34).*

In order to determine the character of the bifurcation and the stability of the bifurcating periodic solutions the linear coordinate transformation

$$\begin{pmatrix} n - n_0 \\ p - p_0 \\ q - q_0 \\ r - r_0 \end{pmatrix} = B \begin{pmatrix} y_1 \\ y_2 \\ y_3 \\ y_4 \end{pmatrix}$$

is performed where

$$B = \begin{pmatrix} 1 & -\left(\frac{1-b}{b}\right)^{1/2} & 1 & 1 \\ \frac{-\beta a(1-b)}{\gamma\alpha b(2b-1)} & \frac{\beta a}{\gamma\alpha(2b-1)}\left(\frac{1-b}{b}\right)^{1/2} & \frac{\beta a}{\gamma\alpha} & 0 \\ b & b\left(\frac{1-b}{b}\right)^{1/2} & -\frac{b(2b-1)}{1-b} & \frac{b(2b-1)^2}{1-b} \\ 1 & 0 & 0 & \frac{-b(2b-1)}{1-b} \end{pmatrix},$$

$$b := \left(\frac{a}{2(K\beta - \gamma)}\right)^{1/2}.$$

This brings system (7.3.27) at $\varepsilon = \varepsilon_0$ to the following canonical form:

$$y' = Ay + F(y),$$

where $y = \mathrm{col}\,[y_1, y_2, y_3, y_4]$,

$$A = \begin{pmatrix} 0 & -\omega & 0 & 0 \\ \omega & 0 & 0 & 0 \\ 0 & 0 & -\frac{\gamma(2b-1)}{2K\beta b(1-b)} & \frac{\gamma(2b-1)}{4K\beta b(1-b)} \\ 0 & 0 & -\frac{\gamma}{2K\beta b} & -\frac{\gamma(2b^2-2b+1)}{2K\beta b(1-b)} \end{pmatrix}$$

and the coordinates of the vector $F(y)$ are quadratic forms of y whose coefficients are fairly complicated expressions of b and γ/a which we have to omit here because of lack of space (see Farkas, A. [1988]). In all four coordinates of $F(y)$ the coefficient of y_2^2 is the opposite of the coefficient of y_1^2. In spite of this feature, which simplified the calculations, these could be performed only by making use of the formula manipulating computer program REDUCE. The Poincaré-Liapunov constant has been determined in Farkas,A. [1988] (see also Farkas,A.-Farkas [1988]). Instead of writing it out in the original parameters we introduce a new parameter θ by

$$\theta = (2a/(K\beta - \gamma))^{1/2}. \tag{7.3.35}$$

It turns out that the sign of the Poincaré-Liapunov constant is equal to the sign of the following expression:

$$\Phi(\theta, \gamma/(K\beta)) = Y_2(\theta)(\gamma/(K\beta))^2 + Y_1(\theta)\gamma/(K\beta) + Y_0(\theta) \tag{7.3.36}$$

where

$$\begin{aligned} Y_2(\theta) &= -2\theta^5 + 23\theta^4 - 86\theta^3 + 134\theta^2 - 90\theta + 20, \\ Y_1(\theta) &= 2\theta^6 - 10\theta^5 - 6\theta^4 + 102\theta^3 - 187\theta^2 + 119\theta - 20, \\ Y_0(\theta) &= 2\theta(\theta - 1)(2 - \theta)(\theta^3 - 3\theta^2 - 4\theta + 10). \end{aligned}$$

Thus, there holds:

Theorem 7.3.4 *(Farkas,A.-Farkas [1988]). If (7.3.29) and (7.3.32) hold, and $\Phi(\theta, \gamma/(K\beta))$ is negative, resp. positive, then the Andronov-Hopf bifurcation at $\varepsilon = \varepsilon_0$ is supercritical, resp. subcritical, i.e. there exists a $\delta > 0$ such that for $\varepsilon \in (\varepsilon_0 - \delta, \varepsilon_0)$, resp. $\varepsilon \in (\varepsilon_0, \varepsilon_0 + \delta)$, system (7.3.27) has exactly one small amplitude periodic orbit in a neighbourhood of (n_0, p_0, q_0, r_0) and this is asymptotically orbitally stable, resp. unstable.*

Introducing new parameters by (7.3.23) similar to the previous case the present situation may be put into a more general context. In the new parameters the condition $\varepsilon = \varepsilon_0$ where ε_0 is given by (7.3.33) assumes the form

$$uw(\sqrt{2} - \sqrt{w(v-1)}) = v\sqrt{2}(\sqrt{2w(v-1)} - 1), \tag{7.3.37}$$

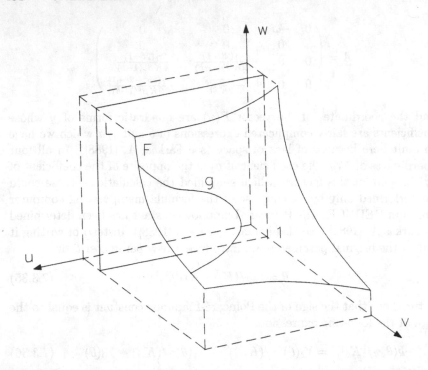

FIGURE 7.3.3. Bifurcation diagram for system (7.3.27) in the three dimensional parameter space u, v, w. Between the vertical plane $L : v = 1$ and the surface F given by (7.3.37), the equilibrium (7.3.28) is asymptotically stable. As F is crossed an Andronov-Hopf bifurcation occurs which is supercritical, resp. subcritical according as the crossing takes place outside or inside the tongue formed by the curve g given by $\Phi(\theta, 1/v) = 0$. (From Farkas,A.-Farkas-Szabó, *J. Math. Biol.* 26 (1988) 93-103).

while (7.3.29) and (7.3.32) are equivalent to

$$1 < v, \qquad 1/2 < w(v-1) < 2, \qquad (7.3.38)$$

respectively.

Corollary 7.3.5 *(Farkas,A.-Farkas-Szabó [1988]). If (7.3.38) holds and the surface F given by (7.3.37) in the parameter space u, v, w is crossed transversally, then the equilibrium (n_0, p_0, q_0, r_0) undergoes an Andronov-Hopf bifurcation; if at the point of crossing $\Phi(\theta, 1/v)$ of (7.3.36) is negative, resp. positive, then the bifurcation is supercritical, resp. subcritical.*

See Figure 7.3.3 where g is the curve on F determined by $\Phi(\theta, 1/v) = 0$. Clearly, $\theta = (1/(w(v-1)))^{1/2}$. The region where $\Phi < 0$, i.e. where the bifurcation is supercritical, is outside the "tongue" formed by g.

One may deduce more practical conditions of supercriticality taking into account the properties of the polynomials $Y_i(\theta)$ $(i = 0, 1, 2)$. First of all, condition (7.3.38) is equivalent to $1 < \theta < 2$, and $Y_1(\theta) < 0$, $Y_2(\theta) < 0$ for $\theta \in (1, 2)$. The third degree polynomial in the factorization of Y_0 has a single root in this interval, namely

$$\theta_0 = 1 + 2(7/3)^{1/2}\cos((4\pi + \cos^{-1}(-2(3/7)^{3/2}))/3)$$
$$\approx 1.6027.$$

Having established these facts an easy calculation yields

Corollary 7.3.6 (*Farkas, A.-Farkas-Szabó [1988]*). *If (7.3.38) holds and at the crossing of F the inequality*

$$1/2 < w(v-1) \leq 2/\theta_0^2 \approx 0.7786$$

holds, then the bifurcation is supercritical; if (7.3.38) holds and

$$0.7786 < w(v-1) < 2,$$

then for v sufficiently near 1 the bifurcation is supercritical, and for large v it is subcritical.

Summing up the results for both cases, both systems (7.3.6) and (7.3.26) have a unique equilibrium in the interior of the positive orthant if and only if $\gamma < K\beta$. This inequality determines the *admissible part* of the positive orthant of the four dimensional parameter space $\varepsilon, \gamma, 1/a, K\beta$. The meaning of the condition $\gamma < K\beta$ is clear: if the mortality of the predator is low, and the carrying capacity and the conversion rate are large, then coexistence is possible. If the converse is true, then the predator dies out. It is interesting to note that in both cases the increase of the intrinsic growth rate of prey and/or the increase of the carrying capacity for the prey *does not increase the prey* coordinate of the equilibrium but does increase the predator. The increase of the prey coordinate of the equilibrium can be achieved by the increase of predator mortality or by the decrease of the conversion rate. In both cases, if the parameter values lie in the admissible set, and, loosely speaking, *the intrinsic growth rate of prey ε and the mortality of predator γ are large, and the delay $1/a$ and the limiting factor $K\beta$ are small, then the equilibrium is asymptotically stable.* We have determined the equation of the bifurcation surface; (7.3.24) in the first case, (7.3.37) in the second. If this surface is crossed by *increasing the delay or the limiting factor, or decreasing the intrinsic birth rate of prey or the mortality of predator* (keeping in all the cases the rest of the parameter constant), *then the equilibrium loses its stability by an Andronov-Hopf bifurcation*, i.e. periodic solutions, closed trajectories occur in the neighbourhood of the equilibrium. The bifurcation surface can be crossed, naturally, by varying two, three or all four parameters, simultaneously leading to the same consequence. We have determined the part of the surface where the bifurcation is supercritical and the part where it is subcritical. Supercriticality means that we have small amplitude, orbitally asymptotically stable, periodic solutions for parameter values near to the surface on the side where the equilibrium is unstable. Subcriticality means that we have small amplitude, unstable periodic solutions near the equilibrium for parameter values near the surface on the side where the equilibrium is still stable, and the system behaves unpredictably on the other side. However, we have obtained computer simulation evidence suggesting that at parameter configurations corresponding to subcritical bifurcation the system has large amplitude stable periodic solutions (cf. Mallet-Paret, Yorke [1982]).

It is worthwhile to observe the *differences between the two cases*. In doing this we are going to consider "γ more fixed than ε which, in turn, will be more fixed than $1/a$", and $K\beta$ will be the easiest variable parameter.

In the first case, if $K\beta/a < 1$, then the equilibrium is asymptotically stable for arbitrary (small) ε and γ. In the second case, the more restrictive $K\beta/a < 1/2$ must hold for the same effect. On the other hand, in the second case, if $(K\beta - \gamma)/a > 2$, then the equilibrium is unstable for arbitrary (large) ε; in the first case we have no such subset, i.e. the increase of ε can always stabilize the equilibrium.

In both cases, for fixed values of $\varepsilon, \gamma, 1/a$, the increase of $K\beta$ destabilizes the equilibrium at a certain well determined value; the larger the delay $1/a$ is, the sooner this destabilization occurs. This is the phenomenon of the "*paradox of enrichment*". However, in the first case there is a *critical delay* under which the bifurcation is supercritical, and above which it is subcritical. In the second case, for small fixed ε the bifurcation is always supercritical; for large fixed ε the bifurcation is supercritical for small and for very large delay; it is subcritical if the delay falls into a certain interval in between.

We note that the same models have been applied to a macro-economic situation by Farkas-Bródy [1987] and Farkas-Kotsis [1992].

The results above could be achieved because the special choice (7.3.3), resp. (7.3.4) of the density functions made it possible to reduce the original integro-differential system to a system of ordinary differential equations. If the density function G is not a polynomial multiplied by an exponential function, then (7.3.2) is to be treated as a "real" functional differential equation (these will be dealt with briefly in Section 5) whose linearization at the equilibrium point has an infinite spectrum. However, if the density function G is near a polynomial multiplied by an exponential function in a natural sense, then as it has been shown in Farkas-Stépán [1992], the Andronov-Hopf bifurcation result remains valid (it is "robust") and it occurs at a parameter value near the one corresponding to the polynomial times exponential case.

There is no point in trying to give a list of applications of the Andronov-Hopf Theorem because it would be so extensive. In order to illustrate the wide range of applications just two will be mentioned. Ghil-Tavantzis [1983] have modelled the variation of the global temperature and the meridional ice sheet extent and have obtained periodic solutions corresponding to records of quaternary glaciation cycles, i.e. alternating cold and warm periods that characterized the last two million years of the Earth's history. Vancsa, Á.S. [1984] and Stépán, G. [1991] have studied the behaviour of wheels of drawn axes which can choose the direction of their rolling, like trolleys, draw-bars, nose-gears of aeroplanes, etc. They have found conditions for the appearance of stable, resp. unstable vibrations (affecting the direction of the rolling of the wheel) and even of chaotic behaviour.

7.4 Zip Bifurcation in Competitive Systems

In this Section certain three dimensional systems will be studied that model the competition of two predator species for a regenerating prey species. First a concrete model will be treated that has drawn considerable attention. The model has grown out of "chemostat type models", the chemostat being a laboratory apparatus used for the production and study of microorganisms. Nutrient is supplied at a constant rate, and the output flow equal to the input carries away cells, waste products and unused nutrients. However, in our case the nutrient for the two predators will be a prey species that is regenerating according to a logistic differential equation in the absence of predation. The model is conveniently exhibiting different kinds of behaviour showing instances of "the competitive exclusion principle" and the competition of an "r-strategist" and a "K-strategist". Under certain conditions Andronov-Hopf bifurcations occur, and also a phenomenon called "zip bifurcation" in Farkas [1982$_b$]. Then the model will be generalized showing that similar phenomena arise in a much wider class of models.

Following the majority of the extensive literature about the subject the quantities of prey and predators 1 and 2 will be denoted by $S(t)$, $x_1(t)$, $x_2(t)$, respectively, at time t. The model is

$$\dot{S} = \gamma S(1 - S/K) - m_1 x_1/(a_1 + S) - m_2 x_2 S/(a_2 + S),$$
$$\dot{x}_1 = m_1 x_1 S/(a_1 + S) - d_1 x_1,$$

$$\dot{x}_2 = m_2 x_2 S/(a_2 + S) - d_2 2 x_2 \tag{7.4.1}$$

where $\gamma > 0$ is the intrinsic birth rate of the prey, $K > 0$ is the carrying capacity of the environment with respect to the prey, $m_i > 0$, $d_i > 0$, $a_i > 0$ are the maximum birth rate, the mortality and the "half saturation constant" of predator i $(i = 1, 2)$ respectively. The meaning of the *half saturation constant* is that at $S = a_i$ the specific birth rate $m_i S/(a_i + S)$ (called also the "functional response") of the i-th predator is equal to half its maximum m_i. Those values of S at which the respective right-hand sides of the equations for the two predator species are zero will be denoted by λ_i $(i = 1, 2)$. Clearly,

$$\lambda_i = a_i d_i/(m_i - d_i) \qquad (i = 1, 2). \tag{7.4.2}$$

$\dot{x}_i$ is positive if and only if $S > \lambda_i$ for $x_i > 0$.

System (7.4.1) has been studied by several authors in its generality. Basic properties of the model have been established by Hsu-Hubbel-Waltman [1978$_a$,1978$_b$]. Butler-Waltman [1981], H.L.Smith [1982] and Keener [1983] have established the existence of periodic solutions too at certain parameter configurations; see also Cheng [1981]. It is easy to see that all solutions with initial values in the positive octant of S, x_1, x_2 space (which is, clearly, invariant) are bounded on $t \in [0, \infty)$. Indeed, if we add the three differential equations, we obtain

$$(S + x_1 + x_2)^\bullet = \gamma S(1 - S/K) - d_1 x_1 - d_2 x_2.$$

$(S + x_1 + x_2)^\bullet < 0$ if either $S > K$ or $0 < S < K$ and $d_1 x_1 + d_2 x_2 > \gamma K$; thus, if $c > 0$ is sufficiently large, then the planes $S + x_1 + x_2 = c$ are crossed inwards by the trajectories in the positive octant. Also one can easily show that predator i *may* survive only if $0 < \lambda_i < K$ (implying that $m_i > d_i$). Indeed, $S = \lambda_i$ is the threshold value of prey above which x_i may grow but $\limsup S(t)$ is, clearly, less than or equal to K as $t \to \infty$. If the limiting long range value of prey were less than the threshold value, then the predator density would go to zero. For the details see Hsu-Hubbel-Waltman [1978$_b$]. Unfortunately, we have no space here for presenting all the results of the general case; we restrict ourselves to the non-generic special case when the two threshold values are equal, i.e. we shall assume from now on that

$$0 < \lambda := \lambda_1 = \lambda_2 < K. \tag{7.4.3}$$

Though this case is non-generic it gives some insight into the dependence of the outcome of competition on the parameter values. This case has been treated by Wilken [1982] and Farkas [1984$_b$, 1985].

We introduce two more constants

$$\beta_i := m_i - d_i, \qquad b_i := m_i/d_i \qquad (i = 1, 2),$$

the former being the *maximal growth rate* (birth rate minus death rate) of predator i ($i = 1, 2$). With these constants

$$\lambda_i = a_i d_i / \beta_i = a_i / (b_i - 1) \qquad (i = 1, 2).$$

Assumption (7.4.3) implies that $\beta_i > 0$, $b_i > 1$ ($i = 1, 2$), and the system is

$$
\begin{aligned}
\dot{S} &= \gamma S(1 - S/K) - m_1 x_1 S/(a_1 + S) - m_2 x_2 S/(a_2 + S), \\
\dot{x}_1 &= \beta_1 x_1 (S - \lambda)/(a_1 + S), \\
\dot{x}_2 &= \beta_2 x_2 (S_\lambda)/(a_2 + S).
\end{aligned}
\tag{7.4.4}
$$

In the positive octant the equilibria of (7.4.4) are $(0,0,0)$, $(K,0,0)$ and the points on the straight line segment

$$
\begin{aligned}
L = \{&(S, x_1, x_2) \in \mathbf{R}^3 : S = \lambda, x_1 \geq 0, x_2 \geq 0, \\
&m_1 x_1/(a_1 + \lambda) + m_2 x_2/(a_2 + \lambda) = \gamma(1 - \lambda/K)\}. \tag{7.4.5}
\end{aligned}
$$

The points on L will be denoted by (λ, ξ_1, ξ_2). The end points of the line segment L in the $x_1 = 0$, resp. $x_2 = 0$ plane are

$$
\begin{aligned}
P_2 : &= (\lambda, 0, \xi_2) = (\lambda, 0, \gamma(a_2 + \lambda)(K - \lambda)/(m_2 K)), \\
P_1 : &= (\lambda_1, \xi_1, 0) = (\lambda, \gamma(a_1 + \lambda)(K - \lambda)/(m_1 K), 0). \tag{7.4.6}
\end{aligned}
$$

It is easy to see that both $(0,0,0)$ and $(K,0,0)$ are unstable. We are going to study the stability of the points on L and of the *set L*.

First the case when also $a := a_1 = a_2$ will be treated. This implies $b := b_1 = m_1/d_1 = b_2 = m_2/d_2$. Set $\rho := d_2/d_1 = m_2/m_1$. Also $\rho = \beta_2/\beta_1$, since $\beta_i = d_i(m_i/d_i - 1) = (b - 1)d_i$. We assume without loss of generality that $\rho \geq 1$. In this case when the half saturation constants are also equal, with the notations just introduced, the system assumes the form

$$
\begin{aligned}
\dot{S} &= \gamma S(1 - S/K) - (x_1 + \rho x_2) m_1 S/(a + S), \\
\dot{x}_1 &= \beta_1 x_1 (S - \lambda)/(a + S), \\
\dot{x}_2 &= \rho \beta_1 x_2 (S - \lambda)/(a + S). \tag{7.4.7}
\end{aligned}
$$

Dividing the third equation with the second one we get that the equation of the trajectories satisfies the differential equation $dx_2/dx_1 = \rho x_2/x_1$. Thus, we get easily that x_2/x_1^ρ is a first integral of system (7.4.7). As a consequence, the parabolic cylinders

$$x_2/x_1^\rho = c, \qquad c \geq 0, \tag{7.4.8}$$

are invariant surfaces of system (7.4.7) for arbitrary non-negative constant c. It is clear that these surfaces foliate completely the positive octant, $S \geq 0$, $x_1 > 0$, $x_2 \geq 0$, i.e. through each point in this octant there passes one and only one surface of the family (7.4.8). Let us fix the value of c and consider the restriction of system (7.4.7) to the invariant manifold (7.4.8) parametrized by S and x_1:

$$\begin{aligned}
\dot{S} &= \gamma S(1 - S/K) - (x_1 + \rho c x_1^\rho) m_1 S/(a + S), \\
\dot{x}_1 &= \beta_1 x_1 (S - \lambda)/(a + S).
\end{aligned} \tag{7.4.9}$$

The equilibria of (7.4.9) are $(S, x_1) = (0, 0)$, $(S, x_1) = (K, 0)$ and the single point at which the line L intersects (7.4.8). However, now the equation of L is (taking into account $a_1 = a_2 = a$)

$$S = \lambda, \quad x_1 + \rho x_2 = \gamma(a + \lambda)(K - \lambda)/(m_1 K), \quad x_1 \geq 0, x_2 \geq 0. \tag{7.4.10}$$

Thus, if the point of intersection is denoted by (λ, ξ_1, ξ_2), then ξ_1 is the unique positive solution of the equation

$$\xi_1 + \rho c \xi_1^\rho = \gamma(a + \lambda)(K - \lambda)/(m_1 K) \tag{7.4.11}$$

(and $\xi_2 = c\xi_1^\rho$). It is easy to see that both equilibria $(0, 0)$ and $(K, 0)$ are unstable.

In studying the stability of the equilibrium (λ, ξ_1) of (7.4.9) where ξ_1 is the positive solution of (7.4.11) the carrying capacity K will be considered as a bifurcation parameter. As $K > \lambda$ is varied, the line (7.4.10) is moving in a parallel way and is cutting the surface (7.4.8) at different points $(\xi_1(c, K), \xi_2(c, K))$ (see Figure 7.4.1); however, if there is no danger of confusion the short notation $\xi_1 = \xi_1(c, K)$ will be used too. The following theorem concerns system (7.4.9) on the manifold defined by (7.4.8).

Theorem 7.4.1 (*Farkas [1984b]*). *If* $\lambda < K < a + 2\lambda$, *then the equilibrium* $(\lambda, \xi_1(c, K))$ *of system (7.4.9) is asymptotically stable with region of attractivity* $\{(S, x_1) : S > 0, x_1 > 0\}$; *at* $K = a + 2\lambda$ *the system undergoes a supercritical Andronov-Hopf bifurcation, i.e. there exists a* $\delta > 0$ *such that for* $a + 2\lambda < K < a + 2\lambda + \delta$, *system (7.4.9) has a single closed path in a neighbourhood of* $(\lambda, \xi_1(c, K))$ *surrounding this equilibrium, and this closed path is orbitally asymptotically stable.*

Note that the bifurcation point $K = a + 2\lambda$ is independent of c, i.e. the bifurcation occurs at the same value of the parameter K in each surface of the family (7.4.8).

Note also that the family (7.4.8) contains the S, x_1 plane ($c = 0$) but does not contain the S, x_2 plane which is also an invariant manifold of system (7.4.7). Nevertheless, the end point of the line segment L in the S, x_2 plane behaves the same way as the rest of the points on L. For $\lambda < K \leq a + 2\lambda$ it is asymptotically stable and at $K = a + 2\lambda$ it undergoes a supercritical Andronov-Hopf bifurcation. This follows from the fact that the restriction of (7.4.7) to the $x_1 = 0$ plane is analogous to (7.4.9) taken at $c = 0$; only

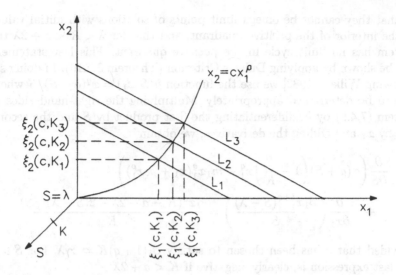

FIGURE 7.4.1. Displacement of the equilibrium in an invariant surface of system (7.4.7) due to the increase of K; L_i is the straight line segment of equilibria of the system corresponding to the value of K_i ($i = 1, 2, 3; \lambda < K_1 < K_2 < K_3$).

m_1 and β_1 are to be replaced by m_2 and β_2, respectively.

Proof. We move the origin into the equilibrium point $(S, x_1) = (\lambda, \xi_1)$ by the transformation $y_1 = S - \lambda$, $y_2 = x_1 - \xi_1$. System (7.4.9) assumes the form

$$\dot{y}_1 = \gamma(y_1 + \lambda)(1 - (y_1 + \lambda)/K)$$
$$- \frac{m_1(y_1 + \lambda)(y_2 + \xi_1 + (y_2 + \xi_1)^p(\gamma(a + \lambda)(1 - \lambda/K) - m_1\xi_1)/(m_1\xi_1^p))}{a + \lambda + y_1}$$
$$\dot{y}_2 = \beta_1(y_2 + \xi_1)y_1/(a + \lambda + y_1) \tag{7.4.12}$$

where (7.4.11) has been applied. Linearizing this system at $(y_1, y_2) = (0, 0)$ the characteristic polynomial turns out to be

$$D(\mu, K) = \mu^2 + \mu\gamma\lambda(a + 2\lambda - K)/(K(a + \lambda))$$
$$+ \lambda m_1\beta_1\xi_1(1 - p + p\gamma(a + \lambda)(1 - \lambda/K)/(m_1\xi_1))/(a + \lambda)^2.$$

Here the constant term is positive as a consequence of (7.4.11). The coefficient of μ is positive if and only if $K < a + 2\lambda$. Thus, the origin is asymptotically stable for $\lambda < K < a + 2\lambda$ and unstable if $K > a + 2\lambda$. The global attractivity (with respect to the interior of the positive quadrant) of the equilibrium (λ, ξ_1) of system (7.4.9) follows from the facts that all the solutions are bounded, $(S, x_1) = (0, 0)$ and $(S, x_1) = (K, 0)$ are both saddles with ingoing trajectories on the x_1-axis and the S-axis, respectively, so that they cannot be omega limit points of solutions with initial values in the interior of the positive quadrant, and that for $\lambda < K < a + 2\lambda$ the system has no limit cycle in the positive quadrant. This last statement can be shown by applying Dulac's Criterion (Theorem 3.5.16). In doing so, following Wilken [1982] we use the function $h(S, x_1) = x_1^q (a + S)/S$ where q is to be determined appropriately. Multiplying the right-hand sides of system (7.4.9) by h, differentiating the first product by S and the second one by x_1 and adding the derivatives, we obtain

$$\frac{\partial}{\partial S}\left(\gamma(a + S)\left(1 - \frac{S}{K}\right)x_1^q - m_1 x_1^q (x_1 + \rho c x_1^\rho)\right)$$

$$+ \frac{\partial}{\partial x_1}\left(\frac{\beta_1 x_1^{1+q}(S - \lambda)}{S}\right) = \frac{\gamma x^q (K - a - 2\lambda - 2(S - \lambda)^2/S)}{K}$$

provided that q has been chosen to satisfy $\beta_1(1 + q)K = 2\gamma\lambda$. For $S > 0$ the last expression is, clearly, negative if $K < a + 2\lambda$.

At $K = a + 2\lambda$ the roots of the characteristic polynomial D are pure imaginary:

$$\mu_{1,2}(a + 2\lambda) = \pm i\omega$$

where

$$\omega = \omega(\xi_1) = (\lambda\beta_1(m_1\xi_1 + \rho(\gamma(a + \lambda)^2/(a + 2\lambda) - m_1\xi_1)))^{1/2}/(a + \lambda). \tag{7.4.13}$$

An easy calculation shows that

$$d\,\mathrm{Re}\,\mu(a + 2\lambda)/dK = \gamma\lambda/(2(a + \lambda)(a + 2\lambda)) > 0;$$

thus, the conditions of the Andronov-Hopf Theorem (Lemma 7.2.5) are satisfied apart from (iii) for the time being.

Developing the right-hand side of (7.4.12) to power series at the critical parameter value $K = a + 2\lambda$ and truncating at the third order terms the following system is obtained:

$$\dot{y}_1 = -\frac{\lambda m_1}{a + \lambda}\left(1 - \rho + \frac{\rho\gamma(a + \lambda)^2}{m_1\xi_1(a + 2\lambda)}\right)y_2$$

$$-y_1^2\frac{\gamma\lambda}{(a + \lambda)(a + 2\lambda)} - y_1 y_2\frac{am_1}{(a + \lambda)^2}\left(1 - \rho + \frac{\rho\gamma(a + \lambda)^2}{m_1\xi_1(a + 2\lambda)}\right)$$

$$-y_2^2\frac{\lambda\rho(\rho - 1)}{(a + \lambda)2\xi_1^2}\left(\frac{\gamma(a + \lambda)^2}{a + 2\lambda} - m_1\xi_1\right)$$

$$-y_1^3\frac{a\gamma}{(a + \lambda)^2(a + 2\lambda)} + y_1^2 y_2\frac{am_1}{(a + \lambda)^3}\left(1 - \rho + \frac{\rho\gamma(a + \lambda)^2}{m_1\xi_1(a + 2\lambda)}\right)$$

$$-y_1 y_2^2\frac{a\rho(\rho - 1)}{(a + \lambda)^2 2\xi_1^2}\left(\frac{\gamma(a + \lambda)^2}{a + 2\lambda} - m_1\xi_1\right)$$

$$-y_2^3 \frac{\lambda\rho(\rho-1)(\rho-2)}{6\xi_1^2}\left(\frac{\gamma(a+\lambda)^2}{a+2\lambda}-m_1\xi_1\right)+\cdots$$

$$\dot{y}_2 = \frac{\beta_1\xi_1}{a+\lambda}y_1-y_1^2\frac{\beta_1\xi_1}{(a+\lambda)^2}+y_1y_2\frac{\beta_1}{a+\lambda}+y_1^3\frac{\beta_1\xi_1}{(a+\lambda)^3}-y_1^2y_2\frac{\beta_1}{(a+\lambda)^2}+\cdots$$

$$(7.4.14)$$

System (7.4.14) is not exactly in canonical form, for the coefficients of y_2 in the first equation and y_1 in the second equation are not equal to $\pm\omega$; still, it serves the purpose well. We assume a Liapunov function V in the form

$$V(y_1,y_2) = y_1^2 + By_2^2 + p_3(y_1,y_2) + p_4(y_1,y_2)$$

where B is a constant and p_3 and p_4 are homogeneous polynomial of degree 3 and 4, respectively. B, p_3 and p_4 are to be determined in such a way that the derivative of V with respect to system (7.4.14) shall be equal to

$$\dot{V}_{(7.4.14)}(y_1,y_2) = G(y_1^2+y_2^2)^2+\cdots.$$

After simple but somewhat lengthy calculations one gets that

$$B = \lambda m_1(1-\rho+\rho\gamma(a+\lambda)^2/(m_1\xi_1(a+2\lambda)))/\beta_1\xi_1 > 0$$

using (7.4.11) with $K = a+2\lambda$, and

$$G\left(2+\frac{3\lambda m_1}{\beta_1\xi_1}\left(1-\rho+\frac{\rho\gamma(a+\lambda)^2}{m_1\xi_1(a+2\lambda)}\right)\right.$$
$$\left.+\frac{3\beta_1\xi_1}{m_1(1-\rho+\rho\gamma(a+\lambda)^2/(m_1\xi_1(a+2\lambda)))}\right) = \frac{-4m_1\lambda\gamma}{(\beta_1\xi_1(a+\lambda)(a+2\lambda))}$$

implying $G < 0$. This result means that the origin is 3-asymptotically stable, and this completes the proof of the Theorem. $\square$

Corollary 7.4.2. *The points of the straight line segment L given by (7.4.10) are stable equilibria in the Liapunov sense of system (7.4.7) for $\lambda < K \le a+2\lambda$ and are unstable for $K > a+2\lambda$.*

In the following proposition, "neighbourhood" of a set A means the intersection of an open set containing A with the octant $\{(S,x_1,x_2) : S \ge 0, x_1 \ge 0, x_2 \ge 0\}$.

Corollary 7.4.3. *For $\lambda < K \le a+2\lambda$ the line segment L given by (7.4.10) is a global attractor of system (7.4.7) with respect to the interior of the positive octant; at $K = a+2\lambda$ the segment L bifurcates into a topological cylinder, i.e. there exists a $\delta > 0$ such that for $a+2\lambda < K < a+2\lambda+\delta$, system (7.4.7) has an invariant topological*

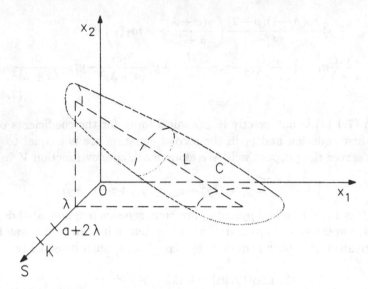

FIGURE 7.4.2. Bifurcation of the line of equilibria into a cylinder of periodic orbits.

cylinder C which is the union of periodic orbits and is an attractor of the system, i.e. it has a "neighbourhood" such that trajectories with initial values in this "neighbourhood" tend to C as t tends to infinity (see Figure 7.4.2).

Secondly, the more interesting case $a_1 > a_2$ will be treated. This assumption implies $b_1 > b_2$ if $\lambda = \lambda_1 = \lambda_2$ is taken into account. As a consequence, predator 1 and 2 can be considered a so-called "*r-strategist*" and a so-called "*K-strategist*", respectively. An r-strategist has a high birth rate compared to death rate (b) and a high half saturation constant (a), i.e. it needs much food. A K-strategist has a relatively low birth rate and half saturation constant, i.e. it can live on at relatively low prey density. Also in this case the carrying capacity K will be considered a bifurcation parameter and our interest lies in the response of the dynamics of the system to the change of K.

We linearize system (7.4.4) at an arbitrary point of L given by (7.4.5), i.e. at (λ, ξ_1, ξ_2) where

$$\frac{m_1\xi_1}{a_1 + \lambda} + \frac{m_2\xi_2}{a_2 + \lambda} = \frac{\gamma(K - \lambda)}{K}, \quad \xi_1 \geq 0, \quad \xi_2 \geq 0. \tag{7.4.15}$$

The coefficient matrix of the linearized system is

$$A = \begin{pmatrix} -\frac{\gamma\lambda}{K} + \lambda\left(\frac{m_1\xi_1}{(a_1+\lambda)^2} + \frac{m_2\xi_2}{(a_2+\lambda)^2}\right) & -\frac{m_1\lambda}{a_1+\lambda} & -\frac{m_2\lambda}{a_2+\lambda} \\ \frac{\beta_1\xi_1}{a_1+\lambda} & 0 & 0 \\ \frac{\beta_2\xi_2}{a_2+\lambda} & 0 & 0 \end{pmatrix}$$

and the characteristic polynomial is

$$D(\mu) = \mu\left[\mu^2 + \mu\lambda\left(\frac{\gamma}{K} - \frac{m_1\xi_1}{(a_1+\lambda)^2} - \frac{m_2\xi_2}{(a_2+\lambda)^2}\right) + \lambda\left(\frac{\beta_1 m_1\xi_1}{(a_1+\lambda)^2} + \frac{\beta_2 m_2\xi_2}{(a_2+\lambda)^2}\right)\right].$$

$$(7.4.16)$$

The quadratic polynomial in brackets is stable if and only if

$$\frac{m_1\xi_1}{(a_1+\lambda)^2} + \frac{m_2\xi_2}{(a_2+\lambda)^2} < \frac{\gamma}{K}. \qquad (7.4.17)$$

Now, if $\lambda < K < a_2 + 2\lambda$, then

$$\frac{m_1\xi_1}{(a_1+\lambda)^2} + \frac{m_2\xi_2}{(a_2+\lambda)^2} \leq \frac{1}{a_2+\lambda}\left(\frac{m_1\xi_1}{a_1+\lambda} + \frac{m_2\xi_2}{a_2+\lambda}\right)$$

$$= \frac{\gamma}{a_2+\lambda}\left(1 - \frac{\lambda}{K}\right) < \frac{\gamma}{a_2+\lambda}\left(1 - \frac{\lambda}{a_2+2\lambda}\right) < \frac{\gamma}{K}.$$

If $K > a_1 + 2\lambda$, then by an analogous estimate we get that

$$\frac{m_1\xi_1}{(a_1+\lambda)^2} + \frac{m_2\xi_2}{(a_2+\lambda)^2} > \frac{\gamma}{K}.$$

Thus, for $\lambda < K < a_2 + 2\lambda$ each equilibrium (λ, ξ_1, ξ_2) on L has a zero eigenvalue and two eigenvalues with negative real parts. This implies in view of Theorem A3.1 and Definition A3.6 that through each equilibrium (λ, ξ_1, ξ_2) on L there passes a two dimensional smooth locally invariant manifold. All trajectories on this surface tend to (λ, ξ_1, ξ_2) as t tends to infinity. On the other hand, if $K > a_1 + 2\lambda$, then all equilibria (λ, ξ_1, ξ_2) on L are unstable.

Now we are going to study the situation when $a_2 + 2\lambda \leq K \leq a_1 + 2\lambda$. It is easy to see that as the point (ξ_1, ξ_2) is moving along the line segment (7.4.15) from $(0, \gamma(a_2+\lambda)(K-\lambda)/(m_2 K))$ to $(\gamma(a_1+\lambda)(K-\lambda)/(m_1 K), 0)$, the left-hand side of (7.4.17) is decreasing. If K is fixed in the interval $(a_2 + 2\lambda, a_1 + 2\lambda)$, then there is a point $(\lambda, \xi_1(K), \xi_2(K))$ on L at which (7.4.17) turns into an equation. This point is easily determined by solving the system consisting of (7.4.15) and the equation

$$\frac{m_1\xi_1}{(a_1+\lambda)^2} + \frac{m_2\xi_2}{(a_2+\lambda)^2} = \frac{\gamma}{K}. \qquad (7.4.18)$$

We obtain

$$(\xi_1(K), \xi_2(K)) = \left(\frac{\gamma(a_1+\lambda)^2(K-a_2-2\lambda)}{Km_1(a_1-a_2)}, \frac{\gamma(a_2+\lambda)^2(a_1+2\lambda-K)}{Km_2(a_1-a_2)}\right).$$

$$(7.4.19)$$

At points of L to the left of this point, system (7.4.4) has a zero eigenvalue and two eigenvalues with positive real parts; at points of L to the right

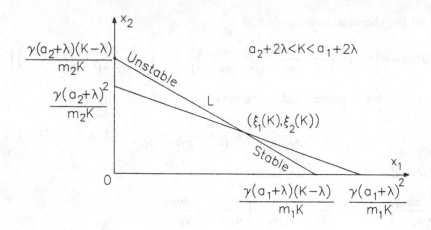

FIGURE 7.4.3. Intersection of lines L (7.4.15) and (7.4.18): point of bifurcation.

of this point, the system has a zero eigenvalue and two eigenvalues with negative real parts. At $(\lambda, \xi_1(K), \xi_2(K))$ one eigenvalue is zero and the other two eigenvalues have real parts equal to zero. Thus, by Hartman's theorem quoted above, through each point of L to the left of the point (7.4.19) there passes a two dimensional unstable manifold and through each point of L to the right of the point (7.4.19) there passes a two dimensional stable manifold. There holds the following Theorem (see Figure 7.4.3).

Theorem 7.4.4. *For any K satisfying $a_2 + 2\lambda \leq K \leq a_1 + 2\lambda$ the point $(\lambda, \xi_1(K), \xi_2(K))$ divides L into two; the equilibria of system (7.4.4) in the set*

$$\{(\lambda, \xi_1, \xi_2) \in L : \xi_1 < \xi_1(K)\}$$

are unstable, and the equilibria in the set

$$L_S = \{(\lambda, \xi_1, \xi_2) \in L : \xi_1 > \xi_1(K)\}$$

are stable in the Liapunov sense (L_S stands for the "stable part of L").

As K is increased from $a_2 + 2\lambda$ up to $a_1 + 2\lambda$ the point $(\lambda, \xi_1(K), \xi_2(K))$ moves along L continuously from $(\lambda, 0, \gamma(a_2 + \lambda)^2/(m_2(a_2 + 2\lambda)))$ to $(\lambda, \gamma(a_1 + \lambda)^2/(m_1(a_1 + 2\lambda)), 0)$ so that the points left behind become unstable. We call this phenomenon a *zip bifurcation*.

Note that as K is varied, the line L itself undergoes a parallel displacement; however, this has no bearing on the qualitative picture.

Proof. The first part of the Theorem is clear because of the position of the eigenvalues of the system linearized at points $(\lambda, \xi_1, \xi_2) \in L$, $\xi_1 < \xi_1(K)$. We have to prove that each point of $L_s, (\lambda, \xi_1, \xi_2) \in L$, with $\xi_1 > \xi_1(K)$ is stable in the Liapunov sense. We know that through each

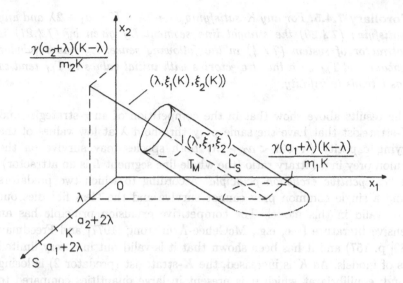

FIGURE 7.4.4. The tubular basin T_M of the attractor $\tilde{L}_S$.

point of L_s there goes a locally invariant two dimensional manifold such that all trajectories on this manifold tend to (λ, ξ_1, ξ_2) exponentially as t tends to infinity. One can show that these stable manifolds fill in a "tubular neighbourhood" of L_S, and this proves the Theorem. The topological details will be omitted (see Farkas [1984b]), but the "tubular neighbourhood" will be defined in the sequel. $\square$

Let $\tilde{\xi}_1$ be such that

$$\xi_1(K) < \tilde{\xi}_1 < \gamma(a_1 + \lambda)(K - \lambda)/(m_1 K), \qquad (7.4.20)$$

and consider the closed proper subset of L_s given by

$$\tilde{L}_s := \{(\lambda, \xi_1, \xi_2) \in L : \xi_1 \geq \tilde{\xi}_1\}. \qquad (7.4.21)$$

Let M be a smooth surface intersecting L transversally at $(\lambda, \tilde{\xi}_1, \tilde{\xi}_2) \in L$. The compact set T_M whose boundary is the part of M inside the surface

$$\{(S, x_1, x_2) \in \mathbf{R}^3 : \text{dist}\,((S, x_1, x_2), \tilde{L}_S) = \rho\} \qquad (7.4.22)$$

where $\rho > 0$, the part of the surface (7.4.22) between M and the S, x_1 plane and the part of the S, x_1 plane inside the surface (7.4.22) is called a *tubular neighbourhood of* $\tilde{L}_S$. Clearly, if $\rho > 0$ is sufficiently small, the intersection of M and the surface (7.4.22) is a simple Jordan curve, $\tilde{L}_S \subset T_M$ and the interior of the straight line segment $\tilde{L}_S$ is in the interior of T_M (see Figure 7.4.4).

Corollary 7.4.5. *For any K satisfying $a_2 + 2\lambda \leq K \leq a_1 + 2\lambda$ and any $\tilde{\xi}_1$ satisfying (7.4.20) the straight line segment $\tilde{L}_S$ given by (7.4.21) is an attractor of system (7.4.4) in the following sense: $\tilde{L}_S$ has a tubular neighbourhood T_M such that trajectories with initial values in T_M tend to $\tilde{L}_S$ as t tends to infinity.*

The results above show that in the competition of an r-strategist and a K-strategist that have the same prey threshold λ at low values of the carrying capacity $(\lambda < K < a_2 + 2\lambda)$ both species may survive on the common prey in arbitrary ratio (the whole line segment L is an attractor). The "*competitive exclusion principle*" according to which two predators having a single common prey cannot survive and the "less fit" dies out is not valid in this model. The competitive exclusion principle has an extensive literature (see, e.g., McGehee-Armstrong [1977] and Freedman [1980] p. 157) and it has been shown that it is valid but in a very limited class of models. As K is increased, the K-strategist (predator 2) is losing ground: equilibria at which it is present in large quantities compared to the r-strategist get destabilized by the zip bifurcation. When K reaches the value $a_1 + 2\lambda$, then no stable equilibrium remains in the interior of the positive octant, only the point P_1 showing the absence of the K-strategist remains stable. Thus, at large quantities of food available the r-strategist outcompetes the K-strategist. If K is increased further, then even P_1 loses its stability by an Andronov-Hopf bifurcation, and what remains is an orbitally asymptotically stable periodic orbit in the S, x_1 plane. This last result is a simple consequence of Theorem 7.4.1 since the two dimensional system governing the dynamics of S and x_1 in the absence of x_2 is, in fact, (7.4.9) with $c = 0$ and $a = a_1$.

We show now that these results are valid in a more general context. System (7.4.1) is generalized the following way:

$$
\begin{aligned}
\dot{S} &= \gamma S g(S, K) - x_1 p(S, a_1) - x_2 p(S, a_2), \\
\dot{x}_1 &= x_1 p(S, a_1) - d_1 x_1, \\
\dot{x}_2 &= x_2 p(S, a_2) - d_2 x_2
\end{aligned}
\tag{7.4.23}
$$

where S, x_1, x_2, $\gamma > 0$, $d_1 > 0$, $d_2 > 0$ have the same meaning as before, and for the rest of the functions and parameters the following conditions hold:

$$g \in C^5((0, \infty) \times (0, \infty), \mathbf{R}), \qquad g \in C^0([0, \infty) \times (0, \infty), \mathbf{R}),$$

$$g(0, K) = 1, \quad g'_S(S, K) < 0 < g''_{SK}(S, K), \quad S \geq 0, \quad K > 0, \quad (7.4.24)$$

$$\lim_{K \to \infty} g'_s(S, K) = 0 \tag{7.4.25}$$

uniformly in $S \in [\delta, S_0]$ for arbitrary $0 < \delta < S_0$; the possibly improper integral $\int_0^{S_0} g'_S(S, K) dS$ is uniformly convergent in $K \in [K_0, \infty)$ for arbitrary $K_0 > 0$, and

$$(K - S)g(S, K) > 0, \qquad S \geq 0, \qquad K > 0, \qquad S \neq K, \qquad (7.4.26)$$

$$p \in C^5((0, \infty) \times (0, \infty), \mathbf{R}), \qquad p \in C^0([0, \infty) \times (0, \infty), \mathbf{R}),$$

$$p(0, a) = 0, \qquad p'_S(S, a) > 0, \qquad S > 0, \qquad a > 0, \qquad (7.4.27)$$

$$p'_S(S, a) < p(S, a)/S, \qquad S > 0, \qquad a > 0, \qquad (7.4.28)$$

$$p'_a(S, a) < 0, \qquad S > 0, \qquad a > 0. \qquad (7.4.29)$$

Fifth order differentiability is supposed in order to make the application of the Andronov-Hopf Bifurcation Theorem 7.2.3 possible. Otherwise the conditions coincide with those of Butler [1983] apart from a genericity condition made by Butler which is unnecessary now. Conditions (7.4.24) mean that the highest specific growth rate of prey is achieved at $S = 0$, $x_1 = x_2 = 0$ and it is $\gamma > 0$, the growth rate decreases with the increase of prey quantity, $|g'_S(S, K)|$ is decreasing with the increase of K, i.e. at high values of K the effect of increasing prey quantity is less significant; in fact, by (7.4.25) at very high values of K this effect is negligible. Relation (7.4.26) means that K is the carrying capacity of the environment with respect to the prey: in the absence of predators prey quantity grows if it is below K and decreases if it is above this value. It is easy to see that these conditions imply

$$\lim_{K \to \infty} g(S, K) = 1, \qquad S \geq 0.$$

Conditions (7.4.27) mean that the birth rate p of the predators (which is equal to the "predation rate") is zero in the absence of prey and is an increasing function of the quantity of prey. Condition (7.4.28) is a "weak concavity" condition sometimes called Krasnoselskij's condition. If p is a strictly concave function of S (for any $a > 0$), then (7.4.28) is implied with the possible exception of isolated points where it holds with an equality sign. Inequality (7.4.29) throws light on the role of the parameter a: the birth rate of the predator is a decreasing function of a, i.e. the higher the value of a, the more food needed to maintain the same birth rate of the specific predator (this is the role of the "half saturation constant" in system (7.4.1)). The conditions imply that $p(S, a) > 0$, $S > 0$, $a > 0$. If p is a bounded function for fixed $a > 0$, then $m_i := \sup p(S, a_i)$, $S > 0$ is the *maximal birth rate* of the i-th predator ($i = 1, 2$). Clearly,

$$\lim_{S \to \infty} p(S, a_i) = \begin{cases} m_i & \text{if } p \text{ is bounded} \\ \infty & \text{if } p \text{ is not bounded} \end{cases} \qquad (7.4.30)$$

for $a = a_i$.

In the second and third terms on the right-hand side of the differential equation describing the growth of the prey population some constants (called "yield factors") used appear in a realistic model. However, these yield factors can be transformed out of the system without loss of generality and they do not affect the qualitative behaviour. Therefore, we have chosen not to introduce them at all.

Omitting the analysis of the less interesting case in which a_1 and a_2 are equal, we shall assume $a_1 > a_2 > 0$, i.e.

$$p(S, a_1) < p(S, a_2) \qquad \text{for} \quad \text{all} \qquad S > 0. \tag{7.4.31}$$

According to this condition, at any given level of prey, the quantity the birth rate of predator 2 is higher than that of predator 1 or, in other words, predator 1 needs a higher quantity of prey to achieve the same birth rate as predator 2. Now, if d_1 is greater than d_2, (7.4.31) implies $p(S, a_1) - d_1 < p(S, a_2) - d_2$, i.e. the net growth rate of predator 2 is higher than that of predator 1. One can prove that in this case predator 2 outcompetes predator 1, provided that the conditions for the survival of the former hold (see (7.4.36)). We assume from now on that

$$d_1 < d_2. \tag{7.4.32}$$

As a consequence, (7.4.31) does *not* imply that the net growth rate of predator 2 also exceeds that of predator 1.

Another important characteristic of the respective predator species is the prey threshold quantity $S = \lambda_i$, above which their growth rate is positive, i.e.

$$p(\lambda_i, a_i) = d_i \qquad (i = 1, 2).$$

Obviously, the lower λ_i, the fitter the predator i. However, we shall assume (non-generically) that $\lambda_1 = \lambda_2$, i.e. the two predator species have equal prey thresholds, although they achieve this by different means. Thus, our assumption will be that there exists a $\lambda > 0$ such that

$$p(\lambda, a_i) = d_i \qquad (i = 1, 2). \tag{7.4.33}$$

We note that, because of condition (7.4.27), equation (7.4.33) has one and only one solution λ if and only if either p is unbounded or m_i is greater than d_i. The real content of (7.4.33) is that the two solutions for $i = 1, 2$ coincide.

The class of models under consideration can be divided into three subclasses according to

Definition 7.4.1. We say that the model (7.4.23) under conditions (7.4.27)-(7.4.29) and (7.4.31)-(7.4.33) is *natural, artificial,* and *degenerate* if

$$\frac{\partial}{\partial S} \left[\frac{p(S, a_2)}{p(S, a_1)} \right]_{S=\lambda} \begin{cases} < 0 \\ > 0 \\ = 0 \end{cases}, \tag{7.4.34}$$

respectively.

The first inequality of (7.4.34) means that, by continuity, the ratio of the birth rates (which is, by (7.4.31), greater than unity) decreases in the neighbourhood of $S = \lambda$, i.e., the advantage of species 2 over species 1 expressed by (7.4.31) decreases as the quantity of prey increases. This is what is usually expected to happen. The second inequality of (7.4.34) means that the same advantage is increasing. The importance of the point $S = \lambda$ will become clear in what follows.

Before turning to the study of the equilibria of system (7.4.23) we note that, obviously, the coordinate planes of S, x_1, x_2 space are invariant manifolds of the system, and that it can be proved by standard methods that all solutions with non-negative initial conditions of the system are defined in $[0, \infty)$, are bounded, and remain non-negative.

System (7.4.23) has the following equilibria:
$Q_1 = (0,0,0)$, $Q_2 = (K,0,0)$, and the points on the straight line segment

$$
\begin{aligned}
L_K \;=\; & \{(S,x_1,x_2) : p(\lambda,a_1)x_1 + p(\lambda,a_2)x_2 = \gamma\lambda g(\lambda,K), \\
& S = \lambda, x_1 \geq 0, x_2 \geq 0\}.
\end{aligned} \tag{7.4.35}
$$

It is easy to see by linearization that Q_1 is unstable, and Q_2 is asymptotically stable for $K < \lambda$ and unstable for $K > \lambda$. Actually, it is known (see Butler [1983]) that

$$
K > \lambda \tag{7.4.36}
$$

is a necessary condition for the survival of each predator. Therefore, (7.4.36) will also be assumed in what follows. Note that, by (7.4.26), if K is less than λ, then L_K is empty, and if $K = \lambda$, then its only point is the origin Q_1.

The analysis and the results are analogous (but somewhat more complicated) to (than) those concerning the special model (7.4.1). Proofs will be omitted (see Farkas [1987]). Linearizing system (7.4.23) at an arbitrary point (λ, ξ_1, ξ_2) of the straight line segment L_K the characteristic polynomial always has 0 as a root. The remaining two eigenvalues either have (both) negative real parts or non-negative real parts. The calculations lead to the point $(\lambda, x_1(K), x_2(K))$ on the straight line containing the segment L_K where

$$
x_i(K) = (-1)^i \gamma \frac{-\lambda g_S'(\lambda,K)p(\lambda,a_{3-i}) + g(\lambda,K)(\lambda p_S'(\lambda,a_{3-i}) - p(\lambda,a_{3-i}))}{p(\lambda,a_2)p_S'(\lambda,a_1) - p_S'(\lambda,a_2)p(\lambda,a_1)}
$$

$$
(i = 1,2), \tag{7.4.37}
$$

provided that the model is not degenerate, i.e. the denominator is different from zero.

Theorem 7.4.6 *(Farkas [1987]). Suppose that system (7.2.23) satisfies conditions (7.4.24)- (7.4.29), (7.4.31)-(7.4.33) and that it is natural; then there exist $\lambda < K_1 < K_2 < \infty$ such that for $K \in (\lambda, K_1)$ all points of segment L_K are stable in the Liapunov sense, and L_K is an attractor of the system; for $K \in (K_2, \infty)$ the system has no stable equilibrium point in the closed positive octant of S, x_1, x_2 space; for $K \in (K_1, K_2)$ the point $(\lambda, x_1(K), x_2(K))$ divides L_K into two parts (one of which may be empty): the points of L_K to the left of this point are unstable, the points to the right are stable in the Liapunov sense, and the part of L_K to the right of this point is an attractor of the system.*

If $g(\lambda, \cdot)$ is a non-decreasing function in $K \in (\lambda, \infty)$, i.e.

$$g'_K(\lambda, K) \geq 0, \qquad K > \lambda, \tag{7.4.38}$$

then $x_1(\cdot)$ is monotone increasing and $x_2(\cdot)$ monotone decreasing (under the conditions of the previous Theorem). As a consequence, if K is increased from K_1 to K_2 the point $(\lambda, x_1(K), x_2(K))$ moves steadily along L_K from the left-hand end to the right-hand end (while the segment L_K undergoes a parallel displacement). In the process the points left behind by $(\lambda, x_1(K), x_2(K))$ become destabilized, i.e. a zip bifurcation occurs. In the more general case when (7.4.38) does not hold, the movement of the point $(\lambda, x_1(K), x_2(K))$ that separates the stable equilibria from the unstable ones is still continuous and yields, at the end, the same result, the "opening up of the zip"; however, in some parts of the interval (K_1, K_2) the zip may move "backwards" (to the left) with increasing K.

It is worthwhile to study the bifurcation of the two end points of segment L_K. This could be done in the general case; however, the picture is clearer if (7.4.38) is assumed, and this we shall do. For $K > \lambda$, system (7.4.23) has an equilibrium point in the interior of the positive quadrant of each coordinate plane S, x_i $(i = 1, 2)$. These are $(\lambda, \gamma \lambda g(\lambda, K)/p(\lambda, a_1), 0)$ and $(\lambda, 0, \gamma \lambda g(\lambda, K)/p(\lambda, a_2))$, respectively. The coordinate planes are invariant manifolds of the system, and the restriction of (7.4.23) to any of them is the two dimensional system

$$\begin{aligned} \dot{S} &= \gamma S g(S, K) - x_i p(S, a_i), \\ \dot{x}_i &= x_i p(s, a_i) - d_i x_i \qquad (i = 1, 2). \end{aligned} \tag{7.4.39}$$

The equilibrium of the latter system inside the positive quadrant of the S, x_i plane is

$$P_i(K) = (\lambda, \gamma \lambda g(\lambda, K)/p(\lambda, a_i)) \qquad (i = 1, 2).$$

The previous results imply that in a natural model $P_i(K)$ is asymptotically stable for $\lambda < K < K_{3-i}$ $(i = 1, 2)$. There holds:

Theorem 7.4.7 *(Butler [1983], Farkas [1987]). If conditions (7.4.24)-(7.4.29), (7.4.31)-(7.4.33) and (7.4.38) hold and (7.4.23) is a natural model, then the equilibrium point $P_i(K)$ of system (7.4.39) undergoes an Andronov-Hopf bifurcation at $K = K_{3-i}$; the bifurcation is supercritical, resp. subcritical according as*

$$G(S) := \left(\frac{f''(S)}{p_i(S)p_i'(S)} \right)' p_i^2(S) p_i'^2(S) + \left(\frac{f(S)p_i'(S)}{p_i''(S)} \right)' p_i''^2(S) \quad (7.4.40)$$

at $S = \lambda$ is negative, resp. positive where $f(S) := Sg(S, K_{3-i})$, $p_i(S) := p(S, a_i)$ $(i = 1, 2)$.

Theorems analogous to Theorems 7.4.6 and 7.4.7 are valid in the case of artificial models, though there are differences in the "direction of the zip" and in the critical values of K at the Andronov-Hopf bifurcation. In the degenerate case the system behaves like system (7.4.7); see Farkas [1987].

We note that the global behaviour of the trajectories of a system exhibiting a zip bifurcation has been studied by Gyökér [1988].

In this chapter we have concentrated on the Andronov-Hopf bifurcation as the "main source" of the appearance of periodic solutions. We have to note that there are other bifurcations that may produce periodic orbits. Unfortunately, there is no space for the treatment of these in detail here. We have to confine ourselves to an example that shows how a "*homoclinic bifurcation*" can produce periodic orbits. A general theory concerning this phenomenon can be found in Andronov-Leontovich-Gordon-Mayer [1967]. The following Example is due to Hale-Kocak [1991] p. 385.

Example 7.4.1 (*Hale-Kocak [1991]*). Consider the two dimensional system

$$\dot{x}_1 = 2x_2, \qquad \dot{x}_2 = 2x_1 - 3x_1^2 - x_2(x_1^2 - x_1^2 + x_2^2 - \mu). \quad (7.4.41)$$

For arbitrary $\mu \in \mathbf{R}$ the system has two equilibria, $(0,0)$ and $(2/3, 0)$. The first one is a saddle, the second one is asymptotically stable for $\mu < -4/27$ and it is unstable for $\mu > -4/27$, as this can easily be seen by linearization. The eigenvalues at $(2/3, 0)$ are

$$\lambda_{1,2}(\mu) = (1/2)(4/27 + \mu \pm ((\mu + 112/27)(\mu - 104/27))^{1/2}).$$

For $\mu \in (-112/27, 104/27)$ they are complex conjugate, and $\operatorname{Re} \lambda_{1,2}(-4/27) = 0$, $d \operatorname{Re} \lambda_1(-4/27)/d\mu = 1/2 > 0$. Thus, the equilibrium $(2/3, 0)$ undergoes an Andronov-Hopf bifurcation at $\mu = -4/27$. Denote the bifurcating periodic solutions for small $|\mu|$ by $(p_1(t, \mu), p_2(t, \mu))$ and their period by $T(\mu) > 0$. Clearly,

$$\lim_{\mu \to -4/27} T(\mu) = 2\pi / \operatorname{Im} \lambda_1(-4/27) = \pi.$$

Now, denoting the vector field on the right-hand side of (7.4.41) by $f(x, \mu)$ we have

$$\operatorname{div} f(x, \mu) = -x_1^3 + x_1^2 - 3x_2^2 + \mu. \tag{7.4.42}$$

But differentiating the "Liapunov function"

$$V(x_1, x_2) := x_1^3 - x_1^2 + x_2^2$$

with respect to the system we obtain

$$\dot{V}_{(7.4.41)}(x_1, x_2, \mu) = -2x_2^2(x_1^3 - x_1^2 + x_2^2 - \mu).$$

V is *not* a first integral but its level curve

$$x_1^3 - x_1^2 + x_2^2 = \mu, \tag{7.4.43}$$

clearly, is an orbit of the system: $V(x_1, x_2) = \mu$ along this curve and $\dot{V}_{(7.4.41)} = 0$. It is easy to see that for $\mu \in (-4/27, 0)$, (7.4.43) is a smooth closed curve around the equilibrium point $(2/3, 0)$ with diameter tending to zero as $\mu \to -4/27 + 0$. Thus, the periodic solutions exist for $\mu > -4/27$, and because of the uniqueness part of the Andronov-Hopf Theorem (Lemma 7.2.5) the bifurcation is supercritical. Indeed, for the periodic solution (p_1, p_2) that satisfies (7.4.43) we have by (7.4.42)

$$\operatorname{div} f(p_1(t, \mu), p_2(t, \mu), \mu) = -2p_2^2(t, \mu);$$

thus, by (5.1.27)

$$\int_0^{T(\mu)} \operatorname{div} f(p_1(t, \mu), p_2(t, \mu), \mu) dt = -2 \int_0^{T(\mu)} p_2^2(t, \mu) dt < 0;$$

hence, the periodic solutions $(p_1(t, \mu), p_2(t, \mu))$ are orbitally asymptotically stable for $\mu \in (-4/27, 0)$. As $\mu \to -0$ the corresponding periodic orbits tend to the curve $x_1^3 - x_1^2 + x_2^2 = 0$, $x_1 \geq 0$. This is still an orbit of the system (7.4.41) with $\mu = 0$ but is no longer a smooth periodic orbit. It is the homoclinic orbit of the solution that tends to the origin as $t \to \pm\infty$. It can be seen that for $\mu = 0$ the trajectories that have initial values in the interior of this homoclinic loop tend to it as $t \to \infty$. (One has to take into consideration that $(2/3, 0)$ is a local minimum of the function V, and for $c \in (-4/27, 0)$ the level curves $V = c$ are crossed outwards by the trajectories of the system (with $\mu = 0$).) Thus, at $\mu = 0$ the periodic orbit of the system disappears, giving place to a homoclinic orbit, and for $\mu > 0$ even this homoclinic orbit disappears (see Figure 7.4.5). This is called a *homoclinic bifurcation*. Naturally, we may start with a positive value of μ and decrease it. As $\mu = 0$ is crossed, large amplitude and large period periodic solutions appear through the homoclinic bifurcation.

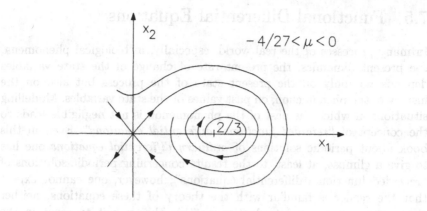

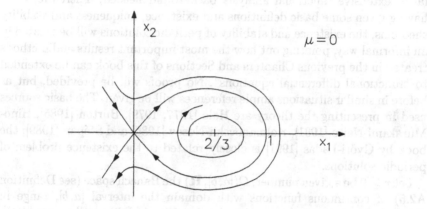

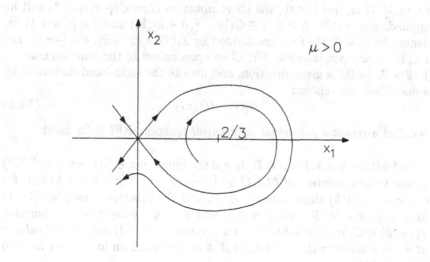

FIGURE 7.4.5. Homoclinic bifurcation in system (7.4.41) at $\mu = 0$.

7.5   Functional Differential Equations

In many processes of the real world, especially, in biological phenomena, the present dynamics, the present rate of change of the state variables depends not only on the present state of the process but also on the history of the phenomenon, on past values of the state variables. Modelling situations in which the past of the phenomenon is not negligible leads to the concept of *"retarded functional differential equations"*. Even in this book about periodic solutions of *ordinary differential equations* one has to give a glimpse, at least, to the results concerning periodic solutions of "retarded functional differential equations"; however, one cannot expect that the reader is familiar with the theory of these equations, neither can one give here an introduction to this theory, not to mention the fairly extensive functional analysis background needed. Therefore, after having given some basic definitions and existence, uniqueness and stability theorems, the existence and stability of periodic solutions will be treated in an informal way, pointing out how the most important results and methods treated in the previous Chapters and Sections of this book can be extended to "functional differential equations". No proofs will be provided, but as before in similar situations ample references will be given. The basic sources used in presenting the theory are Hale [1977, 1979], Burton [1985], Hino-Murakami-Naito [1991], Kolmanovskii-Nosov [1986] and Stépán [1989]; the book by Györi-Ladas [1991] is closely related to the existence problem of periodic solutions.

Let $r \geq 0$ be a given number, $C([a, b], \mathbf{R})$ the Banach space (see Definition A2.5) of continuous functions with domain the interval $[a, b]$, range in $\mathbf{R}^n$ and the supremum norm, i.e. for $\phi \in C([a, b], \mathbf{R}^n) : \|\phi\| := \sup |\phi(t)|$, $t \in [a, b]$. If $[a, b] = [-r, 0]$, the short notation $C := C([-r, 0], \mathbf{R}^n)$ will be applied. For $\sigma \in \mathbf{R}$, $h \geq 0$, $x \in C([\sigma - r, \sigma + h], \mathbf{R})$ and $t \in [\sigma, \sigma + h]$ we denote by $x_t \in C$ the function defined by $x_t(\theta) = x(t + \theta)$, $\theta \in [-r, 0]$. Let $I \subset \mathbf{R}$ be an open interval, $X \subset C$ an open subset of the function class C, $f : I \times X \to \mathbf{R}^n$ a given function, and denote the right-hand derivative by a dot. Then the relation

$$\dot{x}(t) = f(t, x_t) \tag{7.5.1}$$

is called a *retarded functional differential equation*, RFDE for short.

Definition 7.5.1. For $\sigma \in \mathbf{R}$, $h \geq 0$ the function $x \in C([\sigma - r, \sigma + h), \mathbf{R}^n)$ is said to be a *solution of (7.5.1) on* $[\sigma - r, \sigma + h)$ if $[\sigma - r, \sigma + h) \subset I$, for $t \in [\sigma - r, \sigma + h)$ there holds $x_t \in X$, and the function x satisfies (7.5.1) for $t \in [\sigma, \sigma + h)$. For given $\sigma \in I$ and $\phi \in X$ we say that the function $x(\cdot, \sigma, \phi) \in C([\sigma - r, \sigma + h), \mathbf{R}^n)$ is *a solution of (7.5.1) with initial value ϕ at σ* (or a *solution through (σ, ϕ)*) if it is a solution on $[\sigma - r, \sigma + h)$, and $x_\sigma(\cdot, \sigma, \phi) = \phi$.

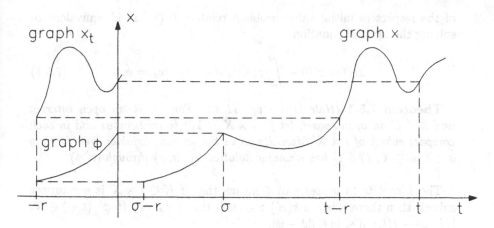

FIGURE 7.5.1. Graphs of a solution x, an initial function ϕ and of x_t related to the RFDE (7.5.1).

According to (7.5.1) the right-hand derivative of the solution x at t (which is an n-vector for fixed t) is determined by t and (through an operator defined on C) by the restriction of x to the interval $[t-r, t]$. Figure 7.5.1 shows the graphs of a solution x, of an "*initial function*" ϕ and of x_t.

The RFDE (7.5.1) includes ordinary differential equations: $r = 0$. In this case, x_t is defined on the interval consisting of the single point 0, and $x_t = x_t(0) = x(t)$. It includes "*differential-difference equations*" of the form

$$\dot{x}(t) = f(t, x(t), x(t-r_1), ..., x(t-r_m)) \qquad (7.5.2)$$

where $0 \le r_i \le r$, $(i = 1, ..., m)$. Here f is a function of $nm + 1$ real variables, and the operator f on the right-hand side of (7.5.1) acts in such a way that for a given t it selects the values assumed by the function x_t at $\theta = 0, -r_1, ..., -r_m$, and makes an n-vector correspond to this set of values. In this case we say that there are m *delays* or *time lags* in the equation each less than $r \ge 0$. The delays r_i may even depend on the time t. Equation (7.5.1) includes also *integro-differential equations* of the form, say,

$$\begin{aligned}
\dot{x}(t) &= \int_{-r}^{0} g(t, x(t), \theta, x_t(\theta)) d\theta \\
&= \int_{-r}^{0} g(t, x(t), \theta, x(t+\theta)) d\theta \qquad (7.5.3)
\end{aligned}$$

where g is a function of $2n + 2$ real variables (cf. the predator-prey model (7.3.2) which is a system with infinite delay $r = \infty$).

It can be shown easily that x_t is a continuous function of $t \in [\sigma, \sigma + h]$, i.e. $x. : [\sigma, \sigma + h] \to C$ is continuous, and that finding the solution $x(\cdot, \sigma, \phi)$

of the respective initial value problem related to (7.5.1) is equivalent to solving the integral equation

$$x(t) = \phi(0) + \int_\sigma^t f(s, x_s)ds, \qquad x_\sigma = \phi. \qquad (7.5.4)$$

Theorem 7.5.1 *(Hale [1977] pp. 41-42). For $I \subset \mathbf{R}$ an open interval and $X \subset C$ an open subset, let $f : I \times X \to \mathbf{R}^n$ be continuous and in each compact subset of $I \times X$ Lipschitzian in the second variable; then for any $\sigma \in I$, $\phi \in X$, (7.5.1) has a unique solution $x(\cdot, \sigma, \phi)$ through (σ, ϕ).*

The *Lipschitzian property* of f means that if $K \subset I \times X$ is a compact subset, then there exists a $k(K) > 0$ such that for every $(t, \phi), (t, \psi) \in K$, $|f(t, \phi) - f(t, \psi)| \le k(K)|\phi - \psi|$.

Theorems about the continuous dependence of the solutions on initial values and on the right-hand side and about the continuation of solutions hold and are more or less analogous to the corresponding theorems for ordinary differential equation. In some cases one has to assume that f is completely continuous. For details, see also Hale [1977].

In order to state a theorem about the differentiability of the solution with respect to the initial function one has to introduce the space $C^1(I \times X, \mathbf{R}^n)$ of functions that have bounded continuous derivatives with respect to $\phi \in X$. Here $I \subset \mathbf{R}$ and $X \subset C$ are as before and "derivative" is meant to be the Fréchet derivative (see Definition A2.11). $C^1(I \times X, \mathbf{R}^n)$ is a Banach space if the norm of its element is chosen to be the sum of the suprema of the function and its derivative.

Theorem 7.5.2 *(Hale [1977] p.46). If $f \in C^1(I \times X, \mathbf{R}^n)$, then the solution $x(\cdot, \sigma, \phi)$ of (7.5.1) through (σ, ϕ) is unique and continuously differentiable with respect to $\phi \in X$; for each $t \ge \sigma$ the derivative of x with respect to ϕ is a linear operator*

$$D_\phi x(t, \sigma, \phi) : C \to \mathbf{R}^n,$$

$D_\phi x(\sigma, \sigma, \phi)$ is the identity, and for each $\psi \in C$ the function $y = D_\phi x(\cdot, \sigma, \phi)\psi$ satisfies the homogeneous linear RFDE

$$\dot{y}(t) = D_\phi f(t, x_t, (\cdot, \sigma, \phi))y_t \qquad (7.5.5)$$

where $D_\phi f$ is the Fréchet derivative of f with respect to its second variable.

System (7.5.5) is called the *variational system* of (7.5.1) with respect to the solution $x(\cdot, \sigma, \phi)$, and the last statement of the Theorem means that $y(t) = D_\phi x(t, \sigma, \phi)\psi$ is the solution of (7.5.5) that satisfies $y_\sigma = \psi$.

Turning now to stability problems it will be assumed that $f : \mathbf{R} \times C \to \mathbf{R}^n$ is completely continuous and it is sufficiently smooth ensuring that the solution $x(t, \sigma, \phi)$ of the RFDE

$$\dot{x}(t) = f(t, x_t) \tag{7.5.6}$$

through $\sigma \in \mathbf{R}$, $\phi \in C$ is unique and continuous in all its arguments in its domain of definition. It will be assumed further that $x = 0$ is a solution, i.e. $f(t, 0) \equiv 0$. We extend now the basic stability concepts to RFDEs.

Definition 7.5.2. The equilibrium solution $x = 0$ of (7.5.6) is said to be *stable in the Liapunov sense* if for any $\varepsilon > 0$ and $\sigma \in \mathbf{R}$ there is a $\delta(\varepsilon, \sigma) > 0$ such that if for $\phi \in C$ we have $||\phi|| < \delta(\varepsilon, \sigma)$, then $||x_t(\cdot, \sigma, \phi)|| < \varepsilon$ for $t \geq \sigma$. The solution $x = 0$ is said to be *uniformly stable* if it is stable in the Liapunov sense and δ does not depend on $\sigma \in \mathbf{R}$. The solution $x = 0$ is *asymptotically stable* if it is stable in the Liapunov sense, and for any $\sigma \in \mathbf{R}$ there is a $\delta_1(\sigma)$ such that $||\phi|| < \delta_1(\sigma)$ implies $x(t, \sigma, \phi) \to 0$ as $t \to \infty$. The solution $x = 0$ is *uniformly asymptotically stable* if it is uniformly stable, and there is $\delta_1 > 0$ such that for every $\varepsilon_1 > 0$ there is a $t_1(\varepsilon_1) > 0$ such that $||\phi|| < \delta_1$ implies $||x_t(\cdot, \sigma, \phi)|| < \varepsilon_1$ for $t \geq \sigma + t_1(\varepsilon_1)$ for every $\sigma \in \mathbf{R}$.

Definition 7.5.3. A solution $x(\cdot, \sigma, \phi)$ of (7.5.6) is said to be *bounded* if there is an $m(\sigma, \phi) > 0$ such that for $t \geq \sigma - r$ there holds $|x(t, \sigma, \phi)| < m(\sigma, \phi)$. The solutions of (7.5.6) are *uniformly bounded* if for any $\delta > 0$ there is an $m(\delta) > 0$ such that for all $\sigma \in \mathbf{R}$, $\phi \in C$, $||\phi|| < \delta$ we have $|x(t, \sigma, \phi)| < m(\delta)$ for $t \geq \sigma$. The solutions are *ultimately bounded* if there is an $m > 0$ such that for any $\sigma \in \mathbf{R}$, $\phi \in C$ there is a $t_1(\sigma, \phi)$ such that $|x(t, \delta, \phi)| < m$ for $t \geq \sigma + t_1(\sigma, \phi)$. The solutions are *uniformly ultimately bounded* if there is an $m > 0$ such that for any $\delta > 0$ there is a $t_1(\delta) > 0$ such that for any $\sigma \in \mathbf{R}$, $\phi \in C$, $||\phi|| < \delta$ we have $|x(t, \sigma, \phi)| < m$ for $t \geq \sigma + t_1(\delta)$.

One can prove (see Hale [1977] p.104) that if (7.5.6) is *autonomous* or *periodic*, i.e. either $f(t, \phi) = f(\phi)$ or there is a $T > 0$ such that $f(t + T, \phi) = f(t, \phi)$ for all $t \in \mathbf{R}$, $\phi \in C$, then Liapunov stability implies uniform stability and asymptotic stability implies uniform asymptotic stability.

Liapunov's direct method (see Section 1.5) can be generalized to RFDEs, but one has to consider "*Liapunov functionals*" instead of Liapunov functions.

Definition 7.5.4. Let $V : \mathbf{R} \times C \to \mathbf{R}$ be a continuous functional and $(\sigma, \phi) \in \mathbf{R} \times C$ arbitrarily given; *the upper right-hand derivative of V with respect to system (7.5.6) at (σ, ϕ)* is by definition

$$\dot{V}_{(7.5.6)}(\sigma, \phi) := \lim_{h \to +0} \sup(V(\delta + h, x_{\sigma+h}(\cdot, \sigma, \phi)) - V(\sigma, \phi))/h.$$

We shall quote a typical theorem from Hale [1977] p.105 which is the analogue of Liapunov's Second Theorem 1.5.2. The function class $\mathcal{H}$

consists of functions $h : \mathbf{R}_+ \to \mathbf{R}_+$ that are strictly increasing, continuous and satisfy $h(0) = 0$ (see Definition 1.5.2).

Theorem 7.5.3. *Suppose that for an arbitrary bounded subset $D \subset C$ the mapping f maps $\mathbf{R} \times D$ into a bounded set of $\mathbf{R}^n$; if there exist a continuous functional $V : \mathbf{R} \times C \to \mathbf{R}$ and $h_1, h_2, h_3 \in \mathcal{H}$ such that for arbitrary $t \in \mathbf{R}$ and $\phi \in C$, the inequalities*

$$h_1(|\phi(0)|) \le V(t, \phi) \le h_2(\|\phi\|),$$

$$\dot{V}_{(7.5.6)}(t, \phi) \le -h_3(|\phi(0)|)$$

hold, then the solution $x = 0$ of (7.5.6) is uniformly asymptotically stable; if $h_1(s) \to \infty$ as $s \to \infty$, then the solutions are uniformly bounded.

The method of Liapunov functionals will be illustrated by

Example 7.5.1 (Hale [1977]). Consider the linear RFDE

$$\dot{x}(t) = Ax(t) + Bx(t - r) \tag{7.5.7}$$

where $r > 0$, A is an $n \times n$ stable matrix and B is an $n \times n$ matrix. Here the right-hand side $f(x_t) := Ax(t) + Bx(t - r)$ is linear and does not depend on t; the function $f : C \to \mathbf{R}^n$ is defined in the following way: for arbitrary $\phi \in C$ one has $f(\phi) := A\phi(0) + B\phi(-r)$. Let D be an arbitrary (symmetric) positive definite matrix; by Theorem 1.5.6 there is a (unique symmetric) positive definite matrix G such that

$$A'G + GA = -D. \tag{7.5.8}$$

Let H be a positive definite matrix, and the functional $V : C \to \mathbf{R}$ be defined by

$$V(\phi) := \phi'(0)G\phi(0) + \int_{-r}^{0} \phi'(\theta)H\phi(\theta)d\theta, \quad \text{for} \quad \phi \in C.$$

This is, clearly, positive definite and satisfies the first inequality in Theorem 7.5.3 with $h_1(s) := \underline{\lambda}_G s^2$, $h_2(s) := (\overline{\lambda}_G + r\overline{\lambda}_H)s^2$ where $\underline{\lambda}_G, \overline{\lambda}_G, \overline{\lambda}_H$ are the smallest, the largest eigenvalues of G and the largest eigenvalue of H, respectively (cf. (2.4.10)). The derivative of V with respect to (7.5.7) is

$$
\begin{aligned}
\dot{V}_{(7.5.7)}(\phi) &= \lim_{h \to +0} (1/h)\big(x'_{\sigma+h}(0, \sigma, \phi)Gx_{\sigma+h}(0, \sigma, \phi) \\
&\quad -\phi'(0)G\phi(0) + \int_{-r}^{0} (x'_{\sigma+h}(\theta, \sigma, \phi)Hx_{\sigma+h}(\theta, \sigma, \phi) \\
&\quad -\phi'(\theta)H\phi(\theta))d\theta\big) \\
&= \lim_{h \to +0} (1/h)((x'_{\sigma+h}(0, \sigma, \phi) - \phi'(0))Gx_{\sigma+h}(0, \sigma, \phi)
\end{aligned}
$$

$$+\phi'(0)G(x_{\sigma+h}(0,\sigma,\phi)-\phi(0)))$$

$$+\lim_{h\to+0}(1/h)\int_{-r}^{0}((x'_{\sigma+h}(\theta,\sigma,\phi)-\phi'(\theta))Hx_{\sigma+h}(\theta,\sigma,\phi)$$

$$+\phi'(\theta)H(x_{\sigma+h}(\theta,\sigma,\phi)-\phi(\theta)))d\theta$$

$$=\dot{x}_{\sigma}(0,\sigma,\phi)Gx_{\sigma}(0,\sigma,\phi)+\phi'(0)G\dot{x}(0,\sigma,\phi)$$

$$+\int_{-r}^{0}(\dot{x}'(\sigma+\theta,\sigma,\phi)Hx_{\sigma}(\theta,\sigma,\phi)+\phi'(0)H\dot{x}(\sigma+\theta,\sigma,\phi))d\theta$$

$$=(A\phi(0)+B\phi(-r))'G\phi(0)+\phi'(0)G(A\phi(0)+B\phi(-r))$$

$$+\int_{-r}^{0}(\phi'(\theta)H\phi(\theta))\cdot d\theta$$

where $x(\cdot,\sigma,\phi)$ denotes the solution of (7.5.7) through $\sigma\in\mathbf{R}$ (arbitrary) and $\phi\in C$, and the identities $x_{\sigma+h}(\theta,\sigma,\phi)=x(\sigma+h+\theta,\sigma,\phi)$, $x_{\sigma}(\theta,\sigma,\phi)=\phi(\theta)$ have been applied. Rearranging the last expression and performing the integration we obtain

$$\dot{V}_{(7.5.7)}(\phi)=-\phi'(0)D\phi(0)+2\phi'(0)GB\phi(-r)$$
$$+\phi'(0)H\phi(0)-\phi'(-r)H\phi(-r).\qquad(7.5.9)$$

Now, this derivative can be considered as a quadratic form

$$\dot{V}_{(7.5.7)}(\phi)=-(\phi_1(0),...,\phi_n(0),\phi_1(-r),...,\phi_n(-r))M\operatorname{col}(\phi_1(0),...,\phi_n(-r))$$

in the $2n$ variables $(\phi_1(0),...,\phi_n(0),\phi_1(-r),...,\phi_n(-r))$ where the $2n\times 2n$ symmetric coefficient matrix is

$$M=\begin{pmatrix}D-H & -(B'G+GB)/2\\ -(B'G+GB)/2 & H\end{pmatrix}.$$

Suppose that the positive definite matrices D and H have been chosen in such a way that $D-H$ is also positive definite. Then for $B=0$, obviously, M is also positive definite, and as a consequence, for sufficiently small B the derivative $\dot{V}_{(7.5.7)}$ is negative definite. Thus, if the influence of the past is not too strong, we may expect asymptotic stability. Indeed, if $\underline{\lambda}$ and λ_H denote the least (positive) eigenvalues of the matrices $D-H$ and H, respectively, then

$$x'(D-H)x\geq\underline{\lambda}|x|^2,\qquad x'Hx\geq\lambda_H|x|^2,\qquad x\in\mathbf{R}^n,$$

and from (7.5.9)

$$\dot{V}_{(7.5.7)}(\phi)\leq -\underline{\lambda}|\phi(0)|^2+2|GB|\cdot|\phi(0)|\cdot|\phi(-r)|-\lambda_H|\phi(-r)|^2$$
$$\leq -\underline{\lambda}|\phi(0)|^2+(\underline{\lambda}\lambda_H)^{1/2}|\phi(0)|\cdot|\phi(-r)|-\lambda_H|\phi(-r)|^2$$

provided that

$$|GB|\leq(\underline{\lambda}\lambda_H/4)^{1/2}.\qquad(7.5.10)$$

Continuing the estimate

$$\dot{V}_{(7.5.7)}(\phi) \leq -(\lambda/2)|\phi(0)|^2 - (\lambda_H/2)|\phi(-r)|^2$$
$$-((\lambda/2)^{1/2}|\phi(0)| - (\lambda_H/2)^{1/2}|\phi(-r)|)^2 \leq -(\lambda/2)|\phi(0)|^2,$$

i.e. the second inequality of Theorem 7.5.3 holds with $h_3(s) := -(\lambda/2)s^2$. Thus, the solution $x = 0$ of (7.5.7) is uniformly asymptotically stable if B satisfies the inequality (7.5.10). Note that the stability of the matrix A and (7.5.10) ensure uniform asymptotic stability for an arbitrary (large) delay $r > 0$.

The stability concepts and criteria can be generalized easily to arbitrary solutions (cf. to what has been said around (1.4.2)). If ξ is a solution of (7.5.6) defined for $t \geq \sigma$, then $y := x - \xi$ satisfies the system

$$\dot{y}(t) = f(t, \xi_t + y_t) - f(t, \xi_t) \tag{7.5.11}$$

provided that x satisfies (7.5.6). By this transformation the stability, asymptotic stability, etc., of ξ is defined as the stability, asymptotic stability, etc., of the corresponding solution $y = 0$ of (7.5.11).

The theory of linear RFDEs has strong analogies with the corresponding theory of ordinary differential equations. Let $L : \mathbf{R} \times C \to \mathbf{R}^n$ be continuous and linear in $\phi \in C$. By the Riesz Representation Theorem (see e.g. Kolmogorov-Fomin [1972] p.347) there exists a matrix function $\eta : \mathbf{R} \times [-r, 0] \to \mathbf{R}^{n^2}$ of bounded variation in $\theta \in [-r, 0]$ such that for $t \in \mathbf{R}$, $\phi \in C$

$$L(t, \phi) = \int_{-r}^{0} [D_\theta \eta(t, \theta)]\phi(\theta) \tag{7.5.12}$$

where D_θ is the derivative with respect to the second variable and the integral is a Riemann-Stieltjes one. We assume also that there exists a locally integrable positive function $m : \mathbf{R} \to \mathbf{R}_+$ such that for all $t \in \mathbf{R}$, $\phi \in C$ there holds

$$|L(t, \phi)| \leq m(t)\|\phi\|. \tag{7.5.13}$$

The relation

$$\dot{x}(t) = L(t, x_t), \qquad t \geq \sigma, \qquad x_\sigma = \phi \in C \tag{7.5.14}$$

is a *homogeneous linear RFDE* with initial condition for some $(\sigma, \phi) \in \mathbf{R} \times C$.

The most common type of such systems occurring in the applications is

$$\dot{x}(t) = \sum_{k=1}^{q} A_k x(t - r_k) + \int_{-r}^{0} A(t, \theta)x(t + \theta)d\theta \tag{7.5.15}$$

where $0 \leq r_k \leq r$, A_k is a constant matrix for $k = 1, 2, ..., q$, the matrix $A(t, \theta)$ is integrable in θ and there is a locally integrable function $m : \mathbf{R} \to \mathbf{R}_+$ such that for all $t \in \mathbf{R}$, $\phi \in C$

$$\left| \int_{-r}^0 A(t + \theta)\phi(\theta)d\theta \right| \leq m(t)\|\phi\|.$$

It can be proved that under the above conditions the initial value problem (7.5.14) has a unique solution $x(\cdot, \sigma, \phi)$ that is defined on $[\sigma - r, \infty]$ and satisfies the RFDE on $[\sigma, \infty]$.

As in the case of homogeneous linear ordinary differential equations, (7.5.14) can be considered also as a matrix differential equation (cf. (1.2.5)). In this case, by an abuse of notation, we consider C the space of continuous matrix functions defined on the interval $t \in [-r, 0]$. The matrix function $\Phi : \mathbf{R} \times \mathbf{R} \to \mathbf{R}^{n^2}$ that satisfies

$$\partial\Phi(t, s)/\partial t = L(t, \Phi_t(\cdot, s))$$

for $t \geq s$ almost everywhere in $s \in \mathbf{R}$ and $t \in \mathbf{R}$ and

$$\Phi(t, s) = \begin{cases} O, & s - r \leq t < s \\ I, & t = s \end{cases}$$

where O and I is the null matrix and the unit matrix of order n, respectively, and

$$\Phi_t(\theta, s) = \Phi(t + \theta, s) \quad \text{for} \quad \theta \in [-r, 0]$$

is called the *fundamental matrix* of (7.5.14).

The space of solutions of (7.5.14) is, in general, not of finite dimension; thus, one may not expect that the fundamental matrix contains *all* the informations about the solutions. In fact, denote the special initial function that is zero over $[-r, 0)$ and equal to $c \in \mathbf{R}^n$ at $t = 0$ by ϕ_c, i.e.

$$\phi_c(t) := \begin{cases} 0, & t \in [-r, 0) \\ c, & t = 0 . \end{cases} \tag{7.5.16}$$

Then, clearly, for arbitrary $\sigma \in \mathbf{R}$ the solution of (7.5.14) through (σ, ϕ_c) is

$$x(t, \sigma, \phi_c) = \Phi(t, \sigma)c.$$

Thus, the fundamental matrix provides us only with the solutions corresponding to very special initial functions. Nevertheless, it has an important role in dealing with the inhomogeneous linear RFDE

$$\dot{z}(t) = L(t, z_t) + b(t) \tag{7.5.17}$$

where $b : \mathbf{R} \to \mathbf{R}^n$ is locally integrable. The solution of (7.5.17) through $(\sigma, \phi) \in \mathbf{R} \times C$ is denoted by $z(\cdot, \sigma, \phi)$. Its existence and uniqueness over

$[\sigma - r, \infty)$ can be proved. There holds the *"variation of constants" formula* (Hale [1977] p.143)

$$z(t, \sigma, \phi) = x(t, \sigma, \phi) + \int_\sigma^t \Phi(t, s)b(s)ds, \quad t \ge \sigma \qquad (7.5.18)$$

which is in complete analogy with (1.2.12).

If the homogeneous linear RFDE (7.5.14) is autonomous, i.e. L does not depend on t, we have the RFDE

$$\dot{x}(t) = L(x_t) \qquad (7.5.19)$$

where L satisfies the conditions above. In this case one can prove easily that the fundamental matrix is of the form

$$\Phi(t - s) := \Phi(t - s, 0) = \Phi(t, s),$$

and the variation of constants formula is

$$z(t, \sigma, \phi) = x(t - \sigma, 0, \phi) + \int_\sigma^t \Phi(t - s)b(s)ds, \quad t \ge \sigma, \qquad (7.5.20)$$

for the solution $z(\cdot, \sigma, \phi)$ through $(\sigma, \phi) \in \mathbf{R} \times C$ of the inhomogeneous linear RFDE

$$\dot{z}(t) = L(z_t) + b(t). \qquad (7.5.21)$$

The stability problems of the autonomous linear RFDE (7.5.19) and also of (7.5.21) can be reduced to an "eigenvalue problem". One arrives at the characteristic equation of (7.5.19) by assuming $x(t) = s\exp(\lambda t)$ as a solution for some $\lambda \in \mathbf{C}$, $s \in \mathbf{C}^n$ and substituting this function into the equation. Assuming that by (7.5.12)

$$L(\phi) = \int_{-r}^0 [D_\theta \eta(\theta)]\phi(\theta)$$

where $\eta : [-r, 0] \to \mathbf{R}^{n^2}$ is a matrix function of bounded variation, we obtain

$$s\lambda \exp(\lambda t) = \int_{-r}^0 [D_\theta \eta(\theta)]s \exp(\lambda(t + \theta)),$$

or, dividing by $\exp(\lambda t)$,

$$\left(\lambda I - \int_{-r}^0 [D_\theta \eta(\theta)] \exp(\lambda\theta) \right) s = 0.$$

The condition of solvability of the last equation for (the "eigenvector") s is the *characteristic equation* of (7.5.19):

$$\det\left(\lambda I - \int_{-r}^0 [D\eta(\theta)] \exp(\lambda\theta) \right) = 0; \qquad (7.5.22)$$

its solutions are the *eigenvalues*.

In the important special case of (7.5.15) when the latter is autonomous, i.e. $A(t, \theta) \equiv A(\theta)$, the characteristic equation assumes the form

$$\det\left(\lambda I - \sum_{k=1}^{a} A_k \exp(-\lambda r_k) + \int_{-r}^{0} A(\theta) \exp(\lambda\theta) d\theta\right) = 0.$$

Clearly, the characteristic equation for RFDEs is no longer a polynomial equation, and as a consequence, generically, the set of eigenvalues (the *spectrum*) is no longer a finite set. Since the position of the eigenvalues in the complex plane is decisive for the stability of an autonomous linear RFDE and also of quasi-linear RFDEs, it has a vast literature. We draw the attention here only to the powerful conditions in Stépán [1989] and the references therein.

We illustrate the relation between the stability of the zero solution and the location of the spectrum by quoting just one theorem for quasi-linear systems. Let $X \subset C$ be a neighbourhood of the zero function in C, $f \in C^1(\mathbf{R} \times X, \mathbf{R}^n)$ (see the paragraph preceding Theorem 7.5.2) and consider the system

$$\dot{x}(t) = L(x_t) + f(t, x_t) \tag{7.5.23}$$

where also $f(t, 0) \equiv 0$, $\partial f(t, 0)/\partial\phi \equiv 0$.

Theorem 7.5.4 *(Hale [1977] p.213). If all the eigenvalues of the linear RFDE (7.5.19) have negative real parts, then the zero solution of (7.5.23) is uniformly asymptotically stable.*

Up to this point we tried to give as short a survey of basic concepts and facts concerning RFDEs as possible in order to be able to present the most important results about the existence and stability of *periodic solutions* in a similarly concise way in the rest of the Section. Besides giving some theorems without proof we shall, mainly, point out the analogies with the corresponding theory of ODEs without going into detail.

First, linear periodic systems are considered, i.e. it is assumed that $L : \mathbf{R} \times C \to \mathbf{R}^n$ satisfies the condition preceding (7.5.14) and there is a $T > 0$ such that for all $(t, \phi) \in \mathbf{R} \times C$ there holds $L(t+T, \phi) = L(t, \phi)$. There is a "Floquet theory" for the *periodic linear system*

$$\dot{x}(t) = L(t, x_t). \tag{7.5.24}$$

For arbitrary $(s, \phi) \in \mathbf{R} \times C$, (7.5.24) has a unique solution $x(\cdot, s, \phi)$ through (s, ϕ) defined on $[s, \infty)$, and $x_t(\cdot, s, \phi)$ is continuous in t, s and ϕ. A continuous linear operator $U(t, s) : C \to C$ can be defined by

$$x_t(\cdot, s, \phi) = U(t, s)\phi$$

for all $t \in \mathbf{R}$, $s \in \mathbf{R}$, $\phi \in C$. The operator U satisfies $U(t,s)U(s,q) = U(t,q)$ for $t \geq s \geq q$ and because of the periodicity of (7.5.24)

$$U(t+T,s) = U(t,s)U(s+T,s), \qquad t \geq s.$$

While the operator $U(t,s)$ plays the role of the fundamental matrix, the role of the principal matrix (cf. (2.2.5)) is played by $U_T := U(T,0)$. Denote the spectrum of the operator U_T by $\sigma(U_T)$. It can be proved that $\sigma(U_T)$ is at most countable and it is a compact subset of the complex plane with the only possible cluster point being zero. If $\mu \neq 0$, $\mu \in \sigma(U_T)$, then there is a $\phi \in C$ such that $U_T\phi = \mu\phi$. The non-zero elements of $\sigma(U_T)$ are called the *characteristic multipliers* of system (7.5.24). If μ is a characteristic multiplier and for some $\lambda \in \mathbf{C}$ there holds $\mu = \exp(\lambda T)$, then λ is called a *characteristic exponent* of (7.5.24).

If μ is a characteristic multiplier, then there are two closed subspaces of C: the *eigensubspace* E_μ of μ and its "complement" K_μ such that E_μ is finite dimensional, $E_\mu \oplus K_\mu = C$ where $\oplus$ denotes the direct sum, both E_μ and K_μ are invariant with respect to the mapping U_T, and the spectrum of the restriction of U_T to E_μ consists of the single element μ, whereas the spectrum of the restriction of U_T to K_μ is $\sigma(U_T)\backslash\{\mu\}$. The dimension of E_μ is called the *multiplicity* of the characteristic multiplier μ. The solutions $x(\cdot,0,\phi)$ of (7.5.24) have a Floquet representation (cf. Theorem 2.2.5) in E_μ, i.e. for $\phi \in E_\mu$, which we are not giving here in detail. In particular this implies

Theorem 7.5.5. $\mu = \exp(\lambda T)$ *is a characteristic multiplier if and only if (7.5.24) has a non-zero solution* $x(t) = p(t)\exp(\lambda t)$ *where* $p(t+T) = p(t)$.

There holds, clearly, the

Corollary 7.5.6. *Equation (7.5.24) has a non-zero T-periodic solution if and only if it has 1 as a characteristic multiplier.*

Although the restriction of (7.5.24) to the eigensubspace of each characteristic multiplier has a "Floquet representation", this is not true for (7.5.24) as a whole in the infinite dimensional space where it operates. For a counterexample, see Hale [1977] p.197.

For the existence of a T-periodic solution of the inhomogeneous linear RFDE

$$\dot{z}(t) = L(t, z_t) + b(t) \tag{7.5.25}$$

where L is the same as in (7.5.24) and b is locally integrable and T-periodic: $b(t+T) = b(t)$. A similar alternative holds like that expressed by Theorem 2.3.1 and Corollary 2.3.2 for ODEs. If (7.5.24) has no T-periodic solution apart from the trivial one, then (7.5.25) has a unique T-periodic solution. We note that by a theorem due to Halanay [1966] the same condition ensures the existence of a T-periodic solution even if $b(t)$ is replaced by a non-linear perturbation term $f(t, x_t)$ where $f(t + T, \phi) = f(t, \phi)$ for all $t \in \mathbf{R}$, $\phi \in C$ and the C^1-norm of f is sufficiently small (see also Hale [1977] p.211). For the stability of the trivial solution of (7.5.24), resp. of the unique periodic solution of (7.5.25) we have, similarly to the ODE case,

Theorem 7.5.7 *The solution $x = 0$ of (7.5.24) is uniformly asymptotically stable if and only if all the characteristic multipliers are in modulus less than 1; the same condition is also necessary and sufficient for the uniform asymptotic stability of the unique T-periodic solution of (7.5.25).*

The critical cases in which the homogeneous system (7.5.24) *has* non-trivial T-periodic solutions can also be treated and conditions can be given for the existence of periodic solutions of the inhomogeneous RFDE (7.5.25) or of the quasilinear one where $b(t)$ is replaced by $f(t, x_t)$ as described above. A recent result in this direction is Hatvani-Krisztin [1992].

We turn now to the study of periodic solutions of the non-linear autonomous RFDE

$$\dot{x}(t) = f(x_t) \tag{7.5.26}$$

where $f : C \to \mathbf{R}^n$ is continuous along with its first and second Fréchet derivative. The first Fréchet derivative of f at $x_t \in C$ is a continuous linear operator mapping C into $\mathbf{R}^n$ and it is denoted by

$$L(t, \cdot) := Df(x_t),$$

i.e. for $\phi \in C$ we have $L(t, \phi) = Df(x_t)\phi$.

Suppose that (7.5.26) admits a non-constant periodic solution $p : \mathbf{R} \to \mathbf{R}^n$ with period $T > 0 : p(t + T) = p(t)$. The *trajectory* (*path*, *orbit*) of p in C is the set $\Gamma := \{p_t \in C : t \in \mathbf{R}\} \subset C$. This is, obviously, a closed curve in C since $p_t = p_{t+T}$. For a non-constant periodic solution p the derivative $\dot{p}$ is nowhere zero. This implies that also $\dot{p}_t \neq 0$, $t \in \mathbf{R}$. The *variational system* of (7.5.26) with respect to the periodic solution p is the T-periodic linear RFDE

$$\dot{y}(t) = L(t, y_t) \equiv Df(p_t)y_t. \tag{7.5.27}$$

From the identity $\dot{p}(t) \equiv f(p_t)$ follows by differentiation that $\ddot{p}(t) \equiv Df(p_t)\dot{p}_t$, i.e. $\dot{p}$ is a non-trivial T-periodic solution of (7.5.27); hence, the number 1 is a characteristic multiplier of (7.5.27).

The problem of isolation of a periodic solution can be treated analogously to how this was done for ODEs in Section 5.2. Similar to Theorem 5.2.3 the condition that 1 be a simple characteristic multiplier of the variational system (7.5.27) ensures the isolatedness of the periodic solution p.

There holds also the analogue of the Andronov-Witt Theorem 5.1.2:

Theorem 7.5.8 *(Stokes [1969]). If 1 is a simple characteristic multiplier of the variational system (7.5.27) and all the other characteristic multipliers are in modulus less than 1, then the periodic solution p is "exponentially asymptotically orbitally stable with asymptotic phase", i.e. its path Γ has a neighbourhood $V \subset C$ such that there are constants $K > 0$, $\delta > 0$, and for $\phi \in V$ there is an $\alpha(\phi) \in [0, T]$ such that*

$$|x(t, 0, \phi) - p(t + \alpha(\phi))| \leq K \|\phi - P_{\alpha(\phi)}\| \exp(-\delta t), \qquad t \geq 0.$$

Establishing the existence of periodic solutions for strongly non-linear periodic or autonomous differential RFDEs presents problems similar to those in the ODE case. The basic methods here are again the fixed point theorems on the one hand and perturbation techniques on the other. Besides these there are certain RFDEs that behave like two dimensional systems for which the Poincaré-Bendixson theory can be applied. For these R.A.Smith [1980$_b$] has worked out a method of proving the existence of periodic solutions (cf. Theorems 5.4.1 and 5.4.3). With reference to the perturbation techniques we note only that the theory of controllably periodic perturbations (see Sections 6.2-6.4) has been extended to RFDEs by Garay [1979$_a$, 1979$_b$].

Schauder's Fixed Point Theorem A2.16 and some other more sophisticated fixed point theorems have been applied by Burton [1985] in order to obtain existence results. We quote a theorem from p.249 of the latter reference concerning the RFDE

$$\dot{x}(t) = f(t, x_t) \tag{7.5.28}$$

where f is continuous on $\mathbf{R} \times C$, takes bounded sets into bounded sets and is periodic in t with period $T > 0$.

Theorem 7.5.9 *(Yoshizawa [1966]). If solutions of (7.5.28) are uniformly bounded and uniformly ultimately bounded, then the RFDE has a T-periodic solution.*

Compare this with Theorem 4.1.2.

The Andronov-Hopf Bifurcation Theorem 7.2.3 is generalized to RFDEs in a straightforward way. For the exact formulation, see Hale [1977, 1979]; and for a practical setting see Hassard-Kazarinoff-Wan [1981] and Stech [1985]. The one parameter family of RFDEs is assumed in the form

$$\dot{x}(t) = L(x_t, \mu) + f(x_t, \mu) \tag{7.5.29}$$

where for each $\mu \in \mathbf{R}$ the operator $L(\cdot, \mu)$ is continuous and linear; it has continuous derivatives with respect to $\phi \in C$ and $\mu \in \mathbf{R}$, f has also continuous derivatives with respect to $\phi \in C$ and $\mu \in \mathbf{R}$, $f(0, \mu) = 0$, $D_\phi f(0, \mu) = 0$, $\mu \in \mathbf{R}$. It is assumed that the spectrum of the RFDE $\dot{y}(t) = L(y_t, 0)$ consists of a pair of pure imaginary eigenvalues $\lambda_1 = i\beta(0) = \bar{\lambda}_2$ where $\beta(0) > 0$, and the rest of the spectrum lies in the open left half plane. As a consequence, for small $|\mu|$ the RFDE $\dot{y}(t) = L(y_t, \mu)$ has a pair of complex eigenvalues $\alpha(\mu) \pm i\beta(\mu)$ where $\alpha(0) = 0$ and the rest of the spectrum lies in the left half plane. It is assumed that $d\alpha(0)/d\mu > 0$. Under these conditions a theorem analogous to the Andronov-Hopf Theorem 7.2.3 can be proved. The main tool of the proof is the infinite dimensional Banach space version of the Centre Manifold Theorem A3.3 whose application reduces the infinite dimensional problem to the finite dimensional centre manifold. A recent treatment of centre manifolds in infinite dimensional spaces can be found in Vanderbauwhede-Iooss [1992].

RFDEs with infinite delay are important and need a special treatment. We say that the RFDE is of infinite delay, loosely speaking, if r, introduced preceding Definition 7.5.1, is infinity. The basic problem arising here is how to select the space of initial functions defined on $(-\infty, 0]$. We have no space here to treat this problem cf. Corduneanu-Lakshmikantham [1980], Burton [1985] and Hino-Murakami-Naito [1991]. Following Burton [1985] p. 279 et seq., the Volterra integro-differential equation

$$\dot{x}(t) = h(t, x) + \int_{-\infty}^{t} q(t, s, x(s))ds \qquad (7.5.30)$$

will be considered where $h \in C^0(\mathbf{R} \times \mathbf{R}^n, \mathbf{R}^n)$, $q \in C^0(\mathbf{R} \times \mathbf{R} \times \mathbf{R}^n, \mathbf{R}^n)$. We assume that there is a decreasing function $g \in C^1((-\infty, 0], (0, \infty))$ such that $g(s) \to \infty$ as $s \to -\infty$, $g(0) = 1$, and for any $\phi \in C^0((-\infty, 0], \mathbf{R}^n)$ which satisfies $|\phi(s)| \le \gamma g(s)$ for some $\gamma > 0$ and for $s \in (-\infty, 0]$, the function defined by $\int_{-\infty}^{0} q(t, s, \phi(s))ds$ is continuous for $t \in [0, \infty)$. With this g the Banach space of continuous functions $\phi : (-\infty, 0] \to \mathbf{R}^n$ for which

$$|\phi|_g := \sup_{t \in (-\infty, 0]} |\phi(t)/g(t)|$$

exists is denoted by X. For $\phi \in X$ a solution of (7.5.30) through $(0, \phi)$ is $x(t, 0, \phi)$ that satisfies (7.5.30) on $t \in [0, \delta)$ for some $\delta > 0$ and $x(t, 0, \phi) \equiv \phi(t)$ on $t \in (-\infty, 0]$. The *g-uniform boundedness* and *g-uniform ultimate boundedness* of solutions of (7.5.30) is defined similarly to Definition 7.5.3 with C replaced by X and $\|\phi\|$ replaced by $|\phi|_g$. For illustration we quote a theorem giving conditions for the existence of a periodic solution.

Theorem 7.5.10. *(Burton [1985] p.292). Suppose that (7.5.30) is periodic with period $T > 0$, i.e. $h(t + T, x) = h(t, x)$, $q(t + T, s + T, x) = q(t, s, x)$, for $\phi \in X$, the solution $x(t, 0, \phi)$ is defined on $[0, \infty)$ and is a continuous function of $\phi \in X$, the solutions are g-uniformly bounded and g-uniformly ultimately bounded with bound $m > 0$ at $\sigma = 0$, and for each*

$H > 0$ *there is an* $M > 0$ *such that* $\phi \in X$, $|\phi(t)| \leq m(g(t - H))^{1/2}$, $t \in [-\infty, 0]$ *implies* $|\dot{x}(t, 0, \phi)| \leq M$, $t \in [0, \infty)$. *Then (7.5.30) has a T-periodic solution.*

In proving this Theorem a fairly sophisticated fixed point theorem, Horn's Theorem, is applied.

Finally, we quote an existence theorem from Burton-Hatvani [1991] concerning the integro-differential equation

$$\dot{x}(t) = Ax(t) + \int_{-\infty}^{\infty} [D_s \eta(t, s)] g(x(t + s)) + f(t, x(t)) \qquad (7.5.31)$$

where A is a constant matrix, the matrix function $\eta : \mathbf{R} \times \mathbf{R} \to \mathbf{R}^{n^2}$ is measurable and continuous to the left (or to the right) in the second variable, the function $\int_{-\infty}^{\infty} |D_s \eta(t, s)|$ is continuous, $g \in C^0(\mathbf{R}^n, \mathbf{R}^n)$, $f \in C^0(\mathbf{R} \times \mathbf{R}^n, \mathbf{R}^n)$, $\eta(t + T, s) = \eta(t, s)$, $f(t + T, x) = f(t, x)$ for some period $T > 0$ and for all $t, s \in \mathbf{R}$, $x \in \mathbf{R}^n$. Alongside of (7.5.31) the one parameter family of systems

$$\dot{x}(t) = Ax(t) + \lambda\left(\int_{-\infty}^{\infty} [D_s \eta(t, s)] g(x(t + s)) + f(t, x(t))\right), \qquad 0 \leq \lambda \leq 1,$$
$$(7.5.32)$$

is considered. As can be seen from this preparation, (7.5.31) will be considered homotopic to $\dot{x} = Ax$, and indeed, in proving the following Theorem, the Leray-Schauder degree is applied (see Definition A2.9 and Theorem A2.15).

Theorem 7.5.11 (*Burton-Hatvani [1991]*). *Suppose that* $2k\pi i/T$ *for* $k = 0, \pm 1, \pm 2, \ldots$ *is not an eigenvalue of matrix* A, *and that there is a bounded open set* D *in the Banach space of continuous and T-periodic functions (from* $\mathbf{R}$ *to* $\mathbf{R}^n$*) such that* $0 \in D$ *and for every* $\lambda \in (0, 1)$ *every T-periodic solution* x_λ *of (7.5.32) satisfies* $x_\lambda \notin \partial D$. *Then (7.5.31) has a T-periodic solution in* D.

Actually, this is only a special case of a more general theorem in the quoted paper.

In order to see how periodic solutions of RFDEs occur in the applications besides the predator prey model treated in Section 7.3 and the several examples in texts already quoted in the present Section, the attention is drawn to Cushing's [1977] book and to the several papers dealing with the *sunflower equation* (see e.g. Somolinos [1978]).

7.6   Through Periodic Motions to Chaos

In this last Section of the Chapter on bifurcations and of the book we try to go beyond the proper scope of our study again (as was done in the previous Section) and describe in an informal way how a system gets into a "chaotic mood" through successive bifurcations of more and more complicated periodic solutions. Chaotic behaviour has been observed long ago in different branches of science, technology, economics, etc., but it withstood mathematical modelling up to the 1970s. The oldest puzzling problem of this sort has been, perhaps, the phenomenon of turbulence in the fluid mechanics of viscous fluids. In dealing with these problems we shall be even more informal than we were in the previous Section. We omit not only the proofs but, mostly, also the formal statement of theorems. The reason for this is not just the lack of space but also the fact that while "chaos" has now an enormous literature, rigorously proven theorems concerning flows in $\mathbf{R}^n$ ($n \geq 2$) are lacking. The state of art can be seen, e.g., from Cvitanovic [1983] and Devaney [1989].

Imagine a flow generated by an autonomous system of differential equations, say, that is modelling a real life process. The simplest easily conceivable situation is when the system has a (linearly) asymptotical stable equilibrium point, so that if initial conditions are realized in the basin of this equilibrium, then the system goes to this rest position exponentially. Suppose that the system depends on a real parameter that is steadily increased, say. As we have seen it in Section 7.2 the consequence of this can be generically that a pair of complex conjugate eigenvalues of the linearization at this equilibrium crosses transversally from the left to the right half plane. The result is an Andronov-Hopf bifurcation. Suppose that this bifurcation occurs at the bifurcation parameter value $\mu = \mu_0$ and it is supercritical, i.e. for $\mu < \mu_0$ the equilibrium is asymptotically stable, and for $\mu > \mu_0$ it is unstable and one has an orbitally asymptotically stable periodic solution for each $\mu > \mu_0$, $\mu - \mu_0 > 0$ sufficiently small. The question is raised now: What may happen to these bifurcating periodic solutions as μ is increased further? This problem has already been mentioned at the end of Section 7.2. There are several possibilities. The bifurcating stable periodic solutions may exist infinitely, i.e. for $\mu \to \infty$. They may disappear in a homoclinic bifurcation at a certain value $\mu_1 > \mu_0$ of the bifurcation parameter, their period tending to infinity as $\mu \to \mu_1 - 0$ (see Example 7.4.1). They may collapse to the equilibrium point again in another "inverse" supercritical Andronov-Hopf bifurcation, that is to say that the stable periodic orbits may exist for $\mu \in (\mu_0, \mu_1)$ and the system may undergo a supercritical Andronov-Hopf bifurcation as μ is *decreased* through μ_1. This way a "snake" of periodic solutions lies between μ_0 and μ_1 (cf. Mallet-Paret, Yorke [1982] and see Figure 7.6.1).

Besides these possibilities there is the possibility that the bifurcating periodic orbits lose their stability because as μ is increased through $\mu_1 > \mu_0$

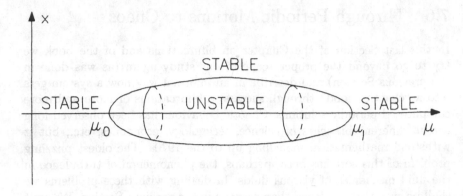

FIGURE 7.6.1. Supercritical Andronov-Hopf bifurcations as μ is increased through μ_0 and decreased through μ_1 with the bifurcating periodic orbits existing in the whole interval (μ_0, μ_1).

a real multiplier or a pair of complex conjugate characteristic multipliers crosses the unit circle from the inside to the outside in a non-degenerate way. If the dimension of the system is n, we know that $n-1$ characteristic multipliers are in modulus less than 1 for small $\mu - \mu_0 > 0$ (see the end of the proof of the Andronov-Hopf Bifurcation Theorem following formula (7.2.54)). This crossing may happen three different ways. First, through the number 1 on the real axis meaning that at $\mu = \mu_1$, the number 1 becomes a double multiplier; secondly through the unit circle off the real axis; thirdly through the point -1 on the real axis. We shall treat all three cases separately. The emphasis will be placed on the third case, the latter being the most important from the point of view of the onset of chaos. The second case is also related to the problem of being the starting point of a theory of turbulence. The first case is not related to chaos but will be treated briefly for the sake of completeness and because of its interest on its own. Thus, we assume that the n dimensional autonomous system depending on the bifurcation parameter μ has an orbitally asymptotically stable periodic orbit for $\mu < \mu_1$ and the conditions of the Andronov-Witt Theorem 5.1.2 hold, i.e. 1 is a simple characteristic multiplier and the rest of the multipliers lie in the interior of the unit circle.

First case: a second (real) characteristic multiplier assumes the value 1 *at* $\mu = \mu_1$. The typical situation in which this happens is the melting together of a family of stable and a family of unstable periodic orbits in a "degenerate", "non-hyperbolic" periodic orbit for which 1 is a double characteristic multiplier. This is called a *saddle-node bifurcation of periodic orbits*: for $\mu < \mu_1$ there are two periodic orbits, a stable and an unstable one; for $\mu > \mu_1$ there is none (cf. Example 7.1.1, see Figure 7.6.2). This is

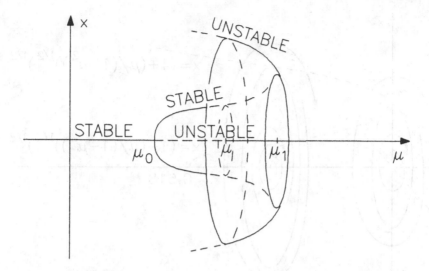

FIGURE 7.6.2. Supercritical Andronov-Hopf bifurcation at μ_0 followed by a saddle-node bifurcation of periodic orbits at μ_1. For $\mu_0 < \mu < \mu_1$ the system has an unstable equilibrium surrounded by a stable periodic orbit surrounded by another unstable periodic orbit.

illustrated by the following:

Example 7.6.1.

$$\dot{x} = -\mu x + (\mu^2 - 1)y + (1 - x^2 - y^2)^2((1 - \mu^2)x - \mu y)$$
$$\dot{y} = (1 - \mu^2)x - \mu y + (1 - x^2 - y^2)^2(\mu x + (1 - \mu^2)y) \quad (7.6.1)$$

where μ is a real parameter, and $|\mu|$ is small. It can be checked easily that the only equilibrium point of this system is $(x, y) = (0, 0)$. Performing the polar transformation $x = r\cos\theta$, $y = r\sin\theta$ the system assumes the form

$$\dot{r} = r((1 - r^2)^2(1 - \mu^2) - \mu),$$
$$\dot{\theta} = (1 - r^2)^2\mu + 1 - \mu^2.$$

For $|\mu| < 1$, $\mu \neq 0$ we have $\dot{\theta} = 0$ iff $(1 - r^2)^2 = (\mu^2 - 1)/\mu$ which has solutions only if $\mu < 0$. On the other hand, $\dot{r} = 0$ iff apart from $r = 0$ we have $(1 - r^2)^2 = \mu/(1 - \mu^2)$ which has solutions only if $\mu \geq 0$. The solutions are $r = (1 \mp (\mu/(1 - \mu^2))^{1/2})^{1/2}$, i.e. for $-1 < \mu < 0$ the system has no periodic orbit, and since $\dot{r} > -\mu r$, all solutions tend to infinity as $t \to \infty$; at $\mu = 0$ the system has a single periodic orbit whose equation is $x^2 + y^2 = 1$; for $0 < \mu$ the system has two periodic orbits whose equations are

$$x^2 + y^2 = 1 - (\mu/(1 - \mu^2))^{1/2}, \qquad x^2 + y^2 = 1 + (\mu/(1 - \mu^2))^{1/2}$$

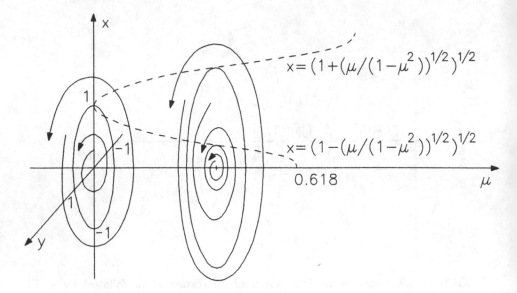

FIGURE 7.6.3. Saddle-node bifurcation of periodic orbits at $\mu = 0$ in system (7.6.1).

provided that the right-hand side of the first equation is positive. This is the case if $0 < \mu < (-1 + \sqrt{5})/2 \approx 0.618$. Checking the eigenvalues of the linearization at $(0,0)$, one obtains easily that the equilibrium is unstable for $-1 < \mu < 0.618$. For the sign of $\dot{r}$ we have that $\dot{r}$ is positive inside the periodic orbit of smaller amplitude, $\dot{r}$ is negative between the two orbits and is again positive outside the orbit of larger amplitude; hence, the smaller orbit is orbitally asymptotically stable, and the larger is unstable. At $\mu = 0$ when the two periodic orbits have melted into the unit circle, the latter is "stable from the inside and unstable from the outside" (see Figure 7.6.3). Thus, at $\mu = 0$ this system exhibits a saddle-node bifurcation of periodic orbits. By the way, at $\mu = 0$ the variational system with respect to the non-constant periodic solution is $\dot{y}_1 = -y_2$, $\dot{y}_2 = y_1$; hence, 1 is a double characteristic multiplier indeed. Observe also that the inner periodic orbit disappears in an "inverse" Andronov-Hopf bifurcation at $\mu = 0.618$. The outer periodic orbit exists in $\mu \in (0,1)$, but its amplitude tends to infinity as $\mu \to 1 - 0$.

Second case: a pair of complex conjugate characteristic multipliers crosses the unit circle off the real axis at $\mu = \mu_1$. We have in this case an orbitally asymptotically stable periodic solution $p(t, \mu)$ for $\mu < \mu_1$: the number 1 is a simple characteristic multiplier and the remaining multipliers are in modulus less than 1. At $\mu = \mu_1$ a complex conjugate pair of the latter reaches the unit circle and crosses it transversally so that for

$\mu > \mu_1$, $\mu - \mu_1$ small, the periodic solution $p(t, \mu)$ is already unstable. In order to see what is going on, we consider the *Poincaré map* in a neighbourhood of the periodic orbit, i.e. we fix a point $p(0, \mu_1)$, say, of the periodic orbit corresponding to the parameter value $\mu = \mu_1$ and assume an $n - 1$ dimensional hyperplane L transversal to the orbit at this point. The Poincaré map P_{μ_1} (see Definition 5.2.1 and Theorem 5.2.1) maps a neighbourhood $U \subset L$ of the point $p(0, \mu_1)$ into L, and $P_{\mu_1} : U \to L$ is a diffeomorphism. The same L and the same neighbourhood will do also for $\mu \neq \mu_1$ provided that $|\mu - \mu_1|$ is small ($p(t, \mu)$ is supposed to be smooth in both variables). The point $p(0, \mu)$ is a fixed point of the Poincaré map P_μ, i.e. $P_\mu(p(0, \mu)) = p(0, \mu)$. For $\mu < \mu_1$, $\mu_1 - \mu$ small, the Poincaré map P_μ is contractive, its spectrum coincides with the set of characteristic multipliers different from 1 (see Theorem 5.2.2). This means that for any $x^0 \in U$ the iterates of P_μ tend to $p(0, \mu) : P_\mu^k(x^0) \to p(0, \mu)$ as $k \to \infty$ (see Figure 5.2.3). As μ is crossing the value μ_1, an "*Andronov-Hopf Bifurcation of the map P_μ*" occurs under some generic conditions. For a rigorous formulation of this theorem see Marsden-McCracken [1976] p.23 and Section 6; for a two-dimensional version see Hale-Kocak [1991] p. 474. We give here but an informal description.

Just like in the case of the Andronov-Hopf Theorem 7.2.3 and Lemma 7.2.5 for flows, the corresponding Theorem for *maps* can be stated easily in two dimensions. The n or infinite dimensional case is then reduced to the former one by a Centre Manifold Theorem for maps (cf. Theorem A3.3). The *Andronov-Hopf Theorem for the one parameter family of* C^k-*diffeomorphisms* $P_\mu : \mathbf{R}^2 \to \mathbf{R}^2$, $k \geq 5$ says that if $P_\mu(0,0) = (0,0)$, $DP_\mu(0,0)$ has a pair of complex conjugate eigenvalues $\lambda(\mu)$, $\overline{\lambda}(\mu)$ (with non-zero imaginary part) such that for $\mu < 0$ there holds $|\lambda(\mu)| < 1$; for $\mu > 0 : |\lambda(\mu)| > 1$, $d|\lambda(0)|/d\mu > 0$, the eigenvalue $\lambda(0)$ is not a j-th root of unity for $j = 1, 2, 3, 4$, and a certain "Poincaré-Liapunov coefficient" is negative. Then for sufficiently small $\mu > 0$ the map P_μ has an attracting invariant closed curve around the fixed point $(0,0)$ while $(0,0)$ is, clearly, repelling. The *Centre Manifold Theorem* says that if P is a C^k-mapping of a neighbourhood of the origin of a Banach space into the Banach space, $P(0) = 0$, the spectrum of $DP(0)$, splits into a part on the unit circle with generalized eigenspace of finite dimension $c \geq 1$ and another part in the interior of the unit circle, then there is a C^k-manifold M of dimension c passing through the origin (the *centre manifold*) that is locally invariant and locally attractive. Now, if $P_\mu : E \to E$ is a one parameter family of C^∞ mappings of the *Banach space* E into itself, $P_\mu(0) = 0$, $DP_\mu(0)$ having a pair of complex conjugate eigenvalues $\lambda(\mu)$, $\overline{\lambda}(\mu)$ such that for $\mu < 0$ there holds $|\lambda(\mu)| < 1$, for $\mu > 0$ we have $|\lambda(\mu)| > 1$, $d|\lambda(0)|/d\mu > 0$ and $\lambda(0)$ is not a j-th root of unity for $j = 1, 2, 3, 4$, the rest of the spectrum lies in the interior of the unit circle at a non-zero distance from the unit circle, and a certain "Poincaré-Liapunov coefficient" is negative; then for $\mu > 0$, $|\mu|$ sufficiently small, P_μ has an attracting invariant closed curve in a neighbourhood of the origin.

If this theory is applied to the Poincaré map P_μ of the paragraph preceding the previous one attached to the periodic solution $p(t, \mu)$ for μ's around the critical value μ_1, then we see that as μ crosses μ_1, the periodic solution $p(t, \mu)$ becomes unstable, and for these $\mu > \mu_1$ the Poincaré map P_μ has an attracting closed curve around the unstable fixed point $p(0, \mu)$. The implication of such an attractive closed curve for the flow is that the union of the orbits with initial value on the attracting closed curve form a two dimensional *torus* T_μ^2 around the unstable periodic orbit of $p(t, \mu)$. The torus T_μ^2 is an attracting invariant manifold of the flow.

According to the Ruelle-Takens [1971] theory of *turbulence* this bifurcation of a stable periodic solution into a stable torus is the starting step of turbulence. On this stable invariant torus it may happen that there is no periodic motion at all (cf. Example 5.2.1). In their paper which was a milestone on the road towards understanding turbulence besides establishing the just quoted version of the Andronov-Hopf Bifurcation Theorem, they have applied this theory to the *Navier-Stokes equation*

$$\partial v / \partial t + (v \cdot \nabla)v - (1/R)\triangle v = - \operatorname{grad} p + F$$

describing the motion of some viscous fluid where $v(t, \cdot) : \mathbf{R}^3 \to \mathbf{R}^3$ is the velocity field, p the pressure, F the external force, R the "Reynolds number", and ∇ and $\triangle$ the Hamilton and the Laplace operators, respectively. This partial differential equation can be considered as a differential equation

$$dv/dt = N(v, R)$$

over the Hilbert space H (see Definition A2.6) of certain fields $v : \mathbf{R}^3 \to \mathbf{R}^3$. A solution $v : \mathbf{R} \to H$ is the velocity field (satisfying naturally certain boundary conditions too), $v(t) = v(t, \cdot)$. For $t \in \mathbf{R}$, $x \in \mathbf{R}^3$ the vector $v(t, x)$ is the velocity at time t and point x. A constant solution corresponds to a stationary flow, a periodic solution to a flow generated by a velocity field which varies periodically in time at each fixed point x. In the Ruelle-Takens theory (see also Marsden-McCracken [1976]) the Reynolds number may play the role of the bifurcation parameter. As it is increased, an originally stationary flow turns into a periodic one at the first critical value. Such a flow may be imagined, for instance, as the *von Karman vortices* behind the profile of the wing of an airplane. At the second critical value the periodic character of the flow disappears and an invariant torus T^2 bifurcates the velocity field, becoming more complicated. According to the Ruelle-Takens theory as the Reynolds number is steadily increased, higher dimensional tori T^k, $k \geq 3$ bifurcate along which the field's variation becomes "*chaotic*" and the flow turbulent.

This theory has induced a long series of papers, improvement and criticism since, no doubt, it was one of the first attempts to describe chaotic behaviour in a deterministic way.

Third case: a real characteristic multiplier crosses the unit circle at the point -1 *at the parameter value* $\mu = \mu_1$. If the periodic solution is denoted by $p(t, \mu)$ as before and the transversal $n - 1$ dimensional hyperplane L at $p(0, \mu_1)$ is considered again, then we know (see Corollary 2.2.3) that the variational system with respect to $p(t, \mu_1)$ has (besides the T-periodic solution $\dot{p}(t, \mu_1)$) a $2T$-periodic solution. Now, there are non-linear systems that have a close to $2T$-periodic solution themselves for $\mu > \mu_1$, and the bifurcation of this $2T$-periodic solution at μ_1 is such that while $p(t, \mu)$ loses its stability for $\mu > \mu_1$ the bifurcating $2T$-periodic solution will be orbitally asymptotically stable. This phenomenon is called *a period doubling bifurcation*. In order to have a deeper look into the situation we reduce the study of the problem again to the Poincaré map P_μ attached to the flow. We know that $p(0, \mu)$ is a fixed point of this Poincaré map and the system has a periodic solution of least period close to $2T$ if and only if the second iterate of the Poincaré map P_μ^2 has a fixed point that is not a fixed point of P_μ (see what has been said following the proof of Theorem 5.2.1 and also Figure 5.2.2). Thus, we assume that the Poincaré map P_μ has an attracting fixed point for $\mu < \mu_1$ which undergoes a period doubling bifurcation at μ_1; the former fixed point is unstable for $\mu > \mu_1$ and a fixed point of P_μ^2 bifurcate from it, which is attractive for $\mu > \mu_1$. We assume further that this phenomenon is repeated infinitely as μ is increasing, i.e. that there is a $\mu_2 > \mu_1$ such that the fixed point of P_μ^2 is attracting for $\mu \in (\mu_1, \mu_2)$ and it undergoes a period doubling bifurcation at μ_2: it is unstable for $\mu > \mu_2$ and a fixed point of $P_\mu^{2^2} = P_\mu^4$ bifurcates from it, which is attractive for $\mu > \mu_2$ and so on. The possibility of such an infinite sequence of period doubling bifurcations and its actual occurrence has been discovered by Feigenbaum [1978] along with some "universal" quantitative properties in the framework of one dimensional maps. Since the theory has been worked out extensively for one dimensional maps, we shall present the most important features also in this setting. Then we shall return to our Poincaré map attached to the n dimensional flow.

Let $F_\mu : \mathbf{R} \to \mathbf{R}$ be a one parameter family of functions, $\mu \in \mathbf{R}$, $F \in C^r(\mathbf{R} \times \mathbf{R}, \mathbf{R})$, $r \geq 3$. For $x \in \mathbf{R}$ the notation $F(\mu, x) \equiv F_\mu(x)$ will also be used. For fixed $\mu \in \mathbf{R}$ the mapping F_μ generates a *discrete dynamical system* or a *discrete flow* on the real line by repetition, i.e. by taking compositions of F_μ with itself: $F_\mu^2 := F_\mu \circ F_\mu, ..., F_\mu^k := F_\mu \circ F_\mu \circ \cdot \circ F_\mu$ (k times, $k = 1, 2, ...$). The point $x \in \mathbf{R}$ is said to be a *fixed point* of F_μ if $F_\mu(x) = x$. The point $x_k \in \mathbf{R}$ is called a *periodic point of F_μ of period k* if it is a fixed point of F_μ^k but is *not* fixed a point of $F_\mu, F_\mu^2, ..., F_\mu^{k-1}$, $k \geq 2$. The sequence of points, "the iterates of x_k" $\{x_k, x_{k1} := F_\mu(x_k), x_{k2} := F_\mu(x_{k1}) = F_\mu^2(x_k), ..., x_{k,k-1} := F_\mu(x_{k,k-2}) = F_\mu^{k-1}(x_k)\}$, is called the *periodic orbit* of the periodic point k. This set consists of k different points, and clearly $F_\mu(x_{k,k-1}) = x_k$ where the orbit closes in. A *fixed point x_1 of F_μ* is said to be *stable*, , resp. *asymptotically stable* if for $\varepsilon > 0$ there is a $\delta > 0$ such that for $|x - x_1| < \delta$ we have $|F_\mu^k(x) - x_1| < \varepsilon$ for all $k \geq 1$, resp. if

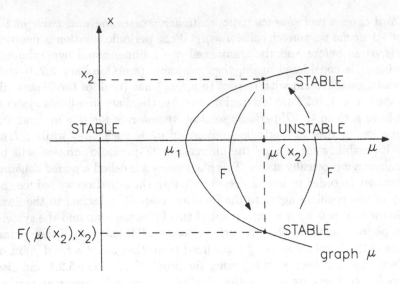

FIGURE 7.6.4. Supercritical period doubling bifurcation of the scalar map F.

besides being stable there is a $\delta_0 > 0$ such that for $|x - x_1| < \delta_0$ we have $F_\mu^k(x) \to x_1$ as $k \to \infty$. The fixed point is *unstable* if it is not stable. It is easy to see that the fixed point x_1 is asymptotically stable, resp. unstable if $|F_\mu'(x_1)| < 1$, resp. $|F_\mu'(x_1)| > 1$. A *periodic point* x_k of period k or its periodic orbit is said to be *stable, asymptotically stable*, resp. *unstable* if it is a stable, asymptotically stable, resp. unstable fixed point of F_μ^k.

In the context of such families of scalar functions the situation leading to a period doubling bifurcation is expressed by the following:

Theorem 7.6.1 *(see Devaney [1989] p.90). If $F(\mu, 0) \equiv 0$ in a neighbourhood of the value $\mu = \mu_1$, $F_x'(\mu_1, 0) = -1$, and $\partial^2 F^2(\mu_1, 0)/\partial \mu \partial x \neq 0$, then $x = 0$ has a neighbourhood U and there exists a C^r function $\mu : U \to \mathbf{R}$ such that $\mu(0) = \mu_1$, and for $x \in U \backslash \{0\}$ we have $F(\mu(x), x) \neq x$ but $F^2(\mu(x), x) = x$.*

Generic conditions can be given (see Hale-Kocak [1991] p.88) that ensure that the function $\mu : U \to \mathbf{R}$ is having either a strict minimum or a maximum at $x = 0$. In the former case, say, the typical situation is that $F_x'(\mu, 0) > -1$ for $\mu < \mu_1$ and this derivative is less than -1 for $\mu > \mu_1$; hence, $x = 0$ is asymptotically stable, resp. unstable for $\mu < \mu_1$, resp. $\mu > \mu_1$, and for $\mu < \mu_1$ there is no periodic orbit of period 2 in the neighbourhood of $x = 0$, but to each $\mu > \mu_1$, $\mu - \mu_1 > 0$ small, there belongs one and only one periodic orbit of period 2 and this is asymptotically stable. This *supercritical period doubling bifurcation* (see Figure 7.6.4) plays the main role in the sequel.

Feigenbaum [1978] has discovered numerically that a large class of one parameter families of scalar maps has the property that as the bifurcation parameter is increased, an infinite sequence of supercritical period doubling bifurcations occurs: first the originally asymptotically stable fixed point loses its stability, giving rise to an asymptotically stable periodic orbit with period 2; at the next critical value of the parameter, this period 2 orbit loses its stability, giving rise to an asymptotically stable periodic orbit with period 4, etc. Moreover the sequence $\mu_1 < \mu_2 < \cdots$ of critical values, being "asymptotically" a geometric series, tends to a finite limit μ_∞. For the details see also Feigenbaum's papers in Cvitanovic [1983] and also Collet-Eckmann [1980] and Lanford [1982]. There are also some "universal" numerical characteristics of this sequence that appear at each family of this class after some rescaling. As $\mu \to \mu_\infty$, the orbits of the map F_μ become more and more "chaotic": there will be a large number of unstable periodic orbits and a single asymptotically stable one with very large period. If $\mu_k < \mu < \mu_{k+1}$, then the period of the latter is 2^k. In the process periodic orbits with period different from any power of 2 occur also.

There are several definitions of "chaos". One of the often used definitions applies the concepts of "topological transitivity" and "sensitive dependence on initial conditions". We say that the map $F_\mu : A \to A$ where $A \subset \mathbf{R}$ is *topologically transitive* if for any pair of relatively open sets $U, V \subset A$ there exists a positive integer k such that $F_\mu^k(U) \cap V \neq \emptyset$. We say that F_μ *depends sensitively on initial conditions* if there exists a $\rho > 0$ such that for any $x \in A$ and any neighbourhood U_x of x there is a $y \in U_x$ and a positive integer k such that $|F_\mu^k(y) - F_\mu^k(x)| > \rho$.

Definition 7.6.1. We say that $F_\mu : A \to A$ is *chaotic (on the set A)* if it is topologically transitive, depends sensitively on initial conditions and its periodic points form a dense set in A.

For scalar maps there is an interesting result yielding some insight into the relation between periodic orbits on the one hand and chaos on the other. This is Sharkovskii's Theorem which is based on the "Sharkovskii ordering" of positive integers:

$$3 \triangleright 5 \triangleright 7 \triangleright \cdots \triangleright 2 \cdot 3 \triangleright 2 \cdot 5 \triangleright 2 \cdot 7 \triangleright \cdots \triangleright 2^2 \cdot 3 \triangleright 2^2 \cdot 5 \triangleright 2^2 \cdot 7 \triangleright \cdots$$

$$\triangleright 2^3 \cdot 3 \triangleright 2^3 \cdot 5 \triangleright \cdots \triangleright 2^3 \triangleright 2^2 \triangleright 2 \triangleright 1.$$

This ordering includes all the positive integers.

Theorem 7.6.2 *(Sharkovskii [1964]). Let $F : \mathbf{R} \to \mathbf{R}$ be a continuous function and suppose that F has a periodic point of period m; then for any $n \triangleleft m$ in the Sharkovskii ordering, F has also a periodic point of period n.*

This implies, obviously, that if F has a periodic point of period 3, then it has periodic points of every period. See also Li-Yorke [1975].

The most often quoted typical representative of the class of mappings exhibiting the "Feigenbaum properties" is the *logistic map*

$$F(\mu, x) := \mu x(1 - x), \qquad \mu > 1.$$

For $\mu \in (1, 4)$ this maps the interval $(0, 1)$ into itself. It has two fixed points, $x = 0$ and $x_\mu := 1 - 1/\mu$. The fixed point $x = 0$ is unstable for all $\mu \in (1, 4)$. Applying the stability condition based on the modulus of the derivative $\partial F/\partial x$ it is easy to see that x_μ is asymptotically stable for $\mu \in (1, 3)$ and it is unstable for $\mu \in (3, 4)$. At $\mu = \mu_1 = 3$ we have $\partial F(3, x_3)/\partial x = -1$, and by Theorem 7.6.1, x_μ undergoes a (supercritical) period doubling bifurcation. The bifurcating 2-periodic orbit is asymptotically stable for $\mu \in (3, 1 + \sqrt{6}) = (3, 3.449)$. At $\mu = \mu_2 = 1 + \sqrt{6}$ another period doubling bifurcation occurs: the 2-periodic orbit loses its stability and a 4-periodic asymptotically stable periodic orbit bifurcates from it (i.e. from the fixed point of F^2), and so on. An infinite sequence of period doubling bifurcations occurs as μ is increased at $\mu_1 = 3$, $\mu_2 = 1 + \sqrt{6} = 3.449$, $\mu_3 = 3.544, ...$. The infinite sequence $\{\mu_k, \; k = 1, 2, 3, ...\}$ is "asymptotically a geometric series", meaning that

$$\lim_{k \to \infty} \frac{\mu_{k+1} - \mu_k}{\mu_k - \mu_{k-1}} = 1/4.669... ,$$

the quotient $1/4.669...$ being a "universal constant" discovered by Feigenbaum. $\lim_{k \to \infty} \mu_k = 3.570$. Increasing μ still further at $\mu = 3.839$ there appears a periodic orbit of period 3, and beyond 3.839 another sequence of period doubling bifurcations takes place, this time starting from the period 3 orbit. By Sharkovskii's Theorem 7.6.2, in this region of the bifurcation parameter, periodic orbits of period equal to any positive integer are present, and the dynamics of the map $F(\mu, x)$ for $\mu > 3.839$ is chaotic. (A more detailed analysis of this example can be found in Hale-Kocak [1991] Section 3.5.)

When one tries to generalize these results concerning period doubling bifurcations of one parameter families of scalar mappings to higher dimensions, one encounters great difficulties. Results concerning this "third case" of destabilization of a fixed point in maps of $\mathbf{R}^n$ to $\mathbf{R}^n$ are very scarce. Sharkovskii's Theorem 7.6.2 is definitely *not* true in $\mathbf{R}^2$. Still there are results according to which under certain conditions relatively large classes of one parameter maps of $\mathbf{C}^n$ into $\mathbf{C}^n$ exhibit an infinite sequence of period doubling bifurcations with some "universal" properties; see, e.g., Collet-Eckmann-Koch [1981]. Thus, returning to the original setting of a one parameter family of flows in $\mathbf{R}^n$ ($n \geq 3$) with a periodic solution $p(t, \mu)$ and reducing the problem of successive bifurcations of p to the bifurcations of the corresponding Poincaré map P_μ of an $n-1$ dimensional hyperplane transversal to the periodic orbit of p, one may encounter an infinite sequence of period doubling bifurcations leading to *chaotic dynamics* around the originally stable periodic orbit.

Cases two and three, treated briefly above, present, perhaps, the most characteristic and typical processes involving successive bifurcations of periodic solutions and tori that lead to chaos. There are, naturally, other ways for the onset of chaos, but this is already a different story.

7.7   Problems

Problem 7.7.1. Show that the right-hand side of the system $\dot{x}_1 = -x_1 - f_1(x_1, x_2)$, $\dot{x}_2 = -x_2 + f_2(x_1, x_2)$ where

$$f_i(x_1, x_2) = \begin{cases} x_i / \ln(x_1^2 + x_2^2), & (x_1, x_2) \neq (0,0) \\ 0, & (x_1, x_2) = (0,0) \end{cases} \qquad (i = 1, 2)$$

is in the C^1 class and that while $(0,0)$ is a proper node of the linearized system, it is a spiral point of the non-linear one. (Hint: use polar coordinates.)

Problem 7.7.2. Determine explicitly the solutions of system (7.1.6) for arbitrary $\mu \in \mathbf{R}$. Show that in some cases the solutions are not defined in $[0, \infty)$ but still the phase pictures are valid as t tends to the upper end point of the maximal interval of definition.

Problem 7.7.3. The system $\dot{x} = y$, $\dot{y} = x - x^2 + \mu y$ has two equilibria, $E_0 = (0,0)$ and $E_1 = (1,0)$. Show that while E_0 is always a saddle, E_1 is a stable, resp. unstable spiral point according as μ is negative, resp. positive. At $\mu = 0$ the system undergoes a "*homoclinic bifurcation*". Show that at $\mu = 0$ the equilibrium E_1 is a centre and there is a homoclinic orbit belonging to E_0. Determine the explicit equation of the homoclinic orbit.

Problem 7.7.4. Determine explicitly the solutions of system (7.2.1).

Problem 7.7.5. (Lepoeter [1983]). Give a complete classification of the different possible stability behaviours of the equilibrium point $(x_1, x_2) = (0,0)$ for the system

$$\begin{aligned} \dot{x}_1 &= -\omega x_2 + a_1 x_1^2 + a_2 x_1 x_2, \\ \dot{x}_2 &= \omega x_1 + c_1 x_1^2 + c_2 x_1 x_2. \end{aligned}$$

Problem 7.7.6. Show that *Rayleigh's equation* $\ddot{u} + (\dot{u}^2 - \mu)\dot{u} + u = 0$ undergoes a supercritical Andronov-Hopf bifurcation at $\mu = 0$ (cf. Farkas-Sparing-Szabó [1986]).

Problem 7.7.7. Prove the analogues of Theorems 7.4.6 and 7.4.7 for artificial models.

Problem 7.7.8. In system (7.4.23) one may choose $g(S, K) = 1 - (S/K)^u$, $0 < u \leq 1$ satisfying all the conditions. Let $A > 0$,

$B \geq 0$, $C \in \mathbf{R}$ be given so that $0 < A/(Ba_1 + C) \leq A/(Ba_2 + c)$ for some $0 < a_2 < a_1$, and

$$p_1(S, a) := (A/(Ba + C))S/(S + a) \qquad \text{(Holling)}$$
$$p_2(S, a) := (A/(Ba + C))(1 - \exp(-S/a)) \qquad \text{(Ivlev)}$$
$$p_3(S, a) := (A/(Ba + C))S^q, \quad 0 < q < 1 \qquad \text{(Rosenzweig)}$$
$$p_4(S, a) := (A/(Ba + C))(\ln(1 + S) + S/(1 + a));$$

show that p_i $(i = 1, 2, 3, 4)$ satisfy conditions (7.4.27)-(7.4.29) and (7.4.31), and that p_1 and p_2 make a natural, p_3 a degenerate and p_4 an artificial model.

Appendix

A1 Matrices

In this Appendix we treat those results of matrix calculus that go beyond a regular introductory course of Linear Algebra and are needed, mainly, in the study of linear systems of differential equations.

The *characteristic polynomial* of a real (a complex) $n \times n$ matrix A will be denoted by

$$D(\lambda) := (-1)^n \det(A - \lambda I) = \lambda^n + d_{n-1}\lambda^{n-1} + \cdots + d_1\lambda + d_0$$

where I is the $n \times n$ unit matrix. We denote the *eigenvalues* of A, the different zeroes of D by $\lambda_1, \lambda_2, ..., \lambda_r$, $\lambda_i \neq \lambda_k$, $i \neq k$ and their respective multiplicities by $n_1, n_2, ..., n_r$, $n_k \geq 1$ $(k = 1, ..., r)$, $\sum_{k=1}^{r} n_k = n$.

The proofs of the following theorems can be found in Gantmacher [1954].

Theorem A1.1 *(Cayley-Hamilton). Every square matrix A satisfies its characteristic equation, i.e.*

$$D(A) = A^n + d_{n-1}A^{n-1} + \cdots + d_1 A + d_0 I = 0.$$

Let $p(\lambda) = p_m\lambda^m + p_{m-1}\lambda^{m-1} + \cdots + p_1\lambda + p_0$ be a polynomial of degree m, and A a square matrix; if $p(A) = p_m A^m + \cdots + p_1 A + p_0 I = 0$, then p is said to be an *annulling polynomial* of A. Thus, by the Cayley-Hamilton Theorem the characteristic polynomial of A is one of its annulling polynomials.

There holds the generalized *Bézout Theorem*:

Theorem A1.2. *If the polynomial $p(\lambda)$ is an annulling polynomial of the matrix A, then the factorization*

$$p(\lambda)I = (A - \lambda I)C(\lambda)$$

is valid where C is a matrix whose elements are polynomials of λ of degree $(m-1)$ at most.

Theorem A1.3. *In the set of annulling polynomials of the matrix A there is one of least degree, and all the annulling polynomials are divisible with this one.*

This theorem, obviously, implies that apart from a constant non-zero factor the annulling polynomial of least degree is defined uniquely. In canonical form, i.e. with coefficient 1 at the leading term, this polynomial is called the *minimal polynomial* of A and will be denoted by $\Delta(\lambda)$. The following theorem provides a practical method for the determination of the minimal polynomial.

Theorem A1.4. *Denote the greatest common divisor of the $(n-1)$-st order minors of the matrix $A - \lambda I$ in canonical form by $D_{n-1}(\lambda)$. Then the minimal polynomial of A is*

$$\Delta(\lambda) = D(\lambda)/D_{n-1}(\lambda). \tag{A1.1}$$

Example A1.1. Let

$$A = \begin{pmatrix} 3 & -3 & 2 \\ -1 & 5 & -2 \\ -1 & 3 & 0 \end{pmatrix}, \quad A - \lambda I = \begin{pmatrix} 3-\lambda & -3 & 2 \\ -1 & 5-\lambda & -2 \\ -1 & 3 & -\lambda \end{pmatrix}.$$

The associate matrix of $A - \lambda I$ whose elements are the cofactors of the transpose is

$$A_\lambda I = \begin{pmatrix} \lambda^2 - 5\lambda + 6 & -3\lambda + 6 & 2\lambda - 4 \\ -\lambda + 2 & \lambda^2 - 3\lambda + 2 & -2\lambda + 4 \\ -\lambda + 2 & 3\lambda - 6 & \lambda^2 - 8\lambda + 12 \end{pmatrix}.$$

The greatest common divisor of its elements is $D_{n-1}(\lambda) = \lambda - 2$. The characteristic polynomial of A is $D(\lambda) = -\det(A - \lambda I) = \lambda^3 - 8\lambda^2 + 20\lambda - 16$. The minimal polynomial is

$$\Delta(\lambda) = D(\lambda)/D_{n-1}(\lambda) = \lambda^2 - 6\lambda + 8.$$

Its roots are 4 and 2 (4 is a simple 2 a double root of D).

Theorem A1.5. λ *is an eigenvalue of A if and only if it is a root of the minimal polynomial $\Delta(\lambda)$.*

Proof. We note first that if λ_0 is an eigenvalue, $s \neq 0$ is a corresponding eigenvector of A, and $p(\lambda)$ is an arbitrary polynomial, then $p(\lambda_0)$ is an eigenvalue of the matrix $p(A)$ with the same eigenvector s, i.e.

$$p(A)s = p(\lambda_0)s. \tag{A1.2}$$

This follows immediately from $A^k s = \lambda_0^k s$ the latter equation being obviously true for arbitrary $k \in \mathbf{N}$. Now, if λ_0 is an eigenvalue of A with eigenvector $s \neq 0$, then

$$0 = \Delta(A)s = \Delta(\lambda_0)s,$$

i.e. λ_0 is a root of Δ. Conversely, if $\Delta(\lambda_0) = 0$, then $\Delta(\lambda) = (\lambda - \lambda_0)d(\lambda)$ where, clearly, $d(\lambda)$ is a polynomial not annulling A, i.e. $d(A) \neq 0$. Therefore there exists a vector $x \neq 0$ such that $s := d(A)x \neq 0$, but

$$0 = \Delta(A)x = (A - \lambda_0 I)d(A)x = (A - \lambda_0 I)s;$$

thus, λ_0 is an eigenvalue of A. $\square$

Now we are able to define the value of an analytic function f on a square matrix A, and prove the existence under reasonable conditions. We are also giving a practical method for the determination of $f(A)$. The respective multiplicities of the eigenvalues $\lambda_1, \lambda_2, ..., \lambda_r$ of A *in the minimal polynomial* $\Delta(\lambda)$ will be denoted by $m_1, m_2, ..., m_r$. Because of the previous theorem, $1 \leq m_k \leq n_k$ (see the beginning of this Appendix), $k = 1, 2, ..., r$, and $\sum_{n=1}^{r} = m \leq n$, m being the degree of Δ.

Definition A1.1. Assume that the spectrum of A lies in the interior of the domain of analyticity of the complex function f; we say that the $(m - 1)$-st degree polynomial

$$h(\lambda) = h_{m-1}\lambda^{m-1} + \cdots + h_1\lambda + h_0$$

is the *Hermite interpolation polynomial of f on the spectrum of A* if

$$h^{(l)}(\lambda_k) = f^{(l)}(\lambda_k) \qquad (k = 1, 2, ..., r; l = 0, 1, ..., m_k - 1). \tag{A1.3}$$

According to this definition the Hermite interpolation polynomial assumes the values of f at the eigenvalues of A, and at the eigenvalues its derivatives are also equal to the derivatives of f up to the order equal to the multiplicity minus one of the respective eigenvalue in the minimal polynomial Δ. If all the roots of the minimal polynomial are simple, then the Hermite interpolation polynomial reduces to the *Lagrange interpolation polynomial* of f corresponding to $\lambda_1, \lambda_2, ..., \lambda_r$. It can be proved (see Smirnov [1955] p.279) that all together the m conditions (A1.3) determine the coefficients $(h_{m-1}, ..., h_1, h_0)$ uniquely.

Definition A1.2. Let $f(\lambda) = \sum_{k=0}^{\infty} c_k \lambda^k$ be a complex function analytic in the interior of a circle with centre in the origin of the complex plane λ, and denote the N-th partial sum of the series of f by

$$s_N(\lambda) := \sum_{k=0}^{N} c_k \lambda^k;$$

then *the value of f at the square matrix A is*

$$f(A) = \sum_{k=0}^{\infty} c_k A^k := \lim_{N \to \infty} s_N(A) \qquad (A1.4)$$

where $A^0 = I$, i.e. $f(A)$ is the limit of the matrix sequence of the N-th partial sums taken at A.

Theorem A1.6. *If the eigenvalues $\lambda_1, ..., \lambda_r$ of the matrix A lie in the interior of the circle of analyticity of the complex function, then $f(A)$ is defined, and*

$$f(A) = h(A)$$

where h is the Hermite interpolation polynomial of f on the spectrum of A.

Proof. Divide the N-th partial sum $s_N(\lambda)$ of the series of f by the minimal polynomial $\Delta(\lambda)$ of A and denote the remainder by $r_N(\lambda)$:

$$s_N(\lambda) = \Delta(\lambda) q_N(\lambda) + r_N(\lambda). \qquad (A1.5)$$

For every $N \in \mathbf{N}$, clearly, r_N is a polynomial of degree $m - 1$, at most (m is the degree of Δ). Since λ_k is an m_k-tuple root of Δ, we have

$$s_N^{(l)}(\lambda_k) = r_N^{(l)}(\lambda_k) \qquad (k = 1, 2, ..., r; l = 0, 1, ..., m_k - 1).$$

Thus, $\lim_{N \to \infty} r_N^{(l)}(\lambda_k) = f^{(l)}(\lambda_k) = h^{(l)}(\lambda_k)$. But $r_N(\lambda)$, $N \in \mathbf{N}$, is a sequence of $(m - 1)$-st degree polynomials, and since the sequences of their m defining data tend to the corresponding defining data of the polynomial h, we have

$$\lim_{N \to \infty} r_N(\lambda) = h(\lambda) \qquad (A1.6)$$

(the sequences of the coefficients tend to the respective coefficients of h). Since Δ is an annulling polynomial of A, (A1.5) implies that $s_N(A) = r_N(A)$, $n \in \mathbf{N}$. Because of (A1.6)

$$\lim_{N \to \infty} s_N(A) = \lim_{N \to \infty} r_N(A) = h(A). \qquad \square$$

Example A1.2. Determine A^{100} where

$$A = \begin{pmatrix} 1 & 2 \\ 12 & 3 \end{pmatrix}.$$

The characteristic polynomial of A is $D(\lambda) = \lambda^2 - 4\lambda - 21$. Its roots are 7 and -3, both simple, so that the minimal polynomial is identical to D. The complex function whose Hermite interpolation polynomial is to be determined is $f(\lambda) = \lambda^{100}$. The Hermite interpolation polynomial is of degree one: $h(\lambda) = h_1\lambda + h_0$. It must be equal to f on the spectrum of A:

$$h_1 7 + h_0 = 7^{100}, -h_1 3 + h_0 = 3^{100}.$$

Hence $h_1 = 0.1(7^{100} - 3^{100})$ and $h_0 = 7^{100}0.3 + 3^{100}0.7$. Thus,

$$\begin{aligned} A^{100} &= h(A) = 0.1(7^{100} - 3^{100})A + (7^{100}0.3 + 3^{100}0.7)I \\ &= 0.1\begin{pmatrix} 4 \cdot 7^{100} + 6 \cdot 3^{100} & 2(7^{100} - 3^{100}) \\ 12(7^{100} - 3^{100}) & 6 \cdot 7^{100} + 4 \cdot 3^{100} \end{pmatrix}. \end{aligned}$$

Corollary A1.7. *For an arbitrary square matrix A and $t \in \mathbf{R}$ the function $\exp(At)$ is defined, and if A is real, then $\exp(At)$ has real entries.*

Proof. Since the function $f(\lambda, t) = \exp(\lambda t)$ is analytic in the open complex λ-plane, the eigenvalues of an arbitrary matrix A fall into the domain of analyticity.

$$\exp(At) = h(A, t) = h_{m-1}(t)A^{m-1} + \cdots + h_1(t)A + h_0(t)I$$

where $h(\lambda, t)$ is the Hermite interpolation polynomial of $\exp(\lambda t)$ on the spectrum of A. Its coefficients are to be determined from conditions (A1.3) which, in our case, assume the form

$$\sum_{j=l}^{m-1} \binom{j}{k} h_j \lambda_k^{j-l} = \frac{1}{l!} t^l e^{\lambda_k t} \qquad (k = 1, 2, ..., r; l = 0, 1, ..., m_k - 1).$$

$$\text{(A1.7)}$$

Now, if A is *real*, and λ_k is a real eigenvalue, then this is a linear equation for the unknown h_js with real coefficients and a real right-hand side. If $\lambda_k = \alpha_k + i\beta_k$, $\beta_k \neq 0$, then introduce the notation

$$\lambda_k^{j-l} = \alpha_{kj} + i\beta_{kj} \qquad (j = l + 1, ..., m - 1; \alpha_{k,l+1} = \alpha_k, \beta_{k,l+1} = \beta_k),$$

and note that the conjugate of λ_k is also a root of the minimal polynomial of the same multiplicity as λ_k; assume without loss of generality that it is

λ_{k+1}, i.e. $\lambda_{k+1} = \alpha_k - i\beta_k$. For each $l = 0, 1, ..., m_k - 1$ consider equation (A1.7) and the corresponding equation with λ_k replaced by λ_{k+1}:

$$\sum_{j=l}^{m-1} \binom{j}{l} h_j(\alpha_{kj} + i\beta_{kj}) = \frac{1}{l!} t^l e^{(\alpha_k + i\beta_k)t},$$

$$\sum_{j=l}^{m-1} \binom{j}{l} h_j(\alpha_{kj} - i\beta_{kj}) = \frac{1}{l!} t^l e^{(\alpha_k - i\beta_k)t}.$$

If first we add these two equations and divide by 2, and then we subtract the second from the first one and divide by $2i$, we get

$$\sum_{j=l}^{m-1} \binom{j}{l} h_j \alpha_{kj} = \frac{1}{l!} e^{\alpha_k t} \cos \beta_k t,$$

$$\sum_{j=l}^{m-1} \binom{j}{l} h_j \beta_{kj} = \frac{1}{l!} t^l e^{\alpha_k t} \sin \beta_k t. \tag{A1.8}$$

With these two equations we may replace the former pair of equations in the system, and these have already real coefficients and real right-hand sides. Performing this operation for every pair of conjugate complex roots and for every l, we see that we get a linear system of equations in which all the coefficients as well as the right-hand sides are real. Thus, the solutions $h_j(t)$ $(j = 0, 1, ..., m - 1)$ are also real valued functions of the real variable t. $\square$

Corollary A1.8. *Assume that the eigenvalues* $\lambda_1, ..., \lambda_r$ *of the matrix A are in the interior of the domain of analyticity of the function f; then the set of eigenvalues of the matrix $f(A)$ is the set* $\{f(\lambda_k) : k = 1, 2, ..., r\}$, *and if s is an eigenvector corresponding to λ_k, then it is one corresponding to $f(\lambda_k)$ too.*

Proof. We are using the notations of Theorem A1.6. First, we show that the eigenvalues of $f(A)$ are among the $f(\lambda_k)$s. Set

$$G(\lambda) := \prod_{k=1}^{r} (\lambda - f(\lambda_k))^{m_k}$$

where m_k is the multiplicity of λ_k in the minimal polynomial Δ of A. We are going to show that G is an annulling polynomial of $f(A)$. Then the minimal polynomial of $f(A)$ is a divisor of G; hence, its roots, the eigenvalues of $f(A)$, are among the $f(\lambda_k)$s. Set

$$G_N(\lambda) := \prod_{k=1}^{r} (r_N(\lambda) - r_N(\lambda_k))^{m_k}$$

(for the notations, cf. (A1.5)). Clearly, λ_k is an m_k-tuple root of G_N; hence, Δ is a divisor of G_N, and this means that G_N is an annulling polynomial of A:

$$G_N(A) = \prod_{k=1}^{r} (r_N(A) - r_N(\lambda_k)I)^{m_k} = 0.$$

If N tends to infinity, then $r_N(\lambda_k) \to f(\lambda_k)$, $r_N(A) \to f(A)$, and, therefore, $G_N(A) \to G(f(A))$. Thus,

$$G(f(A)) = \prod_{k=1}^{r} (f(A) - f(\lambda_k)I)^{m_k} = 0.$$

Secondly, we show that each $f(\lambda_k)$ is an eigenvalue of $f(A)$ indeed. This is easy, since

$$s_N(A)s = s_N(\lambda_k)s$$

where s is an eigenvector corresponding to λ_k (cf. also (A1.2)). If N tends to infinity, $s_N(A)$ tends to $f(A)$, and $s_N(\lambda_k)$ to $f(\lambda_k)$, so $f(A)s = f(\lambda_k)s$. □

Besides the exponential function of a matrix we need the *"logarithm"* of a matrix. While the exponential of an *arbitrary* square matrix is uniquely determined, the logarithm can be defined only for *regular* matrices. This is so because, as a consequence of the previous Corollary, the exponential of a matrix is always a regular matrix (zero cannot occur as an eigenvalue of the exponential). Further, the logarithm is not defined uniquely due to the fact that the exponential function of a complex variable is periodic with period $2\pi i$.

Theorem A1.9. *Let C be a regular $n \times n$ matrix; an $n \times n$ matrix B exists that commutes with C and satisfies*

$$\exp B = C. \tag{A1.9}$$

(This matrix B may be called the *logarithm* of C.)

Proof. Denote the eigenvalues of C by λ_j ($j = 1, 2, ..., r$), $\lambda_j \neq \lambda_k$ if $j \neq k$ and their respective multiplicities in the minimal polynomial of C by m_j ($j = 1, 2, ..., r$). $m_1 + m_2 + \cdots + m_r = m \leq n$ where m is the degree of the minimal polynomial. Since C is regular, no λ_j is equal to zero. Choose a value of $\ln \lambda_j$ and denote it by $\mu_j = \ln \lambda_j$ ($j = 1, 2, ..., r$). In a sufficiently small neighbourhood of each $\lambda_j \neq 0$ we choose the branch of the logarithm function that assumes the value μ_j at λ_j. Denoting this function (which is defined in sufficiently small neighbourhoods of $\lambda_1, ..., \lambda_r$) by $\ln \lambda$, clearly,

$$F(\lambda, \ln \lambda) \equiv 0 \tag{A1.10}$$

where $F(\lambda, \mu) = \exp \mu - \lambda$. In particular $F(\lambda_j, \mu_j) = 0$ $(j = 1, 2, ..., r)$. Differentiating the identity (A1.10) we get $F^{(k)}(\lambda, \ln \lambda) \equiv 0$, $k = 0, 1, 2, ...$. In particular

$$F^{(k)}(\lambda, \ln \lambda)\big|_{\lambda = \lambda_j} = 0 \qquad (j = 1, 2, ..., r; k = 0, 1, ..., m_j - 1). \quad (A1.11)$$

Now determine the Hermite interpolation polynomial $h(\lambda)$ of degree $m - 1$ corresponding to the values $ln^{(k)}\lambda_j$, i.e. let $h(\lambda)$ be the $(m - 1)$-degree polynomial that satisfies

$$h^{(k)}(\lambda_j) = \ln^{(k)} \lambda_j \qquad (j = 1, 2, ..., r; k = 0, 1, ..., m_j - 1). \quad (A1.12)$$

We are going to show that $B := h(C)$ satisfies (A1.9). Let us substitute $\ln \lambda$ with $h(\lambda)$ in F, i.e. consider the function
$H(\lambda) := F(\lambda, h(\lambda)) = \exp h(\lambda) - \lambda$. Because of (A1.11) and (A1.12), $H^{(k)}(\lambda_j) = 0$ $(j = 1, 2, ..., r; k = 0, 1, ..., m_j - 1)$. This means that the Hermite interpolation polynomial of H on the spectrum of C is identical to the Hermite interpolation polynomial of the identically zero function which is, naturally, zero. Thus, $H(C) = F(C, B) = \exp B - C = 0$. Note also that since B is defined as a linear combination of powers of C, it commutes with C. $\square$

In general, even if C is real, the matrix B satisfying (A1.9) may have complex entries. However, if the matrix C of the previous theorem is real and has no negative real eigenvalues, then a matrix with real entries can be chosen for its logarithm.

Theorem A1.10. *Let C be a regular $n \times n$ matrix with real entries, and assume that no eigenvalue of C is a negative (real) number. Then an $n \times n$ matrix B_1, with real entries exists which commutes with C and satisfies*

$$\exp B_1 = C. \qquad (A1.13)$$

Proof. In the proof of the previous Theorem in the case λ_j is real (positive) we choose $\mu_j = \ln \lambda_j$ real; if λ_l is complex, and say, $\lambda_{l+1} = \overline{\lambda_l}$, then we choose $\mu_{l+1} = \ln \lambda_{l+1} = \overline{\mu_l} = \overline{\ln \lambda_l}$. Then starting with equations (A1.12), defining the Hermite polynomial $h(\lambda)$ we may repeat the argument following (A1.7) in the proof of Corollary A1.7. This yields that the coefficients of the Hermite interpolation polynomial are real numbers, and so $B_1 := h(C)$ is a real matrix. $\square$

Corollary A1.11. *If C is a regular $n \times n$ matrix with real entries then a real $n \times n$ matrix B_1 commuting with C exists satisfying*

$$\exp B_1 = C^2. \qquad (A1.14)$$

An extension of Corollary A1.8 will be presented here without proof (for the proof see Demidovich [1967$_a$] p.56).

Theorem A1.12. *Let B be an $n \times n$ matrix, $C = \exp B$, and λ be an eigenvalue of B; the multiplicity of $\exp \lambda$ in the minimal polynomial of C is equal to the multiplicity of λ in the minimal polynomial of B.*

There are fortunate situations (see, e.g., Section 4.3) when the principal matrix of a homogeneous linear system of differential equations can be transformed by similarity transformations into a matrix with positive entries. In such a case the following Theorem due to Perron [1907] may be useful.

Theorem A1.13 *(Perron's Theorem). If all the elements of the $n \times n$ matrix A are positive, then this matrix has a positive eigenvalue that is simple in the characteristic polynomial and greater than the absolute value of any other eigenvalue; to this maximal positive eigenvalue there belongs an eigenvector with positive coordinates.*

For the proof see Gantmacher-Krein [1950] p.101.

In what follows we shall establish a result which may be applied in proving the definiteness of a quadratic form. The proofs of the following theorems will be provided since they are not easily available in textbooks on linear algebra. The presentation is based on Parodi [1952].

Theorem A1.14 *(Hadamard's Theorem). Let $A = [a_{ij}]$ be an $n \times n$ matrix with complex elements; if*

$$|a_{ii}| > \sum_{j=1, j\neq i}^{n} |a_{ij}| \qquad (i = 1, 2, ..., n), \qquad (A1.15)$$

then A is regular.

Proof. We shall give a positive lower estimate to the modulus of det A. Introducing the notations

$$\sigma_i := |a_{ii}| - \sum_{j=1, j\neq i}^{n} |a_{ij}| \qquad (i = 1, 2, ..., n)$$

we shall prove that

$$|\det A| \geq \sigma_1 \sigma_2 \cdots \sigma_n. \qquad (A1.16)$$

Dividing the i-th row of A by σ_i $(i = 1, 2, ..., n)$ we get a matrix $\tilde{A} = [\tilde{a}_{ij}]$ such that

$$\tilde{\sigma}_i := |\tilde{a}_{ii}| - \sum_{j=1, j\neq i}^{n} |\tilde{a}_{ij}| = 1 \qquad (i = 1, 2, ..., n)$$

and
$$\det \tilde{A} = \det A/(\sigma_1 \cdots \sigma_n).$$

We shall prove that $|\det \tilde{A}| \geq 1$, and this, clearly, proves (A1.16). Denote the eigenvalues of $\tilde{A}$ by $\lambda_1, \lambda_2, ..., \lambda_n$. We have

$$|\det \tilde{A}| = |\lambda_1 \lambda_2 ... \lambda_n|.$$

Consider an arbitrary eigenvalue λ_k, let s^k be a corresponding eigenvector, and
$$|s_i^k| = \max_{i \leq j \leq n} |s_j^k|$$

where s_j^k are the coordinates of s^k. The division of the equality

$$\sum_{j=1}^n \tilde{a}_{ij} s_j^k = \lambda_k s_i^k$$

by $|s_i^k| > 0$ yields

$$|\lambda_k| \geq |a_{ii}| - \sum_{j=1, j \neq i}^n |\tilde{a}_{ij}||s_j^k/s_i^k| \geq |\tilde{a}_{ii}| - \sum_{j=1, j \neq i}^n |\tilde{a}_{ij}| = 1.$$

Since the last inequality is valid for every $k = 1, 2, ..., n$, we obtain that $|\det \tilde{A}| \geq 1$ implying (A1.16), and all the σ_is being positive proves the Theorem. $\square$

Note that applying the result to the transpose A' of A yields that condition (A1.15) ensuring the regularity of the matrix A can be replaced by

$$|a_{ii}| > \sum_{j=1, j \neq i}^n |a_{ji}| \qquad (i = 1, 2, ..., n). \qquad (A1.17)$$

The conditions (A1.15) and (A1.17) can be expressed by saying that the element in the main diagonal is *dominant* in its row, resp. column. Hadamard's Theorem has the following consequence.

Theorem A1.15 (*Parodi [1952]*). *Let $A = [a_{ij}]$ be an $n \times n$ matrix with real elements; if*

$$a_{ii} > 0, \qquad a_{ii} > \sum_{j=1, j \neq i}^n |a_{ij}| \qquad (i = 1, 2, ..., n),$$

then $\det A > 0$.

Proof. Applying the notations

$$\sigma_i = a_{ii} - \sum_{j=1, j \neq i}^{n} |a_{ij}| > 0 \qquad (i = 1, 2, ..., n) \tag{A1.18}$$

as before, we may write

$$\det A = \begin{vmatrix} \sum |a_{1j}| + \sigma_1 & a_{12} & \cdots & a_{1n} \\ a_{21} & \sum |a_{2j}| + \sigma_2 & \cdots & a_{2n} \\ \cdot & \cdot & & \cdot \\ a_{n1} & a_{n2} & \cdots & \sum |a_{nj}| + \sigma_n \end{vmatrix}$$

where the summation in the i-th row -i-th column element goes for $j = 1, ..., n$, $j \neq i$. We fix the values σ_i determined by (A1.18) and vary the off-diagonal elements of A, i.e. we consider the function F of $n(n-1)$ variables defined by

$$F(x_{12}, x_{13}, ..., x_{n,n-1})$$
$$:= \begin{vmatrix} \sum |x_{1j}| + \sigma_1 & x_{12} & \cdots & x_{1n} \\ x_{21} & \sum |x_{2j}| + \sigma_2 & \cdots & x_{2n} \\ \cdot & \cdot & & \cdot \\ x_{n1} & x_{n2} & & \sum |x_{nj}| + \sigma_n \end{vmatrix}$$

with the same summation convention as above. Because of Hadamard's Theorem, $F(x_{12}, ..., x_{n,n-1}) \neq 0$ for $(x_{12}, ..., x_{n,n-1}) \in \mathbf{R}^{n(n-1)}$, and $F(0, ..., 0) = \sigma_1 ... \sigma_n > 0$. Since F is continuous,

$$\det A = F(a_{12}, ..., a_{n,n-1}) > 0$$

must hold because of the Intermediate Value Theorem. □

Note that as in the case of Hadamard's Theorem the conditions that the diagonal elements a_{ii} are positive and they are *dominant* over their respective *columns* is also sufficient to the positivity of the determinant. It is also easy to see that condition (A1.18) is not necessary to the positivity of the determinant. In stability investigations the following Corollary is important.

Corollary A1.16. *If $A = [a_{ij}]$ is a symmetric $n \times n$ matrix with real elements and the conditions of the previous Theorem hold, then A is positive definite.*

Proof. The conditions of the previous Theorem, clearly, imply that all the principal minors of A are positive. But it is well known (see Gantmacher [1954] p.248) that the positivity of the elements of the sequence

$$a_{11}, \begin{vmatrix} a_{11} & a_{12} \\ a_{21} & a_{22} \end{vmatrix}, ..., \det A$$

is necessary and sufficient for the positive definiteness of A. $\square$

In several places in this book we make use of the norm of a square matrix. This is supposed to be a familiar concept, and for most cases we do not specify what kind of a norm is to be applied. It is to be remembered that the *norm of a square matrix* (or of a linear operator in a normed linear space, see Appendix 2 around Definitions A2.4-A2.7) A is by definition

$$|A| := \max_{|x|=1} |Ax| = \max |Ax|/|x|. \tag{A1.19}$$

It is clear from the definition that for an arbitrary vector x we have

$$|Ax| \le |A||x|. \tag{A1.20}$$

If the norm of a vector $x \in \mathbf{R}^n$ is defined by

$$|x| := \left(\sum x_i^2 \right)^{1/2} \tag{A1.21}$$

(*"Euclidean norm"*), then the corresponding natural martrix norm is

$$|A| = (\max_i \lambda_i(A'A))^{1/2} \tag{A1.22}$$

where $\lambda_i(A'A)$ $(i = 1, 2, ..., n)$ are the eigenvalues of the positive semidefinite matrix $A'A$ (in case matrices with complex elements are considered A^*A is to be used). Indeed,

$$|Ax| = ((Ax)'Ax)^{1/2} = (x'A'Ax)^{1/2},$$

and if $|x| = 1$, then the maximal value of the right-hand side is achieved when a unit eigenvector of $A'A$ corresponding to the maximal eigenvalue is substituted for x.

If the norm of a vector $x \in \mathbf{R}^n$ is defined by

$$|x| := \max_i |x_i|, \tag{A1.23}$$

then the corresponding natural matrix norm is

$$|A| = \max_i \sum_{k=1}^{n} |a_{ik}| \tag{A1.24}$$

where, naturally, $A = [a_{ik}]$. Indeed,

$$
\begin{aligned}
|Ax| &= \max_i |\sum_{k=1}^{n} a_{ik}x_k| \le \max_i \sum_{k=1}^{n} |a_{ik}| \max_k |x_k| \\
&= \max_i \sum_{k=1}^{n} |a_{ik}||x| = \max_i \sum_{k=1}^{n} |a_{ik}|
\end{aligned}
$$

if $|x| = 1$. The value on the right-hand side is achieved when $x_k = \operatorname{sign} a_{ik}$ $(k = 1, 2, ..., n)$ is substituted for each $i = 1, 2, ..., n$ and from these the maximal value is chosen.

It is assumed that the reader knows that every square matrix A can be brought to *Jordan normal form* by a linear, regular coordinate transformation, i.e. to a form where the matrix consists of "*Jordan blocks*" along the main diagonal

$$J_i = \begin{pmatrix} \lambda_i & 0 & ... & 0 & 0 \\ 1 & \lambda_i & ... & 0 & 0 \\ 0 & 1 & ... & 0 & 0 \\ 0 & 0 & ... & 1 & \lambda_i \end{pmatrix} \qquad (A1.25)$$

where λ_i $(i = 1, 2, ..., r)$ are the eigenvalues of the matrix. Several Jordan blocks may correspond to the same eigenvalue, and a block may consist of just one element: $[\lambda_i]$. If all of the Jordan blocks consist of one element, then the matrix has been (and could be) brought to diagonal form. This is possible even if the characteristic polynomial has multiple eigenvalues, the condition being that all the eigenvalues must be simple in the *minimal polynomial*.

We show now that given an arbitrary $\varepsilon > 0$ the matrix can be brought to a "*Jordan ε-normal form*" where in each block the numbers 1 in the diagonal are replaced by ε:

$$J_i(\varepsilon) = \begin{pmatrix} \lambda_i & 0 & ... & 0 & 0 \\ \varepsilon & \lambda_i & ... & 0 & 0 \\ . & . & ... & . & . \\ 0 & 0 & ... & \varepsilon & \lambda_i \end{pmatrix}. \qquad (A1.26)$$

In order to get (A1.26) from (A1.25) we have to apply the simple scale varying linear coordinate transformation that is given by the diagonal matrix whose part corresponding to the block (A1.26) is written out only (see Hartman [1964] p.68):

$$\begin{bmatrix} \ddots & & & & \\ & 1 & & & & 0 \\ & & 1/\varepsilon & & & \\ & & & 1/\varepsilon^2 & & \\ & & & & \ddots & \\ & 0 & & & & 1/\varepsilon^{h-1} \\ & & & & & & \ddots \end{bmatrix}$$

where the block (A1.26) is supposed to be of dimension $h \times h$. Indeed, if the notation $Q_\varepsilon := \operatorname{diag} [1, 1/\varepsilon, 1/\varepsilon^2, ..., 1/\varepsilon^{h-1}]$ is introduced, it is easy to compute that

$$J_i(\varepsilon) = Q_\varepsilon^{-1} J_i Q_\varepsilon.$$

The Jordan ε-normal form enables us to deduce easily an estimate for the norm of a matrix in terms of the eigenvalues.

Theorem A1.17. *If the eigenvalues of the $n \times n$ square matrix A are denoted by $\lambda_1, ..., \lambda_n$, and $\varepsilon > 0$ is arbitrarily given, then there is a linear coordinate transformation transforming A into $\hat{A}$ where $\hat{A}$ satisfies the inequality*

$$|\hat{A}| \le \varepsilon + \max_i |\lambda_i|. \tag{A1.27}$$

Proof. The norm (A1.24) attached to the vector norm (A1.23) will be used, and A will be transformed into its Jordan ε-normal form denoted by $\hat{A}$. The column vectors will be denoted by $\hat{x} = \text{col} [\hat{x}_1, ..., \hat{x}_n]$ in this new coordinate system. Then for vectors $\hat{x}$ with norm 1, i.e. $|\hat{x}| = \max_i |\hat{x}_i| = 1$ we have

$$
\begin{aligned}
|\hat{A}\hat{x}| &= \max_j |(\hat{A}\hat{x})_j| \\
&= \max(..., |\lambda_i \hat{x}_k|, |\varepsilon \hat{x}_k + \lambda_i \hat{x}_{x+1}|, |\varepsilon \hat{x}_{k+1} + \lambda_i \hat{x}_{k+2}|, ...) \\
&\le \varepsilon + \max_i |\lambda_i|.
\end{aligned}
$$

Thus,

$$|\hat{A}| = \max_{|\hat{x}|=1} |\hat{A}\hat{x}| \le \varepsilon + \max_i |\lambda_i|. \qquad \square$$

A similar estimate is not true for the norm of the original matrix A. "Conversely", there holds the following Theorem which will be used in the proof of Ostrowski's Theorem to be presented right after this one.

Theorem A1.18. *Denote the eigenvalues of the $n \times n$ matrix A by $\lambda_1, ..., \lambda_n$; then*

$$\max_k |\lambda_k| \le |A|. \tag{A1.28}$$

Proof. In (A1.28) any natural matrix norm may be used. Let λ_k be an arbitrary eigenvalue and s a corresponding unit eigenvector: $As = \lambda_k s$, $|s| = 1$. Then

$$|A| = \max_{|x|=1} |Ax| \ge |As| = |\lambda_k s| = |\lambda_k| \qquad (k = 1, 2, ..., n). \square$$

Consider now a system of n linear equations for n unknows $x_1, ..., x_n$:

$$Ax = b \tag{A1.29}$$

where $A = [a_{ik}]$ is an $n \times n$ matrix and $x = [x_1, ..., x_n]$, $b = [b_1, ..., b_n]$ are n dimensional column vectors. If A is a regular matrix, then (A1.29) has one and only one solution. We raise the question: *how much will this solution vary if the coefficient matrix A and the column vector b are perturbed?* More exactly, along with (A1.29) consider the perturbed system of equations

$$(A + \delta A)(x + \delta x) = b + \delta b \tag{A1.30}$$

where $\delta A = [\delta a_{ik}]$, resp. $\delta b = [\delta b_1, ..., \delta b_n]$ is a "small" square matrix, resp. column vector. What is a sufficient condition for $A + \delta A$ remaining regular, and how can $|\delta x|$ be estimated if we have an upper estimate for $|\delta A|$ and $|\delta b|$? For sake of definiteness we shall use the norms (A1.23)-(A1.24). We give a new proof for the following:

Theorem A1.19 (*Ostrowski [1937]*). *If A is regular and $|A^{-1}\delta A| < 1$, then the matrix $A + \delta A$ is regular, and for the difference δx of the solutions of (A1.30) and (A1.29) we have*

$$|\delta x| \le |A^{-1}|(|\delta b| + (|b| + |\delta b|)|A^{-1}||\delta A|/(1 - |A^{-1}\delta A|)). \qquad (A1.31)$$

Since $|A^{-1}\delta A| \le |A^{-1}||\delta A|$, therefore, clearly the condition

$$|\delta A| < 1/|A^{-1}| \qquad (A1.32)$$

implies $|A^{-1}\delta A| < 1$.

Proof.

$$A + \delta A = A(I + A^{-1}\delta A),$$

so that if $I + A^{-1}\delta A$ is regular, then $A + \delta A$ is regular too. But by Hadamard's Theorem A1.14, $I + A^{-1}\delta A$ is regular if

$$\left|1 + \sum_{k=1}^{n} a_{ik}^{-1}\delta a_{ik}\right| > \sum_{l\ne i} \left|\sum_{k=1}^{n} a_{ik}^{-1}\delta a_{kl}\right| \qquad (i = 1, 2, ..., n)$$

holds where $A^{-1} = [a_{ik}^{-1}]$. The last inequality holds if

$$1 > \sum_{l=1}^{n} \left|\sum_{k=1}^{n} a_{ik}^{-1}\delta a_{kl}\right| \qquad (i = 1, 2, ..., n)$$

since

$$\left|1 + \sum_{k=1}^{n} a_{ik}^{-1}\delta a_{ki}\right| \ge 1 - \left|\sum_{k=1}^{n} a_{ik}^{-1}\delta a_{ki}\right|.$$

But

$$\sum_{l=1}^{n} \left|\sum_{k=1}^{n} a_{ik}^{-1}\delta a_{kl}\right| \le |A^{-1}\delta A| < 1 \qquad (i = 1, 2, ..., n)$$

by assumption, so that $A + \delta A$ is regular.

Now, multiply (A1.30) by A^{-1} from the left; simple manipulations yield

$$(I + A^{-1}\delta A)(x + \delta x) = x + A^{-1}\delta b$$

or, since $I + A^{-1}\delta A$ is regular,

$$x + \delta x = (I + A^{-1}\delta A)^{-1}(x + A^{-1}\delta b).$$

Introducing the notation $B := A^{-1}\delta A$ the conditions of the Theorem, Theorem A1.18 and Theorem A1.6 imply that B can be substituted into the analytic function $1/(1 + \lambda)$, and $(I + B)^{-1} = 1/(1 + B) = \sum_{k=0}^{\infty}(-B)^k$. Hence,

$$x + \delta x = x + A^{-1}\delta b + \sum_{k=1}^{\infty}(-B)^k(x + A^{-1}\delta b),$$

that is,

$$
\begin{aligned}
|\delta x| &\leq |A^{-1}\delta b| + (|x| + |A^{-1}\delta b|)\sum_{k=1}^{\infty}|B|^k \\
&= |A^{-1}\delta b| + (|A^{-1}b| + |A^{-1}\delta b|)|B|/(1 - |B|)
\end{aligned}
$$

from where (A1.31) follows. $\square$

Besides the Jordan normal form which has, in general, complex entries it is often useful to transform a matrix with *complex eigenvalues* to a canonical form with *real entries*. We are showing this only in the case of 2×2 matrices.

Lemma A1.20. *Suppose that the real 2×2 matrix A has a pair of complex conjugate eigenvalues $\lambda_{1,2} = \alpha \pm i\omega$, $\alpha, \omega \in \mathbf{R}$, $\omega \neq 0$; then there is a linear coordinate transformation that carries A into the canonical form*

$$A \simeq \begin{pmatrix} \alpha & -\omega \\ \omega & \alpha \end{pmatrix}.$$

Proof. Denote an eigenvector corresponding to $\lambda_1 = \alpha + i\omega$ by $s = u - iv$ where $u = \mathrm{col}\,[u_1, u_2] \in \mathbf{R}^2$ and $v = \mathrm{col}\,[v_1, v_2] \in \mathbf{R}^2$. Then $\bar{s} = u + iv$ is an eigenvector corresponding to $\lambda_2 = \alpha - i\omega$, and since $\lambda_1 \neq \lambda_2$ the eigenvectors s and $\bar{s}$ are linearly independent. But this implies that the real vectors u and v are also linearly independent. Introducing u and v as a new basis, the matrix of the basis transformation is

$$T = \begin{pmatrix} u_1 & v_1 \\ u_2 & v_2 \end{pmatrix}.$$

It is easy to verify, applying

$$Au - iAv = \alpha u + \omega v + i(\omega u - \alpha v),$$

that in the new basis, we obtain

$$\begin{pmatrix} \alpha & -\omega \\ \omega & \alpha \end{pmatrix} = T^{-1}AT.\square$$

A2   Topological Degree and Fixed Point Theorems

In this Appendix a concise survey of degree theory and of fixed point theorems will be presented with an emphasis on those aspects that are most important in establishing existence of periodic solutions of differential systems. Because of lack of space most proofs will be omitted, and only relatively short and instructive proofs will be given. The omitted proofs can be found either in Nagumo [1951$_a$,1952$_b$] or in Cronin [1964]. Useful surveys of degree theory are presented in Rouche-Mawhin [1973] and in Rothe, E.H. [1976] also mainly without proofs. We are treating, first, degree theory in finite dimensional Euclidean spaces, then we turn to Banach spaces. Basic ideas of functional analysis will be supplied; for these see, e.g., Kolmogorov-Fomin [1972] or Hutson-Pym [1980].

Some notations are to be introduced. In the sequel $D \subset \mathbf{R}^n$ will denote an open and bounded subset of the Euclidean space $\mathbf{R}^n$. If $f : \overline{D} \to \mathbf{R}^n$ is a continuous mapping, $f \in C^0(\overline{D}, \mathbf{R}^n)$ ($\overline{D}$: the closure of D), then the *set of the zeroes of f in $\overline{D}$* will be denoted by

$$A_f := \{x \in \overline{D} : f(x) = 0\}$$

or by A if f is fixed. If $f \in C^1(D, \mathbf{R}^n)$, then $f'_x(x)$ denotes the derivative matrix of f whose elements are the first partial derivatives of the coordinates of f, i.e. $f'_x(x) = [f'_{ix_k}(x)]$. The determinant of this matrix is the *Jacobian* of f at $x \in D$, and it is denoted by

$$Jf(x) := \det f'_x(x).$$

A point $x \in D$ where $Jf(x) \neq 0$ is called a *regular point*, a point where $Jf(x) = 0$ a *singular point* of the mapping f. At a singular point the derivative matrix of f is a singular matrix. If f is linearized at a singular point, the resulting linear mapping is degenerate: it maps $\mathbf{R}^n$ into a lower dimensional subspace, or, in other words, its *kernel*, i.e. the subspace mapped into zero is of dimension one at least. (Some texts use the term "critical point" instead of singular point. We do not because we have called the zeroes of a vector field critical points.) The set of the singular points of f will be denoted by

$$B_f := \{x \in D : Jf(x) = 0\}$$

or by B if f is fixed. A *zero of f in D is* called *degenerate*, resp. *non-degenerate* if it is a singular, resp. a regular point of f. If f has no zero on ∂D, the boundary of D, and if every zero of f in D is *non-degenerate*, then f is called *non-degenerate*. Using symbols, f is *non-degenerate* if $f(x) \neq 0$, $x \in \partial D$ and $A_f \cap B_f = \emptyset$, the empty set. We shall need two Lemmata.

Lemma A2.1. *If $f \in C^0(\overline{D})$, $f \in C^1(D)$ and it is non-degenerate, then the set of zeroes of f, the set A_f, is finite.*

Proof. The conditions imply that each zero of f is non-degenerate; hence, because of the Inverse Function Theorem each zero has a neighbourhood where there is no other zero. If there were an infinite number of zeroes in $\overline{D}$, then because of the compactness of $\overline{D}$ they would have a cluster point in $\overline{D}$ which would be a zero too because of the continuity of f and it would not be isolated, a contradiction. □

Lemma A2.2 *(Sard's [1942] Lemma, see also Narasimhan [1968]). Let $f \in C^1(D)$, then the n dimensional volume (the Lebesgue measure) of the image of the singular points is zero, i.e. measure $f(B_f) = 0$.*

Proof. It is well known that the open set $D \subset \mathbf{R}^n$ is the union of a countable set of closed n dimensional cubes. The union of a countable set of sets of measure zero has measure zero. So if we can prove that the measure of the image of the intersection of a closed cube $C \subset D$ and B_f is zero, i.e. measure $f(C \cap B_f) = 0$, then we have proved the Lemma.

Let $C \subset D$ be a closed n dimensional cube. For arbitrary $x^0, x \in C$ we have

$$f(x) - f(x^0) = df(x^0, x - x^0) + r_{x^0}(x - x^0) \qquad \text{(A2.1)}$$

where $df(x^0, \cdot)$ is the differential mapping of f at x^0:

$$df(x^0, h) = f'_x(x^0)h, \qquad h \in \mathbf{R}^n;$$

the remainder r_{x^0} is such that

$$|r_{x^0}(h)|/|h| \to 0, \qquad \text{as} \quad h \to 0 \quad \text{uniformly} \quad \text{in} \quad x^0 \in C.$$

The cube C will be divided into small cubes, and those small cubes will be considered which have a non-empty intersection with the set of singular points B_f. We shall show that the union of the images of these cubes has measure less than an arbitrary positive number. Let the side of C be of length L, its diameter is then $Ln^{1/2}$, and $\varepsilon > 0$ be arbitrary given. There exists a positive integer m such that

$$|r_{x^0}(x - x^0)|/|x - x^0| < \varepsilon \quad \text{if} \quad x, x^0 \in C \quad |x - x^0| < Ln^{1/2}/m. \text{ (A2.2)}$$

Divide C into m^n equal cubes of side L/m, denote those which have a non-empty intersection with B_f by C_i $(i = 1, 2, ..., p \le m^n)$, i.e. it is assumed that $C_i \cap B_f \ne \emptyset$, and the rest of the small cubes do not contain singular points. Choose a singular point x^i in C_i. The restriction of f to C_i can be given by (A2.1) in the following way:

$$f(x) = f(x^i) + f'_x(x^i)(x - x^i) + r_{x^i}(x - x^i), \qquad x \in C^i.$$

The linearization of this mapping will be denoted by $D_i : C_i \to \mathbf{R}^n$,

$$D_i(x) := f(x^i) + f'_x(x^i)(x - x^i), \quad x \in C^i.$$

Introduce the notation

$$M = \max_{x \in C} \sum_{j,k=1}^{n} |f'_{jx_k}(x)|.$$

The diameter of C_i is $Ln^{1/2}/m$, so that the diameter of $D_i(C_i)$ is less than or equal to $MLn^{1/2}/m$. But, since $f'_x(x^i)$ is a singular matrix, $D_i(C_i)$ is contained in a subspace of dimension $n - 1$ at most, so that measure $D_i(C_i) \leq (MLn^{1/2}/m)^{n-1}$. The image of C_i by the non-linear mapping f has points which are of distance $|r_{x^i}(x - x^i)|$ at most from the set $D_i(C_i)$. By (A2.2), $|r_{x^i}(x - x^i)| \leq \varepsilon Ln^{1/2}/m$. Hence,

$$\text{measure } f(C_i) \leq (MLn^{1/2}/m)^{n-1} 2\varepsilon Ln^{1/2}/m.$$

Thus,

$$\text{measure } f(C \cap B_f) \leq \text{measure } f(\cup_{i=1}^{p} C_i) \leq \sum_{i=1}^{p} \text{measure } f(C_i)$$

$$\leq m^n (MLn^{1/2}/m)^{n-1} 2\varepsilon Ln^{1/2}/m = 2\varepsilon M^{n-1} L^n n^{n/2}.$$

Since here the coefficient of ε does not depend on the division of C, this bound is arbitrarily small. Hence, measure $f(C \cap B_f) = 0$, and as a consequence, measure $f(B_f) = 0$. $\square$

We define the topological degree of a mapping with respect to a bounded open set first for non-degenerate mappings then for general C^1 mappings and finally for C^0 mappings. The topological degree is a kind of generalization of the Poincaré index which has been defined for two dimensional vector fields (see Definition 3.5.3, Corollary 3.5.10 and Theorem 3.5.13).

Definition A2.1. Let $f \in C^0(\overline{D})$, $f \in C^1(D)$ be non-degenerate; *the topological degree of f with respect to the set D (and to 0) is the integer*

$$d[f, D, 0] : \; = \sum_{x \in A_f} \text{sign } Jf(x) \quad \text{if} \quad A_f \neq \emptyset,$$

$$d[f, D, 0] : \; = 0 \quad \text{if} \quad A_f = \emptyset.$$

Because of Lemma A2.1 this is a finite sum and its value is well defined. The words in parantheses stand there because the degree can be defined not only for 0 but also for arbitrary $y \in \mathbf{R}^n$ considering those points $x \in D$

where $f(x) = y$. In this case it is assumed that $f(x) \neq y$ if $x \in \partial D$ and that those points of D where f assumes the value y are regular. Under these assumptions the degree with respect to y is

$$d[f, D, y] := d[f - y, D, 0]$$

where $f - y$ is the mapping $x \rightarrow f(x) - y$. However, only the degree with respect to 0 will be used and, therefore, the clause "with respect to 0" will be omitted.

Now, we extend the degree concept to general C^1 mappings. Sard's Lemma A2.2 implies that for a mapping $f \in C^0(\overline{D})$, $f \in C^1(D)$ the interior of the set $f(B_f)$ is empty. If f is degenerate, i.e. $A_f \cap B_f \neq \emptyset$, then $0 \in f(B_f)$ but because of the implication of Sard's Lemma there is a sequence of points $y^k \in \mathbf{R}^n$, $k = 1, 2, 3, ...$, such that $y^k \rightarrow 0$ as $k \rightarrow \infty$, and $y^k \notin f(B_f)$ $(k = 1, 2, ...)$. Define the mapping $f^k : \overline{D} \rightarrow \mathbf{R}^n$, $f^k(x) = f(x) - y^k$ $(k = 1, 2, ...)$. Clearly, $B_{f^k} = B_f$, and $A_{f^k} \cap B_{f^k} = \emptyset$. Furthermore, if $f(x) \neq 0$, $x \in \partial D$, then for sufficiently large ks, $f^k(x) \neq 0$, $x \in \partial D$ since for sufficiently large ks, $|y^k| < m/2$, say, where m is the positive minimum of the continuous function $|f(x)|$ on the compact set ∂D. Thus, for sufficiently large ks the topological degrees $d[f^k, D, 0]$ are defined. It can be proved that the limit of the sequence $d[f^k, D, 0]$, $k = k_0 + 1, k_0 + 2, ...$ (k_0 sufficiently large) consisting of integer values exists and it is independent of the sequence y^k, $k = 1, 2, ...$.

Definition A2.2. Let $f \in C^0(\overline{D})$, $f \in C^1(D)$ be such that $f(x) \neq 0$, $x \in \partial D$; the topological degree of f with respect to the set D (and to 0) is the integer

$$d[f, D, 0] := \lim_{k \to \infty} d[f^k, D, 0].$$

Now we extend the concept of degree to continuous functions. The following theorem can be proved.

Theorem A2.3. Let the mapping $f \in C^0(\overline{D})$ be such that $f(x) \neq 0$, $x \in \partial D$, and $g^k \in C^0(\overline{D})$, $g^k \in C^1(D)$ $(k = 1, 2)$ be arbitrary mappings satisfying the inequality

$$|g^k(x) - f(x)| < \min_{x \in \partial \overline{D}} |f(x)| \qquad (k = 1, 2) \qquad (A2.3)$$

(here, obviously, the right-hand side is positive); then

$$d[g^1, D, 0] = d[g^2, D, 0].$$

This Theorem implies that the following Definition makes sense.

Definition A2.3 Let the mapping $f \in C^0(\overline{D})$ be such that $f(x) \neq 0$, $x \in \partial D$; the topological degree of f with respect to the set D (and to 0) is the integer

$$d[f, D, 0] := d[g, D, 0]$$

where $g \in C^0(\overline{D})$, $g \in C^1(D)$ is an arbitrary mapping satisfying (A2.3). (By the Weierstrass Approximation Theorem, such a g always exists.)

We sum up some simple properties of the topological degree in the next theorem.

Theorem A2.4.

(i) *If f is the identity, i.e. $f(x) = x$, $x \in D$, then*

$$d[f, D, 0] = \begin{cases} 1 & \text{if } 0 \in D \\ 0 & \text{if } 0 \notin \overline{D} . \end{cases}$$

(ii) *If $f \in C^0(\overline{D})$ is such that $f(x) \neq 0$, $x \in \partial D$, and $d[f, D, 0] \neq 0$, then there is a point $x^0 \in D$ such that $f(x^0) = 0$.*

(iii) *If D_1 and D_2 are bounded and open sets such that $D_1 \cup D_2 \subset D$, $D_1 \cap D_2 = \emptyset$, $\overline{D}_1 \cup \overline{D}_2 = \overline{D}$, and $f \in C^0(\overline{D})$, $f(x) \neq 0$ for $x \in \partial D_1 \cup \partial D_2$, then*

$$d[f, D, 0] = d[f, D_1, 0] + d[f, D_2, 0]$$

("additivity").

On the other hand, the following Theorem which establishes the invariance of the degree by homotopy (cf. Definition 3.5.2 and Lemma 3.5.2) needs much preparation.

Theorem A2.5 (*Invariance by homotopy*). *Let the one parameter family of mappings $f \in C^0([0,1] \times \overline{D})$ be such that $f_t(x) := f(t, x) \neq 0$ for $t \in [0,1]$, $x \in \partial D$. Then $d[f_t, D, 0] = $ constant for $t \in [0,1]$.*

This Theorem provides, probably, the strongest tool for determining the degree of a mapping. If we can imbed the function f_1 into a one parameter family of mappings f_t satisfying the conditions of the Theorem, and the degree of f_0 can be determined somehow, this will be also the degree of f_1.

From Theorem A2.5 we may get the following generalization of Lemma 3.5.1.

Corollary A2.6 (*Poincaré-Bohl Theorem*). *If $f, g \in C^0(\overline{D})$, and in no point x of ∂D have the vectors $f(x)$ and $g(x)$ opposite directions, then $d[f, D, 0] = d[g, D, 0]$.*

Proof. Note that the assumptions imply that $f(x) \neq 0 \neq g(x)$, $x \in \partial D$, because the zero vector is parallel and opposite to every vector. The

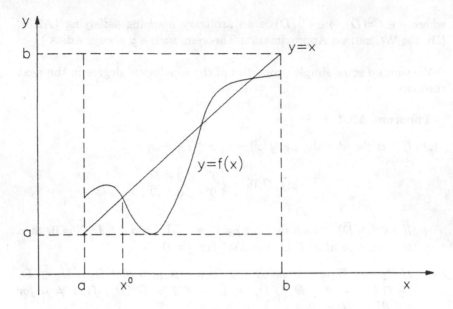

FIGURE A2.1. Illustration of Theorem A2.7: the graph of the continuous function $f : [a, b] \to [a, b]$ must intersect the straight line $y = x$.

assumption that $f(x)$ and $g(x)$ are not opposite on the boundary of D can be expressed by the formula

$$\lambda f(x) + (1 - \lambda)g(x) \neq 0, \quad \lambda \in [0, 1], \quad x \in \partial D. \tag{A2.4}$$

Thus, f and g are homotopic by the family

$$F(\lambda, x) := \lambda f(x) + (1 - \lambda)g(x), \quad f(x) = F(1, x), \quad g(x) = F(0, x). \square$$

The results of degree theory listed above enable us to prove Brouwer's Fixed Point Theorem in a simple way. This theorem is one of the most often applied tools in establishing the existence of a periodic solution of a differential system. It is the generalization of the following elementary theorem:

Theorem A2.7. *Let $f : [a, b] \to [a, b]$ be a continuous function mapping the compact interval $[a, b] \subset \mathbf{R}$ into itself; then this mapping has a fixed point, i.e. there exists an $x^0 \in [a, b]$ such that $f(x^0) = x^0$.*

The statement is obviously true if one looks at Figure A2.1. Its proof is left to the reader as an exercise.

Theorem A2.8 *(Brouwer's Fixed Point Theorem, special case). If f is a continuous mapping of the compact unit ball $\overline{B}$ into itself where*

$$B = B(0, 1) = \{x \in \mathbf{R}^n : |x| < 1\},$$

then there is an $x^0 \in \overline{B}$ such that $f(x^0) = x^0$.

Proof. If there is an $x \in \partial B$ such that $f(x) = x$, then we have a fixed point and the Theorem has been proved. Assume that $x - f(x) \neq 0$, $x \in \partial B$, and consider the family $F(t, x) = x - tf(x)$, $t \in [0, 1]$, $x \in \overline{B}$. We have $F(0, x) = x$, i.e. F_0 is the identity, and $F(1, x) = x - f(x)$. If $0 \leq t < 1$, then $F(t, x) = x - tf(x) \neq 0$ for $x \in \partial B$ because on the boundary $|x| = 1$ and $|tf(x)| = t|f(x)| < 1$. So $x - f(x)$ and the identity mapping I are homotopic, and because of Theorem A2.5

$$d[I - f, B, 0] = d[I, B, 0] = 1$$

by Theorem A2.4 (i). But this implies by Theorem A2.4 (ii) that there is an $x^0 \in B$ such that $x^0 - f(x^0) = 0$. $\square$

Corollary A2.9 (*Brouwer's Fixed Point Theorem, general case*). *Let $H \subset \mathbf{R}^n$ be a closed bounded subset homeomorphic to the closed unit ball $\overline{B}$; if $f \in C^0(H, H)$. Then there exists a point $x^0 \in H$ such that $f(x^0) = x^0$.*

Proof. Let $h : \overline{B} \to H$ be a homeomorphism mapping $\overline{B}$ onto H. Then $h^{-1} \circ f \circ h : \overline{B} \to \overline{B}$ is a continuous mapping, and according to the previous Theorem there is an $y^0 \in \overline{B}$ such that $(h^{-1} \circ f \circ h)(y^0) = y^0$. Denoting $x^0 = h(y^0)$, this implies that $f(x^0) = x^0$. $\square$

We conclude this study of topological degree in $\mathbf{R}^n$ with the computation of the degree in some simple cases. Further examples can be found among the Problems in Section 4.6.

Theorem A2.10. *Let*

$$B(0, r) = \{x \in \mathbf{R}^n : |x| < r\}$$

be the ball of radius $r > 0$ with centre at the origin, and consider the function $f \in C^0(\overline{B}(0, r), \mathbf{R}^n)$ as a vector field on $\overline{B}(0, r)$; if for all $x \in \partial B(0, r) = \{x \in \mathbf{R}^n : |x| = r\}$ the vector $f(x)$ is "directed outward" (resp. "inward"), then $d[f, B(0, r), 0] = 1$ (resp.$= (-1)^n$).

Proof. The expression that $f(x)$ is directed outward means that it forms an acute angle with the outward normal vector of the sphere ∂B which is the vector x at the point $x \in \partial B$. This means that f and the identity mapping I satisfy the conditions of Corollary A2.6 (the Poincaré-Bohl Theorem). Hence, $d[f, B(0, r), 0] = d[I, B(0, r), 0] = 1$ by Theorem A2.4 (i). If $f(x)$ is directed inward everywhere on ∂B, then it forms an acute angle with the vector $-x$ at $x \in \partial B$; thus, f and the mapping $-I$ satisfy the conditions of the Poincaré-Bohl Theorem. Hence, $d[f, B(0, r), 0] = d[-I, B(0, r), 0]$. But the mapping $-I(x) = -x$ has a single zero in $B(0, r)$, the origin, and its Jacobian is the determinant

$$\begin{vmatrix} -1 & 0 & \cdots & 0 \\ 0 & -1 & \cdots & 0 \\ \cdot & \cdot & \cdots & \cdot \\ 0 & 0 & \cdots & -1 \end{vmatrix} = (-1)^n,$$

so that by Definition A2.1 $d[-1, B(0, r), 0] = (-1)^n$. $\square$

Theorem A2.11. *If the mapping $f : \mathbf{R}^n \to \mathbf{R}^n$ is defined by $f(x) = Ax + b$ for every $x \in \mathbf{R}^n$ where A is an $n \times n$ regular matrix, $\det A \neq 0$, $b \in \mathbf{R}^n$, and $D \subset \mathbf{R}^n$ is a bounded open set such that $-A^{-1}b \notin \partial D$, then the topological degree of f with respect to D is defined, and*

$$d[f, D, 0] = \begin{cases} \operatorname{sign} \det A & \text{if } -A^{-1}b \in D \\ 0 & \text{if } -A^{-1}b \in \mathbf{R}^n \setminus \overline{D}. \end{cases}$$

Proof. This is an immediate consequence of Definition A2.1 since $-A^{-1}b$ is the only zero of f, and the Jacobian is $\det A$. $\square$

The proof of the following theorem can be found in Krasnoselskij [1966] p.95.

Theorem A2.12 *(Borsuk's Theorem). Assume that the bounded open set D contains the origin and is symmetrical to it, i.e. $x \in D$ implies that $-x \in D$; if $f : \overline{D} \to \mathbf{R}^n$ is an odd function, i.e. $f(-x) = -f(x)$ for $x \in \overline{D}$, and $f(x) \neq 0$, $x \in \partial D$, then $d[f, D, 0]$ is an odd number (hence non-zero).*

In the rest of this Section we shall extend the concept of degree to mappings of Banach spaces into themselves, i.e. we shall sum up the basic facts of the *Leray-Schauder degree theory* (see Leray-Schauder [1934] and Schauder [1930]), and formulate Schauder's and Banach's Fixed Point Theorems. The proofs of the theorems which follow can be found in the references mentioned at the beginning of this Appendix. Before introducing the degree some basic concepts of functional analysis will be presented. It is assumed that the reader is familiar with the concept of the (finite dimensional) *linear space* and with the concept and properties of *norm*. If in a linear space an arbitrary large number of linearly independent elements (vectors) can be found, then we speak about an *infinite dimensional linear space*. The norm of an element x of the normed linear space E will be denoted by $||x||$. The norm induces a *metric* in the space: the *distance* ρ of $x, y \in E$ is

$$\rho(x, y) := ||x - y||.$$

This way E becomes a *metric space*. The metric satisfies the following conditions: (i) $\rho(x, y) \geq 0$, and the distance is zero if and only if $x = y$;

(ii) $\rho(x,y) = \rho(y,x)$ (symmetricity); (iii) $\rho(x,y) \leq \rho(x,z) + \rho(z,y)$ for $x,y,z \in E$ (triangle inequality).

Definition A2.4. We say that the metric space E is *complete* if every Cauchy sequence of its elements has a limit point in E, i.e. if the sequence $x^k \in E$ ($k = 1,2,...$) has the property that for arbitrary $\varepsilon > 0$ there exists an $N \in \mathbf{N}$ such that $k,j > N$ implies $\rho(x^k, x^j) < \varepsilon$, then there is an $x \in E$ such that $\rho(x^k, x) \to 0$ as $k \to \infty$.

Definition A2.5. A normed linear space that is complete in the metric induced by the norm is called a *Banach space*.

It is assumed that the reader is familiar with the procedure that makes a Euclidean space out of a finite dimensional linear space by the introduction of a scalar product. If E is an arbitrary linear space over the field $\mathbf{C}$ of the complex numbers, then a *scalar product* is a function mapping the Cartesian product $E \times E$ into $\mathbf{C}$, the value of this function at $(x,y) \in E \times E$ is denoted by $<x,y>$. The scalar product must have the following properties: (i) if $x \neq 0$, then $<x,x>\ > 0$; (ii) $<y,x>=\ \overline{<x,y>}$ the complex conjugate; (iii) $<ax + by, z>=\ a <x,z> +b <y,z>$ for $x,y,z \in E$, $a,b \in \mathbf{C}$. The scalar product induces a norm in the space defined by $||x|| =<x,x>^{1/2}$, and this in turn induces a metric.

Definition A2.6. A linear space with a scalar product that is complete in the metric induced by the scalar product is called a *Hilbert space*.

A Hilbert space is said to be *separable* if it has a countable set of elements that is everywhere dense. An infinite dimensional, separable Hilbert space E has a *countable basis*, i.e. a sequence $x^k \in E$ ($k = 1,2,...$) such that for each vector $x \in E$ there is a sequence $c_k \in \mathbf{C}$ ($k = 1,2,...$) such that $||x - \sum_{k=1}^{m} c_k x^k|| \to 0$ as $m \to \infty$. The last relation is expressed by writing $x = \sum_{k=1}^{\infty} c_k x_k$.

A subset K of a normed linear space E is said to be *compact* if every infinite subset of K has a cluster point in K. It is important to note that in an infinite dimensional Banach space the fact that a subset is closed and bounded does *not* imply compactness. It can be shown, for instance, that the closed unit ball $\overline{B} = \{x \in E : ||x|| \leq 1\}$ of an infinite dimensional Banach space E is not compact. Examples can be given showing that one cannot expect a reasonable degree theory and fixed point theorems for mappings of infinite dimensional Banach spaces if they are assumed just to be continuous. The problems that arise are related to the compactness problem. In order to set up the type of mappings or rather *operators*, as they are called in the infinite dimensional case, for which topological degree can be defined, one proceeds in the following way. First "completely continuous operators" are introduced that can be approximated by operators that map a bounded set into a finite dimensional subspace. Then it can be shown that the degree of the difference of the identity and an operator that maps into a finite dimensional subspace can reasonably be defined by the restriction of this difference to the finite dimensional subspace. This way the degree theory can be generalized to a special but important class of operators.

Definition A2.7. Let $N : G \to E$ be an operator defined on the subset G of the Banach space E; the operator N is said to be *completely continuous* (or *compact*) if it is continuous, and if for any bounded subset $D \subset G$ the image $N(D)$ is a subset of a compact set of E.

Theorem A2.13. *Let $G \subset E$ be a bounded subset of the Banach space E and $N : G \to E$ be completely continuous; for every $\varepsilon > 0$ there exists a positive integer m and a completely continuous operator $N_\varepsilon : G \to E^m$ where E^m is an m dimensional subspace of E such that $\|N_\varepsilon x - N x\| \le \varepsilon$ for $x \in G$.*

Thus, according to this Theorem, completely continuous operators defined on bounded sets can be arbitrarily approximated by operators that carry the bounded set into a finite dimensional subspace. Now, in order to make what follows more plausible we return to the degree of continuous operators in finite dimensional spaces and establish one more property in the following:

Theorem A2.14. *Let $0 \le m < n$ be integers, $D \subset \mathbf{R}^n$ a bounded open subset and $f : \overline{D} \to \mathbf{R}^m$ a continuous mapping into the m dimensional subspace of $\mathbf{R}^n$ given by*

$$\mathbf{R}^m = \{x \in \mathbf{R}^n : x_{m+1} = \cdots = x_n = 0\}$$

such that $f(x) \ne x$ for $x \in \partial D$, denote the continuous mapping $I - f$ by g where I is the identity, i.e. $g(x) = x - f(x)$, $x \in \overline{D}$, and let $g|_{\overline{D} \cap \mathbf{R}^m}$ be the restriction of g to $\overline{D} \cap \mathbf{R}^m$. Then

$$d[g, D, 0] = d[g|_{\overline{D} \cap \mathbf{R}^m}, D \cap \mathbf{R}^m, 0].$$

Proof. The conditions imply that g has no zero on the boundary of D and that

$$(g_1(x), ..., g_n(x)) = (x_1 - f_1(x), ..., x_m - f_m(x), x_{m+1}, ..., x_n)$$

because $f_{m+1} = \cdots = f_n = 0$. Hence, all the zeroes of g are in $D \cap \mathbf{R}^m$, and in case f is in the C^1 class, the Jacobian of g in points of $D \cap \mathbf{R}^m$ is equal to the Jacobian of the restriction to $\overline{D} \cap \mathbf{R}^m$.

$$\left. \frac{\partial(g_1, ..., g_n)}{\partial(x_1, ..., x_n)} \right|_{(x_1, ..., x_m, 0, ..., 0) \in D}$$

$$= \begin{vmatrix} 1 - f_{1x_1} & -f_{1x_2} & \cdots & \cdot & \cdot & \cdots & -f_{1x_n} \\ \cdot & \cdot & \cdots & \cdot & \cdot & \cdots & \cdot \\ -f_{mx_1} & -f_{mx_2} & \cdots & 1 - f_{mx_m} & \cdot & \cdots & -f_{mx_n} \\ 0 & 0 & \cdots & 0 & 1 & \cdots & 0 \\ 0 & 0 & \cdots & 0 & 0 & \cdots & 1 \end{vmatrix}$$

where all the f_{ix_k} are to be taken at $(x_1, ..., x_m, 0, ..., 0)$. $\square$

We proceed now by analogy.

Definition A2.8. Let $D \subset E$ be a bounded open set of the Banach space E and $f \in C^0(\overline{D}, E)$ such that $f(\overline{D}) \subset E^m$ a finite dimensional subspace of E; if $x - f(x) \neq 0$ for $x \in \partial D$, then the *topological degree of the operator* $I - f$ with respect to the set D (and to 0) is by definition

$$d[I - f, D, 0] := d[(I - f)|_{D \cap E^m}, D \cap E^m, 0]$$

where the right-hand side is the degree of a continuous mapping of a finite dimensional space.

This way the definition of degree of operators that send their domain into a finite dimensional subspace goes back to the degree concept of mappings in finite dimensional spaces. It is easy to see, applying Theorem A2.14, that this definition is independent of the choice of the subspace E^n which contains $f(\overline{D})$.

Let now $D \subset E$ be a bounded open subset of the Banach space E and

$$N : \overline{D} \to E$$

be a completely continuous operator such that $x - N(x) \neq 0$ for $x \in \partial D$. One may prove that a $\delta > 0$ exists such that $\|x - N(x)\| \geq \delta$ for $x \in \partial D$. According to Theorem A2.13, for arbitrary $0 < \varepsilon \leq \delta/2$ there exists a completely continuous operator N_ε that maps $\overline{D}$ into a finite dimensional subspace, satisfies $\|N_\varepsilon x - Nx\| < \varepsilon$ for $x \in \overline{D}$, and by virtue of $\varepsilon < \delta/2$ and the triangle inequality $x - N_\varepsilon x \neq 0$ for $x \in \partial D$. Applying these assumptions and notations the degree of $I - N$ can be defined in the following way.

Definition A2.9 The *topological degree of the operator $I - N$ with respect to the bounded open subset D of the Banach space E (and to 0)* is

$$d[I - N, D, 0] = \lim_{\varepsilon \to 0} d[I - N_\varepsilon, D, 0].$$

It can be shown that this limit exists, is independent of the appropriate choice of N_ε, and, naturally, is an integer. The degree of $I - N$ where N is a completely continuous operator defined above is sometimes called the *Leray-Schauder degree* in distinction of the "*Brouwer degree*" of continuous mappings in finite dimensional spaces. The Leray-Schauder degree has

properties analogous to the Brouwer degree. In particular it is invariant
with respect to homotopy as expressed by the following:

Theorem A2.15. *Let $D \subset E$ be a bounded, open subset of the Banach
space E and*
$$N : [0, 1] \times \overline{D} \to E$$
*be a completely continuous operator in $[0, 1] \times \overline{D}$ such that $N(\lambda, x) \neq x$ for
$\lambda \in [0, 1]$, $x \in \partial D$; introducing the notation $N_\lambda(x) = N(\lambda, x)$, the degree
$d[I - N_\lambda, D, 0]$ is defined and is independent of $\lambda \in [0, 1]$.*

We have

Theorem   A2.16   (Schauder's   Fixed   Point   Theorem).   Let
$B = \{x \in E : \|x\| < 1\}$ *be the open unit ball of the Banach space E and
$N : \overline{B} \to \overline{B}$ a completely continuous operator mapping $\overline{B}$ into itself; then
N has a fixed point in $\overline{B}$, i.e. there is an $x \in \overline{B}$ such that $Nx = x$.*

Note that the same theorem holds also if $\overline{B}$ is replaced by a bounded
convex closed set or by a set homeomorphic to it.

The computation of the Leray-Schauder degree is difficult. Some results
are summed up in Rouche-Mawhin [1973] II. p.185, see also Schwartz [1964].

We conclude this introduction to fixed point theorems by proving one
more that guarantees the uniqueness of the fixed point too. The price to
be paid for this is the assumption that the operator must be "contractive".

Definition A2.10. Let $N : D \to E$ be a continuous operator mapping a
subset $D \subset E$ into the Banach space E; we say that N is a *contraction
operator* if there exists a number $\alpha \in (0, 1)$ such that for arbitrary
$x^1, x^2 \in D$ the inequality
$$\|N(x^1) - N(x^2)\| \leq \alpha\|x^1 - x^2\|$$
holds.

Theorem A2.17 *(Banach's Fixed Point Theorem). Let $N : D \to E$ be
a contraction operator defined on the closed subset $D \subset E$ of the Banach
space E; if $N(D) \subset D$, then N has a unique fixed point in D.*

Proof. The proof goes by constructing a sequence of "succesive
approximations" that converges to the fixed point. Consider an $x^0 \in D$;
if $x^0 = Nx^0$, then we have found a fixed point; if this is not the case, define
$x^1 := Nx^0, ..., x^{n+1} := Nx^n \in D$ $(n = 0, 1, 2, ...)$. If the procedure stops at
a finite n because $x^n = Nx^n$, then we have found a fixed point. Otherwise,
for the sequence x^n, $n \in \mathbf{N}$, we have
$$\|x^{n+1} - x^n\| = \|Nx^n - Nx^{n-1}\|;$$

hence,

$$\|x^{n+1} - x^n\| \leq \alpha^n \|x^1 - x^0\|.$$

Since $0 < \alpha < 1$, this means that the sequence is a Cauchy sequence and, because the space is complete, has a limit $\tilde{x}$. But D is closed; therefore, $\tilde{x} \in D$. Because of the continuity of N, taking the limit of both sides in $x^{n+1} = Nx^n$ we obtain $\tilde{x} = N\tilde{x}$, i.e. $\tilde{x}$ is a fixed point. If there were another fixed point $\hat{x} \in D$, then

$$\|\tilde{x} - \hat{x}\| = \|N\tilde{x} - N\hat{x}\| \leq \alpha \|\tilde{x} - \hat{x}\|$$

implying $\|\tilde{x} - \hat{x}\| = 0$ because $0 < \alpha < 1$. $\square$

Besides the fixed point theorems above we need the concept of the derivative of an operator in a Banach space. Let E and Y be two Banach spaces, $D \subset E$ an open subset and $N : D \to Y$ an operator mapping D into Y.

Definition A2.11. We say that N is *differentiable* at the point $x \in D$ if there exists a bounded linear operator $L_x : E \to Y$ such that for $h \in E$

$$\|N(x + h) - N(x) - L_x h\| = v(\|h\|)$$

where $x + h \in D$ is assumed; the bounded linear operator L_x (which is uniquely determined if it exists) is called the *Fréchet derivative* of N at x.

One can easily prove that the derivative of a continuous linear operator is itself, that if an operator is differentiable then it is continuous, that there holds the "chain rule", that a finite sum of operators can be differentiated term by term, etc.

A3   Invariant Manifolds

Definition A3.1. The subset M of $\mathbf{R}^n$ is said to be an m *dimensional differentiable manifold of order* C^k (or an m *dimensional* C^k-*manifold*, $1 \leq m < n$, $k \geq 1$) if there exists a finite family of pairs (V_i, v_i) $(i = 1, 2, ..., N)$ where $V_i \subset M \subset \mathbf{R}^n$ is relatively open in M, and v_i is a C^k-diffeomorphism of V_i onto an open set of $\mathbf{R}^m$ such that $M = \cup_{i=1}^{N} V_i$, and for any pair $i, j \in \{1, 2, ..., N\}$ if $V_i \cap V_j \neq \emptyset$, then this intersection is relatively open in M. The set of pairs (V_i, v_i) $(i = 1, 2, ..., N)$ is called an *atlas* of M, and V_i is a *coordinate neighbourhood* of any of its points.

Here a C^k-*diffeomorphism* is a one-to-one map which is k times continuously differentiable along with its inverse. The assumptions imply that if $V_i \cap V_j \neq \emptyset$, then the map

$$v_j \circ v_i^{-1} : v_i(V_i \cap V_j) \to v_j(V_i \cap V_j) \tag{A3.1}$$

is a C^k-*diffeomorphism*. Here $v_i(V_i \cap V_j) \subset \mathbf{R}^m$ is the image of $V_i \cap V_j$ by the mapping v_i, and similarly for v_j.

The concept of the m dimensional C^k-manifold is a generalization of a smooth curve or a smooth surface of the three dimensional Euclidean space. A helix or a non-degenerate conic section is a one dimensional differentiable manifold in $\mathbf{R}^3$ (with the usual metric). Two crossing straight lines is not one because the last requirement of the Definition does not hold at the point of intersection. A paraboloid, a sphere, a cylinder or a Möbius band are two dimensional differentiable manifolds in $\mathbf{R}^3$. A cone is not unless it is deprived of its vertex.

If the conditions of the Definition hold and the v_is are analytic, then M is said to be an *analytic manifold*.

We note that the concept of a C^k-manifold can be introduced abstractly without relying on the "embedding space" $\mathbf{R}^n$ (see Narasimhan [1968]). However, by Whitney's Imbedding Theorem (in Narasimhan [1968]) any abstractly defined m dimensional manifold can be embedded many ways into $\mathbf{R}^n$ if $n \geq 2m + 1$.

$\mathbf{R}^m$ of Definition A3.1 will be called the *coordinate space*. If a point $x \in M$ is an element of a coordinate neighbourhood V_i, then $v_i = v_i(x) \in v_i(V_i) \subset \mathbf{R}^m$ represents its coordinates $u_i = (u_i^1, ..., u_i^m)$ $(i = 1, 2, ..., N)$ (the coordinates of u_i in $\mathbf{R}^m$ are denoted by upper indices). If $x \in V_i \cap V_j$, then (A3.1) represents, actually, the coordinate transformation that determines the coordinates of x considered as a point in V_j from the coordinates of x in V_i. However, in order to avoid an orgy of indices we shall denote V_i by V and V_j by $\tilde{V}$ in the sequel, and accordingly u_i by u and u_j by $\tilde{u}$. The derivative of the k-th coordinate of $v_j \circ v_i^{-1}$ with respect to u^l will be denoted by $\partial \tilde{u}^k / \partial u^l$ $(k, l = 1, 2, ..., m)$. (For the details of the differential geometric background of what follows, see, e.g., Laugwitz [1960] or Stoker [1969]). It will be assumed that the Jacobian of (A3.1) is nowhere zero:

$$\det[\partial \tilde{u}^k / \partial u^l] \neq 0.$$

Definition A3.2. We say that the differentiable manifold M is *orientable* if an atlas can be found such that for arbitrary two coordinate neighbourhoods $V_i = V$ and $V_j = \tilde{V}$ if $V \cap \tilde{V} \neq \emptyset$, then in every point of this intersection

$$\det[\partial \tilde{u}^k / \partial u^l] > 0. \tag{A3.2}$$

A paraboloid, a sphere, and a cylinder are orientable manifolds of dimension two; a Möbius band is not.

Consider a point x that belongs to two coordinate neighbourhoods V and $\tilde{V}$, i.e. $x \in V \cap \tilde{V}$, and consider all the smooth curves in M that pass through x. The tangent vectors of these curves at the point x span an m dimensional linear space T_x that is called the *tangent space* of M at x. Considering x as a point of V, respectively, of $\tilde{V}$, denote the coordinates of a vector of T_x in the natural local base by $p = (p^1, ..., p^m)$, resp., by

$\tilde{p} = (\tilde{p}^1, ..., \tilde{p}^m)$. It is easy to see that the coordinate transformation (A3.1) generates the transformation of the vector coordinates expressed by

$$\tilde{p}^k = \frac{\partial \tilde{u}^k}{\partial u^l} p^l \qquad (A3.3)$$

where the *Einstein convention* is applied, i.e. for an index occurring in a lower and in an upper position in the same term, summation is to be performed from 1 to m. (We note that the coordinates p^l, resp. $\tilde{p}^k$ of the vectors of the tangent space T_x are the so-called *contravariant coordinates*).

We shall introduce now (Riemannian) metric in M. First we define the scalar product of two vectors in the tangent space T_x at $x \in M$.

Let

$$g : \bigcup_{x \in M} \{x\} \times T_x^2 \to \mathbf{R} \qquad (A3.4)$$

be an invariant scalar function which is symmetric and linear in the second and third variables and whose restriction to an arbitrary coordinate neighbourhood V of any atlas belongs to the C^k class. Invariant means that if the atlas or the coordinate system is transformed, the values of g at the same point and for the same pair of vectors stay the same. The requirements imply that if $v \in V$, the coordinates of x in V are $u = (u^1, ..., u^m)$ and $p, q \in T_x$, then

$$g(x, p, q) = g_{ij} p^i q^j \qquad (A3.5)$$

where $g_{ij} = g_{ji}$ and $g_{ij} \in C^k$. We assume, further, that for arbitrary fixed $x \in M$ the quadratic form $g_{ij}(u) p^i p^j$ is positive definite. This implies that $\det[g_{ij}] > 0$. The invariance of the values of g, (A3.3) and (A3.5) imply that if $x \in V \cap \tilde{V}$ where V and $\tilde{V}$ are two intersecting coordinate neighbourhoods of an atlas, and $g_{ij}(u)$ and $\tilde{g}_{kl}(\tilde{u})$ are the *coordinates* of the "*tensor*" g in V and $\tilde{V}$, respectively, where u and $\tilde{u}$ are the coordinate vectors of the same point x in V and $\tilde{V}$, respectively, then the following transformation formula holds:

$$\tilde{g}_{kl}(\tilde{u}) = \frac{\partial u^i}{\partial \tilde{u}^k} \frac{\partial u^j}{\partial \tilde{u}^l} g_{ij}(u) \qquad (A3.6)$$

where, naturally, $u = (v_i \circ v_j^{-1})(\tilde{u})$ is to be substituted (cf. (A3.1)) into the right-hand side.

Definition A3.3. Let m be an m dimensional orientable differentiable manifold and g a "tensor field" satisfying all the conditions in the previous paragraph; we say that (A3.5) is the *scalar product* of the vectors $p, q \in T_x$ for arbitrary $x \in M$, and that by g a *Riemannian metric* has been introduced into M; the manifold M equipped with a Riemannian metric (i.e. the pair (M, g)) is called a *Riemannian manifold* and g its *metric tensor*.

Let V be a coordinate neighbourhood of an atlas of the Riemannian manifold (M, g) and $u : [\alpha, \beta] \to V$ a smooth curve in V, i.e. for $t \in [\alpha, \beta]$, $(\alpha < \beta)$, $u(t) \in V$ and $u \in C^k$. Denote $a = u(\alpha)$, $b = u(\beta)$. The *arc length* s of the curve $u(t)$, $t \in [\alpha, \beta]$ is, by definition,

$$s := \int_\alpha^\beta (g_{ij}(u(t))\dot{u}^i(t)\dot{u}^j(t))^{1/2} dt. \qquad (A3.7)$$

The integrand is, obviously, at each fixed $t \in [\alpha, \beta]$ the square root of the scalar product of the tangent vector to the curve in the given point with itself, i.e. the *norm* of the tangent vector.

We note that if Euclidean metric is introduced in the embedding space $\mathbf{R}^n$, then it *generates* a Riemannian metric on the differentiable manifold M; however, the Riemannian metric introduced by us on M choosing arbitrarily a metric tensor does not coincide necessarily with the metric generated by the metric of the embedding space. On the other hand, if M is an m dimensional differentiable manifold equipped with a Riemannian metric, then it can always be embedded (preserving also the metric) into a $2m + 1$ dimensional Euclidean space.

If $x = (x_1, ..., x_n)$ are considered Cartesian orthogonal coordinates in $\mathbf{R}^n$ where scalar product is defined the usual way, and $x = v^{-1}(u)$ is a parametric representation of a coordinate neighbourhood V of the m dimensional differentiable manifold M, then the Riemannian metric on M generated by the Euclidean metric of $\mathbf{R}^n$ is defined by the metric tensor

$$
\begin{aligned}
g_{ij}(u) : &= \; < v_{u^i}^{-1}(u), v_{u^j}^{-1}(u) > \\
&= \sum_{k=1}^n v_{ku^i}^{-1'}(u) v_{ku^j}^{-1'}(u) \qquad (i, j = 1, 2, ..., m) \qquad (A3.8)
\end{aligned}
$$

where, e.g., $v_{ku^i}^{-1'}$ is the partial derivative of the k-th coordinate of v^{-1} with respect to u^i.

Definition A3.4. A smooth curve of the Riemannian manifold M is called a *geodesic* if it is the shortest curve in M connecting any pair of its sufficiently close points.

This means that if we fix two sufficiently near points on M and consider the smooth curves of M connecting these two points, then the curve with the least arc length (A3.7) is a geodesic. In other words the geodesics arise as the "*extremals*" or rather the "*stationary curves*" of the fixed end points variational problem related to the integral in (A3.7) considered as a functional. A calculation that is routine in variational calculus yields the *Euler-Lagrange differential equations* of this problem (see, e.g., Arnold [1974]). Provided that the curves are given in *natural parametrization*, i.e. *the arc length is introduced as the parameter* on the curves, the Euler-Lagrange equations are

$$\frac{d^2 u^i}{dt^2} + \Gamma^i_{kl}(u)\frac{du^k}{dt}\frac{du^l}{dt} = 0 \qquad (i = 1, 2, ..., m) \qquad (A3.9)$$

where Γ^i_{kl} are the so-called *Christoffel symbols of the second kind*

$$\Gamma^i_{kl} = (g^{ij}/2)(\partial g_{kj}/\partial u^l + \partial g_{jl}/\partial u^k - \partial g_{kl}/\partial u^j); \qquad (A3.10)$$

here $[g^{ij}]$ is the inverse matrix of $[g_{ij}]$. We note that (A3.9) is the differential equation of the geodesics also if the independent variable is a linear function of the arc length measured along the solutions.

We need that much of Riemannian geometry, first of all, for Example 5.3.2. For the study of invariant manifolds related to dynamical systems we need only the concept of a differentiable manifold without the Riemannian structure. Following Definition 1.3.1 we have defined the invariant set of a flow. Dynamical systems will be considered in a neighbourhood of an equilibrium point along with the underlying autonomous system of differential equations. We may assume without loss of generality that the equilibrium is the origin, and the system is of the form

$$\dot{x} = Ax + g(x) \qquad (A3.11)$$

where A is an $n \times n$ constant matrix $g \in C^1(\mathbf{R}^n)$, $g(0) = 0$, $g'_x(0) = 0$, $g'_x(0)$ being the derivative matrix of g at $x = 0$. The flow generated by the system is denoted by φ_t, i.e. $\varphi_t(x) = \varphi(t, x)$ is the solution of (A3.11) that satisfies the initial condition $\varphi(0, x) = x$. We assume as usual in this book that all solution are defined on $t \in \mathbf{R}$.

Definition A3.5. The differentiable manifold M passing through the origin $x = 0$ is called a *local invariant manifold* of system (A3.11) or of the flow φ_t if an $\varepsilon > 0$ exists such that $x \in M, |\varphi(t, x)| < \varepsilon$ implies $\varphi(t, x) \in M$.

The variational system of (A3.11) with respect to the solution $\varphi(t, x)$ is

$$\dot{y} = (A + g'_x(\varphi(t, x)))y. \qquad (A3.12)$$

Denote the fundamental matrix of the latter system that assumes the unit matrix I at $t = 0$ by $\Phi(t, x)$, i.e. $\Phi(0, x) = I$, and

$$\dot{\Phi}(t, x) \equiv (A + g'_x(\varphi(t, x)))\Phi(t, x).$$

Following Theorem 1.1.3 we have shown that $\Phi(t, x) \equiv \varphi'_x(t, x)$. In particular, if $x = 0$, then $\varphi(t, 0) \equiv 0$, also $g'_x(\varphi(t, 0)) \equiv g'_x(0) = 0$, and (A3.12) assumes the form

$$\dot{y} = Ay. \qquad (A3.13)$$

Hence, $\Phi(t,0) = \varphi'_x(t,0) = \exp(At)$. This means that the flow φ_t has the form

$$\varphi_t(x) = \varphi(t,x) = e^{At}x + G(t,x) \qquad (A3.14)$$

where $G \in C^1$, $G(t,0) = 0$, $G'_x(t,0) = 0$.

Assumption (A). We assume without loss of generality that the coefficient matrix A has $s \geq 0$ eigenvalues with negative real parts, $c \geq 0$ eigenvalues with real part zero and $u \geq 0$ eigenvalues with positive real parts counted with multiplicities in the characteristic polynomial: $s + c + u = n$.

Denote the s dimensional, resp. c dimensional, resp u dimensional eigenspace correspondig to the eigenvalues with negative, resp. zero, resp. positive real parts by $\mathbf{R}^s$, $\mathbf{R}^c$, $\mathbf{R}^u$, respectively. Then $\mathbf{R}^n$ is the direct sum of these eigenspaces

$$\mathbf{R}^n = \mathbf{R}^s \oplus \mathbf{R}^c \oplus \mathbf{R}^u$$

which are invariant with respect to the linear system (A3.13). The restrictions or the projections of system (A3.13) to these eigenspaces generate three independent linear systems on these. The one on $\mathbf{R}^s$ is asymptotically stable: all solutions tend to zero exponentially as t tends to infinity. The one on $\mathbf{R}^c$ has some bounded non-trivial solutions and may have also solutions that tend to infinity as t tends to infinity not exponentially but as some positive power of t. The one on $\mathbf{R}^u$ is "*completely unstable*", i.e. all non-trivial solutions tend to infinity exponentially as t tends to infinity (all solutions tend to zero as t tends to $-\infty$). In other words, there is a regular linear coordinate transformation (see what has been written about the Jordan normal form around (A1.25)) that carries system (A3.13) into

$$\dot{\eta} = \tilde{A}\eta \qquad (A3.15)$$

where $\tilde{A} = \text{diag}\,[S,C,U]$, i.e. it is a diagonal "hypermatrix", S being an $s \times s$ stable matrix, C a $c \times c$ matrix whose eigenvalues have real part zero, and U an $u \times u$ matrix whose eigenvalues have positive real part. Considering the vector $\eta \in \mathbf{R}^n$ as the direct sum of the vectors $\eta^s \in \mathbf{R}^s$, $\eta^c \in \mathbf{R}^c$, $\eta^u \in \mathbf{R}^u$:

$$\eta = \eta^s \oplus \eta^c \oplus \eta^u,$$

system (A3.15) can be written in the form

$$\dot{\eta}^s = S\eta^s, \qquad \dot{\eta}^c = C\eta^c, \qquad \dot{\eta}^u = U\eta^u. \qquad (A3.16)$$

It is reasonable to call $\mathbf{R}^s$, $\mathbf{R}^c$ and $\mathbf{R}^u$ the *stable*, the *centre* and the *unstable subspace* of system (A3.13), respectively. The names of $\mathbf{R}^s$ and $\mathbf{R}^u$ are obvious. $\mathbf{R}^c$ is called the centre subspace because if C is diagonizable, i.e. every eigenvalue of A with zero real part has multiplicity 1 in the minimal polynomial, then every solution of system $\dot{\eta}^c = C\eta^c$ is bounded over $\mathbf{R}$ (see Theorems 1.4.5-1.4.6); actually, they are either periodic or linear combinations of periodic solutions (i.e, "*quasi-periodic*"). This means that the origin $\eta^c = 0$ of the subspace $\mathbf{R}^c$ is a centre.

The important question arises whether the non-linear system (A3.11) has a similar inner structure. The positive answer to this question is given in the following two theorems which we present here without proof.

The first theorem (for the proof see Hartman [1964] p.243 and also Kelley [1967]) guarantees the existence of the "stable and the unstable manifolds" of the non-linear system. The existence of a transformation analogous to the one that has carried (A3.13) over to (A3.16) is stated here. We shall deal simultaneously with the system (A3.11) and the flow (A3.14) generated by it.

Theorem A3.1. *Suppose that Assumption (A) holds for system (A3.11); then there exists a neighbourhood V_x of the origin $x = 0$ of $\mathbf{R}^n$ and a regular C^1 coordinate transformation $r : V_x \to V_z$ of this neighbourhood onto a neighbourhood V_z of the origin $z = 0$ such that $r(0) = 0$, the flow (A3.14) is transformed into the flow $\psi_t(z) := (r \circ \varphi_t \circ r^{-1})(z)$ of the form*

$$\psi(t, z) = \psi_t(z) = \mathrm{diag}\,[e^{St}, e^{Ct}, e^{Ut}]z + \tilde{G}(t, z) \qquad (A3.17)$$

where for arbitrary $T > 0$ the matrix $\exp(ST)$ is $s \times s$ with all its eigenvalues in modulus less than 1, $\exp(CT)$ is a $c \times c$ matrix with all its eigenvalues in modulus equal to 1, $\exp(UT)$ is an $u \times u$ matrix with all its eigenvalues in modulus greater than 1, the vector function $\tilde{G}$ is the direct sum of the vector functions $\tilde{G}^s$, $\tilde{G}^c$, $\tilde{G}^u$ of dimension s, c and u, respectively, i.e. $\tilde{G} = \tilde{G}^s \oplus \tilde{G}^c \oplus \tilde{G}^u$ or $\tilde{G} = (\tilde{G}_1, ..., \tilde{G}_n) = (\tilde{G}_1^s, ..., \tilde{G}_s^s, \tilde{G}_1^c, ..., \tilde{G}_c^c, \tilde{G}_1^u, ..., \tilde{G}_u^u)$, $\tilde{G}(t, 0) = 0$, $\tilde{G}_z'(t, 0) = 0$,

$$\tilde{G}^u(t, z^s, 0, 0) \equiv 0, \qquad \tilde{G}^c(t, z^s, 0, 0) \equiv 0 \qquad (A3.18)$$

and

$$\tilde{G}^s(t, 0, 0, z^u) \equiv 0, \qquad \tilde{G}^c(t, 0, 0, z^u) \equiv 0 \qquad (A3.19)$$

where $z = z^s \oplus z^c \oplus z^u$, $z^s \in \mathbf{R}^s$, $z^c \in \mathbf{R}^c$, $z^u \in \mathbf{R}^u$; the same coordinate transformation transforms system (A3.11) into

$$\begin{aligned} \dot{z}^s &= Sz^s + \tilde{g}^s(z^s, z^c, z^u), \\ \dot{z}^c &= Cz^c + \tilde{g}^c(z^s, z^c, z^u), \\ \dot{z}^u &= Uz^u + \tilde{g}^u(z^s, z^c, z^u) \end{aligned} \qquad (A3.20)$$

where S is an $s \times s$ stable matrix, C a $c \times c$ matrix every eigenvalue of which has zero real part, and U an $u \times u$ matrix every eigenvalue of which has positive real part,

$$\begin{aligned} \tilde{g}^u(z^s, 0, 0) &\equiv 0, & \tilde{g}^c(z^s, 0, 0) &\equiv 0, & (A3.21) \\ \tilde{g}^s(0, 0, z^u) &\equiv 0, & \tilde{g}^c(0, 0, z^u) &\equiv 0, & (A3.22) \end{aligned}$$

and $\tilde{g}(z) = (\tilde{g}^s \oplus \tilde{g}^c \oplus \tilde{g}^u)(z) = o(|z|)$ as $z \to 0$.

A few explanatory remarks are appropriate. A coordinate transformation is said to be *regular* if its Jacobian is nowhere zero. In other words it is locally a diffeomorphism. The transformed flow can be written in the form (A3.17), naturally, only until it stays in the neighbourhood V_z, i.e. until $\psi(t, z) \in V_z$. For every $z \in V_z$ there exists an open time interval containing $t = 0$ in which this is true. We may write (A3.17) in the form

$$
\begin{aligned}
\psi^s(t, z) &= e^{St} z^s + \tilde{G}^s(t, z^s, z^c, z^u), \\
\psi^c(t, z) &= e^{Ct} z^c + \tilde{G}^c(t, z^s, z^c, z^u), \\
\psi^u(t, z) &= e^{Ut} z^c + \tilde{G}^u(t, z^s, z^c, z^u).
\end{aligned}
\tag{A3.23}
$$

Here ψ^s represents the first s coordinates, ψ^c the following c coordinates, and ψ^u the last u coordinates of ψ.

The eigenvalues of the inverse of a regular square matrix are the reciprocals of the eigenvalues of the original matrix. Every matrix norm has the property $|PQ| \leq |P||Q|$ where P and Q are arbitrary square matrices of the same dimension. In both matrix norms (A1.22) and (A1.24) the norm of the unit matrix I is 1. Hence, if P is a regular matrix, then $1 = |PP^{-1}| \leq |P||P^{-1}|$, i.e. $|P| \geq 1/|P^{-1}|$. For arbitrary $T > 0$ the properties of the matrices $\exp(ST)$ and $\exp(UT)$ described in the Theorem and Theorems A1.17 and A1.9, respectively, imply that regular linear coordinate transformations can be performed both in $\mathbf{R}^s$ and in $\mathbf{R}^u$ such that denoting the transformed matrices again by $\exp(ST)$ and $\exp(UT)$, respectively, we have

$$
|\exp(ST)| < 1, \qquad |\exp(UT)| > 1.
\tag{A3.24}
$$

The first inequality is obvious, the second follows from $|\exp(UT)| \geq 1/|\exp(-UT)|$ and from the fact that the eigenvalues of $\exp(-UT)$ are in modulus less than 1.

Now, properties (A3.18) of the transformed flow or, alternatively, properties (A3.21) of the transformed system imply obviously that the s dimensional subspace characterized by $z^c = 0$, $z^u = 0$ is invariant with respect to the transformed flow (A3.17). In the original coordinates $x \in \mathbf{R}^n$ this is, naturally, not a linear subspace but an s dimenzional differentiable manifold whose parametric representation is $x = r^{-1}(z^s, 0, 0)$. Similarly (A3.19) or, alternatively, (A3.22) imply that $z^c = 0$, $z^s = 0$ is invariant, i.e. $x = r^{-1}(0, 0, z^u)$ is a locally invariant differentiable manifold of dimension u.

Definition A3.6. The locally invariant manifolds

$$
\begin{aligned}
W_s : &= \{x \in V_x \subset \mathbf{R}^n : x = r^{-1}(z^s, 0, 0), (z^s, 0, 0) \in V_z\}, \\
W_u : &= \{x \in V_x \subset \mathbf{R}^n : x = r^{-1}(0, 0, z^u), (0, 0, z^u) \in V_z\}
\end{aligned}
$$

are called the *stable* and the *unstable manifolds of the equilibrium point*

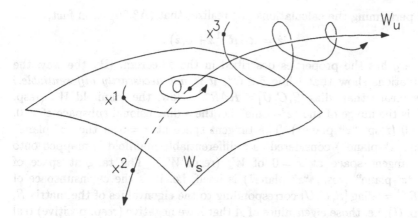

FIGURE A3.1. Here O is a hyperbolic equilibrium point with a positive eigenvalue and a pair of complex conjugate eigenvalues with negative real parts; the stable manifold W_s is two dimensional, the unstable manifold W_u is one dimensional; $x^1 \in W_s$, $x^2 \in W_u$, $x^3 \notin W_s \cup W_u$.

$x = 0$, respectively.

W_s is an s dimensional and W_u a u dimensional manifold. The properties of the matrices S and U imply that if the initial values are $x^1 \in W_s$, resp. $x^2 \in W_u$ then $\varphi(t, x^1) = r^{-1}(\psi(t, r(x^1))) \to 0$ exponentially as $t \to \infty$, and $\varphi(t, x^2) = r^{-1}(\psi(t, r(x^2))) \to 0$ exponentially as $t \to -\infty$, respectively, provided that $\varphi(t, x^1)$ and $\varphi(t, x^2)$ stay in V_x for $t \in [0, \infty)$, $t \in (-\infty, 0]$, respectively. This explains the names. Some authors including Hartman call W_s and W_u the stable and the unstable manifold, respectively, only if $c = 0$ in Assumption (A), i.e. if the coefficient matrix A has no eigenvalues with zero real part. If this is the case, then the equilibrium point $x = 0$ is said to be *hyperbolic*. If $x = 0$ is hyperbolic, then the path of every solution that tends to zero as $t \to \infty$ (resp. $t \to -\infty$) must belong to W_s (resp. to W_u) locally. This is called sometimes the *saddle point property* (see Hale [1969] Chapter 3.3). If $x = 0$ is not hyperbolic, the situation is not so simple; off the stable, resp. the unstable manifold there might be trajectories along which the solutions tend toward the equilibrium as $t \to \infty$ (resp. $t \to -\infty$). Figure A3.1 shows a hyperbolic equilibrium in three dimensions when the coefficient matrix A of the linearization of the system has a positive eigenvalue and a pair of conjugate complex eigenvalues with negative real parts.

The diffeomorphism r occurring in the Theorem can be written in the form $r(x) = Rx + \tilde{r}(x)$ where R is a regular matrix, $\tilde{r} \in C^1(V_x)$, $\tilde{r}(0) = 0$, $\tilde{r}'_x(0) = 0$. Its inverse is, clearly, $r^{-1}(z) = R^{-1}z + \rho(z)$ where $\rho(0) = 0$, $\rho'_z(0) = 0$. This transformation carries the system (A3.11) into (A3.20), and performing the calculations one realizes that (A3.20) is, in fact,

$$\dot{z} = RAR^{-1}z + \tilde{g}(z)$$

where $\tilde{g}$ has the properties described in the Theorem. (By the way, the calculations show that while $\tilde{g} \in C^0$, *it is not necessarily differentiable.*) This means that $\text{diag}[S, C, U] = RAR^{-1}$. Now, the manifold W_s (resp. W_u) is the image of the "z^s-plane" i.e. the s dimensional subspace $z^c = 0$, $z^u = 0$ (resp. "z^u-plane"). The tangent space at $z = 0$ of the "z^s-plane" (resp. "z^u-plane") considered as a differentiable manifold is mapped onto the tangent space at $x = 0$ of W_s (resp. W_u). The tangent space of the "z^s-plane" (resp. "z^u-plane") is itself, i.e. it is the eigensubspace of $RAR^{-1} = \text{diag}[S, C, U]$ corresponding to the eigenvalues of the matrix S, (resp. U), i.e. those eigenvalues of A that have negative (resp. positive) real parts. Hence, denoting again the the eigenspace of A corresponding to the eigenvalues with negative (resp. positive) real parts by $\mathbf{R}^s$ (resp. $\mathbf{R}^u$), we obtain that *the tangent space of the stable manifold W_s (resp. the unstable manifold W_u) is $\mathbf{R}^s$ (resp. $\mathbf{R}^u$).*

We assume from now on that $c \neq 0$, i.e. the matrix A has eigenvalues with zero real parts and turn to the problem of how the "centre subspace" of (A3.16) can be generalized to non-linear systems. The assumptions and notations introduced before are applied; however, since in some applications higher order continuous differentiability is needed, we assume that on the right-hand side of (A3.11) we have

$$g \in C^k(\mathbf{R}^n), \qquad k \geq 1. \tag{A3.25}$$

We recall that $\mathbf{R}^c$ denotes the $c > 0$ dimensional eigensubspace of the matrix A corresponding to the eigenvalues with zero real part. The direct sum of the stable and the unstable subspaces $\mathbf{R}^s \oplus \mathbf{R}^u$ will be called the *hyperbolic subspace* of the linearized system (A3.13). The hyperbolic subspace is the eigensubspace of A corresponding to the eigenvalues with *non-zero* real part; it is $s + u$ dimensional. We assume that the linear coordinate transformation $x \to \eta$ has been performed that carried A into the matrix $\tilde{A} = \text{diag}[S, C, U]$ (see also (A3.15) and (A3.16)). The vectors of $\mathbf{R}^c$ are denoted by η^c. The proof of the following Theorem can be found in Vanderbauwhede [1989] (see also Kelley [1967]).

Theorem A3.2. *Under the assumptions on systems (A3.11) and (A3.25) there exists a mapping $h : \mathbf{R}^c \to \mathbf{R}^s \oplus \mathbf{R}^u$ and a neighbourhood $V \subset \mathbf{R}^n$ of the origin $x = \eta = 0$ such that $h \in C^k(\mathbf{R}^c)$, $h(0) = 0$, $h'_{\eta^c}(0) = 0$; the set*

$$M_h := \{\eta^c \oplus h(\eta^c) \in \mathbf{R}^n : \eta^c \in \mathbf{R}^c\}$$

is a local invariant C^k-manifold for system (A3.11) in V; and if for an $x \in V$ there holds $\varphi(t, x) \in V$ for all $t \in \mathbf{R}$, then $x \in M_h$.

We note that h'_{η^c} is an $(s+u) \times c$ matrix, the set M_h is, in fact, the graph of the function h in the direct sum (or product) of its domain and image set $\mathbf{R}^n = \mathbf{R}^c \oplus (\mathbf{R}^s \oplus \mathbf{R}^u)$, as such, because of its smoothness it is a differentiable manifold indeed. It passes through the origin and the condition $h'_{\eta^c}(0) = 0$ implies that its tangent space is $\mathbf{R}^c$. The requirement that M_h is locally invariant means, naturally, that if $x \in M_h \cap V$, then $\varphi(t, x) \in M_h$ as long as $\varphi(t, x) \in V$. According to the last requirement fulfilled by M_h, every solution that is defined on the whole $\mathbf{R}$ and stays near the origin has its path contained completely in M_h. More vaguely and inaccurately, every solution bounded in the whole future *and* past must have its path contained in M_h. This is fairly intuitive since solutions off M_h have projections on the stable and unstable manifolds, i.e. may tend to infinity as $t \to -$ or $+\infty$.

Definition A3.7. Under the conditions of the previous Theorem the manifold M_h is called a *local centre manifold of system* (A3.11) *or of the flow* (A3.14).

The problem with the local centre manifold is that it may not be unique. This is shown by the following:

Example A3.1 (*Vanderbauwhede* [1989]). Consider the two dimensional system

$$\dot{x} = -x^3, \qquad \dot{y} = -y. \tag{A3.26}$$

The eigenvalues at the origin $(0,0)$ are $\lambda_1 = -1$, $\lambda_2 = 0$. The "one dimensional eigenspace" $\mathbf{R}^s$ corresponding to $\lambda_1 = -1$ is $\mathbf{R}^s = \{(0, y) : y \in \mathbf{R}\}$, and the eigenspace $\mathbf{R}^c$ corresponding to $\lambda_2 = 0$ is $\mathbf{R}^c = \{(x, 0) : x \in \mathbf{R}\}$. Let $a \in \mathbf{R}$, $b \in \mathbf{R}$ be chosen arbitrarily and define the function $h_{ab} : \mathbf{R} \to \mathbf{R}$ by

$$h_{ab}(x) := \begin{cases} a \exp(-1/(2x^2)), & x < 0 \\ 0, & x = 0 \\ b \exp(-1/(2x^2)), & x > 0. \end{cases}$$

It is easy to see that $h_{ab} \in C^\infty$, $h'_{ab}(0) = 0$, and that

$$\begin{aligned} M_{ab} : \quad &= \quad \{(x, 0) + (0, h_{ab}(x)) : x \in \mathbf{R}\} \\ &= \quad \{(x, h_{ab}(x)) : x \in \mathbf{R}\} \end{aligned}$$

is locally invariant, i.e. M_{ab} is a local centre manifold. System (A3.26) is easily integrable. The equation of the path corresponding to the initial condition $x(0) = x_0$, $y(0) = y_0$ is $y = y_0 \exp(1/(2x_0^2)) \exp(-1/(2x^2))$ if $x_0 \neq 0$ and $x = 0$ if $x_0 = 0$. Thus, we see that the graph of the function h_{ab}, i.e. the local centre manifold M_{ab}, is the union of two trajectories (one defined for $x > 0$, the other for $x < 0$) and the point trajectory $(0,0)$. If

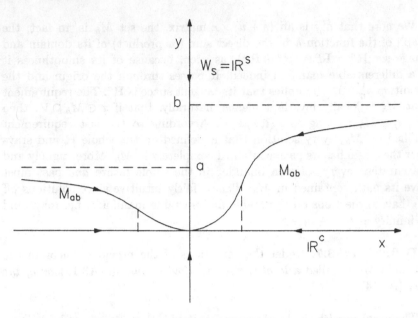

FIGURE A3.2. Phase portrait of system (A3.26) with the stable manifold W_s and a centre manifold M_{ab}, the last one being the union of two trajectories and the point trajectory $(0,0)$.

$(a_1, b_1) \neq (a_2, b_2)$, then $M_{a_1 b_1} \neq M_{a_2 b_2}$ even in small neighbourhoods of the origin. See the phase picture and the stable and centre manifolds in Figure A3.2. The last statement of Theorem A3.2 is not applicable here because no solution with initial values off the stable manifold is defined on the whole $\mathbf{R}$, and as t tends to the left end point of the domain of such a solution, $x(t)$ tends to $+$ or $-$ infinity.

The problem with the function h_{ab} of the previous example is that though it is C^∞, it is not analytic in any neighbourhood of the origin. $h_{ab}^{(k)}(0) = 0$ for $k \in \mathbf{N}$, so that its Taylor series expansion at $x = 0$ is identically zero but this has nothing to do with the function itself (the sum of the Taylor series is not equal to the function). There is a result due to Wan [1977] stating that all local centre manifolds of a system at an equilibrium have coinciding Taylor expansions up to the order but one of their common continuous differentability. This implies that an analytic system (like that of the previous Example) has *at most one analytic* local centre manifold (in the previous Example it is the x-axis.) At the same time it is also true that an analytic, resp. C^∞ system does not have necessarily an analytic, resp. C^∞ centre manifold. More about this and about *global* centre manifolds can be found in Vanderbauwhede [1989].

In most applications in bifurcation and stability theory those situations are the interesting ones in which the coefficient matrix of the linearized system has no eigenvalues with positive real part, i.e. $u = 0$ in Assumption (A) since otherwise we cannot expect stability anyway. The following Theorem concerns this case. Its proof can be found in Vanderbauwhede [1989], see also Hassard, Kazarinoff, Wan [1981].

Theorem A3.3. *Suppose that for system (A3.11) there holds also (A3.25), Assumption (A) holds with $s > 0$, $c > 0$, $u = 0$, i.e. in particular, the coefficient matrix A has no eigenvalues with positive real part, and let M be a local centre manifold; then there exist a neighbourhood $V \subset \mathbf{R}^n$ of the origin and a $0 < \alpha < \min_{1 \le i \le s} | \operatorname{Re} \lambda_i |$ where λ_i $(i = 1, 2, ..., s)$ are the eigenvalues of A with negative real part such that*

(i) *if $x \in M \cap V$, and $\varphi(t, x) \in V$, then $\varphi(t, x) \in M$;*

(ii) *if $x \in V$, and for $t \in (-\infty, 0]$ we have $\varphi(t, x) \in V$, then $x \in M$;*

(iii) *if $x \in V$, and the closure of the positive semitrajectory of x is contained in V, then there exist a $t_0 \ge 0$ and an $m \in M \cap V$ such that*

$$\sup_{t \ge t_0} \exp(\alpha t) | \varphi(t, x) - \varphi(t - t_0, m)| < \infty.$$

We shall refer to this Theorem as the *Centre Manifold Theorem*, though usually the latter is understood to contain also Theorem A3.2. The proposition (i) simply says that V is a neighbourhood in which M is an invariant manifold in the sense of Definition A3.5 and Theorem A3.2. Proposition (ii) states that only solutions corresponding to initial values on the centre manifold can stay near the origin for all t in the past. This is fairly intuitive, since solutions with projection on the stable manifold must tend away from the origin as $t \to -\infty$. Proposition (iii) means that the centre manifold is locally attractive, i.e. solutions which stay near the origin tend to M as $t \to \infty$; moreover, the dist $(\varphi(t, x), M)$ tends to zero exponentially as t tends to infinity, and further there is a solution with path in M such that the difference between $\varphi(t, x)$ and this latter solution tends to zero as $t \to \infty$. One could say that $\varphi(t, x)$ tends to M with "asymptotic phase" (cf. Definition 5.1.3).

Note that in case of Example A3.1 the conclusions of this Theorem hold true for an arbitrarily chosen centre manifold M_{ab}. In fact, every solution with initial value not on *this* centre manifolds tends to it, namely to the origin which is the common point of all the centre manifolds and of the stable manifold.

Up to this point in this Appendix we have dealt with invariant manifolds attached to a single equilibrium point, the origin of system (A3.11). We end this Appendix by quoting two theorems concerning invariant manifolds of autonomous systems which are unions of equilibria. The proofs of these theorems can be found in Aulbach [1984]. Consider the system

$$\dot{x} = f(x) \tag{A3.27}$$

where $f \in C^3(\mathbf{R}^n, \mathbf{R}^n)$, and assume that it has an invariant manifold M of dimension c consisting of equilibria of the system. There holds the following:

Theorem A3.4 *(Aulbach [1984]). Let M be a c dimensional invariant manifold consisting of equilibria of system (A3.27), and $\varphi : [0, \infty) \to \mathbf{R}^n$ an arbitrary solution; if the omega limit set Ω of φ is non-empty, for $\omega \in \Omega$ there exists a neighbourhood $V \subset \mathbf{R}^n$ such that $\Omega \cap V \subset M$ and $n - c$ eigenvalues of $f'_x(\omega)$ have non-zero real parts, then $\varphi(t) \to \omega$ as $t \to \infty$.*

It is easy to see that if $\omega \in M$ is an arbitrary point of M, then the directional derivatives of each coordinate of f in c linearly independent directions in the tangent space of M at ω are zero. Hence, if (A3.27) is linearized at ω, then at least c eigenvalues of the linearized system are equal to zero; thus, M is a submanifold of the local centre manifold. If ω is "as hyperbolic as it can be", i.e. $n - c$ eigenvalues of the linearized system have real parts different from zero, then M is a local centre manifold of ω. According to the Theorem if the omega limit set of a solution is locally contained in M which is a local centre manifold, then the solution cannot "drift" by M but must tend to a point ω of M, i.e. its omega limit set $\Omega = \{\omega\}$. This interpretation of the previous Theorem can be put in a stricter form.

Theorem A3.5 *(Aulbach [1984]). Let M be a c-dimensional compact invariant manifold of system (A3.27) consisting of equilibria, and assume that M is "normally hyperbolic", for every $x \in M$ the derivative matrix $f'_x(x)$ has $n - c$ eigenvalues with non-zero real parts; then M has a neighbourhood V_M such that if the positive semitrajectory of a solution stays in V_M, then it lies on the stable manifold of some point of M (alternatively, if the negative semitrajectory of a solution stays in V_M, then it lies on the unstable manifold of some point of M).*

We note that generalizations of these theorems for non-autonomous systems also hold true (see Aulbach [1984]).

References

Ahlfors, L. V. [1966]: *Complex Analysis*, McGraw Hill, New York.

Akulenko, L. D. [1967]: On oscillations and rotations in resonance in some mechanical systems, *Vestnik Moscow Univ.*, 1. 90–97 (Russian).

Akulenko, L. D.; Volosov, V. M. [1967$_a$]: On the resonance in a rotating system, *Vestnik Moscow Univ.* (Math-Mech), 1. 12–16 (Russian).

Akulenko, L. D.; Volosov, V. M. [1967$_b$]: Rotations in resonance of higher degree, *Vestnik Moscow Univ.* (Math-Mech), 2. 10–14 (Russian).

Alexander, J. C.; Yorke, J. A. [1983]: On the continuability of periodic orbits of parametrized three-dimensional differential equations, *J.Diff.Equ.*, 49. 171–184.

Allée, W. C.; Emerson, A. E. [1949]: *Principles of Animal Ecology*, Saunders, Philadelphia.

Alvarez, C.; Lazer, A. C. [1986]: An application of the topological degree to the periodic competing species problem, *J. Austral. Math. Soc.*, B 28. 202–219.

Amel'kin, V. V. [1977]: On the isochronism of a centre of two dimensional analytic differential systems, *Diff. Equ.*, 13. 971–980 (Russian).

Amel'kin, V. V.; Gaishun, I. V.; Ladis, N. N. [1975]: On periodic solutions under constantly acting perturbations, *Diff. Equ.*, 11. 2115–2120 (Russian).

Amel'kin, V. V.; Lukashevich, N. A. [1974]: Criteria of the existence of a centre and of its isochronism, *Diff. Equ.*, 10. 583–590 (Russian).

Andrès, J. [1987]: Solution with periodic second derivative of a certain third order differential equation, *Math. Slovaca*, 37. 239–245.

Andrès, J. [1990]: Periodic derivative of solutions to nonlinear differential equations, *Czechslovak Math. Journal*, 40. 353–360.

Andronov, A. A.; Leontovich, E. A. [1937]: Some cases of dependence of limit cycles on parameters, *Uchenie Zapiski Gor'kovskovo Gos. Univ.*, 6. 3–24 (Russian).

Andronov, A. A.; Leontovich, E. A.; Gordon, I. I.; Mayer, A.G. [1966]: *Qualitative Theory of Second Order Dynamical Systems*, Nauka, Moscow (Russian).

Andronov, A. A.; Leontovich, E. A.; Gordon, I. I.; Mayer, A. G. [1967]: *Bifurcation Theory of Dynamical Systems on the Plane*, Nauka, Moscow (Russian).

Andronov, A. A.; Pontriagin, L. S. [1937]: Robust systems, *Dokladi Akad. Nauk SSSR* 16., 247–250 (Russian).

Andronov, A.; Witt, A. [1933]: On Liapunov stability, *J. Electr. Techn. Phys.*, 3 (Russian).

Arnold, V. I. [1974]: *The Mathematical Methods of Classical Mechanics*, Nauka, Moscow (Russian).

Arnold, V. I. [1983]: *Geometrical Methods in the Theory of Ordinary Differential Equations*, Springer-Verlag, New York.

Arnold, V. I. [1984]: *Ordinary Differential Equations*, Nauka, Moscow (Russian).

Aulbach, B. [1981]: Behaviour of solutions near manifolds of periodic solutions, *J. Diff. Equ.*, 39. 345–377.

Aulbach, B. [1984]: *Continuous and Discrete Dynamics near Manifolds of Equilibria*, Lecture Notes in Maths 1058, Springer-Verlag, Berlin.

Bajaj, A. K. [1986]: Resonant parametric perturbations of the Hopf bifurcation, *J. Math. Anal.*, 115. 214–224.

Barbashin, E. A. [1967a]: *Introduction to the Theory of Stability*, Nauka, Moscow (Russian).

Barbashin, E. A. [1967$_b$]:   Conditions for the existence of recurrent trajectories in dynamical systems with cylindrical phase space, *Diff. Equ.*, 3. 1627–1633 (Russian).

Barbashin, E. A. [1967$_c$]:   The classification of the trajectories of a dynamical system with cylindrical phase space, *Diff. Equ.*, 3. 2015–2020 (Russian).

Barbashin, E. A. [1968]: On the construction of Liapunov functions, *Diff. Equ.*, 4. 2127–2158 (Russian).

Barbashin, E. A.; Krasovskij, N. N. [1952]: On the stability of motion in the large, *Dokladi Akad. Nauk SSSR* 86., 453–456 (Russian).

Barbashin, E. A.; Tabueva, V. A. [1969]:   *Dynamical Systems with Cylindrical Phase Space*, Nauka, Moscow (Russian).

Bautin, N. N. [1949]:   *The Behaviour of Dynamical Systems near to the Boundary of their Stability Domain*,   Gostechizdat, Moscow (Russian).

Bautin, N. N. [1952]:   On the number of limit cycles appearing by the variation of coefficients from a spiral point or centre type equilibrium, *Mat. Sb.*, 30. 181–196 (Russian).

Bender, C. M.; Orszag, S.A. [1978]: *Advanced Mathematical Methods for Scientists and Engineers*, McGraw-Hill, New York.

Bendixson, I. [1901]:   Sur les courbes définies par des équations différentielle, *Acta Math.*, 24. 1–88.

Besicovitch, A. S. [1954]:   *Almost Periodic Functions*,   Dover Publ., Cambridge.

Birkhoff, G. D. [1927]:   *Dynamical Systems*,   American Mathematical Society, New York.

Blehman, I. I. [1971]: *The Sinchronization of Dynamical Systems*, Nauka, Moscow (Russian).

Blows, T. R.; Lloyd, N. G. [1984]: The number of limit cycles of certain polynomial differential equations, *Proc. Roy. Soc. Edinburgh*, 98A. 215–239.

Bogoliubov, N. N.; Mitropolskij, Yu. A. [1965]: *Asymptotische Methoden in der Theorie der nichtlinearen Schwingungen*, Akademie Verlag, Berlin.

Bunke, H. [1972]:   *Gewöhnliche Differentialgleichungen mit zufälligen Parametern*, Akademie-Verlag, Berlin.

Burton, T. A. [1985]: *Stability and Periodic Solutions of Ordinary and Functional Differential Equations*, Academic Press, Orlando, FL.

Burton, T. A.; Hatvani, L. [1991]: On the existence of periodic solutions of some nonlinear functional differential equations with unbounded delay, *Nonlinear Anal. TMA*, 16. 389–398.

Busenberg, S.; Van den Driessche, P. [1993]: A method for proving the non-existence of limit cycles, *J. Math. Anal. Appl.*, 172. 463–479.

Butler, G. J. [1983]: Competitive predator-prey systems and coexistence, in: *Population Biology Proceedings, Edmonton 1982*, Lecture Notes in Biomath. 52, Springer-Verlag, Berlin, 210–297.

Butler, G. J.; Waltman, P. [1981]: Bifurcation from a limit cycle in a two predator-one prey ecosystem modeled on a chemostat, *J. Math. Biol.*, 12. 295–310.

Cesari, L. [1963]: *Asymptotic Behavior and Stability Problems in Ordinary Differential Equations*, Springer-Verlag, Berlin.

Chang, K. W.; Howes, F. A. [1984]: *Nonlinear Singular Perturbation Phenomena: Theory and Applications*, Springer-Verlag, New York.

Cheng, Kuo-Shung [1981]: Uniqueness of a limit cycle for a predator-prey system, *SIAM J. Math. Anal.*, 12. 541–548.

Chetaev, N. G. [1955]: *Stability of Motion*, Gostechizdat, Moscow (Russian).

Cima, Anna; Llibre, Jaume [1989]: Configurations of fans and nests of limit cycles for polynomial vector fields in the plane, *J. Diff. Equ.*, 82. 71–97.

Coddington, E. A.; Levinson, N. [1955]: *Theory of Ordinary Differential Equations*, McGraw-Hill, New York.

Collet, P.; Eckmann, J. P. [1980]: *Iterated Maps on the Interval as Dynamical Systems*, Progress in Physics vol.1, Birkhäuser, Basel.

Collet, P.; Eckmann, J. P.; Koch, H. [1981]: Period doubling bifurcations for families of maps on $\mathbf{R}^n$, *J. Stat. Phys.*, 25. 1–14 (also in Cvitanovic [1984], 353-366).

Cook, J. S.; Louisell, W. H.; Yocom, W. H. [1958]: Stability of an electron beam on a slalom orbit, *J. Appl. Phys.*, 29. 583–587.

Coppel, W. A. [1966]: A survey of quadratic systems, *J. Diff. Equ.*, 2. 293–304.

Corduneanu, C.; Lakshmikantham, V. [1980]: Equations with unbounded delay: a survey, *Nonlinear Analysis TMA* 4. 831–877.

Cronin, Jane [1964]: *Fixed Points and Topological Degree in Nonlinear Analysis*, American Mathematical Society, Providence R.I.

Cronin, Jane [1974]: Quasilinear equations and equations with large nonlinearities, *Rocky Mountain J. Math.*, 4. 41–63.

Cushing, J. M. [1977]: *Integrodifferential Equations and Delay Models in Population Dynamics*, Lecture Notes in Biomathematics 20, Springer-Verlag, Berlin.

Cushing, J. M. [1980]: Two species competition in a periodic environment, *J. Math. Biol.*, 10. 385–400.

Cvitanovic, P. (ed.) [1983]: *Universality in Chaos*, a reprint selection, Hilger, Bristol.

Dai, Lo-Sheng [1981]: Nonconstant periodic solutions in predator-prey systems with continuous time delay, *Math. Biosci.*, 53. 149–157.

Demidovich, B. P. [1967a]: *Lectures on the Mathematical Theory of Stability*, Nauka, Moscow, (Russian).

Demidovich, B. P. [1967b]: Analogue of the Andronov-Witt theorem, *Soviet Math. Dokl.*, 8. 1230-1232.

Devaney, R. L. [1989]: *An Introduction to Chaotic Dynamical Systems*, Addison-Wesley, Redwood City, CA.

Dieudonné, J. [1960]: *Foundations of Modern Analysis*, Academic Press, New York.

Diliberto, S. P.; Hufford, G. [1956]: Perturbation theorems for non-linear ordinary differential equations, in: *Contrib. to the Theory of Nonlin. Osc. III.*, Princeton Univ. Press, Princeton, NJ, 207–236.

Duffing, G. [1918]: *Erzwungene Schwingungen bei veränderlicher Eigenfrequenz*, Vieweg, Braunschweig.

Duistermaat, J. J. [1970]: Periodic solutions of periodic systems of ordinary differential equations containing a parameter, *Arch. for Rat. Mech. Anal.*, 38. 59–80.

Dulac, H. [1923]: Sur les cycles limites, *Bull. Sci. Math.*, 61. 45–188.

550   References

Écalle, J.; Martinet, J.; Mousse, R.; Ramis, J-P. [1987]: Non-accumulation des cycles limités I-II, *C.R. Acad. Sci. Paris I*, 304. 375–377, 431–434.

Écalle, J. [1993]:   *Introduction aux fonctions analysables et preuve constructive de la conjecture de Dulac*, Actualités mathématiques, Hermann, Paris.

El-Owaidy, H. [1972]: On perturbations of Liénard's equation, PhD thesis, Hungarian Academy of Science, Budapest.

El-Owaidy, H. [1975$_a$]:   Further stability conditions for controllably periodic perturbed solutions, *Studia Sci. Math. Hungar.*, 10. 277–286.

El-Owaidy, H. [1975$_b$]: On perturbations of Liénard's equation, *Studia Sci. Math. Hungar.*, 10. 287–296.

Fargue, D. [1973]: Réductibilité des systèmes héréditaires à des systèmes dinamiques, *C.R. Acad. Sci. Paris B*, 277. 471–473.

Farkas, A. [1988]: Andronov-Hopf bifurcation in a general predator-prey model with delay, PhD thesis, Eötvös Loránd University, Budapest (Hungarian).

Farkas, M. [1964]: *Special Functions with Applications in Engineering and Physics*, Műszaki K., Budapest (Hungarian).

Farkas, M. [1968]:   On stability and geodesics, *Annales Univ. Sci. Budapest, Math.*, 11. 145–159.

Farkas, M. [1970]:   Controllably periodic perturbations of autonomous systems, in: *Congrés International des Mathématiciens, Nice*, p. 228.

Farkas, M. [1971]:   Controllably periodic perturbations of autonomous systems, *Acta Math. Acad. Sci. Hungar.*, 22. 337–348.

Farkas, M. [1972]:   Determination of controllably periodic perturbed solutions by Poincaré's method, *Studia Sci. Math. Hungar.*, 7. 257–266.

Farkas, M. [1975$_a$]: On isolated periodic solutions of differential systems, *Annali di Mat. pura applicata*, 106. 233–243.

Farkas, M. [1975$_b$]: On periodic perturbations of autonomous systems, *Alk. Mat. Lapok*, 1. 197–254 (Hungarian).

Farkas, M. [1978$_a$]:   Estimates on the existence regions of perturbed periodic solutions, *SIAM J. Math. Anal.*, 9. 876–890.

Farkas, M. [1978$_b$]:   On isolatedness of orbits belonging to periodic solutions of differential systems, *Trudi MEI*, 357. 107–108 (Russian).

Farkas, M. [1980]:   The attractor of Duffing's equation under bounded perturbation, *Annali di Mat. pura applicata*, 128. 123-132.

Farkas, M. [1981]: Attractors of systems close to periodic ones, *Nonlinear Analysis TMA*, 5. 845–851.

Farkas, M. [1982]: Attractors of systems close to autonomos ones, *Acta Sci. Math. Szeged*, 44. 329–334.

Farkas, M. [1984$_a$]: Stable oscillations in a predator-prey model with time lag, *J. Math. Anal. Appl.*, 102. 175–188.

Farkas, M. [1984$_b$]: Zip bifurcation in a competition model, *Nonlinear Analysis TMA*, 8. 1295–1309.

Farkas, M. [1985]: A zip bifurcation arising in population dynamics, in: *ICNO X, Varna*, Bulagarian Acad. of Sci., Sofia, 150-155.

Farkas, M. [1987]: Competitive exclusion by zip bifurcation, in: *Dynamical Systems, IIASA Workshop, 1985, Sopron*, Lecture Notes in Econ. and Math. Systems 287, Springer-Verlag, Berlin, 165–178.

Farkas, M. [1990]: On the stability of one-predator two prey systems, *Rocky Mountain J. Math.*, 20. 909-916.

Farkas, M.; Abdel Karim, R. [1972]:   On controllably periodic perturbations of Liénard's equation. *Periodica Polytechnica Electr. Eng.*, 16. 41–45.

Farkas, M.; Bródy, A. [1987]: Forms of economic motion, *Acta Oeconomica Acad. Sci. Hungar.*, 38. 361–370.

Farkas, M.; Farkas, A. [1988]: Stable oscillations in a more realistic predator-prey model with time lag, in: *Asymptotic Methods of Mathematical Physics*, Naukova Dumka, Kiev, 250–256.

Farkas, M.; Farkas, A.; Szabó, G. [1987]:   Bifurcation charts for predator-prey models with memory, in: *Proc. ICNO XI.*, J. Bolyai Math. Soc., Budapest, 808–811.

Farkas, M.; Farkas, A.; Szabó, G. [1988]: Multiparameter bifurcation diagrams in predator-prey models with time lag, *J. Math. Biol.*, 26. 93–103.

Farkas, M.; Farkas, I. [1972]: On perturbations of Van der Pol's equation, *Annales Univ. Sci. Budapest*, Math., 15. 155–164.

552   References

Farkas, M.; Kotsis, M. [1992]:   Modelling predator-prey and wage-employment dynamics, in: *Dynamic Economic Models and Optimal Control*, Elsevier, Amsterdam, 513–526.

Farkas, M.; Pidal, M. [1981]:   *Estabilidad Estructural y Bifurcaciones*, Univ. Central de Venezuela, Caracas.

Farkas, M.; Sparing, L.; Szabó,G. [1986]: On Hopf bifurcation of Rayleigh's equation, *Periodica Polytechnica Mech. Eng.*, 30.263–271.

Farkas, M.; Stépán, G. [1992]: On perturbation of the kernel in infinite delay systems, *Z. angew. Math. Mech.*, 72. 153–156.

Feigenbaum, M. J. [1978]: Quantitative universality for a class of nonlinear transformations, *J. Stat. Phys.*, 19. 25–52.

Floquet, G. [1883]: Sur les équations différentielles linéaires à coefficients périodiques, *Annales École Norm. Sup.*, 12. 47–89.

Frame, J. S. [1974]: Explicit solutions in two species Volterra systems, *J. Theor. Biol.*, 43. 73–81.

Freedman, H. I. [1968]: Estimates on the existence region for periodic solutions of equations involving a small parameter. I: The noncritical case, *SIAM J. Appl. Math.*, 16. 1341–1349.

Freedman, H. I. [1971]: Estimates on the existence region for periodic solutions of equations involving a small parameter. II: Critical cases, *Annali di Mat. pura applicata*, 90. 259–279.

Freedman, H. I. [1980]: *Deterministic Mathematical Models in Population Ecolgy*, Dekker, New York.

Freedman, H. I.; So, J. W-H. [1985]: Global stability and persistence of simple food chains, *Math. Biosci.*, 76. 69–86.

Fricke, R. [1924]: *Lehrbuch der Algebra*, Vieweg, Braunschweig.

Friedrichs, K. O.; Wasow, W. R. [1946]:   Singular perturbations of non-linear oscillations, *Duke Math. J.*, 13. 367–381.

Gantmacher, F. R. [1954]: *The Theory of Matrices*, Gostechizdat, Moscow (Russian).

Gantmacher, F. R.; Krein, M. G. [1950]:   *Oscillation Matrices and Kernels and Small Oscillations of Mechanical Systems*, Gostechizdat, Moscow-Leningrad (Russian).

Garay, B. [1979_a]: On controllably periodic perturbations of autonomous functional differential equations, *Acta Math. Acad. Sci. Hungar.* 34. 317–320.

Garay, B. [1979_b]: Controllably periodic perturbations of autonomous functional differential equations, in: *Qualitative Theory of Differential Equations, Szeged*, North Holland, Amsterdam, 267–276.

Gause, G. F. [1934]: *The Struggle for Existence*, Williams and Wilkins, Baltimore.

Gause, G. F.; Smaragdova, N. P.; Witt, A. A. [1936]: Further studies of interaction between predator and prey, *J. Anim. Ecol.*, 5. 1–18.

Ghil, M.; Tavantzis, J. [1983]: Global Hopf bifurcation in a simple climate model, *SIAM J. Appl. Math.*, 43. 1019-1041.

Golubitsky, M.; Schaeffer, D. G. [1985]: *Singularities and Groups in Bifurcation Theory*, Springer-Verlag, New York.

Gopalsamy, K. [1982]: Exchange of equilibria in two species Lotka-Volterra competition models, *J. Austral. Math. Soc.* B, 24. 160–170.

Gopalsamy, K. [1985]: Global asymptotic stability in a periodic Lotka-Volterra system, *J. Austral. Math. Soc.* B, 27. 66–72.

Gopalsamy, K.; Kulenovic, M. R. S.; Ladas, G. [1990]: Environmental periodicity and time delays in a food limited population model, *J. Mat. Anal. Appl.*, 147. 545–555.

Grasman, J. [1987]: *Asymptotic Methods for Relaxation Oscillations and Applications*, Springer-Verlag, New York.

Grasman, J.; Jansen, M. J. W. [1979]: Mutually synchronized relaxation oscillators as prototypes of oscillating systems in biology, *J. Mat. Biol.*, 7. 171–197.

Grasman, W. [1977]: Periodic solutions of autonomous differential equations in higher dimensional spaces, *Rocky Mountain J. Math.* 7. 457–466.

Guckenheimer, J.; Holmes, P. J. [1983]: *Nonlinear Oscillations, Dynamical Systems and Bifurcations of Vector Fields*, Springer-Verlag, New York.

Gyökér, S. [1988]: *Stability investigation of concrete dynamical systems*, PhD thesis, University of Technology, Budapest.

Győri, I.; Ladas, G. [1991]: *Oscillation Theory of Delay Differential Equations with Applications*, Clarendon, Oxford.

Hagedorn, P. [1988]: *Non-linear Oscillations*, Clarendon, Oxford.

Halanay, A. [1966]: *Differential Equations, Stability, Oscillations, Time Lags*, Academic Press, Orlando, FL.

Hale, J. K. [1963]: *Oscillations in Non-linear Systems*, McGraw-Hill, New York.

Hale, J. K. [1969]: *Ordinary Differential Equations*, Wiley, New York.

Hale, J. K. [1977]: *Theory of Functional Differential Equations*, Springer-Verlag, New York.

Hale, J. K. [1979]: Nonlinear oscillations in equations with delays, *Lectures in Appl. Math.*, 17. 158–185.

Hale, J. K.; Kocak, H. [1991]: *Dynamics and Bifurcations*, Springer-Verlag, New York.

Hale, J. K.; Stokes, A. P. [1960]: Behaviour of solutions near integral manifolds, *Archive Rat. Mech. Anal.*, 6. 133–170.

Hartman, P. [1964]: *Ordinary Differential Equations*, Wiley, New York.

Has'minskij, R. A. [1969]: *The Stability of Differential Systems under Stochastic Perturbations of their Parameters*,   Nauka, Moscow (Russian).

Hassard, B. D.; Kazarinoff, N. D.; Wan, Y.-H. [1981]: *Theory and Applications of Hopf Bifurcation*, Cambridge Univ. Press, Cambridge.

Hastings, S. P.; Murray, J. D. [1975]: The existence of oscillatory solutions in the Field-Noyes model for the Belousov-Zhabotinskij reaction, *SIAM J. Appl. Math.*, 28. 678–688.

Hatvani, L.; Krisztin, T. [1992]: On the existence of periodic solutions for linear inhomogeneous and quasilinear functional differential equations, *J. Diff. Equ.*, 97. 1–15.

Hausrath A. R.; Manasevich, R. F. [1988]:   Periodic solutions of periodically forced non-degenerate systems,   *Rocky Mountain J. Math.*, 18. 49–65.

Hilbert, D. [1902]: Mathematical problems, *Bull. Amer. Math. Soc.*, 8. 437–479.

Hill, G. W. [1886]: On the part of motion of the lunar perigee which is a function of the mean motions of the sun and the moon, *Acta Math.*, 8. 1–36.

Hino, Y.; Murakami, S.; Naito, T. [1991]:   *Functional Differential Equations with Infinite Delay*, Lecture Notes in Math. 1473, Springer-Verlag, Heidelberg.

Hirsch, M. W. [1982]:   Systems of differential equations which are competitive or cooperative I, *SIAM J. Math. Anal.*, 13. 167–179.

Hirsch, M. W. [1985]:   Systems of differential equations that are competitive or cooperative II, *SIAM J. Mat. Anal.*, 16. 423–439.

Hirsch, M. W. [1990]:   Systems of differential equations that are competitive or cooperative IV: Structural stability in three dimensional systems, *SIAM J. Math. Anal.*, 21. 1225–1234.

Hirsch, M. W.; Pugh, C. C.; Shub, M. [1977]: *Invariant Manifolds*, Lecture Notes in Math. 583, Springer-Verlag, Heidelberg.

Hirsch, M. W.; Smale, S. [1974]:   *Differential Equations, Dynamical Systems, and Linear Algebra*, Academic Press, New York.

Hofbauer, J.; Mallet-Paret, J.; Smith, H. L. [1991]:   Stable periodic solutions for the hypercycle system, *J. Dyn. Diff. Equ.*, 3. 423–436.

Hofbauer, J.; Sigmund, K. [1988]: *The Theory of Evolution and Dynamical Systems*, London Math. Soc. Student Texts 7, Cambridge Univ. Press., Cambridge.

Hopf, E. [1942]:   Abzweigung einer periodischen Lösung von einer stationären Lösung eines Differenetialsystems, *Ber. Verh. Sachs. Akad. Wiss.* Leipzig, Math.-Nat., 95. 3–22.

Hoppensteadt, F. [1974]: Asymptotic stability in singular perturbation problems II. Problems having matched asymptotic expansion solutions, *J. Diff. Equ.*, 15. 510–521.

Hsu, C. S. [1974]: On approximating a general linear periodic systems, *J. Math. Anal. Appl.*, 45. 234–251.

Hsu, S. B.; Hubbel, S. P.; Waltman, P. [1978a]: A contribution to the theory of competing predators, *Ecological Monographs*, 48. 337–349.

Hsu, S. B.; Hubbel, S. P.; Waltman, Paul [1978b]: Competing predators, *SIAM J. Appl. Math.*, 35. 617–625.

Hutson, V.; Pym, J.S. [1980]: *Applications of Functional Analysis and Operator Theory*, Academic Press, London.

Ilyashenko, Yu. S. [1984]: Limit cycles of polynomial vector fields with non-degenerate equilibria on the real plane, *Funk. Anal. Priloz.*, 18. (3) 32–42 (Russian).

Ilyashenko, Yu. S. [1985]: Dulac's memoir "On limit cysles" and related problems of the local theory of differential equations, *Uspehi Mat. Nauk* 40. 41–78 (Russian).

Ilyashenko, Yu. S. [1991]: *Finiteness Theorems for Limit Cycles*, Translations of Math. Monographs 94, American Mathematical Society, Providence, RI.

Kamke, E. [1932]: Zur Theorie der Systeme gewöhnlicher Differentialgleichungen II, *Acta Math.*, 58. 57–85.

Kamke, E. [1959]: *Differentialgleichungen, Lösungsmethoden und Lösungen*, Akademische Verlags., Leipzig.

Kayumov, D. Ch.; Rozet, I. G.; Begiev, B. B. [1981]: The existence of four limit cycles for rational-quadratic differential equations, *Dokladi AN Tadzhik. SSR*, 24. 718–720 (Russian).

Keener, J. P. [1983]: Oscillatory coexistence in the chemostat: a codimension two unfolding, *SIAM J. Appl. Math.*, 43., 1005–1018.

Kelley, A. [1967]: The stable, center-stable, center, center-unstable and unstable manifolds, *J. Diff. Equ.*, 3. 546–570.

Kertész, V. [1987]: On stability of approximately periodic solutions, in: *Differential Equations: Qualitative Theory, Szeged*, North-Holland, Amsterdam, 527–541.

Kertész, V. [1988]: Notes on a theorem of M.Farkas, *Z. angew. Math. Mech.*, 68. 315–317.

Kertész, V.; Kooij, R. E. [1991]: Degenerate Hopf bifurcation in two dimensions, *Nonlinear Analysis TMA*, 17. 267–283.

Kolmanovskii, V. B.; Nosov, V. R. [1986]: *Stability of Functional Differential Equations*, Academic Press, London.

Kolmogoroff, A. [1936]: Sulla teoria di Volterra della lotta per l'esistenza, *Giornale dell Instituto Italiano degli Attuari*, 7. 74–80.

Kolmogorov, A. N.; Fomin, S. V. [1972]: *The Elements of Function Theory and of Functional Analysis*, third edition, Nauka, Moscow (Russian).

Kotsis, D. [1976]: On approximate determination of the characteristic multipliers of periodic systems, *Alk. Mat. Lapok*, 2. 269–276 (Hungarian).

Krasnosel'skij, M. A. [1966]: *The Displacement Operator along the Trajectories of Differential Equations*, Nauka, Moscow (Russian).

Krasovskij, N. N. [1963]: *Stability of Motion*, Stanford Univ. Press Stanford, CA.

Kryloff, N.; Bogoliuboff, N. [1949]: *Introduction to Non-linear Mechanics*, Princeton Univ. Press, Princeton, NJ.

Krylov, N. M.; Bogoliubov, N. N. [1934]: *The Application of the Methods of Non-linear Mechanics to the Theory of Stationary Oscillations*, Izd. Vseukr. Akad. Nauk, Kiev (Russian).

Kuo, B.C. [1977]: *Digital Control Systems*, SRL, Champaign, IL.

Lanford, O. E. III [1982]: A computer assisted proof of the Feigenbaum conjectures, *Bull. Amer. Math. Soc.*, 6. 427–434.

Langford, W. F. [1979]: Periodic and steady state interactions lead to tori, *SIAM J. Appl. Math.*, 37. 22–48.

Laugwitz, D. [1960]: *Differentialgeometrie*, Teubner, Stuttgart.

Lazer, A. C. [1975]: Topological degree and symmetric families of periodic solutions of nondissipative second-order systems, *J. Diff. Equ.*, 19. 62–69.

Lazer, A. C.; McKenna, P. J. [1990]: On the existence of stable periodic solutions of differential equations of Duffing type, *Proc. Amer. Math. Soc.*, 110. 125–133.

Lepoeter, K. [1983]: Hopf bifurcation, Research Report, Budapest University of Technology, Budapest.

Leray, J.; Schauder, J. [1934]: Topologie et équations fonctionelles, *Ann. Sci. École Nor. Sup.*, 51. 45–78.

Levi, M. [1981]: *Qualitative Analysis of the Periodically Forced Relaxation Oscillations*, Memoirs of the AMS, no. 244, vol. 32, American Mathematical Society, Providence, RI.

Levinson, N. [1943]: On the existence of periodic solutions for second order differential equations with a forcing term, *J. Math. Phys. Mass. Inst. Techn.*, 22. 41–48.

Levinson, N.; Smith, O. K. [1942]: A general equation for relaxation oscillations, *Duke Math. J.*, 9. 384–403.

Levitan, B. M. [1953]: *Almost Periodic Functions*, Gos. Izd. Tech-teor. Lit., Moscow (Russian).

Lewis, D. C. [1938]: Invariant manifolds near an invariant point of unstable type, *Amer. J. Math.*, 60. 577–587.

Lewis, D. C. [1955$_a$]: On the perturbation of a periodic solution when the variational system has non-trivial periodic solutions, *J. Rat. Mech. Anal.*, 4. 795–815.

Lewis, D. C. [1955$_b$]: Periodic solutions of differential equations containing a parameter, *Duke Math. J.*, 22. 39–56.

Lewis, D. C. [1956]: The role of first integrals in the perturbation of periodic solutions, *Annals of Math.*, 63. 535–548.

Lewis, D. C. [1961]: Autosynartetic solutions of differential equations, *Am. J. Math.*, 83. 1–32.

Li, Bingxi [1981]: Periodic orbits of autonomous ordinary differential equations: theory and applications, *Nonlinear Analysis TMA*, 5. 931–958.

Li, T. Y.; Yorke, J. [1975]: Period three implies chaos, *Amer. Math. Monthly*, 82. 985–992.

Liapunov, A. M. [1892]: *Problème générale de la stabilité de mouvement*, Annals of Mathematics Studies 17, Princeton, NJ, 1947 (originally: Kharkov, 1892, Russian).

Liapunov, A. M. [1902]: Sur une série dans la théorie des équations différentielles linéaires du second ordre à coefficients périodiques, *Zap. Akad. Nauk Fiz.-Mat. Otd.*, 8th series 13. 1–70.

Liénard, A. [1928]: Études des oscillations entretenues, *Rev. gen. Électr.*, 23. 901–946.

Lomov, S. A. [1981]: *Introduction to the General Theory of Singular Perturbations*, Nauka, Moscow (Russian), also: AMS Translations 112, Providence, RI, 1992.

Lotka, A. J. [1924]: *Elements of Mathematical Biology*, Dover, New York, 1956, (originally: 1924).

Loud, W. S. [1959]: Periodic solutions of a perturbed autonomous system, *Annals of Math.*, 70. 490–529.

MacDonald, N. [1977]: Time delay in prey-predator models. II. Bifurcation theory, *Math. Biosci.*, 33. 227–234.

MacDonald, N. [1978]: *Time Lags in Biological Models*, Lecture Notes in Biomathematics 27, Springer-Verlag, Berlin.

Malkin, I. G. [1956]: *Some Problems of the Theory of Non-linear Oscillations*, Gostechizdat, Moscow (Russian).

Mallet-Paret, J.; Smith, H. L. [1990]: The Poincaré-Bendixson Theorem for monotone cyclic feedback systems, *J. Dyn. Diff. Equ.*, 2. 367–421.

Mallet-Paret, J.; Yorke, J. A. [1982]: Snakes: Oriented families of periodic orbits, their sources, sinks, and continuation, *J. Diff. Equ.*, 43. 419–450.

Marsden, J. E.; McCracken, M. [1976]: *The Hopf Bifurcation and its Applications*, Springer-Verlag, New York.

Massera, J. L. [1950]: The existence of periodic solutions of systems of differential equations, *Duke Math. J.*, 17. 457–475.

Mathieu, É. [1868]: Mémoire sur le mouvement vibratoire d'une membrane de forme elliptique, *J. Math. Pure Appl.*, 13. 137–203.

Matrosov, V. M. [1962]: On the stability of motion, *Prikl. Mat. Mech.*, 26. 885–895 (*J. Appl. Math. Mech.*, 26. 1337-1353).

May, R. M.; Leonard, W. J. [1975]: Nonlinear aspects of competition between three species, *SIAM J. Appl. Math.*, 29. 243–253.

McGehee, R.; Armstrong, R. A. [1977]: Some mathematical problems concerning the ecological principle of competitive exclusion, *J. Diff. Equ.*, 23. 30–52.

Meixner, J.; Schäfke, F. W. [1954]: *Mathieusche Funktionen und Sphäroidfunktionen*, Springer-Verlag, Berlin.

Minorsky, N. [1947]: *Introduction to Non-linear Mechanics*, J. W. Edwards, Ann Arbor, MI.

Mishchenko, E. F.; Rozov, N. Kh. [1980]: *Differential Equations with Small Parameters and Relaxation Oscillations*, Plenum Press, New York.

Mitropolskij, Yu. A.; Homa, G. P. [1983]: *The Mathematical Foundation of the Asymptotic Methods of Non-linear Mechanics*, Naukova Dumka, Kiev (Russian).

Mitropolskij, Yu. A.; Lykova, O. B. [1973]: *Integral Manifolds in Nonlinear Mechanics*, Nauka, Moscow (Russian).

Moson P. [1976]: On the isolation of periodic solutions of autonomous systems, *Annales Univ. Sci. Budapest*, Math., 19. 63–67 (Russian).

Moson, P. [1977]: Examples of isolated periodic solutions, *Periodica Polytechnica Mech. Eng.*, 21. 13–23 (Russian).

de Mottoni, P.; Schiaffino, A. [1981]: Competition systems with periodic coefficients: a geometric approach, *J. Math. Biol.*, 11. 319–335.

Murdock, J. [1988]: Qualitative theory of nonlinar resonance by averaging and dynamical systems methods, in: *Dynamics Reported I*, Teubner-Wiley, Stuttgart, 91-172

Nagumo, M. [1951$_a$]: A theory of degree of mapping based on infinitesimal analysis, *Amer. J. Math.*, 73. 485–496.

Nagumo, M. [1951$_b$]: Degree of mapping in convex linear topological spaces, *Amer. J. Math.*, 73. 497–511.

Namachchivaya, N. Sri; Ariaratnam, S. T. [1987]: Periodically perturbed Hopf bifurcation, *SIAM J. Appl. Math.*, 47. 15–39.

Narasimhan, R. [1968]: *Analysis on Real and Complex Manifolds*, Masson and Cie-North-Holland, Paris-Amsterdam.

Nazarov, E. A. [1970]: Conditions for the existence of periodic trajectories in dynamical systems with cylindrical phase space, *Diff. Equ.*, 6. 337–380.

Negrini, P.; Salvadori, L. [1979]: Attractivity and Hopf bifurcation, *Nonlin. Anal. TMA*, 3. 87–99.

Nguyen Van Minh; Tran Van Nhung [1991]: Attractors of systems close to autonomous ones having a stable limit cycle, *Acta Mat. Hung.*, 58. 17–23.

Nguyen Van Su; Tran Van Nhung [1987]: An application of Farkas' estimates on the existence region of periodic solutions to perturbed Liénard's equation, in: *Differential Equations: Qualitative Theory, Szeged 1984*, North-Holland, Amsterdam, 789–797.

Nipp, K. [1988]: An algorithmic approach for solving singularly perturbed initial value problems, in: *Dynamics Reported*, vol I, Wiley-Teubner, Chichester-Stuttgart, 173–263.

Ostrowski, A. [1937]: Sur la détermination des bornes inférieures pour une classe des déterminants, *Bull. Sci. Math.*, 61., 19–32.

Parodi, M. [1952]: *Sur quelques propriétés des valeurs caractéristiques des matrices carrées*, Mémorial des Sciences Mathématiques 118, Gauthier-Villars, Paris.

Peixoto, M. M. [1959]: On structural stability, *Annals of Mathematics*, 69. 199–222.

Peixoto, M. M. [1962]: Structural stability on two dimensional manifolds, *Topology*, 1. 101–120.

Perron, O. [1907]: Jacobischer Kettenbruchalgoritmus, *Math. Annalen*, 64. 1–76.

Petrovskij, I. G.; Ladis, E. M. [1955]: On the number of limit cycles of the equation $dy/dx = P(x, y)/Q(x, y)$ where P and Q are polynomials of the second degree, *Mat. Sb.*, 37. 209–250 (Russian).

Pliss, V. A. [1965]: Families of periodic solutions of dissipative systems of second order differential equations, *Diff. Equ.*, 1. 1428–1448 (Russian).

Pliss, V. A. [1966]: *Nonlocal Problems of the Theory of Oscillations*, Academic Press, New York.

Pliss, V. A. [1977]: *Integral Sets of Periodic Differential Systems*, Nauka, Moscow (Russian).

Poincaré, H. [1886]:   Sur les courbes définies par les équations différentielles, *J. de Math.*, 7. (1881) 375–422, 8. (1882) 251–296, 1. (1885) 167–244, 2. (1886) 151–217.

Poincaré, H. [1890]: Sur le problème des trois corps et les équations de la dynamique, *Acta Math.*, 13. 1–270 (Oeuvres VII, 262–479).

Poincaré, H. [1899]: *Les méthodes nouvelles de la mécanique céleste*, I, II, III, Gauthier-Villar, Paris, 1892, 1893, 1899.

Ponzo, P. J.; Wax, N. [1984]: Periodic solutions of generalized Liénard equations, *J. Math. Anal. Appl.*, 104. 117–127.

Poston, T.; Stewart, I. [1978]: *Catastrophe Theory and its Applications*, Pitman, London.

Proskuryakov, A. P. [1977]: *Poincaré's Method in the Theory of Non-linear Oscillations*, Nauka, Moscow (Russian).

Rand, R. H.; Armbruster, D. [1987]: *Perturbation Methods, Bifurcation Theory and Computer Algebra*, (Applied Math. Sci. 65), Springer-Verlag, New York.

Reijn, J. W. [1989]: *A Bibliography of the Qualitative Theory of Quadratic Systems of Differential Equations in the Plane*, Reports Fac. Techn. Math. Informatics, no. 89-71, Delft.

Rinaldi, S.; Muratori, S. [1992]: Limit cycles in slow-fast forest-pest models, *Theoret.Population Biol.*, 41. 26-43.

Roseau, M. [1966]: *Vibrations non-linéaires et théorie de la stabilité*, Springer-Verlag, Berlin.

Rosenblat, S.; Cohen, D. S. [1980]: Periodically perturbed bifurcation-I. Simple bifurcation, *Studies Appl. Math.*, 63. 1-23.

Rosenblat, S.; Cohen, D. S. [1981]: Periodically perturbed bifurcation-II. Hopf bifurcation, *Studies Appl. Math.*, 64. 143-175.

Rothe, E. H. [1976]: Expository introduction to some aspects of degree theory, in: *Nonlinear Functional Analysis and Differential Equations*, Dekker, New York-Basel, 291-317.

Rothe, F. [1985]: The periods of the Volterra-Lotka system, *J. reine angew. Math.*, 355. 129-138.

Rouche, N.; Mawhin, J. [1973]: *Équations différentielles ordinaires* I. II., Masson, Paris.

Rouche, N.; Habets, P.; Laloy, M. [1977]: *Stability Theory by Liapunov's Direct Method*, Springer-Verlag, New York.

Rudenok, A. E. [1975]: Strong isochronism of a centre. On the periods of the limit cycles of Liénard's system, *Diff. Equ.*, 11. 811-819 (Russian).

Ruelle, D.; Takens, F. [1971]: On the nature of turbulence, *Commun. Math. Phys.*. 20. 167-192.

Samoilenko, A. M.; Ronto, N. I. [1979]: *Numerical-Analytic Methods of Investigating Periodic Solutions*, Mir, Moscow.

Samoilenko, A. M.; Ronto, N. I. [1990]: A modification of the numerical-analytic method of successive approximations for boundary value problems of ordinary differential equations, *Ukr. Mat. J.*, 42. 1107-1116 (Russian).

Sanders, J. A.; Verhulst, F. [1985]: *Averaging Methods in Nonlinear Dynamical Systems*, Springer-Verlag, New York.

Sansone, G.; Conti, R. [1964]: *Non-linear Differential Equations*, Pergamon, New York.

Sard, A. [1942]: The measure of the critical values of differentiable maps, *Bull. Amer. Math. Soc.*, 48. 883–890.

Schapiro, S. M.; Sethna, P. R. [1976]: An estimate for the small parameter in the asymptotic analysis of non-linear systems by the method of averaging, *Int. J. Non-Lin. Mechanics*, 12. 127–140.

Schauder, J. [1930]: Der Fixpunktsatz in Funktionalraumen, *Studia Math.*, 2. 171–180.

Schwartz, J. T. [1964]: *Nonlinear Functional Analysis*, New York University Lecture Notes, New York.

Sharkovskii, A. N. [1964]: Coexistence of cycles of a continuous map of a line into itself, *Ukr. Mat. Zh.*, 16. 61–71 (Russian).

Shi, Songling [1980]: A concrete example of the existence of four limit cycles for quadratic systems, *Sci. Sinica*, 23. 16–21.

Shi, Songling [1981]: A method of constructing cycles without contact around a weak focus, *J. Diff. Equ.*, 41. 301–312.

Smale, S. [1961]: On gradient dynamical systems, *Ann. Math.*, 74. 199–206.

Smale, S. [1966]: Structurally stable systems are not dense, *Amer. J. Math.*, 88. 491–496.

Smale, S. [1967]: Differentiable dynamical systems, *Bull. Amer. Math. Soc.*, 73. 747–817.

Smirnow, W. I. [1955]: *Lehrgang der höheren Mathematik* III. VEB Deutscher Verlag der Wissenschaften, Berlin.

Smith, F. E. [1963]: Population dynamics in Daphnia magna, *Ecology*, 44. 651–663.

Smith, H. L. [1981]: Nonresonant periodic perturbation of the Hopf bifurcation, *Applicable Anal.*, 12. 173–195.

Smith, H. L. [1982]: The interaction of steady state and Hopf bifurcations in a two-predator-one-prey competition model, *SIAM J. Appl. Math.*, 42. 27–43.

Smith, H. L. [1986]: Periodic orbits of competitive and cooperative systems, *J. Diff. Equ.*, 65. 361–373.

Smith, R. A. [1979]: The Poincaré-Bendixson theorem for certain differential equations of higher order, *Proc. Roy. Soc. Edinburgh* A, 83. 63–79.

Smith, R. A. [1980$_a$]: Existence of periodic orbits of autonomous ordinary differential equations, *Proc. Roy. Soc. Edinburgh* A, 85. 153–172.

Smith, R. A. [1980$_b$]: Existence of periodic orbits of autonomous retarded functional differential equations, *Math. Proc. Camb. Phil. Soc.*, 88. 89–109.

Smith, R. A. [1981]: An index theorem and Bendixson's negative criterion for certain differential equations of higher dimension, *Proc. Roy. Soc. Edinburgh* A, 91. 63–77.

Smith, R. A. [1984]: Certain differential equations have only isolated periodic orbits, *Annali di Mat. pura applicata*, 137. 217–244.

Somolinos, A. S. [1978]: Periodic solutions of the sunflower equation: $\ddot{x} + (a/r)\dot{x} + (b/r)\sin x(t-r) = 0$, *Quarterly Appl. Math.*, 35. 465–478.

Stanisic, M. M.; Mlakar, P. F. [1969]: Rotating pendulum under periodic disturbance, *Z. angew. Math. Mech.*, 49. 359–361.

Stech, H. W. [1985]: Hopf bifurcation calculations for functional differential equations, *J. Math. Anal. Appl.*, 109. 472–491.

Stépán, G. [1986]: Great delay in a predator-prey model, *Nonlin. Anal. TMA*, 10. 913–929.

Stépán, G. [1989]: *Retarded Dynamical Systems: Stability and Characteristic Functions*, Longman, Harlow.

Stépán, G. [1991]: Chaotic motion of wheels, *Vehicle Systems Dynamics*, 20. 341–351.

Stépán, G.; Steven, A.; Maunder, L. [1990]: Design principles of digitally controlled robots, *Mech. Mach. Theory*, 25. 515–527.

Stoker, J. J. [1950]: *Nonlinear Vibrations in Mechanical and Electrical Systems*, Interscience, New York.

Stoker, J. J. [1969]: *Differential Geometry*, Wiley-Interscience, New York.

Stokes, A. P. [1969]: Some inplications of orbital stability in Banach spaces, *SIAM J. Appl. Math.*, 17. 1317–1325.

Szabó, G. [1987]: A remark on M. Farkas: Stable oscillations in a predator-prey model with time-lag, *J. Math. Anal. Appl.*, 128. 205–206.

Thom, R. [1972]: *Stabilité structurelle et morphogénèse*, Benjamin, New York.

Tineo, A.; Alvarez, C. [1991]: A different consideration about the globally asymptotically stable solution of the periodic *n*-competing species problem, *J. Math. Anal. Appl.*, 159. 44–50.

Tran Van Nhung [1982]: Über das asymptotische Verhalten von Lösungen gewöhnlicher asymptotisch periodischer Differentialgleichungen mit zufälligen Parametern, *Acta Math. Acad. Sci. Hungar.*, 39. 353–360.

Tran Van Nhung [1983]: Estimates on the existence region of periodic solutions of perturbed Van der Pol's equation, *Annales Univ. Sci. Budapest* Math., 26. 81–91.

Tyson, J. J. [1975]: On the existence of oscillatory solutions in negative feedback cellular control processes, *J. Math. Biol.*, 1. 311–315.

Urabe, M. [1967]: *Nonlinear Autonomous Oscillations*, Academic Press, New York-London.

Vancsa, Ágnes S. [1984]: Application of the Hopf bifurcation theory for the dynamical model of the nose-gear, *Acta Techn. Acad. Sci. Hungar.*, 97. 323–340.

Vanderbauwhede, A. [1989]: Centre manifolds, normal forms and elementary bifurcations, in: *Dynamics Reported 2*, Teubner, Stuttgart, 89–169.

Vanderbauwhede, A.; Iooss, G. [1992]: Center manifold theory in infinite dimensions, in: *Dynamics Reported*, New Series 1, Springer-Verlag, Heidelberg, 125-163.

Van der Pol, B. [1922]: On oscillation hysteresis in a triode generator with two degrees of freedom, *Phil. Mag.*, 1. 700–719.

Van der Pol, B. [1926]: On relaxation oscillations, *Phil. Mag.*, 2. 978–992.

Vasil'eva, A. B.; Butuzov, V. F. [1973]: *Asymptotic Expansions of Solutions of Singularly Perturbed Systems*, Nauka, Moscow (Russian).

Vejvoda, O. [1959]: On the existence and stability of the periodic solution of the second kind of a certain mechanical system, *Czech. Math. J.*, 84. 390–415.

Villari, Gabriele [1982]: Periodic solutions of Liénard's equation, *J. Math. Anal. Appl.*, 86. 379–386.

Volosov, V. M.; Morgunov, B. I. [1971]: *The Averaging Method in the Theory of Nonlinear Oscillating Systems*, Izd. Moscow University, Moscow (Russian).

Volterra, V. [1931]: *Lecons sur la théorie mathématique de la lutte pour la vie*, Gauthier-Villars, Paris.

Vorobev, A. P. [1963$_a$]: On isochronous systems of two differential equations, *Dokladi Akademii Nauk Belorusskoi SSR*, 7. 155–156 (Russian).

Vorobev, A. P. [1963$_b$]: The construction of isochronous systems of two differential equations, *Dokladi Akademii Nauk Belorusskoi SSR*, 7. 513–515 (Russian).

Wan, Y. H. [1977]: On the uniqueness of invariant manifolds, *J. Diff. Equ.*, 24. 268–273.

Ward, J. R. [1978]: Periodic solutions of perturbed conservative systems, *Proc. Amer. Math. Soc.*, 72. 281–285.

Wiggins, S. [1988]: *Global Bifurcations and Chaos: Analytical Methods*, Springer-Verlag, Heidelberg.

Wiggins, S. [1990]: *Introduction to Applied Nonlinear Dynamical Systems and Chaos*, Springer-Verlag, Heidelberg.

Wiggins, S.; Holmes, Ph. [1987]: Periodic orbits in slowly varying oscillators, *SIAM J. Math. Anal.*, 18. 592-611.

Wilken, D. R. [1982]: Some remarks on a competing predators problem, *SIAM J. Appl. Math.*, 42. 895–902.

Wörz-Busekros, Angelika [1978]: Global stability in ecological systems with continuous time delay, *SIAM J. Appl. Math.*, 35. 123–134.

Yakubovich, V. A.; Starzhinskij, V. M. [1972]: *Linear Differential Equations with Periodic Coefficients*, Nauka, Moscow (Russian).

Ye, Yanqian [1982]: Some problems in the qualitative theory of ordinary differential equations, *J. Diff. Equ.*, 46. 153–164.

Ye, Yanqian [1986]: *Theory of Limit Cycles*, Transl. Math. Monographs 66. Amer. Math. Soc., Providence, RI.

Yoshizawa, T. [1966]: *Stability Theory by Liapunov's Second Method*, Mathematics Society of Japan, Tokyo.

Zeeman, E. C. [1988]: Stability of dynamical systems, *Nonlinearity*, 1. 115–155.

Zheng, Zuo-Huan [1990]: Periodic solutions of generalized Liénard equations, *J. Math. Anal. Appl.*, 148. 1–10.

Symbols

$\in$	element of
$\subset$	subset of
$\cup$	union (of sets)
$\cap$	intersection of (sets)
$A \backslash B$	the set consisting of those elements of A which are not elements of B
$A \times B$	the direct product of the sets A and B
$P \oplus Q$	the direct sum of the linear spaces P and Q (the same as the direct product)
$\emptyset$	empty set
Int A	the interior of A
∂A	the boundary of A
$\overline{A}$	the closure of A
dist (A, B)	distance of the sets A and B
$a := b$	defining equality: a is defined by b
$\mathbf{R}$	the set of real numbers
$\mathbf{R}_+$	the set of non-negative real numberas $[0, \infty)$
$\mathbf{R}_-$	the set of non-positive real numbers $(-\infty, 0]$
$[\cdot, \cdot]$	closed interval
$(\cdot, \cdot)$	open interval
$\mathbf{R}^n$	the set of ordered real n-tuples
$\mathbf{R}_+^n$	the set of ordered real n-tuples with non-negative entries
$\mathbf{C}$	the set of complex numbers
$\mathbf{C}^n$	the set of ordered complex n-tuples

$\mathbf{R}^{n^2}$	the set of $n \times n$ matrices with real entries
$\mathbf{C}^{n^2}$	the set of $n \times n$ matrices with complex entries
$\mathbf{N}$	the set of natural numbers: $\{0, 1, 2, 3, ..., \}$
$\mathbf{Z}$	the set of integers
$a \cdot b$	the scalar product of two n dimensional vectors a and b
$< f, g >$	the scalar product of two vectors f and g
$B(a, \delta)$	the ball with centre a and radius $\delta > 0$
$\bar{a}$	the complex conjugate of the complex quantity a
A'	the transpose of the matrix A
A^*	the conjugate transpose of the matrix A with complex entries
$|A|$	the norm of the matrix A
$A < B$	iff $a_{ik} < b_{ik}$ for all, i, k where $A = [a_{ik}]$, $B = [b_{ik}]$ are matrices of the same type
$f : A \rightarrow B$	the function with domain A and values in B
$f \circ g$	the composition of the functions f and g
$\mathcal{H}$	function class, see Definition 1.5.2
$C^n(A)$	the class of n times continuously differentiable functions defined on the set A
$C^n(A, B)$	the class of n times continuously differentiable functions defined on A and with range in B

Index

Applied Mathematical Sciences

(continued from page ii)